T0189210

Lecture Notes in Artificial Intelligence 1085

Subseries of Lecture Notes in Computer Science
Edited by J. G. Carbonell and J. Siekmann

Lecture Notes in Computer Science

Edited by G. Goos, J. Hartmanis and J. van Leeuwen

Springer
Berlin
Heidelberg
New York
Barcelona
Budapest
Hong Kong
London
Milan
Paris
Santa Clara
Singapore
Tokyo

Dov M. Gabbay   Hans Jürgen Ohlbach   (Eds.)

Practical Reasoning

International Conference on Formal and
Applied Practical Reasoning, FAPR'96
Bonn, Germany, June 3-7, 1996
Proceedings

 Springer

Series Editors
Jaime G. Carbonell, Carnegie Mellon University, Pittsburgh, PA, USA
Jörg Siekmann, University of Saarland, Saarbrücken, Germany

Volume Editors

Dov M. Gabbay
Imperial College of Science, Technology and Medicine
Department of Computer Science
180 Queen's Gate, London SW7 2AZ, UK

Hans Jürgen Ohlbach
Max-Planck-Institut für Informatik
Im Stadtwald, D-66123 Saarbrücken, Germany

Cataloging-in-Publication Data applied for

Die Deutsche Bibliothek - CIP-Einheitsaufnahme

Practical reasoning : proceedings / International Conference on
Formal and Applied Practical Reasoning, FAPR '96, Bonn,
Germany, June 3 - 7, 1996. Dov M. Gabbay ; Hans Jürgen
Ohlbach (ed.). - Berlin ; Heidelberg ; New York ; Barcelona ;
Budapest ; Hong Kong ; London ; Milan ; Paris ; Santa Clara ;
Singapore ; Tokyo : Springer, 1996
   (Lecture notes in computer science ; Vol. 1085 : Lecture notes in
   artificial intelligence)
   ISBN 3-540-61313-7
NE: Gabbay, Dov M. [Hrsg.]; International Conference on Formal and
   Applied Practical Reasoning <1996, Bonn>; GT

CR Subject Classification (1991): I.2, F.4.1

ISBN 3-540-61313-7 Springer-Verlag Berlin Heidelberg New York

© Springer-Verlag Berlin Heidelberg 1996
Printed in Germany

Typesetting: Camera ready by author
SPIN 10513089   06/3142 – 5 4 3 2 1 0   Printed on acid-free paper

Preface

Research interest and activity is increasing in artificial intelligence, philosophy, psychology, and linguistics in the analysis and mechanisation of human practical reasoning. Philosophers and linguists, continuing the ancient quest that began with Aristotle, are vigorously seeking to deepen our understanding of human reasoning and argumentation. Significant communities of researchers, faced with the shortcomings of traditional deductive logic in modelling human reasoning and argumentation, are actively engaged in developing new approaches to logic (informal logic, dialogue logic) and argumentation (rhetoric, pragmatic and dialectical) that are better suited to the task. In parallel and with equal dedication and ingenuity, many software engineering and AI researchers are pursuing similar goals. These computer scientists are in urgent need of models of human reasoning and argumentation in order to develop better software tools for aiding or replacing humans or their activities.

A quick look at the research programs of these communities instantly reveals that there is a close conceptual connection and complementary mutual research interest among these diverse communities. Indeed there are strong similarities in aims and case studies between the non-monotonic logic community in AI and the informal logic community in philosophy, between the planning community and the practical reasoning and action community, between the dialogue community and the human computer interaction and user modelling communities, between the fuzzy logic and control community and the many–valued and algebraic logic and philosophy of science communities.

There is moreover a great interest in industry in intelligent systems. The greater family of practical reasoning communities and industry should not fail to collaborate. The main purpose of the FAPR conference was therefore to introduce these communities to each other, to compare the current state of research in these areas, and to make such research available to all researchers involved.

The four invited lectures were given by some of the leading researchers in the various areas: Lotfi Zadeh, Doug Walton, Ray Reiter, and Jaako Hintikka. The presentation of the technical papers included in this volume was accompanied by eight tutorials and three workshops.

We are indebted to the program committee for their effort and thought in organizing the program, to the invited speakers and to the presenters of the tutorials. Our special thanks go to the colleagues and secretaries for their support in organizing this conference: Lydia Rivlin, Janice Londsdale, Ellen Fries, and Christine Kiesel, and in particular to Christine Harms, who has been an invaluable help in ensuring that the event ran smoothly.

March 1996   Dov Gabbay and Hans Jürgen Ohlbach

Program Committee

Lugia Carlucci Aiello	Rome
Johan van Benthem	Amsterdam
Wolfgang Bibel	Darmstadt
Gert Brewka	Leipzig
Luis Fariñas del Cerro	Toulouse
Anthony Cohn	Leeds
Jim Cunningham	London
Robert Demolombe	Toulouse
Didier Dubois	Toulouse
John Fox	London
Dov Gabbay	London
Rob Grootendorst	Amsterdam
David Israel	Menlo Park
Ralph H. Johnson	Windsor
Andrew Jones	Oslo
Robert A. Kowalski	London
Rudolf Kruse	Braunschweig
Alberto Martelli	Turin
Donald Nute	Athens, GA
Hans Jürgen Ohlbach	Saarbrücken
David Pearce	Saarbrücken
David Perkins	Cambridge, MA
Henri Prade	Toulouse
Ian Pratt	Manchester
Uwe Reyle	Stuttgart
Hans Rott	Konstanz
Eric Sandewall	Linkoeping
Michael Scriven	Michigan
Jörg Siekmann	Saarbrücken
Philippe Smets	Bruxelles
Richmond H. Thomason	Pittsburgh
Doug Walton	Winnipeg
John Woods	Lethbridge
Emil Weydert	Saarbrücken
Lotfi Zadeh	Berkeley

Additional Reviewers

We are grateful to all the people who helped referee the papers and run the conference, in particular to

Atzeni, P.	Giordano, L.	Korn, D.	Saffiotti, A.
Bonatti, P.A.	Gordon, T.	Kreitz, C.	Sandner, E.
Cialdea, M.	Herzig, A.	Nonnengart, A.	Thielscher, M.
Finger, M.	Hustadt, U.	Pirri, F.	
Gebhardt, J.	Klawonn, F.	Rath, T.	

Contents

Position Papers and System Descriptions

Schedule

The schedule of the talks in the conference is a compromise between time constraints and a subject oriented grouping.

Wednesday *Track A*

User Modelling and Belief

Legal Reasoning

Track B

Reasoning about Real Systems

Actions and Agents

Miscellaneous Reasoning 2

Integrating Statistical Audit Evidence with Belief Function Theory

Carine Van den Acker, Jan Vanthienen[*]

Katholieke Universiteit Leuven
Naamsestraat 69, B-3000 Leuven
BELGIUM

Abstract

This paper proposes a method for integrating statistical audit evidence (sampling evidence) with belief function theory. Our point of view is the interpretation of belief-function theory as a theory of evidence. The need for integration is urged by the suitability of belief function theory to model the auditor's thought process of collecting and aggregating evidence on the one hand, and the frequent use of both statistical and non-statistical evidence in auditing practice on the other hand. The method we propose is of a general nature and is motivated by the analogy with the way in which sampling evidence is used in auditing practice. Its properties are briefly discussed from the point view of probability theory, though the main part of the paper concentrates on our proposed method and a comparison with Shafer's (Shafer, 1976). Several examples are used to demonstrate both methods' properties.

1 Introduction

The representation of statistical evidence in belief-function theory has always been a problematic issue. Many authors recognized that statistical evidence is not the kind of evidence for which belief-function theory is the most "natural"[1]. In contrast to the witness-and-testimony paradigm, practical applications using belief-function theory as a theory of evidence are confronted with statistical evidence. As long as there is widespread use of statistical techniques for the collection of evidence, there is a need for representing it in the belief-function formalism. According to Dempster (1990), belief-function representation can be useful when it is necessary to combine statistical and nonstatistical evidence. This is definitely the case when using belief-function theory in our context of a financial statements audit, because both statistical and non-statistical evidence are combined to draw a conclusion about the fairness of financial statements. For these purposes, we present a general approach to the problem that is felt to be suited for our application domain. Srivastava and Shafer (1994) developed a method for the particular case of variables sampling in auditing.

Solutions to this representation problem (Dempster, 1968, Shafer, 1976) have received much criticism from proponents of probability theory. We will first formulate the problem in the next section and then review these basic criticisms in the third section, because they will be referred to when discussing Shafer's method and our proposed method. In line with their requirements, Bayesians argue that the only "well-performing" belief-function representation is the Bayesian belief function. We will include this solution for completeness also in the third section. Then we will go into Shafer's method of integrating statistical evidence, because it was the starting point for our alternative method. Next, our proposed approach will be explained and will be compared to Shafer's. Finally, conclusions are summarized.

---

[*] e-mail : Carine.VandenAcker@econ.kuleuven.ac.be, Jan.Vanthienen@econ.kuleuven.ac.be
[1] See e.g. Shafer (1976), Shafer (1990), Dempster (1990), Shafer and Srivastava (1990).

2 Formulation of the Problem

Suppose we perform a random experiment for which we know it is governed by a chance density p_θ on a set of outcomes X. We don't know the true value θ of the density's parameter, but do know the set of its possible values, Θ. Hence, we don't know the true density that governs the experiment, but we do know the set of densities $\{p_\theta\}_{\theta \in \Theta}$ that will contain the true density. The experiment will be performed to get an outcome $x \in X$ that will be evidence as to which p_θ is the correct density. Hence, it will also be evidence as to which θ is the correct value for the unknown parameter[2].

The problem of representing statistical evidence can then be formulated as follows : for each $x \in X$, we would like to determine a mass assignment, m_x over Θ, such that $m_x(A)$, $A \subset \Theta$, is the mass committed to A on the basis of the evidence outcome x.

An example of our application domain may clarify the problem. Suppose we are interested in the number of overdue Accounts Receivable (A/Rs) that are part of a company's A/R general ledger. We know that occurrence of overdue accounts is governed by a Poisson distribution, Poi_θ, but we do not know the true value of its parameter θ. We do know that the number of overdue accounts per 100 accounts on average is an integer number between 3 and 10, meaning $\Theta = \{3, 4, 5, 6, 7, 8, 9, 10\}$. To collect evidence, a random sample of A/Rs will be examined and the number of overdue accounts, x, will be counted. The outcome of this sample will provide support as to which Poi_θ was governing the experiment, and hence, which θ is the true parameter. Determining this support for all possible subsets of Θ on the basis of the sample outcome, is the problem of integrating statistical evidence.

This problem is different from the case in which a probability distribution over the elements of Θ is known[3]. A known distribution over Θ is usually represented as a Bayesian belief function based on Hacking frequency principle (Hacking, 1965)[4]. In that case the probability distribution and the Bayesian belief function share the same values, but not the same meaning (Smets, 1994). It is however not the problem that we aim to solve in this context, because distributions over Θ are not available in our application domain. Moreover, we feel that this situation is quite natural for the typical applications of evidence theory. A related problem that is more complex and in our opinion more fit to use the strengths of belief-function theory, is the representation of a probability distribution that is only partially known. Smets (1994) developed a method to deal with this problem.

3 Integrating Statistical Evidence using Bayesian Belief Functions

A belief function Bel is Bayesian if it obeys the three following rules (Shafer, 1976, p.19) :

   (1) $Bel(\varnothing)=0$
   (2) $Bel(\Theta)=1$
   (3) If $A \cap B=\varnothing$, then $Bel(A \cup B)=Bel(A)+Bel(B)$ **Definition 1**

For our problem, a Bayesian belief function (Bbf) can be constructed by letting $m_x(\{\theta\}) = c.p_\theta(x)$, $c \in \mathbb{R}_0^+$ and restricting total mass committed to one.

Then $m_x(\{\theta\}) = \dfrac{p_\theta(x)}{\sum\limits_{\theta \in \Theta} p_\theta(x)}$ and $Bel_x(A) = \dfrac{\sum\limits_{\theta \in A} p_\theta(x)}{\sum\limits_{\theta \in \Theta} p_\theta(x)}$

A Bbf has two properties which are important (if not essential) to proponents of probability theory[5].

---

[2]   Adapted from Shafer (1976).
[3]   In Bayesian terminology this comes down to the problem of representing priors.
[4]   See e.g. Shafer (1976, p. 16), Smets (1994, p. 1), Halpern and Fagin (1992).
[5]   See e.g. Halpern and Fagin (1992).

The first property says that a Bbf representation of statistical evidence, combined with a prior distribution, results in a new Bbf that is equal to the representation of the posterior distribution. Any belief-function representation for which plausibilities are proportionate to likelihoods, possesses that property. Shafer's method which is included in section 4 has this property, our alternative method of section 5 has not.

The second property is the "correct behavior under combination" stating that the combination of two Bbfs, each representing the distribution one of two independent pieces of evidence, results in a combined Bbf that is the same as the Bbf representation of the joint probability distribution. Halpern and Fagin (1992) proofed that the only representation that can have this property is a Bbf. All solutions that are different from a Bbf have received criticism because the consequence of lacking this property is the difference in results when combining at the aleatory or at the epistemic level (Shafer, 1976). Aleatory combination means that multiple pieces of independent evidence are combined at the chance level (using the joint distribution) and then represented as a belief function. Epistemic combination means that independent pieces of evidence each are represented as a belief function and then combined using Dempster's rule into one belief function.

We will come back to both properties when discussing Shafer's method and the alternative method in section 6.

In our view of belief-function theory as a theory of evidence, the Bbf representation of statistical evidence is not appealing. As argued by Shafer (1976, p. 201-2) "Bbfs represent infinite weights of evidence impugning each of the frame's elements ... they cannot arise from actual, finite evidence". For our purposes, statistical evidence is to be combined with non-statistical evidence. Whatever the outcomes of later evidence collection, combining with a Bbf would lead to conflict, which is not really conflict in an evidence theory interpretation or at least has a different meaning than the conflict we should try to explain instead of normalizing it away. For these reasons, we will prefer consonant belief functions.

4 Integrating Statistical Evidence : Shafer's Method

Shafer (1976) proposed a general method for integrating statistical evidence with the belief-function formalism. His approach is based on two assumptions, being proportionality between probability and plausibility, and consonance. According to Shafer, *"these assumptions must be regarded as conventions for establishing degrees of support and are only justified by their general intuitive appeal and their success in dealing with practical examples"* (Shafer, 1976, p. 239). We will discuss his approach because it served as a starting point for the alternative method that we propose.

4.1 Shafer's Assumptions

The assumption of *proportionality of probability and plausibility* is based on the intuitive appeal of considering likelihoods[6] as a determinant of support for hypotheses. We can say that an outcome $x \in X$ should favor those elements $\theta \in \Theta$ that make x more likely or plausible. In other words, θ should obtain a larger support than θ' whenever the density p_θ makes x more plausible than the density $p_{\theta'}$, i.e. whenever $p_\theta(x) > p_{\theta'}(x)$. This idea is reflected in the belief function when the plausibility of θ and the likelihood of x given θ are proportionate :

$$Pl_x(\{\theta\}) = c \, p_\theta(x) \quad , \quad c \in \mathbb{R}_0^+$$

---

6   Fisher (1950, p. 10.310) wrote "The likelihood that any parameter (or set of parameters) should have any assigned value (or set of values) is proportional to the probability that if this were so, the totality of observations should be that observed"

A belief function Bel is *consonant* when all focal elements are nested, meaning that all focal elements can be arranged in order so that each element is contained in the following one. Stated differently, Bel is consonant if for any two focal elements A, B :
Bel(A∩B) = Min{Bel(A), Bel(B)} and Bel(A∪B) = Max{Bel(A), Bel(B)}. Hence, constructing a consonant belief function from statistical evidence is a possibilistic approach.

Assuming consonance is appealing because a consonant belief function is free of contradiction and represents evidence that is pointing in a single direction[7].

4.2 Belief Obtained from Statistical Evidence

Both assumptions lead to Bel to be derived from observing x as follows :

$$Pl_x(A) = \frac{\max\limits_{\theta \in A} p_\theta(x)}{\max\limits_{\theta \in \Theta} p_\theta(x)}$$

for all non-empty subsets $A \subseteq \Theta$;

and $Bel_x : 2^\Theta \to [0, 1]$:

$$Bel_x(A) = 1 - Pl_x(\overline{A}) = 1 - \frac{\max\limits_{\theta \in \overline{A}} p_\theta(x)}{\max\limits_{\theta \in \Theta} p_\theta(x)}$$

for all proper subsets $A \subset \Theta$. **Definition 2**

5 Integrating Statistical Evidence : Alternative Method

5.1 Alternative Method's Motivation

For our purposes, we agree with the attractiveness of enforcing *consonance* and therefore with Shafer's second assumption.

With respect to the *proportionality between plausibility and likelihood*, we feel more reluctant. At first sight, it agrees to intuition that evidence outcomes render some hypotheses more plausible than others. However, the fact that this verbal explanation uses the word "plausible" does not in itself justify the proportionality of probabilities with plausibilities instead of beliefs (or mass). In our opinion, making $Pl_x(\{\theta\})$ proportionate to $Pl_x(\{\theta\})$ to $p_\theta(x)$ is not in accordance with the formalism's philosophy as a theory of evidence. Plausibility is an indirect measure, that represents the extent to which there is no counter evidence. Relating $Pl_x(\{\theta\})$ to $p_\theta(x)$ means that $Bel_x(\{\overline{\theta}\})$ is related to $1-p_\theta(x)$ which gives the wrong impression that the likelihood contains positive evidence supporting the complement $\{\overline{\theta}\}$ to an extent proportionate with $1-p_\theta(x)$, whereas it is in fact evidence in support of θ itself, not in support of its complement.

5.2 Alternative Method's Assumptions

Resulting from the discussion above, we will keep the *consonance* assumption but reject the proportionality of plausibility and probability for our purposes. Indeed, if one is willing to accept the information rendered by the density $p_\theta(x)$ as evidence to be represented by a belief function, one would expect it to justify a certain amount of belief in θ, not in its complement. Hence, we want *belief* to be *proportionate to probability* or : $Bel_x(\{\theta\}) = c\, p_\theta(x)$, $c \in \mathbb{R}_0^+$

---

7   See Shafer (1976), p. 219.

5.3 Belief Obtained from Statistical Evidence

For a belief function to be consonant, at most one singleton hypothesis can receive a positive amount of mass. If only one hypothesis can be chosen, it is sensible to take the one that makes the evidence outcome the most likely. To select an additional hypothesis in order to construct the next focal element, it is equally appealing to take the one of the remaining hypotheses that renders the evidence outcome the most likely. Proceeding in this way, a complete sequence of focal elements can be constructed gradually by extending each time the largest focal element with the hypothesis (more hypotheses in case of a tie) that is not yet included and makes the evidence outcome the most likely. The result is by definition a consonant belief function. Consonance results in the characteristic that the belief committed to any subset A of Θ equals the belief committed to its largest subset that is also a focal element. This means that all elements θ of A not contained in that largest subset do not have any contribution to the total belief in A. Stated differently, the information contained in $p_\theta(x)$ for these elements is ignored when computing Bel(A).

This reasoning of ignoring information (or selectively withholding information) is inspired by the simplifying way in which statistical evidence is often used in auditing practice : the sample outcome x is used to assess the probability $p_\theta(x)$ for each θ that is a relevant population value or range of population values. Next, only the value of θ for which $p_\theta(x)$ is maximal, is retained as *the* value that supported by the evidence. This corresponds to the situation in which the choice of only one single hypothesis would allowed. Hence, the sample information is reduced to one conclusion (θ); knowledge of the probabilities under other hypotheses is discarded. Adopting this reasoning to obtain beliefs would lead to a simple support function with the most likely singleton hypothesis as the only focal element. In our approach, the number of focal elements is extended gradually, with F_n representing the result of a situation in which the choice of n hypotheses would be allowed. In that way, statistical information that is incorporated in the belief function exceeds information that is often retained in auditing practice.

Masses are normalized using the sum of probabilities.

The resulting method is described formally as[8] :

$Bel_x : 2^\Theta \to [0, 1]$ is a consonant belief function with n focal elements F of the form :

$$F_0 = \bigcup_{p_\theta(x)=\max_{\epsilon\in\Theta}\{p_\epsilon(x)\}}\{\theta\}$$

$$F_j = F_{j-1} \cup \{ \bigcup_{p_\theta(x)=\max_{\epsilon\in\Theta\setminus F_{j-1}}\{p_\epsilon(x)\}}\{\theta\} \}$$

with bpa :

$$m_x(F_0) = \frac{\sum_{\theta\in F_0} p_\theta(x)}{\sum_{\theta\in\Theta} p_\theta(x)}$$

$$m_x(F_j) = \frac{\sum_{\theta\in F_j\setminus F_{j-1}} p_\theta(x)}{\sum_{\theta\in\Theta} p_\theta(x)}$$

and belief for focal elements :

$$Bel_x(F_0) = \frac{\sum_{\theta\in F_0} p_\theta(x)}{\sum_{\theta\in\Theta} p_\theta(x)}$$

$$Bel_x(F_j) = \frac{\sum_{\theta\in F_j} p_\theta(x)}{\sum_{\theta\in\Theta} p_\theta(x)}$$

Definition 3

---

[8] The method is not invariant to a monotone transformation of the scale of Θ, which is not seen as a problem in auditing as we cannot think of any rescaling other than multiplication by a constant.

6 Shafer's Method versus the Alternative Method

To clarify how both methods derive a consonant belief function from statistical evidence, we use a simple numerical example. Then the example is extended to illustrate particular features of both methods. In the remainder of this chapter, the amount of belief is only included for focal elements.

6.1 Simple Example

One ball is taken at random from one of three urns. Each urn contains 20 balls with the following proportions of black and red balls :

	1	2	3
black	5	10	15
red	15	10	5

The ball that was taken is black. We want to know which urn the ball was taken from.

Let Θ be the set of all urns $\{1, 2, 3\}$, Bel_B be the belief function to build on Θ on the basis of the evidence of one B(lack) ball. $P_i(B)$ is the probability of B when i is the urn it was taken from. Then $P_1(B)=1/4$, $P_2(B)=1/2$, $P_3(B)=3/4$.

Using Shafer's formula (**Definition 2**), we obtain : $Bel_B(\{3\})=1/3$, $Bel_B(\{2,3\})=2/3$ and $Bel_B(\{1,2,3\})= 1$. Using the alternative method (**Definition 3**), $Bel_B(\{3\})=1/2$, $Bel_B(\{2,3\})= 5/6$ and $Bel_B(\{1,2,3\})=1$. Computations are illustrated below :

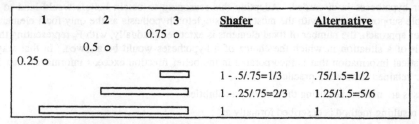

Figure 1 : Computations for Shafer's Method and Alternative Method : Simple Example

Because both methods enforce consonance, they result in the same focal elements but with a different amount of belief. The difference between both methods in assigning mass to subsequent focal elements leads to higher belief values committed to smaller focal elements for the alternative method compared to Shafer's. As a consequence, the former "highlights" the most likely hypotheses more than Shafer's method does. This property is in line with the alternative's method motivation. In **Appendix I** both methods are compared on one of Shafer's original examples, showing that this property can make an important difference in cases of weak evidence.

6.2 Extensions

6.2.1 Version 1

Suppose now there are 10 urns, each with 20 balls and the following proportions of black and red balls :

	1	2	3	4	5	6	7	8	9	10
black	10	10	10	10	10	5	5	5	5	5
red	10	10	10	10	10	15	15	15	15	15

We assume that urns can be distinguished using other attributes than their red/black balls proportion, which are irrelevant to this context. One ball is taken at random out of an urn. The ball is black. We are interested in knowing which urn it was taken from.

Let Θ be the set of all urns $\{1, 2, ... 10\}$, Bel_B be the belief function to build on Θ on the basis of the evidence. $P_i(B)$ is the probability of evidence outcome B(lack ball) when i is the urn it was taken from. Then $P_i(B)=1/2$ for $i=1,..,5$ and $P_i(B)=1/4$ for $i=6,..,10$.

Using Shafer's formula $Bel_B(\{1,2,..,5\})=1/2$ and $Bel_B(\{1,2,..,10\})=1$. Using the alternative method, $Bel_B(\{1,2,..,5\})=2/3$ and $Bel_B(\{1,2,..,10\})=1$.

6.2.2 Version 2

Continuing the same example, suppose now that urns 7 till 10 are taken out of the experiment's setup, leaving us with the frame $\Theta=\{1,2,..,6\}$. Again, one ball is taken at random from one of the urns. The ball is black.

Intuitively, we expect stronger belief in favor of the first five urns in comparison with the sixth, because there is a greater proportion of black balls in that group of urns. Moreover, we would expect a larger amount of belief for subset $\{1,..,5\}$ than in version 2, because of the reduced number of competing urns.

Using Shafer's method, the belief for $\{1,..,5\}$ equals 1/2 which is the same as in the simple example, whereas the alternative methods assigns an increased amount of belief to $\{1,...,5\}$ equal to 5/7.

Shafer's method does not change $Bel_B(\{1,..,5\})$ because there was no change in the probability of the subset's most likely element, nor in the probability of the overall most likely element. Under these conditions, the method does not respond to a change in the number of frame elements with non-zero likelihoods. In contrast, the alternative method does adapt the belief committed to any subset in case of such a change, because there is always a denominator change in **Definition 3**. In our opinion, the alternative method agrees with intuitive expectations in this respect.

6.3 Discussion

Neither Shafer's nor the alternative method correspond to the properties required from a Bayesian point of view (see section 3). Shafer's approach has only the first property, the alternative approach has neither of both.

Missing the first property is not felt to be dramatic for our point of view because in the application of belief-function theory that we pursue, we do not have prior distributions. If we had priors, and evidence were mainly statistical, we would simply have looked for solutions using probability theory.

The second property can only be achieved using a Bayesian belief-function representation for which we explained our objections in section 3. The property of consonance was felt to be better suited. Missing the second property has the consequence that a choice must be made between aleatory and epistemic combination. In our opinion, evidence that is really statistical in nature should be combined at the aleatory level[9]. Aleatory combination is in line with the recognition that "purely statistical problems are not the most important domain of application for belief functions" (Shafer,1990,p.345). Being aware of the fact that representing statistical evidence as a belief function is a necessary but problematic issue, it seems defensible to make this step as late as possible.

---

[9] For the possibly biased coin of Appendix I, it is intuitively more correct to consider H and T from two tosses as evidence for the chance of heads to be 0.5, instead of considering the outcomes separately as conflicting evidence for the chance to be 1 and 0 respectively.

7 Illustrations

This section elaborates on two more complex examples. The first is Newman's Tramcar problem, the second is a coin tossing experiment. Both are taken and adapted from Shafer (1976).

7.1 Newman's Tramcar Problem

7.1.1 Version 1

Problem Description

"A man traveling in a foreign country has to change trains at a junction, and goes into the town, the existence of which he has only just heard. He has no idea of its size. The first thing he sees is a tramcar numbered 100. What can he infer about the number of tramcars in the town ? It may be assumed for the purpose that they are numbered consecutively from 1 upwards."[10]

The problem consists of finding out the true number of tramcars, θ, out of the set of possible numbers $\Theta=\{1, .., N\}$, where N is the maximal number of tramcars. The evidence is the random encounter of tramcar no. 100 (x). Knowing that the random encounter of tramcar x when θ tramcars are available, is governed by a uniform distribution, we can write :

$$p\theta(x)= 1/\theta , \quad \theta \geq x$$
$$0 , \quad \theta < x$$

Shafer's Method

Application of the formula in **Definition 2**, leads to the support and plausibility of **Table 1**.[11]

θ	Pl($\{\theta\}$)	A	S(A)	Pl(A)
99	0	{0,..,99}	0	0
100	1	{100}	1/101	1
150	2/3	{100,..,149}	1/3	1
200	1/2	{100,..,199}	1/2	1
400	1/4	{100,..,399}	3/4	1
2000	1/20	{100,..,1999}	19/20	1
N	100/N	{100,..,N}	1	1
		{101,..,N}	0	100/101

Table 1 : Tramcar Problem - Shafer's Method

Non-focal elements containing 100 also have a positive degree of support. The remaining subsets (not containing 100) all have a zero support. Hence, no singleton, except for {100} has a positive degree of belief. The support in favor of {100} is rather weak. Therefore, according to Shafer, evidence can better be summarized by saying that the number of tramcars is at least 100 and that it is 95 % certain that the number is in [100,2000], 75 % certain that the number is in [100,400] etc.

Alternative Method

In the alternative approach, results depend on the maximum number of tramcars, N. Results are shown for N=2000 (**Table 2**) and for N=8000 (**Table 3**).

---

[10] by H. Jeffreys (1939) in "Theory of Probability" in : Shafer (1976), p. 241.
[11] Taken from Shafer (1976), p. 242.

θ	Pl({θ})	A	S(A)	Pl(A)
99	0	{0,..,99}	0	0
100	1	{100}	0.003333	1
150	0.864332	{100,..,149}	0.135668	1
200	0.768191	{100,..,199}	0.231809	1
400	0.536802	{100,..,399}	0.463198	1
2000	0.000166	{100,..,1999}	0.999834	1
N	0.000166	{100,..,N}	1	1
		{101,..,N}		0.996667

Table 2 : Tramcar Problem - Alternative Method (N=2000)

θ	Pl({θ})	A	S(A)	Pl(A)
99	0	{0,..,99}	0	0
100	1	{100}	0.002279	1
150	0.907197	{100,..,149}	0.092803	1
200	0.841432	{100,..,199}	0.158568	1
400	0.68315	{100,..,399}	0.31685	1
2000	0.316065	{100,..,1999}	0.683935	1
N	0.000029	{100,..,N}	0.999971	1
		{101,..,N}	0	0.997721

Table 3 : Tramcar Problem - Alternative Method (N=8000)

Discussion

Both methods implicitly put forward the most likely parameter value as the only sensible one "to belief in", because no other single parameter value receives any support. This choice is inevitable to build a consonant belief function.

In spite of this similarity between the two methods, there is the same counterintuitive result from Shafer's approach as the one that was shown in the extension of the urns example. Here it comes down to the fact that support can be rendered without knowing the maximum number of tramcars (N). One would expect that the belief in the singleton {100} depends on the number of alternative hypotheses : in case there are only ten alternatives, it should be possible to commit a larger support to this hypothesis than in case there are a thousand alternatives. This dependence is not reflected in Shafer's approach, because support is dependent on the best alternative, being the one closest to 100 that is not an element of the subset of interest.

7.1.2 Version 2

Problem Description

To facilitate further illustration, we assume that the town has either 100, 200 or 300 tramcars. Hence, $\Theta=\{100,200,300\}$. Again, we are interested in knowing the true number of tramcars, knowing that the only observation (x) is again tramcar no. 100.

Results

Computing belief according to Shafer's method results in :
$Bel_x(\{100\})=1/2$, $Bel_x(\{100,200\})=2/3$ and $Bel_x(\{100,200,300\})=1$.
Using the alternative method, $Bel_x(\{100\})=6/11$, $Bel_x(\{100,200\})=9/11$ and $Bel_x(\{100,200,300\})=1$.

7.1.3 Version 3

Problem Description

Continuing with version 2, let us now assume that in addition to tramcar no. 100 (x), tramcar no. 90 (y) has been observed. The evidence will be treated as a joint observation xy[12], hence using : $p_{100}(xy)=(1/100)^2$, $p_{200}(xy)=(1/200)^2$, $p_{300}(xy)=(1/300)^2$. We wonder whether observation y adds to our belief.

Results

Computing belief according to Shafer's method results in :
$Bel_{xy}(\{100\})=3/4$, $Bel_{xy}(\{100,200\})=8/9$ and $Bel_{xy}(\{100,200,300\})=1$.
Using the alternative method, $Bel_{xy}(\{100\})=36/49$, $Bel_{xy}(\{100,200\})=45/49$ and $Bel_{xy}(\{100,200,300\})=1$.

Discussion

Both methods increase their support committed to smaller focal elements. This agrees with our intuition : when one observation of a tramcar with a number lower than any θ makes smaller θ's more likely than larger ones, one would expect a stronger effect when two such observations are made.

7.1.4 Version 4

Problem Description

Continuing with version 3, we assume that tramcar no. 100 (x) and tramcar no. 150 (z) instead of no. 90 have been observed. We then know that $p_{100}(xz)=0$, $p_{200}(xz)=(1/200)^2$, $p_{300}(xz)=(1/300)^2$. Because observation z excludes 100 from the frame, we now wonder how this affects support rendered by both methods.

Results

Computing belief according to Shafer's method results in :
$Bel_{xz}(\{200\})=5/9$ and $Bel_{xz}(\{200,300\})=1$.
Using the alternative method, $Bel_{xz}(\{200\})=9/13$, $Bel_{xz}(\{200,300\})=1$
To compare to the case where only no. 150 would have been observed, we compute also :
$Bel_z(\{200\})=1/3$ and $Bel_z(\{200,300\})=1$ according the Shafer's method, and :
$Bel_z(\{200\})=3/5$ and $Bel_z(\{200,300\})=1$ according to the alternative.

Discussion

The fact that both methods exclude $\theta=100$ thereby depending on the largest observation is rather trivial. Comparing Bel_{xz} with Bel_z shows that support does however not only dependent on the largest observation, but instead the observation of no. 100 adds to the support for smallest possible θ.

7.2 The Coin Tossing Experiment

7.2.1 Version 1 : Epistemic Combination

Problem Description

A coin tossing experiment is performed with a coin that is possibly biased. We know that the chance of H(eads) is 2/10, 5/10 or 8/10. The coin will be tossed twice to find out which of the possible multiples of 1/10 represents the true chance of H. Hence, $\Theta=\{2,5,8\}$. For one toss, we know that $p_\theta(H)=\theta/10$ and $p_\theta(T)=1-\theta/10$. The first toss results in H, the second in

---

[12]   Treating the evidence as a joint observation was called earlier "aleatory combination", as opposed to "epistemic combination". The coin tossing example (7.2) will demonstrate the difference.

T(ails). Using both methods, a consonant belief function will be built for each outcome to be combined into one belief function.

Shafer's Method

On the basis of the outcome H for the first toss, Shafer's approach induces the following belief: $Bel_H(\{8\})=3/8$, $Bel_H(\{5,8\})=6/8$, $Bel_H(\{2,5,8\})=1$.

On the basis of the outcome of the second toss, T : $Bel_T(\{2\})=3/8$, $Bel_T(\{2,5\})=6/8$, $Bel_T(\{2,5,8\})=1$

Combining both leads to $Bel=Bel_H \oplus Bel_T$ with belief :

$Bel(\{2\})=6/37$	$Bel(\{2,5\})=21/37$	$Bel(\{2,5,8\})=1$
$Bel(\{5\})=9/37$	$Bel(\{2,8\})=12/37$	
$Bel(\{8\})=6/37$	$Bel(\{5,8\})=21/37$	

Alternative Method

Using the alternative method, the first outcome (H) leads to : $Bel_H(\{8\})= 8/15$, $Bel_H(\{5,8\})=13/15$ and $Bel_H(\{2,5,8\})=1$,
and for the outcome of the second toss, T : $Bel_T(\{2\})=8/15$, $Bel_T(\{2,5\})=13/15$, $Bel_T(\{2,5,8\})=1$.

Combining both leads to $Bel=Bel_H \oplus Bel_T$:

$Bel(\{2\})=16/81$	$Bel(\{2,5\})=51/81$	$Bel(\{2,5,8\})=1$
$Bel(\{5\})=25/81$	$Bel(\{2,8\})=32/81$	
$Bel(\{8\})=16/81$	$Bel(\{5,8\})=51/81$	

Discussion

Consonance is only enforced when deriving Bel_H and Bel_T separately. When combining two consonant belief functions, consonance is not preserved unless there is perfectly corroborating evidence.

This example of only two evidence outcomes, one H and one T, demonstrates the alternative method's property of stronger highlighting the most likely hypothesis, in this case {5}. Another example of the effect of rather weak evidence on both methods results is included in **Appendix I**.

7.2.2 Version 2 : Aleatory Combination

Problem Description

Consider the same coin tossing experiment as performed in version 2. Hence, we know the coin is possibly biased and that the chance of heads is 2/10, 5/10 or 8/10. Again, the coin will be tossed twice to find out which of the possible multiples of 1/10 represents the true chance of heads. Hence, $\Theta=\{2,5,8\}$. The first toss results in H, the second in T. Both outcomes are treated as one outcome from the joint outcome space {TT,HT,TH,HH}. Therefore, $p_\theta(HH)=(\theta/10)^2$, $p_\theta(TT)=(1 - \theta/10)^2$ and $p_\theta(HT)=p_\theta(TH)=(1 - \theta/10)(\theta/10)$.

Shafer's Method

Applying the formula of **Definition 2** for the joint evidence outcome HT leads to : $Bel_{HT}(\{5\})=9/25$ and $Bel_{HT}(\{2,5,8\})=1$

Alternative Method

On the basis of the outcome HT for the experiment consisting of two tosses of a coin, the alternative method induces : $Bel_{HT}(\{5\})=25/57$, $Bel_{HT}(\{2,5,8\})=1$.

Discussion

Both methods obviously result in a consonant belief function, which is only a trivial result as consonance is enforced. Furthermore, the fact that the belief committed to smaller focal elements is greater when using the alternative method compared to Shafer's, has been extensively shown before.

More interesting is the result that both methods lead to different belief functions when combining at the aleatory level compared to the epistemic level. This difference was already observed by Shafer (1976) and raised many criticisms from probabilists as was discussed in section 3.

7.3.3 Version 3 : Corroborating Evidence

We will briefly present the results for the coin tossing experiment (version 2) in case only corroborating evidence was found. Let us assume H was observed twice.

Combining at the epistemic level, Shafer's method uses $Bel_H(\{8\})=3/8$, $Bel_H(\{5,8\})=6/8$ and $Bel_H(\{2,5,8\})=1$, to compute $Bel=Bel_H \oplus Bel_H$:
$Bel(\{8\})=39/64$, $Bel(\{5,8\})=60/64$ and $Bel(\{2,5,8\})=1$,
whereas the alternative method uses $Bel_H(\{8\})=8/15$, $Bel_H(\{5,8\})=13/15$ and $Bel_H(\{2,5,8\})=1$ to compute $Bel=Bel_H \oplus Bel_H$:
$Bel_H(\{8\})=176/225$, $Bel_H(\{5,8\})=221/225$ and $Bel_H(\{2,5,8\})=1$

Combining at the aleatory level, Shafer's method leads to Bel_{HH} , with :
$BelHH(\{8\})=39/64$, $BelHH(\{5,8\})=60/64$, $BelHH(\{2,5,8\})=1$,

whereas the alternative method leads to Bel_{HH}, with :
$BelHH(\{8\})=64/93$, $BelHH(\{5,8\})=89/93$, $BelHH(\{2,5,8\})=1$.

Using the alternative method, Bel differs from Bel_{HH}, meaning that aleatory and epistemic combination lead to different results. Using Shafer's method Bel and Bel_{HH} are equal. The latter is due to the perfect corroboration of the evidence and is not a general result. Halpern and Fagin (1992) proofed that only Bayesian belief functions lead to the same results under aleatory and epistemic combination under all circumstances.

8 Concluding Remarks

To solve the problem of representing statistical evidence in belief function theory, two methods are presented, being Shafer's method and the alternative method that we propose. The construction of a Bayesian belief function was considered inadequate for our domain of application, mainly because it would inevitably lead to conflict.

Neither of both methods possesses the two properties that are desirable from a probability theory point of view. Shafer's methods obeys the first property. We felt that these properties were not important considering the fact that our application domain does not pose problems of a pure statistical nature. Instead, we chose for other properties that were felt to be more appropriate for our purposes.

Compared to Shafer's method, our alternative method enforces weak evidence and is dependent on the number of elements the frame. The "best" method is the one best suited for the application domain in question.

References

Dempster, A.P. 1990. Construction and local computation aspects of network belief functions, in : Oliver, R.M. ea. *Influence Diagrams, Belief Nets, and Decision Analysis*, Wiley, N.Y.

Fisher, R.A., 1950. Contributions to Mathematical Statistics, Wiley, N.Y.

Hacking, I., 1965. *Logic of Statistical Inference*, Cambridge University Press, U.K.

Halpern, J., Fagin, R. 1992. Two views of belief : belief as generalized probability and belief as evidence. *Artificial Intelligence*, vol. 54, p. 275-317.

Shafer, G. 1976. A Mathematical Theory of Evidence, Princeton University Press.

Shafer, G. 1990. Perspectives on the Theory and Practice of Belief Functions, International Journal of Approximate Reasoning, vol 4, p. 323-362.

Shafer, G., Srivastava, R. 1990.The Bayesian and Belief-Function Formalisms : A General Perspective for Auditing, Auditing : A Journal of Practice & Theory, vol 9 (suppl.), p. 110-148

Smets, P. 1994. Belief Induced by the Knowledge of the Probabilities. Technical Report IRIDIA no. 94-4.1, 8p.

Srivastava, R. and Shafer, G. 1994. Integrating Statistical and Nonstatistical Audit Evidence Using Belief Functions : A Case of Variable Sampling. International Journal of Intelligent Systems. vol. 9. 519-39.

Appendix I : A Possibly Biased Coin

Problem Description

"Consider a coin-tossing experiment in which we only know that the true chance of heads is a multiple of 1/10. Tossing this possibly biased coin will provide evidence as to which is the true chance of heads.[13]"

We are interested to know which is the multiple of 1/10 that represents the true chance of heads. Hence, $\Theta=\{0,1,...10\}$. One toss will result in (H)eads or (T)tails. $p_\theta(H)=\theta/10$ and $p_\theta(T)=1 - \theta/10$.

Consider the case of two independent tosses, one resulting in H, the other in T. Restricting ourselves to combining at the epistemic level, a consonant belief function is constructed for each independent outcome, one for H and one for T. Then these belief functions are combined using Dempster's rule of combination.

Shafer's Method

According to Shafer's method (**Definition 2**) each evidence outcome results in Bel_H and Bel_T respectively. Then $Bel=Bel_H \oplus Bel_T$ is computed. Shafer's results (Shafer, 1967, p. 244) are shown in **Table I.1**. Each entry (i,j) represents Bel({i, ..., j}). For example, Bel({1,2,3})= Bel({1,..., 3})=0.13 (values are rounded to the nearest 0.01)

$i \setminus j$	1	2	3	4	5	6	7	8	9
1	0.02	0.07	0.13	0.22	0.33	0.47	0.62	0.80	1.00
2		0.02	0.07	0.13	0.22	0.33	0.47	0.62	0.80
3			0.02	0.07	0.13	0.22	0.33	0.47	0.62
4				0.02	0.07	0.13	0.22	0.33	0.47
5					0.02	0.07	0.13	0.22	0.33
6						0.02	0.07	0.13	0.22
7							0.02	0.07	0.13
8								0.02	0.07
9									0.02

Table I.1 : Epistemic Combination of one heads and one tails - Shafer's Method

---

[13] Taken from Shafer, 1967, p. 243.

Alternative Method

On the basis of the formula of **Definition 3**, the alternative approach results in two consonant belief functions, which are combined using Dempster's combination rule.

The first toss of the coin resulting in H, leads to the following bpa :

$$m_H(\{10\}) = \frac{\frac{10}{10}}{\frac{1}{10} + \frac{2}{10} + .. + \frac{10}{10}} \cong 0.1818$$

$$m_H(\{9,10\}) = \frac{\frac{9}{10}}{\frac{1}{10} + \frac{2}{10} + .. + \frac{10}{10}} \cong 0.1636$$

$$m_H(\{8,9,10\}) = \frac{\frac{8}{10}}{\frac{1}{10} + \frac{2}{10} + .. + \frac{10}{10}} \cong 0.1455$$

$$m_H(\{7,8,9,10\}) = \frac{\frac{7}{10}}{\frac{1}{10} + \frac{2}{10} + .. + \frac{10}{10}} \cong 0.1273$$

...

A similar belief function is constructed for the second outcome. Combining both with Dempster's rule leads to the results of **Table I.2** (values are rounded to the nearest 0.01).

$i \setminus j$	1	2	3	4	5	6	7	8	9
1	0.02	0.07	0.15	0.27	0.42	0.59	0.76	0.91	1.00
2		0.03	0.10	0.21	0.35	0.51	0.68	0.82	0.91
3			0.04	0.13	0.25	0.39	0.54	0.68	0.76
4				0.05	0.14	0.26	0.38	0.51	0.59
5					0.05	0.14	0.25	0.34	0.42
6						0.05	0.13	0.21	0.26
7							0.04	0.10	0.15
8								0.03	0.07
9									0.02

Table I.2 : Epistemic Combination of one heads and one tails - Alternative Method

Discussion

A comparison of **Table I.1** and **Table I.2** shows that in case of Shafer's approach, the evidence from only two tosses (one H, one T) is too weak to distinguish between hypotheses in terms of different beliefs. On the basis of the evidence, one heads and one tails, one would expect hypothesis {5} and any subset containing it, to get at least a slightly stronger belief than other subsets, even though evidence comes from only two tosses. When instead of two tosses, ten tosses are performed, then Shafer's approach also shows an increase in belief for {5} and "near" hypotheses, as well as for other subsets containing them. In general, the alternative method is more sensitive to evidence than Shafer's by committing larger degrees of belief to smaller subsets. In cases of weak evidence, it may happen that this property "highlights" some hypotheses, where Shafer's method does not.

A Comparative Survey of Default Logic Variants

Grigoris Antoniou

Dept. of Management, The University of Newcastle
Callaghan, NSW 2308, Australia
mgga@alinga.newcastle.edu.au
Phone: +61 49 216848
Fax: +61 49 216911

Abstract. This is an overview paper on default logic and its variants. Default reasoning is one of the most prominent approaches to nonmonotonic reasoning, and allows one to make plausible conjectures when faced with incomplete information about the problem at hand. Default rules prevail in many application domains such as medical diagnosis and leagal reasoning.

Default logic in its original form suffers from some deficiencies, and several variants have been developed in the past years[1]. In this paper we give an overview of the most important of these variants by presenting their motivations and intuitions, and establishing relationships among the approaches. Besides, we give operational models for all logics discussed which allow for a better understanding of the concepts, and make the methods more easily accessbile to a broader audience and practical applications.

Keywords: Nonmonotonic practical reasoning mechanisms, default logic

1 Introduction: Default reasoning

When an intelligent system (either computer–based or a human) tries to solve a problem, it may be able to rely on complete information about this problem, and its main task is to draw the correct conclusions using classical reasoning. In such cases classical predicate logic is sufficient.

In many situations, though, the system has only *incomplete information* at hand, be it because some pieces of information are not available, be it because it has to respond quickly and does not have the time to collect all relevant data. In this case, the system has to make some plausible conjectures. In the case of default reasoning, these conjectures are based on rules of thumb called *defaults*.

When decisions are based on assumptions, these may turn out to be wrong when more information becomes available. The phenomenon of having to revise previous conclusions is called *nonmonotonicity*, and default logic, originally presented by Reiter [14], provides formal methods to support this change of conclusions.

---

[1] Actually, this *family of related default reasoning approaches* just illustrates the strength of Reiter's original idea of default logic.

Default logic is perhaps the most prominent method for nonmonotonic reasoning, basically because of its simplicity and because defaults prevail in many application areas. Nontheless, several deficiencies of the original approach have been identified, and they have led to the development of variations of the initial idea; actually we can talk of a *family of default reasoning methods*, because they all share the same foundations. In this paper we review the most important of these methods. We present their motivations and basic concepts, and compare them systematically both with respect to interconnections and the fullfilment of some properties.

We have put effort to present all methods using *operational models* in addition to the usual fixed–point definitions. The advantages of operational interpretation include the following:

- they make the formalisms easier to learn, thus contributing to making default logic more popular,
- they make the methods more easily accessible to people without a strong formal background (most users fall into this category), and
- they provide a starting point for implementation work.

2 Default Logic

A *default* δ is a string $\frac{\varphi : \psi_1, \ldots, \psi_n}{\chi}$ with closed first–order formulae φ, $\psi_1, \ldots, \psi_n$, χ ($n > 0$). φ is the *prerequisite*, $\psi_1, \ldots, \psi_n$ the *justifications*, and χ the *consequent* of δ. If there are free variables, then the default is interpreted as a default schema, that is, as the set of all ground instances.

Default logic distinguishes between two kinds of knowledge: hard facts (definitely known information) and default rules. Formally, a *default theory* T is a pair (W, D) consisting of a set of formulae W (the set of facts) and a countable set of defaults D. A default of the form $\frac{\varphi : \chi}{\chi}$ is called *normal*. A default of the form $\frac{\varphi : \psi \wedge \chi}{\chi}$ is called *semi–normal*. A default theory T is called *normal* (*semi–normal*) if all defaults in D are normal (semi–normal).

The meaning of default logic is given in terms of so–called *extensions*, 'world views' that are supported by and based on the given information. Here is the formal definition. Let $\delta = \frac{\varphi : \psi_1, \ldots, \psi_n}{\chi}$ be a default, and E a deductively closed set of formulae. We say that δ *is applicable to* E iff φ is included in E, and $\neg\psi_1, \ldots, \neg\psi_n$ are not included in E. For a set D of defaults, we say that E *is closed under D with respect to* E iff, for every default δ in D that is applicable to E, its consequent χ is contained in E.

Let $T = (W, D)$ be a default theory and $\Pi = (\delta_0, \delta_1, \delta_2, \ldots)$ a finite or infinite sequence of defaults from D not containing any repetitions[2] (modelling an application order of defaults from D). We denote by $\Pi[k]$ the initial segment of Π of length k, provided the length of Π is at least k.

---

[2] That means, no default occurs in Π more than once.

- $In(\Pi)$ is $Th(M)$ (the first–order deductive closure of M), where M includes W and the consequents of the defaults occurring in Π.
- $Out(\Pi)$ is the set of the negations of justifications of defaults in Π.
- Π is called a *process of* T iff δ_k is applicable to $In(\Pi[k])$ with respect to $In(\Pi[k])$, for every k such that δ_k occurs in Π.
- Π is *successful* iff $In(\Pi) \cap Out(\Pi) = \emptyset$, otherwise it is *failed*.
- Π is *closed* iff every $\delta \in D$ which is applicable to $In(\Pi)$ with respect to belief set $In(\Pi)$ already occurs in Π.

$In(\Pi)$ collects all formulae in which we believe after application of the defaults in Π, while $Out(\Pi)$ consists of formulae that we should avoid to believe for the sake of consistency. E is an *extension* of T iff there is a closed and successful process Π of T such that $E = In(\Pi)$. This definition is equivalent to the standard fixed–point definition of Reiter, as shown in [1].

3   Representational deficiencies of default logic

- *Default Logic does not guarantee the existence of extensions.*

This is demonstrated by the default theory $(\emptyset, \{\frac{true:p}{\neg p}\})$.

- *Default Logic is not semi–monotonic.*

To see this, consider the default theory consisting of the single default $\frac{true:p}{p}$. This theory has the unique extension $E = Th(\{p\})$. But if we add the new default $\frac{true:q}{\neg q}$ it is easy to see that the enhanced default theory has no extension.

- *Default Logic does not commit to default justifications.*

That means, it is possible to obtain an extension based on contradictory assumptions (joint consistency of justifications is not required). For example, the default theory $(\emptyset, \{\frac{true:p}{q}, \frac{true:\neg p}{r}\})$ has the single extension $Th(\{q, r\})$ which is derived using contradictory assumptions.

If we restrict attention to normal default theories, that is, theories where all defaults are normal, the deficiencies described so far cannot occur. In this sense, normal default theories constitute the well–behaved part of default logic. Actually, when seen on their own, all natural defaults seem to be normal. The problem arises when several defaults are put together to form a theory. What is particularly missing is a means of representing priorities among defaults.

- *Default Logic does not provide an explicit representation of priorities among defaults.*

Defaults may conflict one another. This often leads to different extensions of the default theory under consideration. But in many cases we would like to prefer the application of some default to the application of another. The only way to represent priorities in default logic is by 'patching' this information into the defaults themselves. For example, if we would like to prevent the application of the default rule 'Typically birds fly' in the case of penguins, we can modify the default as follows:

$$\frac{bird(X) : flies(X) \wedge \neg penguin(X)}{flies(X)}.$$

The first drawback of this approach is that we are mixing knowledge with control information, which would make the knowledge base very difficult to maintain. The second problem is that the default is no longer normal; in other words, we have abandoned the well–behaved part of default logic. In the light of this observation, two viable approaches are conceivable, both of which will be followed in subsequent sections:

- Either normal default theories will be enhanced by means of representing priorities among defaults, or
- variants of the full default logic will have to be considered.

Representatives of the latter approach will be introduced in the sections 4, 5 and 6, while approaches along the first line will be discussed in sections 7 and 8.

It should be noted that many formal properties of nonmonotonic logics have been studied in the literature, see, for example, [9,11]. We selected the properties discussed above because they vary from variant to variant and because they have much to do with the original motivation of the approaches discussed in the following sections.

4 Justified default logic

Justified default logic was introduced by Lukaszewicz [10] to overcome one of the problems with default logic, the possibility that a theory has no extensions.

An additional justification of the alternative extension concept we shall present is that it can lead to more intuitive conclusions than the extension concept we have been using so far. For example, let us consider the following piece of information:

Usually I go fishing on Sundays unless I wake up late.
When I am on holidays I usually wake up late.

Suppose it is Sunday and I am on holidays. Then, of course, both default rules cannot be applied in conjunction. Instead we would expect two alternative possibilities, depending on whether I wake up late or not. The information above is expressed by the default theory $T = (W, D)$ with $W = \{holidays, sunday\}$) and D consisting of the defaults

$$\delta_1 = \frac{sunday : goFishing \wedge \neg wakeUpLate}{goFishing}, \ \delta_2 = \frac{holidays : wakeUpLate}{wakeUpLate}$$

It is easily seen that T has only one extension, namely $Th(\{holidays, sunday, wakeUpLate\})$. The other expected outcome, $Th(\{holidays, sunday, goFishing\})$ is not an extension because it is not closed under application of defaults in D: δ_2 can be applied and leads to a failed process. The technical definition presented below aims at avoiding this 'destruction' of a previously constructed In–set.

It is possible to give a simple characterisation of modified extensions using processes. Note that $Out(\Pi)$ contains the set of all justifications of the defaults in Π. Additionally, we should take care that a default is only applied if it does not lead to a failed process. Here is the formal definition:

Let T be a default theory, and let Π and Γ be processes of T. We define $\Pi < \Gamma$ iff the set of defaults occurring in Π is a proper subset of the defaults occurring in Γ. Π is called a *maximal process of T* iff Π is successful, and there is no successful process Γ such that $\Pi < \Gamma$. A set of formulae E is called a *modified extension* of T iff there is a maximal process Π of T such that $E = In(\Pi)$. This definition was proven to be equivalent to Lukaszewicz' original definition in [2].

In the example above $\Pi = (\delta_1)$ is a maximal process: the only process that strictly includes Π is $\Gamma = (\delta_1, \delta_2)$, but it is not successful, so $\Pi \not< \Gamma$.

Among the successful processes of T, maximal processes are maximal w.r.t. the set of defaults they contain. Since every maximal process is successful, it is interesting to determine the relationship between maxiamal processes, and closed and successful processes, i.e. the extensions of T.

1. Theorem *Let T be a default theory. Every closed and successful process of T is also a maximal process of T. Therefore, every extension of T is also a modified extension of T.*

The converse of this result is not true, as shown by the following example. Consider $T = (W, D)$ with $W = \emptyset$ and $D = \{\delta_1 = \frac{true:p}{\neg q}, \delta_2 = \frac{true:q}{r}\}$. There are two closed processes of T, $\Pi_1 = (\delta_1)$ and $\Pi_2 = (\delta_2, \delta_1)$. Π_1 is also successful, so $In(\Pi_1) = Th(\{\neg q\})$ is an extension of T, but Π_2 is failed. On the other hand, there are two maximal processes of T, (δ_1) as predicted by theorem 1, and (δ_2). Therefore T has two modified extensions, $Th(\{\neg q\})$ and $Th(\{r\})$.

It is instructive to look at the 'classical' default theory without an extension, $T = (W, D)$ with $W = \emptyset$ and $D = \{\frac{true:p}{\neg p}\}$. The empty process is maximal (though not closed), because the application of the 'strange default' would lead to a failed process. In general, the following can be shown.

2. Theorem *Every default theory has a modified extension. Justified default logic is semi–monotonic, but does not commit to justifications.*

The latter claim is demonstrated by the default theory $T = (W, D)$ with $W = \emptyset$ and $D = \{\frac{true:p}{q}, \frac{true:\neg p}{r}\}$. T has the single extension $Th(\{q, r\})$ which is supported by the conflicting assumptions p and $\neg p$.

5 Constrained Default Logic

The approach of the previous section avoids running into inconsistencies and can therefore guarantee the existence of (modified) extensions. On the other hand, it does not commit to default justifications. In order to satisfy this property we have to ensure the *joint consistency* of all justifications involved in a process; this is the main idea of *Constrained Default Logic (CDL)* [8]). According to this approach, after the application of the first default in the example above, the second default may not be applied because $\{p, \neg p\}$ is inconsistent.

Furthermore, since the justifications are consistent with each other, we test the consistency of their conjunction with the current knowledge base. In the terminology of processes, we require the consistency of $In(\Pi) \cup \neg Out(\Pi)$.

Finally, we adopt the idea from the previous section, namely that a default may only be applied if it does not lead to a contradiction a posteriori. That means, if $\frac{\varphi : \psi_1 \ldots \psi_n}{\chi}$ is tested for application to a process Π, then $In(\Pi) \cup \neg Out(\Pi) \cup \{\psi_1, \ldots, \psi_n, \chi\}$ must be consistent. We note that the set Out no longer makes sense since we require joint consistency. Instead we have to maintain the set of formulae which consists of W, all consequents *and all justifications* of the defaults that have been applied.

- Given a default theory $T = (W, D)$ and a sequence Π of defaults in D without repetition, we define $Con(\Pi) = Th(W \cup \{\varphi \mid \varphi$ is a consequent or a justification of a default occurring in $\Pi\})$. Sometimes we refer to $Con(\Pi)$ as the *set of constraints*.

For the default theory $T = (W, D)$ with $W = \emptyset$ and $D = \{\delta_1 = \frac{true:p}{q}, \delta_2 = \frac{true:\neg p}{r}\}$ let $\Pi_1 = (\delta_1)$. Then $Con(\Pi_1) = Th(\{p, q\})$.

We say that a default $\delta = \frac{\varphi : \psi_1 \ldots \psi_n}{\chi}$ is *applicable to a pair of deductively closed sets of formulae* (E, C), iff $\varphi \in E$ and $\psi_1 \wedge \ldots \wedge \psi_n \wedge \chi$ is consistent with C. A pair (E, C) of deductively closed sets of formulae is called *closed under* D if, for every default $\frac{\varphi : \psi_1 \ldots \psi_n}{\chi} \in D$ which is applicable to (E, C), $\chi \in E$ and $\{\psi_1, \ldots, \psi_n, \chi\} \subseteq C$.

In the example above, δ_2 is not applicable to $(In(\Pi_1, Con(\Pi_1)) = (Th(\{q\}), Th(\{p, q\}))$ wrt $Th(\{p, q\})$ because $\{\neg p \wedge r\} \cup Th(\{p, q\})$ is inconsistent.

Let $\Pi = (\delta_0, \delta_1, \ldots)$ be a sequence of defaults in D without repetitions.

- Π is a *constrained process* of the default theory $T = (W, D)$ iff, for all k such that $\Pi[k]$ is defined, δ_k is applicable to $(In(\Pi[k]), Con(\Pi[k]))$.
- A *closed constrained process* Π is a constrained process such that every default δ which is applicable to $(In(\Pi), Con(\Pi))$, already occurs in Π.
- A pair of sets of formulae (E, C) is a *constrained extension of* T iff there is a closed constrained process of T such that $(E, C) = (In(\Pi), Con(\Pi))$. This definition was shown to be equivalent to the original one in [4].

Note that we do not need a concept of success here because of the definition of default applicability we adopted: δ is only applicable to (E, C) if it does not lead to a contradiction.

3. Theorem *Constrained default logic guarantees the existence of constrained extensions. Furthermore, it is semi–monotonic and commits to default justifications.*

Let us now establish some relationships between constrained extensions and the other extension concepts we have seen so far.

4. Theorem *Let T be a default theory and $E = In(\Pi)$ an extension of T, where Π is a closed and successful process of T. If $E \cup \neg Out(\Pi)$ is consistent, then $(E, Con(\Pi))$ is a constrained extension of T.*

The converse does not hold since the existence of an extension is not guaranteed.

5. Theorem *Let T be a default theory and $E = In(\Pi)$ a modified extension of T, where Π is a maximal process of T. If $E \cup \neg Out(\Pi)$ is consistent, then $(E, Con(\Pi))$ is a constrained extension of T.*

The example $T = (W, D)$ with $W = \emptyset$ and $D = \{\frac{true:p}{q}, \frac{true:\neg p}{r}\}$ shows that we cannot expect the first component of a constrained extension to be a modified extension: T has the single modified extension $Th(\{q, r\})$, but possesses two constrained extensions, $(Th(\{q\}), Th(\{p, q\}))$ and $(Th(\{r\}), Th(\{\neg p, r\}))$. As the following result demonstrates, it is not accidental that the first component of both constrained extensions is included in the modified extension (for a proof see [8]).

6. Theorem *Let T be a default theory and (E, C) a constrained extension of T. Then there is a modified extension F of T such that $E \subseteq F$.*

6 Rational Default Logic

Constrained Default Logic requires joint consistency of the justifications of the defaults that contribute to an extension, but goes one step further by requiring that the consequent of a default is consistent with the current Con–set. Rational Default Logic [12] does not require this.

Technically, a default $\frac{\varphi:\psi_1,\dots,\psi_n}{\chi}$ is *rationally applicable* to a pair of deductively closed sets of formulae (E, C) with respect to a set of formulae C' iff $\varphi \in E$ and $\{\psi_1, \dots, \psi_n\} \cup C'$ is consistent.

Let $\Pi = (\delta_0, \delta_1, \dots)$ be a sequence of defaults in D without repetitions. We define:

- Π is a *rational process* of the default theory $T = (W, D)$ iff, for all k such that $\Pi[k]$ is defined, δ_k is rationally applicable to $(In(\Pi[k]), Con(\Pi[k]))$ with respect to $Con(\Pi[k])$.

- A *closed rational process* Π is a rational process such that every default δ which is rationally applicable to $(In(\Pi), Con(\Pi))$ with respect to $Con(\Pi)$, already occurs in Π.
- Π is *successful* iff $Con(\Pi)$ is consistent.
- A set of formulae E is a *rational extension of T* iff there is a closed and successful rational process of T such that $E = In(\Pi)$. This definition is equivalent to the original one [12].

As an example, consider the default theory $T = (W, D)$ with $W = \emptyset$ and $D = \{\frac{true:b}{c}, \frac{true:\neg b}{d}, \frac{true:\neg c}{e}, \frac{true:\neg d}{f}\}$. T has one extension, $Th(\{c, d\})$, two rational extensions, $Th(\{c, f\})$ and $Th(\{d, e\})$, and three constrained extensions,

$$(Th(\{e, f\}), Th(\{e, \neg c, f, \neg d\}))$$
$$(Th(\{c, f\}), Th(\{c, b, f, \neg d\}))$$
$$(Th(\{d, e\}), Th(\{d, \neg b, e, \neg c\})).$$

The default theory $T = (W, D)$ with $W = \emptyset$ and $D = \{\frac{true:p}{\neg p}\}$ has no rational extension; this should not be surprising since consistency of the default consequent with the justifications is not required in the applicability test. So we have the following result:

7. Theorem *Rational Default Logic does not guarantee the existence of rational extensions and is not semi-monotonic. On the other hand, it commits to default justifications.*

Obviously, the concepts of rational and constrained extensions are equivalent for semi-normal default theories. Also, if E is a rational extension, then there exists a constrained extension with E as its first component; the converse is not true, as the example in the beginning of this subsection shows.

8. Theorem *Let $E = In(\Pi)$ be a rational extension of a default theory T, where Π is a closed and successful rational process of T. Then Π is a closed constrained process of T and $(E, Con(\Pi))$ is a constrained extension of T. If T is semi-normal, then the converse is also true: if (E, C) is a constrained extension of T, then E is a rational extension of T.*

Finally we note that all approaches discussed so far are identical for normal default theories; in other words, the variants coincide on the well-behaved portion of default logic, but extend this in different ways.

9. Theorem *Let T be a normal default theory. E is an extension of $T \Longleftrightarrow E$ is a modified extension of $T \Longleftrightarrow E$ is a rational extension of $T \Longleftrightarrow$ there is a set of formulae C such that (E, C) is a constrained extension of T.*

7 PDL: Prioritized Default Logic

In *PDL* [6] the user *explicitly gives the priority order* in which defaults are to be applied in situations where more than one default is applicable. In the simplest case a strict well–ordering is used (that means, an irreflexive total order in which every subset of D has a smallest element). Let us consider the default theory $T = (W, D)$ with $W = \{bird, penguin\}$ and D consisting of the defaults

$$\delta_1 = \frac{penguin : \neg flies}{\neg flies}, \quad \delta_2 = \frac{bird : flies}{flies}$$

Suppose a total order $\ll$ is given according to which δ_1 is preferred over δ_2, i.e. $\delta_1 \ll \delta_2$. Then T together with $\ll$ should only admit the extension $Th(\{bird, penguin, \neg flies\})$.

Let $T = (W, D)$ be a normal default theory[3], and $\ll$ a strict well–ordering on D. A process $\Pi = (\delta_0, \delta_1, \ldots)$ of T is *generated by* $\ll$ iff, for all i such that $\Pi[i]$ is defined, δ_i is the $\ll$–minimal default that is applicable to $In(\Pi[i])$. Clearly, Π is closed; it is also successful since T is a normal default theory. We say that E is *generated by* $\ll$ iff $E = In(\Pi)$ for a process Π that is generated by $\ll$.

Π and E as defined above are uniquely determined. In the previous example, $\ll$ is defined by $\delta_1 \ll \delta_2$. (δ_1) is the process generated by $\ll$, and $Th(\{bird, penguin, \neg flies\})$ is the extension generated by $\ll$.

Of course, we cannot expect the defaults in all problem domains to be ordered in a total way (for examples demonstrating this, see [5]). Often defaults should be left incomparable with one another. In other words, requiring a total ordering of defaults will often be a kind of overspecification. Let us look at a simple example from politics.

Technically speaking, the priority information will be given in the form of a *strict partial order* $<$ on the set of defaults. The definition of the semantics will make use of all total orders that are compatible with $<$. This approach resembles the treatment of concurrency in computer science, where meaning is assigned by considering all possible linearizations.

Definition $T = (W, D, <)$ is a *prioritized default theory* if (W, D) is a normal default theory, and $<$ a strict partial order on D. E is a *PDL–extension of* T iff there is a strict well–ordering $\ll$ on D which contains $<$ and generates E. —

By definition of a PDL–extension E, $E = In(\Pi)$ for a closed and successful process Π. Therefore we make the following observation:

10. Theorem *If E is a PDL–extension of the prioritized default theory $T = (W, D, <)$ then E is an extension of (W, D).*

---

[3] As explained at the end of section 3, we consider normal default theories with priorities; this class is widely regarded by the field leaders to be suitable for most practical applications.

The following result follows from the observation that given a finite set M, every strict partial order on M is included in a strict well–ordering on M.

11. Theorem *A prioritized default theory $T = (W, D, <)$ always has an extension if D is finite. Furthermore, PDL is semi–monotonic (in the case of a finite set of defaults) and commits to default justifications.*

Note that PDL allows the user to specify priorities among defaults, but only in a rigid, predefined form. In contrast to this, the logic we shall present in the following section allows for far greater flexibility in specifying and reasoning about priorities. We call that logic $PRDL$ for 'Priority Reasoning in Default Logic'.

8 PRDL: Reasoning about priorities

In this section we present an alternative approach to priorities in default reasoning, introduced by Brewka in [7]. Whereas in PDL the priority information was *extralogical*, that is, it was given outside the usual default logic, here priority information will be part of the logical language. The information 'δ_1 has higher priority than δ_2' will be a formula as every other assertion and included in the default theory given. For the sake of technical simplicity we restrict our presentation to finite sets D of defaults.

To be able to reason about default priorities, defaults are augmented by a (unique) name; so, named defaults have the form $d \equiv \frac{\varphi:\psi}{\psi}$. Whenever confusion does not arise, we shall assume that d_i is the name of δ_i, and that δ_i is the default with name d_i. Please notice the difference between d_i and δ_i: d_i is a default name *within the logical language*, whereas δ_i is used *on the meta–level*, that means outside the logical language, to allow us to refer to a default without having to write down its details.

Apart from naming defaults, a special symbol $\prec$ representing default priorities and acting on default names is introduced. $d_1 \prec d_2$ is to be read as 'give the default (with name) d_1 priority over (the default with name) d_2'.

A *named default theory* T is a tuple (W, D) consisting of a set of first order formulae W and a finite set D of named normal defaults $d \equiv \frac{\varphi:\psi}{\psi}$. It is implicitly assumed that W contains axioms for the transitivity and asymmetry of $\prec$:

$$\forall X \forall Y \forall Z (X \prec Y \land Y \prec Z \to X \prec Z), \quad \forall X \neg X \prec X$$

Therefore, if both assertions $d_1 \prec d_2$ and $d_2 \prec d_1$ are included in a set E containing W, E will be inconsistent. For a named default theory T, let T' denote the unnamed version of T (in which the defaults are as defined in section 2). A *DL–extension of T* is defined as an extension of T'.

So far there is no semantic difference between priority statements and other formulae. But, of course, priority statements must be taken into consideration, essentially by ruling out the application of a default if there exists another applicable default with higher priority. Let us look at an example to illustrate this point. We use the well-known birds domain, and consider the named default theory $T = (W, D)$ with $W = \{penguin, bird, d_2 \prec d_1\}$ and D consisting of the defaults

$$d_1 \equiv \frac{bird : flies}{flies}, \quad d_2 \equiv \frac{penguin : \neg flies}{\neg flies}$$

The unnamed default theory T' has two extensions: $E_1 = Th(\{penguin, bird, d_2 \prec d_1, flies\})$, and $E_2 = Th(\{penguin, bird, d_2 \prec d_1, \neg flies\})$. Intuitively, only E_2 should be a priority extension of T, since deriving E_1 (d_1 is preferred over d_2) is not consistent with the priority information in W (that means, $d_2 \prec d_1$).

In the example above all priority information was given in W, in which case it is quite easy to deal with it. The situation becomes more complicated when *different processes (and thus extensions) contain different priority information.* Consider, for example, the named default theory $T = (W, D)$ with $W = \emptyset$ and W consisting of the defaults

$$d_1 \equiv \frac{true : d_2 \prec d_1}{d_2 \prec d_1}, \quad d_2 \equiv \frac{true : d_1 \prec d_2}{d_1 \prec d_2}$$

There are two closed and successful processes of T, $\Pi_1 = (\delta_1)$ and $\Pi_2 = (\delta_2)$ (they are closed because of the implicitly given asymmetry of $\prec$). Note that $In(\Pi_1) = Th(\{d_2 \prec d_1\})$ and $In(\Pi_2) = Th(\{d_1 \prec d_2\})$ contain mutually inconsistent priority information. Furthermore, if the first default is applied then it turns out that the second default should have been applied instead, and vice versa. Therefore we wouldn't expect T to have any priority extensions.

The interpretation of $PRDL$ we present in the following is very similar to the process model of default logic discussed in section 2. In fact, the only thing we must additionally take care of is the priority information. If a default δ has been just applied instead of some others, it should not turn out that δ should not have been applied at that stage. Stated another way: when a default is applied, we decide to give it higher priority than all other defaults that are applicable at the time. This decision should remain consistent even after application of further defaults.

Since we restrict attention to *normal* defaults augmented by priority information, the *Out*-set is unnecessary (because all processes of a normal default theory are successful). Instead, we use a set Pri to record the decisions regarding priorities among defaults that were made during the construction of the current knowledge base. Figure 1 shows an overview of the processes for the example from the previous section. The left-hand branch leads to a failed situation because $In \cup Pri$ is inconsistent (it includes both $d_1 \prec d_2$ and $d_2 \prec d_1$, thus contradicting the asymmetry of $\prec$). Therefore T has a unique priority extension.

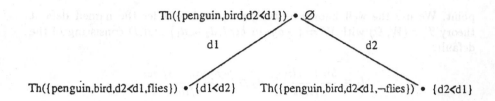

Figure 1

The formal definition below concerns the entire process tree and not one process as in section 2. The reason is simple: whereas in default logic we may analyze a process on its own, in $PRDL$ we must take alternative defaults into consideration. This means that at each node, we have to determine all applicable defaults before proceeding along one choice.

Given a named default theory $T = (W, D)$, the *process tree* $proc(T)$ of T consists of nodes n and edges e. With each node n, we associate sets of formulae $In(n)$ and $Pri(n)$. Also, with each edge e we associate a default $def(e)$. The root of $proc(T)$ is the node n_0 with $In(n_0) = Th(W)$ and $Pri(n_0) = \emptyset$.

Let n be any node that has already been constructed. The tree can be expanded at n in the following way: let $\delta_1, \ldots, \delta_k$ be the unnamed defaults that are applicable to $In(n)$ w.r.t. $In(n)$, and that have not been used as labels of edges on the path connecting n with the root. Then, the process tree is expanded at the node n which becomes the father of the new nodes $n_1, \ldots, n_k$. For each of these nodes n_i we define the In and Pri–set as follows:

$$In(n_i) = Th(In(n) \cup \{cons(\delta_i)\})$$
$$Pri(n_i) = Pri(n) \cup \{d_i \prec d_j \mid j \neq i\}.$$

n is *closed* if it is a leaf, i.e. if $k = 0$. n is *successful* if $In(n) \cup Pri(n)$ is consistent, otherwise it is *failed*. Finally, one notational convention: we denote the path within $proc(T)$ from the root to a node n_m by $\Pi = (n_0, < \delta_1, n_1 >, \ldots, < \delta_m, n_m >)$, where δ_i is $def(e_i)$ for the edge e_i connecting n_{i-1} with n_i. Note that all paths are finite since the set D of defaults in named default theories are finite by definition.

E is a *priority extension* of T iff $E = In(n)$ for a closed and successful node n in $proc(T)$. This definition is equivalent to Brewka's original one, as shown in [4].

As pointed out in section 3, normal default theories have some very desirable properties such as the existence of extensions and semi–monotonicity. Unfortunately, these properties are not preserved when reasoning about priorities is added. Consider the theory $T = (W, D)$ with $W = \emptyset$ and $D = \{d_1 \equiv \frac{true:d_2 \prec d_1}{d_2 \prec d_1}, d_2 \equiv \frac{true:d_1 \prec d_2}{d_1 \prec d_2}\}$. It is easy to check using the definition from the previous subsection that T has no priority extensions: In the beginning both defaults can be applied. If we decide to apply d_1, then $d_1 \prec d_2$ becomes part of the Pri–set, and the default consequent $d_2 \prec d_1$ part of the In–set, so their union is inconsistent (because of the transitivity and asymmetry of $prec$). The same happens if we decide to apply d_2.

Here is an example which shows that $PRDL$ is not semi–monotonic. Consider the theory $T_1 = (W, D_1)$ with $W = \emptyset$ and $D_1 = \{d_1 \equiv \frac{true:a}{a}, d_2 \equiv \frac{true:\neg a}{\neg a}\}$. T_1 has two priority extensions, $E_1 = Th(\{\neg a\})$ and $E_2 = Th(\{a\})$. Now expand D_1 as follows: $D_2 = D_1 \cup \{d_3 \equiv \frac{true:d_1 \prec d_2}{d_1 \prec d_2}\}$. The named default theory $T_2 = (W, D_2)$ has only the extension $E = Th(\{a, d_1 \prec d_2\})$. The priority extension E_1 of (W, D) is not included in a priority extension of (W, D_2).

9 Conclusion

In this paper we reviewed the most important variants of default logic, compared them, and showed how their concepts can be applied to concrete examples by providing operational models.

In fact, these models are the starting point for implementation work conducted within the CIN Project (Change and Information) [3] at the Universities of Newcastle and Sydney. The aim is to develop a toolkit for default reasoning which provides the variants presented in this paper. The reason why we favor a toolkit is that, in our opinion, there is not *the* 'right' default logic approach, but rather the most appropriate for the specific application at hand.

We would like to conclude with a remark regarding the fulfillment of abstract properties. One view is that the nonmonotonic reasoning systems should be safe in their use, so, for example, they should guarantee the existence of extensions. The alternative view would be that the user should be given more choices, which in some cases may lead to anomalous situations.

To illustrate this point, consider the default theory consisting of the default $\frac{true:p}{\neg p}$. According to the first view, we should be using a method such as Justified or Constrained Default Logic which would assign a meaning to the theory. The alternative view is that this theory is *anomalous*, and such anomalies should be detected using *verification and validation* techniques [13]; indeed, it is most likely that the 'anomalous default' is due to an entry error and not really desired by the user.

Nevertheless, we adopt the view that it is usually preferable to use safe nonmonotonic reasoning systems, and this due to the current trend of integrating information from different sources. In this process anomalous situations are very likely to arise even though the separate pieces of information seem to be natural.

In such cases it is difficult to determine the source of error, so we would prefer formal systems with reasonable behaviour.

References

1. G. Antoniou and V. Sperschneider. Operational Models of Nonmonotonic Logics. Part 1: Default Logic. *Artificial Intelligence Review* 8(1994): 3–16.
2. G. Antoniou. An operational interpretation of justified default logic. In *Proc. 8th Australian Joint Conference on Artificial Intelligence*, World Scientific 1995.
3. G. Antoniou and M.A. Williams. Reasoning with Incomplete and Changing Information: The CIN Project. In *Proc. 2nd Joint Conference on Information Sciences*, 1995.
4. G. Antoniou. Operational characterization of extensions in some logics for default reasoning. *Information Sciences* (forthcoming).
5. G. Brewka. *Nonmonotonic reasoning: logical foundations of commonsense*. Cambridge University Press 1991.
6. G. Brewka. Adding priorities and specificity to default logic. *Technical Report, GMD* 1993.
7. G. Brewka. Reasoning About Priorities in Default Logic. In *Proc. 12th National Conference on Artificial Intelligence (AAAI-94)*, MIT Press 1994.
8. J.P. Delgrande, T. Schaub and W.K. Jackson. Alternative approaches to default logic. *Artificial Intelligence* 70(1994): 167–237.
9. S. Kraus, D. Lehmann and M. Magidor. Nonmonotonic Reasoning, Preferential Models and Cumulative Logics. *Artificial Intelligence* 44 (1990): 167–207.
10. W. Lukaszewicz. Considerations on default logic: an alternative approach. *Computational Intelligence* 4,1(1988): 1–16.
11. D. Makinson. General patterns in nonmonotonic reasoning. In *Handbook of Logic in Artificial Intelligence and Logic Programming* Vol. 3, Oxford University Press 1994.
12. A. Mikitiuk and M. Truszczynski. Constrained and rational default logics. In *Proceedings International Joint Conference on Artificial Intelligence 1995*.
13. A.D. Preece and R. Shinghal. Foundation and Application of Knowledge Base Verification. *International Journal of Intelligent Systems* 9(1994): 683–701.
14. R. Reiter. A logic for default reasoning. *Artificial Intelligence* 13(1980): 81–132.

Modal Logics with Relative Accessibility Relations

Philippe BALBIANI

Laboratoire d'informatique de Paris-Nord, Institut Galilée, Université Paris-Nord,
Avenue Jean-Baptiste Clément, F-93430 Villetaneuse, email:
balbiani@ura1507.univ-paris13.fr

Abstract. This paper[1] presents a systematic study of the logics based
on Kripke models with relative accessibility relations as well as a general
method for proving their completeness.

1 Introduction

The modal logics with relative accessibility relations have been introduced by
Orlowska [10] in the context of the modal logics for similarity relations in knowledge representation systems. They naturally appear in dynamic logic [8] — where
the modal operators are parametrized by programs, epistemic logic for reasoning
about knowledge in a distributed environment [5] [6] — where the modal operators are parametrized by sets of epistemic agents, deontic logic for reasoning
about obligation and permission in a hierarchical universe [1] — where the modal
operators are parametrized by sets of deontic agents. Let PAR be a nonempty
set of parameters.

As for the epistemic logic with distributed knowledge [5] [6], for every subset P of PAR, the modal operator $[P]$ — "it is distributed knowledge among
the agents in P" — is added to the standard propositional formalism. A standard model for this language is a structure of the form (W, R, m) where W
is a nonempty set of "knowledge states" and R is a mapping on the set of
the subsets of PAR to the set of the equivalence relations on W such that
$R(P \cup Q) = R(P) \cap R(Q)$. The method presented in the section 4.2 of this paper
and in its extended version [2] can be applied for the proof of the completeness of
this logic. Let it be remarked that the proof of the completeness of the epistemic
logic with distributed knowledge presented by Fagin, Halpern and Vardi [6] —
who use the technics of the unravelling introduced by Sahlqvist — is extremely
complicated.

As for the deontic logic with distributed obligation and permission [1], for
every subset P, S of PAR, the modal operators $[P \mid S]$ — "some agent in P
makes it obligatory for every agent in S that" — and $[P \mid S]'$ — "some agent in
P makes it obligatory for some agent in S that" — are added to the standard
propositional formalism. A standard model for this language is a structure of
the form (W, R, R', m) where W is a nonempty set of "normative states", R is a

---

[1] The extended version of this paper [2] will appear under the title "Axiomatization
of logics based on Kripke models with relative accessibility relations" in the book
entitled "Reasoning with incomplete information" edited by Ewa Orlowska.

mapping on the set of the subsets of PAR to the set of the binary relations on W such that $R(P \cup Q, S) = R(P, S) \cap R(Q, S)$ and $R(P, S \cap T) = R(P, S) \cap R(P, T)$ and R' is a mapping on the set of the subsets of PAR to the set of the binary relations on W such that $R'(P \cup Q, S) = R'(P, S) \cap R'(Q, S)$ and $R'(P, S \cup T) = R'(P, S) \cap R'(P, T)$. The method presented in the section 4.2 of this paper and in its extended version [2] can be applied for the proof of the completeness of this logic. Let it be remarked that Bailhache [1] does not even conjecture the completeness of the deontic logic with distributed obligation and permission.

In this paper, the language of the modal logics with relative accessibility relations contains the modal operator $[P]$, for every cofinite subset P of PAR. A standard model for this language is a structure of the form (W, R, m) where W is a nonempty set of possible worlds and R is a mapping on the set of the cofinite subsets of PAR to the set of the binary relations on W such that $R(P \cup Q) = R(P) \cap R(Q)$ and $R(P \cap Q) = R(P) \cup R(Q)$, for every cofinite subset P, Q of PAR. The class of all standard models defines a set of valid formulas.

We prove that this set is finitely axiomatizable and that its finite axiomatization is the smallest normal set of formulas containing the schemata $[P]A \vee [Q]A \rightarrow [P \cup Q]A$ and $[P]A \wedge [Q]A \rightarrow [P \cap Q]A$, for every cofinite subset P, Q of PAR. Our completeness proof is in two steps. Firstly, we prove that our finite axiomatization is complete for the class of all models such that $R(P \cup Q) \subseteq R(P) \cap R(Q)$ and $R(P \cap Q) = R(P) \cup R(Q)$, for every cofinite subset P, Q of PAR. Secondly, we use copying to transform any model into a standard model satisfying the same formulas.

Our proof of the completeness of RK — the finite axiomatization of the set of the formulas valid in every model — for the class of all standard models shows evidence of the interest of our method for the study of the modal logics with relative accessibility relations. It solves an open problem brought up by Orlowska [10] in the context of the modal logics for knowledge representation systems.

2 Language

The linguistic basis of the modal logics with relative accessibility relations is the propositional calculus. Let VAR be the set of its *atomic formulas*. Let PAR be a nonempty set of *parameters*. Let $\mathcal{P}(PAR)^\circ$ be the set of the cofinite subsets of PAR. For every parameter $a \in PAR$, let $\underline{a} = PAR \setminus \{a\}$. For every cofinite subset P of PAR, let $\underline{P} = PAR \setminus P$. For every cofinite subset P of PAR, the modal operator $[P]$ is added to the standard propositional formalism.

3 Semantics

The semantics of the modal logics with relative accessibility relations comes from modal logic: a set W of possible worlds together with an interpretation of every cofinite subset P of PAR as a binary relation $R(P)$ on W. A satisfiability relation between possible worlds and formulas is defined such that the truth value of a formula at a possible world depends on the truth values of its subformulas at other possible worlds.

3.1 Satisfiability

A *frame* is a pair $\mathcal{F} = (W, R)$ where W is a nonempty set of *possible worlds* and R is a mapping on the set of the cofinite subsets of PAR to the set of the binary relations on W such that, for every cofinite subset P, Q of PAR:

- $R(P \cup Q) \subseteq R(P) \cap R(Q)$,
- $R(P \cap Q) \subseteq R(P) \cup R(Q)$.

Direct calculations would lead to the conclusion that:

Lemma 1 *Let $\mathcal{F} = (W, R)$ be a frame. Then, for every cofinite subset P, Q of PAR, $R(P \cap Q) = R(P) \cup R(Q)$.*

Lemma 2 *Let $\mathcal{F} = (W, R)$ be a frame. Then, for every cofinite subset P of PAR, if $P \neq PAR$ then $R(P) = \bigcup_{a \in \underline{P}} R(\underline{a})$.*

$\mathcal{F}$ is *standard* when, for every cofinite subset P, Q of PAR:

- $R(P \cup Q) = R(P) \cap R(Q)$.

A mapping m on the set of the possible worlds to the set of the subsets of VAR is called *assignment* on $\mathcal{F}$. A *(standard) model* is a pair $\mathcal{M} = (\mathcal{F}, m)$ where $\mathcal{F}$ is a (standard) frame and m is an assignment on $\mathcal{F}$. The *satisfiability relation* in $\mathcal{M}$ between a formula A and a possible world x is defined in the following way:

- $x \models_{\mathcal{M}} A$ iff $A \in m(x)$, A atomic formula,
- $x \models_{\mathcal{M}} \neg A$ iff $x \not\models_{\mathcal{M}} A$,
- $x \models_{\mathcal{M}} A \wedge B$ iff $x \models_{\mathcal{M}} A$ and $x \models_{\mathcal{M}} B$,
- $x \models_{\mathcal{M}} [P]A$ iff, for every possible world y, $x\, R(P)\, y$ only if $y \models_{\mathcal{M}} A$, for every cofinite subset P of PAR.

A formula is *valid* in a model when it is satisfied in every possible world of this model. A schema is *valid* in a frame when every instance of the schema is valid in every model of the frame.

3.2 Copying

Let $\mathcal{F} = (W, R)$ be a frame. Let $\mathcal{F}' = (W', R')$ be a frame and I be a set of mappings on W to W'. I is a *copying* from $\mathcal{F}$ into $\mathcal{F}'$ when:

- for every mapping $f, g \in I$ and for every possible world $x, y \in W$, $f(x) = g(y)$ only if $x = y$,
- for every possible world $x' \in W'$, there exists a mapping $f \in I$ and a possible world $x \in W$ such that $f(x) = x'$,

- for every mapping $f \in I$ and for every possible world $x, y \in W$, $x\ R(P)\ y$ only if there exists a mapping $g \in I$ such that $f(x)\ R'(P)\ g(y)$, for every cofinite subset P of PAR,
- for every mapping $f, g \in I$ and for every possible world $x, y \in W$, $f(x)\ R'(P)$ $g(y)$ only if $x\ R(P)\ y$, for every cofinite subset P of PAR.

Copying preserves the satisfiability of a formula according to the following lemma:

Lemma 3 *Let $\mathcal{F} = (W, R)$, $\mathcal{F}' = (W', R')$ be two frames and m be an assignment on $\mathcal{F}$. Let I be a copying from $\mathcal{F}$ into $\mathcal{F}'$ and m' be the assignment on $\mathcal{F}'$ defined in the following way:*

- *for every mapping $f \in I$ and for every possible world $x \in W$, $m'(f(x)) = m(x)$.*

Let $\mathcal{M} = (\mathcal{F}, m)$, $\mathcal{M}' = (\mathcal{F}', m')$. Then, for every formula A, for every mapping $f \in I$ and for every possible world $x \in W$, $x \models_{\mathcal{M}} A$ iff $f(x) \models_{\mathcal{M}'} A$.

Its proof can be done by induction on the complexity of A. The technics of the copying have been introduced by Vakarelov [11] in the context of the modal logics for knowledge representation systems. They are mainly used for proving that a modal logic which is characterized by a certain class of frames is also characterized by another class.

4 RK

We present the finite axiomatization RK of the set of the formulas valid in every model. We prove that this axiomatization is complete for the class of all standard models.

4.1 Axioms

Together with the classical tautologies, all the instances of the following schemata are axioms of RK, for every cofinite subset P, Q of PAR:

- $[P](A \to B) \to ([P]A \to [P]B)$,
- $[P]A \vee [Q]A \to [P \cup Q]A$,
- $[P]A \wedge [Q]A \to [P \cap Q]A$.

Together with the *modus ponens*, the following schema is a rule of RK, for every cofinite subset P of PAR:

- if $\vdash_{RK} A$ then $\vdash_{RK} [P]A$.

Direct calculations would lead to the conclusion that:

Theorem 1 *The theorems of RK are valid in every model.*

Let $\mathcal{M} = (W, R, m)$ be the canonical model of RK. Direct calculations would lead to the conclusion that $\mathcal{M}$ is a model. Therefore:

Theorem 2 *RK is complete for the class of all models, that is to say: the formulas valid in every model are theorems of RK.*

The technics of the canonical model have been introduced in the Lemmon notes[2]. They are mainly used for proving that a modal logic is complete for a certain class of models, that is to say: for proving that the formulas valid in every model of a certain class are theorems of a modal logic [9].

4.2 Completeness

This section presents the proof of the completeness of RK for the class of all standard models. Let $\mathcal{F} = (W, R)$ be a frame. We use copying to transform $\mathcal{F}$ into a standard frame $\mathcal{F}' = (W', R')$ validating the same schemata.

$\boldsymbol{W'}$ Let I be the set of the mappings on $\mathcal{P}(PAR)^{\circ} \times PAR$ to $\mathcal{P}(W)$. Let us notice that $\mathcal{B} = \mathcal{P}(W)$ is a boolean ring where:

- $0_{\mathcal{B}} = \emptyset$,
- $1_{\mathcal{B}} = W$,
- $A +_{\mathcal{B}} B = (A - B) \cup (B - A)$,
- $A \times_{\mathcal{B}} B = A \cap B$.

Let $W' = W \times I$.

$\boldsymbol{R'(PAR)}$ Let $R'(PAR)$ be the binary relation on W' defined by $(x, f)\, R'(PAR)$ (y, g) iff:

- $x\, R(PAR)\, y$,
- for every parameter $\alpha \in PAR$ and for every cofinite subset P of PAR, if $\alpha \in P$ then $f(P, \alpha) + g(P, \alpha) = \emptyset$.

$\boldsymbol{R'(\underline{a})}$ For every cofinite subset P of PAR, let $\pi(P)$ be a mapping on $W \times W$ to $\mathcal{P}(W)$ such that $\pi(P)(x, y) = \emptyset$ iff $x\, R(P)\, y$. An example of such a mapping is $\pi(P)(x, y) =$ if $x\, R(P)\, y$ then $\emptyset$ else W. For every parameter $a \in PAR$, let $R'(\underline{a})$ be the binary relation on W' defined by $(x, f)\, R'(\underline{a})\, (y, g)$ iff:

- $x\, R(\underline{a})\, y$,
- for every parameter $\alpha \in \underline{a}$ and for every cofinite subset P of PAR, if $\alpha \in P$ then $f(P, \alpha) + g(P, \alpha) = \emptyset$,
- for every cofinite subset P of PAR, if $a \in P$ then $f(P, a) + g(P, a) = \pi(P)(x, y)$.

Direct calculations would lead to the conclusion that:

Lemma 4 $R'(PAR) \subseteq R'(\underline{a})$, *for every parameter* $a \in PAR$.

---

[2] The "Lemmon Notes" have been edited by K. Segerberg: *An Introduction to Modal Logic.* Blackwell, 1977.

$R'(P)$ For every cofinite subset P of PAR, if $P \neq PAR$ then let $R'(P) = \bigcup_{a \in \underline{P}} R'(\underline{a})$.

"$\mathcal{F}' = (W', R')$ is a frame" Let it be proved that $R'(P \cup Q) \subseteq R'(P) \cap R'(Q)$, for every cofinite subset P, Q of PAR.

First case: $P \cup Q \neq PAR$. If $(x, f) \; R'(P \cup Q) \; (y, g)$ then there exists $c \in \underline{P \cup Q}$ such that $(x, f) \; R'(\underline{c}) \; (y, g)$. Therefore, there exists $a \in \underline{P}$ such that $\overline{(x, f)}$ $R'(\underline{a}) \; (y, g)$ and there exists $b \in \underline{Q}$ such that $(x, f) \; R'(\underline{b}) \; (y, g)$. Consequently, $(x, f) \; R'(P) \; (y, g)$ and $(x, f) \; R'\overline{(Q)} \; (y, g)$.
Second case: $P \cup Q = PAR$. Direct consequence of lemma 4.

Therefore:

Lemma 5 $R'(P \cup Q) \subseteq R'(P) \cap R'(Q)$, *for every cofinite subset* P, Q *of* PAR.

Let it be proved that $R'(P \cap Q) \subseteq R'(P) \cup R'(Q)$, for every cofinite subset P, Q of PAR.

First case: $P \cap Q \neq PAR$. If $(x, f) \; R'(P \cap Q) \; (y, g)$ then there exists $c \in \underline{P \cap Q}$ such that $(x, f) \; R'(\underline{c}) \; (y, g)$. Therefore, there exists $a \in \underline{P}$ such that $\overline{(x, f)}$ $R'(\underline{a}) \; (y, g)$ or there exists $b \in \underline{Q}$ such that $(x, f) \; R'(\underline{b}) \; (y, g)$. Consequently, $(x, f) \; R'(P) \; (y, g)$ or $(x, f) \; R'\overline{(Q)} \; (y, g)$.
Second case: $P \cap Q = PAR$. Direct calculations.

Therefore:

Lemma 6 $R'(P \cap Q) \subseteq R'(P) \cup R'(Q)$, *for every cofinite subset* P, Q *of* PAR.

"I is a copying from $\mathcal{F}$ into $\mathcal{F}'$" Let it be proved that, for every mapping $f \in I$ and for every possible world $x, y \in W$, $x \; R(PAR) \; y$ only if there exists a mapping $g \in I$ such that $(x, f) \; R'(PAR) \; (y, g)$. If $x \; R(PAR) \; y$ then let g be a mapping on $\mathcal{P}(PAR)^\circ \times PAR$ to $\mathcal{P}(W)$ such that:

- for every parameter $\alpha \in PAR$ and for every cofinite subset P of PAR, if $\alpha \in P$ then $g(P, \alpha) = f(P, \alpha)$.

Direct calculations would lead to the conclusion that $(x, f) \; R'(PAR) \; (y, g)$. Consequently:

Lemma 7 *For every mapping* $f \in I$ *and for every possible world* $x, y \in W$, x $R(PAR) \; y$ *only if there exists a mapping* $g \in I$ *such that* $(x, f) \; R'(PAR) \; (y, g)$.

Let it be proved that, for every parameter $a \in PAR$, for every mapping $f \in I$ and for every possible world $x, y \in W$, $x \; R(\underline{a}) \; y$ only if there exists a mapping $g \in I$ such that $(x, f) \; R'(\underline{a}) \; (y, g)$. If $x \; R(\underline{a}) \; y$ then let g be a mapping on $\mathcal{P}(PAR)^\circ \times PAR$ to $\mathcal{P}(W)$ such that:

- for every parameter $\alpha \in \underline{a}$ and for every cofinite subset P of PAR, if $\alpha \in P$ then $g(P, \alpha) = f(P, \alpha)$,

- for every cofinite subset P of PAR, if $a \in P$ then $g(P,a) = f(P,a) + \pi(P)(x,y)$.

Direct calculations would lead to the conclusion that (x,f) $R'(\underline{a})$ (y,g). Consequently:

Lemma 8 *For every parameter $a \in PAR$, for every mapping $f \in I$ and for every possible world $x,y \in W$, x $R(\underline{a})$ y only if there exists a mapping $g \in I$ such that (x,f) $R'(\underline{a})$ (y,g).*

Let it be proved that, for every cofinite subset P of PAR, if $P \neq PAR$ then, for every mapping $f \in I$ and for every possible world $x,y \in W$, x $R(P)$ y only if there exists a mapping $g \in I$ such that (x,f) $R'(P)$ (y,g). If x $R(P)$ y then, according to lemma 2, there exists $a \in \underline{P}$ such that x $R(\underline{a})$ y. Therefore, according to lemma 8, there exists a mapping $g \in I$ such that (x,f) $R'(\underline{a})$ (y,g). Consequently, there exists a mapping $g \in I$ such that (x,f) $R'(P)$ (y,g). Therefore:

Lemma 9 *For every cofinite subset P of PAR, if $P \neq PAR$ then, for every mapping $f \in I$ and for every possible world $x,y \in W$, x $R(P)$ y only if there exists a mapping $g \in I$ such that (x,f) $R'(P)$ (y,g).*

Every mapping $f \in I$ can equally be considered as the mapping on W to W' defined in the following way:

- $f(x) = (x,f)$.

Consequently:

Lemma 10 *I is a copying from $\mathcal{F}$ into $\mathcal{F}'$.*

"$\mathcal{F}'$ is standard" Let it be proved that, for every parameter $a,b \in PAR$ such that $a \neq b$, $R'(\underline{a}) \cap R'(\underline{b}) \subseteq R'(PAR)$. If (x,f) $R'(\underline{a})$ (y,g) and (x,f) $R'(\underline{b})$ (y,g) then, for every parameter $\alpha \in PAR$, either $\alpha \in \underline{a}$ or $\alpha \in \underline{b}$ and, for every cofinite subset P of PAR, if $\alpha \in P$ then $f(P,\alpha) + g(P,\alpha) = \emptyset$. Moreover, $f(PAR,a) + g(PAR,a) = \pi(PAR)(x,y)$ and $f(PAR,b) + g(PAR,b) = \pi(PAR)(x,y)$. Therefore, $\pi(PAR)(x,y) = \emptyset$ and x $R(PAR)$ y. Consequently, (x,f) $R'(PAR)$ (y,g). Therefore:

Lemma 11 *For every parameter $a,b \in PAR$ such that $a \neq b$, $R'(\underline{a}) \cap R'(\underline{b}) \subseteq R'(PAR)$.*

Let it be proved that $R'(P \cup Q) = R'(P) \cap R'(Q)$, for every cofinite subset P,Q of PAR. If $P \neq PAR$ and $Q \neq PAR$ then (x,f) $R'(P)$ (y,g) and (x,f) $R'(Q)$ (y,g) only if there exists $a \in \underline{P}$ such that (x,f) $R'(\underline{a})$ (y,g) and there exists $b \in \underline{Q}$ such that (x,f) $R'(\underline{b})$ (y,g).

First case: $a = b$. Then, there exists $c \in \underline{P \cup Q}$ such that (x,f) $R'(\underline{c})$ (y,g). Consequently, (x,f) $R'(P \cup Q)$ (y,g).

Second case: $a \neq b$. Then, according to lemma 11, $(x, f)\ R'(PAR)\ (y, g)$ and, according to lemma 4, $(x, f)\ R'(P \cup Q)\ (y, g)$.

If $P = PAR$ or $Q = PAR$ then $P \cup Q = PAR$. Therefore:

Lemma 12 $R'(P \cup Q) = R'(P) \cap R'(Q)$, *for every cofinite subset* P, Q *of* PAR.

Consequently:

Lemma 13 *For every frame* $\mathcal{F}$, *there exists a standard frame* $\mathcal{F}'$ *and a copying* I *from* $\mathcal{F}$ *into* $\mathcal{F}'$.

Completeness of RK Let A be a consistent formula of RK. According to theorem 2, A is satisfied in a possible world of some model. According to lemmas 3 and 13, A is satisfied in a possible world of some standard model. Therefore:

Theorem 3 RK *is complete for the class of all standard models, that is to say: the formulas valid in every standard model are theorems of* RK.

Let us remark that, for every cofinite subset P of PAR, the only condition on $\pi(P)$ is $\pi(P)(x, y) = \emptyset$ iff $x\ R(P)\ y$.

5 Conclusion

Our proof of the completeness of RK — the finite axiomatization of the set of the formulas valid in every model — for the class of all standard models constitutes our main result. It uses copying to transform a frame into a standard frame validating the same schemata.

Its extension to the proof of the completeness of RS — the finite axiomatization of the set of the formulas valid in every model of similarity where $R(P)$ is reflexive and symmetrical, for every cofinite subset P of PAR — for the class of all standard models of similarity[3] shows evidence of our method for the study of the modal logics with relative accessibility relations.

Such logics have been introduced by Fariñas del Cerro and Orlowska [7] [10] in the context of the modal logics for the systems of the representation of knowledge. Axiomatization of the logic of data analysis [7] is even then an open problem [10] [4]. We believe that our method can be applied so that the proof of the completeness of other modal logics with relative accessibility relations is overcome:

- Replace the condition $R(P \cap Q) = R(P) \cup R(Q)$ of the modal logics with relative accessibility relations by the condition $R(P \cap Q) = R(P) \cup^* R(Q)$.
- Let PAR be a nonempty set of parameters. The linguistic basis of the modal logics with growing accessibility relations contains the modal operator $[P]$, for every finite subset P of PAR. A standard model for this language is

---

[3] See annex A.

a structure of the form (W, R, m) where W is a nonempty set of possible worlds and R is a mapping on the set of the finite subsets of PAR to the set of the binary relations on W such that $R(P \cup Q) = R(P) \cup R(Q)$ and $R(P \cap Q) = R(P) \cap R(Q)$, for every finite subset P, Q of PAR. The class of all standard models defines a set of valid formulas. We prove in annex B that this set is finitely axiomatizable and that its finite axiomatization is the smallest normal set of formulas containing the schemata $[P]A \wedge [Q]A \to [P \cup Q]A$ and $[P]A \vee [Q]A \to [P \cap Q]A$, for every finite subset P, Q of PAR.
- Replace the condition $R(P \cup Q) = R(P) \cup R(Q)$ of the modal logics with growing accessibility relations by the condition $R(P \cup Q) = R(P) \cup^* R(Q)$.

Acknowledgement

The author wishes to express his thanks and his gratitude to Ewa Orlowska, Dimiter Vakarelov as well as to his former colleagues of the *Institut de recherche en informatique de Toulouse* for their valuable comments about the preliminary draft of this paper.

References

1. **P. Bailhache**. *Essai de logique déontique*. Vrin, 1991.
2. **P. Balbiani**. *Axiomatization of logics based on Kripke models with relative accessibility relations*. E. Orlowska (editor), Reasoning with Incomplete Information. To appear.
3. **B. Chellas**. *Modal Logic: an Introduction*. Cambridge University Press, 1980.
4. **S. Demri** and **E. Orlowska**. *Logical analysis of indiscernability*. E. Orlowska (editor), Reasoning with Incomplete Information. To appear.
5. **R. Fagin, J. Halpern, Y. Moses** and **M. Vardi**. *Reasoning About Knowledge*. MIT Press, 1995.
6. **R. Fagin, J. Halpern** and **M. Vardi**. *What can machines know ? On the properties of knowledge in distributed systems*. Journal of the ACM, Volume 39, 328-376, 1992.
7. **L. Fariñas del Cerro** and **E. Orlowska**. *DAL - a logic for data analysis*. Theoretical Computer Science, Volume 36, 251-264, 1985.
8. **D. Harel**. *Dynamic logic*. D. Gabbay and F. Guenthner (editors), Handbook of Philosophical Logic, Volume 2, Extensions of Classical Logic. 497-604, Reidel, 1984.
9. **G. Hughes** and **M. Cresswell**. *A Companion to Modal Logic*. Methuen, 1984.
10. **E. Orlowska**. *Kripke semantics for knowledge representation logics*. Studia Logica, Volume 49, 255-272, 1990.
11. **D. Vakarelov**. *Modal logics for knowledge representation systems*. Theoretical Computer Science, Volume 90, 433-456, 1991.

6 Annex A

This annex presents the extension of the results of section 4 to the following classes of frames:

- serial frames,
- reflexive frames,
- symmetrical frames,
- frames of similarity.

6.1 Seriality

Semantics Let $\mathcal{F} = (W, R)$ be a frame. $\mathcal{F}$ is *serial* if $R(P)$ is serial, for every cofinite subset P of PAR.

Axioms Together with the axioms of RK, all the instances of the following schema are axioms of RQ, for every cofinite subset P of PAR:

- $[P]A \to\; < P > A$.

Direct calculations would lead to the conclusion that the canonical model of RQ is a serial model.

Completeness Let $\mathcal{F} = (W, R)$ be a serial frame. Let $\mathcal{F}'$ be the structure defined as in section 4.2. Direct calculations would lead to the conclusion that $\mathcal{F}'$ is a serial frame. Therefore:

Theorem 4 *RQ is complete for the class of all standard serial models.*

6.2 Reflexivity

Semantics Let $\mathcal{F} = (W, R)$ be a frame. $\mathcal{F}$ is *reflexive* if $R(P)$ is reflexive, for every cofinite subset P of PAR.

Axioms Together with the axioms of RK, all the instances of the following schema are axioms of RT, for every cofinite subset P of PAR:

- $[P]A \to A$.

Direct calculations would lead to the conclusion that the canonical model of RT is a reflexive model.

Completeness Let $\mathcal{F} = (W, R)$ be a reflexive frame. Let $\mathcal{F}'$ be the structure defined as in section 4.2. Direct calculations would lead to the conclusion that $\mathcal{F}'$ is a reflexive frame. Therefore:

Theorem 5 *RT is complete for the class of all standard reflexive models.*

6.3 Symmetry

Semantics Let $\mathcal{F} = (W, R)$ be a frame. $\mathcal{F}$ is *symmetrical* if $R(P)$ is symmetrical, for every cofinite subset P of PAR.

Axioms Together with the axioms of RK, all the instances of the following schema are axioms of RB, for every cofinite subset P of PAR:

- $A \rightarrow [P] < P > A$.

Direct calculations would lead to the conclusion that the canonical model of RB is a symmetrical model.

Completeness Let $\mathcal{F} = (W, R)$ be a symmetrical frame. For every cofinite subset P of PAR, let $\pi(P)$ be a mapping on $W \times W$ to $\mathcal{P}(W)$ such that $\pi(P)(x, y) = \emptyset$ iff x $R(P)$ y, $\pi(P)(x, y) = \pi(P)(y, x)$. An example of such a mapping is $\pi(P)(x, y) = $ if x $R(P)$ y then $\emptyset$ else W. Let $\mathcal{F}'$ be the structure defined as in section 4.2. Direct calculations would lead to the conclusion that $\mathcal{F}'$ is a symmetrical frame. Therefore:

Theorem 6 *RB is complete for the class of all standard symmetrical models.*

6.4 Similarity

Semantics Let $\mathcal{F} = (W, R)$ be a frame. $\mathcal{F}$ is a frame of *similarity* if $R(P)$ is reflexive and symmetrical, for every cofinite subset P of PAR.

Axioms Together with the axioms of RK, all the instances of the following schema are axioms of RS, for every cofinite subset P of PAR:

- $[P]A \rightarrow A$,
- $A \rightarrow [P] < P > A$.

Direct calculations would lead to the conclusion that the canonical model of RS is a model of similarity.

Completeness Let $\mathcal{F} = (W, R)$ be a frame of similarity. For every cofinite subset P of PAR, let $\pi(P)$ be a mapping on $W \times W$ to $\mathcal{P}(W)$ such that $\pi(P)(x, y) = \emptyset$ iff x $R(P)$ y, $\pi(P)(x, y) = \pi(P)(y, x)$. An example of such a mapping is $\pi(P)(x, y) = $ if x $R(P)$ y then $\emptyset$ else W. Let $\mathcal{F}'$ be the structure defined as in section 4.2. Direct calculations would lead to the conclusion that $\mathcal{F}'$ is a frame of similarity. Therefore:

Theorem 7 *RS is complete for the class of all standard models of similarity.*

6.5 Indiscernability

Semantics Let $\mathcal{F} = (W, R)$ be a frame. $\mathcal{F}$ is a frame of *indiscernability* if $R(P)$ is reflexive, symmetrical and transitive, for every cofinite subset P of PAR. Direct calculations would lead to the conclusion that:

Lemma 14 *Let* $\mathcal{F} = (W, R)$ *be a frame of indiscernability. Then, for every possible world* $x \in W$, *either* $R(P)(x) \subseteq R(Q)(x)$ *or* $R(Q)(x) \subseteq R(P)(x)$, *for every cofinite subset* P, Q *of* PAR.

Consequently:

Lemma 15 *Let* $\mathcal{F} = (W, R)$ *be a frame of indiscernability. Then,* $\mathcal{F}$ *is standard iff, for every possible world* $x \in W$, *either* $R(P)(x) \subseteq R(P \cup Q)(x)$ *or* $R(Q)(x) \subseteq R(P \cup Q)(x)$, *for every cofinite subset* P, Q *of* PAR.

Therefore:

Lemma 16 *Let* $\mathcal{F} = (W, R)$ *be a frame of indiscernability. Then,* $\mathcal{F}$ *is standard iff the schema* $[P \cup Q]A \to [P]A \vee [Q]A$ *is valid in* $\mathcal{F}$.

Axioms Together with the axioms of RK, all the instances of the following schema are axioms of RI, for every cofinite subset P, Q of PAR:

- $[P]A \to A$,
- $A \to [P] < P > A$,
- $[P]A \to [P][P]A$,
- $[P \cup Q]A \to [P]A \vee [Q]A$.

Direct calculations would lead to the conclusion that the canonical model of RI is a standard model of indiscernability.

Completeness Therefore:

Theorem 8 *RI is complete for the class of all standard models of indiscernability.*

7 Annex *B*

This annex presents the modal logics with growing accessibility relations.

7.1 Language

Let VAR be a nonempty set of *atomic formulas*. Let PAR be a nonempty set of *parameters*. Let $\mathcal{P}(PAR)^\bullet$ be the set of the finite subsets of PAR. For every finite subset P of PAR, the modal operator $[P]$ is added to the standard propositional formalism.

7.2 Semantics

A *growing frame* is a pair $\mathcal{F} = (W, R)$ where W is a nonempty set of *possible worlds* and R is a mapping on the set of the finite subsets of PAR to the set of the binary relations on W such that, for every finite subset P, Q of PAR:

- $R(P \cup Q) \subseteq R(P) \cup R(Q)$,
- $R(P \cap Q) \subseteq R(P) \cap R(Q)$.

Direct calculations would lead to the conclusion that:

Lemma 17 *Let $\mathcal{F} = (W, R)$ be a growing frame. Then, for every finite subset P, Q of PAR, $R(P \cup Q) = R(P) \cup R(Q)$.*

Lemma 18 *Let $\mathcal{F} = (W, R)$ be a growing frame. Then, for every finite subset P of PAR, if $P \neq \emptyset$ then $R(P) = \bigcup_{a \in P} R(\{a\})$.*

$\mathcal{F}$ is *standard* when, for every finite subset P, Q of PAR:

- $R(P \cap Q) = R(P) \cap R(Q)$.

7.3 GK

We present the finite axiomatization GK of the set of the formulas valid in every growing model. We prove that this axiomatization is complete for the class of all standard growing models.

Axioms Together with the classical tautologies, all the instances of the following schemata are axioms of GK, for every finite subset P, Q of PAR:

- $[P](A \rightarrow B) \rightarrow ([P]A \rightarrow [P]B)$,
- $[P]A \wedge [Q]A \rightarrow [P \cup Q]A$,
- $[P]A \vee [Q]A \rightarrow [P \cap Q]A$.

Together with the *modus ponens*, the following schema is a rule of GK, for every finite subset P of PAR:

- if $\vdash_{GK} A$ then $\vdash_{GK} [P]A$.

Direct calculations would lead to the conclusion that:

Theorem 9 *The theorems of GK are valid in every growing model.*

Let $\mathcal{M} = (W, R, m)$ be the canonical model of GK. Direct calculations would lead to the conclusion that $\mathcal{M}$ is a growing model. Therefore:

Theorem 10 *GK is complete for the class of all growing models.*

Completeness This section presents the proof of the completeness of GK for the class of all standard growing models. Let $\mathcal{F} = (W, R)$ be a growing frame. We use copying to transform $\mathcal{F}$ into a standard growing frame $\mathcal{F}' = (W', R')$ validating the same schemata. Let I be the set of the mappings on $\mathcal{P}(PAR)^\bullet \times PAR$ to $\mathcal{P}(W)$. Let $W' = W \times I$. Let us remark that if PAR is finite then W is finite only if W' is finite. Let $R'(\emptyset)$ be the binary relation on W' defined by $(x, f)\ R'(\emptyset)\ (y, g)$ iff:

- $x\ R(\emptyset)\ y$,
- for every parameter $\alpha \in PAR$ and for every finite subset P of PAR, if $\alpha \notin P$ then $f(P, \alpha) + g(P, \alpha) = \emptyset$.

For every finite subset P of PAR, let $\pi(P)$ be a mapping on $W \times W$ to $\mathcal{P}(W)$ such that $\pi(P)(x, y) = \emptyset$ iff $x\ R(P)\ y$. An example of such a mapping is $\pi(P)(x, y) =$ if $x\ R(P)\ y$ then $\emptyset$ else W. For every parameter $a \in PAR$, let $R'(\{a\})$ be the binary relation on W' defined by $(x, f)\ R'(\{a\})\ (y, g)$ iff:

- $x\ R(\{a\})\ y$,
- for every parameter $\alpha \in \underline{a}$ and for every finite subset P of PAR, if $\alpha \notin P$ then $f(P, \alpha) + g(P, \alpha) = \emptyset$,
- for every finite subset P of PAR, if $a \notin P$ then $f(P, a) + g(P, a) = \pi(P)(x, y)$.

For every finite subset P of PAR, if $P \neq \emptyset$ then let $R'(P) = \bigcup_{a \in P} R'(\{a\})$. Direct calculations would lead to the conclusion that: $\mathcal{F}' = (W', R')$ is a growing frame, I is a copying from $\mathcal{F}$ into $\mathcal{F}'$ and $\mathcal{F}'$ is standard. Therefore:

Theorem 11 *GK is complete for the class of all standard growing models.*

Geometrical Structures and Modal Logic

Philippe BALBIANI, Luis FARIÑAS DEL CERRO,
Tinko TINCHEV and Dimiter VAKARELOV

[1] Laboratoire d'informatique de Paris-Nord, Institut Galilée, Université Paris-Nord, Avenue Jean-Baptiste Clément, F-93430 Villetaneuse, email: balbiani@ura1507.univ-paris13.fr
[2] Institut de recherche en informatique de Toulouse, Université Paul Sabatier, 118 route de Narbonne, F-31062 Toulouse Cedex, email: farinas@irit.fr
[3] Department of Mathematical Logic with Laboratory for Applied Logic, Faculty of Mathematics and Informatics, Sofia University, Boul. James Bouchier 5, 1126 Sofia, Bulgaria, email: dvak@ipx.fmi.uni-sofia.bg

Abstract. Although, in natural language, space modalities are used as frequently as time modalities, the logic of time is a well-established branch of modal logic whereas the same cannot be said of the logic of space. The reason is probably in the more simple mathematical structure of time: a set of moments together with a relation of precedence. Such a relational structure is suited to a modal treatment. The structure of space is more complex: several sorts of geometrical beings as points and lines together with binary relations as incidence or orthogonality, or only one sort of geometrical beings as points but ternary relations as collinearity or betweeness. In this paper, we define a general framework for axiomatizing modal logics which Kripke semantics is based on geometrical structures: structures of collinearity, projective structures, orthogonal structures.

1 Introduction

Although, in natural language, space modalities are used as frequently as time modalities, the logic of time is a well-established branch of modal logic [5] whereas the same cannot be said of the logic of space. The reason is probably in the more simple mathematical structure of time: a set of moments together with a relation of precedence. Such a relational structure is suited to a modal treatment. The structure of space is more complex: several sorts of geometrical beings as points and lines together with binary relations as incidence [7] [8] or orthogonality [6], or only one sort of geometrical beings as points but ternary relations as collinearity or betweeness [11]. Such relational structures are not suited to a modal treatment.

Structures of collinearity, first example, consist of a set of points together with a ternary relation of collinearity between points. They cannot constitute the standard semantics of the modal logic of collinearity. To overcome this problem, a frame of collinearity is associated to every structure of collinearity in the following way. Let $\underline{S} = (P, C)$ be a structure of collinearity, with P the set of points and C the ternary relation of collinearity between points. Let W be the graph of the relation C. For every $i, j \in \{1, 2, 3\}$, let $\equiv_{ij}$ be the binary relation on

W defined by $(X_1, X_2, X_3) \equiv_{ij} (Y_1, Y_2, Y_3)$ iff $X_i = Y_j$. Intuitively, each element (X, Y, Z) of W can be considered either as X, Y or Z. Let $u = (X_1, X_2, X_3)$, $v = (Y_1, Y_2, Y_3)$ and $w = (Z_1, Z_2, Z_3)$. The expression $(\exists x \in W)(u \equiv_{i1} x \wedge v \equiv_{j2} x \wedge w \equiv_{k3} x)$ is equivalent to $C(X_i, Y_j, Z_k)$. Therefore, $W(\underline{S}) = (W, \equiv_{ij})$ contains in some sense the whole information of $\underline{S}$. $W(\underline{S})$ is the frame of collinearity over $\underline{S}$. It satisfies the properties detailed in section 4.2. A frame of collinearity is a relational structure of the form $(W, \equiv_{ij})$ that satisfies the same properties. It can be proved that any frame of collinearity is isomorphic to a frame of collinearity over some structure of collinearity and that the set of all structures of collinearity is categorically equivalent to the set of all frames of collinearity. Therefore, frames of collinearity can constitute the standard semantics of the modal logic of collinearity.

Projective structures, second example, consist of a set of points and a set of lines together with a binary relation of incidence between points and lines. They cannot constitute the standard semantics of the modal logic of projective geometry. To overcome this problem, a projective frame is associated to every projective structure in the following way. Let $\underline{S} = (P, L, in)$ be a projective structure, with P the set of points, L the set of lines and in the binary relation of incidence between points and lines. Let W be the graph of the relation in. Let $\equiv_{11}$ and $\equiv_{22}$ be the binary relations on W defined by $(X, x) \equiv_{11} (Y, y)$ iff $X = Y$ and $(X, x) \equiv_{22} (Y, y)$ iff $x = y$. Intuitively, each element (X, x) of W can be considered either as X or x. Let $u = (X, x)$ and $v = (Y, y)$. The expression $u \equiv_{11} \circ \equiv_{22} v$ is equivalent to X in y. Therefore, $W(\underline{S}) = (W, \equiv_{11}, \equiv_{22})$ contains in some sense the whole information of $\underline{S}$. $W(\underline{S})$ is the projective frame over $\underline{S}$. It satisfies the properties detailed in section 7.1.2. A projective frame is a relational structure of the form $(W, \equiv_{11}, \equiv_{22})$ that satisfies the same properties. It can be proved that any projective frame is isomorphic to a projective frame over some projective structure and that the set of all projective structures is categorically equivalent to the set of all projective frames. Therefore, projective frames can constitute the standard semantics of the modal logic of projective geometry.

Section 2 introduces point n-frames and n-arrow frames and gives the proof of their categorial equivalence. Section 3 extends Vakarelov's basic arrow logic [12] [13]. Its standard semantics is the set of all n-arrow frames. Section 4 describes an example of point 3-frames: the structures of collinearity, proves its categorial equivalence with the associated example of 3-arrow frames: frames of collinearity, and identifies the modal logic with standard semantics in the set of all frames of collinearity. Sections 5 and 6 extend the results of sections 2 and 3 to sorted point n-frames and sorted n-arrow frames. Section 7.1 describes an example of sorted point 2-frames: the projective structures, proves its categorial equivalence with the associated example of sorted 2-arrow frames: projective frames, and identifies the modal logic with standard semantics in the set of all projective frames. Section 7.2 describes an example of sorted point 3-frames: the orthogonal structures, proves its categorial equivalence with the associated example of sorted 3-arrow frames: orthogonal frames, and identifies the modal logic with standard semantics in the set of all orthogonal frames.

2 Point n-frames and n-arrow frames

Let $n \geq 2$ and $(n) = \{1, \ldots, n\}$. This section is devoted to the proof of the categorial equivalence between point n-frames and n-arrow frames.

2.1 Point n-frames

A *point n-frame* consists of a non-empty set S together with a n-ary relation R on S such that, for every $X \in S$, there exists $X_1, \ldots, X_n \in S$ and there exists $i \in (n)$ such that $R(X_1, \ldots, X_n)$ and $X_i = X$. The class of all point n-frames is denoted by Σ_n and is considered as a category with morphisms the usual homomorphisms between relational structures. Namely, let $\underline{S} = (S, R)$ and $\underline{S}' = (S', R')$ be point n-frames. Then f is called a *homomorphism* from $\underline{S}$ into $\underline{S}'$ if, for every $X_1, \ldots, X_n \in S$, $R(X_1, \ldots, X_n)$ only if $R'(f(X_1), \ldots, f(X_n))$. A one-to-one f is called an *isomorphism* from $\underline{S}$ into $\underline{S}'$ if, for every $X_1, \ldots, X_n \in S$, $R'(f(X_1), \ldots, f(X_n))$ only if $R(X_1, \ldots, X_n)$.

2.2 n-arrow frames

A *n-arrow frame* consists of a non-empty set W of *tips* together with n^2 binary relations $\equiv_{ij}$ on W such that, for every $i, j, k \in (n)$:

- for every $u \in W$, $u \equiv_{ii} u$,
- for every $u, v \in W$, $u \equiv_{ij} v$ only if $v \equiv_{ji} u$,
- for every $u, v, w \in W$, $u \equiv_{ij} v$ and $v \equiv_{jk} w$ only if $u \equiv_{ik} w$.

The n-arrow frame $\underline{W} = (W, \equiv_{ij})$ is *normal* if:

- for every $u, v \in W$, $u \equiv_{ii} v$, for every $i \in (n)$, only if $u = v$.

The class of all normal n-arrow frames is denoted by Φ_n and is considered as a category with morphisms the usual homomorphisms between relational structures.

2.3 From point n-frames to n-arrow frames

Let $\underline{S} = (S, R)$ be a point n-frame. Let $W = \{(X_1, \ldots, X_n): X_1, \ldots, X_n \in S$ and $R(X_1, \ldots, X_n)\}$. For every $i, j \in (n)$, let $\equiv_{ij}$ be the binary relation on W defined by $(X_1, \ldots, X_n) \equiv_{ij} (Y_1, \ldots, Y_n)$ iff $X_i = Y_j$.

Lemma 1 $W(\underline{S}) = (W, \equiv_{ij})$ *is a normal n-arrow frame.*

Example 1 *Suppose $n = 3$. It can be proved that if, for every $X, Y \in S$, $R(X, Y, X)$ then, for every $i, j \in (3)$ and for every $x, y \in W$, there exists $u \in W$ such that $x \equiv_{i1} u$, $y \equiv_{j2} u$ and $x \equiv_{i3} u$.*

2.4 From n-arrow frames to point n-frames

Let $\underline{W} = (W, \equiv_{ij})$ be a normal n-arrow frame. For every $\alpha_1, \ldots, \alpha_n \in \mathcal{P}(W)$, if:

  - for every $i, j \in (n)$ and for every $u, v \in W$, $u \in \alpha_i$ and $v \in \alpha_j$ only if $u \equiv_{ij} v$,
  - for every $i, j \in (n)$ and for every $u, v \in W$, $u \in \alpha_i$ and $u \equiv_{ij} v$ only if $v \in \alpha_j$,
  - $\alpha_1 \cup \ldots \cup \alpha_n \neq \emptyset$

then $(\alpha_1, \ldots, \alpha_n)$ is a *generalized point* of $\underline{W}$. For every $u \in W$ and for every $i \in (n)$, let $i(u) = (\equiv_{i1}(u), \ldots, \equiv_{in}(u))$. Direct calculations would lead to the conclusion that:

Lemma 2 *For every $u, v \in W$ and for every $i, j \in (n)$, $i(u) = j(v)$ iff $u \equiv_{ij} v$.*

Lemma 3 *For every generalized point $(\alpha_1, \ldots, \alpha_n)$ of $\underline{W}$, there exists $u \in W$ and $i \in (n)$ such that $i(u) = (\alpha_1, \ldots, \alpha_n)$.*

Let S be the set of the generalized points of $\underline{W}$. Let R be the n-ary relation on S defined by $R(X_1, \ldots, X_n)$ iff there exists $u \in W$ such that, for every $i \in (n)$, $X_i = i(u)$.

Lemma 4 $S(\underline{W}) = (S, R)$ *is a point n-frame.*

Example 2 *Suppose $n = 3$. It can be proved that if, for every $i, j \in (3)$ and for every $x, y \in W$, there exists $u \in W$ such that $x \equiv_{i1} u$, $y \equiv_{j2} u$ and $x \equiv_{i3} u$ then, for every $X, Y \in S$, $R(X, Y, X)$.*

2.5 Representation theorems

Let $\underline{W} = (W, \equiv_{ij})$ be a normal n-arrow frame and $\underline{W}' = W(S(\underline{W}))$. For every $u \in W$, let $g(u) = (1(u), \ldots, n(u))$. Direct calculations would lead to the conclusion that g is an isomorphism from $\underline{W}$ into $\underline{W}'$. Therefore:

Lemma 5 $\underline{W}$ *and* $\underline{W}'$ *are isomorphic.*

Let $\underline{S} = (S, R)$ be a point n-frame and $\underline{S}' = S(W(\underline{S}))$. For every $X \in S$, let $f(X) = (1(X), \ldots, n(X))$ where $i(X) = \{(X_1, \ldots, X_n): X_1, \ldots, X_n \in S, R(X_1, \ldots, X_n) \text{ and } X_i = X\}$. Direct calculations would lead to the conclusion that f is an isomorphism from $\underline{S}$ into $\underline{S}'$. Consequently:

Lemma 6 $\underline{S}$ *and* $\underline{S}'$ *are isomorphic.*

Direct calculations would lead to the conclusion that the mapping $S \colon \underline{W} \to S(\underline{W})$ is a functor from Φ_n into Σ_n and that the mapping $W \colon \underline{S} \to W(\underline{S})$ is a functor from Σ_n into Φ_n. Therefore:

Theorem 1 *The categories Σ_n and Φ_n are equivalent.*

3 Basic arrow logic

Let $n \geq 2$. This section introduces a modal logic with standard semantics in the class of all normal n-arrow frames.

3.1 Language

The linguistic basis of basic arrow logic is the propositional calculus. Let VAR be the set of its *atomic formulas*. For every $i,j \in (n)$, the modal operator $[\equiv_{ij}]$ is added to the standard propositional formalism and, for every $i \in (n)$, the modal operator $[\not\equiv_{ii}]$ is added to the standard propositional formalism. Let $[\neq]A = [\not\equiv_{11}]A \wedge \ldots \wedge [\not\equiv_{nn}]A$ and $[U]A = A \wedge [\neq]A$.

3.2 Semantics

A *general n-arrow frame* consists of a non-empty set W together with, for every $i,j \in (n)$, a binary relation $\equiv_{ij}$ on W such that $(W, \equiv_{ij})$ is a n-arrow frame and, for every $i \in (n)$, a binary relation $\not\equiv_{ii}$ on W. A general n-arrow frame $\underline{W} = (W, \equiv_{ij}, \not\equiv_{ii})$ is $\neq$-*standard* if, for every $u, v \in W$, $u \neq v$ iff there exists $i \in (n)$ such that $u \not\equiv_{ii} v$. A $\neq$-standard frame $\underline{W} = (W, \equiv_{ij}, \not\equiv_{ii})$ is *quasi-standard* if, for every $i \in (n)$, $\not\equiv_{ii}$ is the complement of $\equiv_{ii}$.

Lemma 7 *If the general n-arrow frame* $\underline{W} = (W, \equiv_{ij}, \not\equiv_{ii})$ *is quasi-standard then the n-arrow frame* $(W, \equiv_{ij})$ *is normal.*

Let $\underline{W} = (W, \equiv_{ij}, \not\equiv_{ii})$ be a general n-arrow frame. A *valuation* on $\underline{W}$ is a mapping which assigns a subset of W to every atomic formula. A *($\neq$-standard, quasi-standard) general n-arrow model* is a structure of the form $\mathcal{M} = (W, \equiv_{ij}, \not\equiv_{ii}, m)$ where $\underline{W} = (W, \equiv_{ij}, \not\equiv_{ii})$ is a ($\neq$-standard, quasi-standard) general n-arrow frame and m is a valuation on $\underline{W}$. The satisfiability relation in $\mathcal{M}$ between a formula A and a possible world $u \in W$ is defined in the following way:

- $u \models_{\mathcal{M}} A$ iff $u \in m(A)$, A atomic formula,
- $u \models_{\mathcal{M}} \neg A$ iff $u \not\models_{\mathcal{M}} A$,
- $u \models_{\mathcal{M}} A \wedge B$ iff $u \models_{\mathcal{M}} A$ and $u \models_{\mathcal{M}} B$,
- for every $i,j \in (n)$, $u \models_{\mathcal{M}} [\equiv_{ij}]A$ iff, for every possible world $v \in W$, $u \equiv_{ij} v$ only if $v \models_{\mathcal{M}} A$,
- for every $i \in (n)$, $u \models_{\mathcal{M}} [\not\equiv_{ii}]A$ iff, for every possible world $v \in W$, $u \not\equiv_{ii} v$ only if $v \models_{\mathcal{M}} A$.

If $\mathcal{M}$ is $\neq$-standard then:

Lemma 8 *For every formula A and for every possible world $u \in W$:*

- $u \models_{\mathcal{M}} [\neq]A$ *iff, for every possible world $v \in W$, $u \neq v$ only if $v \models_{\mathcal{M}} A$,*
- $u \models_{\mathcal{M}} [U]A$ *iff, for every possible world $v \in W$, $v \models_{\mathcal{M}} A$.*

A formula is *valid* in a general n-arrow model when it is satisfied in every possible world of this model. A schema is *valid* in a general n-arrow frame if every instance of the schema is valid in every model on this frame. A schema is *valid* in a class of general n-arrow frames if it is valid in every frame of this class. Let Σ and Σ' be two classes of general n-arrow frames. Σ is *modally definable* in Σ' by a schema A if, for every frame $\underline{W} \in \Sigma'$, A is valid in $\underline{W}$ iff $\underline{W} \in \Sigma$.

Lemma 9 *The quasi-standard n-arrow frames are modally definable in the class of all $\neq$-standard n-arrow frames by the conjunction of the following schemata:*

- *for every $i \in (n)$, $[\equiv_{ii}]A \wedge [\not\equiv_{ii}]A \to [U]A$,*
- *for every $i \in (n)$, $<\equiv_{ii}> [\neq]A \to [\not\equiv_{ii}]A$.*

3.3 Axiomatics

Together with the classical tautologies, all the instances of the following schemata are axioms of BAL_n:

- for every $i, j \in (n)$, $[\equiv_{ij}](A \to B) \to ([\equiv_{ij}]A \to [\equiv_{ij}]B)$,
- for every $i \in (n)$, $[\neq_{ii}](A \to B) \to ([\neq_{ii}]A \to [\neq_{ii}]B)$,
- $A \to [\neq] <\neq> A$,
- $[U]A \to [\neq][\neq]A$,
- for every $i, j \in (n)$, $[U]A \to [\equiv_{ij}]A$,
- for every $i \in (n)$, $[\equiv_{ii}]A \wedge [\neq_{ii}]A \to [U]A$,
- for every $i \in (n)$, $<\equiv_{ii}> [\neq]A \to [\neq_{ii}]A$,
- for every $i \in (n)$, $[\equiv_{ii}]A \to A$,
- for every $i, j \in (n)$, $A \to [\equiv_{ij}] <\equiv_{ji}> A$,
- for every $i, j, k \in (n)$, $[\equiv_{ik}]A \to [\equiv_{ij}][\equiv_{jk}]A$.

Together with the *modus ponens*, the following schemata are inference rules of BAL_n:

- for every $i, j \in (n)$, the $[\equiv_{ij}]$-necessitation rule is: if $\vdash_{BAL_n} A$ then $\vdash_{BAL_n} [\equiv_{ij}]A$,
- for every $i \in (n)$, the $[\neq_{ii}]$-necessitation rule is: if $\vdash_{BAL_n} A$ then $\vdash_{BAL_n} [\neq_{ii}]A$,
- the irreflexivity rule is: if B is an atomic formula not in A and $\vdash_{BAL_n} ([\neq]B \to B) \vee A$ then $\vdash_{BAL_n} A$.

Theorem 2 *The theorems of BAL_n are valid in every quasi-standard n-arrow model.*

The $\neq$-standard n-arrow frames are not modally definable in the class of all general n-arrow frames. Nevertheless, $\neq$-standard n-arrow frames can be characterized by an inference rule: the irreflexivity rule. The irreflexivity rule have been introduced by Gabbay [4] and studied by de Rijke [10]. Usually, the axiomatization of modal logic does not contain any rule of this kind [9]. If the accessibility relation associated to a modal operator is required to be irreflexive then the irreflexivity rule makes the completeness proof easier.

3.4 Completeness

This section is devoted to the proof of the completeness of BAL_n for the class of all quasi-standard n-arrow models. A formula A is *consistent* when $\nvdash_{BAL_n} \neg A$. A finite set $\{A_1, \ldots, A_n\}$ of formulas is *consistent* when the formula $A_1 \wedge \ldots \wedge A_n$ is consistent. An infinite set of formulas is *consistent* when every of its finite subset is consistent. A set of formulas is *maximal* when, for every formula A, either A or $\neg A$ belongs to the set. A set of formulas is a $\neq$-*theory* if there exists an atomic formula B such that the formula $\neg([\neq]B \to B)$ belongs to the set. Complicated calculations would lead to the conclusion that every consistent set of formulas is a subset of a maximal consistent $\neq$-theory (see [2] for details). Let W be the set of all the maximal consistent $\neq$-theories. For every $i, j \in (n)$, let

$\equiv_{ij}$ be the binary relation on W defined by $\Gamma \equiv_{ij} \Delta$ iff $\{A: [\equiv_{ij}]A \in \Gamma\} \subseteq \Delta$. Direct calculations would lead to the conclusion that, for every $i, j, k \in (n)$:

- for every $\Gamma \in W$, $\Gamma \equiv_{ii} \Gamma$,
- for every $\Gamma, \Delta \in W$, $\Gamma \equiv_{ij} \Delta$ only if $\Delta \equiv_{ji} \Gamma$,
- for every $\Gamma, \Delta, \Phi \in W$, $\Gamma \equiv_{ij} \Delta$ and $\Delta \equiv_{jk} \Phi$ only if $\Gamma \equiv_{ik} \Phi$.

Consequently, $(W, \equiv_{ij})$ is a n-arrow frame. For every $i \in (n)$, let $\not\equiv_{ii}$ be the binary relation on W defined by $\Gamma \not\equiv_{ii} \Delta$ iff $\{A: [\not\equiv_{ii}]A \in \Gamma\} \subseteq \Delta$. The following three lemmas imply that, for every $\Gamma, \Delta, \Phi \in W$ and for every sequence $(R_1, \ldots, R_k), (S_1, \ldots, S_l)$ of elements of $\{\equiv_{ij}, \not\equiv_{ii}\}$, if $\Gamma R_1 \circ \ldots \circ R_k \Delta$ and $\Gamma S_1 \circ \ldots \circ S_l \Phi$ then either $\Delta = \Phi$ or there exists $m \in (n)$ such that $\Delta \not\equiv_{mm} \Phi$.

Lemma 10 *For every $\Gamma, \Delta, \Phi \in W$ and for every $i, j \in (n)$, if $\Gamma \not\equiv_{ii} \Delta \not\equiv_{jj} \Phi$ then either $\Gamma = \Phi$ or there exists $k \in (n)$ such that $\Gamma \not\equiv_{kk} \Phi$.*

Lemma 11 *For every $\Gamma, \Delta \in W$ and for every $i, j \in (n)$, if $\Gamma \equiv_{ij} \Delta$ then either $\Gamma = \Delta$ or there exists $k \in (n)$ such that $\Gamma \not\equiv_{kk} \Delta$.*

Lemma 12 *For every $\Gamma, \Delta \in W$ and for every $i \in (n)$, if $\Gamma \not\equiv_{ii} \Delta$ then there exists $j \in (n)$ such that $\Delta \not\equiv_{jj} \Gamma$.*

Let $\underline{W} = (W, \equiv_{ij}, \not\equiv_{ii})$. Let m be the valuation on $\underline{W}$ defined by $m(A) = \{\Gamma: \Gamma \in W$ and $A \in \Gamma\}$, A atomic formula. Let $\mathcal{M} = (\underline{W}, m)$. Direct calculations would lead to the conclusion that, for every $\Gamma \in W$ and for every formula A, $\Gamma \models_{\mathcal{M}} A$ iff $A \in \Gamma$. $\mathcal{M}$ is the *canonical model* of BAL_n. Let A be a consistent formula. There exists $\Gamma \in W$ such that $A \in \Gamma$ and $\Gamma \models_{\mathcal{M}} A$. Let $W^\circ = \{\Delta: \Delta \in W$ and there exists a sequence $(R_1, \ldots, R_k)$ of elements of $\{\equiv_{ij}, \not\equiv_{ii}\}$ such that $\Gamma R_1 \circ \ldots \circ R_k \Delta\}$. Let $\equiv_{ij}^\circ, \not\equiv_{ii}^\circ$ be the restrictions of $\equiv_{ij}, \not\equiv_{ii}$ to W°. Direct calculations would lead to the conclusion that $\underline{W}^\circ = (W^\circ, \equiv_{ij}^\circ, \not\equiv_{ii}^\circ)$ is a general n-arrow frame. Let m° be the restriction of m to W°. Let $\mathcal{M}^\circ = (\underline{W}^\circ, m^\circ)$. Direct calculations would lead to the conclusion that $\Gamma \models_{\mathcal{M}^\circ} A$. Moreover, for every $\Delta, \Phi \in W^\circ$, either $\Delta = \Phi$ or there exists $i \in (n)$ such that $\Delta \not\equiv_{ii} \Phi$. Let $\Delta \in W^\circ$. Since Δ is a $\neq$-theory, then there exists an atomic formula B such that $\neg([\neq]B \to B) \in \Delta$. Therefore, $[\neq]B, \neg B \in \Delta$. Consequently, for every $\Phi \in W^\circ$, if there exists $i \in (n)$ such that $\Delta \not\equiv_{ii} \Phi$ then $B \in \Phi$, $\neg B \notin \Phi$ and $\Delta \neq \Phi$. Therefore, for every $\Delta, \Phi \in W^\circ$, $\Delta \neq \Phi$ iff there exists $i \in (n)$ such that $\Delta \not\equiv_{ii} \Phi$. Consequently, $\underline{W}^\circ$ is $\neq$-standard. Let it be proved that, for every $\Delta, \Phi \in W^\circ$ and for every $i \in (n)$, either $\Delta \equiv_{ii} \Phi$ or $\Delta \not\equiv_{ii} \Phi$. If neither $\Delta \equiv_{ii} \Phi$ nor $\Delta \not\equiv_{ii} \Phi$ then there exists a formula $[\equiv_{ii}]A \in \Delta$ such that $A \notin \Phi$ and there exists a formula $[\not\equiv_{ii}]B \in \Delta$ such that $B \notin \Phi$. Consequently, $[\equiv_{ii}](A \lor B) \land [\not\equiv_{ii}](A \lor B) \in \Delta$ and $\neg(A \lor B) \in \Phi$. Therefore, $[U](A \lor B) \in \Delta$ and $A \lor B \in \Phi$, a contradiction. Similarly, direct calculations would lead to the conclusion that, for every $\Delta, \Phi \in W^\circ$ and for every $i \in (n)$, either $\neg\Delta \equiv_{ii} \Phi$ or $\neg\Delta \not\equiv_{ii} \Phi$. Consequently, $\underline{W}^\circ$ is quasi-standard. Therefore:

Theorem 3 BAL_n *is complete for the class of all quasi-standard n-arrow models, that is to say: the formulas valid in every quasi-standard n-arrow model are theorems of BAL_n.*

4 Collinearity

Let $n = 3$. Collinearity is one of the basic ternary relations between the points of a geometrical structure.

4.1 Structures of collinearity

A *structure of collinearity* is a point 3-frame $\underline{S} = (P, C)$ such that:

- for every $X, Y \in P$, $C(X, Y, X)$,
- for every $X, Y, Z \in P$, $C(X, Y, Z)$ only if $C(Y, X, Z)$,
- for every $X, Y, Z, T \in P$, $C(X, Y, Z)$ and $C(X, Y, T)$ only if $X = Y$ or $C(X, Z, T)$.

The class of all structures of collinearity is denoted by Σ_3^C. The elements of P are called *points* and are denoted by capital letters. C is the relation of *collinearity* between the points of the relational structure.

Example 3 *The affine geometries axiomatized by Szczerba and Tarski [11] are structures of collinearity.*

4.2 Frames of collinearity

A *frame of collinearity* is a 3-arrow frame $\underline{W} = (W, \equiv_{ij})$ such that, for every $i, j, k, l \in (3)$:

- for every $x, y \in W$, there exists $u \in W$ such that $x \equiv_{i1} u$, $y \equiv_{j2} u$ and $x \equiv_{i3} u$,
- for every $x, y, z \in W$, there exists $u \in W$ such that $x \equiv_{i1} u$, $y \equiv_{j2} u$ and $z \equiv_{k3} u$ only if there exists $v \in W$ such that $y \equiv_{j1} v$, $x \equiv_{i2} v$ and $z \equiv_{k3} v$,
- for every $x, y, z, t \in W$, there exists $u \in W$ such that $x \equiv_{i1} u$, $y \equiv_{j2} u$ and $z \equiv_{k3} u$ and there exists $v \in W$ such that $x \equiv_{i1} v$, $y \equiv_{j2} v$ and $t \equiv_{l3} v$ only if either $x \equiv_{ij} y$ or there exists $w \in W$ such that $x \equiv_{i1} w$, $z \equiv_{k2} w$ and $t \equiv_{l3} w$.

The class of all normal frames of collinearity is denoted by Φ_3^C.

4.3 Structures and frames of collinearity

Direct calculations would lead to the conclusion that:

Lemma 13 *Let $\underline{S}$ be a structure of collinearity. Then $W(\underline{S})$ is a normal frame of collinearity.*

Lemma 14 *Let $\underline{W}$ be a normal frame of collinearity. Then $S(\underline{W})$ is a structure of collinearity.*

Therefore:

Theorem 4 *The categories Σ_3^C and Φ_3^C are equivalent.*

4.4 Modal logic of collinearity

This section introduces a modal logic with standard semantics in the class of all normal frames of collinearity. A *general frame of collinearity* consists of a non-empty set W together with, for every $i, j \in (3)$, a binary relation $\equiv_{ij}$ on W such that $(W, \equiv_{ij})$ is a frame of collinearity and, for every $i \in (3)$, a binary relation $\not\equiv_{ii}$ on W.

Lemma 15 *The quasi-standard frames of collinearity are modally definable in the class of all quasi-standard 3-arrow frames by the conjunction of the following schemata:*

- *for every $i, j \in (3)$, $< U > A \wedge B \rightarrow <\equiv_{i1}> (<\equiv_{2j}> A \wedge <\equiv_{3i}> B)$,*
- *for every $i, j, k \in (3)$, $<\equiv_{i1}> (<\equiv_{2j}> A \wedge <\equiv_{3k}> B) \rightarrow$*
  $<\equiv_{i2}> (<\equiv_{1j}> A \wedge <\equiv_{3k}> B)$,
- *for every $i, j, k, l \in (3)$, $<\equiv_{i1}> (<\equiv_{2j}> (A \wedge$*
  $<\equiv_{j2}> (<\equiv_{1i}> B \wedge <\equiv_{3k}> C)) \wedge <\equiv_{3l}> D)$
  $\rightarrow <\equiv_{ij}> A \vee <\neq> B \vee <\equiv_{i1}> (<\equiv_{2k}> C \wedge <\equiv_{3l}> D)$.

Together with the axioms of BAL_3, all the instances of the previous schemata are axioms of BAL_3^C.

Theorem 5 BAL_3^C *is complete for the class of all quasi-standard models of collinearity.*

5 Sorted point n-frames and sorted n-arrow frames

Let $n \geq 2$. This section is devoted to the proof of the categorial equivalence between sorted point n-frames and sorted n-arrow frames.

5.1 Sorted point n-frames

A *sorted point n-frame* consists, for every $i \in (n)$, of a non-empty set S_i together with, for every $i, j \in (n)$ such that $i \neq j$, a binary relation R_{ij} between S_i and S_j such that:

- for every $i, j \in (n)$ such that $i \neq j$ and for every $X \in S_i$, there exists $Y \in S_j$ such that $X R_{ij} Y$,
- for every $i, j \in (n)$ such that $i \neq j$ and for every $X \in S_i, Y \in S_j$, if $X R_{ij} Y$ then there exists $Z_1 \in S_1, \ldots, Z_n \in S_n$ such that $X = Z_i$, $Y = Z_j$ and, for every $k, l \in (n)$ such that $k \neq l$, $X_k R_{kl} X_l$.

Lemma 16 *For every $i, j \in (n)$ such that $i \neq j$ and for every $X \in S_i, Y \in S_j$, if $X R_{ij} Y$ then $Y R_{ji} X$.*

The class of all sorted point n-frames is denoted by Σ^n and is considered as a category with morphisms the usual homomorphisms between relational structures.

5.2 Sorted n-arrow frames

The n-arrow frame $\underline{W} = (W, \equiv_{ij})$ is *sorted* if:

– for every $u_1, \ldots, u_n \in W$, if, for every $i, j \in (n)$ such that $i \neq j$, $u_i \equiv_{ii} \circ \equiv_{jj}$ u_j then there exists $u \in W$ such that, for every $i \in (n)$, $u_i \equiv_{ii} u$.

The class of all normal sorted n-arrow frames is denoted by Φ^n.

5.3 From sorted point n-frames to sorted n-arrow frames

Let $\underline{S} = (S_i, R_{ij})$ be a sorted point n-frame. Let $W = \{(X_1, \ldots, X_n): X_1 \in S_1, \ldots, X_n \in S_n$ and, for every $i, j \in (n)$ such that $i \neq j$, $X_i \; R_{ij} \; X_j\}$. For every $i, j \in (n)$, let $\equiv_{ij}$ be the binary relation on W defined by $(X_1, \ldots, X_n) \equiv_{ij}$ $(Y_1, \ldots, Y_n)$ iff $X_i = Y_j$.

Lemma 17 $W(\underline{S}) = (W, \equiv_{ij})$ *is a normal sorted n-arrow frame.*

Example 4 *It can be proved, for every $i, j \in (n)$ such that $i \neq j$, that if $S_i \cap S_j = \emptyset$ then $\equiv_{ij} = \emptyset$ and that if $S_i \subseteq S_j$ then $\equiv_{ij}$ is serial.*

5.4 From sorted n-arrow frames to sorted point n-frames

Let $\underline{W} = (W, \equiv_{ij})$ be a normal sorted n-arrow frame. For every $i \in (n)$, let $S_i = \{i(u): u \in W\}$. For every $i, j \in (n)$ such that $i \neq j$, let R_{ij} be the binary relation between S_i and S_j defined by $i(u) \; R_{ij} \; j(v)$ iff $u \equiv_{ii} \circ \equiv_{jj} v$. Direct calculations would lead to the conclusion that:

Lemma 18 $S(\underline{W}) = (S_i, R_{ij})$ *is a sorted point n-frame.*

Example 5 *It can be proved, for every $i, j \in (n)$ such that $i \neq j$, that if $\equiv_{ij} = \emptyset$ then $S_i \cap S_j = \emptyset$ and that if $\equiv_{ij}$ is serial then $S_i \subseteq S_j$.*

5.5 Representation theorems

Let $\underline{W} = (W, \equiv_{ij})$ be a normal sorted n-arrow frame and $\underline{W}' = W(S(\underline{W}))$. For every $u \in W$, let $g(u) = (1(u), \ldots, n(u))$. Direct calculations would lead to the conclusion that g is an isomorphism from $\underline{W}$ into $\underline{W}'$. Therefore:

Lemma 19 $\underline{W}$ *and* $\underline{W}'$ *are isomorphic.*

Let $\underline{S} = (S_i, R_{ij})$ be a sorted point n-frame and $\underline{S}' = S(W(\underline{S}))$. For every $X \in S_i$, let $f(X) = \{(X_1, \ldots, X_n): X_1 \in S_1, \ldots, X_n \in S_n, X = X_i$ and, for every $j, k \in (n)$ such that $j \neq k$, $X_j \; R_{jk} \; X_k\}$. Direct calculations would lead to the conclusion that f is an isomorphism from $\underline{S}$ into $\underline{S}'$. Consequently:

Lemma 20 $\underline{S}$ *and* $\underline{S}'$ *are isomorphic.*

Therefore:

Theorem 6 *The categories Σ^n and Φ^n are equivalent.*

6 Sorted arrow logic

This section introduces a modal logic with standard semantics in the class of all normal sorted n-arrow frames.

6.1 Semantics

Let $A_1, \ldots, A_n$ be formulas. Let $B_1 = true$ and, for every $k \geq 1$, $B_{k+1} = <\equiv_{(k+1)(k+1)}><\equiv_{kk}> ([\neq]A_k \wedge B_k) \wedge \bigwedge_{i \in (k)} <\equiv_{(k+1)(k+1)}><\equiv_{ii}> \neg A_i$. Let $\mathcal{M} = (W, \equiv_{ij}, \neq_{ii}, m)$ be a quasi-standard n-arrow model. Direct calculations would lead to the conclusion that, for every $k \in (n)$ and for every $x \in W$, $x \models_{\mathcal{M}} B_k$ iff there exists $x_1, \ldots, x_k \in W$ such that $x_k = x$, $x_i \models_{\mathcal{M}} [\neq]A_i \wedge B_i$, for every $i \in (k-1)$ and $x_i \equiv_{ii} \circ \equiv_{jj} x_j$, for every $i, j \in (k)$ such that $i \neq j$. Consequently:

Lemma 21 *The quasi-standard sorted n-arrow frames are modally definable in the class of all quasi-standard n-arrow frames by the following schema:*

$$- C_n = B_n \rightarrow <\equiv_{nn}> \bigwedge_{i \in (n-1)} <\equiv_{ii}> \neg A_i.$$

Example 6 B_2 *is logically equivalent to the schema* $<\equiv_{22}><\equiv_{11}> [\neq]A_1 \wedge <\equiv_{22}><\equiv_{11}> \neg A_1$. *The schema* B_3 *is equivalent to* $<\equiv_{33}><\equiv_{22}> ([\neq]A_2 \wedge B_2) \wedge <\equiv_{33}><\equiv_{11}> \neg A_1 \wedge <\equiv_{33}><\equiv_{22}> \neg A_2$.

Together with the axioms of BAL_n, all the instances of the previous schema are axioms of BAL^n.

Theorem 7 BAL^n *is complete for the class of all quasi-standard sorted n-arrow models.*

7 Geometrical sorted point n-frames

Projective structures and orthogonal structures are examples of sorted point n-frames.

7.1 Projective geometry

Incidence is one of the basic binary relations between the points and the lines of a geometrical structure.

Projective structures A *projective structure* is a relational structure $\underline{S} = (P, L, in)$ such that:

- $P \cap L = \emptyset$,
- for every $X, Y \in P$, there exists $x \in L$ such that X *in* x and Y *in* x,
- for every $x, y \in L$, there exists $X \in P$ such that X *in* x and X *in* y,

– for every $X, Y \in P$ and for every $x, y \in L$, X *in* x, Y *in* x, X *in* y and Y *in* y only if $X = Y$ or $x = y$.

The elements of L are called *lines* and are denoted by lower case letters. *in* is the relation of incidence between the points and the lines of the relational structure. The class of all projective structures is denoted by Σ_P^2.

Example 7 *The projective geometries axiomatized by Heyting [7] are projective structures.*

A projective structure (P, L, in) is considered as a sorted point 2-frame $(S_1, S_2, R_{12}, R_{21})$ where $S_1 = P$, $S_2 = L$, $R_{12} = in$ and $R_{21} = in^{-1}$.

Projective frames A *projective frame* is a sorted 2-arrow frame $\underline{W} = (W, \equiv_{ij})$ such that:

– $\equiv_{12} = \emptyset$,
– for every $x, y \in W$, $x \equiv_{11} \circ \equiv_{22} \circ \equiv_{11} y$,
– for every $x, y \in W$, $x \equiv_{22} \circ \equiv_{11} \circ \equiv_{22} y$,
– for every $x, y, z, t \in W$, $x \equiv_{11} \circ \equiv_{22} z$, $y \equiv_{11} \circ \equiv_{22} z$, $x \equiv_{11} \circ \equiv_{22} t$ and $y \equiv_{11} \circ \equiv_{22} t$ only if $x \equiv_{11} y$ or $z \equiv_{22} t$.

The class of all normal projective frames is denoted by Φ_P^2.

Projective structures and projective frames Direct calculations would lead to the conclusion that:

Lemma 22 *Let $\underline{S}$ be a projective structure. Then $W(\underline{S})$ is a normal projective frame.*

Lemma 23 *Let $\underline{W}$ be a normal projective frame. Then $S(\underline{W})$ is a projective structure.*

Therefore:

Theorem 8 *The categories Σ_P^2 and Φ_P^2 are equivalent.*

Projective modal logic This section introduces a modal logic with standard semantics in the class of all normal projective frames. A *general projective frame* consists of a non-empty set W together with, for every $i, j \in (2)$, a binary relation $\equiv_{ij}$ on W such that $(W, \equiv_{ij})$ is a projective frame and, for every $i \in (2)$, a binary relation $\not\equiv_{ii}$ on W.

Lemma 24 *The quasi-standard projective frames are modally definable in the class of all quasi-standard sorted 2-arrow frames by the conjunction of the following schemata:*

– $[\equiv_{12}] false$,
– $[\equiv_{11}][\equiv_{22}][\equiv_{11}]A \rightarrow [U]A$,

- $[\equiv_{22}][\equiv_{11}][\equiv_{22}]A \rightarrow [U]A$,
- $<\equiv_{11}><\equiv_{22}> (A\wedge <\equiv_{22}><\equiv_{11}> ([\neq]B \wedge C)) \rightarrow$
  $[\equiv_{11}][\equiv_{22}](<\equiv_{22}> A \vee [\equiv_{22}][\equiv_{11}]B)\vee <\equiv_{11}> C.$

Together with the axioms of BAL^2, all the instances of the previous schemata are axioms of BAL_P^2.

Theorem 9 BAL_P^2 *is complete for the class of all quasi-standard projective models.*

7.2 Orthogonal geometry

Orthogonality is one of the basic binary relations between the lines of a geometrical structure.

Orthogonal structures An *orthogonal structure* is a relational structure $\underline{S} = (P, L, in, \perp)$ such that:

- $P \cap L = \emptyset$,
- for every $X, Y \in P$, there exists $x \in L$ such that X in x and Y in x,
- for every $X, Y \in P$ and for every $x, y \in L$, X in x, Y in x, X in y and Y in y only if $X = Y$ or $x = y$,
- for every $X \in P$ and for every $x \in L$, there exists $y \in L$ such that X in y and $x \perp y$,
- for every $X \in P$ and for every $x, y, z \in L$, X in y, $x \perp y$, X in z and $x \perp z$ only if $y = z$,
- for every $x, y \in L$, $x \perp y$ only if $y \perp x$.

$\perp$ is the relation of orthogonality between the lines of the relational structure. The class of all orthogonal structures is denoted by Σ_O^3.

Example 8 *The orthogonal geometries axiomatized by Goldblatt [6] are orthogonal structures.*

An orthogonal structure $(P, L, in, \perp)$ is considered as a sorted point 3-frame $(S_1, S_2, S_3, R_{12}, R_{13}, R_{21}, R_{23}, R_{31}, R_{32})$ where $S_1 = P$, $S_2 = L$, $S_3 = L$, $R_{12} = in$, $R_{13} = in$, $R_{21} = in^{-1}$, $R_{23} = \perp$, $R_{31} = in^{-1}$ and $R_{32} = \perp^{-1}$.

Orthogonal frames An *orthogonal frame* is a sorted 3-arrow frame $\underline{W} = (W, \equiv_{ij})$ such that, for every $i, j, k \in \{2, 3\}$:

- $\equiv_{12} = \emptyset$,
- $\equiv_{23}$ is serial,
- $\equiv_{32}$ is serial,
- for every $x, y \in W$, $x \equiv_{11} \circ \equiv_{23} y$ iff $x \equiv_{11} \circ \equiv_{33} y$,
- for every $x, y \in W$, $x \equiv_{11} \circ \equiv_{32} y$ iff $x \equiv_{11} \circ \equiv_{22} y$,
- for every $x, y \in W$, $x \equiv_{11} \circ \equiv_{ij} \circ \equiv_{11} y$,
- for every $x, y, z, t \in W$, $x \equiv_{11} \circ \equiv_{ii} z$, $y \equiv_{11} \circ \equiv_{ii} z$, $x \equiv_{11} \circ \equiv_{jj} t$ and $y \equiv_{11} \circ \equiv_{jj} t$ only if $x \equiv_{11} y$ or $z \equiv_{ij} t$,

- for every $x, y \in W$, $x \equiv_{11} \circ \equiv_{i2} \circ \equiv_{3j} y$,
- for every $x, y, z, t \in W$, $x \equiv_{11} \circ \equiv_{jj} z$, $y \equiv_{i2} \circ \equiv_{3j} z$, $x \equiv_{11} \circ \equiv_{kk} t$ and $y \equiv_{i2} \circ \equiv_{3k} t$ only if $z \equiv_{jk} t$,
- for every $x, y \in W$, $x \equiv_{i2} \circ \equiv_{3j} y$ only if $x \equiv_{i3} \circ \equiv_{2j} y$.

The class of all normal orthogonal frames is denoted by Φ_O^3.

Orthogonal structures and orthogonal frames Direct calculations would lead to the conclusion that:

Lemma 25 *Let $\underline{S}$ be an orthogonal structure. Then $W(\underline{S})$ is a normal orthogonal frame.*

Lemma 26 *Let $\underline{W}$ be a normal orthogonal frame. Then $S(\underline{W})$ is an orthogonal structure.*

Therefore:

Theorem 10 *The categories Σ_O^3 and Φ_O^3 are equivalent.*

Orthogonal modal logic A *general orthogonal frame* consists of a non-empty set W together with, for every $i, j \in \{1, 2, 3\}$, a binary relation $\equiv_{ij}$ on W such that $(W, \equiv_{ij})$ is an orthogonal frame and, for every $i \in (3)$, a binary relation $\neq_{ii}$ on W.

Lemma 27 *The quasi-standard orthogonal frames are modally definable in the class of all quasi-standard sorted 3-arrow frames by the conjunction of the following schemata:*

- $[\equiv_{12}]false$,
- $<\equiv_{23}> true$,
- $<\equiv_{32}> true$,
- $[\equiv_{11}][\equiv_{33}]A \leftrightarrow [\equiv_{11}][\equiv_{23}]A$,
- $[\equiv_{11}][\equiv_{22}]A \leftrightarrow [\equiv_{11}][\equiv_{32}]A$,
- $[\equiv_{11}][\equiv_{ij}][\equiv_{11}]A \rightarrow [U]A$,
- $<\equiv_{11}><\equiv_{ii}> (A \wedge <\equiv_{ii}><\equiv_{11}> ([\neq]B \wedge C)) \rightarrow [\equiv_{11}][\equiv_{jj}](<\equiv_{ji}> A \vee [\equiv_{jj}][\equiv_{11}]B) \vee <\equiv_{11}> C$,
- $[\equiv_{11}][\equiv_{i2}][\equiv_{3j}]A \rightarrow [U]A$,
- $<\equiv_{11}><\equiv_{jj}> (A \wedge <\equiv_{j3}><\equiv_{2i}> [\neq]B) \rightarrow [\equiv_{11}][\equiv_{kk}](<\equiv_{kj}> A \vee [\equiv_{k3}][\equiv_{2i}]B)$,
- $[\equiv_{i3}][\equiv_{2j}]A \rightarrow [\equiv_{i2}][\equiv_{3j}]A$.

Together with the axioms of BAL^3, all the instances of the previous schemata are axioms of BAL_O^3:

Theorem 11 BAL_O^3 *is complete for the class of all quasi-standard orthogonal models.*

8 Conclusion

The methodology presented in this paper for a modal treatment of structures of collinearity, projective structures and orthogonal structures can be applied to other geometrical structures as well:

- Structures of betweeness consist of a set of points together with a ternary relation of betweeness between points [11].
- Projective structures of space consist of a set of points, a set of lines and a set of planes together with three binary relations of incidence between points, lines and planes.

However, many questions remain unsolved:

- Proof of the completeness or the incompleteness of BAL_3^C, BAL_P^2 and BAL_O^3 without the irreflexivity rule.
- Proof of the decidability or the undecidability of BAL_3^C, BAL_P^2 and BAL_O^3.

References

1. **P. Balbiani, V. Dugat, L. Fariñas del Cerro** and **A. Lopez.** *Eléments de géométrie mécanique.* Hermès, 1994.
2. **P. Balbiani, L. Fariñas del Cerro, T. Tinchev** and **D. Vakarelov.** *Modal logics for incidence geometries.* Journal of Logic and Computation, to appear.
3. **J. van Benthem.** *The Logic of Time.* Reidel, 1983.
4. **D. Gabbay.** *An irreflexivity lemma with applications to axiomatizations of conditions on tense frames.* U. Mönnich (editor), Aspects of Philosophical Logic. 67-89, Reidel, 1981.
5. **D. Gabbay, I. Hodkinson** and **M. Reynolds.** *Temporal Logic: Mathematical Foundations and Computational Aspects.* Volume I, Oxford University Press, 1994.
6. **R. Goldblatt.** *Orthogonality and Spacetime Geometry.* Springer-Verlag, 1987.
7. **Heyting.** *Axiomatic Projective Geometries.* North-Holland, 1963.
8. **D. Hilbert.** *Foundations of Geometry.* Second english edition, Open Court, 1971.
9. **G. Hughes** and **M. Cresswell.** *A Companion to Modal Logic.* Methuen, 1984.
10. **M. de Rijke.** *The modal logic of inequality.* Journal of Symbolic Logic, Volume 57, Number 2, 566-584, 1992.
11. **L. Szczerba** and **A. Tarski.** *Metamathematical properties of some affine geometries.* Y. Bar-Hillel (editor), Logic, Methodology and Philosophy of Science. 166-178, North-Holland, 1972.
12. **D. Vakarelov.** *A modal theory of arrows. Arrow logics I.* D. Pearce and G. Wagner (editors), Logics in AI, European Workshop JELIA '92, Berlin, Germany, September 1992, Proceedings. Lecture Notes in Artificial Intelligence 633, 1-24, Springer-Verlag, 1992.
13. **D. Vakarelov.** *Many-dimensional arrow structures. Arrow logics II.* Journal of Applied Non-Classical Logics, to appear.

A Unified Framework for Hypothetical and Practical Reasoning (1): Theoretical Foundations

S. K. Das[1], J. Fox[2], P. Krause[2]

[1] William Penney Laboratory, Imperial College, London SW7 2AZ
[2] Advanced Computation Laboratory, ICRF, London WC2A 3PX

Abstract. We describe here a general and flexible framework for decision making which embodies the concepts of beliefs, goals, options, arguments and commitments. We have employed these concepts to build a generic decision support system which has been successfully applied in a number of areas in clinical medicine. In this paper, we present the formalisation of the decision making architecture within a framework of modal propositional logics. A possible-world semantics of the logic is developed and the soundness and completeness result is also established.

1 Introduction

A *decision* is a choice between two or more competing hypotheses about some world or possible courses of action in the world. A *decision support system* is a computerised system which helps decision makers by utilizing knowledge about the world to recommend beliefs or actions [1]. Such a system when built on *symbolic theory* [4] offers a general and flexible framework for decision making. The theory embodies the concepts of beliefs and goals of decision makers, represents options to satisfy a goal, argues for and against for every option and commits to a suitable option. We have employed this theory to build a generic decision support system [9, 13]. A paper focussing on applications of the theory to clinical medicine has also been submitted to this conference [12].

The scope of our decision framework is illustrated by the simple medical example in Figure 1. We should emphasise at this stage that the medical content of the examples in this paper has been simplified to avoid the need to explain complex medical terminology. Suppose that a patient presents complaining of serious and unexplained loss of weight. As an abnormal and potentially serious condition a decision has to be taken as to the most likely cause of the complaint. In the graphical notation in the figure circles are used to represent decisions; this diagnosis decision is on the left of the figure. To make the decision we have to identify all potentially relevant causes of weight loss (such as cancer and peptic ulcer) and then generate and evaluate arguments for and against each of these candidates. Arguments will be based on relevant information sources such as the patient's age, history, symptoms and so on. Suppose, after evaluating the arguments, we take the decision that cancer is the most likely cause of the weight loss - i.e. we commit to acting on this conclusion.

Now we must take another decision, about the appropriate therapy. Suppose the possible candidates for this are chemotherapy and surgery. As before, arguments for and against these options need to be considered, taking into account such information as the patient's age (if a patient is elderly this may argue against more aggressive forms of treatment), the likely efficacy, cost and side-effects of each therapy etc. Finally, suppose that after weighing up the arguments we conclude that chemotherapy is most appropriate for the patient. Once again we must be careful about committing to this option since once an action is taken it cannot be reversed.

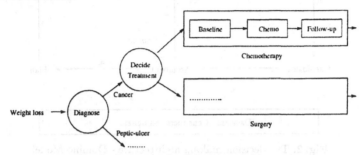

Fig. 1. An example of decision making.

A system which implements this model operates as follows (refer to Figure 2). First it maintains a database of beliefs about a particular situation; in a medical context this may include a set of clinical data about a patient for example. Certain beliefs (e.g. unexplained weight loss) cause the system to raise problem goals (e.g. to explain the weight loss). Such goals lead to problem solving to find candidate solutions (e.g. the weight loss may be caused by cancer or peptic ulcer) and arguments are constructed for and against the candidates, perhaps by instantiating general argument schemas with specific patient data and specific knowledge and beliefs about cancer, ulcers and so forth. As additional patient data are acquired a point may arise where an assessment of the various arguments for and against the various candidates permits the system to commit to a single hypothesis (e.g. cancer). This is adopted as a new belief which, while the belief is held, guides further problem solving and action. Since the belief concerns an undesirable - indeed life threatening - condition a new goal is raised, to decide on the best therapy for the patient. This initiates a further cycle of reasoning summarised in the left half of the figure. As before, candidate decision options are proposed (surgery, chemotherapy etc) and arguments are generated for and against the alternatives. In due course a commitment is made to a single therapy (e.g. chemotherapy).

Many clinical therapies, such as chemotherapy, are in fact complex procedures executed over time. Such therapies can usually be modelled as hierarchical plans that decompose into atomic actions (e.g. administer a drug) and subplans (e.g. take baseline measurements, administer several cycles of therapy, and then follow up the patient for a period after the last cycle of treatment). Our framework acknowledges this by providing formalisms for representing plans and for specifying the control processes required during plan execution. In particular the atomic actions of a plan must be scheduled with respect to any other actions which have been previously scheduled as a result of problem-solving or decision processes required for achieving the goal, or other goals raised by the system.

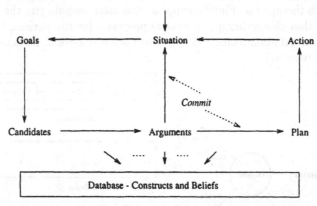

Fig. 2. The decision making architecture - Domino Model

We have proposed a high-level and expressive language called R^2L (RED Representational Language) [8] to be used by knowledge engineers who wish to encode knowledge and beliefs of a particular application domain for decision making based on the above scheme. R^2L explicitly supports the key concepts required in our framework, such as beliefs, goals, arguments and commitments. Our approach to formalisation of the above decision making scheme is by providing a sound translation mechanism of R^2L structures into a lower level but more general language which is the language of the logic LR^2L. The logic LR^2L enriches the propositional calculus by the introduction of a number of appropriately specialized modal operators and its semantics are well defined. This kind of modal formalism has the major advantage that its possible-world semantics [2, 16, 17] reflects the dynamic nature of applications such as medicine. In the implementation, decision making and scheduling processes are performed by an LR^2L theorem prover.

The paper is organized as follows. We present the R^2L language in the following section 2. This incorporates a model of argumentation in which arguments may be annotated with some qualifier indicating the confidence in a line of argument. These qualifiers may be drawn from one of a number of numeric or symbolic dictionaries, and then aggregated to provide an overall indication of the support for a given proposition. Section 3 describes the aggregation of arguments. The LR^2L syntax for representing theories of properties and of actions is presented in section 4 which is followed by the section describing the translation mechanism from R^2L to LR^2L syntax. The axioms and possible-world semantics of LR^2L are presented in sections 6 and 7, respectively. To illustrate and motivate the abstract model LR^2L, we provide a medical example in section 8. The soundness and completeness result is established in section 9. The proofs of the propositions and theorems can be found in the longer version of the paper.

2 R^2L

A decision schema in R^2L corresponding to "Diagnose" in figure 1 is represented as follows:

```
decision:: weight_loss_diagnosis
  situation   weight_loss
  goal        weight_loss_diagnosed
  candidates  cancer;
              peptic_ulcer
  arguments   elderly => support(cancer, d1);
              smoker => support(cancer, d2);
              positive_biopsy => support(cancer, d3);
              pain_in_upper_abdomen => support(peptic_ulcer, d4);
              young => support(~cancer & ~peptic_ulcer, d2);
              pain_after_meals=>support(cancer & peptic_ulcer, d5);
  commits     netsupport(X, M) & exceed_threshold(M) => add(X).
```

In this example the decision weight_loss_diagnosis is considered for activation when the *situation* weight_loss is observed. A *goal* is considered as a property to be brought about. An *argument* schema is like an ordinary rule with support(F, <degree>) as its consequent, where <degree> is drawn from a dictionary of qualitative or quantitative measures of support, and indicates the support conferred on this candidate by the argument [15].

A commitment rule will often, though not necessarily, make use of the meta-predicate netsupport such as above. This meta-predicate has the general form netsupport(F, <support>). It computes the total support for the specified candidate using an aggregation algorithm (discussed in the following section) selected from a library of such algorithms [15]. When a commitment rule involves the netsupport as in the above example, the computation of its truth value involves meta-level reasoning whose formalisation is beyond the scope of this paper.

A *dictionary* is a set of symbols which can be used to label a proposition. In general, a dictionary will be a semi-lattice with the partial order relation $\leq$. For simplicity, we shall consider a dictionary as a chain with one distinguished element $\triangle$ known as the *top element*. Let $d1$ be an element from some dictionary. Then the argument elderly => support(cancer, d1) specifies that if a person is believed to be elderly then this argument confers evidence level $d1$ on the candidate cancer. We might consider $d1$ as a member of the quantitative dictionary of probabilities $dict(Prob) =_{def} [0, 1]$. However, there is no requirement that we should restrict dictionaries to $dict(Prob)$. Among the obvious dictionaries we may consider is $dict(Qual) =_{def} \{+, ++\}$. As mentioned, a dictionary has always a top element to represent the highest support for arguments. For example, elements $++$ and 1 are the top elements of the two dictionaries $dict(Qual)$ and $dict(Prob)$ respectively. A number of different dictionaries for reasoning under uncertainty have been discussed in [15], together with their mathematical foundations and their relation to classical probability and other uncertainty formalisms.

A *commitment rule* is like an ordinary rule with one of add(<property>) or schedule(<plan>) as its consequent. The former adds a new belief to the database and the latter causes a plan to be scheduled as in the following construct:

```
decision:: cancer_treatment
  situation  cancer
  goal       cancer_cured
  candidates chemotherapy;
             surgery
  arguments  elderly => support(chemotherapy, +);
             young => support(surgery, +)
  commits    netsupport(X, M) & netsupport(Y, N) & M > N =>
             schedule(X).
```

When we are committed to scheduling a plan, the plan involves executing constituent subplans and actions in a certain order. For example, if we schedule chemotherapy then a *plan construct* [8] of R^2L guides us to achieve the goal of carrying out the chemotherapy plan. Formalisation of this aspect involving actions and temporal reasoning is presented elsewhere [9] and is out of the scope of this paper.

3 Argumentation and Aggregation

In classical logic an argument is a sequence of inferences leading to a conclusion. The usual interest of the logician is in procedures by which arguments may be used to establish the validity (truth or falsity) of a formula. In LA, a logic of argument [15], arguments do not necessarily prove formulae but may merely indicate support for (or doubt about) them. Also in classical logic, so long as we can construct *one* argument (proof) for F, any further arguments for F are of no interest. In our system all distinct arguments of candidates are of interest (intuitively, the more arguments we have for F the greater is our knowledge about the validity of F). We therefore distinguish distinct arguments by identifying the unique grounds of each (essentially a normalised proof term in LA [15]) and a sign drawn from some dictionary which indicates the support provided to F by the argument. An example of an R^2L argument is elderly => support(cancer, d1), where F is cancer, the *ground* is elderly => cancer and the support is $d1$.

Suppose a decision maker has a set of arguments for and against a set of mutually exclusive decision options ("candidates", that is, alternative beliefs or plans under consideration) whose signs are drawn from a single dictionary. The decision maker can *aggregate* these arguments to yield a sign representing the decision maker's overall confidence in each of the candidates. Every dictionary has a characteristic aggregation function for aggregating arguments. Consider the argument presented above and positive_biopsy => support(cancer, d3).

Considering the dictionary as $dict(Prob)$, the two arguments can be aggregated by using a special case of Dempster's epistemic probability giving the value $d1 + d3 - d1 \times d3$. This formula can be generalised incrementally if there are more than two arguments for the candidate **cancer**.

In general, suppose a decision maker has a set of arguments for and against a set of mutually exclusive decision options, C, (candidates, that is, alternative beliefs or plans under consideration) whose signs are drawn from a single dictionary D. The decision maker can *aggregate* these arguments to yield a sign drawn from D' which represents the decision maker's overall confidence in each C. The general form of an aggregation function is as $\mathcal{A} : \Pi(C \times G \times D) \to C \times D'$, where Π stands for "power set" and G is the set of all grounds. The simple **netsupport** predicate in R^2L implements the function $\mathcal{A}$. If D is $dict(Qual)$ then D' is the set of non-negative integers whereas D' is D itself when we consider $dict(Prob)$ as D. In the former case, $\mathcal{A}$ assigns an aggregation number to each decision option, giving a total preference ordering over the options. This suggests a simple rule for taking a decision; choose the alternative which maximises this value.

If we allow both F and ˜F to occur in the support then by applying our usual aggregation algorithm we compute total evidence for F (say, $d1$) and ˜F (say, $d2$) separately. If we have used the dictionary $\{+, ++\}$ then we have the following four cases:
- total evidence for F is $d1 - d2$ if $d1 > d2$;
- the total evidence for ˜F is $d2 - d1$ if $d2 > d1$;
- dilemma if $d1 = d2$
- inconsistent if $d1 = d2$.

If we have used the dictionary $dict(Prob)$ then we have the following cases:
- total evidence for F is $d1(1 - d2) \div (1 - d1 \times d2)$;
- total evidence for ˜F is $d2(1 - d1) \div (1 - d1 \times d2)$;
- dilemma if $d1 = d2$;
- inconsistent if $d1 \times d2 = 1$.

4 Syntax of LR^2L

The language of LR^2L is the usual propositional language extended with a few modal operators introduced as follows. Suppose $\mathcal{P}$ is the set of all propositions which includes the special symbol $\top$ (true). Suppose D is an arbitrary dictionary with the the top element Δ. The modal operators of LR^2L corresponding to belief and goal are $\langle bel \rangle$ and $\langle goal \rangle$ respectively. In addition, for each dictionary symbol $d \in D$, we have a modal operator $\langle sup_d \rangle$ for support. The *formulae* (or *assertions*) of LR^2L are as follows:
- propositions are formulae.
- $\langle bel \rangle F$ and $\langle goal \rangle F$ are formulae, where F is a formula.
- $\langle sup_d \rangle F$ is a formula, where F is a formula and d is in the dictionary D.
- $\neg F$ and $F \wedge G$ are formulae, where F and G are formulae.

We take $\perp$ (false) to be an abbreviation of $\neg\top$. Other logical connectives are defined using '$\neg$' and '$\wedge$' in the usual manner.

5 Translating R^2L to LR^2L

This section details how R^2L constructs can be translated into sentences of the base language LR^2L. First of all, if the situation in a decision construct is believed then the corresponding goal is raised. Thus the situation and goal portion in decision *weight_loss_diagnosis* is translated to the rule

$\langle bel \rangle weight\_loss \rightarrow \langle goal \rangle weight\_loss\_diagnosed$

For any particular situation a raised goal is considered as achieved if it is true. The raised goal from a decision construct is true if any possible situation for the candidates is believed. In the context of the decision *weight_loss_diagnosis*, this is reflected in the following formulae:

$\langle bel \rangle (cancer \wedge \neg peptic\_ulcer) \rightarrow weight\_loss\_diagnosed$

$\langle bel \rangle (peptic\_ulcer \wedge \neg cancer) \rightarrow weight\_loss\_diagnosed$

$\langle bel \rangle (\neg cancer \wedge \neg peptic\_ulcer) \rightarrow weight\_loss\_diagnosed$

$\langle bel \rangle (cancer \wedge peptic\_ulcer) \rightarrow weight\_loss\_diagnosed$

Note that $\langle bel \rangle weight\_loss\_diagnosed$ can be derived from the first of the above four, where $\langle bel \rangle cancer$ and $\langle bel \rangle \neg peptic\_ulcer$ are true. The equivalent LR^2L representations of all the arguments in decision *weight_loss_diagnosis* are given below:

$\langle bel \rangle (weight\_loss \wedge elderly) \rightarrow \langle sup_{d1} \rangle cancer$

$\langle bel \rangle (weight\_loss \wedge smoker) \rightarrow \langle sup_{d2} \rangle cancer$

$\langle bel \rangle (weight\_loss \wedge positive\_biopsy) \rightarrow \langle sup_{d3} \rangle cancer$

$\langle bel \rangle (weight\_loss \wedge pain\_in\_upper\_abdomen) \rightarrow \langle sup_{d4} \rangle peptic\_ulcer$

$\langle bel \rangle (weight\_loss \wedge young) \rightarrow \langle sup_{d3} \rangle (\neg cancer \wedge \neg peptic\_ulcer)$

$\langle bel \rangle (weight\_loss \wedge pain\_after\_meals) \rightarrow \langle sup_{d5} \rangle (cancer \wedge peptic\_ulcer)$

If `support(cancer & peptic_ulcer, d)` holds then there is support d for each of `cancer` and `peptic_ulcer`. The converse is not necessarily true.

The version of LR^2L presented here excludes temporal reasoning and reasoning with actions; otherwise, the set of propositional symbols would have been divided into properties (e.g. the patient is elderly) and actions (e.g. the patient is given chemotherapy). A plan construct in R^2L would have been transformed to a set of temporal rules of LR^2L [9] involving action symbols. Reasoning with time and actions is out of scope of the current, as we are concentrating primarily on the process of decision making.

6 Axioms of LR^2L

We consider every instance of a propositional tautology to be an axiom. Instances of propositional tautologies may involve any number of modal operators, for example, $\langle bel \rangle p \rightarrow \langle bel \rangle p$. We have the modus ponens inference rule and adopt a set of standard axioms of beliefs which can be found in [5, 11, 14, 18]:

$$\neg \langle bel \rangle \perp \tag{1}$$

$$\langle bel \rangle F \wedge \langle bel \rangle (F \rightarrow G) \rightarrow \langle bel \rangle G \tag{2}$$

$$\langle bel \rangle F \rightarrow \langle bel \rangle \langle bel \rangle F \tag{3}$$

$$\neg \langle bel \rangle F \rightarrow \langle bel \rangle \neg \langle bel \rangle F \tag{4}$$

Axiom (1) expresses that an inconsistency is not believable by a decision maker. The derivation of the symbol ⊥ from the database implies inconsistency. Axiom (2) states that a decision maker believes all the logical consequences of its beliefs, that is, a decision maker's beliefs are closed under logical deduction. The two facts that a decision maker believes that s/he believes in something and a decision maker believes that s/he does not believe in something are expressed by axioms (3) and (4) respectively. We also have the rule of necessitation for beliefs:

$$if \vdash F \ then \vdash \langle bel \rangle F \qquad (5)$$

Proposition 1. *The following are theorems of LR^2L:*
$\langle bel \rangle (F \wedge G) \leftrightarrow \langle bel \rangle F \wedge \langle bel \rangle G$
$\langle bel \rangle F \vee \langle bel \rangle G \rightarrow \langle bel \rangle (F \vee G)$
$\langle bel \rangle F \rightarrow \neg \langle bel \rangle \neg F$

There is no support for an inconsistency and the following axiom reflect this property:

$$\neg \langle sup_d \rangle \perp, \ for \ every \ d \in D \qquad (6)$$

Support is closed under tautological implications by preserving degrees. In other words, if F has a support d and $F \rightarrow G$ is an LR^2L tautology then G too has a support d:

$$if \vdash F \rightarrow G \ then \vdash \langle sup_d \rangle F \rightarrow \langle sup_d \rangle G, \ for \ every \ d \in D \qquad (7)$$

If an observation in the real world generates support d for F and if $F \rightarrow G$ is a decision maker's belief then it is unreasonable to conclude that d is also a support for G. This prevents us from considering supports closed under believed implications. The following rule of inference states that an LR^2L tautology has always the highest support:

$$if \vdash F \ then \vdash \langle sup_\triangle \rangle F \qquad (8)$$

Proposition 2. *For every d in D, the following is a theorem of LR^2L:*
$\langle sup_\triangle \rangle \top$
$\langle sup_d \rangle (F \wedge G) \rightarrow \langle sup_d \rangle F \wedge \langle sup_d \rangle G$

A *rational decision maker* believes in something which has support with the top element of the dictionary. Thus, the axiom $\langle sup_\triangle \rangle F \rightarrow \langle bel \rangle F$. should be considered for a rational decision maker. This axiom, of course, derives that an assertion and its negation are not simultaneously derivable with the top element as support, that is, an integrity constraint [6] of the form $\langle sup_\triangle \rangle F \wedge \langle sup_\triangle \rangle \neg F \rightarrow \perp$. It is difficult to maintain consistency of a database in the presence of this axiom, particularly when the database is constructed from different sources; mutual inconsistency and mistakes sometimes need to be tolerated. In these circumstances, it might be left to the decision maker to arbitrate over which to believe or not believe.

A decision maker might believe in something even if the database derives no support for it. We call a decision maker who does not believe in something unless there is support with the top element a *strict decision maker*. If a decision maker is both rational and strict then the concepts of believability and support with the top element coincide. In other words, $\langle sup_\Delta \rangle F \leftrightarrow \langle bel \rangle F$. Note that we do not consider $\langle sup_{d1} \rangle F \rightarrow \langle sup_{d2} \rangle F$ (where $d2 \leq d1$) as an axiom which says that certain evidence for an assertion also implies every evidence for the assertion lower than the evidence. The reason for exclusion will be given in the context of model definition. The exclusion also avoids the unnecessary contributions to the aggregation process for F.

We adopt the following two standard axioms of goals [5, 19]:

$$\neg \langle goal \rangle \perp \tag{9}$$
$$\langle goal \rangle F \wedge \langle goal \rangle (F \rightarrow G) \rightarrow \langle goal \rangle G \tag{10}$$

Axiom (9) says that something that is impossible to achieve cannot be a goal of a decision maker. Axiom (10) states that all the logical consequences of a decision maker's goal are goals themselves.

Proposition 3. *The following are theorems of LR^2L:*
$\langle goal \rangle F \wedge \langle bel \rangle (F \rightarrow G) \rightarrow \langle goal \rangle G$
$\langle goal \rangle (F \wedge G) \leftrightarrow \langle goal \rangle F \wedge \langle goal \rangle G$
$\langle goal \rangle F \vee \langle goal \rangle G \rightarrow \langle goal \rangle (F \vee G)$
$\langle goal \rangle F \rightarrow \neg \langle goal \rangle \neg F$
$\langle goal \rangle F \rightarrow \neg \langle bel \rangle \neg F$

According to [5], worlds compatible with a decision maker's goals must be included in those compatible with the decision maker's beliefs. This is summarised in the following axiom:

$$\langle bel \rangle F \rightarrow \langle goal \rangle F \tag{11}$$

A database is full of decision maker's belief. Consequently, many redundant goals will be generated due to the presence of the above axiom. A goal will be considered *achieved* (resp. *active*) in a state if it is derivable (resp. not derivable) in the state.

7   Semantics of LR^2L

A *model* of LR^2L is a tuple $\langle W, V, R_b, R_s, R_g \rangle$ in which W is a set of all possible worlds and V is a valuation which associates a world to a set of propositions which are true in that world. In other words, $V : W \rightarrow \Pi(P)$, where P is the set of propositions and $\Pi(P)$ is the power set of P. Some additional restrictions have to be placed on the relations of the model corresponding to some of the axioms presented in section 6.

The relation R_b relates a world w to a set of worlds considered possible by the decision maker from w. If there are n candidates in a decision construct which is active in a world w then the size of such set of possible worlds will be 2^n. An assertion is said to be a *belief* of the decision maker at a world w if and only if it is provable in every possible world accessible from w by the accessibility relation R_b. The presence of axiom $\neg\langle bel \rangle \perp$ (axiom (1)) in our system guarantees the existence of a world in which an assertion is true if it is believed in the current state. The accessibility relation R_b has the following set of properties due to axioms (1), (3) and (4):

(A) R_b is serial, transitive and euclidean.

The relation R_s is a *hyperelation* which is a subset of $W \times D \times \Pi(W)$. Semantically, if $\langle w, d, W' \rangle \in R_s$ then there is an amount of support d for moving to one of the worlds in W' from the world w, where W' is non-empty. In other words, the support d is for the set of assertions uniquely characterised by the set of worlds W'. Given the current world w, a decision maker either continues to stay in the current world or to move to one of the possible worlds accessible from w by R_b. In either case, the changed world always belongs to W. This states that there is always the highest support for moving to one of the worlds in W from any world. The restrictions on R_s are now summarised as the following set of properties:

(B) for every w in W and d in D, if $\langle w, d, W' \rangle \in R_s$ then $W' \neq \emptyset$, and for every w in W, $\langle w, \triangle, W \rangle \in R_s$.

Suppose, $\langle w, 0.7, \{w_1, w_2\} \rangle$ is in R_s, where the support 0.7 gets distributed as 0.4 and 0.3 to w_1 and w_2 respectively. In addition, $\langle w, 0.6, \{w_1, w_2\} \rangle$ is also in R_s, where the support 0.6 gets distributed as 0.5 and 0.1 to w_1 and w_2 respectively. Although, 0.6 is less than 0.7, having $\langle w, 0.7, \{w_1, w_2\} \rangle$ in R_s does not imply $\langle w, 0.6, \{w_1, w_2\} \rangle$ is in R_s. This demonstrates that, in general, $\langle w, d1, W' \rangle \in R_s$ does not necessarily mean $\langle w, d2, W' \rangle \in R_s$, for $d2 \leq d1$.

Aggregation of arguments introduces a hierarchy of preferences [7] among the set of all possible worlds accessible from w by the relation R_b. The maximal elements and possibly some elements from the top of the hierarchy of this preference structure will be called *goal worlds*. The relation R_g, which is a subset of R_b, relates the current world to the set of goal worlds. An assertion is a *goal* in a world w if and only if it is provable in every goal world accessible from w by the accessibility relation R_g. Axiom (9) introduces the seriality property on the accessibility relation R_g and axiom (11) restricts R_g to a subset of R_b. Thus we have the following set of properties of R_g:

(C) R_g is serial and $R_g \subseteq R_b$.

To keep our development practical and simple we have excluded a number of axioms related to goals. Two such axioms concerned with goals [19] are (a) if a decision maker has a goal of having a goal then s/he has this goal and the converse (b) if a decision maker has a goal of not having a goal then s/he has not got this goal and vice versa. If we had considered these axioms this would have introduced some extra properties on the accessibility relation R_g. Only one of the goal worlds is committed to move from the current world and this world will be called the *committed world*.

Given a model $\mathcal{M} = \langle W, V, R_b, R_s, R_g \rangle$, truth values of formulae with respect to a world w are determined by the rules given below:

$\models_{\mathcal{M}}^{w} \top$

$\models_{\mathcal{M}}^{w} p$ iff $p \in V(w)$.

$\models_{\mathcal{M}}^{w} \langle sup_d \rangle F$ iff there exists $\langle w, d, W' \rangle$ in R_s such that $\models_{\mathcal{M}}^{w'} F$, for every $w' \in W'$

$\models_{\mathcal{M}}^{w} \langle bel \rangle F$ iff for every w' in W such that $wR_b w'$, $\models_{\mathcal{M}}^{w'} F$

$\models_{\mathcal{M}}^{w} \langle goal \rangle F$ iff for every w' in W such that $wR_g w'$, $\models_{\mathcal{M}}^{w'} F$

$\models_{\mathcal{M}}^{w} \neg F$ iff $\not\models_{\mathcal{M}}^{w} F$

$\models_{\mathcal{M}}^{w} F \wedge G$ iff $\models_{\mathcal{M}}^{w} F$ and $\models_{\mathcal{M}}^{w} G$

A formula F is said to be *true* in model $\mathcal{M}$, written as $\models_{\mathcal{M}} F$, if and only if $\models_{\mathcal{M}}^{w} F$, for every world w in W. A formula F is said to be *valid*, written as $\models F$, if F is true in every model.

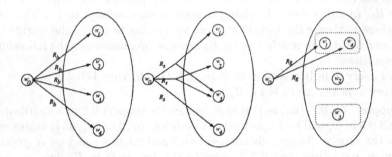

Fig. 3. Relation between the current world and possible worlds.

8 Medical Example

This section provides an example which illustrates the semantics presented in the previous section. First of all, we consider the dictionary D as *dict(Prob)* and D' is D itself. Suppose the current world w_0 is described by a database consisting of the formulae in section 5 (which are translated from the decision constructs presented in section 2) as hypotheses and the following set as *knowledge* ($\equiv F \wedge \langle bel \rangle F$): {*young, smoker, pain_in_upper_abdomen, weight_loss*}

The valuation V on w_0 is defined as follows:

$V(w_0) = \{$*young, smoker, pain_in_upper_abdomen, weight_loss*$\}$

Since there are 2 candidates in the *weight_loss_diagnosis* decision construct, there will be 2^2, that is, four worlds w_1, w_2, w_3 and w_4 considered possible by the decision maker whose valuations are as follows (see Figure 3):

$V(w_1) = V(w_0) \cup \{$*cancer, weight_loss_diagnosed*$\}$

$V(w_2) = V(w_0) \cup \{$*peptic_ulcer, weight_loss_diagnosed*$\}$

$V(w_3) = V(w_0) \cup \{$*cancer, peptic_ulcer, weight_loss_diagnosed*$\}$

$V(w_4) = V(w_0) \cup \{$*weight_loss_diagnosed*$\}$

The relations R_b and R_s in the model definition are defined as follows:

$R_b = \{\langle w_0, w_1 \rangle, \langle w_0, w_2 \rangle, \langle w_0, w_3 \rangle, \langle w_0, w_4 \rangle\}$

$R_s = \{\langle w_0, d2, \{w_1, w_3\}\rangle, \langle w_0, d4, \{w_2, w_3\}\rangle, \langle w_0, d2, \{w_4\}\rangle\}$

Note that *weight_loss_diagnosed* is true in each of the possible worlds and therefore this is a goal as the set of goal worlds is a subset of the the set of possible worlds. The goal corresponds to the provability of $\langle goal\rangle weight\_loss\_diagnosed$ in the current world using $\langle bel\rangle weight\_loss$ in conjunction with the formula $\langle bel\rangle weight\_loss \rightarrow \langle goal\rangle weight\_loss\_diagnosed$

The goal is active in w_0. We are, of course, assuming that the theorem prover of LR^2L is able to derive the negation of $\langle bel\rangle weight\_loss\_diagnosed$ from the current world by a mechanism similar to negation by failure. The total supports for the mutually exclusive possibilities are computed by the aggregation process (using the domain knowledge that *cancer* and *peptic_ulcer* are almost mutually exclusive candidates) as follows:

- support for $C_1(cancer \wedge \neg peptic\_ulcer) = \mathcal{A}(\{\langle C_1, G_1, d2\rangle\}) = d2$
- support for $C_2(\neg cancer \wedge peptic\_ulcer) = \mathcal{A}(\{\langle C_2, G_2, d4\rangle\}) = d4$
- support for $C_3(cancer \wedge peptic\_ulcer) = \mathcal{A}(\{\langle C_3, G_1, d2\rangle, \langle C_3, G_2, d4\rangle\}) = 0$
- support for $C_4(\neg cancer \wedge \neg peptic\_ulcer) = \mathcal{A}(\{\langle C_4, G_4, d4\rangle\}) = d2$

where each *di* is drawn from $dict(Prob)$ and the grounds G_1, G_2 and G_4 are:

$G_1 = weight\_loss \wedge smoker \rightarrow cancer$
$G_2 = weight\_loss \wedge pain\_in\_upper\_abdomen \rightarrow peptic\_ulcer$
$G_4 = weight\_loss \wedge young \rightarrow \neg cancer \wedge \neg peptic\_ulcer$

Assuming that *d4* is less than *d2*, the preference relation $\prec$ among the set of possible worlds is derived as $w_3 \prec w_2$, $w_2 \prec w_1$ and $w_2 \prec w_4$. The maximally preferred possible worlds are w_1 and w_2. The relation R_g in the model definition is now defined as follows:

$R_g = \{\langle w_0, w_1\rangle, \langle w_0, w_4\rangle\}$

This yields a dilemma. In case the decision maker cannot gather any more evidence, s/he may commit to w_4 by preferring w_4 to w_1. This involves adding $\neg cancer$ and $\neg peptic\_ulcer$ to the current state of the database as either beliefs or knowledge depending on the strength of support and decision maker's confidence. The goal *weight_loss_diagnosis* in the new situation will no longer be active due to the presence of

$\langle bel\rangle(\neg cancer \wedge \neg peptic\_ulcer) \rightarrow weight\_loss\_diagnosed$

Note that we add only beliefs to keep the belief revision option open in case of wrong diagnosis. Alternatively, if we now add $\langle bel\rangle positive\_biopsy$ as an additional evidence into the database that would increase the total support for C_1 as follows:

- total support for $C_1 = \mathcal{A}(\{\langle C_1, G_1, d2\rangle, \langle C_1, G'_1, d3\rangle\}) = d2 + d3 - d2 * d3$

where the additional ground G'_1 for C_1 is the following:

$G'_1 = weight\_loss \wedge positive\_biopsy \rightarrow cancer$

The revised valuation on each w_i will be as before except *positive_biopsy* changes its truth value. The relations R_s and R_g will be redefined as follows:

$R_s = \{\langle w_0, d2, \{w_1, w_3\}\rangle, \langle w_0, d4, \{w_2, w_3\}\rangle, \langle w_0, d2, \{w_4\}\rangle, \langle w_0, d3, \{w_1, w_3\}\rangle\}$
$R_g = \{\langle w_0, w_1\rangle\}$

Since w_1 is the only goal world, the decision maker considers w_1 as the committed world. Changing to the committed world from the current world involves adding *cancer* and ¬*peptic_ulcer* to the database as decision maker's beliefs. Adding ⟨*bel*⟩*cancer* to the database will trigger the decision for cancer treatment and the decision making process continues as before.

If $d2 = d3$ then the two members $\langle w_0, d2, \{w_1, w_3\}\rangle$ and $\langle w_0, d3, \{w_1, w_3\}\rangle$ of R_s are indistinguishable although they correspond to two different arguments. The relation in a more accurate model takes the following form:
$R_s = \{\langle w_0, d2, W_1\rangle, \langle w_0, d4, W_2\rangle, \langle w_0, d2, W_4\rangle, \langle w_0, d3, W_5\rangle\}$
where W_1 (resp. W_2, W_4 and W_5) is the set of possible worlds uniquely characterised by the set {*smoker, cancer*} (resp. {*pain_in_upper_abdomen, peptic_ulcer*}, {*young*} and {*positive_biopsy, cancer*}) which includes $\{w_1, w_3\}$ (resp. $\{w_2, w_3\}$, $\{w_4\}$ and $\{w_1, w_3\}$). In fact, the distribution of support $d2$ (resp. $d4$, $d2$ and $d3$) among the members of W_1 (resp. W_2, W_4 and W_5) is confined to the subset $\{w_1, w_3\}$ (resp. $\{w_2, w_3\}$, $\{w_4\}$ and $\{w_1, w_3\}$) as any other member is not considered possible by the decision maker.

9 Soundness and Completeness

The technique adopted in this section for establishing soundness and completeness of LR^2L is similar to the one for the class of normal logics in [3]. First of all, the following two propositions prove that the validity in a class of models is preserved by the use of the rule of inference and the axioms of LR^2L.

Proposition 4. *For every formulae F and G the following hold:*
- *if* $\models F$ *then* $\models \langle bel\rangle F$
- *if* $\models F \to G$ *then* $\models \langle sup_d\rangle F \to \langle sup_d\rangle G$
- *if* $\models F$ *then* $\models \langle sup_\triangle\rangle F$

Proposition 5. *For every formulae F and G the axioms (1)-(11) are valid in* LR^2L.

Propositions 4 and 5 establishes the basis of the soundness result. In order to prove the completeness result, the following class of models is relevant.

Definition 6. A model $\mathcal{M} = \langle W, V, R_b, R_s, R_g\rangle$ of LR^2L is called a *canonical model*, written as $\mathcal{M}_c$, if and only if
- $W = \{w: w$ is a maximal consistent set in logic $LR^2L\}$.
- For every w, $\langle bel\rangle F \in w$ iff for every w' in W such that wR_bw', $F \in w'$.
- For every w, d and W', $\langle sup_d\rangle F \in w$ iff there exists $\langle w, d, W'\rangle \in R_s$ such that $F \in w'$, for every w' in W'.
- For every w, $\langle goal\rangle F \in w$ iff for every w' in W such that wR_gw', $F \in w'$.
- For each proposition p, $\models^w_{\mathcal{M}_c} p$ iff $p \in w$.

Proposition 7. *Let* $\mathcal{M}_c = \langle W, V, R_b, R_s, R_g\rangle$ *be a canonical model of* LR^2L. *Then, for every w in W,* $\models^w_{\mathcal{M}_c} F$ *if and only if $F \in w$.*

Therefore, the worlds in a canonical model for LR^2L will always verify just those sentences they contain. In other words, the sentences which are true in such a model are precisely the theorems of LR^2L.

Theorem 8. *Let* $\mathcal{M}_c = \langle W, V, R_b, R_s, R_g \rangle$ *be a canonical model of* LR^2L. *Then* $\vdash F$ *if and only if* $\models_{\mathcal{M}_c} F$, *for every formula* F.

Existence of a canonical model for LR^2L is shown by the existence of a proper canonical model defined as follows.

Definition 9. A model $\mathcal{M} = \langle W, V, R_b, R_s, R_g \rangle$ of LR^2L is called a *proper canonical model*, written as $\mathcal{M}_{pc}$, if and only if
- $W = \{w : w$ is a maximal consistent set in logic $LR^2L\}$.
- For every w and w', wR_bw' iff $\{F : \langle bel \rangle F \in w\} \subseteq w'$.
- For every w, d and W', $\langle w, d, W' \rangle \in R_s$ iff $\{\langle sup_d \rangle F : F \in \cap W'\} \subseteq w$.
- For every w and w', wR_gw' iff $\{F : \langle goal \rangle F \in w\} \subseteq w'$.
- For each proposition p, $\models_{\mathcal{M}_{pc}}^w p$ iff $p \in w$.

By definition, a proper canonical model exists and the following proposition establishes that a proper canonical model is a canonical model.

Proposition 10. *Suppose* $\mathcal{M}_{pc}$ *is a proper canonical model of* LR^2L *as defined above. Then* $\mathcal{M}_{pc}$ *is also a canonical model.*

Theorem 11. *If* $\mathcal{M}_{pc}$ *is a proper canonical model of* LR^2L *then the model satisfies properties (A), (B) and (C).*

Suppose Γ is the class of all models satisfying the properties (A), (B) and (C). Then the following soundness and completeness theorem establishes the fact that LR^2L is determined by Γ.

Theorem 12. *For every formula F in LR^2L, $\vdash F$ if and only if $\models F$.*

10 Discussion

We have presented a modal formalisation of a very general framework for the design of knowledge-based decision support systems. As well as the conventional evaluation of decision options, we are able to handle within our framework the proposal of decision candidates and reasoning about actions once a decision has been made. Consequently, the model presented in this paper provides a more comprehensive account of decision making than classical decision theory. Furthermore, we are not constrained by requirements for comprehensive probability tables to enable a decision to be made; qualitative reasoning can be used if complete probability tables are unavailable. A discussion detailing the advantage of our approach over the traditional statistical decision theory (e.g. expected utility theory) and knowledge-based (expert) systems can be found in [9].

Although we have focussed on the decision making component of our model in this paper, the theory can be expanded into a general formalism incorporating actions and temporal reasoning. This aspect has been described elsewhere [9, 10].

Our formal model of decision making has already been successfully used in decision support applications in the domains of cancer and asthma protocol management [12]. In these applications, a human decision maker has over-riding control of the management process. However, we envisage that this extended model will provide a promising framework for the design of autonomous agents and multi-agent systems.

References

1. R. H. Bonczek, C. W. Holsapple, and A. B. Whinston. Development in decision support systems. *Advances in Computers*, 23:123–154, 1984.
2. R. Bradley and N. Swartz. *Possible Worlds*. Basil Blackwell, 1979.
3. B. Chellas. *Modal Logic*. Cambridge University Press, 1980.
4. D. A. Clark, J. Fox, A. J. Glowinski, and M. J. O'Neil. Symbolic reasoning for decision making. In K. Borcherding, O. I. Larichev, and D. M. Messick, editors, *Contemporary Issues in Decision Making*, pages 57–75. Elsevier Science Publishers B. V. (North-Holland), 1990.
5. P. R. Cohen and H. Levesque. Intention is choice with commitment. *Artificial Intelligence*, 42, 1990.
6. S. K. Das. *Deductive Databases and Logic Programming*. Addison-Wesley, 1992.
7. S. K. Das. A logical reasoning with preference. *Decision Support Systems*, 15:19–25, 1995.
8. S. K. Das and D. Elsdon. R^2L. Technical Report RED/ QMW/ WP/ 740/ 1/ 4, QMW, University of London, London, 1994.
9. S. K. Das, J. Fox, P. Hammond, and D. Elsdon. A flexible architecture for autonomous agents. *revised version is being considered by JETAI*, 1995.
10. S. K. Das and P. Hammond. Managing tasks using an interval-based temporal logic. *Journal of Applied Intelligence*, in press.
11. R. Fagin and J. Y. Halpern. Belief, awareness and limited reasoning. *Artificial Intelligence*, 34:39–76, 1988.
12. J. Fox and S. K. Das. A unified framework for hypothetical and practical reasoning (2): lessons from medical applications. In *Proceeding of FAPR*, June 1996.
13. J. Fox, S. K. Das, and D. Elsdon. Decision making and planning in autonomous systems: theory, technology and applications. In *Proceedings of the ECAI Workshop on Decision Theory for DAI Applications*, 1994.
14. J. Y. Halpern and Y. O. Moses. A guide to the modal logics of knowledge and belief. In *Proceedings of the 9th International Joint Conference on Artificial Intelligence*, pages 480–490, 1985.
15. P. J. Krause, S. J. Ambler, M. Elvang-Goransson, and J. Fox. A logic of argumentation for uncertain reasoning. *Computational Intelligence*, 11, 1995.
16. S. A. Kripke. Semantical analysis of modal logic I: normal modal propositional calculi. *ZMLGM*, 9:67–96, 1963.
17. E. J. Lemmon. *An Introduction to Modal Logic*. Basil Blackwell, 1977.
18. J.-J. Ch. Meyer, W. van der Hoek, and G. A. W. Vreeswijk. Epistemic logic for computer science: a tutorial (part one). *EATCS*, 44:242–270, 1991.
19. J. Wainer. Yet another semantics of goals and goal priorities. In *Proceedings of the 11th European Conference on Artificial Intelligence*, pages 269–273, August 1994.

A Unified Framework for Hypothetical and Practical Reasoning (2): Lessons from Medical Applications[1]

John Fox[*], Subrata Das[**]

[*] Imperial Cancer Research Fund, London
[**] William Penney Laboratory, Imperial College, London

Abstract . A general theory of decision making which unifies hypothetical reasoning (reasoning about beliefs) and practical reasoning (reasoning about actions and plans) has been reported elsewhere (e.g. Fox et al, 1988; 1990). The theory is grounded in research on non-standard logics, and particularly our proposals for LA, a logic of argumentation (Fox et al, 1992; Krause et al, 1995). A model-theoretic semantics for the theory is presented in a companion paper (Das et al, 1995). In this paper we illustrate the utility of the approach in medical decision making and management of clinical procedures (e.g. diagnosis; selection of investigations and therapies; drug prescribing, execution of care plans). The theory provides foundations for an application development methodology based on a formal language for representing medical knowledge. The paper discusses some lessons learned from these applications and identifies some important problems which theoreticians might valuably address.

1. Introduction

Medical informaticians have been developing computer systems to help doctors make clinical decisions for well over thirty years. However, there has been relatively little impact on routine clinical practice. Among the reasons for this have been a lack of convincing demonstrations that computers can help to improve care; unacceptable demands on busy clinical staff; inflexibility of user interfaces, and perhaps a certain unwillingness by doctors to accept that they might benefit from the use of computers.

This situation is now changing rapidly. Evidence is becoming available that decision support systems can significantly improve medical care (e.g. Johnstone et al, 1994), and doctors are recognising that rapid growth of medical knowledge, combined with increasing economic pressures and the implications of ageing populations, mean that it will be increasingly difficult to continue to deliver high quality care unaided. In the coming years the main obstacle to the successful introduction of computerised decision support systems will not be professional doubts, but limitations of the technologies and concerns about safety.

---

[1] Many people have been involved in work described in this paper, though the views presented are only those of the authors. We particularly benefited from contributions by: Paul Krause, Peter Johnson, Paul Taylor, Jean-Louis Renaud-Salis, Paul Ferguson, Saki Hajnal and Peter Hammond. We also wish to thank Jun Huang, David Elsdon, Claude Gierl, Richard Thomson, Andrew Jackson-Smale, Colin Gordon, Philippe Lagouarde, Ian Herbert, Robert Walton, Andrzej Glowinski, Mike O'Neil, Ahmed Benlamkaddem, Pierre Robles and Zeljko Ilic, all of whom have contributed to the applications described in the paper.

The term "decision support system" (DSS) is used to describe many different kinds of clinical software. Some authors include general information sources, such as drug databases, electronic texts and multimedia documents in the category, since they provide general information that can be relevant to a clinical decision. Others restrict use of the term to systems which provide assistance that is specific to the care of a particular patient (Johnstone et al, op cit). Examples of this kind of assistance include *reminders* (e.g. about topics which are relevant to specific diagnosis or treatment decisions), *alerts* (e.g. drug interaction warnings) and *recommendations* (e.g about the most likely diagnosis or most preferred therapy). Recently the term has also come to include systems that help in planning and scheduling clinical procedures (e.g. complex therapy plans such as cancer chemotherapy).

This paper describes a framework for supporting all the above aspects of decision making in a unified way; one which is intended to be flexible and intuitive for doctors to use without sacrificing theoretical soundness or practical safety.

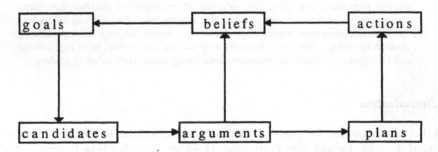

Fig 1: The "DOMINO" model of decision making and plan management

The framework is summarised in figure 1, the "DOMINO model". This consists of a set of reasoning elements which work together in an integrated way. The model is grounded in research on non-standard logics and logic programming.

The DOMINO model has emerged from a decade of work on applying computational logic to the problem of reasoning under uncertainty. This has been motivated by a desire to build practical aids for use in a variety of clinical settings. Our work has addressed general practice (e.g. Fox et al, 1987; Walton and Gierl, 1995), cancer care (e.g. Renaud-Salis and Taylor, 1990; Shortliffe, 1990), radiology (e.g. Taylor, 1995), clinical trials (e.g. Hammond et al, 1994) and multi-disciplinary shared care (Huang et al, 1994).

Many of the systems presented below share functionality with other knowledge-based systems reported in the literature, including diagnostic systems, and systems for therapy management (e.g. Shortliffe, 1990). However, the approach outlined here is believed to differ significantly from other work in a number of respects.

1. Knowledge based DSSs are flexible but have not been based on a well understood theory of decision making. In contrast, systems based on statistical decision theory draw on a "normative", well understood framework, but are inflexible and difficult to use. The DOMINO model is believed to offer a significant advance on these older traditions by providing practical flexibility while preserving theoretical soundness.

2. The model integrates a number of practical reasoning techniques. It can be used to support a range of tasks in many areas of medicine (and perhaps in non-medical fields as well). The main objectives of the present paper are to demonstrate this generality, and to argue that the framework acquires its versatility from a property of "compositionality": different kinds of application can be constructed from different subsets of reasoning elements.

3. The model has strong links with research on autonomous agents (Fox and Krause, 1991) and, in particular, with the class of intelligent agents known as Beliefs-Desires-Intentions or BDI agents (Wooldridge and Jennings, 1995).

4. The DOMINO model is embodied in a well-defined knowledge representation language, R^2L. This is intended to support application designers in clearly specifying the knowledge required for decision making and planning in the application domain. A formal semantics for most of R^2L has been developed; this is presented in our companion paper (Das et al, 1995).

5. Whether used only to support decision making or incorporated in autonomous agents, decision models raise important issues concerning operational safety. We advocate using a number of techniques for maximising safety (Fox, 1994), ranging from formal specification and verification of knowledge bases to more novel methods by which systems reason explicitly from general "safety axioms" (Hammond et al 1994). We believe the work is distinctive in addressing the challenges of safety-critical applications.

Overview

The next section provides a summary of the history and an intuitive presentation of the DOMINO model. Section 3 presents a number of applications. Section 4 summarises the R^2L language and outlines a methodology for developing applications. Finally, section 5 discusses general lessons learned from the medical domain and theoretical challenges arising from it.

2. The DOMINO model: history and informal explanation

The origins of the model are in the *Oxford System of Medicine* (OSM), a decision support system for general practitioners (Fox et al, 1987). When the OSM was conceived, medical DSS research was generally concerned with decision making in specialist medicine, such as statistical methods for diagnosis of abdominal pain or expert systems for diagnosis and treatment of bacterial infections. General practice has requirements which do not arise in specialist applications and which in our view demanded a new approach. Among these requirements were the following:

(1) It is necessary to support many different aspects of care. Patient care involves many different types of decision for a patient, not just diagnosis but also decisions about investigation, therapy selection, drug prescribing, risk assessment, referral to specialists etc..

(2) The DSS should permit a doctor to use it in a variety of ways. Doctors will wish to choose the type and level of support required rather than have a technologist's view imposed upon them. An experienced doctor will often only want quick access to up to date information, or reminders about topics relevant to a specific decision, and perhaps only occasionally will require a "second opinion" about diagnosis or treatment.

(3) General practice covers the whole of medicine; it should incorporate sufficient medical knowledge to address a wide range of common problems seen by the practitioner.

(4) The system should be able to provide practical help using only qualitative information (because reliable statistical data are only available for a small fraction of medicine).

The OSM design addressed these requirements using the layered architecture shown in figure 2. Here the outer layers correspond to general knowledge which is used to interpret more specific, and particularly patient-specific knowledge. Knowledge was encoded declaratively as facts which instantiated a decision procedure encoded as first-order rules (Fox et al, 1990; Huang et al, 1993). These rules embodied *ad hoc* inference techniques, though these were later developed formally.

The *Oxford System of Medicine* provided support for the general practitioner in making several types of decision including diagnosis, investigation selection, and treatment decisions. The OSM demonstrated a range of queries on dynamic patient data and static medical knowledge, and offered various kinds of help to the user in focussing on specific clinical goals and structuring the decision to be taken (Fox et al, 1987; 1990). For example, the user could identify weight loss as a clinical focus in the patient record, and select "diagnosis", "investigation" or "treatment" as a goal. In response the OSM applied an appropriate strategy for decision making in order to:

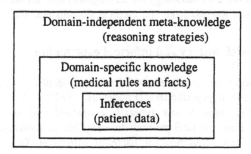

Fig 2: layered architecture of the Oxford System of Medicine

(1) generate possible decision candidates (e.g. diagnoses, treatments);

(2) construct arguments for and against the different decision options, and

(3) aggregate the arguments to yield a preference ordering on the various diagnostic or treatment options.

The OSM was implemented as a deductive database (in Prolog). This permitted the declarative encoding of first-order rules and meta-level knowledge, and simplified the provision of a range of user queries on the static knowledge base (for information retrieval) and dynamic patient database (for patient-specific reminders, explanations etc.).

A second source of requirements which influenced the development of the DOMINO model was a system designed to support decision making in oncology (cancer care). The *Bordeaux Oncology Support System* (BOSS1) was developed to meet many of the same requirements as the OSM, but in addition it was designed to support the execution of complex therapy plans, as pioneered in the ONCOCIN system at Stanford University. BOSS applied decision making functions to the diagnosis of breast cancer, assessment of the stage of the disease, and selection of a therapy plan (often referred to as a "protocol" in medicine), represented as a logic database. The protocol was used to guide the execution of medical tasks: the acquisition of clinical data, and sequencing of decisions and clinical acts etc (Renaud-Salis and Taylor, 1990). BOSS exploited a fairly complete, though still *ad hoc*, version of the DOMINO.

The DOMINO model as a decision procedure

The DOMINO can also be viewed as an architecture for an "agent" which maintains a set of beliefs about a world, and responds to situations which arise in that world by formulating problems, taking decisions about how to solve them, and carrying out plans and actions to implement its decisions.

In the context of clinical medicine the agent is a software system which maintains a set of *beliefs* about a patient (consisting of a set of facts about a patient, recorded over time and stored in a patient record system). Suppose data are added to the patient record which indicate that the patient has lost weight. The agent may infer that this is abnormal and, consequently, adopt a *goal* to diagnose the cause of the abnormaliy. From its general medical knowledge it deduces that possible causes of weight loss include, say, cancer and peptic ulcer and records these as *candidate solutions* for the diagnosis problem. Using meta-knowledge about the kinds of knowledge which are relevant in making diagnosis decisions, and additional facts in the patient record (e.g. the patient is elderly), the agent constructs *arguments* for and against each candidate diagnosis. This is carried out using meta-knowledge about the facts and/or rules which are relevant in generating candidates and constructing arguments for each type of problem. To illustrate, the OSM knowledge base included first-order meta-rules such as "any condition which could *cause* the presenting problem is a possible diagnosis" or "when looking for arguments in favour of a disease consider whether any symptoms or signs in the patient record are statistically associated with each hypothetical disease".

At some point enough information may be available about the patient to *commit* to one diagnostic candidate or another, say cancer. This decision is added to the patient record as a new belief.

The commitment to a diagnosis causes another goal to be raised: to *treat* the cancer. As before, a set of candidate solutions is inferred from the medical knowledge base, but this time the candidates are *plans* and/or *actions* rather than hypotheses: chemotherapy, radiotherapy or surgery, for example. Once again, the pros and cons of these alternatives are *argued* for the specific patient and, in due course, the agent takes a decision to *commit* to a particular therapy. This decision puts in train a plan, a sequence of actions, recursive subplans and decisions, for treating the patient. In principle, therapy plans may be constructed dynamically, but in the applications described here we only choose among standardised protocols.

Execution of a clinical plan or protocol will typically involve a number of steps or *tasks*. These will need to be scheduled for execution in a particular order and/or at particular times. For example, we may need to establish various baseline measurements on the patient, administer several cycles of treatment, monitor the effects of treatment and follow up the patient's condition at intervals following therapy. Scheduling must be done using general knowledge of temporal and logical constraints, and situation-specific factors like the patient's well-being and availability of resources. Some tasks will be obligatory, some optional and some based on decisions. Tasks can be specified as collections of ground facts in the agent's knowledge base.

3. Demonstrations of the practicality of the model

Figure 3 summarises the operations which the OSM carried out in terms of the DOMINO model. The goal of decision making was established manually, by the user, in this system but reasoning about possible candidate decision options and the arguments pro and con each alternative was carried out automatically. Arriving at a decision (about the preferred diagnosis or treatment) was carried out by *aggregating* the arguments. This yielded a preference ordering over the candidates; the system could in principle make the decision autonomously, though in practice we have adopted the policy that a trained doctor must actually take (commit to) any decision.

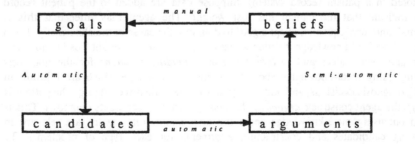

Fig 3: schematic representation of reasoning in the Oxford System of Medicine and CAPSULE prescribing system

An almost identical pattern of decision making was implemented in the CAPSULE drug prescribing system (Walton and Gierl, 1995). CAPSULE was designed to explore the use of logical argumentation in routine prescribing. Figure 4 shows a typical screen from the CAPSULE system. The top half of the display (behind the inset box) contains a view of a simple patient record. It shows the patient's main problem (that which requires medication) and additional information about associated problems, relevant past history, other current drugs etc. These data are all potentially relevant to formulating arguments about the best treatment for the main problem.

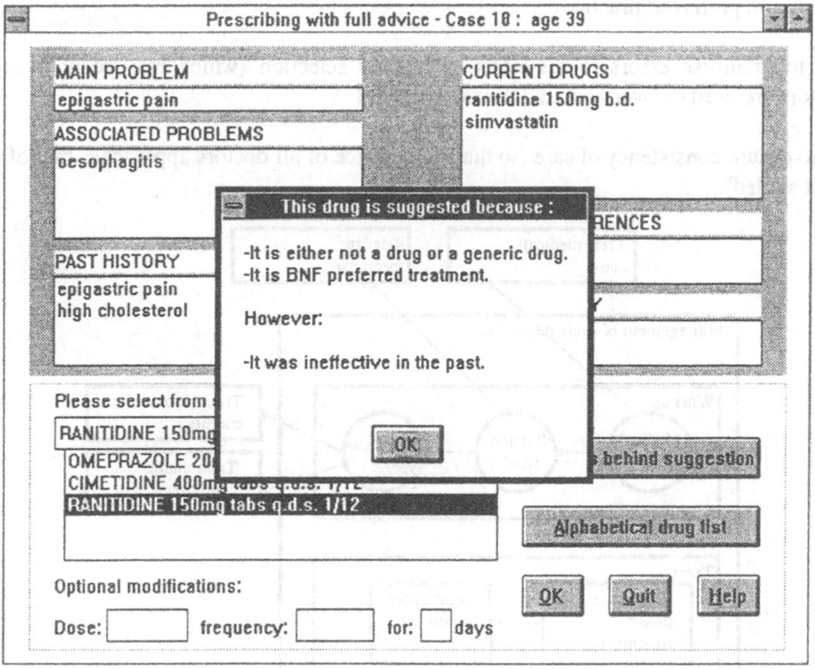

Fig 4: CAPSULE, a system for help in prescribing for common conditions. The system examines data in a patient record (top, back) and proposes a set of possible medications in order of preference (on the left near the bottom of the figure). On request CAPSULE provides a summary of the arguments for and against a selected candidate medication (central window). CAPSULE was, developed by Robert Walton and Claude Gierl, who have shown it can significantly improve the speed and quality of GP prescribing.

CAPSULE and the OSM only support individual decisions. This may be sufficient for many situations but often it would be desirable to provide support for more complex patient management processes. For example decisions may be chained together (e.g. a diagnosis decision followed by a treatment decision). In addition, routine patient management typically involves tasks which do not entail significant uncertainty: collecting and recording information; ordering tests; making appointments; scheduling hospital admissions and, increasingly in modern medical practice, paying attention to issues of resource utilisation and the maintenance of administrative records.

Plans and protocols

An important and growing topic in medicine is the use of standardised care plans ("guidelines" and "protocols" in medical terminology). BOSS used a chemotherapy protocol for the treatment of breast cancer, for example. Among the purposes of standardised guidelines are:

(1) to ensure use of state-of-the-art disease management strategies (e.g. in the management of cancer or chronic diseases, where medical research is leading to frequent changes in preferred practice),

(2) to minimise errors in diagnosis or therapy selection (which may occur because doctors are tired or overworked, for example), and

(3) to ensure consistency of care (so that the practice of all doctors approaches that of the most skilled).

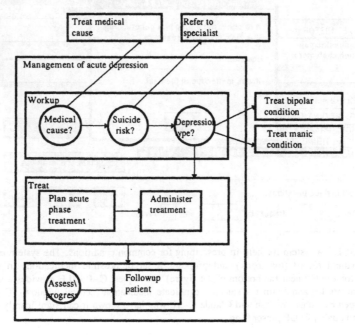

Fig 5: Guideline for the management of acute depression. Decisions are represented by circles, plans and actions by boxes, and arrows are temporal or logical scheduling constraints.

A typical guideline consists of a number of tasks. These include information collection, decision making, therapy planning and administration tasks. In primary care, for example, the management of chronic conditions like asthma, diabetes and hypertension will typically require a detailed "history" and characterisation of the problem, careful selection of therapy and continued monitoring and reassessment of the patient. In specialist care, such as cancer chemotherapy, protocols are often complex and include many obligatory rather than optional procedures. If protocols are presented in the form of conventional documents they can be hard to follow and use in busy clinical situations. It has been recognised for some time that computers might be used to help doctors comply with guidelines and protocols more successfully.

Fig 6: The Synergy system, developed under the EC Health Telematics Project DILEMMA, demonstrates the integration of "intelligent" guidelines for common conditions with conventional general practice management software. The figure shows a summary of the patient record (back window); an interactive guideline reminds the GP of topics which are relevant to the patient's problems (middle) and an example of decision support, in this case showing a suggested diagnosis with the most likely diagnosis highlighted together with the reasons for the recommendation (front). Synergy was developed by Andrew Jackson-Smale and Peter Johnson.

Consider the management of acute depression in general practice (see figure 5). Although this is a common problem, it is often poorly managed because preferred practice demands extensive data collection and scheduling of procedures, as well as several significant decisions. The figure gives a schematic representation of part of a guideline for the management of acute depressesion published by the American College of Psychiatrists. However, although it involves many steps it is not a particularly complex procedure in comparison with many used routinely in some fields of medicine (such as oncology, see Hammond, 1994 for examples).

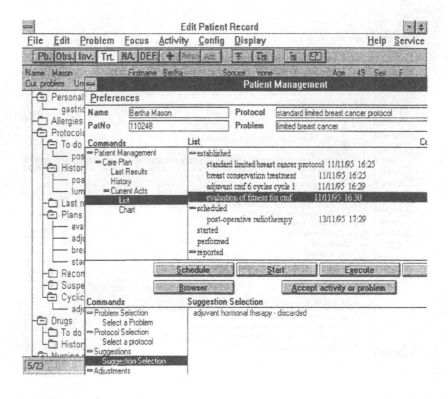

Fig 7: A screen from a cancer management system (BOSS2, developed under the EC's Healthcare Telematics Programme). The application is specifically concerned with managing a combined surgery, radiotherapy and chemotherapy protocol, and provides help with patient data recording, decision making and the scheduling of clinical procedures. The application is accessed via an electronic patient record system at the "back" of the figure (developed by collaborators at the Institut Bergonié, Bordeaux) from which members of medical staff call up the patient management assistant, in the front of the figure. The main panel in the top half of the patient management window summarises the current care plan, with facilities for reviewing and updating it. The panel at the bottom provides decision support (e.g. for diagnosis and staging, protocol selection and suggestions for action).

We have developed several applications which embody the capability to execute guidelines like those in Figure 5. The SYNERGY system is designed to support use of guidelines by general practitioners (Figure 6). BOSS2 provides facilities for supporting protocol-based cancer care (Figure 7). As with BOSS1 outlined earlier it is concerned with chemotherapy for breast cancer, but it is considerably more developed than the earlier version. A system to support the management of clinical trials (CREW), and a radiology workstation which combines decision support and protocol management functions with both manual and automated image interpretation (CADMIUM) are illustrated in figures 8 and 9.

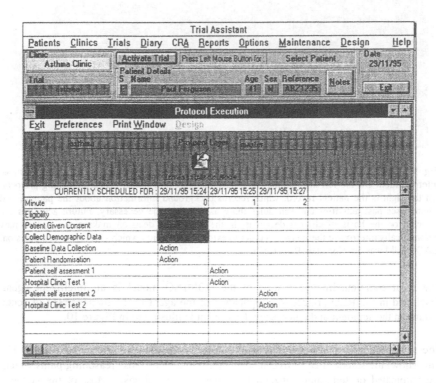

Fig 8: The CREW clinical trials management system. A clinical trial of a new treatment requires precise compliance with a research protocol which sets out the criteria for including patients in the trial (establish medical eligibility, patient consent), defines the data to be collected (e.g. demographic data and baseline clinical data) and the therapeutic procedure to be followed. The primary function of CREW is to schedule and manage these tasks over time, under the control of the clinical research assistant who interacts with the task chart in the bottom window. CREW was developed by Paul Ferguson, now at MedWay S.A., Bordeaux, and Jonathan Gee, now at the Hammersmith Hospital, London.

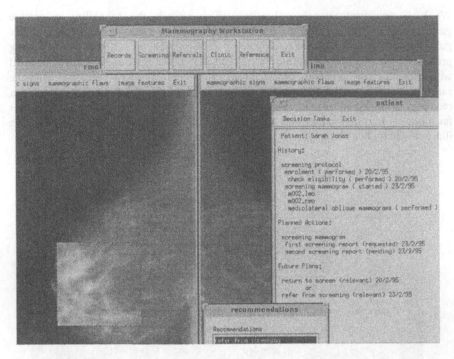

Fig 9: CADMIUM, a radiology workstation which combines decision support with image interpretation. On the left are mammograms taken as part of a breast cancer screening protocol. On the right is the patient record, which records background information about the patient, and summarises past activities and schedules those as they become necessary. Future activities are selected on the basis of clinical decisions. In the process of decision making the system may automatically interrogate the images in order to extract information which is relevant to a decision. CADMIUM was developed by Paul Taylor.

The most complete implementation of the DOMINO model is an intelligent agent developed in the RED project. An application of the agent to the management of acute asthma in the accident and emergency department is illustrated in Figure 10.

The relationships between these systems can be understood by referring to figure 11. This shows the standard DOMINO model but also indicates those reasoning paths which are automated or semi-automated in the different applications. Most of the systems deduce decision candidates, construct arguments and schedule tasks entirely automatically. The clinical trials system CREW, on the other hand, only automates the scheduling function. The CADMIUM radiology workstation alone is capable of automatically interrogating medical images and only the RED asthma management system is capable of generating decision-making goals without human intervention. As remarked earlier, the processes of commitment - to new beliefs or to plans or actions, could be carried out automatically in any of these systems, but again in practice preferences are only offered as recommendations; the decision should only be taken by a trained doctor.

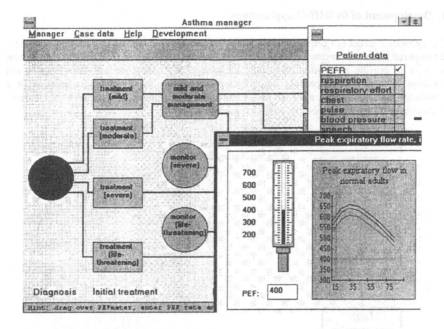

Fig 10: Decision support and guideline management technology used in an accident and emergency application: the management of acute asthma in adults. The decisions and actions recommended by the guideline are shown as a graphical workflow chart which reminds clinical staff of the tasks that need to be carried out. The panels on the right prompt for relevant patient information using easily understood forms. Colour and other kinds of coding show the need for specific decisions or actions or flag hazardous conditions which may arise. The application is based on a guideline published by the British Thoracic Society; it includes a hypertext document providing access to the original text and figures. The application was developed in the RED project (DTI/SERC, ITD 4/1/9053) by Claude Gierl, Paul Ferguson and David Elsdon of Integral Solutions Ltd, Basingstoke, UK.

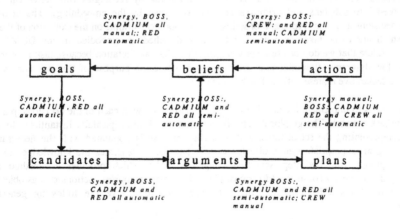

Figure 11. Relationship between SYNERGY, BOSS, CREW, RED and the DOMINO

4. Development of DOMINO applications

The DOMINO model has been embodied in a general purpose reasoning engine which provides the required inference functions. As with an expert system shell, in order to build an application we have to provide an application-specific knowledge base. Most medical knowledge is not available in anything approaching the necessary formalism, however, so the creation of a knowledge base has to be carried out in a number of steps. Figure 12 summarises the current methodology for development of DOMINO applications.

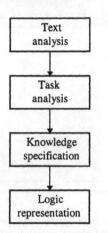

Fig 12: application
development methodology

The first step is concerned with acquiring and structuring relevant sources of knowledge (textbooks and other published documents, records of interviews etc) using text processing tools to extract, organise and index relevant material.

The second step is to define the overall structure of the application, in terms of the procedures, plans, decisions and other components required for the knowledge base. The task network for the management of acute depression shown in figure 5 is an example. In the graphical notation we use decisions are normally indicated by circles and plans by rectangles (the latter can be decomposed into subplans and primitive actions, as shown by the embedding). The arrows linking decisions and plans represent ordering and temporal constraints on the execution of these tasks which are to be taken into account by the task scheduler embodied in the DOMINO software. (Note that we do not use a conventional flow diagram notation because task networks are not flow diagrams; flow diagrams are procedural while the purpose of this notation is to provide a declarative representation of task structure.)

The next step is to implement the detailed knowledge associated with each of the decisions, plans etc in the task network. The graphical network is transformed into a partially instantiated logic database representing the set of tasks, and then the database is populated with the necessary details using an appropriate syntax directed editor. The editor generates the knowledge in a formal language, the Red Representation Language (R^2L), which provides a method of specifying the necessary constructs (decisions, plans, information sources, actions etc) as objects with a well-defined *object-attribute-value* structure. Decisions have the following general structure:

decision <decision ID>

    situation <set of beliefs>   d1

    problem <goal>   d2

    candidates <set of candidates>   d3

    oracles <set of information sources>   d4

    arguments <set of first-order argumentation rules>   d5

    commitments <set of first-order commitment rules>   d6

This structure reflects the entities and inference rules on the decision making (left) side of the DOMINO. Decisions need to be taken in particular situations which are considered to be problems in the application domain (d1,d2: e.g. an abnormal medical observation requires diagnosis and, once the diagnosis is known, therapy). Given a decision problem there may be one or more possible solutions (candidates, d3) and one or more relevant information sources (oracles, d4) which are relevant to deciding among the candidates. The decision procedure applies an argumentation theorem prover to the argumentation rules (d5) together with information in the situation data (d1) in order to derive pros and cons of the different decision candidates. Finally, when one or more of the commitment rules associated with a decision is satisfied (d6) the decision is taken. This results of decisions, such as a diagnosis of the abnormality, or an intension to carry out a treatment plan, are added to the database of beliefs. Plans have the following structure:

plan <plan ID>

    preconditions <set of beliefs>   p1

    components <set of subplans or actions (subtasks)>   p2

    scheduling <set of ordinal or temporal constraints on subtasks> p3

    termination conditions <set of beliefs>   p4

Once plans are committed for execution the scheduler will schedule the plan as and when the pre-conditions of the plan become satisfied (p1). Once a plan is scheduled all the subplans and actions (p2) are scheduled, taking into account any constraints on the order or timing indicated in the plan definition (p3). This is a recursive process which bottoms out in atomic actions, typically involving messages to the user interface or to other agents. The scheduler is sensitive to situation updates; it can terminate plan execution at any time if termination conditions become satisfied (p4). For example if a hazard of some sort arises, such as a patient having an allergic response to a medication, then the plan should be aborted and any scheduled actions cancelled.

The final step in the development methodology is to translate the R^2L definition into an executable knowledge base. The executable language (LR^2L) supports all the basic types of logical inference shown in figure 1, incorporating a sub-language for specifying temporal conditions, integrity and safety constraints and obligations. LR^2L does not have primitive constructs for the R^2L concepts of decisions, plans etc, but these are supported with more primitive notions. LR^2L is therefore a lower level but more expressive language than R^2L; in principle we believe it could be used to implement different architectures than the one discussed here. LR^2L is interpreted by a specialised theorem prover written in Prolog (Das et al, 1995).

5. Lessons learned

We believe that the body of experience gained in building the applications described in this paper has much to say to researchers in formal reasoning. We need their help in return. Having committed much effort to providing a formal foundation for our work we have become aware how difficult some of the underlying issues are (including logical, computational and epistemological issues). In contrast to the common expectations of medical informaticians we believe that the development of successful decision support systems is not just a matter of good practice in software implementation and testing. Important as these matters are, we also require a deep understanding of many theoretical questions. Medical decision support systems are safety-critical. Consequently we would strongly encourage the FAPR community to become actively involved in the medical field and support clinical technologists in their desire to build safe and sound applications.

Medicine is one of the richest fields of human practical reasoning so there is much to be gained in return for such an effort, even for the very theoretically minded. In the remainder of this paper, therefore, we shall try to underline some of the theoretical challenges medicine raises by outlining some of the lessons learned that impact on conference themes, together with some important outstanding questions. The themes that our work addresses include: theories of argumentation and aggregation; reasoning about beliefs, actions and plans; formal models of reasoning and integrated reasoning mechanisms; temporal reasoning; agent theory and logic programming.

Theories of argumentation and aggregation.

Medical problems frequently involve high levels of uncertainty. This uncertainty may be about what to believe (e.g. what is wrong with somebody) or uncertainty about what to do (e.g. what would be the best treatment). The orthodox approach to the management of uncertainty in decision support systems is to quantify the uncertainty in statistical or other terms and use standard probabilistic, fuzzy, or other methods to compute degrees of belief (e.g. in diagnostic propositions) or expected utilities (e.g. for alternative actions) in order to establish a preference ordering over the candidates. The challenge raised by medicine is that there is frequently no reliable basis for establishing objective quantitative parameters.[2] Nevertheless we must deal with the uncertainty we face in the practical world, how may we do this?

Classical logic helps us very little, of course, and even non-monotonic logics are of limited value because they provide no notion of *degrees* of belief, *relative amounts* of evidence etc.

For these reasons we have developed the intuitive idea of *argumentation* into a formal theory for reasoning about uncertain propositions in the absence of quantitative information (Fox, 1986) and the *aggregation* of collections of arguments in order to establish preferences over competing propositions (Fox et al, 1988). As remarked earlier, however, these ideas were somewhat ad hoc when first introduced so we have since put considerable effort into formalising them. The result is the logic LA, a variant of intuitionistic logic which provides a sound axiomatisation of argumentation as a general method for decision making under uncertainty (Fox et al, 1992; Krause et al, 1995; Ambler, 1996; Das et al, 1996).We believe that LA provides a very general

---

[2] Some advocate the use of subjective opinion as a substitute for objective probability but we know from research in psychology that such estimates are unreliable, due to well-known biases in human judgement. In any case there is no scientific justification for interpreting subjective levels of confidence as mathematical probabilities.

framework for practical reasoning under uncertainty which is more flexible than, but potentially compatible with, orthodox probability theory.

Argumentation, however, is a complex subject which raises many issues. Among the questions which arise in medicine are the following. When is it safe to act on a diagnosis (what makes an argument for or against a proposition sound and persuasive?). How should we handle evidence which may not be reliable or exploit promising medical theories which are not yet proven? (how do we deal with arguments which have some *force* but are not *conclusive*?). How do we decide whether or not a partcular body of medical knowledge is relevant to a particular case? (when is an argument relevant to a proposition and when irrelevant?). A common feature of many such questions is that they involve *meta-level reasoning about arguments*, for which at present we have little theoretical understanding but which formal analysis may shed light on.

Reasoning about beliefs, actions and plans.

The formulation and aggregation of a set of arguments to support some clinical conclusion, such as "the patient is suffering from cancer" or "chemotherapy is an appropriate course of action", may establish a preference over the alternatives (the patient does not have cancer; chemotherapy is not appropriate). In some cases the arguments may be conclusive (as when an imaging technique visualises the presence of a tumour, or chemotherapy is the only available therapy). In others there is residual doubt. Under this condition we require safe and sound rules for taking action. That is to say we must be able to prove that, given what we know at the time, our choice is the *best* choice before we make a commitment that could prove to be wrong, and possibly disastrous.

In practical medicine a wide range of policies may be adopted in. A *risk-averse* clinician, for example, might say "we should obtain all relevant information about a patient before committing to action". A more *resource-aware* doctor may say "there is no point in spending time, effort and money on further investigations if none of the possible remaining tests could result in a different preference ordering". Frequently general practitioners appear to adopt apparently ad hoc policies like "if there is any possibility of my patient suffering from a life-threatening condition, however small, I will refer the patient to an appropriate specialist". Or alternatively, "I know enough about this aspect of medicine; I can take treat the patient without a second opinion".

What theory might sanction or refute such policies? How may we formalise risk in logical terms? How do we provide a semantics for the concepts of subjective values or professional ethics? When is it reasonable for an agent to act upon a possibility, as distinct from a probability? and how may an agent reason soundly about what it knows and doesn't know?

Formal models of reasoning and the integration of reasoning models.

The computer science literature, such as that on medical expert systems and logic programming, tends to make the assumption that medical reasoning is just diagnostic reasoning. Clearly it is much more than that. First of all it involves reasoning about goals, which is to say that an agent has a problem if it finds itself in some situation that challenges its goals, or threatens to challenge them in the future. Practical medical reasoning must therefore embody a theory of goals and an understanding of different ways of achieving and maintaining goal states. Adaptive behaviour in the face of clinical problems will sometimes only require simple actions (e.g. prescribe some medication) but often we shall have to put plans in place for long term therapy, perhaps involving many individuals from different medical disciplines over many years. We have mentioned the example of cancer chemotherapy many times, but this is also true of the management of chronic conditions such as arthritis, diabetes, hypertension and many neurological conditions. All this has to be done in the face of considerable uncertainty about the state of the world and the efficacy of our actions for achieving our goals in the world.

To address problems of this complexity we need to combine many different types of reasoning. To be confident of them they must be formalised in some well understood manner, as a set of logics or in some other way. The DOMINO model is our own attempt to meet this objective; it provides a framework in which we can reason about goals and problems; possible solutions to problems; arguments for beliefs, plans and actions; reasoning about time, scheduling and obligations. We hope that the intuitive model we have presented here, and our formalisation in the accompanying paper, will prove to be of interest to theoreticians.

We are aware, however, that our framework is neither complete nor unique.

It is incomplete in several ways. We have provided no theory of dynamic planning, only a theory for selecting among alternative plans (protocols or guidelines). Furthermore the safe execution of plans involves constantly reasoning about the consequences and side-effects of plans (Fox 1993; Hammond et al, 1994) so we need some theory of knowledge-based forecasting. Also we need to develop some theoretical framework for perception of the environment (e.g. regarding the detection of hazardous events in the clinical environment, or abnormal structures in images).

The framework is not unique. Actions and guidelines/protocols are equivalent to *operators* and *skeletal plans* which are discussed in the AI literature on planning. There is a considerable body of work on goal-based reasoning, logics of belief, knowledge, planning and acting etc. Perhaps there are many "dominos", composed from different logics, that interact in different ways to yield different behaviour in the same conditions. While different, these behaviours may be equally "rational" in that they satisfy normative axioms but represent different tradeoffs for different kinds of environment (recall our risk-averse and resource-aware doctors - who is right?). Our conclusion from this is that we require a meta-theory of practical reasoning which acknowledges the possibility of many different instantiations of rationality, and some well understood rules for composing different logics for different purposes.

Agent theory

The DOMINO can be viewed as a collection of logics, or alternatively as an architecture for an agent which unites a set of practical reasoning methods. As an agent theory the model is similar to the BDI agent concept (Wooldridge and Jennings, 1995): they share mentalistic ideas like beliefs and "desires" (goals), and guidelines and protocols seem similar in spirit to the "intentions" of the BDI agent.

The main contribution of our work to agent theory is the proposal embodied in the DOMINO model for the integration of plan management with functions for reasoning about goals and unpredictable events and making decisions under uncertainty. Generally, goals will require hypothetical and/or practical reasoning but many problems may be solved in more than one way. To deal with this we introduced the idea that agents may construct arguments for and against different candidate solutions. Decision taking may involve the aggregation of arguments to induce a preference ordering over the candidates, committing to one or other of the options based on these preferences or on other decision policies which offer different tradeoffs for the agent.

An important area of future development is in multi-agent planning. As remarked, the care of an ill person often requires several different clinical disciplines and hence communication between and coordination of individuals in different places. Huang et al (1995) have explored ways of supporting interactions in healthcare teams and between their decision support systems which will ensure that decision making, planning and acting take place in a coordinated way. To do this Huang et al built a communication and coordination layer on top of an agent implementation; this provided a set of communication primitives for informing other agents of needs or data, requesting information or actions, and establishing and maintaining responsibilities for different tasks be members of the care team. This language is practically promising but, as with our earlier work, their practical solution may be theoretically ad hoc; we look forward to working with formalists to establish a firmer foundation for a technology which is likely to be of growing importance.

6. Conclusions

We have developed a general model for knowledge based decison making and plan management based on an integration of non-standard logics. The model has been validated on a range of medical applications. It shows practical promise and addresses a number of theoretical problems for formal and applied practical reasoning. These include issues concerning argumentation and the management of uncertainty; reasoning about actions and plans; integration of reasoning systems, and agent architectures and theories. We urge computational logicians and other theorists to address problems in complex fields such as medicine, in part because they raise many important problems and challenges to theory, and in part because technologists in the field need more formal tools if they are to successfully develop technologies which can be trusted.

7. References

Ambler, S "A categorical approach to the semantics of argumentation" *Mathematical structures in computer science* (in press).

Das S, Fox J and Krause P "A unified framework for mathematical reasoning (2): theoretical foundations (this volume), 1995a.

Das S, Fox J, Elsdon D and Hammond P "A flexible architecture for intelligent agents" (submitted) 1995b

Fox J "Three arguments for extending the framework of probability" *In Proc. 1st Conference on Uncertainty in Artificial Intelligence*, Los Angeles; Machine intelligence and pattern recognition 4, Amsterdam: North Holland, 1986.

Fox J "Decision theory and autonomous systems" in MG Singh and L Trave-Massuyes (ed) *Proc IMACS International Conference on Decision Support Systems and Qualitative Reasoning*, Amsterdam: Elsevier, 1991.

Fox J "On the soundness and safety of expert systems" *Artificial Intelligence in Medicine*, 5, 159-179,1993

Fox J, Glowinski A J, O'Neil M "The Oxford System of Medicine: a prototype information system for primary care" *Lecture notes in Medical Informatics*, Berlin: Springer, 213-226. 1987.

Fox J "Decision making as a logical process" In B Kelly and A Rector (eds) *Research and development in expert systems*, Cambridge: Cambridge University Press, 1988.

Fox J, Clark D A, Glowinski A J and O'Neil M "Using predicate logic to integrate qualitative reasoning and classical decision theory" *IEEE Trans. Systems Man and Cybernetics*, 20(2), 347-357, 1990.

Fox J, Krause P and Ambler S "Arguments, contradictions and practical reasoning" in B Neumann (ed) *ECAI 92: Proc. 10th Eur. Conf. on AI*, 623-627, 1992.

Fox J and Krause P "Symbolic decision theory and autonomous systems" in D'Ambrosio, B Smets P and Bonissone P (eds) *Proc. 7th Conference on Uncertainty in AI*, 103-110, Morgan Kaufman, 1991.

Hammond P, Harris A L, Das S K and Wyatt J "Safety and decision support in oncology" *Methods of Information in Medicine*, 33 (4), 371-381, 1994.

Huang J, Fox J, Gordon C and Smale A "Symbolic decision support in Medical Care" *Artificial Intelligence in Medicine*, 5 (5), 415-430, 1993.

Huang J, Jennings N R and Fox J "An agent based approach to healthcare management" *Int. J Applied Artificial Intelligence*, 9, 4, 1995

Johnstone M, Langton K B, Haynes B and Matheu A "Effects of computer-based clinical decision support systems on clinician performance and patient outcome, a critical appraisal of research" *Annals of Internal Medicine*, 120, 135-142, 1994.

Krause P, Ambler S, Elvang-Goransson M and Fox J, "A logic of argumentation for reasoning under uncertainty" *Computational Intelligence*, 11 (1), 1995.

Musen M, Tu S, Das A K and Shahar, Y "A component based architecture for automation of protocol-directed therapy", Stanford University Medical School, 1995.

Renaud-Salis J L and Taylor P "The Bordeaux Oncology Support System: Knowledge Representation and prototype" Project LEMMA technical report, Imperial Cancer Research Fund, February 1990

Shortliffe E S "An integrated oncologist's workstation" *National Cancer Institute*, 1990.

Taylor P "Decision support for image interpretation: A mammography workstation" in Bizais, Barillot and Di Paola (eds) *Image processing and medical imaging*, Dordrecht: Kluwer, 1995.

Walton R and Gierl C in Jackson-Smale et al "Evaluation of primary health care decision support systems" DILEMMA project deliverable D11, Imperial Cancer Research Fund, 1995

Wooldridge M and Jennings N "Intelligent agents: theory and practice" *The Knowledge Engineering Review*, 10 (2), 115-152, 1995

General Domain Circumscription and its First-Order Reduction

Patrick Doherty[1]*, Witold Łukaszewicz[2]**, Andrzej Szałas[2]***

[1] Department of Computer and Information Science,
Linköping University, S-581 83 Linköping, Sweden,
e-mail: patdo@ida.liu.se
[2] Institute of Informatics, Warsaw University,
ul. Banacha 2, 02-097 Warsaw, Poland,
e-mail: witlu,szalas@mimuw.edu.pl.

Abstract. We first define *general domain circumscription* (GDC) and provide it with a semantics. GDC subsumes existing domain circumscription proposals in that it allows varying of arbitrary predicates, functions, or constants, to maximize the minimization of the domain of a theory. We then show that for the class of *semi-universal* theories without function symbols, that the domain circumscription of such theories can be constructively reduced to logically equivalent first-order theories by using an extension of the DLS algorithm, previously proposed by the authors for reducing second-order formulas. We also isolate a class of domain circumscribed theories, such that any arbitrary second-order circumscription policy applied to these theories is guaranteed to be reducible to a logically equivalent first-order theory. In the case of semi-universal theories with functions and arbitrary theories which are not *separated*, we provide additional results, which although not guaranteed to provide reductions in all cases, do provide reductions in some cases. These results are based on the use of fixpoint reductions.

1 Introduction

In many common-sense reasoning scenarios, we are given a theory T specifying general laws and domain specific facts about the set of phenomena under investigation. In addition, one provides a number of closure axioms circumscribing the domain of individuals and certain properties and relations among individuals. The closure machinery normally involves the use of non-monotonic rules of inference, or in the case of circumscription, a second-order axiom. In order for a circumscribed theory to be useful, it is necessary to find a means of computing inferences from the circumscribed theory in an efficient manner. Unfortunately, the second-order nature of circumscription axioms creates an obstacle towards doing this.

---

* Supported in part by the Swedish Council for Engineering Sciences (TFR).
** Supported in part by KBN grant 3 P406 019 06.
*** Supported in part by KBN grant 3 P406 019 06.

In previous work [3], we proposed the use of an algorithm (DLS) which when given a second-order formula as input would terminate with failure, or output a logically equivalent first-order formula. Since circumscription axioms are simply second-order formulas, we showed that the DLS algorithm could be used as a basis for efficiently computing inferences for a broad class of circumscribed theories by first reducing the circumscription axiom to a logically equivalent first-order formula and then using classical theorem proving techniques to compute inferences from the original theory augmented with the output of the algorithm. In [4], the DLS algorithm was generalized using a reduction theorem from [11]. It was shown that a broad subset of second-order logic can be reduced into fixpoint logic. Moreover, a class of fixpoint formulas was characterized which can be reduced into their first-order equivalents.

In this paper, we extend the previous work in three ways:

1. We define a general form of domain circumscription which subsumes existing domain circumscription proposals in the literature ([10], [2], [6], [7], and [8]). We call the generalization *general domain circumscription* (GDC). GDC distinguishes itself from other proposals in the following manner. When circumscribing the domain of a theory T, it is permitted to vary arbitrary predicates, functions, or constants, to maximize the minimization of the domain of individuals.
2. We characterize a class of theories which when circumscribed using GDC are guaranteed to be reducible to equivalent first-order theories which are constructively generated as output from extended versions of the original DLS algorithm. Included in this class are theories for which both McCarthy's original domain circumscription [10] and Hintikka's mini-consequence [7] are always reducible to first-order logic.
3. We characterize a class of theories which, when first circumscribed using GDC and then circumscribed using an arbitrary circumscription policy, are guaranteed to be reducible to equivalent first-order theories which are constructively generated as output from the extended versions of the original DLS algorithm mentioned in the previous item.

We approach the characterization and reduction problems in the following manner.

- Given a theory T, we show that if the domain closure axiom is entailed by the domain circumscribed theory $Circ_D(T)$, then $Circ_D(T)$ is always reducible to a logically equivalent first-order theory.
- We then characterize a class of theories where the domain closure axiom is not only entailed by the domain circumscribed theory, but can be automatically generated and used in the extended algorithm to reduce theories from this class to their corresponding first-order equivalents.
- Given a theory in the class characterized above and an arbitrary circumscription policy applied to that theory, we show that the extended version of the DLS algorithm will always generate a first-order theory logically equivalent to the second-order circumscribed theory.

The key to the approach is determining when a domain circumscribed theory $Circ_D(T)$, entails it's domain closure axiom. Semantically, a possible answer is when the cardinalities of all minimal models of the domain circumscribed theory have the same finite upper bound. Syntactically, we can characterize two classes of theories that provide such constraints when minimized:

1. Universal theories without function symbols, where the general domain circumscription policy can include arbitrary constants and predicates that vary.
2. Semi-universal theories without function symbols, where the general domain circumscription policy can include arbitrary constants and predicates that vary.

The class of semi-universal theories is a broad class of theories much more expressive than universal theories which have previously been studied in the context of restricted forms of domain circumscription. In the case of universal and semi-universal theories with function symbols, where the general domain circumscription policy can include arbitrary constants, predicates and functions that vary, reducible classes of theories are difficult to characterize. In this case, we provide additional results which guarantee reduction non-constructively and additional methods which, although not guaranteed to provide first-order reductions in all cases, do provide reductions in some cases.

The paper is organized as follows. Section 2 consists of preliminary definitions and notation. In Section 3, general domain circumscription is introduced together with it model-preferential semantics. In Section 4, the original DLS algorithm is briefly described together with two limitations associated with the basic algorithm. In Section 5, two generalizations of the basic DLS algorithm are described which deal with the limitations previously described. In Section 6, reducibility results concerning different specializations of general domain circumscription are presented together with a number of concrete examples. In Section 7, we consider the potential for reducing a larger class of arbitrarily circumscribed theories which are first circumscribed using general domain circumscription.

We refer the reader to [5], for an extended version of this paper which includes all proofs, additional methods for reduction based on fixpoint methods, and additional examples.

2 Preliminaries

In this paper, the term *theory* always refers to a finite set of sentences of first-order logic. Since each such set is equivalent to the conjunction of its members, a theory may be always viewed as a single first-order sentence. In the sequel, we shall never distinguish between a theory T and the sentence being the conjunction of all members of T. Unless stated otherwise, the term *function symbol* refers to a function symbol of arity n, where $n > 0$.

2.1 Notation

An n-ary *predicate expression* is any expression of the form $\lambda \overline{x}.\ A(\overline{x})$, where $\overline{x}$ is a tuple of n individual variables and $A(\overline{x})$ is any formula of first-order classical logic. If U is an n-ary predicate expression of the form $\lambda \overline{x}.\ A(\overline{x})$ and $\overline{\alpha}$ is a tuple of n terms, then $U(\overline{\alpha})$ stands for $A(\overline{\alpha})$. As usual, a predicate constant P is identified with the predicate expression $\lambda \overline{x}.\ P(\overline{x})$. Similarly, a predicate variable Φ is identified with the predicate expression $\lambda \overline{x}.\ \Phi(\overline{x})$.

An n-ary *function expression* is any expression of the form $\lambda \overline{x}.\ \tau(\overline{x})$, where $\overline{x}$ is a tuple of n individual variables and $\tau(\overline{x})$ is any term of first-order classical logic. If u is an n-ary function expression of the form $\lambda \overline{x}.\ \tau(\overline{x})$ and $\overline{t}$ is a tuple of n terms, then $u(\overline{t})$ stands for $\tau(\overline{t})$. An n-ary $(n \geq 0)$ function constant f is identified with the function expression $\lambda \overline{x}.\ f(\overline{x})$. An n-ary $(n \geq 0)$ function variable ϕ is identified with the function expression $\lambda \overline{x}.\ \phi(\overline{x})$. Note that $0 - ary$ function variables are simply individual variables.

Let $\overline{U} = (U_1, \ldots, U_n)$ and $\overline{V} = (V_1, \ldots, V_n)$ (resp. $\overline{u} = (u_1, \ldots, u_n)$ and $\overline{v} = (v_1, \ldots, v_n)$) be tuples of predicate (resp. function) expressions. $\overline{U}$ and $\overline{V}$ (resp. $\overline{u}$ and $\overline{v}$) are said to be *similar* iff, for each i $(1 \leq i \leq n)$, U_i and V_i (resp. u_i and v_i) are predicate (resp. function) expressions of the same arity.

Truth values *true* and *false* are denoted by $\top$ and $\bot$, respectively.

If U and V are predicate expressions of the same arity, then $U \leq V$ stands for $\forall \overline{x}\ U(\overline{x}) \supset V(\overline{x})$. If $\overline{U} = (U_1, \ldots, U_n)$ and $\overline{V} = (V_1, \ldots, V_n)$ are similar tuples of predicate expressions, then $\overline{U} \leq \overline{V}$ is an abbreviation for $\bigwedge_{i=1}^{n}[U_i \leq V_i]$.

If A is a formula, $\overline{\sigma} = (\sigma_1, \ldots, \sigma_n)$ and $\overline{\delta} = (\delta_1, \ldots, \delta_n)$ are tuples of any expressions, then $A(\overline{\sigma} \leftarrow \overline{\delta})$ stands for the formula obtained from A by simultaneously replacing *each* occurrence of σ_i by δ_i $(1 \leq i \leq n)$. For any tuple $\overline{x} = (x_1, \ldots x_n)$ of individual variables and any tuple $\overline{t} = (t_1, \ldots t_n)$ of terms, we write $\overline{x} = \overline{t}$ to denote the formula $x_1 = t_1 \wedge \cdots \wedge x_n = t_n$. We write $\overline{x} \neq \overline{t}$ as an abbreviation for $\neg(\overline{x} = \overline{t})$.

2.2 Definitions

Definition 1. A theory T is said to be *existential* (*universal*) iff all of its axioms are of the form $\exists \overline{x}\ T_1$ (resp. $\forall \overline{x}\ T_1$), where T_1 is quantifier free.

Definition 2. A theory is called *semi-universal* if its axioms do not contain existential quantifiers in the scope of universal quantifiers.

Definition 3. Let T be a theory without function symbols and suppose that, for $n > 0$, $c_1, \ldots c_n$ are all the individual constants occurring in T. The *domain closure axiom for T*, written $DCA(T)$, is the sentence

$$\forall x.\ x = c_1 \vee \cdots \vee x = c_n.$$

Let $\overline{c}$ be a tuple of individual constants. By $DCA^{-\overline{c}}(T)$ we shall denote the sentence $\forall x\ x = c_1 \vee \ldots \vee x = c_n$, where $c_1, \ldots, c_n$ are all the individual constants of T excluding constants from $\bar{c}$. For $k \in \omega$, by $DCA^{+k}(T)$ we shall denote the sentence

$$\exists z_1 \cdots \exists z_k \forall x \; x = z_1 \vee \cdots \vee x = z_k \vee x = c_1 \vee \ldots \vee x = c_n,$$

where $c_1, \ldots, c_n$ are all the individual constants of T. We also use notation $DCA^{-\bar{c}+k}(T)$ as a combination of the above.

Definition 4. A predicate variable Φ occurs *positively* (resp. *negatively*) in a formula A if the conjunctive normal form of A contains a subformula of the form $\Phi(\bar{t})$ (resp. $\neg\Phi(\bar{t})$). A formula A is said to be *positive* (resp. *negative*) w.r.t. Φ iff all occurrences of Φ in A are positive (resp. negative).

Definition 5. Let Φ be either a predicate constant or a predicate variable and $\overline{\Phi}$ be a tuple of predicate constants or a tuple of predicate variables. Then a formula $T(\Phi)$ is said to be *separated* w.r.t. Φ iff it is of the form $T_1(\Phi) \wedge T_2(\Phi)$ where $T_1(\Phi)$ is positive w.r.t. Φ and T_2 is negative w.r.t. Φ.

3 General Domain Circumscription

In this section, we provide a definition of general domain circumscription (GDC) and its model-preferential semantics. GDC subsumes both McCarthy's original domain circumscription, introduced in [10], studied in [2], and substantially improved in [6], and Hintikka's mini-consequence, formulated in [7] and studied in [8].

Definition 6. Let $\overline{P} = (P_1, \ldots, P_n)$ be a tuple of different predicate constants, $\overline{f} = (f_1, \ldots, f_k)$ be a tuple of different function constants (including, perhaps, individual constants), $T(\overline{P}, \overline{f})$ be a theory and let Φ be a one-place predicate variable, $\overline{\Psi}$ be a tuple of predicate variables similar to $\overline{P}$, and $\overline{\psi}$ be a tuple of function variables similar to $\overline{f}$. By $Axiom(\Phi, \overline{P}, \overline{f})$, sometimes abbreviated by $Axiom(\Phi)$, we shall mean the conjunction of:

- $\Phi(a)$, for each individual constant a in T not occurring in $\overline{f}$,
- $\Phi(\psi_i)$, for each individual constant a in T such that a is f_i,
- $\forall x_1 \ldots x_n [\Phi(x_1) \wedge \cdots \wedge \Phi(x_n) \supset \Phi(f(x_1, \ldots x_n))]$, for each n-ary $(n > 0)$ function constant f in T not occurring in $\overline{f}$, and
- $\forall x_1 \ldots x_n [\Phi(x_1) \wedge \cdots \wedge \Phi(x_n) \supset \Phi(\psi_i(x_1, \ldots x_n))]$, for each n-ary $(n > 0)$ function constant f in T such that f is f_i.

T^{Φ} stands for the result of rewriting $T(\overline{\Psi}, \overline{\psi})$, replacing each occurrence of $\forall x$ and $\exists x$ in $T(\overline{\Psi}, \overline{\psi})$ with "$\forall x \; \Phi(x) \supset$" and "$\exists x \; \Phi(x) \wedge$", respectively.

Definition 7. Let $\overline{P} = (P_1, \ldots, P_n)$, $\overline{f} = (f_1, \ldots, f_k)$ and $T(\overline{P}, \overline{f})$ be as in Definition 6. The *general domain circumscription for* $T(\overline{P}, \overline{f})$ *with variable* $\overline{P}$ *and* $\overline{f}$, written $CIRC_D(T; \overline{P}; \overline{f})$, is the following sentence of second-order logic:

$$T(\overline{P}, \overline{f}) \wedge \forall\Phi\forall\overline{\Psi}\forall\overline{\psi}[(\exists x \Phi(x) \wedge Axiom(\Phi, \overline{P}, \overline{f}) \wedge T^{\Phi}) \supset \forall x \Phi(x)]. \quad (1)$$

A formula α is said to be a *consequence of* $CIRC_D(T; \overline{P}; \overline{f})$ iff $CIRC_D(T; \overline{P}; \overline{f})$ $\models \alpha$, where "$\models$" denotes the entailment relation of classical second-order logic.

The second conjunct of the sentence (1) is called the *domain circumscription axiom*.

It is not difficult to see that (1) asserts that the domain of discourse (represented by Φ) is minimal with respect to T, where $\overline{P}$ and $\overline{f}$ are allowed to vary during the minimization.

We shall write $CIRC_D(T)$ as an abbreviation for $CIRC_D(T; (); ())$, i.e. if neither predicate nor function constants are allowed to vary. This simplest form of domain minimization corresponds closely to McCarthy's original domain circumscription [10] with the augmentation described in [6].[3]

We shall write $CIRC_D(T; \overline{P})$ as an abbreviation for $CIRC_D(T; \overline{P}; ())$, i.e. if some predicate constants, but not function constants, are allowed to vary. If $\overline{P}$ includes all predicate constants occurring in a theory T, then $CIRC_D(T; \overline{P})$ is exactly *mini-consequence*, introduced in [7] and improved in [8]. Following [8], this form of minimization will be referred to as *variable domain circumscription*.[4]

Example 1. Consider a theory T consisting of $\forall x \ P(x) \wedge Q(x) \wedge P(a) \wedge Q(b)$. We shall minimize the domain of T without varying predicate or function constants. $CIRC_D(T)$ is given by

$$T \wedge \forall \Phi[(\exists x \Phi(x) \wedge \Phi(a) \wedge \Phi(b) \wedge \forall x (\Phi(x) \supset (P(x) \wedge Q(x) \wedge P(a) \wedge Q(b)))) \supset \forall x \Phi(x)]. \tag{2}$$

Substituting $\lambda x.x = a \vee x = b$ for Φ, we get

$$T \wedge [\exists x (x = a \vee x = b) \wedge (a = a \vee a = b) \wedge (b = a \vee b = b) \wedge \tag{3}$$
$$\forall x[(x = a \vee x = b) \supset (P(x) \wedge Q(x) \wedge P(a) \wedge Q(b))] \supset$$
$$\forall x (x = a \vee x = b)].$$

Since (3) is equivalent to $T \wedge (\forall x \ x = a \vee x = b)$, we conclude that the domain closure axiom for T, i.e. the sentence $\forall x \ x = a \vee x = b$ is a consequence of $CIRC_D(T)$.

Example 2. Let T consist of $P(a) \wedge P(b)$. We minimize the domain of T with the constant a allowed to vary during the minimization. $CIRC_D(T; (), (a))$ is given by[5]

$$T \wedge \forall \Phi \forall x_a[\exists x \Phi(x) \wedge \Phi(x_a) \wedge \Phi(b) \wedge P(x_a) \wedge P(b) \supset \forall x \Phi(x)]. \tag{4}$$

Substituting $\lambda x.x = b$ for Φ and b for x_a, one easily calculates that (4) implies $T \wedge \forall x \ x = b$. Accordingly, we conclude that the domain of T consists of one object, referred to by both a and b.

---

[3] In fact, $CIRC_D(T)$ is slightly stronger in that it is based on a second-order axiom rather than on a first-order schema.

[4] Note that in variable domain circumscription all predicate constants, but no function constants, are allowed to vary during the minimization process.

[5] Since a is an individual constant, the variable corresponding to a is, in fact, an individual variable. Accordingly, we denote it by x_a, rather than by ψ.

We now proceed to give a semantics for general domain circumscription. Given a model M, we shall write $\mid M \mid$ to denote the domain of M.

Definition 8. Let $\overline{P}, \overline{f}$ and $T(\overline{P}, \overline{f})$ be as in Definition 6. Let M_1 and M_2 be models of T. We say that M_1 is a $(\overline{P}; \overline{f})$-submodel of M_2, written $M_1 \leq M_2$, iff $\mid M_1 \mid \subseteq \mid M_2 \mid$, and for each predicate, function or individual constant C, occurring neither in $\overline{P}$ nor in $\overline{f}$, the interpretation of C in M_1 is the restriction of the corresponding interpretation in M_2 to $\mid M_1 \mid$. A model is said to be $(\overline{P}; \overline{f})$-minimal iff it has no proper $(\overline{P}; \overline{f})$-submodels.

Theorem 9. Let $\overline{P}, \overline{f}$ and $T(\overline{P}, \overline{f})$ be as in Definition 6. A formula A is a consequence of $CIRC_D(T; \overline{P}; \overline{f})$ iff A is true in all $(\overline{P}; \overline{f})$-minimal models of T.

3.1 An Optimization Technique

In this section, we propose a technique that allows one to reduce the size of a domain circumscription axiom. The technique allows one to sometimes remove counterparts of universal formulas from the axiom. More precisely, let theory T consist of axioms, including universal axioms of the form $\forall x_1 \cdots \forall x_n\, A(x_1, \cdots, x_n)$, and suppose that all predicate and/or function constants occurring in $A(x_1, \cdots, x_n)$ are not allowed to vary during the minimization. Then each such axiom reappears in (1) as a part of T^ϕ, equivalent to the formula

$$\forall x_1 \cdots \forall x_n\, \neg\Phi(x_1) \vee \cdots \vee \neg\Phi(x_n) \vee A(x_1, \cdots, x_n). \tag{5}$$

Since (5), together with the corresponding axiom of T, reduces to $\top$, it can be removed from T^ϕ. We thus have the following principle:

Remove a counterpart of every universal axiom A of T from T^ϕ in the domain circumscription axiom, provided that none of the predicate and/or function constants occurring in A are allowed to vary. If B is the resulting formula, then $T \wedge B$ is equivalent to $CIRC_D(T; \overline{P}; \overline{f})$.

Observe also that one can remove formula $\exists x \Phi(x)$ from (1) whenever T contains a constant symbol. This follows from the fact that for each constant symbol, say a, one has $\Phi(a)$ as a conjunct of $Axiom(\Phi)$. Thus $\exists x \Phi(x)$ follows from $Axiom(\Phi)$ and can be removed.

It should be emphasized that the DLS algorithm works successfully without the above mentioned optimizations. However, as shall be seen in the examples in Section 6, they usually considerably decrease the complexity of the reduced formula.

4 DLS Algorithm

4.1 The Basic DLS Algorithm

In this section, we briefly describe the DLS algorithm mentioned in the introduction. Its complete formulation can be found in [3]. The algorithm was originally formulated in a weaker form in [13], in the context of modal logics. It is based on Ackermann's techniques developed in connection with the elimination problem (see [1]).The DLS algorithm is based on the following lemma, proved by Ackermann in 1934 (see [1]). The proof can also be found in [13].

Lemma 10 Ackermann Lemma. *Let Φ be a predicate variable and $A(\bar{x}, \bar{z})$, $B(\Phi)$ be formulas without second-order quantification. Let $B(\Phi)$ be positive w.r.t. Φ and let A contain no occurrences of Φ at all. Then the following equivalences hold:*

$$\exists\Phi\forall\bar{x}[\Phi(\bar{x}) \vee A(\bar{x}, \bar{z})] \wedge B(\Phi \leftarrow \neg\Phi) \equiv B(\Phi \leftarrow A(\bar{x}, \bar{z})) \tag{6}$$

$$\exists\Phi\forall\bar{x}[\neg\Phi(\bar{x}) \vee A(\bar{x}, \bar{z})] \wedge B(\Phi) \equiv B(\Phi \leftarrow A(\bar{x}, \bar{z})) \tag{7}$$

where in the righthand formulas the arguments $\bar{x}$ of A are each time substituted by the respective actual arguments of Φ (renaming the bound variables whenever necessary).

The DLS algorithm is based on eliminating second-order quantifiers of the input formula using a combination of applications of Lemma 10 together with various syntactic transformations which preserve equivalence.

4.2 Problems with the Basic DLS Algorithm

There are two weaknesses associated with the basic DLS algorithm which cause it to terminate with failure:

1. Non-separated input problem.
2. Unskolemization problem.

In order for the DLS algorithm to reduce an input formula, it must be possible for the formula to be transformed into separated form. If the input formula consists of clauses which contain both positive and negative occurrences of the predicate variable being eliminated, then the basic DLS algorithm will return with failure.

Another limitation of the basic DLS algorithm involves unskolemization. Skolemization is sometimes required either due to the original form of the input formula, or to one of the phases in the algorithm which may introduce new existential quantifiers. When applying Ackermann's Lemma, all existentially quantified individual variables have to be removed from the prefix of the formula being reduced. For this purpose, Skolemization is performed using the equivalence,

$$\forall\bar{x}\exists y \, A(\bar{x}, y) \equiv \exists f\forall\bar{x} \, A(\bar{x}, y \leftarrow f(\bar{x})), \tag{8}$$

where f is a new function variable. After application of Ackermann's Lemma, one tries to remove the newly introduced function variables using equivalence (8) in the other direction. Unfortunately, unskolemization is not always successful.

5   Extending the DLS Algorithm

There are two generalizations of the basic DLS algorithm that extend the class of input formulas that can be successfully reduced to include non-separated input formulas and formulas which would normally fail to be reduced due to unskolemization problems.

The first method appeals to the observation that for a particular class of theories whose domain closure axiom is entailed by the corresponding general domain circumscribed theory, both the non-separated input and unskolemization problems can be avoided by combining the basic DLS algorithm with the additional constraints contributed by the domain closure axiom associated with the input theory. Although this method can be used for a particular class of input formulas, it can not be used for all non-separated input formulas.

The second method generalizes Ackermann's Lemma (10) by transforming an input formula into a (possibly) non-separated form which can be shown to be logically equivalent to a fixpoint formula in a fixpoint calculus. In the case where the fixpoint formula is bounded, the non-separated input formula can be reduced to a logically equivalent first-order formula.

Due to page limitations, we will concentrate on the first method whose formal justification is described in Section 5.1. We refer the reader to [5], for a detailed description of both methods.

5.1   DLS Algorithm with the Domain Closure Axiom

As mentioned before, the DLS algorithm may fail due to non-separatedness and unskolemizaton problems. On the other hand, whenever it is known that the domain closure axiom (DCA) follows from the theory considered, the non-separatedness and unskolemization problems are always solvable. This is particularly important in cases when one combines domain circumscription with other second-order formalisms, like e.g. second-order circumscription.

Assume that T is a theory. Then, for each formula A, $DCA(T)$ implies:

$$\exists x \, A(x) \; \equiv \; (A(c_1) \vee \cdots \vee A(c_n)) \tag{9}$$

and

$$\forall x \, A(x) \; \equiv \; (A(c_1) \wedge \cdots \wedge A(c_n)). \tag{10}$$

The following example illustrates the use of equivalences (9) and (10).

Example 3. Assume that $\forall x \; x = a \vee x = b$ holds. An application of equivalence (9) to formula $\forall y \exists z \, P(y, z)$ results in $\forall y (P(y, a) \vee P(y, b))$. An application of equivalence (10) to this formula results in $(P(a, a) \vee P(a, b)) \wedge (P(b, a) \vee P(b, b))$.

Using equivalence (9) one can remove existential quantifiers that would require Skolemization. This solves the unskolemization problem associated with the DLS algorithm. Observe that in order to make the DLS algorithm work one could also use equivalence (10) in order to remove universal quantifiers preceding the existential quantifiers, whenever necessary.

The second reason the DLS algorithm fails is when formulas cannot be separated w.r.t. predicate Φ. In the canonical case, this occurs when a universally quantified clause contains both positive and negative occurrences of Φ. Using equivalence (10), one can remove the universal quantifiers from the clause prefix. This, together with certain distributions across subformulas, is guaranteed to transform the initially non-separated formula into a separated formula.

Of course, the above technique can easily be modified if it is known that $DCA^{-\overline{c}}(T)$, $DCA^{+k}(T)$ or $DCA^{-\overline{c}+k}(T)$ is entailed from T. Before we introduce this modification, consider the following simple example.

Example 4. Assume that $\exists z \forall x\ x = z \lor x = a$ holds. An application of equivalence (10) to formula $\forall y\ P(y)$ results in $\exists z[\forall x(x = z \lor x = a) \land P(z) \land P(a)]$.

Observe that, unlike Example (3), the DCA reappears in the result. This is due to the existential quantifier $\exists z$ that has to bind both the DCA and the resulting formula. ∎

The following theorem justifies the technique.

Theorem 11. *Assume that for a given second-order theory T,*

$$T \models DCA^{-\overline{c}+k}(T).$$

Then T is equivalent to a first-order formula.

Since $CIRC_D(T, \overline{P}, \overline{f})$ is a second-order sentence, we have the following corollary.

Corollary 12. *Assume that $CIRC_D(T, \overline{P}, \overline{f}) \models DCA^{-\overline{c}+k}(T)$. Then $CIRC_D(T, \overline{P}, \overline{f})$ is equivalent to a first-order formula.*

Theorem 11 allows us to modify the DLS algorithm in such a way that whenever there is a Skolemization or separatedness problem, one applies formulas (9) and (10), or their generalizations, respectively. We will denote this modification of the DLS algorithm by DLS^*.

The following example illustrates the use of the DLS^* algorithm.

Example 5. Assume that the DCA is of the form $\exists z \forall x\ x = z \lor x = a$, and let T be the second-order formula

$$\exists \Phi \forall x \forall y[(\Phi(x) \lor \neg\Phi(y) \lor R(a)) \land \exists z \neg\Phi(z) \land \exists u \Phi(u)] \land (\exists z \forall x\ x = z \lor x = a). \quad (11)$$

We first Skolemize DCA and obtain $\forall x(x = b \lor x = a)$. We then try to eliminate the quantifier $\exists \Phi$ from formula (11) using the DLS algorithm. In this case, one first Skolemizes one of the first-order existential quantifiers. Whichever is chosen, we are then faced with a non-separated formula. Due to this, the DLS algorithm fails. If instead one uses the DLS^* algorithm, we first eliminate one of the existential quantifiers, say $\exists u$, by applying equivalence (9) and obtain

$$\exists b[DCA \land \exists \Phi \forall x \forall y((\Phi(x) \lor \neg\Phi(y) \lor R(a)) \land \exists z \neg\Phi(z) \land (\Phi(b) \lor \Phi(a)))]. \quad (12)$$

Formula (12) is not separated. We thus apply equivalence (10) to quantifier $\forall y$ and obtain:

$$\exists b[DCA \wedge \exists \Phi \forall x((\Phi(x) \vee \neg\Phi(b) \vee R(a)) \wedge (\Phi(x) \vee \neg\Phi(a) \vee R(a))) \qquad (13)$$
$$\wedge \exists z \neg\Phi(z) \wedge (\Phi(b) \vee \Phi(a)))].$$

(13) is equivalent to

$$\exists b[DCA \wedge [\exists \Phi(\exists z \neg\Phi(z) \wedge \forall x \Phi(x) \wedge (\Phi(b) \vee \Phi(a)))] \vee \qquad (14)$$
$$[\exists \Phi(\exists z \neg\Phi(z) \wedge (\neg\Phi(b) \vee R(a)) \wedge (\neg\Phi(a) \vee R(a)) \wedge (\Phi(b) \vee \Phi(a)))]$$

It is easily observed that each disjunct is in separated form, and no additional skolemization is necessary, so application of the basic DLS algorithm results in a first-order formula equivalent to (14).

6 Reducing General Domain Circumscription

In this section we provide some reducibility results concerning various variants of general domain circumscription. In what follows, we assume that theories under consideration contain at least one individual constant symbol.

6.1 Fixed GDC

Universal Theories In Example 1, we saw that domain circumscription may allow the derivation of the domain closure axiom. It turns out that for universal theories without function constants this is always the case. Moreover, as the next theorem shows, if T is a theory of that type, then the domain circumscription of T is equivalent to $T \wedge DCA(T)$.

Theorem 13. *Let T be a universal theory without function symbols. Then $CIRC_D(T)$ is always reducible into first-order logic using the DLS algorithm. Moreover, if A is the resulting formula, then A is equivalent to $T \wedge DCA(T)$.*

Observe that according to our assumption, we consider only theories that contain at least one individual constant. This is only a technical assumption. If T has no individual constant symbols then

$$CIRC_D(T) \equiv T \wedge \forall \Phi[\exists x \, \Phi(x) \supset \forall z \, \Phi(z)].$$

After negating the second conjunct of this formula we obtain

$$\exists \Phi[\exists x \, \Phi(x) \wedge \exists z \neg\Phi(z)],$$

which, after applying the DLS algorithm, results in the equivalent $\exists x \exists z[z \neq x]$. Thus $CIRC_D(T) \equiv T \wedge \forall x \forall z(x = z)$.

Semi-Universal Theories As regards semi-universal theories without function symbols we have the following theorem:

Theorem 14. *Let T be a semi-universal theory without function symbols. Then $CIRC_D(T)$ is always reducible into first-order logic using the DLS algorithm. Moreover, if A is the resulting formula, then A implies $DCA^{+k}(T)$, where k is the number of existential quantifiers of T.*

6.2 Variable GDC

For universal and semi-universal theories, we have the following counterparts of Theorems 13 and 14:

Universal Theories

Theorem 15. *Let T be a universal theory without function symbols and suppose that $\overline{P}$ is a tuple of predicate symbols occurring in T. Then $CIRC_D(T;(\overline{P}))$ is always reducible into first-order logic using the DLS* algorithm. Moreover, if A is the resulting formula, then A implies $DCA(T)$.*

For theories with varied individual constants the following theorem holds.

Theorem 16. *Let T be a universal theory without function symbols. Let $\overline{P}$ be a tuple of predicate symbols and $\overline{c}$ be a tuple of individual constants occurring in T. Then $CIRC_D(T;\overline{P};\overline{c})$ is always reducible into first-order logic using the DLS* algorithm. Moreover, if A is the resulting formula, then A implies $DCA^{-\overline{c}}(T)$.*

The following example varies an individual constant.

Example 6. Consider the theory T given by

A1. $S(c) \wedge S(d)$
A2. $\forall x\ R(x,c) \equiv R(x,d)$
A3. $\forall x \neg R(x,c) \wedge \forall y \neg R(y,d) \wedge \forall z \neg R(z,z)$.

This example is taken from [12]. Here $S(x)$, $R(x,y)$, c and d are to be read "the evidence says that x saw the victim alive", "the evidence says that x saw the victim alive after y saw her alive for the last time", "murderer" and "suspect", respectively. Suppose further that the police try to find all individuals who satisfy exactly those formulas that the (unknown) murderer c does, by comparing what is provable about the murderer with what is provable about a particular individual. To formalize this type of procedure, we should minimize the domain under consideration with all constant symbols fixed, except that referring to the murderer which is allowed to vary. In our case, we minimize the domain of T with variable c. The intended conclusion is $d = c$.

The second-order part of $Circ_D(T, (), (c))$, after simplifications, is equivalent to

$$\forall x_c \forall \Phi[\Phi(x_c) \wedge \Phi(d) \wedge S(x_c) \wedge \forall x(\Phi(x) \supset ((\neg R(x, x_c) \vee R(x, d)) \quad (15)$$
$$\wedge (R(x, x_c) \vee \neg R(x, d))) \wedge \forall x(\Phi(x) \supset \neg R(x, x_c)) \supset \forall s \Phi(s)].$$

Negating (15), we obtain

$$\exists x_c \exists \Phi[\Phi(x_c) \wedge \Phi(d) \wedge S(x_c) \wedge \forall x(\neg \Phi(x) \vee \neg R(x, x_c) \vee R(x, d)) \wedge \quad (16)$$
$$(\neg \Phi(x) \vee R(x, x_c) \vee \neg R(x, d))) \wedge \forall x(\neg \Phi(x) \vee \neg R(x, x_c)) \wedge \exists s \neg \Phi(s)]$$

which is transformed by the DLS algorithm to

$$\exists s \exists x_c \exists \Phi \forall x[(\Phi(x) \vee (x \neq x_c \wedge x \neq d)) \wedge S(x_c) \wedge \forall x(\neg \Phi(x) \vee \neg R(x, x_c)) \wedge \neg \Phi(s)]$$

and then, after the application of Ackermann's Lemma, to

$$\exists s \exists x_c[S(x_c) \wedge \forall x((x \neq x_c \wedge x \neq d) \vee \neg R(x, x_c) \wedge (s \neq x_c \wedge s \neq d)]. \quad (17)$$

Negating (17), we obtain

$$\forall s \forall x_c[S(x_c) \wedge \forall x((x \neq x_c \wedge x \neq d) \vee \neg R(x, x_c) \supset (s \neq x_c \wedge s \neq d)]$$

and so, $Circ_D(T, (), (c)) \equiv$

$$T \wedge \forall s \forall x_c[S(x_c) \wedge \forall x((x \neq x_c \wedge x \neq d) \vee \neg R(x, x_c) \supset (s \neq x_c \wedge s \neq d)]$$

It is easily seen, substituting d for x_c, that $Circ_D(T, (), (c)) \models d = c$.

Semi-Universal Theories

Theorem 17. *Let T be a semi-universal theory without function symbols and suppose that $\overline{P}$ is a tuple of predicate symbols occurring in T. Then $CIRC_D(T; \overline{P})$ is always reducible into first-order logic using the DLS* algorithm. Moreover, if A is the resulting formula, then A implies $DCA^k(T)$, where k is the number of existential quantifiers of T.*

The following theorem generalizes Theorem 16 to semi-universal theories.

Theorem 18. *Let T be a semi-universal theory without function symbols. Let $\overline{P}$ be a tuple of predicate symbols and $\overline{c}$ be a tuple of individual constants occurring in T. Then $CIRC_D(T; \overline{P}; \overline{c})$ is always reducible into first-order logic using the DLS* algorithm. Moreover, if A is the resulting formula, then A implies $DCA^{-\overline{c}+k}(T)$, where k is the number of existential quantifiers of T.*

A reduction of variable domain circumscription for a semi-universal theory is illustrated below.

Example 7. Let T consist of $\exists x\ Q(x) \wedge [Q(a) \supset \exists y\ y \neq a]$. This example is taken from [8]. The intended conclusion of domain circumscription with Q allowed to vary is $\exists y \forall x\ x = y \vee x = a$.

We reduce $CIRC_D(T; (Q); ())$. The second-order part of $CIRC_D(T; (Q); ())$ (after removing $\exists x \Phi(x)$) is

$$\forall \Psi \forall \Phi [[\Phi(a) \wedge \exists x(\Phi(x) \wedge \Psi(x)) \wedge (\Psi(a) \supset \exists y(\Phi(y) \wedge y \neq a))] \supset \forall z \Phi(z)]. \quad (18)$$

Negating (18), we obtain

$$\exists \Psi \exists \Phi [\Phi(a) \wedge \exists x(\Phi(x) \wedge \Psi(x)) \wedge (\Psi(a) \supset \exists y(\Phi(y) \wedge y \neq a))] \wedge \exists z \neg \Phi(z),$$

which is equivalent to

$$\exists x \exists y \exists z \exists \Psi \exists \Phi [\Phi(a) \wedge \Phi(x) \wedge \Psi(x) \wedge (\neg \Psi(a) \vee (\Phi(y) \wedge y \neq a)) \wedge \neg \Phi(z)],$$

which is equivalent to

$$\exists x \exists y \exists z \exists \Psi \exists \Phi [\Phi(a) \wedge \Phi(x) \wedge \Psi(x) \wedge (\neg \Psi(a) \vee (\Phi(y) \wedge y \neq a)) \wedge \forall t(\neg \Phi(t) \vee t \neq z)].$$
$$(19)$$

Applying Ackermann's Lemma to (19), we eliminate Φ and obtain

$$\exists x \exists y \exists z \exists \Psi [a \neq z \wedge x \neq z \wedge \Psi(x) \wedge (\neg \Psi(a) \vee (y \neq z \wedge y \neq a))],$$

which is equivalent to

$$\exists x \exists y \exists z \exists \Psi [a \neq z \wedge x \neq z \wedge \forall t(\Psi(t) \vee t \neq x) \wedge (\neg \Psi(a) \vee (y \neq z \wedge y \neq a))]. \quad (20)$$

Applying Ackermann's Lemma to (20), we eliminate Ψ and obtain

$$\exists x \exists y \exists z [a \neq z \wedge (a \neq x \vee (y \neq z \wedge y \neq a))]. \quad (21)$$

The negation of (21) is

$$\forall x \forall y \forall z [a = z \vee (a = x \wedge (y = z \vee y = a))].$$

Thus,

$$CIRC_D(T; (Q); ()) \equiv T \wedge \forall x \forall y \forall z [a = z \vee (a = x \wedge (y = z \vee y = a))].$$

It is easily verified that $CIRC_D(T; (Q); ())$ implies $\exists y \forall x\ x = y \vee x = a$.

6.3   Arbitrary Theories and Functions

A fixpoint generalization of the basic DLS algorithm (DLS^{fix}) is described in [11], applied in [4] to the class of semi-Horn formulas, and applied in [5] to general domain circumscription. Due to page limitations, we can only briefly describe the latter results:

- For arbitrary domain circumscribed theories without functions, neither the DLS algorithm nor its fixpoint generalization DLS^{fix}, guarantee a reduction to classical first-order logic. However reductions can sometimes be obtained using DLS^{fix}.
- For arbitrary domain circumscribed theories with functions, the DLS algorithm always fails. However, the following results apply.

Theorem 19. *Let T be a semi-universal theory. Then $CIRC_D(T)$ is always reducible into fixpoint logic using the DLS^{fix} algorithm.*

If one can show that the fixpoint formula output by the DLS^{fix} algorithm is bounded, then the input formula is reducible to classical first-order logic.

Theorem 20. *Let T be a theory and let $\overline{P}$ be the tuple of all predicate symbols occurring in T. Suppose further that T has a $(\overline{P}, \overline{f})$-minimal model and cardinalities of all such models have the same finite common upper bound. Then $CIRC_D(T; \overline{P}; \overline{f})$ can be reduced into an equivalent first-order sentence using the DLS^* algorithm.*

7   Combining Domain Circumscription with Arbitrary Circumscriptions

Theorem 11 states that given a second-order theory T, if one can show that the domain closure axiom is entailed by T, then T is reducible to a first-order formula using DLS^*. There is a direct connection between this result and the reduction of arbitrary circumscriptive policies applied to a certain class of domain circumscribed theories. The connection works as follows:

1. We know that given a semi-universal theory T, the domain circumscription of T, $Circ_{DC}(T; \bar{P}; \bar{f})^6$, can be reduced to its first-order equivalent using DLS^*. In addition, the DCA used in the DLS^* algorithm can be constructively generated.
2. Suppose the result of $Circ_{DC}(T; \bar{P}, \bar{f})$ is T'. Observe that for any arbitrary circumscription $Circ_{SO}(T'; \bar{P}; \bar{f})$, applied to T' that

$$Circ_{SO}(T'; \bar{P}; \bar{f}) \equiv Circ_{SO}(T' \wedge DCA; \bar{P}; \bar{f}).$$

---

[6] $\bar{f}$ is restricted to individual constants.

3. Since the DLS algorithm can only fail when unskolemization or non-separatedness occur, and we have shown how to avoid these problems for theories which entail the DCA, it follows that $Circ_{SO}(T'; \bar{P}; \bar{f})$ is always reducible to a first-order formula using DLS^* with the DCA.

In summary, we have the following result.

> For any semi-universal first-order theory without functions and any arbitrary circumscription policy applied to the theory, the DLS^* algorithm will always reduce the circumscribed theory to a logically equivalent first-order formula, provided that the theory is first circumscribed using domain circumscription.

The reduction process is achieved as follows.

1. Given a semi-universal theory T, constructively generate the DCA for T using the procedure described in the proof of Theorem 18.
2. Apply DLS^* to T resulting in the output T'.
3. Apply DLS^* to the arbitrary circumscriptive policy applied to T' using the previously generated DCA. This results in T'', a first-order formula logically equivalent to the latter arbitrary circumscription.

The following example illustrates the technique. To save space, we will first domain circumscribe the following theory and then apply a particular circumscription to the original theory in conjunction with the generated DCA.

Example 8. Let T consist of

A1. $Ab(a)$
A2. $\forall x \forall y Ab(x) \wedge S(y, x) \supset Ab(y)$
A3. $\exists x \forall y S(y, x)$

where Ab and S stand for *Abnormal* and *Son-of*, respectively.

The DCA entailed by the domain circumscription $CIRC_D(T)$ is[7]

$$\exists z \forall x (x = z \vee x = a) \tag{22}$$

(22) can be constructively generated using the procedure described in the proof of Theorem 18.

In the next phase, we would like to minimize the predicate Ab relative to $T \wedge$ DCA. Let T' denote the first-order formula output by application of the DLS^* algorithm to $CIRC_D(T)$. Since $T' \models DCA$ and $CIRC_{SO}(T; Ab) \equiv CIRC_{SO}(T \wedge DCA; Ab)$, the DLS^* algorithm can be applied to $CIRC_{SO}(T; Ab)$ using the DCA with a guarantee that the output of DLS^* will be a formula in classical first-order logic, logically equivalent to $CIRC_{SO}(T; Ab) \wedge DCA$. In fact, the output of DLS^* is

$$\exists z [\ \forall x(x = z \vee x = a)$$
$$\wedge \forall x(\neg Ab(x) \vee -S(z, x) \vee Ab(z)) \wedge \exists x \forall y S(y, x) \wedge Ab(a)$$
$$\wedge (S(z, a) \vee S(a, a) \vee \forall d \exists c \neg S(c, d) \vee \forall e(a = e \vee \neg Ab(e)))].$$

---

[7] In this example, by DCA, we mean DCA^{+1}.

References

1. Ackermann, W. (1935) *Untersuchungen über das Eliminationsproblem der mathematischen Logik*, Mathematische Annalen, 110, 390-413.
2. Davis, M. (1980) *The Mathematics of Non-Monotonic Reasoning*, Artificial Intelligence J., **13**, 73-80.
3. Doherty, P., Lukaszewicz, W., Szałas, A. (1994) *Computing Circumscription Revisited. A Reduction Algorithm*, Technical Report LiTH-IDA-R-94-42, Linköping University, 1994. Also in, Proc. 14th IJCAI, 1995, Montreal, Canada.[8] Full version to appear in Journal of Automated Reasoning.[9]
4. Doherty, P., Lukaszewicz, W., Szałas, A. (1995) *A Reduction Result for Circumscribed Semi-Horn Formulas*, To appear in Fundamenta Informaticae, 1996.
5. Doherty, P., Lukaszewicz, W., Szałas, A. (1995) *General Domain Circumscription and its First-Order Reduction*, Technical Report LiTH-IDA-R-96-01, Linköping University.
6. Etherington, D. W., Mercer, R. (1987) *Domain Circumscription: A Revaluation*, Computational Intelligence, **3**, 94-99.
7. Hintikka, J. (1988) *Model Minimization - An Alternative to Circumscription*, Journal of Automated Reasoning, 4,1-13.
8. Lorenz, S. (1994) *A Tableau Prover for Domain Minimization*, Journal of Automated Reasoning, **13**, 375-390.
9. Lukaszewicz, W. (1990) *Non-Monotonic Reasoning - Formalization of Commonsense Reasoning*, Ellis Horwood Series in Artificial Intelligence. Ellis Horwood, 1990.
10. McCarthy, J. (1977) *Epistemological Problems of Artificial Intelligence*, in: Proc. 5th IJCAI, Cambridge, MA, 1977, 1038-1044.
11. Nonnengart, A., Szałas, A. (1995) *A Fixpoint Approach to Second-Order Quantifier Elimination with Applications to Correspondence Theory*, Report of Max-Planck-Institut für Informatik, MPI-I-95-2-007, Saarbrücken, Germany.
12. Suchanek, M. A. (1993) *First-Order Syntactic Characterizations of Minimal Entailment, Domain-Minimal Entailment, and Herbrand Entailment*, Journal of Automated Reasoning, **10**, 237-263.
13. Szałas, A. (1993) *On the Correspondence Between Modal and Classical Logic: an Automated Approach*, Journal of Logic and Computation, **3**, 605-620.

---

[8] To try out the algorithm: http://www.ida.liu.se/labs/kplab/projects/dls/ .
[9] ftp://ftp.ida.liu.se/pub/labs/kplab/people/patdo/jal-final.ps.gz.

Reasoning about Rational, but not Logically Omniscient Agents (Extended Abstract)

Ho Ngoc Duc

Institute of Informatics, University of Leipzig
P.O. Box 920, D-04009 Leipzig, Germany
phone: +49-341-97 32283, fax: +49-341-97 32209
e-mail: duc@informatik.uni-leipzig.de

Epistemic logic, or the logic of the concepts of knowledge and belief, has become a major tool for knowledge representation recently. However, it is a very controversial matter whether epistemic logic is suitable for this purpose. Among the main problems of epistemic logic is the so-called Logical Omniscience Problem. This and related problems are examined, and some recent proposals to solve them are reviewed. Almost all attempts in the literature to solve this problem consist in weakening the standard epistemic systems: weaker systems are considered where the agents do not possess the full reasoning capacities of ideal reasoners. It is argued that this solution is not satisfactory: in this way omniscience can be avoided, but many intuitions about the concepts of knowledge and belief get lost. The main problems of epistemic logic cannot be solved satisfactorily within a static framework. To solve these problems we should take the reasoning activities of the agents into account. Therefore, we need dynamic epistemic logic for modeling knowledge. We shall show that axioms for epistemic logics must have the following form: if the agent knows all premises of a valid inference rule, and if she thinks hard enough, then she will know the conclusion. To formalize such an idea, we propose to "dynamize" epistemic logic, that is, to introduce a dynamic component into the language. Some sample systems of dynamic epistemic logic are then developed within the framework. It will be shown that these logics are suited to formalize resource bounded agents.

Specification of Nonmonotonic Reasoning*

Joeri Engelfriet and Jan Treur

Free University Amsterdam,
Department of Mathematics and Computer Science,
De Boelelaan 1081a,
1081 HV Amsterdam, The Netherlands,
tel. +31 20 4447776, fax. +31 20 4447653
email: {joeri, treur}@cs.vu.nl

Abstract. Two levels of description of nonmonotonic reasoning are distinguished. For these levels semantical formalizations are given. The first level is defined semantically by the notion of belief state frame, the second level by the notion of reasoning frame. We introduce two specification languages to describe nonmonotonic reasoning at each of the levels: (1) a specification language for level 1, with formal semantics based on belief state frames, (2) a fragment of infinitary temporal logic as a general specification language for level 2, with formal semantics based on reasoning frames. In our framework every level 2 description can be abstracted to level 1, and for every level 1 description there are level 2 descriptions which are a specialization of it.

keywords: Nonmonotonic reasoning, temporal logic, specification

1 Introduction

Nonmonotonic reasoning systems address applications where an agent reasoning about the world wants to draw conclusions that are not logically entailed by its (incomplete) knowledge about the world. Under such circumstances it is only possible to build a set of (additional) beliefs of hypothetical nature. Such a set of beliefs represents a hypothetical view on the world. In general it is not unique: multiple views are possible; an agent may (temporarily) commit itself to one view and switch its commitment to another one later. Such a view does not necessarily give a complete world description either. Each view leaves open a number of possible complete world descriptions. However, the additional knowledge defining the view an agent is committing to, may be sufficient for the agent to draw the required (defeasible) conclusions (within the context of that view).

One may focus on the intersection of the different possible sets of beliefs for the agent; this could be described by a nonmonotonic inference operator, e.g. as in [KLM90]. A disadvantage of this (sceptical) approach may be that hypothetical conclusions that are possible within one of the belief sets may be lost due to the

---

* Part of this work has been supported by SKBS and the ESPRIT III Basic Research project 6156 DRUMS II.

restriction to the common beliefs. So, the agent may not be able to draw the required conclusions, only taking into account the beliefs common in all views. Therefore we will concentrate on multiple belief sets rather than on their intersection. The non-multiple view will be incorporated as the special case of a single belief set.

A set of beliefs is usually not available to an agent immediately: usually for a given set X of world knowledge a set of beliefs is constructed by applying some type of reasoning (e.g., inference steps defined by default rules). As this type of reasoning should allow different alternative belief sets as conclusion sets (that may be mutually inconsistent), the type of reasoning used is essentially context-dependent: depending on a chosen context one of the alternative belief sets is generated.

In this paper we distinguish two levels of abstraction at which nonmonotonic reasoning can be specified:

1. *Specification of a set of intended multiple belief sets*
Specification of the possible belief sets for the agent abstracting from the specific reasoning patterns that lead to them.
2. *Specification of a set of intended reasoning patterns*
Specification of the reasoning patterns that lead to the intended possible belief states.

In this paper we will first semantically define the two levels. The first level is defined semantically by the notion of belief state frame (Section 2), the second level by the notion of reasoning frame (Section 3). In Section 4 the semantic connections between the two levels are described. Moreover, we introduce two specification languages to describe nonmonotonic reasoning at each of the levels. In Section 5 a specification language is defined for level 1, with formal semantics based on belief state frames. In Section 6 we introduce a fragment of infinitary (linear time) temporal logic as a general specification language for level 2, with formal semantics based on reasoning frames (interpreted as linear time temporal models).

Thus a specification framework is obtained by which any nonmonotonic system aiming at defining multiple belief sets (or reasoning traces leading to them) can be specified. Such a specification can be written either at the level of the possible alternative sets of beliefs that are the intended outcomes, or at the level of reasoning patterns to construct these outcomes (or both). The first type of specifications at least covers approaches such as preferential semantics (cf. [Sho87]) and the notion of S-expansions for modal nonmonotonic logics (cf. [MT93]). The second type at least covers approaches like default logic (cf. [Rei80]). The general specification framework (and its semantics) introduced here provides a unifying perspective on these and other well-known approaches. For example, the general connections between the two levels developed in the current paper provide connections between preferential semantics and default logic (cf. [Eth87], [Voo93]) as a special case.

Apart from the two levels of abstraction considered here, [EHT95] distinguishes three more levels of abstraction. Sections 2, 3 and 4 review and extend some of the material presented there.

2 Belief state frames

Classical (propositional) logic lies at the basis of most nonmonotonic formalisms, so we will assume we have a propositional language, L, its corresponding set of models, Mod, and the (semantic) consequence relation $\models \subseteq Mod \times L$. Furthermore, for a set $X \subseteq L$, we define the models of X: $Mod(X) = \{ m \in Mod \mid m \models \varphi \text{ for all } \varphi \in X\}$; the consequences of a set $X \subseteq L$ are defined by $Cn(X) = \{ \varphi \in L \mid Mod(X) \subseteq Mod(\varphi)\}$. For a subset K of Mod, the *theory* of K is defined by $Th(K) = \{ \varphi \mid m \models \varphi \text{ for all } m \in K \}$. A set of models K is called *closed* if $K = Mod(Th(K))$, or equivalently, if K is the set of models of a theory.

Semantically, nonmonotonic reasoning could be described by model operators which assign to a set of formulae X, the initial facts, a set of models, the *intended models*, which should be a subset of the set of classical models of X. These models are intended in the sense that these are the worlds the agent considers plausible on the basis of X. Often, however, a reasoning agent does not have just one set of beliefs, but there are (many) alternative belief sets depending on which further assumptions the agent wants to make. For example, in the case of default reasoning, given initial facts and a set of default rules, there may be more than one Reiter extension. To formalize this semantically we introduce the notion of an information state and a belief state frame.

Definition 2.1 (Information state)

a) An *information state* M is a non-empty closed set of propositional models, that is, there is a consistent theory of which it is the model class. The truth of a propositional formula α in such a state is defined by:

$$M \models \alpha \quad \Leftrightarrow \quad m \models \alpha \text{ for each } m \in M$$

b) The *refinement ordering* $\leq$ on information states is defined by:

$$M_1 \leq M_2 \quad \Leftrightarrow \quad M_2 \subseteq M_1$$

c) The *set of all information states* is denoted by IS.

An information state represents a possible view on the world. The information the agent has in a state M is $Th(M)$, which is closed under propositional consequence. Although in practice this may not always be the case, we assume the agent is in principle capable of deriving any propositional consequence of its information. The condition that an information state is closed is merely technical, it improves readability of definitions and proofs, but is not essential. The refinement ordering on information states expresses degree of information: if $M_1 \leq M_2$ then $Th(M_1) \subseteq Th(M_2)$, so the agent has more information in M_2 than it has in M_1.

Definition 2.2 (Belief state operator and belief state frame)

a) A *belief state operator* Γ is a function $\Gamma : \mathcal{P}(L) \to \mathcal{P}(IS)$ satisfying the following conditions for every $X \subseteq L$:

(i) $K \subseteq Mod(X)$ for every $K \in \Gamma(X)$;

(ii) $\forall J, K \in \Gamma(X)$ $J \subseteq K \Rightarrow J = K$ (*noninclusiveness*).

The tuple $\mathcal{SB} = (L, Mod, \models, \Gamma)$ is said to be a *belief state frame*.

b) A belief state operator Γ is called *invariant* if $\Gamma(X) = \Gamma(Cn(X))$ for all $X \subseteq L$.

The first condition expresses conservativity: it means that a possible view on the world at least satisfies the initial facts. In our opinion this is an essential feature of nonmonotonic reasoning (about a single world situation; if the world changes, the agent may have to adapt its initial facts): it gives a method of extending partial information. The third condition is probably most subject to discussion. The idea is that if an agent has two possible views, one of which is included in the other, the agent will want to retain only the most informative one. Most nonmonotonic formalisms (though not all) satisfy this requirement; we may want to drop this condition in further work. The property of invariance is related to the fact that an information state is closed under propositional consequence: the syntactical form in which the initial facts are given, is not important.

Level 1 semantically consists of descriptions of nonmonotonic reasoning using belief state frames. As an example we consider preferential semantics (cf. [Sho87]). Let a preference relation $<$ on **Mod** be given. A belief state operator $\Gamma_<$ (with single belief states) can be defined in the following manner: for each $X \subseteq L$

$$\Gamma_<(X) = \{ \{ m \in \text{Mod} \mid m \text{ is } <\text{-minimal in Mod}(X) \} \}$$

Preferential semantics essentially provides a level 1 description, abstracting from lower levels. Also approaches with non-singleton belief states specified by preference relations exist ([KLM90], [Voo93]).

3 Reasoning frames

We have seen in Section 2 how (nonmonotonic) reasoning can be described by assigning to each set X of initial facts a set $\Gamma(X)$ of belief states, abstracting from the way in which the conclusions of these states have been reached. On a less abstract level, one would also like to be able to describe types of reasoning by specifying not only the conclusions of the reasoning process, but also the reasoning path leading from the initial facts to the conclusions. To do this we will first formalize the notion of such a path, which we call a reasoning trace. After giving a formal ("algebraic") description of these traces, in Section 6 we will introduce a specification language for these traces and investigate the power of this language. We will look at the links between the previous level and the current one in Section 4.

Intuitively, the path from initial set of formulae to final conclusions can be seen as the behaviour of a reasoning process which starts with the initial formulae, then makes some inferences to arrive at a new state, again makes some inferences, et cetera, possibly ad infinitum. The union of all conclusions drawn at all stages of such a process can be seen as the set of final conclusions of the process. A formalization of such reasoning behaviour would have to describe which formulae have been derived at each stage. We will formalize this semantically by the notion of a reasoning trace:

Definition 3.1 (Reasoning trace and limit model)
    a) A *reasoning trace* $\mathfrak{M}$ is a function from the set of natural numbers ($\mathbb{N}$) to **IS** such that for all $s \in \mathbb{N}$:
        (i) $\mathfrak{M}_s \leq \mathfrak{M}_{s+1}$
        (ii) $\mathfrak{M}_s = \mathfrak{M}_{s+1} \quad\quad \Rightarrow \quad\quad \mathfrak{M}_s = \mathfrak{M}_t$ for all $t \geq s$.

b) The *refinement ordering* $\leq$ on reasoning traces is defined by:

$$\mathfrak{M} \leq \mathfrak{N} \qquad \Leftrightarrow \qquad \mathfrak{M}_s \leq \mathfrak{N}_s \text{ for all } s \in \mathbb{N}$$

c) The *limit model*, $\lim \mathfrak{M}$ of a reasoning trace $\mathfrak{M}$ is defined by

$$\lim \mathfrak{M} = \bigcap_{s=0}^{\infty} \mathfrak{M}_s$$

d) A reasoning trace $\mathfrak{M}$ is sometimes denoted by $(\mathfrak{M}_s)_{s \in \mathbb{N}}$.

e) A reasoning trace is called *finitely generated* if each $\mathbf{Th}(\mathfrak{M}_s)$ is finitely generated over $\mathbf{Th}(\mathfrak{M}_0)$, i.e., if $\mathbf{Th}(\mathfrak{M}_s) = \mathbf{Cn}(\mathbf{Th}(\mathfrak{M}_0) \cup \{\alpha_s\})$ for some formula α_s.

We will motivate the two conditions. The first condition expresses conservativity: the information states become more informative over time. This means that the conclusions the agent draws at a certain point in time, remain valid in the future. Of course we do not wish to claim that backtracking never occurs in an *implementation* of a reasoning process; it will. However, we would like to abstract away from particular implementations, and look only at the increase of information of the agent over time: the resulting valid reasoning patterns which may be generated by a backtracking implementation. The second condition states that once the agent does not reach any new conclusions in a reasoning step, it has finished reasoning. This relates on the one hand to the assumption that the agent is reasoning about a particular fixed world situation (it cannot perform any observations during the reasoning), and on the other hand to the above mentioned motivation: although in an implementation the agent may be idle for some time, we are interested in the essentials of a reasoning process. The reason that a trace is always infinite, is again technical and not essential. We want to be able to model infinite reasoning processes (in the case of for instance default logic with an infinite set of defaults), and finite ones. In order to simplify definitions and proofs we model these finite processes as infinite traces which stabilize (become constant) at a certain point in time, instead of as finite traces.

Since we assume a countable language L, each theory can be approximated by a chain of finitely generated theories. Therefore:

Proposition 3.2
   a) For any reasoning trace its limit is an information state.
   b) Any information state is the limit model of a finitely generated reasoning trace.

A (nonmonotonic) type of reasoning can now be described by giving its intended reasoning traces. Given a set of initial formulae, there may of course be several traces leading to different conclusion sets. We do, however, assume that the reasoning is deterministic in the sense that given the set of initial formulae *and* the final conclusion set, the trace between them is uniquely determined. This can be explained in the sense that at each stage of the reasoning process all conclusions that possibly can be drawn, actually are drawn in the next step. (Although most nonmonotonic formalisms satisfy this requirement, we may want to drop this assumption in further work.) Moreover, we do not allow two distinct traces with the same initial facts leading to limit models of which one is a refinement of the other (non-inclusiveness of traces), in analogy with the requirement of non-inclusiveness for belief state operators.

Definition 3.3 (Reasoning frame)

a) A *reasoning frame* is a tuple $(L, \mathbf{Mod}, \vDash, \mathfrak{J})$ with $\mathfrak{J}$ a set of reasoning traces such that for all $\mathfrak{M}$ and $\mathfrak{N}$ in $\mathfrak{J}$: if $\mathfrak{M}_0 = \mathfrak{N}_0$ and $\lim \mathfrak{M} \leq \lim \mathfrak{N}$ then $\mathfrak{M} = \mathfrak{N}$. For shortness, sometimes we also call $\mathfrak{J}$ by itself a reasoning frame.

b) If for all sets of formulae X there exists a trace $\mathfrak{M}$ in $\mathfrak{J}$ such that $\mathrm{Th}(\mathfrak{M}_0) = \mathrm{Cn}(X)$ then $\mathfrak{J}$ is called a *complete* reasoning frame. Otherwise it is called *partial*.

This provides a semantical formalization of level 2. As an example we consider default logic. Let D be a set of defaults. For $X \subseteq L$ let $\mathbb{E}(\langle X, D \rangle)$ denote the set of (Reiter) extensions of the default theory $\langle X, D \rangle$. The following belief state operator can be defined for $X \subseteq L$:

$$\Gamma_D(X) = \{ \mathrm{Mod}(E) \mid E \in \mathbb{E}(\langle X, D \rangle) \}$$

For a given X and $E \in \mathbb{E}(\langle X, D \rangle)$ the following reasoning trace $\mathfrak{M}$ can be associated in a canonical manner:

$$\mathfrak{M}_i = \mathrm{Mod}(E_i)$$

with $E_0 = \mathrm{Cn}(X)$, and for all $i \geq 0$

$$E_{i+1} = \mathrm{Cn}(E_i \cup \{ \omega \mid (\alpha : \beta_1, \ldots, \beta_n) / \omega \in D, \ \alpha \in E_i \text{ and } \neg\beta_1 \notin E, \ldots, \neg\beta_n \notin E\})$$

Note that this is a trace definition based on the given set of defaults D.

4 Connections between belief state frames and reasoning frames

Belief state frames and reasoning frames both provide a means of defining (nonmonotonic) types of reasoning. A specification of a belief state frame (level 1) is more abstract whereas a specification of a reasoning frame (level 2) provides more details of the reasoning process. But can they describe the same types of reasoning, or put differently: are there clear connections between the two levels? We would like every level 1 specification to be "implementable" (by specialisation) by a level 2 specification. On the other hand, for every specification on level 2 it should be possible to find an abstraction of it on level 1 of which it is an "implementation". These issues will be addressed in this section.

4.1 Abstraction

Level 1 descriptions give the final conclusion sets of a type of reasoning given the initial formulae, abstracting from the reasoning process. So if we want to abstract from a level 2 specification, given a trace we should look at the initial formulae and the final outcome, that is the limit model. If we have a reasoning frame of level 2, we can define an invariant belief state operator in a straightforward way.

Definition 4.1 (Belief state operator of a reasoning frame)
   a) Given a complete reasoning frame $\mathfrak{J}$ the *associated belief state operator* $\Gamma_{\mathfrak{J}}$ is defined as follows: for any set $X \subseteq L$,
$$\Gamma_{\mathfrak{J}}(X) = \{ \lim \mathfrak{M} \mid \mathfrak{M} \in \mathfrak{J}, \text{Th}(\mathfrak{M}_0) = \text{Cn}(X) \}$$
   b) For a given invariant belief state operator Γ we define the set $\overline{\mathfrak{J}}(\Gamma)$ of reasoning frames with Γ as their associated belief state operator:
$$\overline{\mathfrak{J}}(\Gamma) = \{ \mathfrak{J} \mid \mathfrak{J} \text{ is a reasoning frame with } \Gamma_{\mathfrak{J}} = \Gamma \}$$

It is easy to see that for a given reasoning frame $\mathfrak{J}$ the defined $\Gamma_{\mathfrak{J}}$ is an invariant belief state operator.

4.2 Specialisation

Many reasoning frames can yield the same associated belief state operator, so we want to analyse the set $\overline{\mathfrak{J}}(\Gamma)$ of possible reasoning frames connected with a belief state operator Γ by defining a parametrization of $\overline{\mathfrak{J}}(\Gamma)$.

If we want to specify, for an invariant belief state operator Γ, an associated reasoning frame, what we have to do is, for all X and $M \in \Gamma(X)$, specify a trace from the information state related to the set of initial formulae X to the belief state $M \in \Gamma(X)$. Therefore a parameter should specify when the formulae of $\text{Th}(M) \backslash \text{Cn}(X)$ can and have to be added during the reasoning. We can do this by assuming that each formula may depend on some other formulae and can only be added if the formulae it depends on have already been added in earlier stages. So, for each X and $M \in \Gamma(X)$ such a dependency ordering (or information ordering) $\prec$ between propositional formulae has to be specified, where $\psi \prec \varphi$ means that φ depends on ψ. Therefore we have chosen a parametrization of $\overline{\mathfrak{J}}(\Gamma)$ by means of functions p which assign to each pair of theories X, Y, where X consists of the initial facts and Y is the theory of a belief state $M \in \Gamma(X)$ an ordering on the propositional formulae. In the reasoning trace one has to make sure the formulae are added to X respecting this order. Doing this, for each X and $M \in \Gamma(X)$, the traces are specified unambiguously.

Definition 4.2 (Reasoning trace parameters)
   a) A *reasoning trace parameter* is a partial order $(L, \prec)$ on propositional formulae such that for each $\varphi \in L$ there is an $n \in \mathbb{N}$ such that $\{\psi \mid \psi \prec \varphi\}$ does not contain a chain of length more than n. A reasoning trace parameter is called *finitely grounded* if for each $\varphi \in L$ the set $\{\psi \mid \psi \prec \varphi\}$ is finite.
   b) Let $\mathbb{TH} := \{X \subseteq L \mid \text{Cn}(X) = X\}$ be the set of theories, and let $\mathbb{BE} = \{ (X, Y) \mid X, Y \in \mathbb{TH}, X \subseteq Y\}$ be the set of pairs of possible starting points and endpoints. Given a trace parameter $\prec$ and $(X, Y) \in \mathbb{BE}$ we define a chain of sets of formulae $(S_t)_{t \in \mathbb{N}}$ as follows:
$$S_0 \quad = \quad X$$
$$S_{t+1} \quad = \quad \text{Cn}(S_t \cup \{ \varphi \in Y \mid \{\psi \in Y \mid \psi \prec \varphi\} \subseteq S_t \})$$

Now we define a reasoning trace $(\mathcal{M}_t)_{t \in \mathbb{N}}$ by $\mathcal{M}_t = \mathrm{Mod}(S_t)$. This trace will be denoted by $\mathcal{M}(\prec, \mathbf{X}, \mathbf{Y})$.

It is easy to see that this indeed defines a reasoning trace with $\mathrm{Th}(\mathcal{M}_0) = \mathbf{X}$ and $\mathrm{Th}(\lim \mathcal{M}(\prec, \mathbf{X}, \mathbf{Y})) = \mathbf{Y}$.

Definition 4.3 (Parametrized reasoning frame of a belief state operator)

Let $\mathbb{TF}$ be the set of all reasoning trace parameters. A function $p : \mathbb{B} \to \mathbb{TF}$ with $\mathbb{B} \subseteq \mathbb{BE}$ such that if $(\mathbf{X}, \mathbf{Y}), (\mathbf{X}, \mathbf{Y}') \in \mathbb{B}$ and $\mathbf{Y} \subseteq \mathbf{Y}'$ then $\mathbf{Y} = \mathbf{Y}'$ is called a *reasoning frame parameter*. For a reasoning frame parameter p let $\mathcal{J}_p$ be the following reasoning frame:

$$\mathcal{J}_p = \{ \mathcal{M}(p(\mathbf{X},\mathbf{Y}), \mathbf{X}, \mathbf{Y})) \mid (\mathbf{X},\mathbf{Y}) \in \mathbb{B}\}$$

A reasoning frame parameter p is *suited for* an invariant belief state operator Γ if $\mathbb{B} = \{(\mathbf{X}, \mathbf{Y}) \mid \mathbf{X} \in \mathbb{TH}, \mathbf{Y} = \mathrm{Th}(M) \text{ for some } M \in \Gamma(\mathbf{X})\}$.

It is easy to verify that in case all $\mathbf{X} \in \mathbb{TH}$ occur as first coordinate in a pair in $\mathbb{B}$, this defines a complete reasoning frame and that all reasoning frames are parametrized by these parameters:

Theorem 4.4

Let an invariant belief state operator Γ and a complete reasoning frame $\mathcal{J}$ be given. The following conditions are equivalent:

(i) $\Gamma_{\mathcal{J}} = \Gamma$

(ii) There exists a reasoning frame parameter p suited for Γ such that $\mathcal{J}_p = \mathcal{J}$.

The reasoning frame parameter can be taken finitely grounded.

5 A specification language for belief state frames

Having introduced a semantical foundation of the level of multiple belief sets based on the notion of a belief state frame, a natural next question is how to specify such a belief state operator. In other words, can a standard language be defined in which a specific belief state frame can be described? To this end we introduce the following specification language.

Definition 5.1 (Belief state frame specification)

a) A *belief state frame expression* is a tuple $< \mathbf{A}, \mathbf{B}, \mathbf{C}, \gamma >$ where $\mathbf{A}, \mathbf{B}$ and $\mathbf{C}$ are sets of formulae and γ is a formula in $\mathbf{L}$. This expression is called *finitary* if $\mathbf{B}$ is finite; otherwise it is called *infinitary*.

b) A *belief state frame theory* or *specification* is a set $\mathbf{S}$ of belief state frame expressions.

c) A belief state frame specification **S** *specifies* the belief state frame $\mathfrak{SB}$ if for any set of formula **X** and any closed set of models **K** the following are equivalent:

    (i) $K \in \Gamma(X)$

    (ii) $K = Mod(X \cup \{ \gamma \mid \exists < A, B, C, \gamma > \in S$ such that $A \subseteq X, B \subseteq L \backslash X$
                           and for all $\beta \in C$: not $K \vDash \neg\beta$ $\})$

In a belief state frame expression $< A, B, \gamma >$ the set **A** can be seen as the preconditions, the set **B** contains formulae which can be considered as "anti-preconditions" (the formulae in **B** should *not* be known initially in order for the rule to be applicable), and the set **C** is a kind of consistency check (as is also used in default logic). Examples as in [MTT96] show that sometimes we really need an infinite **C**.

Definition 5.2

    Let $\mathfrak{SB} = (L, Mod, \vDash, \Gamma)$ be a belief state frame. Then the belief state frame specification $S^{\mathfrak{SB}}$ is defined by the following set of belief state frame expressions for each set of formulae $X : < X, L \backslash X, C, \gamma >$ where γ ranges over a set of generators of $Th(K)$ over X, for each $K \in \Gamma(X)$ and $B = \{ \neg\alpha_{J,K} \mid J \in \Gamma(X), J \neq K \}$ with $\alpha_{J,K}$ defined as follows: as $Th(J) \backslash Th(K) \neq \varnothing$ (non-inclusiveness), we choose $\alpha_{J,K}$ in $Th(J) \backslash Th(K)$.

Notice that $S^{\mathfrak{SB}}$ can be taken finitary if there is only a finite number of atomic proposition symbols. The following theorem shows that any belief state frame $\mathfrak{SB}$ is specified by a belief state frame specification and, conversely, that for every belief state frame specification an associated belief state frame can be defined that is specified by it.

Theorem 5.3

    a) Let **S** be a belief state frame specification. The operator Γ^S defined by
    $\Gamma^S(X) = \{ K \mid K = Mod(X \cup \{ \gamma \mid \exists < A, B, C, \gamma > \in S$ such that $A \subseteq X, B \subseteq L \backslash X$
                          and for all $\beta \in C$: not $K \vDash \neg\beta$ $\})\}$
    for every set of formulae **X**, is a belief state operator and $\mathfrak{SB}^S = (L, Mod, \vDash, \Gamma^S)$ is the belief state frame specified by **S**.

    b) Let $\mathfrak{SB} = (L, Mod, \vDash, \Gamma)$ be a belief state frame. Then the belief state frame specification $S^{\mathfrak{SB}}$ specifies $\mathfrak{SB}$. If Γ is invariant and there is only a finite number of atomic proposition symbols, then we can restrict $S^{\mathfrak{SB}}$ to a finite set of finitary expressions.

6 A temporal specification language for reasoning frames

A simple observation allows us to find a natural description language for reasoning traces: the steps in a reasoning trace can be viewed as temporal steps. This means that the transition from an information state to the next one (as the result of a number of inference steps) can be seen as a temporal one. In this view a trace is a temporal model based on the set of natural numbers as the flow of time. An obvious candidate for describing these models is temporal logic. However, the full (tense) logic will turn out to be not completely appropriate: on the one hand it can describe models which are not traces but on the other hand it is not powerful enough. Therefore we will introduce a limited fragment of infinitary tense logic (based on our essentially three-valued information states).

Definition 6.1 (Temporal model)
   i) A *temporal model* is a function $\mathcal{M}: \mathbb{N} \to$ IS.
   ii) A temporal model $\mathcal{M}$ is *conservative* if $\mathcal{M}_s \le \mathcal{M}_{s+1}$ for all $s \in \mathbb{N}$.
   iii) The *refinement ordering* $\le$ on temporal models is defined by:
   $$\mathcal{M} \le \mathcal{N} \iff \text{for all } s: \ \mathcal{M}_s \le \mathcal{N}_s \text{ and } \mathcal{M}_0 = \mathcal{N}_0.$$
   iv) The *limit model*, $\lim \mathcal{M}$, of a temporal model $\mathcal{M}$ is the information state defined by
   $$\lim \mathcal{M} = \bigcap_{s=0}^{\infty} \mathcal{M}_s$$
   v) A temporal model $\mathcal{M}$ is sometimes denoted by $(\mathcal{M}_s)_{s \in N}$.

Note that the notion of a temporal model as such is close to the notion of a reasoning trace: any reasoning trace can be considered a conservative temporal model with the property that if it stabilises one step, then it stabilises forever. However, in a temporal model temporal operators and temporal formulae are interpreted. The temporal language we will use is built from temporal atoms of the form $\mathbf{O}\varphi$ (with $\mathbf{O} \in \{ \mathbf{F}, \mathbf{G}, \mathbf{C}, \mathbf{H_0} \}$ and φ a propositional formula) using negation and infinite conjunctions (we do not need nesting of temporal operators here, although the definition can easily be extended to include them). The truth of a formula α in a temporal model $\mathcal{M}$ at time point i, denoted as $(\mathcal{M}, i) \vDash \alpha$, is defined inductively:

Definition 6.2 (Temporal interpretation)
   a) For a propositional formulae φ:

$(\mathcal{M}, s) \vDash \mathbf{F}\varphi$	$\iff$	there exists $t \in \mathbb{N}$, $t > s$ such that $\mathcal{M}_t \vDash \varphi$
$(\mathcal{M}, s) \vDash \mathbf{G}\varphi$	$\iff$	for all $t \in \mathbb{N}$ with $t > s$: $\mathcal{M}_t \vDash \varphi$
$(\mathcal{M}, s) \vDash \mathbf{C}\varphi$	$\iff$	$\mathcal{M}_s \vDash \varphi$
$(\mathcal{M}, s) \vDash \mathbf{H_0}\varphi$	$\iff$	$\mathcal{M}_0 \vDash \varphi$

   b) For a temporal formula α:
   $$(\mathcal{M}, s) \vDash \neg\alpha \iff \text{it is not the case that } (\mathcal{M}, s) \vDash \alpha$$
   c) For a set A of temporal formula:
   $$(\mathcal{M}, s) \vDash \wedge A \iff \text{for all } \varphi \in A: (\mathcal{M}, s) \vDash \varphi$$
   d) A formula φ is true in a model $\mathcal{M}$, denoted $\mathcal{M} \vDash \varphi$, if for all $s \in \mathbb{N}$: $(\mathcal{M}, s) \vDash \varphi$
   e) A set of formulae T is true in a model $\mathcal{M}$, denoted $\mathcal{M} \vDash T$, if for all $\varphi \in T$, $\mathcal{M} \vDash \varphi$. We call $\mathcal{M}$ a model of T.

Furthermore the connectives $\vee$ and $\rightarrow$ are introduced as the usual abbreviations.

The temporal language we have just introduced is still too powerful: we want to use only a fragment to describe models which can be seen as reasoning traces. So the question is: which fragment is appropriate for reasoning? As steps in a reasoning process are taken whenever a number of (nonmonotonic) inference steps is used, it seems that temporal rules should prescribe taking the equivalent of inference steps in the temporal model. So the next question is what the nature is of a generalized nonmonotonic inference step that a reasoning process can execute. A general format of (temporal) inference rules is $\alpha \rightarrow G\beta$ where α is a condition for the inference, and β is its conclusion: if the condition α is fulfilled, the conclusion β can be drawn, and will be true henceforth. The condition α may of course include reference to the initial facts and the facts which have been derived earlier (and therefore are still true at the present moment). But in nonmonotonic reasoning there is often also a kind of global consistency check. In default logic for instance, a rule is applicable if certain formulae, called the justifications, are consistent with the final outcome of the reasoning process (called an extension), which means they should be consistent throughout the entire reasoning process. Consistency of a formula usually means that its negation should not be true. Therefore we also allow conditions which state that a certain formula should never (in the future of the reasoning process) be true.

Definition 6.3 (Reasoning theories)
    a) A formula is called a *(nonmonotonic) reasoning formula* if it is of the form
        $\alpha \wedge \beta \wedge \varphi \wedge \psi \rightarrow G\gamma$, where
            $\alpha = \wedge\{ H_0\varepsilon \mid \varepsilon \in A \}$ for a set of propositional formulae A.
            $\beta = \wedge\{ \neg H_0\delta \mid \delta \in B \}$ for a set of propositional formulae B.
            $\varphi = \wedge\{ \neg F\theta \mid \theta \in C \}$ for a set of propositional formulae C.
            $\psi = \wedge\{ C\zeta \mid \zeta \in D \}$ for a set of propositional formulae D.
            γ is a propositional formula.
    A reasoning formula is called *finitary* if all sets of formulae involved are finite; otherwise it is called *infinitary*.
    b) A set **Th** of reasoning formulae is called a *theory of reasoning*. It is called *finitary* if all its elements are; otherwise it is called *infinitary*.

So a reasoning formula prescribes the truth of a formula in the future based on knowledge of initial facts, truth of current facts and consistency of facts in the future (if $\neg F\theta$ is true, then θ is never true in the future, so it is always either false or unknown).

Definition 6.4 (Conservativity)
    The theory $Cons = \{ C\alpha \rightarrow G\alpha \mid \alpha$ **a propositional formula** $\}$ is a theory of reasoning expressing conservativity of temporal models.

A theory of reasoning prescribes truth of facts in the future, analogously to inference steps. But what about facts which become true at a point in time spontaneously, that is without any inference rule prescribing their truth? We should have a way to make sure that this does not happen: we want the models to have minimal information in the sense that nothing becomes true if there are no rules saying so. This leads to the following notion of minimal models:

Definition 6.5 (Minimal temporal models)

A temporal model $\mathcal{M}$ is called a *minimal model* of a theory **Th** if it is a model of **Th** and for any model $\mathcal{N}$ of **Th**, if $\mathcal{N} \le \mathcal{M}$ then $\mathcal{N} = \mathcal{M}$.

A minimal model of a theory is a model for which there are no smaller models of the theory, so they contain a minimum of information.

Given the fragment of temporal logic we have defined, a natural property to investigate is whether this fragment is suited for describing reasoning traces.

From now on we will assume that any theory includes the theory **Cons** described in Definition 6.4. The first result is that all minimal models of any theory are reasoning traces. On the other hand, any reasoning frame is the set of minimal models of a theory of reasoning:

Theorem 6.6

a) For any theory of reasoning **Th** its minimal models constitute a (partial) reasoning frame.

b) For any (partial) reasoning frame $\mathcal{T}$ there exists a theory of reasoning whose minimal models are exactly $\mathcal{T}$. If there is only a finite number of atomic proposition symbols, then such a theory of reasoning can be taken finite and finitary.

Given a belief state operator Γ and a reasoning frame parameter p suited for Γ, we have defined the corresponding reasoning frame $\mathcal{T}_p$. As this is a reasoning frame, by Theorem 6.6 we can find a theory of reasoning whose minimal models are exactly $\mathcal{T}_p$. However, we can find a more intuitive and direct definition of a theory for reasoning for Γ and p yielding the corresponding reasoning frame.

Definition 6.7

Let $p : \mathbb{B} \to \mathbb{TP}$ be a reasoning frame parameter. Take a pair $(X, Y) \in \mathbb{B}$, and a formula $\gamma \in Y\backslash X$, and denote $p(X, Y)$ as $\prec$. Then we define the rule for γ as:
$$\alpha \wedge \beta \wedge \varphi \wedge \psi \to G\gamma, \text{ with}$$
$$\alpha = \wedge \{ H_0 g \mid g \in X \},$$
$$\beta = \wedge \{ \neg H_0 \, \delta \mid \delta \in L\backslash X \},$$
$$\varphi = \wedge \{ \neg F \, \alpha_{Y,Y'}(X) \mid Y' \ne Y, (X, Y') \in \mathbb{B} \},$$
$$\psi = \wedge \{ C\varepsilon \mid \varepsilon \in Y, \varepsilon \prec \gamma \}.$$

In the formula φ, the $\alpha_{Y,Y'}$ are as follows: since $Y'\backslash Y \ne \varnothing$ (noninclusiveness) we choose $\alpha_{Y,Y'}$ in $Y'\backslash Y$. This ensures that γ is added only if we are "heading" for the right limit, Y. Let $\mathbf{Th}_{(X,Y)}$ denote the set of these rules for all $\gamma \in Y\backslash X$.

For any theory X such that there is no pair $(X, Y) \in \mathbb{B}$ we define its rule:
$$\alpha \wedge \beta \to G\bot$$
with α and β as above. Let $\mathbf{Th}_N$ be the set of these rules for all such X.

Then a *theory of (nonmonotonic) reasoning for p*, $\mathbf{Th}_p$ is defined by:

$$\mathbf{Th}_p = \bigcup_{(X,Y) \in \Xi} \mathbf{Th}_{(X,Y)} \cup \mathbf{Th}_N \cup C$$

Note that the formulae $\alpha_{Y,Y'}$ are defined here differently from Definition 5.2.

Theorem 6.8

Let $p : \Xi \to \Xi\Xi$ be a reasoning parameter, then the set of minimal models of $\mathbf{Th}_p$ is exactly $\mathfrak{S}_p$. If there is only a finite number of atomic proposition symbols, then we can restrict $\mathbf{Th}_p$ to a finite and finitary theory.

The logic we described is infinitary. On the one hand, this is needed to get the result that any reasoning frame can be described by a theory of reasoning (see [MTT96]). On the other hand, it is possible to work with infinitary logic if the temporal models have a finite representation and if the truth of an infinite conjunction in such a model is decidable (this may depend on the theory of reasoning).

For the example of default logic a temporal translation was already introduced in [ET93]. Here a default rule $(\alpha : \beta) / \gamma$ is translated into the temporal rule $C\alpha \wedge \neg F \neg \beta \to G\gamma$, which indeed is a reasoning formula in the sense defined above (notice that translated default rules form a strict subset of the temporal language - Default Logic is strictly less expressive than the temporal language, even when only finite conjunctions are allowed).

7 Conclusions and further perspectives

In this paper two levels of specification of nonmonotonic reasoning are distinguished. The notions of belief state frame and reasoning frame were introduced and used as a semantical basis for these levels. Moreover, the semantical connections between the levels were identified.

A specification language for level 1 was introduced and a fragment of infinitary temporal logic was introduced and proposed as a general specification language for level 2. It was shown that the logic is suited for describing reasoning frames and that for a finite number of atomic proposition symbols only finitary formulae can be used. On the other hand, for a countable language examples exist where finitary formulae are insufficient; see [MTT96] for an example with countably many belief states that cannot be expressed by finitary means. Furthermore a correct translation of a reasoning frame parameter to a theory in the logic was given.

Our framework allows a unified way of looking at nonmonotonic reasoning at two different levels of abstraction. Abstract properties of nonmonotonic reasoning at these levels can be studied in a way analogous to for instance in [KLM90] (see [EHT95]). Also specification of reasoning can be done in a unified way. Logical properties of the specification language for level 2 are studied in [En95].

Although many nonmonotonic formalisms can be described and specified in our framework, we cannot claim that any reasoning process falls in its scope. We would like to relax some of the conditions for belief state operators and reasoning frames. Since the specification languages we proposed exactly capture these (restricted) notions of belief state frames and reasoning frames, it means we would also have to adapt the specification languages (for instance by allowing a more general form of formulae, or by dropping the conservativity axioms).

In [ET93], [ET94] it is shown for a number of examples how the language of temporal logic can be used to specify sets of intended reasoning patterns. The dualism between multiple outcomes and multiple reasoning traces of a nonmonotonic reasoning process is also studied in the context of default logic, leading to representation theory: for which set of outcomes can a default theory be found with these outcomes (for a number of results in this area, see [MTT96]).

Acknowledgements

Part of this work has been supported by SKBS and the ESPRIT III Basic Research project 6156 DRUMS II. Discussions with Heinrich Herre have played a stimulating role in developing part of the material in this paper. The paper was read and commented on by Rineke Verbrugge.

References

[Eth87] Etherington, D.W. : A Semantics for Default Logic, *Proc. IJCAI-87*, pp. 495-498; see also in: D.W. Etherington, Reasoning with Incomplete Information, Morgan Kaufmann, 1988

[EHT95] Engelfriet, J., Herre, H., Treur, J.: Nonmonotonic Belief State Frames and Reasoning Frames, in: C. Froidevaux, J. Kohlas (eds.), *Symbolic and Quantitative Approaches to Reasoning and Uncertainty*, Proc. ECSQARU'95, LNAI 946, Springer-Verlag, 1995, pp. 189-196

[En95] Engelfriet, J.: Minimal Temporal Epistemic Logic, Technical Report IR-388, Vrije Universiteit Amsterdam, Department of Mathematics and Computer Science, 1995

[ET93] Engelfriet, J., Treur, J.: A Temporal Model Theory for Default Logic, in: M. Clarke, R. Kruse, S. Moral (eds.), *Proc. 2nd European Conference on Symbolic and Quantitative Approaches to Reasoning and Uncertainty*, ECSQARU '93, Springer Verlag, 1993, pp. 91-96. Extended version: Report IR-334, Vrije Universiteit Amsterdam, Department of Mathematics and Computer Science, 1993

[ET94] Engelfriet, J., Treur, J. : Temporal Theories of Reasoning. In: C. MacNish, D. Pearce, L.M. Pereira (eds.), *Logics in Artificial Intelligence*, Proceedings of the 4th European Workshop on Logics in Artificial Intelligence, JELIA '94, Springer Verlag, pp. 279-299. Also in: *Journal of Applied Non-Classical Logics* 5 (2), 1995, pp. 239-261

[Her94] Herre, H.: Compactness Properties of nonmonotonic Inference Operations. In: C. MacNish, D. Pearce, L.M. Pereira (eds.), *Logics in Artificial Intelligence*, Proceedings of the 4th European Workshop on Logics in Artificial Intelligence, JELIA '94, Springer Verlag, 1994, pp. 19-33. Also in: *Journal of Applied Non- Classical Logics 5* (1), Special Issue with selected papers from JELIA'94, 1995, pp. 121-135

[KLM90] Kraus, S., D. Lehmann, M. Magidor: Nonmonotonic Reasoning, Preferential models and cumulative logics; *A.I.* 44 (1990), 167 - 207

[MT93] Marek, V., M. Truszczynski: *Nonmonotonic Logic*, Springer-Verlag, 1993

[MTT96] Marek, V., J. Treur, M. Truszczynski: Representation Theory for Default Logic, Proceedings of the Fourth International Symposium on Artificial Intelligence and Mathematics, 1996

[Rei80] Reiter, R.: A logic for default reasoning, *Artificial Intelligence* 13, 1980, pp. 81-132

[Sho87] Shoham, Y.: Nonmonotonic Logics: Meaning and Utility, In: J. McDermott (ed.), *Proc. 10th International Joint Conference on Artificial Intelligence*, IJCAI-87, Morgan Kaufmann, 1987, pp. 388-393

[Voo93] Voorbraak, F.: Preference-based Semantics for Nonmonotonic Logics, in: Bajcsy, R. (ed.), *Proc. 13th International Joint Conference on Artificial Intelligence*, IJCAI-93, Morgan Kaufmann, 1993, pp. 584-589

Intelligent Agents in the Situation Calculus: an Application to User Modelling

Bruno Errico[1] and Luigia Carlucci Aiello[1]

Dipartimento di Informatica e Sistemistica, Università di Roma "La Sapienza"
via Salaria 113, 00198 Roma, Italy. e-mail:errico,aiello@dis.uniroma1.it

Abstract. We propose an agent based language, aimed at representing the epistemic states of users of interactive systems, with the goal to devise domain independent tools for building, updating and maintaining user models. The need of coping with change and update motivates the choice to define our language within the Situation Calculus. We refer to a multi agent Situation Calculus with clusters of alternative situations. On top of it we introduce an ontology that captures most of the knowledge needed by Intelligent Interactive Systems. We discuss the adequacy of our proposal and compare it with other agent based languages proposed in the literature.

1 Introduction

Intelligent agents have recently received much attention as a paradigm in the solution of many AI problems. The concept of agent has assumed different meanings in different contexts. An agent sometimes indicates an entity acting on behalf of somebody. For instance, software agents designed to accomplish given tasks. In this paper we consider agents as entities whose description in terms of mental notions such as knowledge, beliefs, wants and desires is relevant to the aims of the systems they belong to.

The aim of our work is to model agents' mental states, with particular attention to the attitudes of users of interactive systems. In fact, modelling users is crucial for adaptive interfaces. Adaptiveness can be realized by exploiting information concerning users' knowledge, beliefs, goals, and preferences, usually acquired during interactive sessions and then stored and updated in a data structure called user model.

Interactions between a system and its user(s) may be viewed in a multi agent scenario, where system and users are agents who communicate with each other with the aim to solve a particular problem or, more in general, to exchange some information. However, there is a distinction between a robot agent, a software agent (softbot) and an agent modelling a user of an interactive system. The formers can be viewed as entities that sense the state of an external environment (for instance, the physical world or a distributed data base) and intensionally perform some action that can affect this state (for instance, moving an object or performing a transaction). Actions performed by agents of interactive systems are also communicative acts aimed to affect other agents' internal state. So, an action can be a request of information of a user agent to the system agent, possibly determining an update of the beliefs of the system about the user's interests; or, conversely, a system may inform a user about a particular subject, involving the update of her knowledge about it.

A problem arising with many empiric approaches to user modelling is the lack of semantics for the concepts being modelled. This often prevents from completely understanding the meaning of the information dealt with. As a consequence, also the behaviour of some existing systems becomes vague and hard to understand. These ambiguities become particularly harmful in the specification of general user modelling components and shells. Besides, it becomes a hard task to show some properties of models and to compare different models. Analogously, many agent based applications lack a formal semantics for epistemic notions such as beliefs and goals, that make the applications somehow brittle. Hence, resorting to an agent based approach to user modelling does not provide, per sé, a solution to the above mentioned problems.

In this paper we present a formal framework for modelling users, based on intelligent agents. We first analyze the kind of knowledge involved in the construction of agents, with particular reference to agents that model users of interactive systems, and present it as a taxonomy. This is set within a *formal* and *executable* framework, appealing to the Situation Calculus, as introduced by McCarthy and Hayes in their seminal paper [9] and extended by Reiter in [12]. Thus, mental notions expressed in the language are provided with a formal semantics, ruling out possible ambiguities arising with more empirical approaches.

Furthermore, our agents should in general account for incomplete, changing and noisy information. In fact, agents acquire their knowledge through a dynamic process. In user modelling, usually systems build a model of a user according to the evolution of the current interaction. Possibly models are initialized by ascribing some default assumption to users and then, when new evidence is available, models are revised and updated. This has motivated our choice of a formalism explicitly dealing with change and updates such as the Situation Calculus.

In the same spirit as Shoham in [14] and Lespérance et al. in [7], we devise an agent based language, aimed at representing the epistemic states of users of interactive systems. The mental notions introduced in the language are characterized by a set of suitable properties that are formally expressed within the language.

The presentation is organized as follows. Section 2 shortly introduces the formal system adopted, namely the Situation Calculus; Section 3 is devoted to the presentation and discussion of our language for modelling agents. In Section 4 we show how to introduce in our language suitable properties of the expressible mental notions. Finally a discussion concludes the paper along with a comparison of relevant literature and some hints on future work.

2 Agents in the Situation Calculus

The language we consider is a many sorted first order one, built on the following ingredients. Four sorts: *agent*, *sit*, *action* and *object*, respectively, for agents, situations, actions and anything else. A binary function $do(ag, a, s)$ from *agent* × *action* × *sit* to *sit*, denoting the situation resulting from agent ag performing action a in situation s. A binary predicate $Poss(ag, a, s)$ defined on *agent* × *action* × *sit*, stating whether action a is possible for agent ag in situation s or not. A finite number of functions and predicates. Predicates containing a situation argument are called *fluents* and are used to define properties changing from one situation to another. A *pseudo fluent* is an extra logical predicate corresponding to a fluent with the last argument of sort *sit* suppressed. In this case, we use the notation that $p[s]$ is the fluent corresponding to p.

A description of the state of the world in a given situation s is simply obtained by considering the truth values assumed by the fluents on argument s. We make the hypothesis that situations evolve if and only if an action is performed by an agent. The evolution of the world state is thus described by changes of the truth values of fluents in the new situation resulting from the action that has been performed, according to a branching time model. We start from an initial situation s_0, whose properties may be stated through fluents having s_0 as argument. When an action a is performed by an agent ag, a new situation $s_1 = do(ag, a, s_0)$ occurs, where possibly different properties hold. These properties are again described by fluents holding on argument s_1, which may have changed w.r.t. s_0.

Note that situations are terms of our language and thus appear directly in formulae, unlike states in modal logics that are semantic entities. Note also that situations do not bear any information about a particular state of the world. It is only a representation choice to model the evolutions of the world through a sequence of *frames* where a situation may be simply thought of as a name for a given frame, and fluents holding on it give the corresponding description.

In Reiter's proposal, where a solution to the frame problem is provided, the effects of actions on the world state may be specified by defining a logical theory with two kinds of axioms: *Action precondition axioms* for actions, specifying the conditions under which an action can be performed; *Successor state axioms*, stating necessary and sufficient conditions under which actions affect the truth value of fluents. In general, some extra axioms are stated in order to properly define how situations evolve. In particular, a set of *foundational axioms* eliminate finite cycles and merging among situations. Finally, unique name assumptions are usually postulated for situations and actions.

Agents can thus be modelled by the evolutions of a Knowledge Base containing facts and properties describing them. In an initial situation, some hypotheses are made, e.g., by a selection among a set of possible cases, represented by a set of fluents that are logically valid. As the interaction goes on, each action, i.e., each communicative act among agents, determines the evolution to a next situation, where new facts might hold. The axiomatic system, i.e., precondition and successor state axioms along with a set of general axioms, allows the system to deductively infer which facts are valid in a new situation.

3 Agents for Interactive Systems

We now introduce a taxonomy of the knowledge that a large class of Adaptive Interactive Systems (AISs) should or could address, in order to successfully tailor the interaction to users' characteristics. The concepts representing users mental notions are formally introduced, by means of *epistemic modalities* that behave as fluents of the Situation Calculus. Thus we can express suitable properties about users' mental notions within our logical theory. However, in the sequel we show that these modalities need not be added to our Situation Calculus, but could be considered as abbreviations of suitable formulae.

We distinguish among two classes of modalities describing *beliefs* and *goals* of agents. Beliefs concern facts that are supposed true by an agent in the current situation, according to the evidence provided. Goals, instead, describe those facts that agents wish true in the current situation. Unlike previous works on agents' goals, like that of Cohen and Levesque [2], where a notion of rationality is expressed through a commitment for agents to their goals, in the rest of this work we do not appeal to any such notion. In fact, our agent modelling is mainly concerned with the problem of capturing human agents' behaviour (that hopefully is a rational one), according to the evidences provided, e.g., users' action on an interactive system. Thus, we do not really need to constrain a rational behaviour: actions may be thought of as an input of our problem, rather than an output, as it is instead the case with planning problems.

We now present a number of epistemic modalities belonging to the two classes considered.

Definition 1. (BELIEF MODALITIES)
$knows(ag, p, s)$: agent ag knows that p holds in situation s;
$believes(ag, p, s)$: agent ag believes that p holds in situation s;
$kWhether(ag, p, s)$: agent ag knows whether p holds in situation s or not;
$doubts(ag, p, s)$: agent ag is uncertain whether p holds in situation s or not.

The fact p appearing as an argument inside an epistemic modality is a sentence of our language, i.e., a state independent predicate, or a *pseudo fluent*. The first modality, *knows*, accounts for knowledge of the truth of facts, for which agents have no uncertainties. In general, knowledge is stable for an agent, unless it concerns a fact that is changed in the actual world and which she is aware of. Modality *believes* accounts for situations where agents are inclined to accept facts as true even if they might turn out to be false. This second modality concerns defeasible facts, i.e., knowledge whose truth value can change in future situations, if simply more information is provided. The third modality, *kWhether*, still captures facts whose truth or falsity is known by agents. Anyway, it provides less information than the first one in that it does not specify whether the fact is known to be true or false. It may be relevant before Y/N questions [4], when an agent wants to assess if another one is able to provide a correct answer. This information could be profitably exploited by AISs adopting a related interaction strategy. The last modality, *doubts*, again concerns defeasible beliefs; it gives a way to explicitly reason about agents ignorance. It provides the lowest degree of information among the operators discussed here, in that it only specifies that agents have no idea about the truth value of a fact. This may be important for systems where the interaction is adapted to users' expertise on the application domain.

Definition 2. (GOAL MODALITIES)

$wants(ag, p, s)$: agent ag wants p in situation s;

$isIndifferent(ag, p, s)$: agent ag has no preferences about p in situation s;

$wantsToKnow(ag, p, s)$: agent ag wants to know whether p holds in situation s or not.

The first modality, *wants*, deals with agents' objectives; these are a central issue for systems tailoring the interaction to users' needs. The second modality, *isIndifferent*, accounts for agent's lack of preferences: even this information can be useful to plan system's actions. The last modality, *wantsToKnow* accounts for knowledge that users are interested in, thus allowing systems to explicitly single out the information to provide. Even though it can be easily expressed in terms of the other modalities, it has been included here because of its importance.

3.1 Semantics

One way to provide a semantics for a given formalism is to translate it into another formalism whose semantics is defined. In the sequel, we present a translation of the notions of our taxonomy into the Situation Calculus, thus indirectly providing them with a (first order logic) semantics. We refer to a possible worlds setting for defining the epistemic modalities previously introduced. Our semantics extends Moore's proposal for Knowledge [10] and has been adopted also by Scherl and Levesque in [13]. Anyway, unlike these works where only knowledge is dealt with, our proposal considers more than one modality.

The intuitive idea behind the introduction of possible worlds in the Situation Calculus is the following one. In order to express epistemic properties, instead of considering single situations, we consider sets of alternative (and contemporary) situations representing alternative conceivable states of affairs. We call them *clusters*. Note that given two situations s and s', if one of them is the result of applying a non empty sequence of actions to the other one, then no one can be an alternative to the other, i.e., they do not belong to the same cluster.

We introduce three accessibility relations among situations as particular fluents having one more argument of sort *sit*. These fluents are called *cognitive fluents* because they describe cognitive states of agents through the relations among situations they represent. Note that, unlike accessibility relations among possible worlds in classical modal logic, these fluents are expressed at the syntactic level, i.e., in the language, rather than at the semantical level. Note also that the notation we adopt here for accessibility relations prescribes the two arguments of sort *sit* to appear in reverse order with respect to classical modal logics.

Definition 3. (COGNITIVE FLUENTS)

$K(ag, s', s)$: agent ag has situation s' as a possible alternative to situation s;
$B(ag, s', s)$: agent ag has situation s' as a plausible alternative to situation s;
$G(ag, s', s)$: agent ag has situation s' as a desirable alternative to situation s.

The first cognitive fluent, K, expresses that, owing to the uncertainties of ag about the state of the world, situation s' is a possible alternative, or a *possible situation*, when the actual situation she is in is s. This amounts to saying that, when agent ag is in s, she is not fully aware of the actual state of the world, as she does not know every property that characterizes it, and so s' is one of the alternatives. The second cognitive fluent, B, expresses that, owing to suppositions or bias of agent ag, s' could be a *plausible situation* the world is in when the actual situation is s. Thus, it expresses an inner awareness of agent ag towards situation s' as a plausible alternative. The third cognitive fluent, G, expresses that situation s' is a desirable alternative or a *desirable situation* from situation s for agent ag. Thus, this fluent describes, for each agent ag, a preference relation among situations that derives from her consideration of some state of affairs being more desirable than others. In our approach the possible worlds model is meshed with the Situation Calculus. Relevant features of the world in the given actual situation are thus expressed through truth values of fluents on such situation and relations among situations are expressed by the epistemic fluents. After performing an action, the state of the world changes and the function *do* determines a new cluster of situations, where fluents may change their truth value and thus different relations among situations may hold. The evolution of the cognitive fluents determines an evolution of the epistemic state of an agent.

Knowledge, beliefs and wants about a fact, that we refer to as *basic epistemic modalities*, can be represented considering the truth values of the fact in the alternative situations. So, we express that an agent ag *knows* that a fact to be true in a situation s, if there is no possible situation where the fact is false. Thus known facts are all those holding in all (possible) situation s' that make fluent $K(ag, s', s)$ true.

We can give a similar definition for beliefs. Anyway, now the situations considered are all those which agents deem as plausible alternatives. So, an agent ag, in a situation s believes all those facts that are true in the situations that are plausible from s, i.e., all situations s' which make fluent $B(ag, s', s)$ true.

The definition is analogous for *wants*. An agent ag wants a fact to be true in a situation s if this is true in all the situations that are desirable from situation s, i.e., all situations s' which make fluent $G(ag, s', s)$ true.

By means of these definitions, we are able to give a semantics to all the epistemic modalities presented in the previous section. We stress the fact that, in the same style as Scherl and Levesque [13], the definitions of the epistemic modalities are not given as new fluents within the language. Instead, they are defined as abbreviations, or macros, of formulae involving the cognitive fluents. These are in fact the only fluents introduced to characterize epistemic states. Anyway, in the sequel , we allow epistemic modalities to appear within formulae of the Situation Calculus. Thus we speak indifferently of epistemic modalities or fluents with the understanding that they stand for the particular formulae they abbreviate.

- $knows(ag, p, s)$ means p true in all situations that are possible when ag is in s. That is: $knows(ag, p, s) \doteq \forall s' K(ag, s', s) \to p[s']$;
- $believes(ag, p, s)$ means p true in all situations that are plausible for ag from s. That is: $believes(ag, p, s) \doteq \forall s' B(ag, s', s) \to p[s']$;
- $wants(ag, p, s)$ means p true in all situations that are more desirable than s, for ag. That is: $wants(ag, p, s) \doteq \forall s' G(ag, s', s) \to p[s']$;
- $kWhether(ag, p, s)$ can be expressed in terms of the fluent $knows$: $kWhether(ag, p, s) \doteq knows(ag, p, s) \lor knows(ag, \neg p, s)$;
- $doubts(ag, p, s)$ can be described by the absence of any belief about a fact in a given situation: $doubts(ag, p, s) \doteq \neg believes(ag, p, s) \land \neg believes(ag, \neg p, s)$;
- $isIndifferent(ag, p, s)$ corresponds to the lack of preference: $isIndifferent(ag, p, s) \doteq \neg wants(ag, p, s) \land \neg wants(ag, \neg p, s)$;
- $wantsToKnow(ag, p, s)$ is defined in terms of $wants$ and $kWhether$: $wantsToKnow(ag, p, s) \doteq wants(ag, kWhether(ag, p), s))$.

4 Characterizations of Epistemic Fluents

In order to illustrate the features and abilities expressed by our logical framework, we need to specify some kind of hypothesis about the epistemic attitudes defined. We can consider two main kinds of properties. First, it makes sense to single out structural properties for the accessibility relations, in order to adequately model the epistemic concepts built upon them. This correspondence resembles that between properties of frames and axiom schemata in modal logic. Anyway, in our case as (cognitive) relations are first class elements and (epistemic) relations are not, we need work in the reverse direction. That is, we need state (first order) properties for our relations such that corresponding desired properties are fulfilled by our epistemic fluents.

Here we characterize the three accessibility relations introduced by the cognitive fluents by means of suitable axioms that capture an (idealized) meaning of the epistemic fluents built on top of them.

We can choose classical properties characterizing knowledge for relation K, i.e., transitivity, reflexivity and euclideaness, corresponding to the standard modal logic for knowledge S5. These are expressed by Axioms 1, 2 and 3 below. As for relation B, we choose that every plausible situation is also possible; this is expressed by Axiom 4. Other properties that characterize relation B are chosen to be transitivity, euclideaness, corresponding to standard modal logic for belief K45; however these two properties are not explicitly axiomatized as they can be derived from Axiom 4 and the axioms given for K. G is chosen to be transitive; see Axiom 5. Finally, our characterization involves that agents are aware of their beliefs and goals; see Axioms 6 and 7.

1. $K(ag, s, s)$;
2. $K(ag, s_2, s_1) \wedge K(ag, s_3, s_2) \to K(ag, s_3, s_1)$;
3. $K(ag, s_2, s_1) \wedge K(ag, s_3, s_1) \to K(ag, s_3, s_2)$;
4. $B(ag, s_2, s_1) \to K(ag, s_2, s_1)$;
5. $G(ag, s_2, s_1) \wedge G(ag, s_3, s_1) \to G(ag, s_3, s_2)$;
6. $K(ag, s_2, s_1) \to B(ag, s_3, s_1) \to B(ag, s_3, s_2)$;
7. $K(ag, s_2, s_1) \to G(ag, s_3, s_1) \to G(ag, s_3, s_2)$.

Note that all these properties are expressed by formulae of the Situation Calculus. In the spirit of the Correspondence Theory of van Benthem [15] between classical first order and modal logic, the above properties and the derivable theorems of the cognitive fluents determine suitable characterizations of the epistemic fluents *knowledge*, *believes* and *wants* that behave like modal operators, for instance we have:

1. $knows(ag, p, s) \to p[s]$;
2. $knows(ag, p \to q, s) \to (knows(ag, p, s) \to knows(ag, q, s))$;
3. $knows(ag, p, s) \to knows(ag, knows(ag, p), s)$;
4. $\neg knows(ag, p, s) \to knows(ag, \neg knows(ag, p), s)$;
5. $believes(ag, p \to q, s) \to (believes(ag, p, s) \to believes(ag, q, s))$;
6. $believes(ag, p, s) \to believes(ag, believes(ag, p), s)$;
7. $believes(ag, p, s) \to believes(ag, \neg believes(ag, p), s)$;
8. $wants(ag, p \to q, s) \to (wants(ag, p, s) \to wants(ag, q, s))$;
9. $wants(ag, p, s) \to wants(ag, wants(ag, p), s)$;
10. $knows(ag, p, s) \to believes(ag, p, s)$;
11. $believes(ag, p, s) \to knows(ag, believes(ag, p), s)$;
12. $wants(ag, p, s) \to knows(ag, wants(ag, p), s)$.

Property 1 corresponds to the *knowledge axiom*. Properties 2, 3, and 4 assert *distribution*, *positive introspection*, and *negative introspection* for knowledge. Properties 5, 6, and 7 assert the same for belief. Properties 8 and 9 state distribution and positive introspection for wants. Property 10 asserts that knowledge *implies* belief. Finally, Properties 11 and 12 state *awareness* for belief and want. All these properties are derivable by a straightforward application of the above mentioned Correspondence Theory, under the abbreviations made for the epistemic modalities, but in the cases of distribution (Properties 2, 5 and 8) that are derivable from first order tautologies.

We now state that two situations are accessible via relations K or G only if they belong to the same cluster, i.e., if they result from the same sequence of actions. To this end, we introduce a fluent $init(ag, s)$ stating that a situation s belongs to the starting cluster for an agent ag. Note that, for all s_0^i in the starting cluster – in particular the initial situation s_0^0, i.e., the real world – $init(ag, s_0^i)$ is postulated. Furthermore, we have that no initial situation can be the result of any action performed by any agent, i.e., $init(ag, s) \equiv (s \neq do(ag, a, s'))$. Then, we introduce a fluent $alternative(ag, s', s)$, determining an equivalence relation among all situations belonging to a cluster of situations resulting from the same sequence of actions. We assume in general that different agents can be related to different initial situations and then different alternative situations, so that a cluster of situations may indeed be thought of as the union of all sets of situations related by the fluent $alternative$ for any agent.

- $init(ag, s) \wedge init(ag, s') \rightarrow alternative(ag, s', s)$;
- $init(ag, s) \wedge alternative(ag, s', s) \rightarrow init(ag, s')$;
- $alternative(ag'', do(ag', a', s'), do(ag, a, s)) \equiv alternative(ag'', s', s) \wedge ag' = ag \wedge a' = a$.

These axioms state that the initial cluster of alternative situations relative to an agent is composed by all and only the situations that are postulated to be *initial* and that the other following clusters are composed, inductively, by situations coming from the initial cluster by applying the same sequence of actions performed by the same agents. By means of this fluent we state that accessibility via relation K (and hence also B) and G is restricted to alternative situations:

- $K(ag, s', s) \rightarrow alternative(ag, s's)$;
- $G(ag, s', s) \rightarrow alternative(ag, s's)$.

This choice is a strong one, as it determines that agents are aware of the actions performed by others, like, for instance, in some systems where agents explicitly try to collaborate to perform some task. Note that no set inclusion relation is postulated between K and G, i.e., between possible and desirable situations. This relation would amount to a *realism* hypothesis stating that only possible worlds can be desirable ones. In fact, as we implicitly assume in a following example, the set of possible and desirable situations are, in general, disjoint, i.e., $K(ag, s', s) \rightarrow \neg G(ag, s', s)$, unless some kind of interdependence need be explicitly modelled between known facts and desirable ones. This is motivated by the fact that we want to model situations where agents are not necessary aware of the feasibility of their goals.

A classical problem with a possible worlds approach is the side effect problem, i.e., the closure of goals under belief implication, which is a non intuitive property, as shown – for instance – by Cohen and Levesque in [2]. In our framework this is expressed by the following sentence that cannot be derived.

$$(wants(ag, p, s) \wedge believes(ag, p \rightarrow q, s)) \rightarrow wants(ag, q, s)$$

In fact, it can be shown, as pointed out by Rao and Georgeff in [11] that the choice of dropping the hypothesis of realism, as we do, prevents the side effect problem.

4.1 Example

In this section we show, by means of a simple example, how it is possible to express the change in agents' mental states. Consider a user U issuing commands to an operating system. We represent the world with a set of situations S_i^j. We start with a cluster of initial situations S_0^j (i.e., $init(S_0^j)$ holds for all of them), where S_0^0 represents the real world. In particular, consider the simple case where U moves a file Foo from a folder $Archive$ to a folder $InProgress$. We start from an initial configuration where files Foo and Fie are in $Archive$, expressed by fluents $isIn(Foo, Archive, S_0^0)$ and $isIn(Fie, Archive, S_0^0)$. We suppose that $file$ can be moved to $folder$, i.e., action $moveTo(file, folder)$ can be performed in a situation s iff the file is readable by the agent ag and the folder is not protected for ag in s, i.e., fluents $readable(ag, file, s)$ and $protected(ag, folder, s)$ are, respectively, true and false. This is expressed by the following action precondition axiom:

$$Poss(ag, moveTo(file, folder), s) \equiv$$
$$(readable(ag, file, s) \wedge \neg protected(ag, folder, s)).$$

Then we assume the following successor state axiom:

$$Poss(ag, a, s_i^j) \rightarrow$$
$$isIn(file, folder, do(ag, a, s_i^j)) \equiv$$
$$[K(ag, s_i^j, s_O^j) \wedge$$
$$(a = moveTo(file, folder) \vee$$
$$(isIn(file, folder, s_i^j) \wedge a \neq moveTo(file, folder'))] \vee$$
$$[G(ag, s_i^j, s_O^j) \wedge isIn(file, folder, s_i^j)].$$

This axiom states that, given a possible action a, $isIn(file, folder, s_i^j)$ is true after ag performs a iff: s_i^j results from a situation accessible via K to ag and either a moves $file$ to $folder$ or the fluent was already true in the starting situation and a has not moved $file$ to a different folder; alternatively, s_i^j results from a situation accessible via G to ag and the fluent was already true in the starting situation.

Note that, for the sake of simplicity, we are assuming that moving a file from a folder to another one is the only action that can insert a file into the destination folder and remove it from the origin folder. In addition, we rule out, for the time being, epistemic actions (i.e., actions directly affecting agents' internal state by providing them with some information). We only deal with actions affecting the external environment and assume that an agent gets to know the effect of her actions. Furthermore, we do not take into account state constraints such as the fact that a file cannot be in two different folders at the same time.

We assume that U knows that Foo is in $Archive$ and she wants it to be in $InProgress$, in the starting situation. U does not know that Fie is in $Archive$ and does not care about it being in $InProgress$ or not. These assumptions about

U are represented by an initial cluster of four situations s.t. S_0^0 and S_0^1 are accessible to U from S_0^0 via relation K and S_0^2 and S_0^3 via relation G. We have that in S_0^0 and S_0^1 $isIn(Foo, Archive, s)$ holds; in S_0^2 and S_0^3 $isIn(Foo, InProgress, s)$ holds; finally fluents $isIn(Fie, Archive, S_0^0)$ and $isIn(Fie, InProgress, S_0^2)$ hold, while $\neg isIn(Fie, Archive, S_0^1)$ and $\neg isIn(Fie, InProgress, S_0^3)$ hold.

This situation is depicted in the left part of the following figure, where $inProgress$ has been abbreviated as I and $Archive$ has been abbreviated as A, and $X(U, S_i^0)$ represents the set of situations accessible via $X, X \in \{K, G\}$ to U from S_i^0.

It can be easily seen that, by our definitions, these assumptions involve that the following sentences hold:

$$knows(U, isIn(Foo, Archive), S_0^0)$$
$$doubts(U, isIn(Fie, Archive), S_0^0)$$
$$wants(U, isIn(Foo, InProgress), S_0^0)$$
$$isIndifferent(U, isIn(Fie, InProgress), S_0^0).$$

Now, let us consider the case where in the initial situation U performs the action $moveTo(Foo, InProgress)$.
The new state of affairs is then represented by a new cluster $S_1^0, \cdots, S_1^3$ and is illustrated in the right part of the figure. We have that:

$$S_1^j = do(U, moveTo(Foo, InProgress), S_0^j).$$

Moreover, by the above successor state axiom, we can derive that $isIn(Foo, Archive, s_i^j)$ now holds in S_1^0, S_1^1, S_1^2 and S_1^3. This amounts to saying that after performing the moving action both these formulae hold:

$$knows(U, isIn(Foo, InProgress), S_1^0)$$
$$wants(U, isIn(Foo, InProgress), S_1^0).$$

It can be noticed that the truth values of the fluents involving Fie in S_1^j are unchanged with respect to the corresponding values in S_0^j. Hence, also these formulae still hold:

$$doubts(U, isIn(Foo, InProgress), S_1^0)$$
$$isIndifferent(U, isIn(Fie, InProgress), S_1^0).$$

The example sketches, in a simple case, how it is possible to represent a set of possibly changing assumptions about users that are relevant for a system to devise an interactive strategy tailored to them.

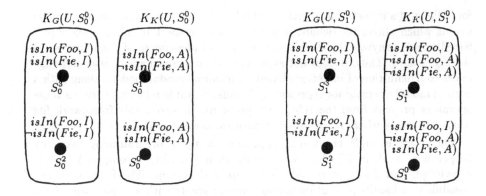

5 Discussion and Further Research

We have introduced a logical framework for representing agents beliefs and goals. This seems epistemologically powerful to capture the mental notions relevant to user modelling for a large class of interactive systems. The adequacy of our proposal, though hard to be proven in general, may be highlighted by a comparison with an empirical analysis carried out in [4] by Kobsa. In this work a general taxonomy of goals and beliefs is presented and discussed. The taxonomy is shown to be expressive enough to capture most of the relevant mental notions dealt with in many dialog systems. It can be easily shown that our logical framework allows us to represent (and reason) about the mental notions surveyed in that work.

We do not insist here on the benefits coming from a formal approach to user modelling. For instance, the possibility to give a clear and unambiguous specification of the behaviour of the user modelling components for system developers; also, implementors of user modelling components are provided with a powerful and expressive means via the translation into a logic programming language. GOLOG, the language being developed at Toronto (see [8]) has strongly inspired our work.

Other logic based approaches to user modelling have appeared in the literature. For instance, Kok in [5] illustrates a modal logic framework for representing users' interests when presenting information in data retrieval applications. A formal approach to student modelling can be found in [1], which mainly concentrates on the diagnostic problem in Intelligent Tutoring Systems. Another example of a formal approach to student modelling is given by Huang [3], who attempts to model the fact that inconsistencies can hold in students' beliefs, whenever they do not recognize them.

The problem of defining a suitable semantics for agents' beliefs and goals has been widely tackled within the AI community and there are many proposals for representing epistemic notions, based on a possible worlds semantics: Wainer [16] and Konolige and Pollack [6] among them. Wainer considers a modal temporal framework with an accessibility relation underlying a classical KD45 definition for belief and a subset of it, a preference relation, determining a preorder among worlds which allows the definition of wishes with respect to all maximally preferred worlds. Anyway this definition of wishes is independent of the actual world the agent is in. Thus, it does not easily extend to a multi agent case. Konolige refers to a definition of intentions based on minimal modal models, along with a normal modal operator for representing beliefs. Suitable relations between these operators prevent from the side effect problem. However, this framework fits static situations where no dynamic acquisition is dealt with.

Our work extends the logical approach to agent programming taken by Lespérance et al. in [7]. This work deals with (possibly incomplete) agents' knowledge, and does not consider other mental notions. The focus is in the possibility to handle perceptual actions among agents with an application for a meeting schedule problem. On the contrary, we have disregarded, for the time being, perceptual actions and we stress the possibility to represent several relevant agents' epistemic features within the extended version of the Situation Calculus adopted.

One of the limitations of the framework proposed here is that it does not allow to represent temporal relations between facts used as arguments to an epistemic modality. For instance, it is not possible to state that an agent has a certain goal only after another goal has been achieved. In this preliminary work we have deemed more interesting to develop means to allow a system representing and reasoning about agents' goals and beliefs. We defer temporal aspects to future extensions.

Furthermore, we plan to extend our framework in several other directions. Work is in progress for introducing a set of actions for modelling communication among agents. Our aim is to express communicative actions in terms of basic actions. Besides, we are currently working on a proposal able to cope with change of agents' epistemic state. On the one side, our idea is concerned with updates of the epistemic state, occurring when an agent is aware of some action changing the underlying domain state. On the other side, we deal with revisions of the epistemic state that occur when agents are provided with more information about an unchanged state of the world.

A major goal of ours is the definition of reasoning abilities for the agents modelled by our framework. So, besides reconsidering procedures like projection and legality testing, we are currently working on the possibility of introducing abductive reasoning abilities, in order for an agent to diagnose the causes of another agent's answers and to infer plans, based on the observations of a sequence of actions.

The choice of referring to a first order language for defining our framework determines many complexity problems for the reasoning procedures. Therefore, another interesting direction for research is the possibility of defining suitable restrictions yielding more manageable subsets of our language.

Furthermore, the problems arising when considering a multi agent framework need further investigation, e.g., problems arising in presence of *exogenous* actions, i.e., actions which an agent is not aware of.

Finally, other directions for research aim at coping with mental notions like intentions, capabilities or common beliefs among (groups of) agents.

6 Acknowledgments

We thank Ray Reiter for helping us rediscover the Situation Calculus, and to Fiora Pirri for many useful discussions and for reading drafts of this paper.

The research reported here has been partly funded by MURST and ASI. Bruno Errico is supported by a grant from ENEA.

References

1. L. Aiello, M. Cialdea, and D. Nardi. Reasoning about student knowledge and reasoning. In *Proceedings of the Twelfth International Joint Conference on Artificial Intelligence (IJCAI-91)*. Morgan Kaufmann, 1991.
2. P. R. Cohen and Levesque H. J. Intention is choice with commitment. *Artificial Intelligence Journal*, 42:213–261, 1990.
3. X. Huang. Modelling a student's inconsistent beliefs and attention. In G. E. Greer and G. I. McCalla, editors, *Student Modelling: The Key to Individualized Knowledge-Based Instruction*, volume 125 of *NATO ASI Series F*, pages 267–280. Springer-Verlag, 1991.
4. A. Kobsa. A Taxonomy of Beliefs and Goals for User Models in Dialog Systems. In A. Kobsa and W. Wahlster, editors, *User Models in Dialog Systems*, pages 52–68. Springer-Verlag, 1989.
5. A. J. Kok. A formal approach to user models in data retrieval. *International Journal of Man-Machine Studies*, 30:675–693, 1991.
6. K. Konolige and M. E. Pollack. A representationalist theory of intention. In *Proceedings of the Thirteenth International Joint Conference on Artificial Intelligence (IJCAI-93)*, pages 390–395. Morgan Kaufmann, 1993.
7. Y. Lespérance, H. J. Levesque, Lin F., D. Marcu, R. Reiter, and R. B. Scherl. Foundations of a logical approach to agent programming. In *IJCAI-95 Workshop on Agent Theories, Architectures, and Languages*, 1995.
8. H. J. Levesque, R. Reiter, Lin F., and R. B. Scherl. GOLOG: A logic programming language for dynamic domains. *Artificial Intelligence*, 1995. submitted.
9. J. McCarthy and P. Hayes. Some philosophical problem from the standpoint of artificial intelligence. In B. Meltzer and D. Michie, editors, *Machine Intelligence*, pages 463–502. Edinburgh University Press, 1969.
10. R. C. Moore. A formal theory of knowledge and action. In J. R. Hobbs and R. C. Moore, editors, *Formal Theories of the Commonsense World*, pages 319–358. Norwood, 1985.
11. A. S. Rao and M. P. Georgeff. Modelling rational agents within a BDI architecture. In *Proceedings of the Second International Conference on the Principles of Knowledge Representation and Reasoning (KR-91)*, pages 473–484. Morgan Kaufmann, 1991.
12. R. Reiter. The frame problem in the situation calculus: A simple solution (sometimes) and a completeness result for goal regression. In V. Lifshitz, editor, *Artificial Intelligence and Mathematical Theory of Computation: Papers in Honor of John McCarthy*, pages 359–380. Academic Press, 1991.

13. R. Scherl and H. J. Levesque. The frame Problem and Knowledge Producing Actions. *Artificial Intelligence*, 1994. submitted.
14. Y. Shoham. Agent-oriented programming. *Artificial Intelligence*, 60(1):51–92, 1993.
15. J. van Benthem. Correspondence theory. In D. Gabbay and Guenthner F., editors, *Handbook of Philosophical Logic II*, pages 167–247. D. Reidel Publishing Company, 1984.
16. J. Wainer. Yet another Semantics of Goals and Goal Priorities. In *Proceedings of the Eleventh European Conference on Artificial Intelligence (ECAI-94)*, pages 269–273. Wiley, 1994.

Talkin'bout Consistency, or:
When Logically Possible Becomes Possible

Luis Fariñas del Cerro, Antonio Frias Delgado*, Andreas Herzig

IRIT, Université Paul Sabatier
118 route de Narbonne, F-31062 Toulouse Cédex, France
E-mail: herzig@irit.fr
Tel.: (+33)-6155.6344, Fax: (+33)-6155.8325

1 Introduction

In the research on nonmonotonic reasoning, "normally A" has often been identified with "*A if there is no contradiction*".

Following such lines of thought, a lot of formal systems have been devised that appeal to consistency in order to establish new theorems. One of the first and perhaps the most prominent of such systems is Reiter's default logic (Reiter 1980). There, a default rule such as

$$\frac{\text{bird : flies}}{\text{flies}}$$

has the intuitive reading "if something is a bird and it is consistent with our knowledge to suppose that it flies then it flies".

Default logic as well as other formal systems such as the modal nonmonotonic logic of (McDermott and Doyle 1980) and autoepistemic logic of (Moore 1985) have to face important formal difficulties. They are due to the fact that in these systems it is attempted to define deduction via a notion of inference which itself refers to consistency, i.e. non-deducibility.

In this paper, we study a logical system where such a circularity is avoided by a two-level architecture, where the lower level is classical logic and the top level is a nonmonotonic extension of the modal logic S5. Precisely, there is a default rule refering to classical consistency w.r.t. a given classical theory T.

On the semantical side, it comes without surprise that our system axiomatizes the set of classical models of T. It is natural to see this set as a possible worlds model: for each classical formula that is consistent with T there is some possible world in the model satisfying it. Imposing the universal accessibility relation, consistency of a formula A can be interpreted as existence of an accessible world satisfying A. Hence semantics is in terms of a particular S5-model, where the possible worlds are all those worlds which are *logically possible* w.r.t. the theory T.

The idea of putting every logically possible world in the model is at the base of several systems that have been proposed in the literature, e.g. Levesque's

---

* On leave from the University of Cadiz (Spain).

logic of only-knowing (Levesque 1990), Winslett's Possible Models Approach for reasoning about actions (Winslett 1988, Fariñas and Herzig 1994), and Katsuno and Mendelzon's semantics for updates (Katsuno and Mendelzon 1991).

It is well-known since Gödel that trying to speak about consistency of a formal system within that system is more difficult than one naively would expect: generally, one cannot capture the notion of consistency within the system itself. Therefore it is the most important feature of our system that its operator of possibility captures exactly its notion of consistency.

In the rest of the paper, after some preliminaries (section 2) we axiomatize our nonmonotonic extension of S5 (section 3). Then we prove the central theorem, saying that for every formula A, either A or $\Diamond \neg A$ can be deduced (section 4). Having given a semantics (section 5), this theorem is used to prove completeness (section 6). We also give a decision procedure based on normal forming, which reduces theorem proving for our system to classical theorem proving (section 7). Finally (section 8) we discuss some more properties of our system. In particular we give a strong modal completeness theorem, and we show that we exactly capture classical consistency as well as consistency within the system itself.

2 Preliminaries

We suppose a language built on a set of propositional constants or atoms $ATOMS = \{p, q, r, \ldots\}$. Formulas are built using the classical connectives $\wedge$, $\vee$ and $\neg$ and the modal operator $\Diamond$. They are denoted by $A, B, C, \ldots$. We shall consider the formula $\Box A$ to be an abbreviation of $\neg \Diamond \neg A$.

We call a formula *classical* if it does not contain any modal operator. A *classical theory* (or theory for short) is a set of classical formulas that is consistent in classical logic (or 'classically consistent' for short).

A formula is *fully modalized* if each of the atoms occurring in it is in the scope of at least one modal operator. In other words,

- $\neg A$ is fully modalized if A is fully modalized,
- $A \wedge B$ is fully modalized if both A and B are fully modalized,
- $A \vee B$ is fully modalized if both A and B are fully modalized,
- $\Box A$ is fully modalized
- $\Diamond A$ is fully modalized

E.g. $\Box(p \vee q) \wedge \Diamond r$ is fully modalized, and $\Box(p \vee q) \wedge r$ is not.

Inductions on formulas require the standard notion of *modal depth* of a formula A (see e.g. Hughes and Cresswell 1968), which is defined inductively by

- $mdepth(A) = 0$ if A is an atom
- $mdepth(\neg A) = mdepth(A)$
- $mdepth(A \wedge B) = max(mdepth(A), mdepth(B))$
- $mdepth(A \vee B) = max(mdepth(A), mdepth(B))$
- $mdepth(\Box A) = mdepth(A) + 1$
- $mdepth(\Diamond A) = mdepth(A) + 1$

Hence classical formulas are of modal depth 0, and fully modalized formulas have modal depth at least 1.

For every theory T, $cmodels(T)$ is the set of classical models of T. We suppose classical models to be represented by sets of atoms. Given a (classical) theory T and a classical formula A, we say that A is *classically T-consistent* if $T \cup \{A\}$ is classically consistent.

Our logic will be be an extension of S5. We give here an axiomatics in terms of inference rule schemas (Bull 1966, Fariñas and Raggio 1983). It nicely illustrates the close similarity between the modal operators of S5 and the classical quantifiers.

1. A where A is a theorem of classical logic
2. $\dfrac{A, A \to B}{B}$
3. $\dfrac{A \to B}{\Box A \to B}$
4. $\dfrac{A \to B}{A \to \Box B}$
   where A is fully modalized
5. $\dfrac{A \to B}{\Diamond A \to B}$
   where B is fully modalized
6. $\dfrac{A \to B}{A \to \Diamond B}$

The rule of substitution of provable equivalents and the necessitation rule

$$\frac{A}{\Box A}$$

can be derived from this presentation of S5. The completeness proof will use as well the following derivable S5 equivalences (see e.g. Hughes and Cresswell 1968).

- $\neg \Box A \leftrightarrow \Diamond \neg A$
- $\Box(A \wedge B) \leftrightarrow (\Box A \wedge \Box B)$
- $\Diamond(\Box A \wedge B) \leftrightarrow (\Box A \wedge \Diamond B)$
- $\Diamond(\Diamond A \wedge B) \leftrightarrow (\Diamond A \wedge \Diamond B)$
- $\Diamond \neg \Box A \leftrightarrow \neg \Box A$

3 Nonmonotonic deduction

Given a classical theory T, we shall define nonmonotonic deduction from T. We use the following rule inference T-CON:

7. $\dfrac{A \text{ classically } T\text{-consistent}}{\Diamond A}$
   for every classical formula A.

The rule $\mathcal{T}$-CON makes sense only because there is no rule of uniform substitution. Working with schemas enables us to avoid that rule.

A *nonmonotonic deduction* of a formula A from a theory $\mathcal{T}$ is a tree labeled by formulas such that the following conditions hold:

- every node has at most two children,
- A labels the root,
- every leaf is labeled by
  - a classical theorem, or
  - an element of $\mathcal{T}$, or
  - an instance of $\mathcal{T}$-CON,
- for every node n having two children n_1 and n_2, the label of n is obtained from the labels of n_1 and n_2 by Modus Ponens,
- for every node n having one child n_1, the label of n is obtained from the label of n_1 by one of the inference rules of S5.

Thus leaves are labelled with classical formulas, which are either classically deduced from $\mathcal{T}$ or classically consistent with $\mathcal{T}$. We write $\mathcal{T} \mathrel{\vert\!\sim} A$ if A can be deduced nonmonotonically from $\mathcal{T}$. As Modus Ponens is an inference rule, and as all classical theorems are leaves of the tree, our nonmonotonic deduction relation $\mathrel{\vert\!\sim}$ contains the classical deduction relation. In other words, if a classical formula is classically deducible from $\mathcal{T}$ then $\mathcal{T} \mathrel{\vert\!\sim} A$.

Our notion of deduction is nonmonotonic in the following sense: it may be the case that $\mathcal{T}_1 \subseteq \mathcal{T}_2$, but we do not have that $\mathcal{T}_1 \mathrel{\vert\!\sim} A$ implies $\mathcal{T}_2 \mathrel{\vert\!\sim} A$. The following examples illustrate that.

Example 1. $\emptyset \mathrel{\vert\!\sim} \Diamond p$ and $\emptyset \mathrel{\vert\!\sim} \Diamond \neg p$ for every atom p.

Example 2. Let $p \in ATOMS$. Then $\{p\} \mathrel{\vert\!\sim} \Diamond q$ and $\{p\} \mathrel{\vert\!\sim} \Diamond \neg q$ for every atom q different from p, and $\{p\} \mathrel{\vert\!\sim} p$ and $\{p\} \mathrel{\vert\!\sim} \Box p$, but $\{p\} \mathrel{\not\vert\!\sim} \neg p$ and $\{p\} \mathrel{\not\vert\!\sim} \Diamond \neg p$.

Example 3. Let $p, q \in ATOMS$. Then $\{p \vee q\} \mathrel{\vert\!\sim} \Diamond \neg p$ and $\{p \vee q\} \mathrel{\vert\!\sim} \Diamond \neg q$, but $\{p \vee q\} \mathrel{\not\vert\!\sim} \Diamond(\neg p \wedge \neg q)$.

4 The central theorem

In this section we shall establish that for every formula A, either $\mathcal{T} \mathrel{\vert\!\sim} \neg A$ or $\mathcal{T} \mathrel{\vert\!\sim} \Diamond A$. This central theorem will enable us to give simple completeness proofs.

But first of all we need some normal forming machinery.

4.1 Normal Forms

We define the notions of conjunctive and disjunctive normal form by mutual indution.

- A formula is in *disjunctive normal form* if it is of the form

$$C_1 \vee \ldots \vee C_k$$

  and every C_i is in conjunctive normal form.
- A formula is in *conjunctive normal form* if it is of the form

$$\Box D_1 \wedge \ldots \wedge \Box D_k \wedge \Diamond C_1 \wedge \ldots \wedge \Diamond C_l \wedge E$$

  and
  - every D_i is in disjunctive normal form,
  - every C_j is in conjunctive normal form, and
  - E is classical.

We say that a formula is in *normal form* if it is in disjunctive or in conjunctive normal form.

Theorem 1. *There is a procedure which transforms every formula into an equivalent formula in disjunctive normal form.*

Proof. See e.g. (Fariñas 1982, Marek and Truszczynski 1991).

4.2 The theorem

Theorem 2. $T \vdash \neg A$ *or* $T \vdash \Diamond A$.

Proof. First, by Theorem 1 it is sufficient to consider only formulas in normal form. Let B be such a formula. We proceed by induction on the modal depth of B. If the modal depth of B is 0, then B is classical, and we apply T-CON. Let B be a consistent formula, and let the modal depth of B be $N + 1$. There are two cases.

- If B is in disjunctive normal form, then B is of the form

$$C_1 \vee \ldots \vee C_k$$

  As B is consistent, there is some C_i which is consistent. C_i is in conjunctive normal form. This is treated as our second case:
- If B is in conjunctive normal form, then B is of the form

$$\Box D_1 \wedge \ldots \wedge \Box D_k \wedge \Diamond C_1 \wedge \ldots \wedge \Diamond C_l \wedge E$$

  We distinguish three subcases.
  - If $k = l = 0$ then B is a classical formula, and we apply T-CON.

- If $k > 0$ and $l = 0$ then B is of the form $\Box D_1 \wedge \ldots \wedge \Box D_k \wedge E$. As B is consistent, we must have $T \not\hspace{-0.3em}\sim \Diamond \neg D_i$ for each D_i.

  Now we put each $\neg D_i$ in disjunctive normal form by "pushing down" negations. $\neg D_i$ being already in disjunctive normal form, this requires de Morgan equivalences together with the modal equivalences $\neg \Box A \leftrightarrow \Diamond \neg A$ and $\neg \Diamond A \leftrightarrow \Box \neg A$. We obtain formulas D_i' such that D_i' is in disjunctive normal form, $D_i' \leftrightarrow \neg D_i$, and D_i' is of the same modal depth as D_i.

  Hence D_i' is of the form $D_{i_1}' \vee \ldots \vee D_{i_m}'$, and $\Diamond D_{i_1}' \vee \ldots \vee \Diamond D_{i_m}'$ is in disjunctive normal form. As $T \not\hspace{-0.3em}\sim \Diamond \neg D_i$, we have $T \not\hspace{-0.3em}\sim \Diamond D_{i_j}'$ for every D_{i_j}. The modal depth of every D_{i_j}' being strictly smaller than that of B, we can apply m times the counterpositive of the induction hypothesis and get that all the D_{i_j}' are inconsistent, i.e. $T \hspace{0.2em}\vdash\hspace{-0.5em}\sim \neg D_{i_j}'$. Therefore, $T \hspace{0.2em}\vdash\hspace{-0.5em}\sim \neg(D_{i_1}' \vee \ldots \vee D_{i_m}')$, which means $T \hspace{0.2em}\vdash\hspace{-0.5em}\sim D_i$ for all D_i. Hence $\hspace{0.2em}\vdash\hspace{-0.5em}\sim D_1 \wedge \ldots \wedge D_k$, and by the derived rule of necessitation we get $T \hspace{0.2em}\vdash\hspace{-0.5em}\sim \Box(D_1 \wedge \ldots \wedge D_k)$. On the other hand, we have $\hspace{0.2em}\vdash\hspace{-0.5em}\sim \Diamond E$ by T-CON. Putting both together we have

$$T \hspace{0.2em}\vdash\hspace{-0.5em}\sim \Box(D_1 \wedge \ldots \wedge D_k) \wedge \Diamond E$$

  Applying the S5 equivalences $\Box D \wedge \Diamond E \leftrightarrow \Diamond(\Box D \wedge E)$ and $\Box(D_i \wedge D_j) \leftrightarrow (\Box D_i \wedge \Box D_j)$ we finally end up with $\Diamond B$.

- If $l > 0$ then suppose $T \not\hspace{-0.3em}\sim \neg B$. For every C_i, the formula $\Box D_1 \wedge \ldots \wedge \Box D_k \wedge C_i$ is consistent, and its modal depth strictly smaller than that of B. Therefore the induction hypothesis applies and gives us $T \hspace{0.2em}\vdash\hspace{-0.5em}\sim \Diamond(\Box D_1 \wedge \ldots \wedge \Box D_k \wedge C_i)$. Applying the S5 equivalence $\Diamond(\Box D \wedge C_i) \leftrightarrow \Box D \wedge \Diamond C_i$ we get $T \hspace{0.2em}\vdash\hspace{-0.5em}\sim \Box D_1 \wedge \ldots \wedge \Box D_k \wedge \Diamond C_i$ for every C_i. On the other hand, we have $T \hspace{0.2em}\vdash\hspace{-0.5em}\sim \Diamond E$ by T-CON. Putting things together we have

$$T \hspace{0.2em}\vdash\hspace{-0.5em}\sim \Box D_1 \wedge \ldots \wedge \Box D_k \wedge \Diamond C_1 \wedge \ldots \wedge \Diamond C_l \wedge \Diamond E$$

  Applying the S5 equivalences $\Box D \wedge \Diamond E \leftrightarrow \Diamond(\Box D \wedge E)$ and $\Diamond C_i \wedge \Diamond E \leftrightarrow \Diamond(\Diamond C_i \wedge E)$ we finally end up with $\Diamond B$.

To prove completeness we must know what the semantics is. Here it comes:

5 Semantics

Suppose given a classical theory T. Semantics of nonmonotonic deduction is in terms of a particular possible worlds model: for a given world, every logically possible world satisfying T is accessible.

The (unique) possible worlds model is just the set of classical models of T, i.e. $cmodels(T)$. Thus we identify worlds and the associated interpretations. (Hence there are no 'copies' of worlds possessing the same interpretation.)

The truth condition is the same as for S5. We only give the case of an atom and that of the modal operator. Let $w \in cmodels(T)$.

- $w \models A$ if $A \in ATOMS$ and $A \in w$.
- $w \models \Diamond A$ if there is $w' \in cmodels(T)$ such that $w' \models A$.

A formula is the (nonmonotonic) semantical consequence of T (noted $T \mathrel{\vDash\hspace{-0.6em}\sim} A$) if $w \models A$ for every $w \in cmodels(T)$.

6 Soundness and completeness

Now we shall show that our axiomatics is sound and complete w.r.t. our semantics.

6.1 Soundness

Theorem 3. *Let T be a classical theory. If $T \vdash A$ then $T \mathrel{\vDash\hspace{-0.6em}\sim} A$.*

Proof. First, the inference rule schemas preserve validity. Second, if $A \in T$ or if A is an axiom of S5 then $T \mathrel{\vDash\hspace{-0.6em}\sim} A$. It remains to prove the validity of the inference rule T-CON.

Let A be a classical formula that is classically T-consistent. Hence there is $w \in cmodels(T)$ such that w (viewed as a classical model) classically satisfies A. Our semantics guarantees that w appears as a possible world in our model. Hence by the truth condition for the modal operator, we have that $w' \models \Diamond A$ for every classical interpretation w', which means that $T \mathrel{\vDash\hspace{-0.6em}\sim} \Diamond A$.

6.2 Completeness

Theorem 4. *Let T be a classical theory. If $T \mathrel{\vDash\hspace{-0.6em}\sim} A$ then $T \vdash A$.*

Proof. Suppose $T \not\vdash A$. By the central theorem, we have $T \vdash \Diamond \neg A$. By the soundness of nonmonotonic deduction (that we have just shown), we have $T \mathrel{\vDash\hspace{-0.6em}\sim} \Diamond \neg A$.

As $T \not\vdash A$, the theory T must be classically consistent (because our nonmonotonic deduction relation $\vdash\hspace{-0.55em}\sim$ contains classical deduction), i.e. the set of its (classical) models must be nonempty. Hence there is a world $w \in cmodels(T)$ such that $w \models \Diamond \neg A$. Now the truth condition tells us that there is some world w' in $cmodels(T)$ such that $w' \models \neg A$. Consequently we cannot have $T \mathrel{\vDash\hspace{-0.6em}\sim} A$.

The completeness we have just proved is called *weak* in (Hughes and Cresswell 1968).

7 Decidability

In this section we prove decidability of our logic by giving a decision algorithm.

W.l.o.g. we suppose here that no $\Box$-operator occurs in formulas: all abbreviations of the form $\Box A$ have been replaced by $\neg \Diamond \neg A$.

Starting with some innermost occurrence of a $\Diamond$-operator, the procedure checks (classical) T-consistency of the subformula in the scope of $\Diamond$.

function DECIDE(A).
Input: A formula A.
Output: DECIDE returns **true** if $T \vdash A$, and **false**, otherwise.
begin
while A is not a classical formula **do**

take an occurrence of some subformula $\Diamond B$ in A such that B is a classical
formula;
if B is classically T-consistent
   then replace that occurrence of $\Diamond B$ in A by $\top$
   else replace that occurrence of $\Diamond B$ in A by $\bot$
endwhile;

if A follows classically from T **then return** {**true**} **else return** {**false**}
end

The algorithm terminates because each step eliminates one modal operator.
It is sound and complete: if B is consistent then the replacement of a subformula
$\Diamond B$ by $\top$ is guaranteed by the rule T-CON, and if B is inconsistent then the
replacement of $\Diamond B$ by $\bot$ is guaranteed by the necessitation rule, again together
with the derived rule of substitution of provable equivalents. At the end of the
"while"-loop we have an equivalent classical formula (whose theoremhood is
checked by some algorithm for classical logic).

8 Discussion

Now we shall give some more properties of our system. First we give a strong
modal completeness theorem. Then we show that we exactly capture classical
consistency as well as consistency within our system itself. For these two results
we suppose that the classical theory T is classically consistent.

8.1 Strong modal completeness

We shall establish a particular strong completeness theorem that holds for fully
modalized formulas.

Theorem 5. *Let A be a fully modalized formula. Then either $T \mathrel{|\!\sim} A$ or $T \mathrel{|\!\sim} \neg A$.*

Proof. Let A be a fully modalized formula. The central theorem tells us that
$T \mathrel{|\!\sim} \Box A$ or $T \mathrel{|\!\sim} \Diamond \neg \Box A$. Using the S5 equivalence $\Diamond \neg \Box A \leftrightarrow \neg \Box A$ we get either
$T \mathrel{|\!\sim} \Box A$ or $T \mathrel{|\!\sim} \neg \Box A$. As A is fully modalized, applying the S5 equivalences we
obtain $A \leftrightarrow \Box A$. Therefore, we have $T \mathrel{|\!\sim} A$ or $T \mathrel{|\!\sim} \neg A$. We cannot have both,
because else our system would be inconsistent, in contradiction with the above
soundness result. Consequently we have either $T \mathrel{|\!\sim} A$ or $T \mathrel{|\!\sim} \neg A$.

8.2 Possibility and classical consistency

We can show that for classical formulas, our modal operators exactly capture
classical consistency as well as classical deducibility.

Theorem 6. *Let T be a classically consistent classical theory. Let A be a classical formula. Then $T \mathrel{|\!\sim} \Diamond A$ iff A is classically T-consistent.*

Proof. From the right to the left, the proof follows from T-CON.

From the left to the right, suppose $T \mathrel{\vdash\hspace{-0.5em}\sim} \Diamond A$. By soundness of nonmonotonic deduction, $T \mathrel{\not\approx} \Diamond A$. As $cmodels(T) \neq \emptyset$ there is a world $w \in cmodels(T)$ such that $w \models A$. As $w \in cmodels(T)$, $T \cup \{A\}$ is classically satisfiable. By the completeness of classical logic, A is classically consistent with T.

Theorem 7. *Let A be a classical formula. Then $T \mathrel{\vdash\hspace{-0.5em}\sim} \Box A$ iff A can be deduced from T in classical logic.*

Proof. From the right to the left, the proof follows immediately from the fact that the axioms and rules for classical deduction are contained in those for nonmonotonic deduction: having thus $T \mathrel{\vdash\hspace{-0.5em}\sim} A$, the derived rule of necessitation gives us $T \mathrel{\vdash\hspace{-0.5em}\sim} \Box A$.

From the left to the right, suppose $T \mathrel{\vdash\hspace{-0.5em}\sim} \Box A$. By soundness of nonmonotonic deduction, $T \mathrel{\approx} \Box A$. Hence for every $w \in cmodels(T)$ we have $w \models A$. Therefore A is a classical semantical consequence of T. By the completeness of classical logic, A can be classically deduced from T.

8.3 Possibility and consistency

Now we show that our modal operators exactly capture nonmonotonic deduction and nonmonotonic consistency.

Theorem 8. *Let T be a classically consistent classical theory. Let A be any formula. Then $T \mathrel{\vdash\hspace{-0.5em}\sim} \Diamond A$ iff A is nonmonotonically consistent with T.*

Proof. From the right to the left, the proof follows from the central theorem.

From the left to the right, suppose that $T \mathrel{\vdash\hspace{-0.5em}\sim} \Diamond A$, and that A is inconsistent with T, i.e. $T \mathrel{\vdash\hspace{-0.5em}\sim} \neg A$. By the derived rule of necessitation we get $T \mathrel{\vdash\hspace{-0.5em}\sim} \Box \neg A$, i.e. $T \mathrel{\vdash\hspace{-0.5em}\sim} \neg \Diamond A$. By soundness, we have $T \mathrel{\approx} \Diamond A$, and $T \mathrel{\approx} \neg \Diamond A$. Hence $cmodels(T)$ must be empty, contradicting our hypothesis that T is classically consistent.

Theorem 9. *Let A be a classical formula. Then $T \mathrel{\vdash\hspace{-0.5em}\sim} \Box A$ iff $T \mathrel{\vdash\hspace{-0.5em}\sim} A$.*

Proof. From the right to the left, the proof comes with the derived rule of necessitation.

From the left to the right, the proof follows from the T-axiom $\Box A \rightarrow A$, which is a theorem of S5. Hence it can be nonmonotonically deduced from every T.

8.4 Other notions of consistency and completeness

In fact, there are other notions of consistency around. In (Hughes and Cresswell 1968), a definition is mentioned where a system is consistent "iff no wff consisting of a single propositional variable is a thesis" (p. 20). Given that whenever A is an atom we have that $T \mathrel{\vdash\hspace{-0.5em}\sim} \Diamond A$ is an instance of T-CON, if we replaced 'propositional variable' by 'atom' then our system would be in some sense 'modally inconsistent'.

There are different notions of completeness around as well. Our (weak) completeness is that of (Hughes and Cresswell 1968). There, an axiom system is strongly complete "if it cannot have any more theses than it has without falling into inconsistency" (p. 19). Note that the axiomatizations there use the rule of uniform substitution, whereas we use schemas. Therefore, e.g. propositional calculus is strongly complete in the framework of (Hughes and Cresswell 1968), but not in ours. The reason is that working with atoms - i.e., propositional constants - we could add e.g. the atom p as an axiom (under the hypothesis that p is consistent with T). This would not the case if we worked with propositional variables.

We have used the term strong modal completeness because we have that property only for fully modalized formulas.

References

R. A. Bull (1966), MIPC as the formalisation of an intuitionist concept of modality. JSL 31, 4, pp 609-616.

L. Fariñas del Cerro (1982), A simple deduction method for modal logic. IPL 14, 2, pp 49-51.

L. Fariñas del Cerro and A. Herzig (1994), A conditional logic for updating in the Possible Models Approach. Proc. German Conf. on AI (KI-94), eds. B. Nebel and L. Dreschler-Fischer, Springer Verlag, LNAI 861, pp 237-247.

L. Fariñas del Cerro and A. Raggio (1983), Some results in intuitionistic modal logic. Logique et Analyse 102, pp 219-224.

G. Hughes, M. J. Cresswell (1968), An introduction to modal logic. Methuen & Co. Ltd.

H. Katsuno and A. O. Mendelzon (1991), On the differenece between updation a knowledge base and revising it. Proc. Int. Conf. on Principles of Knowledge Representation and Reasoning (KR-91), Morgan Kaufmann.

H. Levesque (1990), All I know: A study in autoepistemic logic. Artificial Intelligence 42, pp 263-309.

V. W. Marek, M. Truszczynski (1991), Nonmonotonic logic. Springer Verlag.

D. McDermott and J. Doyle (1980), Non-monotonic logic I. Artificial Intelligence 25, pp 41-72.

R. C. Moore (1985), Semantical considerations on non-monotonic logic. Artificial Intelligence 25, pp 75-94.

R. Reiter (1980), A logic for default reasoning. Artificial Intelligence 13, pp 81-132.

M. Winslett (1988), Reasoning about actions. Proc. Seventh Nat. Conf. on AI (AAAI-88), Minneapolis, pp 89-93.

M. Winslett (1994), Updating logical databases. In: Handbook of Logic in Artificial Intelligence (eds. D. Gabbay, A. Galton, C. Hogger), Vol. 3, Oxford University Press, 1994.

The Analysis and Evaluation of Legal Argumentation from a Pragma-Dialectical Perspective

E.T. Feteris

Department of Speech Communication
University of Amsterdam
Netherlands

This paper shows how a theory of legal argumentation can be developed from a specific dialogical approach, a pragma-dialectical approach. It demonstrates how ideas from pragma-dialectical theory on the analysis and evaluation of legal argumentation can be combined with ideas taken from legal theory. It describes how a model for the analysis and evaluation of legal argumentation can be developed and it specifies a research programme for legal argumentation from a pragma-dialectical perspective.

1. Introduction

Until recently, in research of legal argumentation logical and rhetorical approaches have been the predominant research traditions. In a logical approach, argumentation is analyzed as a form of reasoning consisting of premises which lead to a certain conclusion, the legal decision.[1] The evaluation concentrates on the question whether there is a formally valid argument underlying the argumentation. In a rhetorical approach we find in legal theory and in speech communication, argumentation is analyzed as an attempt to convince a certain audience.[2] The analysis and evaluation concentrate on the specific audience-related criteria of legal rationality and on the techniques used for convincing this audience.

Recently, in argumentation theory as well as in legal theory, a new approach to legal argumentation has been developed. In this approach, legal argumentation is considered as part of a discussion procedure in which a legal standpoint is defended according to certain rules for rational discussion. In such approaches, which can be called *dialectical*, legal argumentation is considered as part of a dialogue about the acceptability of a legal standpoint. The evaluation centres around the question whether the standpoint has been defended successfully according to commonly shared starting

---

1. For representatives of a logical approach see for instance Klug (1951), Soeteman (1989), and Weinberger (1970).

2. For representatives of a rhetorical approach see for instance Toulmin (1958) and Perelman (1976).

points and whether certain standards for rational discussion have been met. The advantage of such a dialectical approach is that argumentation can be evaluated on the basis of both general criteria for a formal discussion procedure ànd field and audience related material grounds.

The aim of this paper is to show how a theory of legal argumentation can be developed from a specific dialogical approach, a pragma-dialectical approach. I will demonstrate how ideas from pragma-dialectical theory on the analysis and evaluation of legal argumentation can be combined with ideas taken from legal theory. Section 2 describes a pragma-dialectical perspective on legal argumentation, and section 3 describes a model for the analysis of legal argumentation. Section 4 specifies the norms for the evaluation of legal argumentation. Finally, section 5 presents an overview of a research programme for legal argumentation from a pragma-dialectical perspective.

2. A Pragma-dialectical Perspective on Legal Discussions

In a pragma-dialectical approach legal argumentation is considered as part of a communication process, called a *critical discussion*, aimed at the resolution of a dispute. Argumentation is considered as part of a critical exchange of speech acts in which the acceptability of a standpoint is tested in view of critical reactions and counter-arguments. The acceptability of a standpoint is dependent on the question whether certain forms of critical doubt put forward by an antagonist have been taken away by the protagonist and whether certain rules for rational discussion have been observed.[3)]

The behaviour of the parties and the judge is viewed as an attempt to resolve a difference of opinion. In a legal process (a civil process or a criminal process) between two parties and a judge the argumentation is part of an explicit or implicit discussion. The parties react to or anticipate certain forms of critical doubt. Feteris (1989) describes the various discussions which can be distinguished in a legal process, the discussion roles the parties and the judge can fulfil, and the contributions which play a role in a rational resolution of the dispute.

In a legal process, various discussions can be distinguished. In the discussion between the parties, it is tested whether the claim of the protagonist (the plaintiff in a civil process/the public prosecutor in a criminal process) can be defended against the critical reactions of the antagonist (the defendant in a civil process/the accused in a criminal process). A specific characteristic of a legal process is that apart from the

---

3. For an extensive description of a pragma-dialectical approach of argumentation see van Eemeren and Grootendorst (1992).

discussion between the parties, there is an (implicit) discussion between the parties and the judge, which is aimed at checking whether the claim of the protagonist can be defended against the critical reactions the judge puts forward as an institutional antagonist in his official capacity. The judge must both check whether the claim is acceptable in the light of the critical reactions of the other party and whether the claim is acceptable in the light of certain legal starting points and evaluation rules which must be taken into account when evaluating arguments in a legal process. In the defence of their standpoints, the parties anticipate the possible critical reactions of the other party and the judge.[4]

When the judge presents his decision, this decision is submitted to a critical test by the audience it is addressed to. This multiple audience consists of the parties, higher judges, other lawyers, and the legal community as a whole. Therefore, the judge must present arguments in support of his decision, he must justify his decision.[5] He has to specify which facts and which legal rule(s) are underlying his decision. From a pragma-dialectical perspective, the justification forms a part of the discussion between the judge and possible antagonists: the party which wants to appeal the decision and the judge in appeal. In his justification the judge anticipates the possible forms of critical doubt which may be brought forward by these antagonists.

The resolution process in a legal process can be considered as a critical discussion in which the four stages which have to be passed through in a pragma-dialectical critical discussion, are represented.[6] These stages are the confrontation stage, the opening stage, the argumentation stage and the concluding stage.

The first stage of a legal process in which the parties advance their point of view, can be characterized as the *confrontation stage* of the process. In this stage the judge remains passive. The only thing he has to do is see to it that the parties present their standpoints in accordance with the rules of procedure.

The second stage, the *opening stage*, in which the participants reach agreement on commonly shared starting points and discussion rules, remains for the main part implicit in a legal process. The opening stage can be represented by the institutional-ized system of rules and starting points laid down in the Codes of Procedure (in the Netherlands the Code of Civil Procedure and the Code of Criminal Procedure, the

---

4. For an extensive description of the critical reactions of a judge in a criminal process see Feteris (1995).

5. In some legal systems, there are statutory provisions which require justification. For example, in the Netherlands section 121 of the Constitution, in Germany s. 313 (1) of the Code of Civil Procedure (ZPO). For a description of conventions and styles of justifying legal decisions in various countries see MacCormick and Summers (1991).

6. For a more extensive account of the analysis of a legal process in terms of a critical discussion see Feteris (1990, 1991,1993,1995).

Civil Code and the Criminal Code). Because it is unlikely that the parties will reach agreement on common rules and starting points among themselves, the legal system provides an institutionalized system of rules and starting points which functions as such an agreement, and thus guarantees that there are rules available for legal conflict resolution. So, for reasons of legal certainty, the opening stage with respect to the agreement on rules and starting points is passed through prior to the discussion.

In the third stage, the *argumentation stage*, the party who has asked the judge for a decision has to defend his standpoint and the other party can put forward his counter-arguments. In the argumentation stage the judge also evaluates the argumentation. This part of the argumentation stage in legal proceedings in the Netherlands and other continental law countries differs from the argumentation stage in the Anglo-American system. In the continental system the decision about the force and weight of the evidence and the answer to the question whether the facts lead to the required legal consequence is taken by the judge and not by a jury. It is the judge who decides on factual and legal matters.

In the fourth and final stage of the process, which can be considered as the *concluding stage*, the judge has to decide whether the claim is defended successfully against the critical counter arguments put forward. If the facts put forward can be considered as established facts and the judge has decided that there is a legal rule which connects the claim to these facts, the judge will grant the claim. If the facts cannot be considered as an established fact, or if there is no legal rule applicable, the judge will reject the claim.

We could say that the stages of a pragma-dialectical critical discussion are all represented in a legal process and that the way the discussion is conducted can be considered as a process of critically testing a standpoint which leads to a resolution of the dispute. However, there are some important differences which need some attention.

First, in a critical discussion the parties jointly see to it that the discussion rules are being observed and they jointly decide about the result of the evaluation and the outcome of the discussion. In a legal process, for reasons of impartiality, it is the task of the judge to see to it that the rules of procedure are observed. It is also the task of the judge to evaluate the argumentation and to give a decision about the final outcome. So, in a legal process the judge does what the parties to a critical discussion do jointly.

Second, in a legal process the idealized conditions which are the prerequisites for a critical discussion are not always fulfilled. Contrary to the participants in a critical discussion who are required to have a reasonable discussion attitude, the parties to a legal process are not expected to be prepared to reach agreement on certain matters and they are not expected to give each other optimal opportunities to put forward their point of view and their criticism. The parties to a legal process cannot be expected to be cooperative.

To guarantee that the result of a legal process is in accordance with the requirements of a rational discussion, certain additional rules are supplied. For instance, there are rules which require that the participants live up to the requirements of rational discussion behaviour by obliging them to defend their claim or counter claim by specifying which factual and legal grounds must be put forward as defence. Also, in order to enable the other party to prepare his defence, there are rules prescribing that a party has to present his standpoint and his defence within a certain time limit. In order to compensate for inequalities in knowledge and power, there are rules guaranteeing the right of legal assistance.

Because of certain specific legal goals such as legal security and equity, in law there are some procedures and rules which differ in certain respects from the rules and procedures of a critical discussion. These rules and procedures must guarantee that the conflict can be resolved within a certain time limit by a neutral third party.

3. The Analysis of Legal Argumentation

As has been demonstrated, legal argumentation forms part of a discussion between various participants: the parties in dispute and the judge. To decide whether a legal standpoint is acceptable, the judge has to decide whether it has been defended successfully against the critical reactions put forward by the other party and the critical reactions which must be put forward by the judge. In order to do this, the judge must first decide which arguments have been put forward and to which critical forms of doubt these arguments respond. This implies that an analysis must be made of the elements which are important to the evaluation of the argumentation. The question is whether the defended standpoint withstands rational critique. Such an analysis is also called a *rational reconstruction*.[7] In the evaluation of the argumentation it must be tested whether the contents of the arguments is acceptable and whether the discussion has been conducted in accordance with the rules for a rational critical discussion.

This section describes how such a model for a rational reconstruction of legal argumentation can be developed. In doing this, ideas from pragma-dialectical argumentation theory with respect to a model for the analysis of legal argumentation are combined with ideas taken from legal theory.

---

7. See for example MacCormick and Summers (1991:21-23).

In the reconstruction, a pragma-dialectical approach distinguishes between various forms of argumentation.[8] In the most simple case, called a *single* argument, the argumentation consists of an argument describing the facts of the case (1.1) and an argument describing the legal rule (1.1'). The justification implies that the decision (1) is defended by showing that the facts (1.1) can be considered as a concrete implementation of the conditions which are required for applying the legal rule (1.1'). In schema:

1
legal decision

↑

1.1   &   1.1'
facts   legal rule

Figure 1

Often the argumentation is more complex, for example if a further justification of an argument is required. For example, if a legal rule must be interpreted and the interpretation, in its turn, must be defended. In the famous 'Electricity case' in 1918 a dentist in The Hague bypassed his electricity meter so that he was able to get free electricity. The dentist was caught and subsequently prosecuted for theft of electricity. In the end, the Supreme Court had to decide whether taking electricity constitutes the criminal offence of theft of 'a good', for which a penalty is prescribed in clause 310 of the Dutch Criminal Code. The Supreme Court (HR 23-5-1921, NJ 1921, 564) decided that taking electricity is considered to be taking a good. The Supreme Court states that clause 310 aims at securing the property of individuals and for that reason makes taking 'a good' punishable under the described circumstances. According to the Supreme Court, this clause applies to electricity because of the properties of electricity. One property of electricity is that it has a certain value, because someone has to incur expenses and make some effort to obtain it, and because someone can use it for their own benefit or can sell it to others for money. Thus, electricity is considered to be a property.

---

8. For an extensive description of the various forms of argumentation see van Eemeren and Grootendorst (1992 chapter 7).

The analysis of the argument is as follows:[9]

1
The accused must be convicted and imprisoned for three months
(legal decision)

↑
1.1   &   1.1'

| The accused has taken a good that, wholly or partly, belongs to someone else with the the intention of appropriating it. | If someone takes a good that, wholly or partly, belongs to someone else with the intention of appropriating it, he or she should be convicted of theft and imprisoned for a maximum term of four years. |
| (legal qualification of the facts) | (legal rule) |

ARGUMENT A

↑
(1.1.1)   &   1.1.1'
The accused has taken a property   If someone takes a property, he or she is
                                                   taking a good

ARGUMENT B

↑
(1.1.1.1)   &   1.1.1.1'
The accused has taken something that   If someone takes something that has a certain
has a certain value   value, he or she is taking a property

ARGUMENT C

↑
(1.1.1.1.1)   &   1.1.1.1.1'
The accused has taken electricity   If someone takes electricity, he or she is
                                                   taking something that has a certain value

ARGUMENT D

Figure 2

---

9. The notation with numbers such as (1.1) etc. is the pragma-dialectical notation. The notation with letters and logical symbols such as $(p \rightarrow Oq)$ is the logical notation.

To sustain that clause 310 of the Dutch Criminal Code should be applied to the facts of this concrete case, it must be shown that the facts (1.1.1.1.1) form a concrete implementation of the conditions for application of the legal rule of clause 310, the legal rule (1.1'). To defend this claim, a chain of subordinate arguments is required containing a step-by-step justification. First, it is shown that electricity is something that has a certain value (argument D); second, that something that has a certain value is a property (argument C); and, finally, that a property is a good in the sense of clause 310 (argument B).

This reconstruction also makes clear which arguments have remained implicit and must be made explicit. To complete all single arguments in the chain, the arguments (1.1'), (1.1.1), (1.1.1.1) must be made explicit.[10]

In legal theory Aarnio, Alexy, and Peczenik also develop models for the reconstruction of legal argumentation.[11] In the reconstruction they make a distinction between *clear cases* and *hard cases*. In clear cases, the facts and the legal rule are not disputed and the judge can put forward what is in pragma-dialectical terms called a *single* argument. In hard cases in which the facts or the legal rule are disputed, a further justification by means of a chain of what is in pragma-dialectical terms called a chain of *subordinate* arguments is required. An example of a hard case is the Electricity case discussed above.

In terms of this model the following reconstruction of the argumentation of the Supreme court may be made:[12]

. (1) (x) (Tx → ORx)
. (2) (x) (M1x → Tx)
. (3) (x) (M2x → M1x)
.
. (4) (x) (Sx → Mnx)
. (5) Sa
　(6) ORa　　　　　　　　　　　　　　　　　　　　　　　((1)-(5))

Figure 3

10. See Plug (1994,1995) for a more extensive description of the various forms of complex argumentation in law.

11. The interconnections between these theories become apparent from the fact that there are several joint publications: Aarnio, Alexy, Peczenik (1981), Alexy and Peczenik (1990) and the contributions of Aarnio, Alexy, MacCormick and Peczenik in MacCormick and Summers (1991).

12. In this reconstruction the analysis of Alexy described in Alexy (1980) is taken as an example.

In this schema, (1) might be the universal norm of clause 310 of the Dutch Criminal Code, which states that if someone (x) takes a good that belongs to someone else with the intention of appropriating it (T), he should be punished with imprisonment for a maximum term of four years (ORx); (2) is the statement that if someone takes a property (M1), he or she is taking a good (T); (3) is the statement that if someone takes something that has a certain value (M2), he or she is taking a property (M1); (4) is the statement that if someone is taking electricity (S), he or she takes something that has a certain value (Mn) (n is a variable representing a certain M, depending on the number of steps required, in this case M2); (5) is the statement that Mr. A has taken electricity belonging to the city of The Hague, and (6) is the normative statement (ORa).[13]

In most cases the argumentation of the judge is incomplete because various elements remain implicit. The pragma-dialectical reconstruction given above clarifies which missing premisses must be added.

In legal theory the authors pay little attention to the question of reconstructing missing premisses. Alexy only says that a legal decision must follow from at least one universal norm together with other statements, but he does not specify how hidden assumptions must be made explicit. From Alexy's description it can be guessed that if the universal rule is missing, it must be made explicit. In Alexy's analysis the universal rule is formulated as '(x) M2x $\rightarrow$ M1x', but he leaves open the question of how other implicit elements must be filled in. None of the legal authors specifies how missing elements must be formulated on the various levels of the chain of arguments defending an interpretation of a legal rule.

In pragma-dialectical theory an instrument is developed for making explicit missing premisses in an adequate way. On the basis of a logical analysis first the *logical minimum* is formulated that is required to make the argument complete and logically valid. Because the logical minimum does not suffice in most cases, a *pragmatic optimum* must be formulated. This is the premiss to which the arguer can be held, given the broader verbal and non-verbal context of the argument.[14]

To analyze legal arguments adequately, an *analytical* model should be developed which can be used as a *heuristic* tool for a rational reconstruction of the justification of legal decisions and interpretations. Such a model should present the relevant

---

13. MacCormick calls such a chain of subordinate arguments *second-order justification* and Aarnio, following Wróblewski, *external justification*. Alexy calls the whole chain of arguments the *Interne Rechtfertigung* and uses the term *Externe Rechtfertigung* for the justification of the content of the arguments.

14. See van Eemeren and Grootendorst (1992 chapter 6) for a description of the procedure for making explicit implicit premisses.

options which must be taken into account when reconstructing legal arguments. The relevant options are dependent on the criteria used in the evaluation. The aim of the analysis is to produce an analytical overview which forms an adequate basis for the evaluation.

The basic form of such an analytical model could be the schema described in the previous section for clear cases in which a justification consists of a description of the facts and the legal rule. The basic model should be elaborated for complex cases in which qualification of the facts, an interpretation of the legal rule, or a choice between two or more rules is required. For each case it should be specified which types of argument can occur in certain 'slots' of the model for hard cases. For a justification, for example, it should be specified on which levels the concrete facts, the legal qualification, the legal rule, and the justification of the interpretation should be situated. In the exemplary analysis of the 'Electricity case', such a reconstruction is performed for an interpretation of a statutory rule. A similar reconstruction may be performed for the justification of a qualification or a choice between various rules.

For various forms of legal argument such as analogy arguments, *a contrario* arguments, arguments in defence of a grammatical interpretation, a teleological interpretation, etc. it must be specified which arguments are required for a successful justification of the interpretation.

4. Norms for the Evaluation of the Argumentation

As has been described in the previous section, the analysis establishes which arguments constitute the justification. The evaluation must determine whether the content of the arguments is acceptable and whether the discussion has been conducted in accordance with the rules for a rational discussion. In this section I distinguish between norms for the evaluation of the *content* of the argumentation and norms for the evaluation of the discussion *procedure*.

Norms for the evaluation of the content

In a pragma-dialectical approach it is first checked whether an argument is identical to a common starting point. Such an evaluation procedure is called the *identification procedure*. If an argument is not identical to a common starting point, the next procedure to follow is the *testing procedure* which checks whether the argument can be considered acceptable according to a common testing method.

When evaluating factual statements in legal argumentation the first thing a judge in Dutch law does is deciding whether the facts are generally known. If this is not the case, he decides whether the facts can be considered proven according to legal rules of proof.

When evaluating the legal arguments, in continental law systems the judge first decides whether the legal rule can be considered a rule of valid law according to generally accepted legal sources (such as statutes etc.). Rules of valid law can be considered as a specific form of common starting points. In some cases, to decide which rule is to be preferred to another rule, a rule of preference must be used. Examples of these rules are *lex posterior derogat legi priori*, which states that an earlier norm is incompatible with a later one, *lex specialis derogat legi generali*, which allows application of a more general norm only in cases not covered by an incompatible, less general norm, and *lex superior derogat legi inferiori*, which states that when a higher norm is incompatible with a norm of a lower standing, one must apply the higher norm.[15] When interpreting a legal rule, the judge uses an interpretation method (for example the grammatical interpretation method in which reference is made to the meaning of a term in everyday language).

When evaluating the content of the argumentation, in pragma-dialectical terms the judge must also check whether the relation between the premises and the conclusion is acceptable: whether the *argumentation schema* is correctly chosen and applied correctly. There are various argumentation schemata such as analogy argumentation, causal argumentation which are used for defending the acceptability of the interpretation of a legal rule. In pragma-dialectical theory it is investigated in which cases the argumentation schemata are correctly chosen (for example in Dutch criminal law it is not allowed to interpret statutory rules analogically) and in which cases they are applied correctly (for example if an analogy does not relate to relevant similarities the analogy is not applied correctly).[16]

Norms for the evaluation of the discussion procedure

With respect to the evaluation of the *procedural* aspects of the argumentation, it must be determined whether the discussion has been conducted in accordance with the rules for rational discussion. In pragma-dialectical theory a system of ten basic rules for rational discussion are formulated. These rules apply to discussions in which the participants behave as rational discussants and aim at a rational resolution of the dispute. The rules relate to the right to put forward standpoints, to the obligation to defend a standpoint which is in dispute, to the relevance of defences and attacks, to the commitment to implicit arguments, the commitment to common starting points and

---

15. Alexy (1989) and Peczenik (1983) formulate rules of preference for the use of argumentation schemas.

16. See Kloosterhuis (1994, 1995) for a description of of model for the analysis and evaluation of arguments based on analogy or *a-contrario* reasoning.

evaluation methods, and to the rules for a successful defence and attack of a stand-point.

Feteris (1989) describes in which respects the general rules for rational discussion apply in the Dutch civil process and criminal process. She describes which general pragma-dialectical rules do not apply in a legal process, which pragma-dialectical rules take on a specific form in a legal context, and which additional rules are required in order to resolve the dispute in a rational way.

According to Aarnio, Alexy, and Peczenik, the basic principles of a system of rules for rational discussion are the principles of consistency, efficiency, testability, coherence, generalizability, and sincerity. These principles are formulated by Alexy and developed into a system of rules for general practical discussions, which is, in turn, elaborated for legal discussions.

The procedural rules also contain the rules for the formal and material evaluation of the justification. Rules which are specific for the discussion procedure are the rules which guarantee the right to participate in discussions, the sincerity rules, the rules concerning the burden of proof, the rules concerning the relevance of the contribu-tions, and the rules for a common use of language.

The authors do not pay attention to the question whether all these rules apply to all forms of legal discussions. Alexy argues that in a legal process the discussion rules differ from the rules for a discussion between legal scholars, but does not specify to what extent the general rules are applicable in various types of legal discussions.

Some authors such as Aarnio, MacCormick, and Peczenik distinguish a separate component in the evaluation, in which it is determined whether the *result* of the justification process (in Aarnio's and MacCormick's terms, the interpretation, in Peczenik's terms, the legal decision) is in accordance with the norms and values of a certain legal community. In Aarnio's theory, an interpretation must be coherent with the norms and values which are shared within a certain legal community, a specific audience. In MacCormick's theory, the interpretation must be coherent with certain legal principles, and must be consistent with relevant legal rules and precedents. In Peczenik's theory, a decision must be in accordance with the legal ideology, the whole of must, should and may sources, source norms, interpretation norms, conflict norms, and the *Grundnorm*.

Alexy does not distinguish a separate evaluation component for the result of the discussion. In his opinion, the rationality of the result depends on the question whether the discussion has been conducted in accordance with the rules for rational discussions. Because the discussion rules already contain the requirement that the argumentation must be acceptable according to common starting points, it ensures that the final result is coherent with the starting points and values which are shared within the legal community.

To evaluate legal argumentation in an adequate way, an *evaluation* model should be developed that may be used as a *critical* tool to establish whether the argumentation is acceptable. In the model it should be specified how common starting points and evaluation standards are to be used. For the use of common starting points, it should be specified for various legal fields which statements can be used as an argument in a legal justification. For example, it should be specified what the role is of legal rules, legal principles, etcetera. For the use of evaluation standards, it should be specified which types of legal argumentation schemata, such as reasoning from analogy, etc. must be distinguished. For various legal systems, it should be specified which legal argumentation schemata must, should and may be used in the justification of a legal decision. Because the correct application of an argumentation schema depends on the question whether certain critical questions can be answered positively, the relevant critical questions must be formulated for the various argumentation schemata.

For an adequate evaluation it should also be specified which discussion rules apply in a concrete case. For various types of legal discussions (discussions in a legal process, discussions in legal science) it should be specified which general and which specific legal rules are relevant for conducting a rational legal discussion.

5. A model for the Analysis and Evaluation of Legal Argumentation: Five Components

In the previous sections I have described how ideas concerning the analysis and evaluation of legal argumentation developed in pragma-dialectical theory can be combined with ideas taken from legal theory. I have also specified which additions and specifications are required in further research.

By way of conclusion, in this section I describe how a research programme of legal argumentation from a pragma-dialectical perspective could be developed. Such a research programme encompasses various research components: a philosophical, a theoretical, an analytical, an empirical and a practical component.

The *philosophical* component should link ideas developed in legal theory about the rationality of legal argumentation with ideas developed in argumentation theory about the rationality of argumentation in general. If one adopts a dialectical approach and takes legal argumentation to be a part of a critical discussion, it should be specified how a legal discussion is to be conducted in order to resolve a dispute in a rational way.

The *theoretical* component should develop a model for a rational reconstruction of legal argumentation. If one adopts a dialectical approach, several theoretical descriptions should be given. First, the stages of a legal discussion and the contributions which are relevant in these stages should be described. Second, the structure, the levels and elements of a legal justification should be specified. Third, the formal and material standards of rationality should be formulated. In the theoretical component, legal ideas about legal standards of acceptability such as legal principles, rules of procedure, legal interpretation methods, rules for the use of legal sources, etc. should be combined with ideas developed in argumentation theory and logic about ideal norms for rational argumentation.

The *reconstruction* component should investigate how a rational reconstruction of legal argumentation can be performed with the aid of the theoretical model. For example, how should a legal interpretation be reconstructed, and which general and which specific legal background knowledge is required to give an adequate reconstruction? How should implicit elements be made explicit?

The *empirical* component should investigate how legal practice relates to the theoretical model. In which respects does legal practice differ from the legal ideal model, what are the reasons to depart from the model, and how can the difference be justified? Which argumentative strategies appear to be successful in legal practice in convincing an audience?

Finally, to be able to give practical recommendations for the analysis and evaluation of legal argumentation, it should be established how the theoretical, analytical, and empirical ideas may be combined to develop methods for improving argumentative skills in legal education. The *practical* component should determine which methods may be used to improves skills in analyzing, evaluating, and writing legal argumentation.

Bibliography

Aarnio, A. (1977). *On legal reasoning.* Turku: Turun Yliopisto.
Aarnio, A. (1987). *The rational as reasonable. A treatise of legal justification.* Dordrecht etc.: Reidel.
Aarnio, A, R. Alexy, A. Peczenik (1981). 'The foundation of legal reasoning'. *Rechtstheorie*, Band 21, Nr. 2, p. 133-158, nr. 3, p. 257-279, nr. 4, p. 423-448.
Alexy, R. (1980). 'Die logische Analyse juristischer Entscheidungen'. In: Hassemer et al (Hrsg.), *Argumentation und Recht. Archiv für Rechts- und Sozialphilosophie*, Beiheft Neue Folge No. 14, Wiesbaden: F. Steiner, p. 181-212.
Alexy, R. (1989). *A theory of legal argumentation. The theory of rational discourse as theory of legal justification.* Oxford: Clarendon Press. (Translation of: *Theorie der juristischen Argumentation. Die Theorie des Rationalen Diskurses als Theorie der juristischen Begründung.* Frankfurt a.M.: Suhrkamp, 1978, Second edition 1991 with a reaction to critics).

Alexy, R., A. Peczenik (1990). 'The concept of coherence and its significance for discursive rationality'. *Ratio Juris*, Vol. 3, nr. 1, p. 130-147.

Eemeren, F.H. van, (1987). 'Argumentation studies' five estates'. In: J.W. Wenzel (ed.), *Argument and critical practices*. Annandale: Speech Communication Association, p. 9-24.

Eemeren, F.H. van, R. Grootendorst (1992). *Argumentation, communication, and fallacies. A pragma-dialectical perspective*. Hillsdale NJ: Erlbaum.

Feteris, E.T. (1989). *Discussieregels in het recht. Een pragma-dialectische analyse van het burgerlijk proces en het strafproces*. (Rules for discussion in law. A pragma-dialectical analysis of the Dutch civil process and criminal process) Dissertation University of Amsterdam. Dordrecht: Foris.

Feteris, E.T. (1990). 'Conditions and rules for rational discussion in a legal process: A pragma-dialectical perspective'. *Argumentation and Advocacy. Journal of the American Forensic Association*. Vol. 26, No. 3, p. 108-117.

Feteris, E.T. (1991). 'Normative reconstruction of legal discussions'. *Proceedings of the Second International Conference on Argumentation, June 19-22 1990*. Amsterdam: Sic Sat, p. 768-775.

Feteris, E.T. (1993). 'Rationality in legal discussions: A pragma-dialectical perspective'. *Informal Logic*, Vol. 15, No. 3, p. 179-188.

Feteris, E.T. (1995). 'The analysis and evaluation of legal argumentation from a pragma-dialectical perspective'. In: F.H. van Eemeren, R. Grootendorst, J.A. Blair, Ch.A. Willard (eds.), *Proceedings of the Third ISSA Conference on Argumentation*, Vol. IV, Amsterdam: Sic Sat, p. 42-51.

Kloosterhuis, H. (1994). 'Analysing analogy argumentation in judicial decisions'. In: F.H. van Eemeren and R. Grootendorst (eds.), *Studies in pragma-dialectics*. Amsterdam: Sic Sat, p. 238-246.

Kloosterhuis, H. (1995). 'The study of analogy argumentation in law: four pragma-dialectical starting points'. In: F.H. van Eemeren, R. Grootendorst, J.A. Blair, Ch.A. Willard (eds.), *Proceedings of the Third ISSA Conference on Argumentation. Special Fields and Cases*, Vol. IV, Amsterdam: Sic Sat, p. 138-145.

Klug, U. (1951). *Juristische Logik*. Berlin: Springer.

MacCormick, N. (1978). *Legal reasoning and legal theory*. Oxford: Oxford University Press.

MacCormick, N. (1992). 'Legal deduction, legal predicates and expert systems'. *International Journal for the Semiotics of Law*, Vol. V, No. 14, p. 181-202.

MacCormick, D.N., R.S. Summers (eds.) (1991). *Interpreting statutes. A comparative study*. Aldershot etc.: Dartmouth.

Peczenik, A. (1983). The basis of legal justification. Lund.

Peczenik, A. (1989). *On law and reason*. Dordrecht etc.: Reidel.

Perelman, Ch., L. Olbrechts-Tyteca (1958). *La nouvelle rhétorique. Traité de l'argumentation*. Bruxelles: l'Université de Bruxelles.

Perelman, Ch. (1976). *Logique juridique. Nouvelle rhétorique.* Paris: Dalloz.

Plug, H.J. (1994). 'Reconstructing complex argumentation in judicial decisions'. In: F.H. van Eemeren and R. Grootendorst (eds.), *Studies in pragma-dialectics.* Amsterdam: Sic Sat, p. 246-255.

Plug, H.J. (1995). 'The rational reconstruction of additional considerations in judicial decisions'. In: F.H. van Eemeren, R. Grootendorst, J.A. Blair, Ch.A. Willard (eds.), *Proceedings of the Third ISSA Conference on Argumentation. Vol. IV Special Fields and Cases,* Amsterdam: Sic Sat, p. 61-72.

Soeteman, A. (1989). *Logic in law.* Dordrecht: Reidel.

Toulmin, S.E. (1958). *The uses of argument.* Cambridge: Cambridge University Press.

Weinberger, O. (1970). *Rechtslogik. Versuch einer Anwendung moderner Logik auf das juristische Denken.* Wien etc.: Springer.

Wróblewski, J. (1974). 'Legal syllogism and rationality of judicial decision'. *Rechtstheorie,* Band 14, Nr. 5, p. 33-46.

Reasoning About Reasoning

Maurice A. Finocchiaro

University of Nevada, Las Vegas

Abstract. Several varieties of metareasoning are discussed. The prototypical case is argument analysis, namely the interpretation and/or evaluation of arguments. A second special case is self-reflective argumentation. A third case is methodological reflection, namely the formulation, interpretation, evaluation, and application of methodological principles; these are inexact and fallible rules stipulating useful procedures in the search for truth. A fourth case is informal logic, conceived as the formulation, testing, systematization, and application of concepts and principles for the interpretation, evaluation, and practice of argument. Two other varieties of theory of argument are characterized without elaboration, "argumentation theory" and "formal logic."

1 Introduction

The purpose of this paper is to discuss several forms which reflection upon practical reasoning may take. That is, it aims to define, distinguish, and interrelate several forms of reasoning about practical reasoning. Reasoning about practical reasoning may be labeled "metareasoning," and so this paper discusses some forms and applications of metareasoning.

By practical reasoning I mean reasoning about any subject matter, as long as we understand it to be a mental process going on in the human mind. Thus I take practical reasoning to be synonymous to actual or real reasoning. The point of the adjective "practical" is not to contrast practical reasoning to some other type, but to stress that we are talking about the human practice of reasoning; about reasoning as it exists in human practice, not reasoning as we may reconstruct it when we begin to theorize, argue, and reason about it.

In other words, the most basic distinction in this context is perhaps the distinction between object-level reasoning and metalevel reasoning. Object-level reasoning is reasoning about such objects as mathematical entities, natural phenomena, historical facts, and practical problems, whereas metalevel reasoning is reasoning about reasoning.

I shall also assume that the most fundamental case of practical reasoning is argumentation. This is not to deny that there are other types, such as problem solving and decision making; rather I am limiting the subject of discussion for the purpose of this paper, and I am suggesting that perhaps argumentation is primary in the sense that problem solving and decision making are reducible to argumentation in a way in which it is not reducible to them.

My next assumption is that argumentation is the mental process of or the linguistic expression of supporting conclusions with reasons. Once again, this is meant to be a minimalist definition consistent with other possible definitions. For example (Blair and Johnson 1987, Eemeren and Grootendorst 1992), it may very well be true that argumentation has an essentially dialectical nature; that it always occurs in a context of controversy; and that it is to be distinguished from the notion of proof. However, all these points could be incorporated into the notion of support that occurs in the minimalist definition. Moreover, this definition calls attention to what may very well be an inescapable fact, namely that the smallest possible unit of practical reasoning and actual argumentation is a conclusion supported by a reason.

As an example of argumentation or practical reasoning, I would like to quote an argument that was widely discussed by physicists, astronomers, and philosophers in the 16th and 17th centuries during the Copernican controversy; this was the controversy which began in 1543 with the publication of Copernicus's book *On the Revolution of the Heavenly Spheres* and climaxed in 1687 with the publication of Newton's *Mathematical Principles of Natural Philosophy*. A key issue was whether the earth stands still at the center of the universe, or instead is a planet spinning daily on its axis and orbiting the sun once a year. The former was, of course, the geostatic, geocentric world view, which for thousands of years had been almost universally accepted on the basis of what I would regard as very plausible and convincing supporting arguments. One of these involved the alleged experiment of dropping a rock from the top of the mast of a ship on two different occasions, when the ship is docked and anchored motionless in a harbor and when it is advancing forward traveling over the water. This is what it claimed:

> because when the ship stands still the rock falls at the foot of the mast, and when the ship is in motion it falls away from the foot, therefore, inverting, from the rock falling at the foot one infers the ship to be standing still, and from its falling away one argues to the ship being in motion; but what happens to the ship must likewise happen to the terrestrial globe; hence, from the rock falling at the foot of the tower one necessarily infers the immobility of the terrestrial globe. [Galilei 1632, 169-70]

2 Argument Analysis

With these ideas in the background, we may say that one type of metareasoning is the reasoned interpretation or evaluation of an argument. Note, however, that this really involves two special cases, the reasoned interpretation of arguments and the reasoned evaluation of arguments; for interpretation and evaluation should be distinguished along obvious lines which I shall not elaborate for lack of space; on the other hand, if we want a single handy label to refer to one or the other or both, we may speak of argument analysis, or alternatively of "critical thinking" (cf. Scriven and Fisher, in press). Similarly, note that I speak of the reasoned interpretation or evaluation and not of mere interpretation or evaluation; that is, in these two special cases of metareasoning we must have not only an analytical claim about an argument, but also a justification of the claim by means of reasons.

For example, let us consider some interpretive claims about the argument just quoted. One might claim that the final conclusion of the argument is the proposition that the terrestrial globe is immobile; that the argument's last step is some kind of conditional argument using the premise that the rock dropped from the tower lands at its base (and not some distance to the west); that another step or subargument is an argument from analogy, beginning with a premise stating differential results for the ship experiment, drawing an analogy between a mast on an advancing ship and a tower on a rotating earth, and implicitly concluding that the rock dropped from the tower would land at its base if and only if the earth is motionless.

Such interpretive claims obviously have to do with the understanding of the argument, and their justification or criticism would involve argumentation about the argument. But it should be equally obvious that such claims are not evaluative and do not attempt to assess the strength or validity of various aspects of the argument. Examples of evaluative claims might be the following (cf. Galilei 1632, 167-75). One might question the strength of the analogy by arguing that there are significant dissimilarities between the case of an advancing ship and that of a rotating earth: (1) on a rotating earth the rock would have an inherent natural tendency to move along with the tower, but on an advancing ship the rock would not have an inherent natural tendency to move along with the mast and ship; and (2) on a rotating earth the air surrounding the tower and the rock would be moving eastward carried by the earth's rotation, whereas on an advancing ship the surrounding air does not share its motion. Another evaluative claim is that the argument's key experimental premise is false: on an advancing ship the rock dropped from the top of the mast lands at the foot of the mast, in the same place where it lands when the ship is motionless. During the Copernican controversy this was a controversial issue whose resolution required not only observation and well designed experimentation, but also argumentation interpreting such empirical data.

3 Informal Logic

These two types of metareasoning need now to be distinguished from and interrelated with other types. The first of these involves the problem of the relationship between argument analysis and informal logic. For the latter too aims at the interpretation and evaluation of arguments. What then is the difference? I believe the difference is one of generality, systematicity, and conceptual explicitness. If we take the latter to be elements of the mental activity called theorizing, then I can express my point by saying that informal logic in the theory of argument or reasoning. It is obvious, however, that this theory is not value-free or purely interpretive, but also normative; thus we may think of informal logic as the normative theory of argument, as long as we do not go to the other extreme and act as if its normative character deprives it of any descriptive or empirical content, and of an analytical or interpretive aim. This point can be made more explicit by defining informal logic as the formulation, testing, systematization, and application of concepts and principles for the interpretation, evaluation, and practice of argument or reasoning (cf. Finocchiaro 1984, and Johnson and Blair 1985). This definition also makes more vivid the problem of distinguishing argument analysis from informal logic.

I have already said that the difference lies along the dimensions of generality, systematicity, and conceptual explicitness. In regard to generality, if one is engaged in the critical analysis of a particular argument and reaching a conclusion merely about that argument, then one is engaged in argument analysis; whereas if one is engaged in the critical analysis of a whole class of arguments or reaching a conclusion about arguments in general, then one is doing informal logic. However, this dimension of generality makes it clear that the difference is a quantitative one of degree rather than a qualitative one of kind.

In regard to systematicity, what readily comes to mind is the Euclidean type of axiomatization. However, just as readily, I would want to add that it is both unrealistic and anachronistic to expect such systematicity in informal logic. But then the challenge becomes that of articulating a different kind of systematicity.

This problem may be expressed differently. Once one has formulated some principles of interpretation or evaluation, one may want to explore the logical relationships among them, logical in the sense of standard formal logic. Certainly the informal logician will want to do some of that. But is that the only kind of systematization one can undertake? If so, we would have the ironical situation that if the informal logician wants to claim to do something above and beyond the argument analyst, he has to engage in formal logic; if not, the question is what is this other kind of systematization.

At any rate, in regard to the difference between argument analysis and informal logic from the point of view of systematicity, it would seem that on the one hand systematicity is a matter of degree, but that on the other hand if a critical interpretation of an argument becomes systematic to any extent then one is taking a qualitative step (however small) away from argument analysis as such and toward informal logic.

Conceptual explicitness was the third distinguishing characteristic. It too is a matter of degree. Even the argument analyst can hardly avoid using such terms as argument, premise, conclusion, serial structure, intermediate propositions, linked versus independent reasons, interpretation, evaluation, criticism, objection, counterargument, fallaciousness, and so on. However, in the critical interpretation of a particular argument, many of the subtler ones of these concepts may be only implicit. On the other hand, I take it that part of the business of informal logic is to render more explicit concepts that are implicit in the practice of argument analysis.

There are probably other distinguishing characteristics, besides generality, systematicity, and conceptual explicitness; and even these three require more clarification and analysis than I have provided. Nevertheless, given the aforementioned definitions of argument analysis and of informal logic, we may say that while they are both meta-argumentative activities concerned with the interpretation and evaluation of arguments, what distinguishes them is that informal logic has a higher degree of generality, systematicity, and conceptual explicitness. In terms of a slogan, we might say that informal logic is generalized or systematized argument analysis, and the latter is applied informal logic.

4 Argumentation Theory and Formal Logic

At this point, however, a few words need to be said about the relationship between informal logic and two other enterprises, argumentation theory and formal logic, although lack of space will force me to briefly state my view without elaboration and justification (but cf. Finocchiaro 1989, 1991, 1994; Freeman 1991). Here argumentation theory may be ostensively defined by referring to the work of the Amsterdam School (cf. Eemeren and Grootendorst 1992). Since I have defined informal logic as the theory of argument, it ought to come as no surprise that there is a large overlap between informal logic and argumentation theory. This overlap is, however, only partial. My impression is that argumentation theory tends to be more empirical and interpretive oriented, whereas informal logic tends to me more a priori and normative oriented; but this does not mean that argumentation theory is, or can be, or ought to be completely a posteriori and value-free; similarly, informal logic is not, should not, and cannot be completely a priori and evaluative; the difference is one of degree and of emphasis. In regard to formal logic, my impression is that it tends to be even more a priori and normative than informal logic; moreover, formal logic tends to theorize about a special class of arguments (usually, mathematical proofs), about special interpretive aspects of their structure (such as the internal microstructural properties of propositions), and about a special kind of evaluation (namely, deductive validity or invalidity), whereas informal logic tends to have a more general scope; finally, formal logic tends to me more systematic that informal logic.

5 Self-Reflective Argumentation

So far, the first case of metareasoning (argument analysis) involves the critical interpretation of already constructed arguments advanced by others. However, sometimes it is proper and advisable to engage in some interpretation and/or evaluation of an argument that one is constructing and advancing. If we subsume self-interpretation and self-evaluation under the notion of self-reflection, then we may say that at such times one is engaged in the self-reflective formulation or construction of one's own argument.

In other words, argument analysis occurs when the subject matter of our argument is itself an argument. However, the subject matter of most arguments is not arguments, but rather numbers, atoms, human affairs, social institutions, historical events, and so on. In such cases, one may engage in argument in a more or less self-reflective manner.

What is required for reasoning to be self-reflective, and how do we distinguish reasoning which is self-reflective from reasoning which is not? We should not expect that, when we are self-reflectively constructing our own argument, we will analyze and evaluate it with the same degree of explicitness and formality as when we are critically analyzing the arguments of others. But the degree of self-interpretation or self-evaluation cannot be too low, otherwise we would have an instance of mere object level reasoning and not of metareasoning.

Sensitivity to self-interpretation is normally shown by careful attention to such questions as what are our conclusions and what are our reasons, whether we are advancing just one reason or more than one, and how our reasons are meant to connect with our conclusions. Sensitivity to self-evaluation is normally shown by careful attention to such questions as whether we are advancing a conclusive or very strong or moderately strong or weak argument; this involves the degree of support the reasons lend to the conclusion. It is also shown by paying attention to possible criticism, objections, and counterarguments against our own argument, and to ways of rebutting these.

The expansion of metareasoning from the case of argument analysis to informal logic and to self-reflective argumentation involves relatively small steps since what is involved is being willing to include our own arguments as well as the arguments of others, and classes of arguments as well as particular arguments, when we speak of the interpretation and evaluation of arguments. The next type of metareasoning involves a bigger step.

6 Methodological Reflection

In fact, there is an important type of thinking consisting of reflections on what one is doing when engaged in inquiry, the search for truth, or the quest for knowledge; that is, thinking aimed at understanding and evaluating the aims, presuppositions, and procedures of knowledge. Some call this metacognition (cf. Kitchener 1983, Meichenbaum 1986); I call it methodological reflection. Methodological reflection is another special case of metareasoning. Let me explain.

Any inquiry, search for truth, or quest for knowledge about nature and physical reality eventually leads to questions about the nature of inquiry, truth, and knowledge; the focus then temporarily shifts from natural science to methodological reflection. The same happens with any inquiry into any other topic; so these shifts occur in other fields, whether they study numbers, electrons, life, human nature, history, society, or the supernatural.

There are many causes for such methodological pauses. Sometimes investigators want to gain a deeper understanding of what they are doing, what their aims are, what procedures they follow, what rules they accept, and what their presuppositions are. Sometimes they are challenged by a critic about one of these things and want to determine whether the criticism is valid. Sometimes important differences arise with fellow practitioners, and their resolution requires that such things be identified, analyzed, and evaluated.

Methodological reflection is concerned with the identification, description, interpretation, and evaluation of the aims, procedures, rules, and presuppositions of inquiry, truth-seeking, or knowledge-gathering. As defined, methodological reflection is essentially context dependent and practically oriented; it arises in the course of inquiry about other topics and ends when further reflection is no longer relevant. However, it is obvious that it can also be undertaken in a systematic, general, conceptually explicit, and theoretical manner, independent of the contextual and practical origin it had initially; indeed, it has become professionalized into a branch of technical philosophy. We may refer to it as systematic methodology; other may wish to call it systematic epistemology, or just epistemology.

This relationship between methodological reflection and systematic methodology is analogous to that between argument analysis and informal logic.

This may be taken to provide another, formal reason for subsuming both argument analysis and methodological reflection under the notion of metareasoning, for they share a metalevel dimension. However, this formal analogy must be seen in the light of the substantive difference between their respective subject matters (arguments versus methodological principles).

A methodological principle is a general rule about the conduct of inquiry, search for truth, or quest for knowledge. Whereas an argument must have at least two propositions one of which is based on the other, a methodological rule is a single proposition. Thus, argument analysis involves reasoning more directly than does methodological reflection.

Methodological principles are special kinds of propositions, defined partly by their content and partly by their form. I just described their content explicitly by reference to inquiry, truth, and knowledge; this content gives them special importance. I also described their form when I said that they are rules and general statements. To say that they are rules means that they are prescriptions about what we should do or what is desirable to do if we want to arrive at the truth, acquire knowledge, and conduct inquiry properly; this in turn means that they may be obeyed or disobeyed, followed or violated, acted upon or disregarded. Thus, another important point is the *application* of methodological principles, and this element should be added explicitly to the definition: methodological reflection aims at the formulation, interpretation, evaluation, *and* application of rules for the conduct of inquiry. To say that methodological principles are general means that they convey information which can be applied to more than one particular case; it does *not* mean that they are necessarily universal and categorical, or that they convey information about each and every situation.

Let me give some examples of methodological principles. In regard to the role of observation in inquiry, some hold the principle that a theory should be rejected if it conflicts with observation; others hold that if there is a conflict between a theory and an observation, then one should either revise the theory or reinterpret the observation. Another principle states that, if two theories are otherwise equivalent, we ought to prefer the one which can explain more observed facts. The principle of simplicity would state that, other things being equal, a simpler theory is to be preferred to one which is less simple.

Besides contrasting methodological reflection with argument analysis and systematic methodology, it is useful to contrast it with method, or at least with one meaning of this word. Method is sometimes conceived as a set of rules which are exact, infallible, unchanging, and mechanically applicable, so that use of the method guarantees that we will arrive at the truth or solve the problem at hand. As conceived here, methodology is *not* meant to have this connotation. Instead a methodological principle is an inexact and fallible rule open to reformulation and requiring judgment for its application.

This talk of procedure leads to the distinction between results and method or methodology. A method is a procedure used to arrive at certain results. The results are the conclusions, ideas, hypotheses, theories, or beliefs at which we arrive at the end of a particular investigation, or which we test by means of such an investigation. Thus, a methodological rule is a procedural rule, a rule about a procedure conducive toward truth.

Another important distinction is that between methodological reflection and methodological practice. This is the distinction between what investigators say about the procedures they follow or should follow in their research and what they actually do when engaged in research. This is also called the distinction between theory and practice, between reflective pronouncements and practical involvement, and between words and deeds.

It should be stressed that methodological reflection is a type of metareasoning not, or not merely, or not primarily, insofar as one can argue for and against various methodological principles, and thus engage in the critical interpretation and the self-reflective formulation of methodological arguments. Given the controversial nature of methodology, this is both common and necessary. However, to do this is to engage in argument analysis and self-reflective argumentation.

The point to stress is that in methodological reflection there may be reasoning even when there may be no argumentation to any significant degree. Since the evaluation of methodological principles clearly and inevitably involves methodological arguments, and self-reflective ones at that, the aspects of methodological reflection that need emphasis are those that involve the identification, formulation, interpretation, and application of methodological principles.

7 An Example Involving Open-Mindedness

I want to conclude with an example of methodological reflection which is relatively short and nontechnical. It is taken from Galileo's *Dialogue on the Two Chief World Systems, Ptolemaic and Copernican* of 1632. This book contains many examples of critical interpretations of physical arguments and of self-reflective arguments about physical reality, which I have elaborated elsewhere and need not concern us here. The passage I have in mind is primarily a comment on open-mindedness and the fact that the Copernicans were more open-minded than the Ptolemaics. It is in the form of a story related by one the three speakers in the dialogue.

> I must take this opportunity to relate to you some things which have happened to me since I began hearing about this opinion. When I was a young man and had just completed the study of philosophy (which I then abandoned to apply myself to other business), it happened that a man from Rostock beyond the Alps (whose name I believe was Christian Wursteisen) came into these parts and gave two or three lectures on this subject at an academy; he was a follower of Copernicus and had a large audience, I believe more for the novelty of the subject than anything else. However, I did not go, having acquired the distinct impression that this opinion could be nothing but solemn madness. When I asked some who had attended, they all made fun of it, except one who told me that this business was not

altogether ridiculous. Since I regarded him as a very intelligent and very prudent man, I regretted not having gone. From that time on, wheneverI met someone who held the Copernican opinion, I began asking whether he had always held it; although I have asked many persons, I have not found a single one who failed to tell me that for a long time he believed the contrary opinion, but that he switched to this one due to the strength of the reasons supporting it; moreover, I examined each one of them to see how well he understood the reasons for the other side, and I found everyone had them at his fingertips; thus, I cannot say that they accepted this opinion out of ignorance, or vanity, or (as it were) to show off. On the other hand, out of curiosity I also asked many Peripatetics and Ptolemaics how well they had studied Copernicus's book, and I found very few who had seen it and none who (in my view) had understood it; I also tried to learn from the same followers of the Peripatetic doctrine whether any of them had ever held the other opinion, and similarly I found none who had. Now, let us consider these findings: that everyone who follows Copernicus's opinion had earlier held the contrary one and is very well-informed about the reasons of Aristotle and Ptolemy; and that, on the contrary, no one who follows Aristotle and Ptolemy has in the past held Copernicus's opinion and abandoned it to accept Aristotle's. Having considered these findings, I began to believe that when someone abandons an opinion imbibed with mother's milk and accepted by infinitely many persons, and he does this in order to switch to another one accepted by very few and denied by all the schools (and such that it really does seem a very great paradox), he must be necessarily moved (not to say forced) by stronger reasons. Therefore, I have become most curious to go, as it were, to the bottom of this business, and I regard myself very fortunate to have met the two of you; without any great effort I can hear from you all that has been said (and perhaps all that can be said) on this subject, and I am sure that by virtue of your arguments I will lose my doubts and acquire certainty. [Galilei 1632, 154-55]

One key claim here is that all Copernicans had previously been Ptolemaics, but no Ptolemaics had previously been Copernicans. Another is that the Copernicans knew the pro-Ptolemaic arguments, but the Ptolemaics did not know the pro-Copernican arguments. Let us take open-mindedness to include at least the following two skills: (1) the ability to know, understand, and learn from the arguments against one's own views, and (2) the ability to reject previously held views and accept opposite views; then these claims can be reformulated by saying either that the Copernicans were open-minded but the Ptolemaics were not, or at least that the Copernicans were more open-minded than the Ptolemaics. A third claim in the passage is that, whenever one abandons a long-held and generally accepted view and adopts a new and unpopular opinion, the arguments for the new view are probably stronger. A fourth claim is not explicit but obviously suggested; that is, the instantiation of this generalization to the case of the Copernicans yields the claim that the Copernican arguments are stronger.

176

The three explicit claims are intriguing and controversial, but unsupported in the passage. Moreover, there is no suggestion that they constitute an argument. Nor it is obvious what conclusion is to be drawn from the first two claims. When the book was first published, Church authorities received the complaint that this passage contained the argument that Copernicanism is true because the Copernicans are open-minded; but this would be an uncharitable interpretation, since the alleged conclusion is stated more strongly than it needs to be. On the other hand, it would be relatively plausible to argue that Copernicanism was worth pursuing as a research program because the Copernicans were more open-minded than the Ptolemaics.

At any rate, my main conclusion is that, despite its lack of (explicit) argumentation, this passage contains metareasoning of the methodological-reflection variety.

8 Epilogue

In this paper I have discussed several varieties of metareasoning. The prototypical case is argument analysis, namely the interpretation and/or evaluation of arguments. A second special case is that of self-reflective argumentation. A third case is methodological reflection, namely the formulation, interpretation, evaluation, and application of methodological principles; these are inexact and fallible rules stipulating useful procedures in the search for truth. A fourth case is informal logic, conceived as the formulation, testing, systematization, and application of concepts and principles for the interpretation, evaluation, and practice of argument. Two other varieties of theory of argument which I have briefly characterized without elaboration are "argumentation theory" and "formal logic."

References

Blair, J. A., and R. H. Johnson. 1987. "Argumentation as Dialectical." *Argumentation* 1: 41-56.

Eemeren, F. van, and R. Grootendorst. 1992. *Argumentation, Communication, and Fallacies*. Hillsdale/London: Lawrence Erlbaum Associates.

Finocchiaro, M. A. 1980. *Galileo and the Art of Reasoning*. Boston: Reidel.

Finocchiaro, M. A. 1984. "Informal Logic and the Theory of Reasoning." *Informal Logic*, vol. vi, no. 2, pp. 3-8.

Finocchiaro, M. A. 1989. "Methodological Problems in Empirical Logic." *Communication and Cognition* 22: 313-35.

Finocchiaro, M. A. 1991. "Induction and Intuition in the Normative Study of Reasoning." In *Probability and Rationality*, ed. E. Eells and T. Maruszewski, pp. 81-95. Amsterdam/Atlanta: Rodopi.

Finocchiaro, M. A. 1994. "Two Empirical Approaches to the Study of Reasoning." *Informal Logic* 16: 1-21.

Freeman, J. B. 1991. *Dialectics and the Macrostructure of Arguments*. New York: Foris.

Galilei, G. 1632. *Dialogo ...*, in *Le opere di Galileo Galilei*, vol. 7, National Edition by A. Favaro *et al.* Florence: Barbera, 1890-1909.

177

Johnson, R. H., & J. A. Blair. 1985. "Informal Logic: The Past Five Years 1978-1983." *American Philosophical Quarterly* 22: 181-96.
Kitchener, K. S. 1983. "Cognition, Metacognition, and Epistemic Cognition." *Human Development* 26:222-32.
Meichenbaum, D. 1986. "Metacognitive Methods of Instruction." In *Facilitating Cognitive Development*, ed. M. Schwebel & C. A. Maher, pp. 23-32. New York: Haworth.
Scriven, M., & A. Fisher. In press. *Critical Thinking*. Newbury Park: Sage.

A Resolution-Based Proof Method for
Temporal Logics of Knowledge and Belief*

Michael Fisher   Michael Wooldridge   Clare Dixon

Department of Computing
Manchester Metropolitan University
Manchester M1 5GD
United Kingdom

{M.Fisher,M.Wooldridge,C.Dixon}@doc.mmu.ac.uk
http://www.doc.mmu.ac.uk/

Abstract. In this paper we define two logics, KL_n and BL_n, and present resolution-based proof methods for both. KL_n is a *temporal logic of knowledge*. Thus, in addition to the usual connectives of linear discrete temporal logic, it contains a set of unary modal connectives for representing the *knowledge* possessed by *agents*. The logic BL_n is somewhat similar: it is a temporal logic that contains connectives for representing the *beliefs* of agents. The proof methods we present for these logics involve two key steps. First, a formula to be tested for unsatisfiability is translated into a normal form. Secondly, a family of resolution rules are used, to deal with the interactions between the various operators of the logics. In addition to a description of the normal form and the proof methods, we present some short worked examples and proposals for future work.

1 Introduction

This paper presents two logics, called KL_n and BL_n respectively, and gives resolution-based proof methods for both. The logic KL_n is a *temporal logic of knowledge*. That is, in addition to the usual connectives of linear discrete temporal logic [4], KL_n contains an indexed set of unary modal connectives that allow us to represent the information possessed by a group of agents. These connectives satisfy analogues of the axioms of the modal system S5 [2], which is widely recognized as a logic of *idealized knowledge* [10]. It is for this reason that we call KL_n a temporal logic of knowledge. (The properties of KL_n-like logics are studied in [11, 5], where a version of KL_n is used that contains only future time connectives.) Syntactically, the logic BL_n is identical to KL_n. It is also a temporal logic that contains connectives for representing the information possessed by a group of agents. However, the *agent modalities* in BL_n satisfy analogues of the modal axioms KD45 — a system widely accepted as a logic of *idealized belief* [10]. For this reason, we say that BL_n is a *temporal logic of belief*.

Logics such as KL_n and BL_n have been studied in both AI and mainstream computer science for some time (see, e.g., [9, 12, 11, 6, 5]). However, very little effort has been directed at developing proof methods for such logics [14]. This is perhaps

---

* Work partially supported under EPSRC Research Grant GR/K57282.

because of the complexity of the problem: it is proved in [11, 5] that even for comparatively simple temporal logics of knowledge, the decision problem for validity is PSPACE complete. For more complex variants, the problem is undecidable even in the propositional case. However, recent advances in proof methods for the underlying temporal logic (for which the decision problem is also PSPACE complete) indicate that practical theorem provers for such complex logics may be possible [7, 3]. In this paper, we extend the proof method for purely temporal logics described in [7] to deal with KL_n and BL_n. Specifically, we present a clausal resolution method for BL_n, which we show to be sound and refutation complete. A simple extension to this method gives a sound and refutation complete proof method for KL_n. This work represents the first attempt to extend this resolution method to multi-modal logics, and is the first resolution method for temporal logics of knowledge and belief.

The remainder of this paper is structured as follows. Section 2 gives complete formal definitions of the two logics. Section 3 defines SNF*, a normal form for KL_n and BL_n. The proof method itself is presented in section 4. Worked examples, illustrating the proof method, are presented in section 5, and some concluding remarks appear in section 6.

Notation: If $\mathcal{L}$ is a logical language, then we write $Form(\mathcal{L})$ for the set of (well-formed) formulae of $\mathcal{L}$. We use the lowercase Greek letters φ, ψ, and χ as meta-variables ranging over formulae of the logical languages we consider.

2 Temporal Logics of Knowledge and Belief

In this section, we formally present the syntax and semantics of two logics: BL_n is a *temporal logic of belief*, and KL_n is a *temporal logic of knowledge*. These logics actually share a common syntax, which we shall call the language $\mathcal{L}$. Note that due to space restrictions, our presentation of the syntax and semantics of the language, though complete, is of necessity somewhat terse.

First, note that $\mathcal{L}$ is not a quantified language. We shall thus build formulae from a set $\Phi = \{p, q, r, \ldots\}$ of *primitive propositions*. In fact, the language $\mathcal{L}$ generalizes classical propositional logic, and thus it contains the standard propositional connectives $\neg$ (not) and $\vee$ (or); the remaining connectives ($\wedge$ (and), $\Rightarrow$, (implies), and $\Leftrightarrow$ (if, and only if)) are assumed to be introduced as abbreviations in the usual way. We use temporal connectives that can refer both to the *past* and to the *future*. With respect to the future, we use two basic connectives: $\bigcirc$ (for 'next'), and $\mathcal{U}$ (for 'until'). With respect to the past, we use '$\bullet$' (for 'last') and '$\mathcal{S}$' (for 'since'). We explain these connectives in detail below. The temporal connectives are interpreted over a *flow of time* that is linear, discrete, bounded in the past, and infinite in the future. An obvious choice for such a flow of time is $(I\!N, <)$, i.e., the natural numbers ordered by the usual 'less than' relation.

With respect to belief/knowledge connectives, we assume a set $Ag = \{1, \ldots, n\}$ of *agents*. We then build an indexed set of unary modal connectives $\{[i] \mid i \in Ag\}$, where a formula $[i]\varphi$ is to be read (in BL_n) as 'agent i believes that φ', or (in KL_n) as 'agent i knows that φ'. In both cases, $\varphi \in Form(\mathcal{L})$.

2.1 Syntax

Definition 1. The set $Form(\mathcal{L})$ of (well-formed) formulae of $\mathcal{L}$ is defined by the following rules:

1. (Primitive propositions are formulae): if $p \in \Phi$ then $p \in Form(\mathcal{L})$;
2. (Nullary connectives): **false** $\in Form(\mathcal{L})$, **true** $\in Form(\mathcal{L})$;
3. (Unary connectives): if $\varphi \in Form(\mathcal{L})$ then $\neg\varphi \in Form(\mathcal{L})$, $\bigcirc\varphi \in Form(\mathcal{L})$, $\bullet\varphi \in Form(\mathcal{L})$, and $(\varphi) \in Form(\mathcal{L})$;
4. (Binary connectives): if $\varphi, \psi \in Form(\mathcal{L})$, then $\varphi \vee \psi \in Form(\mathcal{L})$, $\varphi \mathcal{U} \psi \in Form(\mathcal{L})$, and $\varphi \mathcal{S} \psi \in Form(\mathcal{L})$;
5. (Agent modalities): if $\varphi \in Form(\mathcal{L})$ and $i \in Ag$ then $[i]\varphi \in Form(\mathcal{L})$.

Definition 2. If $p \in \Phi$ then both p and $\neg p$ are *literals*. If l is a literal and $i \in Ag$, then $[i]l$ and $\neg[i]l$ are *agent literals*.

2.2 Semantics

Definition 3. It is assumed that the world may be in any of a set S of *states*. We generally use s (with annotations, e.g., s_0, s', ...) to denote a state.

Definition 4. A *timeline*, l, is an infinitely long, linear, discrete sequence of states, indexed by the natural numbers. For convenience, we define a timeline l to be a function $l : I\!N \rightarrow S$. Let *TLines* be the set of all timelines.

Note that timelines correspond to the *runs* of Halpern *et al* [11, 5].

Definition 5. A point, p, is a pair $p = (l, u)$, where $l \in TLines$ is a time line and $u \in I\!N$ is a temporal index into l. Let the set of all points (over S) be *Points*.

Definition 6. A *valuation*, π, is a function $\pi : Points \times \Phi \rightarrow \{T, F\}$.

Definition 7. A *model*, M, for $\mathcal{L}$ is a structure $M = \langle TL, R_1, \ldots, R_n, \pi \rangle$, where: $TL \subseteq TLines$ is a set of timelines; R_i, for all $i \in Ag$, is an agent accessibility relation over *Points*, i.e., $R_i \subseteq Points \times Points$; and $\pi : Points \times \Phi \rightarrow \{T, F\}$ is a valuation.

As usual, we define the semantics of the language via the satisfaction relation '$\models$'. For $\mathcal{L}$, this relation holds between pairs of the form $\langle M, p \rangle$ (where M is a model and $p \in Points$), and $\mathcal{L}$-formulae. The rules defining the satisfaction relation are given in Figure 1. Satisfiability and validity in $\mathcal{L}$ are defined in the usual way.

Knowledge and belief models: We shall now define two classes of $\mathcal{L}$-models: KL_n-models are models of *knowledge*, and BL_n-models are models of *belief*.

Definition 8. An $\mathcal{L}$-model $M = \langle TL, R_1, \ldots, R_n, \pi \rangle$ is a KL_n-model iff R_i is an equivalence relation, for all $i \in Ag$.

$$\langle M, (l, u) \rangle \models \textbf{true}$$

$\langle M, (l, u) \rangle \models p$ iff $\pi((l, u), p) = T$ (where $p \in \Phi$)

$\langle M, (l, u) \rangle \models \neg\varphi$ iff $\langle M, (l, u) \rangle \not\models \varphi$

$\langle M, (l, u) \rangle \models \varphi \lor \psi$ iff $\langle M, (l, u) \rangle \models \varphi$ or $\langle M, (l, u) \rangle \models \psi$

$\langle M, (l, u) \rangle \models [i]\varphi$ iff $\forall l' \in TL, \forall v \in I\!N$, if $((l, u), (l', v)) \in R_i$, then $\langle M, (l', v) \rangle \models \varphi$

$\langle M, (l, u) \rangle \models \bigcirc\varphi$ iff $\langle M, (l, u + 1) \rangle \models \varphi$

$\langle M, (l, u) \rangle \models \bullet\varphi$ iff $(u > 0)$ and $\langle M, (l, u - 1) \rangle \models \varphi$

$\langle M, (l, u) \rangle \models \varphi \mathcal{U} \psi$ iff $\exists v \in I\!N$ such that $(v \geq u)$ and $\langle M, (l, v) \rangle \models \psi$, and $\forall w \in I\!N$, if $(u \leq w < v)$ then $\langle M, (l, w) \rangle \models \varphi$

$\langle M, (l, u) \rangle \models \varphi \mathcal{S} \psi$ iff $\exists v \in I\!N$ such that $(v < u)$ and $\langle M, (l, v) \rangle \models \psi$, and $\forall w \in I\!N$, if $(v < w < u)$ then $\langle M, (l, w) \rangle \models \varphi$

Fig. 1. Semantics of $\mathcal{L}$

It should be clear that as agent accessibility relations in KL_n models are equivalence relations, the axioms of the normal modal system S5 are valid in the class of KL_n models.

Theorem 9.

$$\models_{KL_n} [i](\varphi \Rightarrow \psi) \Rightarrow ([i]\varphi \Rightarrow [i]\psi) \tag{1}$$

$$\models_{KL_n} [i]\varphi \Rightarrow \neg[i]\neg\varphi \tag{2}$$

$$\models_{KL_n} [i]\varphi \Rightarrow \varphi \tag{3}$$

$$\models_{KL_n} [i]\varphi \Rightarrow [i][i]\varphi \tag{4}$$

$$\models_{KL_n} \neg[i]\varphi \Rightarrow [i]\neg[i]\varphi \tag{5}$$

These axioms are called K, D, T, 4, and 5, respectively. The system S5 is widely recognized as the logic of idealized *knowledge*, and for this reason we say KL_n is a *temporal logic of knowledge*. (The future-time component of KL_n corresponds to Halpern and Vardi's logic $KL_{(m)}$ [11], also known as $S5_n^U$ in [5, p283], where a complete axiomatization is given.) We now define *belief models*.

Definition 10. An $\mathcal{L}$-model $M = \langle TL, R_1, \ldots, R_n, \pi \rangle$ is a BL_n-model iff for all $i \in Ag$, we have:

1. (Euclidean:) $\forall p, p', p'' \in Points$, if $(p, p') \in R_i$ and $(p, p'') \in R_i$, then $(p', p'') \in R_i$;
2. (Serial:) $\forall p \in Points, \exists p' \in Points$ such that $(p, p') \in R_i$; and
3. (Transitive:) $\forall p, p', p'' \in Points$, if $(p, p') \in R_i$ and $(p', p'') \in R_i$, then $(p, p'') \in R_i$.

It is well-known that the axioms K, D, 4, and 5 from normal modal logic are valid in models whose accessibility relations satisfy properties (1)–(3) of Definition 10. However, axiom T (formula 3 above) is not. Axiom T is generally taken to be the axiom that distinguishes knowledge from belief: it says that if an agent knows φ, then φ is true. As this axiom is not BL_n-valid, we say that BL_n is a *temporal logic of belief*.

Derived Temporal Connectives Other standard temporal connectives are introduced as abbreviations, in terms of S, U and $\bullet$:

$$\Diamond\varphi \stackrel{\text{def}}{=} \mathbf{true}\,U\,\varphi \qquad\qquad \blacklozenge\,\varphi \stackrel{\text{def}}{=} \mathbf{true}\,S\,\psi$$

$$\Box\varphi \stackrel{\text{def}}{=} \neg\Diamond\neg\varphi \qquad\qquad \blacksquare\varphi \stackrel{\text{def}}{=} \neg\blacklozenge\neg\varphi$$

$$\varphi\,W\,\psi \stackrel{\text{def}}{=} \varphi\,U\,\psi \vee \Box\varphi \qquad \varphi\,Z\,\psi \stackrel{\text{def}}{=} \varphi\,S\,\psi \vee \blacksquare\varphi$$

$$\bullet\varphi \stackrel{\text{def}}{=} \neg\,\mathbf{O}\,\neg\varphi \qquad\qquad \mathbf{start} \stackrel{\text{def}}{=} \bullet\,\mathbf{false}$$

We now informally consider the meaning of the temporal connectives. First, consider the two basic future-time connectives: $\bigcirc$ and U. The $\bigcirc$ connective means 'at the next time'. Thus $\bigcirc\varphi$ will be satisfied at some time if φ is satisfied at the *next* time. The U connective means 'until'. Thus $\varphi\,U\,\psi$ will be satisfied at some time if ψ is satisfied at that time or some time in the future, and φ is satisfied at all times until the time that ψ is satisfied. Of the derived connectives, $\Diamond$ means 'either now, or at some time in the future'. Thus $\Diamond\varphi$ will be satisfied at some time if either φ is satisfied at that time, or some later time. The $\Box$ connective means 'now, and at all future times'. Thus $\Box\varphi$ will be satisfied at some time if φ is satisfied at that time and at all later times. The binary W connective means 'unless'. Thus $\varphi\,W\,\psi$ will be satisfied at some time if either φ is satisfied until such time as ψ is satisfied, or else φ is always satisfied. Note that W is similar to, but weaker than, the U connective; for this reason it is sometimes called 'weak until'.

The past-time connectives are similar: $\mathbf{O}$ and $\bullet$ are true at some moment if their arguments were true at the previous moment. The difference between them is that, since the model of time underlying the logic is bounded in the past, the beginning of time is a special case: $\mathbf{O}\varphi$ will always be false when interpreted at the beginning of time, whereas $\bullet\varphi$ will always be true at the beginning of time. The $\blacklozenge$ connective is a past-time version of $\Diamond$. Thus $\blacklozenge\varphi$ will be true at some time if φ was true at some previous moment in time. The $\blacksquare$ connective is a past-time version of $\Box$. Thus $\blacksquare\varphi$ will be true at some time if φ was true at all previous moments in time. The S connective mirrors U, and so $\varphi\,S\,\psi$ will be true now if ψ was true at some previous moment in time, and φ has been true since then; Z is the same, but allowing for the possibility that the second argument was never true. Thus Z mirrors W. Finally, a temporal operator that takes no arguments can be defined which is true only at the first moment in time: this operator is '**start**'.

3   A Normal Form for KL_n and BL_n

The proof methods we present in section 4 depend on formulae being rewritten into a normal form, which we call SNF*(after the *separated normal form* of [7, 8]). In this section, we define SNF*, and outline the procedure by which an arbitrary $\mathcal{L}$-formula may be rewritten in this form. The translation depends upon the use of *renaming* [13] to simplify formulae. We therefore begin, in the following section, by describing how renaming works.

3.1   Renaming

The basic idea of renaming is to simplify a formula φ by replacing sub-formulae of φ by new proposition symbols that act, in effect, as abbreviations for the sub-formulae they replace. In order to preserve satisfiability during this process, we must link the truth-value of a new proposition to the truth-value of the sub-formula it replaced. Enforcing this link in modal logic is complicated somewhat by the fact that a formula can be interpreted in many different states: we must ensure that the link is maintained in all states that can play a part in the interpretation of a formula. In temporal logic, we can do this by carrying out the renaming within the scope of a '$\square$' connective. For example, in temporal logic, renaming can be used to replace a formulae such as $\Diamond(\varphi\,\mathcal{U}\,\psi)$ by $\Diamond(\varphi\,\mathcal{U}\,p) \wedge \square(p \Leftrightarrow \psi)$, where p is a new proposition symbol. In temporal logic, the operator '$\square$' accesses all reachable states, thus p is defined as an abbreviation for ψ in every state, and so satisfiability is preserved.

Unfortunately, in our logics, the situation is complicated yet further by the presence of two kinds of modal links: temporal ones, and those given by each agent's accessibility relation R_i. We must therefore introduce a derived operator $\square^*$, such that $\square^*\varphi$ means φ is satisfied in every *reachable* state — intuitively, every state that can play a part in interpreting a formula. Renaming is then carried out within the context of the $\square^*$ operator, and thus the link between a new proposition and the sub-formula it replaces is forced across all reachable states.

In order to define the operator $\square^*$, we must first define an operator C to capture the notion of *common knowledge* (or, in BL_n, mutual belief). This, in turn requires an operator E to capture the idea of *every agent* knowing (believing) a formula. We define E by

$$E\varphi \Leftrightarrow \bigwedge_{i \in Ag} [i]\varphi\,.$$

The common knowledge operator, C, is then defined as the maximal fixpoint of the formula

$$C\varphi \Leftrightarrow E(\varphi \wedge C\varphi)\,.$$

Finally, the $\square^*$ operator is defined as the maximal fixpoint of

$$\square^*\varphi \Leftrightarrow \square(\varphi \wedge C\square^*\varphi).$$

To illustrate the properties of this operator, we must formalise the notion of *reachability*.

Definition 11. Let M be an $\mathcal{L}$-model and $(l, u), (l', v)$ be points in M. Then (l', v) is *reachable* from (l, u) iff either: (i) $l = l'$ and $v \geq u$; (ii) $((l, u), (l', v)) \in R_i$ for some agent $i \in Ag$; or (iii) there exists some point (l'', w) in M such that (l'', w) is reachable from (l, u) and (l', v) is reachable from (l'', w).

The important property of the $\square^*$ operator can now be stated.

Theorem 12. *Let M be an $\mathcal{L}$-model and p, p' be points in M such that $\langle M, p \rangle \models \square^* \varphi$. Then $\langle M, p' \rangle \models \varphi$ if p' is reachable from p.*

Now, renaming of a formula such as $\Diamond(\varphi \mathcal{U} \psi)$ produces $\Diamond(\varphi \mathcal{U} p) \wedge \square^*(p \Leftrightarrow \psi)$. This theorem therefore guarantees that renaming carried out within the context of the $\square^*$ operator will preserve satisfiability.

3.2 SNF*: A Normal Form for $\mathcal{L}$

We now describe SNF*, the normal form that we use in the proof methods described in Section 4. An $\mathcal{L}$-formula in SNF* is of the form:

$$\square^* \bigwedge_{i=1}^{n} (\varphi_i \Rightarrow \psi_i)$$

where each of the '$\varphi_i \Rightarrow \psi_i$' (called *rules*) is one of the following

$$\text{start} \quad \Rightarrow \bigvee_{k=1}^{r} l_k \quad \text{(an } initial \; \square\text{-rule)}$$

$$\text{\bf O} \bigwedge_{j=1}^{q} m_j \Rightarrow \bigvee_{k=1}^{r} l_k \quad \text{(a } global \; \square\text{-rule)}$$

$$\text{true} \quad \Rightarrow \bigvee_{k=1}^{r} l_k \quad \text{(a } global \text{ renaming rule)}$$

$$\text{start} \quad \Rightarrow \Diamond l \quad \text{(an } initial \; \Diamond\text{-rule)}$$

$$\text{\bf O} \bigwedge_{j=1}^{q} m_j \Rightarrow \Diamond l \quad \text{(a } global \; \Diamond\text{-rule)}$$

where each m_j, or l is a literal, and l_k is either a literal or an agent literal.

Theorem 13. *There exists a translation function $\tau : Form(\mathcal{L}) \rightarrow Form(SNF^*)$ such that for any $\varphi \in Form(\mathcal{L})$, we have $\tau(\varphi)$ is satisfiable just in case φ is satisfiable.*

Proof. Full details of the translation process are rather complex, and so we simply sketch the main steps. Note that the translation is similar to that described in [8], where more details can be found. The main steps are:

1. 'Pushing' negations (converting into negation normal form).
   This involves applying transformations such as

   $$\neg(\varphi \wedge \psi) \longrightarrow \neg\varphi \vee \neg\psi$$
   $$\neg\bigcirc\varphi \longrightarrow \bigcirc\neg\varphi$$
   $$\neg\square\varphi \longrightarrow \Diamond\neg\varphi$$

   both inside and outside (but not *across*) agent modalities. This operation preserves validity.

2. Dealing with axiom K:

   $$[i](\varphi \vee \psi) \longrightarrow \neg[i]\neg\varphi \vee [i]\psi$$

   Preservation of validity is obvious.

3. Removing derived temporal operators.
   This simply involves the replacement of various temporal operators by their definitions, for example

   $$\varphi \mathcal{W} \psi \longrightarrow \varphi \mathcal{U} \psi \vee \square\varphi.$$

   Again, this operation preserves validity.

4. Renaming embedded non-literal formulae.
   Here, the renaming procedure described above is applied exhaustively to removed embedded formulae, for example

   $$\varphi \mathcal{U} \psi \longrightarrow x \mathcal{U} y \wedge \square^*(x \Leftrightarrow \varphi) \wedge \square^*(y \Leftrightarrow \psi)$$
   $$[i]\varphi \longrightarrow [i]z \wedge \square^*(z \Leftrightarrow \varphi).$$

   where φ and ψ are non-literal formulae.
   Once such transformations have been applied, each literal is guaranteed to be within the scope of at most one modal or temporal operator (apart from '$\square^*$').

5. Removing maximal fixpoint operators, e.g. '$\square$'.
   This involves unwinding fixpoint operators defined in temporal logic, for example

   $$\square\varphi \longrightarrow \varphi \wedge x \wedge \square^*(\bullet x \Leftrightarrow (\varphi \wedge x))$$

   (For more detail, see [8].) Again, this step preserves satisfiability.

6. Rewriting into SNF* (effectively an extension of CNF).
   This final phase involves the use of classical transformations, analogous to those used to produce clausal form, to produce a set of SNF* rules from the formula produced so far.
   This step is a variation on the standard transformation into CNF, and so preserves validity.

Thus, the proof reduces to showing that the renaming process itself preserves satisfiability, which can be shown by observing that

- the new propositions introduced are defined in every state (see Theorem 12);
- the original formula is satisfied iff the renamed formula is satisfied (which follows from the structure of the formula resulting from renaming).

In order to ensure that renaming rules are available everywhere, formulae such as

$$\Box^*(x \Leftrightarrow \varphi)$$

are represented by SNF* rules

$$\textbf{true} \Rightarrow \neg x \vee \varphi$$
$$\textbf{true} \Rightarrow \neg\varphi \vee x$$

which again appear within the context of an ' $\Box$*' operator. Note that such rules (with '**true**' on their left-hand side) are only used for renaming in this way.

4 Resolution-Based Proof Methods for KL_n and BL_n

Before describing the resolution method in detail, we give an overview of our approach. First, recall the two basic problems associated with extending resolution to modal and multi-modal logics such as those considered in this paper:

1. Literals cannot generally be moved in and out of modal or temporal contexts. In particular, if M is a modal or temporal operator, p and $M\neg p$ cannot be resolved. Thus, the only inferences that can be made occur in particular modal or temporal contexts. For example, both p and $\neg p$ can be resolved, as, for certain types of modal operator, can Mp and $M\neg p$.
2. In many non-classical logics, particularly multi-modal logics, the operators interact in complex ways. For example, the logics considered here have transitive accessibility relations and so the axiom $M\varphi \Rightarrow MM\varphi$ holds. Thus, in addition to the problem of reasoning in the presence of modal and temporal contexts, the proof process must take into account this interaction between operators.

The resolution method described in this paper tackles these problems by:

- using the normal form SNF*, which separates out complex formulae from their contexts through the use of renaming (as described above); and
- utilizing additional translation, resolution, and simplification rules in order to represent the interaction between operators.

The translation of a formula into a normal form is particularly important. In removing formulae from their contexts and replacing them by new propositions (effectively, abbreviations), we are able to manipulate these formulae using what is essentially classical resolution, only returning the results to their contexts under specific conditions.

We can now describe the proof method. This method extends that of the underlying temporal logic [7] and consists of the following steps. To determine whether a formula $\varphi \in Form(\mathcal{L})$ is unsatisfiable:

1. Rewrite φ into SNF*.
2. Repeat
   (a) apply step resolution (effectively a form of classical resolution)

(b) rewrite new resolvents into SNF*
(c) apply simplification and subsumption
(d) apply temporal resolution
(e) rewrite new resolvents into SNF*

until either **false** is derived or no more rules can be applied.

The process of applying temporal and step resolution rules to a set of formulae in SNF* eventually terminates. On termination, either **false** has been derived (showing that the formula is unsatisfiable) or no more rules can be applied (showing that the formula is satisfiable). As in classical resolution, simplification and subsumption procedures are applied throughout the process.

Since agent literals can never occur either on the left-hand side of an SNF* rule, or within a $\Diamond$-formula, rules containing such literals can not directly participate in temporal resolution steps. Thus, in extending the temporal resolution method of [7] to BL_n and KL_n, we need only consider the additional step resolution rules and simplification rules for handling formulae containing modal operators.

4.1 Step Resolution

The step resolution rule is simply a version of the classical resolution rule rewritten as follows.

$$\frac{\begin{array}{c}\varphi_1 \Rightarrow \psi_1 \vee l \\ \varphi_2 \Rightarrow \psi_2 \vee \neg l\end{array}}{\varphi_1 \wedge \varphi_2 \Rightarrow \psi_1 \vee \psi_2} \quad \text{(SRES)}$$

Now, however, the 'l' above is now an atom. In particular, we can resolve agent literals via the following rule:

$$\frac{\begin{array}{c}\varphi_1 \Rightarrow \psi_1 \vee [i]m \\ \varphi_2 \Rightarrow \psi_2 \vee \neg[i]m\end{array}}{\varphi_1 \wedge \varphi_2 \Rightarrow \psi_1 \vee \psi_2} \quad \text{(SRESa)}$$

In addition to this general rule, we add a more specific resolution rule for resolving within modal contexts:

$$\frac{\begin{array}{c}\varphi_1 \Rightarrow \psi_1 \vee [i]m \\ \varphi_2 \Rightarrow \psi_2 \vee [i]\neg m\end{array}}{\varphi_1 \wedge \varphi_2 \Rightarrow \psi_1 \vee \psi_2} \quad \text{(MRES)}$$

This resolution rule derives from axiom (2) of the logic, (which in turn corresponds to axiom D of normal modal logic).

4.2 Simplification

The simplification rules used are similar to the temporal case, which are, in turn, similar to the classical case, consisting of both simplification and subsumption rules. The major rule required in the temporal resolution method is used when a contradiction in a state is produced:

$$\frac{\bigcirc \varphi \Rightarrow \textbf{false}}{\textbf{true} \Rightarrow \neg \varphi} \qquad \text{(SIMP1)}$$

From the properties of the '$\Box^*$' operator, which implicitly surrounds all SNF* rules, we are able to derive the following rule:

$$\frac{\textbf{true} \Rightarrow \varphi}{\textbf{true} \Rightarrow [i]\varphi} \qquad \text{(SIMP2)}$$

where $[i]\varphi$ must again be rewritten into SNF*.

Additional simplification rules are provided by axioms (3), (4), and (5), corresponding to normal modal axioms T, 4, and 5 respectively. The first, (*SIMP3*), is derived from axiom T.

$$\frac{\psi_1 \Rightarrow \psi_2 \ \vee \ [i]\varphi}{\psi_1 \Rightarrow \psi_2 \ \vee \ \varphi} \qquad \text{(SIMP3)}$$

Note that this rule is not used in the BL_n proof method, as axiom (3) is not valid in BL_n: it is *only* used in KL_n. The use of this rule represents the only difference between the proof method for KL_n and that for BL_n. The next simplification rule is derived from axiom (4). It can be used at most once per rule within SNF*.

$$\frac{\psi_1 \Rightarrow \psi_2 \ \vee \ [i]\varphi}{\substack{\psi_1 \Rightarrow \psi_2 \ \vee \ [i]x \\ x \Leftrightarrow [i]\varphi}} \qquad \text{(SIMP4)}$$

This final rule, (*SIMP5*), is derived from axiom 5. Like (*SIMP4*), this rule is also used at most once per formula in SNF*.

$$\frac{\psi_1 \Rightarrow \psi_2 \ \vee \ \neg[i]\varphi}{\substack{\psi_1 \Rightarrow \psi_2 \ \vee \ [i]y \\ y \Leftrightarrow \neg[i]\varphi}} \qquad \text{(SIMP5)}$$

4.3 Temporal Resolution

Rather than describe the temporal resolution rule in detail, we refer the interested reader to [7]. The basic idea is to resolve one $\Diamond$-rule with a *set* of $\Box$-rules as follows[2].

---

[2] A similar rule resolving with a $\Diamond$-rule of the form **start** $\Rightarrow \Diamond l$ is also used.

$$\left.\begin{array}{c} \bullet\,\varphi_1 \Rightarrow \psi_1 \\ \bullet\,\varphi_2 \Rightarrow \psi_2 \\ \vdots \\ \bullet\,\varphi_n \Rightarrow \psi_n \\ \dfrac{\bullet\,\chi \Rightarrow \Diamond l}{\mathbf{true} \Rightarrow \neg(\bigvee\limits_{i=1}^{n} \varphi_i \wedge \chi)} \\ \bullet\,\chi \Rightarrow (\neg\bigvee\limits_{i=1}^{n} \varphi_i)\,\mathcal{W}\,l \end{array}\right\} \quad \text{where } \bigvee\limits_{i=1}^{n} \varphi_i \Rightarrow \Box\neg l \qquad \text{(TRES)}$$

The resolvents produced must again be translated into SNF*.

4.4 Termination

Finally, if any of the following rules are produced, the original formula is unsatisfiable and the resolution process terminates.

$$\mathbf{start} \Rightarrow \mathbf{false} \qquad\qquad \text{(NULL1)}$$

$$\bullet\,\mathbf{true} \Rightarrow \mathbf{false} \qquad\qquad \text{(NULL2)}$$

$$\mathbf{true} \Rightarrow \mathbf{false} \qquad\qquad \text{(NULL3)}$$

4.5 Correctness

Theorem 14. *The resolution method is sound, i.e., if a contradiction is derived using the resolution method, then the original formula is unsatisfiable.*

Theorem 15. *The resolution method is refutation complete, i.e., if a formula is unsatisfiable, then the resolution method will eventually derive a contradiction when applied to that formula.*

5 Worked Examples

In this section, we illustrate the proof methods for KL_n and BL_n through a number of short worked examples. (Note that, for brevity, we omit many of the rules derived but not crucial to the refutation.)

Example 1. Consider the formula

$$([1]\,\Box p) \wedge \Box\neg p$$

which is BL_n-satisfiable, but KL_n-unsatisfiable. We can transform this into SNF*, giving:

1. **start** $\Rightarrow [1]a$
2. **true** $\Rightarrow \neg a \vee p$
3. **true** $\Rightarrow \neg a \vee b$
4. $\bullet b \Rightarrow p$
5. $\bullet b \Rightarrow b$
6. **start** $\Rightarrow n$
7. **start** $\Rightarrow \neg p$
8. $\bullet n \Rightarrow n$
9. $\bullet n \Rightarrow \neg p$

Given this, resolution can proceed as follows.

10. **start** $\Rightarrow \neg a$ [2, 7, SRES]

At this point the refutation stalls for BL_n, as no further resolution or simplification rules can be applied. Thus the formula is BL_n-satisfiable. However, in KL_n, (*SIMP3*) can be used as a simplification rule, producing:

11. **start** $\Rightarrow a$ [1, SIMP3]
12. **start** $\Rightarrow$ **false**   [10, 11, SRES]

As (*NULL1*) has been derived, representing **false**, the formula is KL_n-unsatisfiable.

Example 2. Consider the following formula, illustrating the inductive nature of the underlying temporal logic. This formula is valid in both BL_n and KL_n.

$$(\Diamond [i]a \wedge \Box([i]a \Rightarrow \bigcirc [i]a)) \Rightarrow \Diamond \Box [i]a$$

If we negate this, and transform it into a set of SNF* rules, we get the following (note that m is an abbreviation for $[i]a$).

1.   **start** $\Rightarrow \Diamond m$
2. $\bullet m \Rightarrow m$
3.   **start** $\Rightarrow s$
4. $\bullet s \Rightarrow s$
5. $\bullet(s \wedge m) \Rightarrow \Diamond \neg m$

The resolution process then proceeds as follows.

6. $\bullet(s \wedge m) \Rightarrow \neg m$ [2, 5, TRES]
7.   **true** $\Rightarrow \neg s \vee \neg m$ [2, 5, TRES]
8. $\bullet s \Rightarrow \neg m$ [4, 7, SRES]
9.   **start** $\Rightarrow m \vee \neg s$ [1, 4, 8, TRES]
10.   **start** $\Rightarrow \neg s$ [7, 9, SRES]
11.   **start** $\Rightarrow$ **false**   [3, 10, SRES]

Example 3. To show

$$[i]\bigcirc p \wedge [i]\bigcirc \neg p$$

is BL_n-unsatisfiable, we translate into SNF*, giving:

1. **start** $\Rightarrow [i]x$
2. **start** $\Rightarrow [i]y$
3. **true** $\Rightarrow \neg x \lor a$
4. **true** $\Rightarrow \neg y \lor b$
5. $\bigcirc a \Rightarrow p$
6. $\bigcirc b \Rightarrow \neg p$

A contradiction can be derived as follows.

7.	$\bigcirc (a \land b) \Rightarrow$ **false**	[5, 6, SRES]
8.	**true** $\Rightarrow \neg a \lor \neg b$	[7, SIMP1]
9.	**true** $\Rightarrow \neg x \lor \neg b$	[3, 8, SRES]
10.	**true** $\Rightarrow \neg x \lor \neg y$	[4, 9, SRES]
11.	**true** $\Rightarrow \neg [i]x \lor [i]\neg y$	[10, SIMP2]
12.	**start** $\Rightarrow [i]\neg y$	[1, 11, SRESa]
13.	**start** $\Rightarrow$ **false**	[2, 12, MRES]

Example 4. Finally, we consider a purely modal example:

$$\neg [i]\neg [i]p \land \neg p$$

which is unsatisfiable in KL_n. We first translate this into SNF*, giving

1. **start** $\Rightarrow \neg p$
2. **start** $\Rightarrow \neg [i]x$
3. **true** $\Rightarrow \neg x \lor \neg [i]p$
4. **true** $\Rightarrow x \lor [i]p$

A contradiction can be derived as follows.

5.	**true** $\Rightarrow x \lor p$	[4, SIMP3]
6.	**start** $\Rightarrow x$	[1, 5, SRES]
7.	**start** $\Rightarrow \neg [i]p$	[3, 6, SRES]
8.	**start** $\Rightarrow [i]y$	[7, SIMP5]
9.	**true** $\Rightarrow \neg y \lor \neg [i]p$	[7, SIMP5]
10.	**true** $\Rightarrow y \lor [i]p$	[7, SIMP5]
11.	**true** $\Rightarrow \neg y \lor x$	[4, 9, SRES]
12.	**true** $\Rightarrow \neg [i]y \lor [i]x$	[11, SIMP2]
13.	**start** $\Rightarrow [i]x$	[8, 12, SRESa]
14.	**start** $\Rightarrow$ **false**	[2, 13, SRESa]

Space precludes the presentation of further larger examples.

6 Summary

While the complexity of the decision problem for the temporal logics of knowledge and belief presented here has been studied [11], few proof methods have been developed for these logics [14]. Proof methods for multi-modal logics (e.g., [1]) have been developed, but have generally been based on tableaux methods and, moreover, have not been extended to modal *and* temporal combinations. Our work therefore represents an important step towards the mechanization of a class of logics with a wide variety of applications in distributed systems and distributed AI.

In the future, we hope to extend the proof method to deal with (limited) quantification, to implement the method, and to apply the method to the specification and verification of distributed intelligent systems. We also hope to evaluate the resolution method against tableaux proof methods for such logics [14].

References

1. L. Catach. TABLEAUX: A general theorem prover for modal logics. *Journal of Automated Reasoning*, 7:489–510, 1991.
2. B. Chellas. *Modal Logic: An Introduction*. Cambridge University Press: Cambridge, England, 1980.
3. C. Dixon, M. Fisher, and H. Barringer. A graph-based approach to temporal resolution. In D. M. Gabbay and H. J. Ohlbach, editors, *Temporal Logic — Proceedings of the First International Conference (LNAI Volume 827)*, pages 415–429, July 1994.
4. E. A. Emerson. Temporal and modal logic. In J. van Leeuwen, editor, *Handbook of Theoretical Computer Science*, pages 996–1072. Elsevier Science Publishers B.V.: Amsterdam, The Netherlands, 1990.
5. R. Fagin, J. Y. Halpern, Y. Moses, and M. Y. Vardi. *Reasoning About Knowledge*. The MIT Press: Cambridge, MA, 1995.
6. R. Fagin, J. Y. Halpern, and M. Y. Vardi. What can machines know? on the properties of knowledge in distributed systems. *Journal of the ACM*, 39(2):328–376, 1992.
7. M. Fisher. A resolution method for temporal logic. In *Proceedings of the Twelfth International Joint Conference on Artificial Intelligence (IJCAI)*, Sydney, Australia, August 1991.
8. M. Fisher. A normal form for first-order temporal formulae. In *Proceedings of Eleventh International Conference on Automated Deduction (CADE)*. Springer-Verlag: Heidelberg, Germany, June 1992.
9. J. Y. Halpern. Using reasoning about knowledge to analyze distributed systems. *Annual Review of Computer Science*, 2:37–68, 1987.
10. J. Y. Halpern and Y. Moses. A guide to completeness and complexity for modal logics of knowledge and belief. *Artificial Intelligence*, 54:319–379, 1992.
11. J. Y. Halpern and M. Y. Vardi. The complexity of reasoning about knowledge and time. I: Lower bounds. *Journal of Computer and System Sciences*, 38:195–237, 1989.
12. S. Kraus and D. Lehmann. Knowledge, belief and time. *Theoretical Computer Science*, 58:155–174, 1988.
13. D. A. Plaisted and S. A. Greenbaum. A structure-preserving clause form translation. *Journal of Symbolic Computation*, 2(3):293–304, September 1986.
14. M. Wooldridge and M. Fisher. A decision procedure for a temporal belief logic. In D. M. Gabbay and H. J. Ohlbach, editors, *Temporal Logic — Proceedings of the First International Conference (LNAI Volume 827)*, pages 317–331. Springer-Verlag: Heidelberg, Germany, July 1994.

A Methodology for Iterated Theory Change

Dov Gabbay[1] and Odinaldo Rodrigues[2]

[1] Department of Computing, Imperial College
180 Queen's Gate, London, SW7 2BZ, UK
e-mail: dg@doc.ic.ac.uk
[2] Department of Computing, Imperial College
180 Queen's Gate, London, SW7 2BZ, UK
e-mail: otr@doc.ic.ac.uk

Abstract. In this work, we propose some operators for theory change. We consider Belief Revision and Updates. The operators support combined iterations of revisions and updates and can be efficiently implemented. We show that the revision operator verifies the AGM postulates for Belief Revision and that the update one verifies the postulates for Updates proposed by Katsuno and Mendelzon [7].

1 Introduction

Many aspects of theory change share a number of similarities and can be modelled upon the same concept of minimality by simply changing the way the information available is perceived [9].

In this paper we explore this view and provide a representation of formulae upon which different kinds of information change can be efficiently implemented. We will be concerned initially with two well-known approaches to information change: Belief Revision and Updates.

For Belief Revision, we propose a revision operator through which revisions, expansions and contractions complying with the AGM postulates for Belief Revision can be easily accomplished [5]. We also propose another operator that can be used to model some effects of the execution of actions and complies with the postulates for Updates proposed in [7]. Both operators are defined syntactically but are model-theoretically motivated. The consistency check, usually very expensive in such operations [2, 13, 1], is reduced to checking the consistency of disjuncts of propositional logic.

The degree of change in information to accept the new information in the operations is evaluated by a function to measure distance between interpretations. This function is a metric on the space of interpretations of propositional logic. We also provide a way to compute it by analysing syntactically the sentences involved in the operation.

2 Evaluating the degree of change

When new information is to be accomodated into a belief state, one is usually interested in keeping as much as possible of the information already available.

This involves some sort of measurement of the degree of change in information.

For the case of propositional logic this change can be measured by analysing the truth-values of propositional variables of interpretations. One possibility is to consider the "distance" between two interpretations as the number of propositional variables with different truth-values.

Remark. We assume that **L** is the language of propositional logic over some finite set of propositional variables $\mathcal{P}^3$. A *literal* is a propositional variable or the negation of a propositional variable. The satisfaction and derivability relations (⊩ and ⊢, resp.) for **L** are assumed to have their usual meanings. An interpretation is a function from $\mathcal{P}$ to $\{\mathbf{tt}, \mathbf{ff}\}$, and $\mathcal{I}$ represents the set of all interpretations of **L**.

Definition 1. Let M and N be two elements of $\mathcal{I}$. The *distance d between M and N* is the number of propositional variables p_i, for which $M(p_i) \neq N(p_i)$.

Proposition 2. *The function d is a metric on $\mathcal{I}$.*

Proof. Omitted. See [10] for details.

The function d represents essentially the same distance notion defined in [2]. We have also extended it to classes of interpretations:

Definition 3. Let $\mathcal{M}$ and $\mathcal{N}$, be two classes of interpretations of L. The *distance D between $\mathcal{M}$ and $\mathcal{N}$*, $D(\mathcal{M}, \mathcal{N})$, is defined as

$$D(\mathcal{M}, \mathcal{N}) = \begin{cases} \infty, & \text{if either } \mathcal{M} \text{ or } \mathcal{N} \text{ is empty (or both)} \\ \min\{d(M, N) \mid M \in \mathcal{M}, \ N \in \mathcal{N}\}, & \text{otherwise} \end{cases}$$

Remark. $\mathrm{mod}(\varphi)$ will denote the set of interpretations of $\mathcal{I}$ which satisfy φ.

As expected, the distance between formulae can be introduced as the distance between their set of models.

Definition 4. Let φ and ψ be formulae of propositional logic. The semantic distance between φ and ψ, $dist_{\mathcal{I}}(\varphi, \psi)$ is defined as $D(\mathrm{mod}(\varphi), \mathrm{mod}(\psi))$.

Proposition 5. *Let φ and ψ be two formulae of propositional logic.*

$$dist_{\mathcal{I}}(\varphi, \psi) = 0 \ \textit{iff} \ \mathrm{mod}(\varphi) \cap \mathrm{mod}(\psi) \neq \emptyset$$

If the formulae are only disjuncts of propositional logic, $dist_{\mathcal{I}}$ can be more easily computed:

Definition 6. Let P and Q bet two disjuncts of propositional logic. The syntactical distance between P and Q, $dist(P, Q)$, is defined as:

$dist(P, Q) =$
$$\begin{cases} \infty, & \text{if either } P \text{ or } Q \text{ is contradictory (or both)} \\ \text{number of prop. var. } p_i \text{ s.t. both } p_i \text{ and } \neg p_i \text{ appear in } P \text{ and } Q, & \text{otherwise} \end{cases}$$

Proposition 7. *Let P and Q bet two disjuncts of propositional logic.*

$$dist(P, Q) = dist_{\mathcal{I}}(P, Q).$$

---

[3] This is required by [7], whose results we use.

3 Bands

Now that we have a notion of distance formally defined, we need a representation of formulae upon which changes of information can be easily performed. We will call this representation a *band*, because we perceive it as a "layer" of information to be added to the belief base.

Remark. We will use capital letters from the middle of the alphabet, $P, Q, \ldots$ to denote disjuncts of propositional logic and capital greek letters Δ, Φ, Γ to denote bands (to be defined next).

Definition 8. A *band* is a non-empty finite set of disjuncts of propositional logic of the form $\{P_1, P_2, \ldots, P_k\}$, and it is to be understood as a representation of the DNF formula $P_1 \vee P_2 \vee \ldots \vee P_k$.

Representing DNF as sets makes the definition of the operations simpler. Even though there is the extra cost of converting the sentences to DNF, this does not reduce the expressive power, since given any sentence φ of propositional logic there is a tautologically equivalent sentence φ' in DNF [4, Corollary 15C].

Definition 9 Models of a Band. The set of models of a band Δ, $\mathrm{mod}^*(\Delta)$, is the set $\mathrm{mod}^*(\Delta) = \{M \in \mathcal{I} \mid M \Vdash P, \text{ for some disjunct } P \in \Delta\}$.

Proposition 10. *Suppose Δ is the band $\{P_1, P_2, \ldots, P_k\}$. $\mathrm{mod}^*(\Delta) = \mathrm{mod}(P_1 \vee P_2 \vee \ldots \vee P_k)$.*

Remark. Let φ be a sentence of propositional logic. We will use the expression $\mathrm{band}(\varphi)$ to denote a band Δ such that $\mathrm{mod}^*(\Delta) = \mathrm{mod}(\varphi)$.

4 Revisions of Bands

The operators we propose are defined by selecting some combinations of disjuncts from two bands, namely those with minimum distance. The way the disjuncts are combined changes slightly depending on the particular kind of information change being performed. This is reasonable in face of the differences between the processes. Just to mention an example, whereas revisions by consistent sentences always result in consistent belief states, this does not necessarily happen in an update.

Definition 11 Prioritised Revisions of Disjuncts. Let $P = \bigwedge_i p_i$ and $Q = \bigwedge_j q_j$ be two disjuncts of propositional logic. The *prioritised revision of P by Q*, denoted $\widehat{PQ}$, is a new disjunct R, such that

$$R = \begin{cases} Q, & \text{if } P \text{ is unsatisfiable} \\ \bigwedge_j q_j \bigwedge \{p_i | p_i, \neg p_i \text{ doesn't appear among the } q_j\text{'s}\}, & \text{otherwise} \end{cases}$$

Notice that the first condition in Definition 11 is the only consistency check necessary to achieve (consistent) revisions. It can be easily performed since it amounts to checking whether a disjunct have both positive and negative literals of the same propositional variable.

Proposition 12. $\widehat{PQ} \vdash Q$.

Proof. It follows immediately from Definition 11, since all the literals in Q are always preserved.

Proposition 13. $\widehat{PQ}$ is unsatisfiable iff Q is unsatisfiable.

Proof. It follows immediately from Definition 11.

Definition 14 Revisions of Bands. The *revision of band Δ by band Φ*, $\Delta \triangleleft \Phi$, is a new band Γ such that:

$$\Gamma = \{\widehat{PQ}^* \mid P \in \Delta, Q \in \Phi, \text{ and } dist(P,Q) \text{ is minimum}\}$$

Proposition 15. $\mathrm{mod}^*(\Delta \triangleleft \Phi) \subseteq \mathrm{mod}^*(\Phi)$.

Proof. By Definition 14, the elements of $\mathrm{mod}^*(\Delta \triangleleft \Phi)$ will be disjuncts of the form $\widehat{PQ}^*$, where $Q \in \Phi$. By Proposition 12, $\widehat{PQ}^* \vdash Q$, and then $\widehat{PQ}^* \Vdash Q$. Therefore, if $M \in \mathrm{mod}^*(\Delta \triangleleft \Phi)$, then $M \Vdash R$, for some $R \in \Delta \triangleleft \Phi$, and hence $M \Vdash Q$, for some $Q \in \mathrm{mod}^*(\Phi)$. It follows that $M \in \mathrm{mod}^*(\Phi)$.

Proposition 16. $\mathrm{mod}^*(\Delta \triangleleft \Phi) = \emptyset$ iff $\mathrm{mod}^*(\Phi) = \emptyset$.

From the definition of Revisions of bands and the properties of *dist*, we can prove the following properties of the operator $\triangleleft$.

Proposition 17. *Suppose* $M \in \mathrm{mod}(\varphi)$. $M \in \mathrm{mod}^*(\mathrm{band}(\psi) \triangleleft \mathrm{band}(\varphi))$ *iff* $\forall N \in \mathrm{mod}(\varphi)$, $\forall I' \in \mathrm{mod}(\psi)$, $\exists I \in \mathrm{mod}(\psi)$ *such that* $d(M,I) \leq d(N,I')$.

Proof. ($\Rightarrow$) Suppose that $\forall N \in \mathrm{mod}(\varphi)$, $\forall I' \in \mathrm{mod}(\psi)$, $\exists I \in \mathrm{mod}(\psi)$ s.t.

$$d(M,I) \leq d(N,I') \tag{1}$$

but $M \notin \mathrm{mod}^*(\mathrm{band}(\psi) \triangleleft \mathrm{band}(\varphi))$.

By Proposition 16, $\mathrm{mod}^*(\mathrm{band}(\psi) \triangleleft \mathrm{band}(\varphi)) \neq \emptyset$, and then $\exists O \Vdash \widehat{PQ}^*$, for some $P \in \mathrm{band}(\psi)$, $Q \in \mathrm{band}(\varphi)$. It follows that $dist(P,Q) < dist(P',Q')$, for all $P' \in \mathrm{band}(\psi)$, $Q' \in \mathrm{band}(\varphi)$, such that $M \Vdash \widehat{P'Q'}$. Therefore, $\exists J \in \mathrm{mod}(P)$ such that for all $I'' \in \mathrm{mod}(\psi)$

$$d(O,J) < d(M,I'') \tag{2}$$

But this is a contradiction, since $O \Vdash \varphi$, $J \Vdash \psi$ and then by (1), $d(M,I) \leq d(O,J)$ (take $I'' = I$ in (2)).

($\Leftarrow$) If $M \in \text{mod}^*(\text{band}(\psi)\triangleleft\text{band}(\varphi))$, then $M \Vdash \widehat{PQ}^*$ for some $P \in \text{band}(\psi)$, $Q \in \text{band}(\varphi)$, such that $dist(P,Q)$ is minimum. Now, suppose $\exists N \in \text{mod}$ (φ) and $\exists I' \in \text{mod}(\psi)$ such that $\forall I \in \text{mod}(\psi)$, $d(N,I') < d(M,I)$. $N \Vdash$ P' and $I' \Vdash Q'$, for some $P' \in \text{band}(\varphi)$ and $Q' \in \text{band}(\psi)$. It follows that $dist(P',Q') < dist(P,Q'')$ for all $Q'' \in \text{band}(\psi)$, a contradiction, since we assumed that $dist(P,Q)$ was minimum.

Corollary 18. *Suppose* $M \in \text{mod}(\varphi)$. *If* $M \notin \text{mod}^*(\text{band}(\psi)\triangleleft\text{band}(\varphi))$, *then* $\exists N \in \text{mod}(\varphi)$, $\exists I' \in \text{mod}(\psi)$, *such that* $\forall I \in \text{mod}(\psi)$, $d(N,I') < d(M,I)$.

4.1 Revisions of Bands and the AGM Postulates for Belief Revision

The AGM postulates for Belief Revision were initially proposed by Alchourrón, Gärdenfors and Makinson as basic requirements an operator performing a rational change of belief should satisfy.

Later on, Katsuno et. al. [7, 6], provided a semantical characterisation for all revision operators defined in terms of pre-orders satisfying the postulates. We will use this characterisation to show that our revision operator satisfies the AGM postulates for Belief Revision.

We recap them here briefly. In the postulates below, $K \circ A$ stands for the revision of K by A.

(R1) $K \circ A$ implies A
(R2)   If $K \wedge A$ is satisfiable, then $K \circ A \leftrightarrow K \wedge A$
(R3)   If A is satisfiable, then $K \circ A$ is also satisfiable
(R4)   If $K_1 \leftrightarrow K_2$ and $A_1 \leftrightarrow A_2$, then $K_1 \circ A_1 \leftrightarrow K_2 \circ A_2$
(R5) $(K \circ A) \wedge B \rightarrow K \circ (A \wedge B)$
(R6)   If $(K \circ A) \wedge B$ is satisfiable, then $K \circ (A \wedge B)$ implies $(K \circ A) \wedge B$

Definition 19 Faithful Assignment for Belief Revision - [7]. A *faithful assignment for belief revision* is a function mapping each propositional formula ψ to a pre-order $\leq_\psi$ on $\mathcal{I}$, such that

1. If $I, I' \in \text{mod}(\psi)$, then $I <_\psi I'$ does not hold.
2. If $I \in \text{mod}(\psi)$ and $I' \notin \text{mod}(\psi)$, then $I <_\psi I'$ holds.
3. If $\psi \leftrightarrow \varphi$, then $\leq_\psi = \leq_\varphi$.

Remark. Let $\leq$ be a pre-order on a set S, and M a subset of S. $\min(M, \leq)$ is the set of elements in M which are minimal w.r.t. ordering $\leq$.

Theorem 20. *[7] A revision operator $\circ$ satisfies the AGM postulates for Belief Revision iff there exists a faithful assignment that maps each knowledge base ψ to a total pre-order $\leq_\psi$, such that* $\text{mod}(\psi \circ \varphi) = \min(\text{mod}(\varphi), \leq_\psi)$.

The next step is to define a faithful assignment based on our distance function. For this, we need a function χ which assigns every propositional formula ψ to a pre-order $\leq_\psi$ based on d.

Definition 21. For each formula ψ of propositional logic, $\chi(\psi) = \{\leq_{\psi}^{\lhd}, \mathcal{I}\}$, where for any $M, N \in \mathcal{I}$ and satisfiable ψ

$$M \leq_{\psi}^{\lhd} N \text{ iff } M = N \text{ or } \exists I \in \mathrm{mod}(\psi) \text{ s.t. } \forall I' \in \mathrm{mod}(\psi) \, d(M, I) \leq d(N, I')^4$$

If ψ is unsatisfiable, we define $M \leq_{\psi}^{\lhd} N$, for all $M, N \in \mathcal{I}$.

Proposition 22. *For each propositional formula ψ, $\leq_{\psi}^{\lhd}$ is a pre-order.*

Proof. The interesting case is transitivity when ψ is satisfiable: suppose $M \leq_{\psi}^{\lhd} N$, and $N \leq_{\psi}^{\lhd} O$. If $M \leq_{\psi}^{\lhd} N$, then $\exists I_1 \in \mathrm{mod}(\psi)$, such that $\forall I' \in \mathrm{mod}(\psi) \, d(M, I_1) \leq d(N, I')$. If $N \leq_{\psi}^{\lhd} O$, then $\exists I_2 \in \mathrm{mod}(\psi)$ such that $\forall I'' \in \mathrm{mod}(\psi) \, d(N, I_2) \leq d(O, I'')$. $I_2 \in \mathrm{mod}(\psi)$. Thus, in particular, $d(M, I_1) \leq d(N, I_2)$, and hence $M \leq_{\psi}^{\lhd} O$.

Proposition 23. *χ is a faithful assignment.*

Proof. By Proposition 22, for each propositional formula ψ, $\leq_{\psi}^{\lhd}$ is a pre-order. It can be easily seen that $\leq_{\psi}^{\lhd}$ is also total.

1. If $M, N \in \mathrm{mod}(\psi)$, then $M <_{\psi}^{\lhd} N$ does not hold. If $M \in \mathrm{mod}(\psi)$, then $M \leq_{\psi}^{\lhd} N$, because $\forall I' \in \mathrm{mod}(\psi) \, d(M, M) \leq d(N, I')$. Also, $N \leq_{\psi}^{\lhd} M$, for the same reason, since $N \in \mathrm{mod}(\psi)$, and hence $M \equiv_{\psi}^{\lhd} N$.

2. If $M \in \mathrm{mod}(\psi)$ and $N \notin \mathrm{mod}(\psi)$, then $M \leq_{\psi}^{\lhd} N$. Clearly, $M \leq_{\psi}^{\lhd} N$ (see previous item). $N \not\leq_{\psi}^{\lhd} M$, because since $N \notin \mathrm{mod}(\psi)$ and $d(M, M) = 0$, $\neg \exists I \in \mathrm{mod}(\psi)$ such that $d(N, I) \leq d(M, M)$.

3. If $\psi \leftrightarrow \varphi$, then $\leq_{\psi} = \leq_{\varphi}$. This follows directly, since d and $\leq_{\psi}^{\lhd}$ are defined semantically. □

Theorem 24. *The revision operation defined by $\lhd$ verifies the AGM postulates for Belief Revision.*

Proof. We are left to show that $\mathrm{mod}^*(\mathrm{band}(\psi) \lhd \mathrm{band}(\varphi)) = \min(\mathrm{mod}(\varphi), \leq_{\psi}^{\lhd})$.

($\subseteq$) *i)* Suppose $\mathrm{mod}(\psi) = \emptyset$. It follows that all disjuncts in $\mathrm{band}(\psi)$ are unsatisfiable, and then $\mathrm{band}(\psi) \lhd \mathrm{band}(\varphi) = \mathrm{band}(\varphi)$. But if $\mathrm{mod}(\psi) = \emptyset$, then all interpretations are incomparable w.r.t. to $\leq_{\psi}^{\lhd}$, and hence $\min(\mathrm{mod}(\varphi), \leq_{\psi}^{\lhd}) = \mathrm{mod}(\varphi)$. *ii)* Suppose $\mathrm{mod}(\psi) \neq \emptyset$, and that $M \in \mathrm{mod}^*(\mathrm{band}(\psi) \lhd \mathrm{band}(\varphi))$, but $M \notin \min(\mathrm{mod}(\varphi), \leq_{\psi}^{\lhd})$. By proposition 15, $M \in \mathrm{mod}^*(\mathrm{band}(\varphi))$, and then $M \in \mathrm{mod}(\varphi)$. By Proposition 17, $\forall N \in \mathrm{mod}(\varphi)$, $\forall I' \in \mathrm{mod}(\psi)$, $\exists I \in \mathrm{mod}(\psi)$, such that

$$d(M, I) \leq d(N, I') \tag{3}$$

---

[4] When $\mathrm{mod}(\psi) \neq \emptyset$, the ordering $\leq_{\psi}^{\lhd}$ may be regarded as an extension of Lewis' systems of spheres [8]. The inner most sphere is consisted by $\mathrm{mod}(\psi)$, the next one contains all the interpretations which have distance at most 1 w.r.t. to *some* model of ψ, and so on.

$M \notin \min(\mathrm{mod}(\varphi), \leq_\psi^\lhd)$ and $\mathrm{mod}(\varphi) \neq \emptyset$, therefore $\exists O \in \mathrm{mod}(\varphi)$, such that $O <_\psi^\lhd M \therefore \exists J \in \mathrm{mod}(\psi)$ such that $\forall J' \in \mathrm{mod}(\psi)$

$$d(O, J) < d(M, J') \tag{4}$$

To see that this is a contradiction, take $J' = I$ in (4), and $N = O$ and $I' = J$ in (3).

($\supseteq$) If $M \in \min(\mathrm{mod}(\varphi), \leq_\psi^\lhd)$, then $\neg\exists N \in \mathrm{mod}(\varphi)$, such that $N <_\psi^\lhd M$ holds. That is, $\neg\exists N \in \mathrm{mod}(\varphi)$, $\exists I \in \mathrm{mod}(\psi)$ such that

$$\forall I' \in \mathrm{mod}(\psi) \; d(N, I) <_\psi^\lhd d(M, I')$$

Therefore, $\forall N \in \mathrm{mod}(\varphi)$, $\forall I \in \mathrm{mod}(\psi)$, $\exists I' \in \mathrm{mod}(\psi)$, such that $d(M, I') \leq_\psi^\lhd d(N, I)$. By Proposition 17, $M \in \mathrm{mod}^*(\mathrm{band}(\psi)\lhd\mathrm{band}(\varphi))$.

Proposition 25. *If* $\mathrm{mod}^*(\Delta) = \emptyset$, *then* $\mathrm{mod}^*(\Delta\lhd\Phi) = \mathrm{mod}^*(\Phi)$

It could be argued that the revision operator is then too restrictive in the cases where the previous knowledge base ψ is inconsistent, but we believe that without extra structural information about ψ, no information can be reasonably "recovered" from it.

Corollary 26. *If* $\mathrm{mod}^*(\Delta) \cap \mathrm{mod}^*(\Phi) \neq \emptyset$, *then* $\mathrm{mod}^*(\Delta\lhd\Phi) = \mathrm{mod}^*(\Delta\lhd\Phi)$.

Proof. By **(R2)**, $\mathrm{mod}^*(\Delta\lhd\Phi) = \mathrm{mod}^*(\Delta) \cap \mathrm{mod}^*(\Phi) = \mathrm{mod}^*(\Phi\lhd\Delta)$.

Proposition 27. *Suppose that* $\mathrm{mod}^*(\Delta)$ *and* $\mathrm{mod}^*(\Phi)$ *are both non-empty. If* $\mathrm{mod}^*(\Delta\lhd\Phi) = \mathrm{mod}^*(\Phi\lhd\Delta)$, *then* $\mathrm{mod}^*(\Delta) \cap \mathrm{mod}^*(\Phi) \neq \emptyset$.

Proof. If $\mathrm{mod}^*(\Delta) \neq \emptyset$, then by (R3) $\exists M \in \mathrm{mod}^*(\Phi\lhd\Delta)$. By (R1) $M \in \mathrm{mod}^*(\Delta)$. Since $\mathrm{mod}^*(\Phi\lhd\Delta) \subseteq \mathrm{mod}^*(\Delta\lhd\Phi)$, it follows that $M \in \mathrm{mod}^*(\Delta\lhd\Phi)$. By (R1) again, $M \in \mathrm{mod}^*(\Phi)$, and hence $M \in \mathrm{mod}^*(\Delta) \cap \mathrm{mod}^*(\Phi)$.

The two previous results establish the cases when $\lhd$ is commutative. Any set of sentences Γ, such that $\mathrm{mod}(\Gamma) \neq \emptyset$ can be represented as a series of revisions of its elements in any order. The models of the series of revisions will be exactly those of Γ.

Example 1 Revisions of bands. ① When the sentences are consistent with each other, the order of the revision is irrelevant. *Revision of* $\{p\}$ *by* $\{\neg p, q\}$ *and revision of* $\{\neg p, q\}$ *by* $\{p\}$:

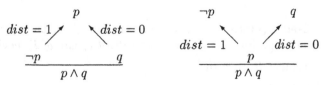

② If the sentences are inconsistent with each other, then the sentence which revises has priority over the revised one.

Revision of $\{\neg p\}$ by $\{p \wedge q\}$ and revision of $\{p \wedge q\}$ by $\{\neg p\}$:

$$\begin{array}{c} \neg p \\ dist = 1 \Big\uparrow \\ \hline \begin{array}{c} p \wedge q \\ p \wedge q \end{array} \end{array} \qquad\qquad \begin{array}{c} p \wedge q \\ dist = 1 \Big\uparrow \\ \hline \begin{array}{c} \neg p \\ \neg p \wedge q \end{array} \end{array}$$

③ *Revision of $p \wedge q$ by $\neg p \vee \neg q$ ($\{p \wedge q\} \triangleleft \{\neg p, \neg q\}$):*

$$\begin{array}{c} p \wedge q \\ dist = 1 \nearrow \qquad \nwarrow dist = 1 \\ \hline \begin{array}{cc} \neg p & \neg q \end{array} \\ (\neg p \wedge q) \vee (p \wedge \neg q) \end{array}$$

In this case, each possible combination of disjuncts of the two sentences have the same distance. Thus, both of them are kept as the result of the revision. This corresponds to the two intuitively correct attitudes: either to accept $\neg p$, rejecting the belief in p but keeping the belief in q, or to accept $\neg q$, keep the belief in p and retract the belief in q: *fairness* according to [2].

Definition 28. Let ψ and φ be two formulae of propositional logic, and $\mathrm{band}(\psi)$ and $\mathrm{band}(\neg\varphi)$, two bands representing ψ and $\neg\varphi$, respectively. The contraction of $\mathrm{band}(\psi)$ by $\mathrm{band}(\varphi)$, $\mathrm{band}(\psi) \blacktriangleleft \mathrm{band}(\varphi)$, is the band

$$\mathrm{band}(\psi) \blacktriangleleft \mathrm{band}(\varphi) = \mathrm{band}(\psi) \cup (\mathrm{band}(\psi) \triangleleft \mathrm{band}(\neg\varphi)).$$

It can be verified that $\blacktriangleleft$ verifies the AGM postulates for Contractions.

5 Updates of Bands

The basic difference between the combination of disjuncts described in the previous section and the one we present now is that the former ignores the disjunct with less priority when it is contradictory, whereas the latter ignores the one with higher priority. This reflects a property of Updates described by the following postulate (see [7]): **(U2)** If K implies A, then $K \diamond A$ is equivalent to K.

(U2) is based on the assumption that if the description of the world is inconsistent, then there is no point in "guessing" how the world looks like after the execution of an action. That is, not by using Updates, as was very well pointed out in [7].

Definition 29 Prioritised Update of Disjuncts. Let $P = \bigwedge_i p_i$ and $Q = \bigwedge_j q_j$ be two disjuncts of propositional logic. The *prioritised update of P by Q*, denoted $P \widehat{\diamond} Q$, is a new disjunct R, such that

$$R = \begin{cases} P, & \text{if } P \text{ is unsatisfiable} \\ \bigwedge_j q_j \bigwedge \{p_i | p_i, \neg p_i \text{ doesn't appear among the } q_j\text{'s}\}, & \text{otherwise} \end{cases}$$

Proposition 30. $\widehat{PQ} \vdash Q$.

The revision of bands could be straightforwardly defined because in Belief Revision one is concerned with minimising the distance between any two given models of the two sentences. However, in an Update of ψ by φ, every model of ψ represents a possible state of the world that has to be taken into account in the operation. This assumption was captured by one of the postulates, called the *disjunction rule*: **(U8)** $(K_1 \vee K_2) \diamond A \leftrightarrow (K_1 \diamond A) \vee (K_2 \diamond A)$.

It is not possible to perform this syntactically without first finding a suitable representation of the classes of models of ψ relevant to the operation.

Remark. We shall use pq to denote a model which satisfies p and q; $\bar{p}q$, a model which satisfies $\neg p$ and q, and so forth.

Example 2 Updating bands.

In the Update of ψ by φ above, we cannot simply consider the distance between a disjunct in band(ψ) and the disjuncts in band(φ), because each disjunct in band(ψ) represents possibly many models of ψ. Suppose **L** over $\{p, q\}$. The disjunct p in ψ represents two models, one satisfying $p \wedge q$ and another one satisfying $p \wedge \neg q$. The only model of $\neg p \vee q$ with minimum distance w.r.t. to pq is pq itself. However, the models of $\neg p \vee q$ with minimum distance w.r.t. $p\bar{q}$ are $\overline{pq}$ and pq ($dist = 1$). The result of the update should then be the union of all such models. That is, $\{\overline{pq}, pq\}$. A simple combination of all the disjuncts in band(ψ) with the disjuncts in band(φ) with minimum distance would result just in the model pq, since $dist(p, q) = 0$. But then we would be neglecting the model $p\bar{q}$, implicitly represented in the disjunct p.

In order to perform this operation syntactically, we need to represent explicitly some classes of models of each of the disjuncts in band(ψ) relevant to compute the update. This can be accomplished by "complementing" band(ψ) with the missing propositional variables appearing in φ.

Definition 31 Complement of a disjunct w.r.t. a band. Let P be a disjunct and Δ a band, and $Q = \{q_1, \ldots, q_k\}$ an enumeration of the propositional variables appearing in Δ but not in P. We define

$$\mathbb{P}_0^\Delta = \{P\}$$
$$\mathbb{P}_i^\Delta = \{P' \wedge q_i, P' \wedge \neg q_i\}, \text{ such that } P' \in \mathbb{P}_{i-1}^\Delta \text{ and } q_i \in Q$$

The complement of P w.r.t. Δ and enumeration Q, $\mathbb{C}_\Delta^Q(P)$ is the set $\mathbb{P}_k^\Delta$, as defined above, or $\mathbb{P}_0^\Delta$, if Q is empty.

Proposition 32. *Let P be a disjunct and Δ a band, and $\mathbf{C}^{\mathcal{Q}}_{\Delta}(P)$ the complement of P w.r.t. Δ and some enumeration $\mathcal{Q} = \{q_1, \ldots, q_k\}$ of the propositional variables appearing in Δ but not in P.* $\mathrm{mod}^*(\mathbf{C}^{\mathcal{Q}}_{\Delta}(P)) = \mathrm{mod}(P)$.

Remark. Since the particular enumeration $\mathcal{Q}$ chosen to compute the complement of a disjunct w.r.t. a band is not relevant to the properties we are interested in, we assume there exists such a suitable enumeration and the reference to it will be dropped in the remaining of the paper.

Definition 33 Complement of a band w.r.t. another band. Let Φ and Δ be two bands. The complement of Δ w.r.t. Φ is the band

$$\mathbf{C}_{\Phi}(\Delta) = \bigcup_{P \in \Delta} \mathbf{C}_{\Phi}(P)$$

where each $\mathbf{C}_{\Phi}(P)$ is the complement of P w.r.t. Φ.

Proposition 34. $\mathrm{mod}^*(\mathbf{C}_{\Phi}(\Delta)) = \mathrm{mod}^*(\Delta)$.

Definition 35 Updates of Bands. Let Δ and Φ be two bands. The *update of Δ by Φ, $\Delta \ominus \Phi$*, is a new band Γ such that:

$$\Gamma = \{\widehat{PQ} \mid P \in \mathbf{C}_{\Phi}(\Delta),\ Q \in \Phi,\ \text{and}\ dist(P,Q)\ \text{is minimum}\}$$

From the definition of Updates of bands we can prove the following properties of the operator $\ominus$.

Proposition 36. *Suppose $M \in \mathrm{mod}(\varphi)$. $M \in \mathrm{mod}^*(\mathrm{band}(\psi) \ominus \mathrm{band}(\varphi))$ iff $\exists I \in \mathrm{mod}(P)$, for some $P \in \mathbf{C}_{\mathrm{band}(\varphi)}(\mathrm{band}(\psi))$, such that $\forall I' \in \mathrm{mod}(\varphi)$, $\forall I'' \in \mathrm{mod}(P)$, $d(M,I) \leq d(I',I'')$.*

Proof. ($\Rightarrow$) If $M \in \mathrm{mod}(\varphi)$, $M \Vdash Q$, for some $Q \in \mathrm{band}(\varphi)$. Let I be such that $I \Vdash P$ for some $P \in \mathbf{C}_{\mathrm{band}(\varphi)}(\mathrm{band}(\psi))$ and $\forall I' \in \mathrm{mod}(\varphi)$, $\forall I'' \in \mathrm{mod}(P)$, $d(M,I) \leq d(I',I'')$, it follows that $dist(P,Q) \leq dist(P,Q')$, for all $Q' \in \mathrm{band}(\varphi)$. Thus, $\widehat{PQ} \in \mathrm{band}(\psi) \ominus \mathrm{band}(\varphi)$, and hence $M \in \mathrm{mod}^*(\mathrm{band}(\psi) \ominus \mathrm{band}(\varphi))$.
($\Leftarrow$) If $M \in \mathrm{mod}^*(\mathrm{band}(\psi) \ominus \mathrm{band}(\varphi))$, then

$$M \Vdash \widehat{PQ},\ \text{for some}\ P \in \mathbf{C}_{\mathrm{band}(\varphi)}(\mathrm{band}(\psi)),\ Q \in \mathrm{band}(\varphi)$$

such that $dist(P,Q)$ is minimum. This Q is such that it provides the combination with that particular P with minimum distance between models of P and models of all $Q \in \mathrm{band}(\varphi)$, and hence of models of P and φ. $M \Vdash Q$. Thus, $\exists I \in \mathrm{mod}(P)$ such that $\forall I' \in \mathrm{mod}(\varphi)$ and $\forall I'' \in \mathrm{mod}(P)$, $dist(M,I) \leq dist(I',I'')$.

Corollary 37. *Suppose $\mathrm{mod}^*(\mathrm{band}(\psi) \ominus \mathrm{band}(\varphi)) \neq \emptyset$ and that $M \in \mathrm{mod}(\varphi)$. If $M \notin \mathrm{mod}^*(\mathrm{band}(\psi) \ominus \mathrm{band}(\varphi))$, then $\forall P \in \mathbf{C}_{\mathrm{band}(\varphi)}(\mathrm{band}(\psi))$, $\exists I \in \mathrm{mod}(P)$ and $\exists N \in \mathrm{mod}(\varphi)$, such that $\forall I' \in \mathrm{mod}(P)$, $d(N,I) < d(M,I')$.*

5.1 Updates of Bands and the Postulates for Updates

In [7], the distinction between Belief Revision and Updates is formally stated and some postulates for Updates are proposed. As for Revisions, a semantical characterisation of all update operators defined in terms of partial orders (or pre-orders) satisfying these postulates is given.

Definition 38 Faithful Assignment for Updates - [7]. A *faithful assignment for updates* is a function mapping each interpretation I to a partial pre-order $\leq_I$, such that for any $I, I' \in \mathcal{I}$, if $I \neq I'$, then $I <_I I'$.

Theorem 39. *[7] An update operator $\diamond$ satisfies the postulates for Updates if there exists a faithful assignment that maps each interpretation I to a partial pre-order $\leq_I$, such that $\mathrm{mod}(K \diamond A) = \bigcup_{I \in \, \mathrm{mod}\,(K)} \min(\mathrm{mod}(A), \leq_I)$.*

Our objective in the next section will be to define a faithful assignment based on our distance function and prove that our update operator verifies the postulates for Updates via Theorem 39.

We need first to define a function ξ which assigns every interpretation I to a partial pre-order $\leq_I$, using our distance function:

Definition 40. For each $I \in \mathcal{I}$, $\xi(I) = \{\leq_I, \mathcal{I}\}$, where for any $M, N \in \mathcal{I}$

$$M \leq_I N \text{ iff } d(M, I) \leq d(N, I)$$

Proposition 41. ξ *is a faithful assignment for updates.*

Proof. Clearly, $\leq_I$ is a pre-order. Now, notice that if $N \neq I$, then $d(N, I) > 0$. On the other hand, $d(I, I) = 0$, and hence $I <_I N$, for any $N \in \mathcal{I}$, such that $N \neq I$. □

Theorem 42. *The update operation defined by $\otimes$ verifies the postulates for Updates defined in [7].*

Proof. We are left to show that

$$\mathrm{mod}^*(\mathrm{band}(\psi) \otimes \mathrm{band}(\varphi)) = \bigcup_{I \in \, \mathrm{mod}\,(\psi)} \min(\mathrm{mod}(\varphi), \leq_I).$$

($\subseteq$) This part is easy. By Proposition 36, If $M \in \mathrm{mod}^*(\mathrm{band}(\psi) \otimes \mathrm{band}(\varphi))$, then $\exists P \in \mathfrak{C}_{\mathrm{band}(\varphi)}(\mathrm{band}(\psi))$ and $\exists I \in \mathrm{mod}(P)$, such that $\forall N \in \mathrm{mod}(\varphi)$ and $\forall I' \in \mathrm{mod}(P)$ $d(M, I) \leq d(N, I')$. In particular, $\forall N \in \mathrm{mod}(\varphi)$ $d(M, I) \leq d(N, I)$, and thus $M \in \min(\mathrm{mod}(\varphi), \leq_I)$.

($\supseteq$) Suppose $M \in \bigcup_{I \in \, \mathrm{mod}\,(\psi)} \min(\mathrm{mod}(\varphi), \leq_I)$, but $M \notin \mathrm{mod}^*(\mathrm{band}(\psi) \otimes \mathrm{band}(\varphi))$. By assumption, $M \in \min(\mathrm{mod}(\varphi), \leq_I)$, for some $I \in \mathrm{mod}(\psi)$, and then $\exists I \Vdash P$, for some $P \in \mathfrak{C}_{\mathrm{band}(\varphi)}(\mathrm{band}(\psi))$. By Definition 40,

$$\forall N \in \mathrm{mod}(\varphi) \ d(M, I) \leq d(N, I) \tag{5}$$

Since $M \notin \text{mod}^*(\text{band}(\psi) \ominus \text{band}(\varphi))$, then $\forall Q \in \mathbb{C}_{\text{band}(\varphi)}(\text{band}(\psi))$, $\exists J \in \text{mod}(Q)$ and $\exists N \in \text{mod}(\varphi)$, such that $\forall J' \in \text{mod}(Q)$

$$d(N, J) < d(M, J') \tag{6}$$

Consider the particular P of (5), and take $J' = I$. From (6), $\exists J \in \text{mod}(P)$, and $\exists N \in \text{mod}(\varphi)$, such that

$$d(N, J) < d(M, I) \tag{7}$$

But $P \in \mathbb{C}_{\text{band}(\varphi)}(\text{band}(\psi))$, thus J, I must agree in every truth-value of the propositional variables in P (which include the propositional variables in φ). It is easy to see that $d(N, J) = d(N, I)$, and then $d(N, I) < d(M, I)$, a contradiction.

Example 3 Updates of bands. ① We start with an example from Winslett's scenario in [13]. Suppose we have a room with exactly two objects: a magazine and a newspaper. All we know is that either the magazine or the newspaper is on the floor. We then send a robot to the room and ask him to pick up the magazine from the floor.

We use m to represent the fact that the magazine is on the floor and n to represent the fact that the newspaper is on the floor. The scenario before the execution of the action can then be represented by the sentence $m \vee n$ and the post-condition of the action by $\neg m$.

$$
\begin{array}{cc}
m & n \\
\end{array}
$$

Update of $m \vee n$ by $\neg m$: $\quad dist = 1 \diagdown \quad \diagup dist = 0$

$$\frac{\neg m}{\neg m \vee (\neg m \wedge n)}$$

All we are supposed to know after the execution of the action is that the magazine is not on the floor; the newspaper may or may not be. Notice that had we used revisions, the result would have been $\neg m \wedge n$, since $m \vee n$ is consistent with $\neg m$. **(U8)** ensures that each possible world is given equal consideration.

② Following the reasoning, all we know now is that the magazine is not on the floor ($\neg m$). We then tell the robot to put either the magazine or the newspaper on the floor ($m \vee n$):

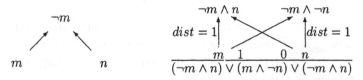

In this case, we need to complement the band representing the first sentence w.r.t. the band representing the second one. The results can be seen on the top of right-hand side of the figure above. In both worlds, the magazine is not on the floor. However, they include the two possible locations for the newspaper w.r.t. the floor.

In the first world, nothing has to be done, since the post-condition of the action is already satisfied[5] (the newspaper was already on the floor anyway). The second one represents the world in which neither the magazine nor the newspaper are on the floor, one of them is then put on the floor (but not both as this would imply distance 2 w.r.t. this state of the world).

6 Related Work

In this work we have presented a formalism to accomplish Belief Revision and Updates syntactically and in a uniform way. We will now make some comments about related work found in the literature.

In [3], Del Val and Shoham present a framework to reason about Belief Revision and Updates. The formalism uses the language of situation calculus to explicitly represent time and circumscription to capture the notion of minimality of change. That work and ours have little in common, apart from the fact of combining the two kinds of belief change in a single formalism.

Another related work is Winslett's Possible Models Approach (PMA). It is well known that Winslett's PMA verifies the postulates for Updates. In fact, it is one of the first works to point out the importance of considering each model of the belief state independently when reasoning about actions. PMA is defined semantically. Moreover, the update operation defined here provides different results, for it uses a different distance notion to evaluate change. PMA's $Diff$ results in a partial order on the set of interpretations w.r.t. to a given world ψ, whereas our $dist$ provides a total ordering. If $Diff$ is used in our approach instead of $dist$, we would have to either modify the revision operator or give up some of the postulates for Belief Revision.

Dalal's revision operation [2] and ours are based on the same distance function and give the same results provided the database is consistent[6]. The cost of Dalal's procedure is increased by the fact that it requires a consistency check in every step of the generalization, whereas in ours this is reduced to a single check of the consistency of disjuncts. Apart from the cost of converting the formulae to DNF, the cost of our procedure for revision is linear both in the sizes of the database and the revise formula.

In [12], Del Val presents an operator for computing updates based on the PMA approach. The database can be either in Disjunctive, Conjunctive or Negation Normal forms, but the algorithm for the update itself receives formulae in DNF as inputs. He also provides an analysis of improvements that can be made if the database is stored in Conjunctive or Negation normal forms, by pointing out that only part of the database is relevant for the update. The same ideas are valid here. Our procedure for update is conceptually simpler, and even though it does not compute the update in the sense of PMA, it could be easily modified to

---

[5] This is what is implied by (U2), and corresponds to what Brewka-Hertzberg called *laziness* in [1]. In some cases, *laziness* may not be adequate.

[6] If the database formula is contradictory, then Dalal's operator doesn't provide the expected results

provide such results. First analysis on the complexity of the update defined here suggests that it is linear in the size of the database and exponential in the size of the update formula. However, further investigation on how to achieve more efficient algorithms remains to be done.

7   Conclusions and Future Work

In this work, we proposed two operators for belief change. One of the operators performs belief revision in the AGM sense. This operator can also be used to obtain Contractions. Since Expansions are a special kind of Revisions, the three basic epistemological changes can be accomplished. We have also defined an Update operator which can be used to model some effects of the execution of actions.

One of the main advantages of the methodology lies on its simplicity. The operations are based on simple superimposition of literals and on a distance function used to evaluate the degree of change caused. The consistency check is reduced to checking the consistency of disjuncts of propositional logic in both cases. The operators can be iterated and combined in a single framework for information change.

Other operations have also been investigated and defined, even though not included in this paper, due to limitation in space. They can be found in [10]. One example is an operator to model the effects of execution of ambiguous actions [1], which are actions with non-deterministic post-conditions, such as "flipping a coin". The effects of ambiguous actions cannot be modelled by ordinary update operators complying with the postulates for Updates.

We do not believe that a whole belief state of an agent can be simplified into a single sentence without any extra information on epistemic entrenchment and yet provide a rational change of belief. However, we do believe that the postulates for revisions provide reasonable guidelines on what to expect from rational revisions between sentences. The operators defined here are meant to be used in a more complex framework in which extra-logical information can also be represented. This information can be used, for instance, to represent priorities among sentences or indicate the time when the change occurred. If the belief base is seen as a list of sentences where sentences later in list have priority over those appearing earlier (e.g., as in [11]), revisions (or updates) of the belief state could be obtained by simply appending the new sentence to the list. Since the operators verify the postulates, this guarantees that new information *can* be accepted in the new belief state (if so desired). This also provides a rather trivial way of generating a new epistemic entrenchment ordering from the original one.

More sophisticated changes of belief could be obtained by defining different procedures according to the extra-logical information available. The "belief state" would be the result of a number of operations on a (structured) set of bands.

Acknowledgments

We would like to thank Mark Ryan for his comments on an early version of this paper. Odinaldo Rodrigues is supported by CNPq – Conselho Nacional de Desenvolvimento Científico e Tecnológico/Brasil. Dov Gabbay is supported by EPSRC, GR/K57268 (Proof Methods for Temporal Logics of Knowledge and Belief).

References

1. Gerhard Brewka and Joachim Hertzberg. How to do things with worlds: on formalizing actions and plans. *Journal of Logic and Computation*, 3(5):517–532, 1993.
2. M. Dalal. Investigations into a theory of knowledge base revision: Preliminary report. *Proceedings of the 7th National Conference on Artificial Intelligence*, pages 475–479, 1988.
3. A. del Val and Y. Shoham. A unified view of belief revision and update. *Journal of Logic and Computation*, 4(5):797–810, 1994.
4. H. B. Enderton. *A Mathematical Introduction to Logic*. Academic Press, New York, 1972.
5. Peter Gärdenfors. *Knowledge in Flux: Modeling the Dynamics of Epistemic States*. A Bradford Book - The MIT Press, Cambridge, Massachusetts - London, England, 1988.
6. Hirofumi Katsuno and Alberto O. Mendelzon. Propositional knowledge base revision and minimal change. *Artificial Intelligence*, 52:263–294, 1991.
7. Hirofumi Katsuno and Alberto O. Mendelzon. On the difference between updating a knowledge base and revising it. *Belief Revision*, pages 183–203, 1992.
8. David K. Lewis. *Counterfactuals*. Harvard University Press, 1973.
9. David Makinson. Five faces of minimality. *Studia Logica*, 52:339–379, 1993.
10. Odinaldo Rodrigues. Transfer report: Mphil/phd. Imperial College of Science, Technology and Medicine, February 1996. Department of Computing.
11. Mark Ryan. *Ordered Presentation of Theories - Default Reasoning and Belief Revision*. PhD thesis, Department of Computing, Imperial College, U.K., 1992.
12. Alvaro Del Val. Computing knowledge base updates. In B. Nebel, C. Rich, and W. Swartout, editors, *Proc. Third International Conference on Principles of Knowledge Representation and Reasoning (KR '92)*, pages 740–750. Morgan Kaufmann, San Francisco, CA, 1992.
13. Marianne Winslett. Reasoning about action using a possible models approach. In *Proceedings of AAAI-88, St Pauls, MS*, pages 89–93, San Mateo, CA, 1988. Morgan Kaufmann.

A Formal Framework for Causal Modeling and Argumentation

Hector Geffner*

Depto. de Computación
Universidad Simón Bolívar
Aptdo 89000, Caracas 1080-A
Venezuela

Abstract. We develop a framework for causal modeling that features a language based on causal defeasible rules, a semantics based on order-of-magnitude probabilities, and a proof-theory based on the interaction of arguments. The framework extends and integrates logical, probabilistic and procedural modeling languages such as logic programs with negation as failure, Qualitative Bayesian Networks and Axelrod's cognitive maps.

1 Introduction

People are good at reasoning with causal assertions like:

1. If I insert a dollar in the dispenser machine, I'll get a drink
2. If I turn the heater on, the room will be warm
3. If demand for a good increases, the price for the good will increase
4. If I put sugar in my coffee, it will taste good
5. If I scratch a match, it will light

We often draw predictions, construct explanations and make decisions based on assertions of this type. In this paper, we will be interested in developing a formal system with these capabilities. The goal is twofold: to understand how people reason with causal assertions and to understand how to construct systems that reason that way. In both cases, we need precise representation languages for describing what is *explicitly* known, precise semantics for describing what is *implicit* in those descriptions, and inference procedures to make explicit what is implicit. A framework of this sort will be *adequate* if the representation language allows us to express what is known in a natural way, the semantics captures closely the inferences that people deem valid, and the algorithms capture closely the way people arrive at those inferences.

In this paper we will develop a representational framework that addresses each one of these aspects. The main difficulty to be faced is that neither the languages based on logic nor the languages based on probabilities are adequate

---

* Mailing address from US and Europe: H. Geffner, Bamco CCS 144-00, P.O.Box 02-5322, Miami Fla 33102-5322, USA. Email: hector@usb.ve.

for the task. The rules above cannot be modeled as logical implications because, first of all, such rules are all *defeasible:* i.e., their antecedents may be true, but their consequents may be false. Indeed, we want to be able to *refine* those models by *adding* new rules indicating contexts where the older rules do not apply, e.g.,

1. If the dispenser machine is broken, I'll not get a drink
2. If the the windows are wide open, the room will not get warm
3. If prices are state controlled, prices will not go up
4. If I put oil in my coffee, it will taste bad
5. If I scratch a wet match, it will not light

Yet defeasibility is not the only difficulty to be dealt with. The other one is *causality.* Indeed, the causal directionality of rules is often crucial for determining whether the rules should be applicable or not. As Pearl argued in [14], chaining one causal rule after another is often reasonable (e.g., the sprinkler was on, therefore the grass is wet, therefore the grass is shining); yet chaining an *evidential* rule after a causal rule often is not (e.g., sprinkler was on, therefore the grass is wet, therefore it rained).

These difficulties suggest that something more than logic is needed to understand the meaning and use of causal rules. The standard alternative is to use probabilities; yet probabilities involve numbers while the rules above involve no numbers at all, and similarly, the rules provide a reason to *accept* the consequents, while in probabilities beliefs are accepted only to a certain *degree.*

Thus neither logic nor probabilities provide good models for reasoning with causal defeasible rules. This of course, does not imply that logic and probabilities are of no use; quite the opposite. As we will show below, the most fruitful interpretations of defeasible causal rules have all been based in either logic or probabilities, but logic and probabilities understood in a broader sense than usual. In this paper we develop a framework that extends and integrates these interpretations.

We will start with a review of three basic interpretations of defeasible rules: preferential logics, the logic of high probabilities, and Spohn's κ functions. These three interpretations provide equivalent frameworks for reasoning with defeasible rules yet none of them addresses the causal aspects of those rules. An extension that addresses these aspects was developed by Goldszmidt and Pearl in [8]. In their interpretation, the causal rules are understood as defining qualitative analogues of Probabilistic Bayesian Network (Sections 4), with probabilities replaced by Spohn's κ measures (Section 5). Here we refine this proposal (Section 6), contrast the result with other accounts of causal reasoning, and develop a suitable inference algorithm in the form of an argumentation system (Section 8). We also discuss the use of the proposed framework for modeling (Section 9) and possible extensions (Section 10).

2   High Probabilities and Model Preference

Two ways for interpreting *defeasible* rules like the ones above rely on the assumption that worlds are ordered by either a 'preference' relation[2] or probability weights. In both cases, rules of the form $p \to q$ are understood as constraining such orderings in such a way that q must true in the preferred or most probable worlds that satisfy p. Since the set of preferred or most probable worlds may shift when an additional piece of information r is obtained, it's perfectly possible for q to be a consequence of p alone, and yet for $\neg q$ to be a consequence of p and r. This is the *non-monotonicity* property that distinguishes these logics from classical ones.

To formalize these interpretations, let us first define the formal objects we will be dealing with. First, we will assume that $p, q, r, \ldots$ all refer to formulas in a finite propositional language $\mathcal{L}$. We will also assume that the descriptions or theories are encoded by means of two types of expressions: a set W of formulas and a set D of (defeasible) rules. Each rule will be represented by an expression of the form $p \to q$ where p and q are formulas.

The *semantic problem* is the problem of defining a precise and meaningful characterization of the acceptable consequences of such theories. With no defeasible rules, the standard definition in logic suffices: the formulas entailed are the formulas that are true in *all* models of W. With defeasible rules, preferential logics replace the notion of 'entailment' in all models by the weaker notion of entailment in all *preferred* models [19, 9, 13]:

Definition 1. A theory $T = \langle W, D \rangle$ *preferentially entails* p iff p is true in all preferred models of W relative to any preference ordering on worlds compatible with D.

In this definition, the *preferred models* refer to the models that are not less preferred than any other model, and the preference orderings compatible with D refer to the orderings in which the consequent of any rule $p \to q$ in D is true in the preferred models of p.

The probabilistic interpretation of defeasible theories is defined in an analogous way with probability distributions playing the role of preference orderings [1, 15, 6]:

Definition 2. A proposition p is *probabilistically entailed* by a theory $T = \langle W, D \rangle$ when for any $\delta > 0$ there exists an $\epsilon > 0$, such that $P(p|W) > 1 - \delta$ holds whenever $P(q|p) > 1 - \epsilon$ holds for every rule $p \to q$ in D.

As shown by Adams [1] and Lehmann and Magidor [10], preferential and probabilistic entailment coincide, and both accept an elegant and simple proof-theory made out of a few qualitative rules of inference [9, 6].

These two interpretations capture important aspects of rules like the implicit preference for rules that are based on more specific information [6]. On the other

---

[2] A preference relation is an irreflexive and transitive binary relation.

hand, they embed important limitations as neither one captures independence assumptions. For example, they will not chain two rules $p \to q$ and $q \to r$ to derive from r from p because they are unable to determine that p does not affect r when q is given.

3 Qualitative Probabilities

A particular type of preference relation results from *assigning a non-negative integer to each world* so that worlds with lower numbers are preferred to worlds with larger numbers. Those numbers can be understood as some type of *plausibility measures* with lower numbers assigned to more plausible worlds and higher numbers assigned to less plausible worlds. Interestingly, if we focus on such numerical preference relations, the resulting semantics is equivalent to the two interpretations above [10].

Let κ denote the mapping from worlds to their plausibility measures. Then, following Spohn [20], the prior and conditional *plausibilities of formulas* can be defined as:[3]

$$\kappa(p) = \min_{w \models p} \kappa(w) \;,\;\; \kappa(p \vee \neg p) = 0 \;,\;\; \kappa(p|q) = \kappa(p \wedge q) - \kappa(q)$$

These definitions are similar the ones defining the calculus of probabilities:

$$P(p) = \sum_{w \models p} P(w) \;,\;\; P(p \vee \neg p) = 1 \;,\;\; P(p|q) = P(p \wedge q) / P(q)$$

with products replaced by sums and sums replaced by minimizations. Indeed, Spohn [20] shows that the κ-calculus can be understood as an order-of-magnitude approximation of the probability calculus.[4] More precisely, if the order-of-magnitude approximation of a probability measure x is defined as the integer i such that $\epsilon^{i+1} \leq x < \epsilon^i$, then *in the limit,* when ϵ goes to zero, $\kappa(p)$ and $\kappa(p|q)$ will be order-of-magnitude approximations of $P(p)$ and $P(p|q)$ respectively, when for each world w, $\kappa(w)$ is an order-of-magnitude approximation of $P(w)$ [20].

Goldszmidt and Pearl refer to the κ measures as *qualitative* probabilities and show that some of the limitations of the interpretations above are overcome when the systems of causal rules are understood as the *qualitative* analogue of Probabilistic Bayesian Networks, with probabilities replaced by κ measures. To understand this proposal, we take a quick look first at Bayesian Networks.

4 Bayesian Networks

Bayesian Networks are a graphical language for representing probabilistic knowledge comprised of three elements [15]: a set of variables, each having a finite set of

---

[3] We also include ∞ in the range of the function κ and define $\kappa(q) = \infty$ and $\kappa(p|q) = 0$ when q is unsatisfiable.

[4] The κ calculus is also in close correspondence with possibility theory; see [4].

possible values, a directed acyclic graph displaying the causal influences among those variables,[5] and probability matrices encoding the conditional probability of each variable given its parents in the graph (its direct causes).

For example, the Bayesian Network displayed in Fig. 1 involves the five variables A, B, C, D, and F, each having two values, *true* and *false*. The full specification of that network requires the probabilities $P(A)$, $P(B|A)$, $P(C|A)$, $P(D|B,C)$ and $P(F|C)$ for each of the values of the variables. Since probabilities of complementary propositions have to add up to one, it is actually sufficient to provide the probability of each variable taking the value *true* given each of the possible value combinations of its parents. That makes for $1 + 2 + 2 + 4 + 2 = 11$ probabilities.

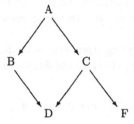

Fig. 1. A Bayesian Network

Once a Bayesian Network is so defined, it has all the information that is required to answer *any* query involving its variables; e.g., what is the probability of $A = true$ given that $B = true$ and $F = false$; what is the probability that both A and B are *true*, given $D = false$, etc. This is because the graph and the numbers together define the unique joint probability distribution over all the variables in the net.

Surprisingly, the textbook approach to specify a joint probability distribution among N binary variables requires $2^N - 1$ probabilities, which for $N = 5$ means 31 numbers.

This raises the question: how can a Bayesian Network with 11 parameters represent a probability distribution that requires 31 parameters?

The answer to this question is that the 20 parameters that are not explicitly in the net, are conveniently encoded in the *structure* of the net. The structure of the network encodes what in probability theory are called *independence assertions*; namely, assertions stating that the probability of certain variable X given information about variables Y_1, Y_2, ..., is not affected when information about other variables Z_1, Z_2, ... is given. In that case, it is said that variable X

---

[5] There is an alternative way for constructing Bayesian Networks which does not rest on causal considerations (see [15]). This alternative method, in which the graph is constructed incrementally by adding one variable at a time, is seldom used in practice as it requires too many parameters and too many independence judgements.

is *independent* of Z_1, Z_2, ... given Y_1, Y_2, ..., and it is written as:

$$P(X|Y_1, Y_2, \ldots) = P(X|Y_1, Y_2, \ldots, Z_1, Z_2, \ldots)$$

These independence assertions together with the probabilities that are explicitly provided are sufficient to completely determine all the numbers that define the joint probability distribution.

There are three different but equivalent criteria for uncovering these independences (see [15] for more details).

Criterion 1. The network determines the formula that defines the joint probability distribution. If the variables are named X_1, X_2, ..., X_n and we let $pa(X)$ stand for the parents of X in the net, then the formula is:[6]

$$P(X_1, X_2, \ldots, X_n) = \prod_{i=1}^{n} P(X_i | pa(X_i))$$

Criterion 2. The criterion above does not explicitly tells us which are the independences in the graph, even though it provides us with all the necessary information to find that out. A more explicit answer is that *the parents $pa(X)$ of each variable X in the net contain all the information that is needed to estimate the probability of X when no effects of X have been observed.* In the literature on Bayesian Networks this is normally phrased as saying that each variable X is independent of its non-descendants given its parents, written as:[7]

$$P(X | pa(X)) = P(X | pa(X), non\text{-}desc(X))$$

Criterion 3. A still more explicit answer is provided by the graphical criterion named *d-separation*. Consider the paths that connect two variables X and Y in the graph (ignoring the directions of arrows). X is *independent* of Y given a set of variables S if every such path is *blocked*, where a path is blocked when three consecutive variables A, B and C along the path 1) are connected by a subpath of the form $A \rightarrow B \leftarrow C$ and S gives no evidence about B,[8] or 2) are connected by a different type of subpath and B belongs to S.

Applying this last criterion to the graph in Fig. 1 it is simple to see that B is not independent of F a priori, yet B is independent of F given A, and B is not independent of F given both A and D.

The three criteria can all be shown to be sound and complete when the network is constructed by the incremental method mentioned above. This method, however, is seldom used as it is not practical. Most of the time, the network is constructed to reflect the causal influences among the variables, and in such case, the three criteria above amount to a *convention for reading independence information from causal information.* This convention, however, has been found very useful in practice as the independences so extracted are often in good correspondence with the independences that the user has in mind.

---

[6] When X has not parents, $P(X|pa(X))$ stands for the prior probability $P(X)$.

[7] A variable Y is a *descendant* of variable X if there is a directed path from X to Y.

[8] S gives no evidence about B when neither B nor a descendant of B belongs to S.

5 Qualitative Bayesian Network

Goldszmidt's and Pearl's [8] proposal for interpreting causal rules is an extension of the logic of high-probabilities that incorporates the above convention for reading independence information from causal information.

More precisely, the causal rules are assumed to determine a causal network that is defined by associating with each propositional symbol x a boolean variable X and by drawing a link from a variable X to a variable Y whenever there is a causal rule that involves the symbol x in the antecedent and the symbol y in the consequent. (causal rule systems that give rise to cycles are excluded).

Provided with this network, they consider among the κ functions that are compatible with the rules (i.e., the κ's that make $\kappa(\neg q|p) > 0$ for each rule $p \to q$ in D), only those that comply with the qualitative analogue of Criteria 1 above:[9]

$$\kappa(X_1, X_2, \ldots, X_n) = \sum_{i=1}^{n} \kappa(X_i \,|\, pa(X_i))$$

The plausibility functions that comply with this condition are called *stratified*. Such functions thus encode qualitative analogues of Bayesian Networks with the conditional probabilities replaced by the 'qualitative probabilities' $\kappa(X|pa(X))$. For this reason, the pairs formed by a causal graph and a stratified plausibility function are called Qualitative Bayesian Networks or QBN's [7].[10]

The *QBN semantics* defines the consequences of a causal theory $T = \langle W, D \rangle$ as the formulas p for which $\kappa(\neg p|W) > 0$ is true in all the *stratified* plausibility functions κ that are compatible with D. Since there are always many, indeed, an infinite number of such functions, a causal theory stands for a family of QBN's, and an inference is valid if it is valid in all such QBN's.

6 The Proposed Framework

The QBN semantics for causal rules is both clean and elegant, and solves many of the difficulties encountered by other approaches. Yet it has also some difficulties of its own.

First, the semantics is not entirely constructive. Even though we may have a very small language, there may be too many plausibility functions to check in order to accept or reject a particular formula. Some of the inference can be tested by making use of the d-separation criterion; yet, in other cases the d-separation criterion does not help. For example, it's not a simple matter to check whether b is a consequence of a and c given the two rules $a \to b$ and $c \to b$.

---

[9] Note that this equation is an abbreviation of all the equations that result from replacing each variable X_i by each of its possible 'values'; in this case, the literals x_i and $\neg x_i$.

[10] QBN's are not to be confused with Wellman's [23] QPN's which represent a different qualitative abstraction of Bayesian Networks.

Second, no reasonable algorithms have yet been developed for deriving semantically valid conclusions, even in the context of the simple theories just mentioned. Actually, the only sound and complete algorithms [3] are for Qualitative Bayesian Networks (QBN) not causal *theories* (which stand for families of QBN's). This is in contrast with the algorithms used in other models of causal reasoning such as inheritance nets, logic programs, and cognitive maps.

Last but not least, the semantics is too weak. In the example above, it turns out that b does *not* follow from $a \rightarrow b$ and $c \rightarrow b$ when *both* a and c have been observed. The reason is that variable B is independent of variable C given $A = true$, yet this is not an independence that can be extracted from the graph.

Here we will develop a refinement of the QBN semantics that addresses these limitations. The resulting semantics will actually bridge the QBN semantics with more standard, but also more ad-hoc models for causal reasoning. combines the best features of the logical, probabilistic and procedural worlds: a logical language, a probabilistic semantics, and an argumentation-based proof-theory. The fundamental idea is very simple though: while the QBN semantics maps systems of rules to a large family of QBN's, the proposed refinement will map the rules into a *canonical QBN* which is unique up to a certain parameter which can be provided by the user or can be assumed by default. In this canonical QBN, the so-called *gates* $\kappa(X|pa(X))$ are determined locally from the rules that relate the variable X to its parents as detailed below.

6.1 Rule Gates

Let us first make precise the rule language we will use. We assume first that rules are of the form $A \rightarrow L$ where A is a conjunction of literals and L is a literal. As before, we associate a boolean variable X with each propositional symbol x, so that x stands for $X = true$ and $\neg x$ stands for $X = false$.

We will also assume that rules have a *priority* or *strength measure* represented by a non-negative integer: the higher the number, the higher the priority. The idea is that when two rules say different things about the same variable, the higher priority rule will prevail. For example, the rule 'on holidays, people do not go to work' has priority over the rule 'on weekdays, people go to work'. Unless stated otherwise, we will be assume all rules to have priority 0.

Consider now a variable X and a particular valuation $pa_i(X)$ for its parent variables $pa(X)$. These are the variables that occur in the antecedents of the rules whose consequent is either x or $\neg x$. The valuation $pa_i(X)$ thus gives a truth value to all such antecedents. Among the rules whose antecedents are true, let us select the ones that have the highest priority and let us refer to the them as the rules for X that are *active* given $pa_i(X)$.

We then say that the valuation $pa_i(X)$ *supports* a particular value of X if $pa_i(X)$ activates a rule that asserts that value.

For example, from the rules $a \rightarrow b$ and $c \rightarrow \neg b$ with the same priorities, we obtain that $pa_1(B) = \{a, \neg c\}$ supports b, $pa_2(B) = \{\neg a, c\}$ supports $\neg b$,

$pa_3(B) = \{a, c\}$ supports both b and $\neg b$, while $pa_4(B) = \{\neg a, \neg b\}$ supports neither b nor $\neg b$.

With these notions, we define the RULE gates so that a supported literal whose complement is not supported is very likely, while a supported literal whose complement is also supported is plausible:

$$\kappa(X|pa_i(X)) = \begin{cases} 0 & \text{when value of } X \text{ supported by } pa_i(X) \\ \infty & \text{when } only \text{ a different value of } X \text{ is supported by } pa_i(X) \\ \pi(X) & \text{when } no \text{ value of } X \text{ supported by } pa_i(X) \end{cases}$$

Here $\pi(X)$ is a function that determines the plausibility of the different values of X *when the causes of X do not provide reasons to believe in either one of them.* We call π the *completion* function. For each value of X, $\pi(X)$ must be a non-negative integer, and for some value of X, $\pi(X)$ must be zero.

To understand how these gates work, consider the pair of rules $p \to q$ and $r \to \neg q$ with the same priority. The parents $pa(Q)$ of variable Q are P and R. The model says that when P is true and R is false, $\neg q$ will be very unlikely, $\kappa(\neg q|p, \neg r) = \infty$, as q will be supported but $\neg q$ will not. The opposite is true when P is false and R is true. On the other hand, when *both* P and R are true, there will be reasons for both q and $\neg q$ so the model sets both $\kappa(q|p, r)$ and $\kappa(\neg q|p, r)$ to 0, making both q and $\neg q$ plausible. Finally, when both p and r are false, neither q nor $\neg q$ will be supported, and thus, the model says that their plausibility will be determined by the completion function $\pi(Q)$. In particular, *if we want to assume that q is false by default,* we will choose a completion function such that $\pi(\neg q) = 0$ and $\pi(q) > 0$. On the other hand, if we want to make no assumptions at all, we will choose $\pi(\neg q) = \pi(q) = 0$.

6.2 The Framework: Causal Rule Systems

It's clear that once a function $\pi(X)$ is chosen for every variable X in the theory, the set of causal rules defines a *single* stratified plausibility function κ and thus a single QBN all of whose gates are RULE gates. We call such particular QBN, the *canonical* QBN. The resulting semantics is such that q is a consequence of an observation p if $\kappa(\neg q|p) > 0$ for such canonical plausibility function.

We will assume throughout the rest of the paper that the completion function is specified by the user or is assumed by default. Actually we will show that some of the familiar systems in the literature can be understood as embedding some standard completion functions π. Two such functions are the *grounded* completion function π_g, *which makes atoms false by default,* defined as $\pi_g(\neg a) = 0$ and $\pi_g(a) = 1$ for every atom a, and the *uniform* completion function π_u, that makes no assumptions, defined as $\pi_u(\neg a) = \pi_u(a) = 0$.

We will refer to the language of causal rules, plausibility functions and observations, interpreted by their canonical QBN's, as the Causal Rule System framework (CRS).

7 Properties

We will show some properties of causal rule systems by showing how they relate to some familiar systems developed in AI. Readers not familiar with such systems may want to skip this section.

Logic Programs

Consider a propositional acyclic logic program with negation as failure [11]. Take each rule in the program as a causal rule and consider the *grounded* completion function π_g above which makes atoms false by default.

Theorem 3. *The canonical model of the program corresponds to the single most plausible world of its corresponding causal rule system.*

Inheritance Networks

Consider an acyclic inheritance network Γ and the causal theory T_Γ that results from mapping each link $p \to q$ in Γ ($p \not\to q$) into a causal rule of the form $p \to q$ ($p \to \neg q$). Moreover, let's set the priorities of the rules in such a way that 'more specific' rules have higher priority.[11]

Consider now a standard inductive definition of the inheritance paths that are *supported* in Γ,[12] where $p \longrightarrow q$ stands for a positive path between p and q:

1. the empty path is *supported*
2. the path $p \longrightarrow s \to q$ is *supported* if the path $p \longrightarrow s$ is supported and for every path of the form $p \longrightarrow r \not\to q$, either $p \to q$ is 'more specific' than $r \not\to q$, or there is a supported path of the form $p \longrightarrow t \not\to r$.

Theorem 4 (Soundness). *If $p \longrightarrow q$ is a supported path in Γ, then q is a consequence of p in the causal rule system T_Γ, provided the grounded completion function π_g above.*

Theorem 5 (Completeness). *If every time two links are in conflict, one is more specific than the other, then if q is a consequence of p in T_Γ, $p \longrightarrow q$ will be a supported path.*

Evidential Reasoning

In logic programs and inheritance hierarchies, reasoning proceeds along the direction of the rules only. Causal reasoning, however, often involves reasoning in two directions: from causes to effects, in what is called *predictive* reasoning, and from effects to causes, in what is called *evidential* or *abductive* reasoning. For example, if cold and allergy are the two known causes of sneezing, from observing 'sneezing' we want to infer that either cold or allergy are present. This scenario can be modeled by means of two rules *cold* $\to$ *sneezing* and *allergy* $\to$ *sneezing* and the function π_g above. More generally:

---

[11] Determining 'specificity' in inheritance networks is a not a simple issue; e.g., see [21].

[12] Paths in Inheritance Networks mean directed paths in which the links of the form $p \not\to q$ are not followed by any other link. Positive paths mean paths where such links do not occur at all. The empty path connects a node to itself.

Theorem 6. *When all rules are of the form* $h \to m$ *for hypotheses* h *and manifestations* m, *and* π *is such that all hypotheses are equally unlikely, e.g.,* $\pi(h) = 1$, *and all manifestations without causes are impossible,* $\pi(m) = \infty$, *then the resulting worlds, having observed a set of manifestations, correspond to the* minimal cardinality diagnoses [17].

This theories could be refined in several way, e.g., to reflect that a cold is more likely than an allergy, by asserting $\pi(cold) = 1 < \pi(allergy) = 2$. In that case the resulting diagnoses would no longer be minimal in the sense above.

8   Causal Argumentation

To model how people reason it is not enough 'to get the conclusions right'; we need to get them in a reasonable way. Here we will consider a valid inference algorithm that has the form of an *argumentation systems*. Argumentation systems cast inference as an argumentation process, where arguments for or against propositions are constructed, and arguments and counterarguments may rebut, defeat and cancel each other [16, 12].

The standard inductive definition of the supported paths in inheritance networks can be seen as an argumentation system where every time we want to extend a supported path with another link $p \to q$, we need to consider the rebuttals $p \longrightarrow r \not\to q$ and show that such rebuttals can be defeated.

We will actually use a similar definition within a more interesting language. First, we allow conjunctions of literals in the antecedent of the rules and single literals in their consequents. Second, and more interestingly, we can accommodate completion functions π that are not necessarily grounded. Namely, we will accommodate both atoms that we want to assume to be false by default (*abnormal* atoms) and atoms for which we want to make no assumptions (*normal* atoms). For the first, we have $\pi(\neg a) = 0$ and $\pi(a) > 0$; and for the second, $\pi(\neg a) = \pi(a) = 0$. For example, in the Yale Shooting scenario it makes sense to treat *shoot* and *loaded* as abnormal and normal atoms respectively. We will refer to literals L for which $\pi(L) > 0$ as *exceptional* literals.

Like in the causal encodings of Logic Programs, we will assume that we have no 'observations' but only rules. That is, observations like p or $\neg p$ will be treated as rules $true \to p$ and $true \to \neg p$.[13]

Following a standard format, we call the literals that can be believed, the *warranted* literals, and define them inductively as follows:

Definition 7. A literal L is *warranted* if it is supported by an argument which is not rebutted or all of whose rebuttals are defeated.

The notions of arguments, rebuttals and defeaters are defined in turn as:

- A rule $A \to L$ is an *argument* for L if A is the symbol *true* or if every literal in A is warranted.

---

[13] The reason for encoding observations as rules is not simple. We will discuss this problem in deep elsewhere.

- A rule $B \to \sim L$ *rebuts* an argument $A \to L$ if it has no lower priority and every exceptional literal in B is warranted
- A rebuttal $B \to \sim L$ is *defeated* if the complement $\sim L'$ of some non-exceptional literal in B is warranted

Let us finally say that a system of rules is *deterministic* when every pair of conflicting rules $A \to L$ and $B \to \sim L$ with the same priority are such that A and B contain at least one complementary literal, and that the system is *normal* when no rule antecedent includes negated abnormal atoms. Then,

Theorem 8 (Soundness). *The argumentation system is sound for normal theories that are deterministic.*

That is, every warranted literal follows in such case from the rules. The argumentation system, however, is not complete, as it lacks that ability to *reason by cases* (e.g., concluding q from two rules $p \to q$ and $\neg p \to q$).

9 Causal Modeling: Cognitive Maps

We want to use and extend this framework to model reasoning involving conditions, events, decisions and preferences. For an illustration of what's needed, consider the following fragment of an editorial by M. Friedman taken from [2]:

> Recent protectionist measures ... have disappointed us. Voluntary limits on Japanese automobiles ... are bad for the nation. They do not promote the long-run health of the industry ... The problem of these industries ......is [high wages] ... Far from saving jobs, the limitations on imports will cost jobs. If we import less, foreign countries will earn fewer dollars. They will have less to spend on American exports. The result will be fewer jobs in export industries.

There is a lot of causal reasoning in arguments of this type and we want to be able to model it. A useful language for this task, is the language of *cognitive maps* developed by political scientists to model the belief systems of political elites and to analyze how they make, predict and understand policy decisions (e.g., whether to pull out from the Middle East, etc.; see Axelrod's [5]).

Cognitive maps are very simple. They are made of variables connected by either positive or negative links. A positive link $A \overset{+}{\to} B$ means that, other things equal, an increase in the value of A will produce an increase in the value of B, and that a decrease in the value of A will produce a decrease in the value of B. A negative link $A \overset{-}{\to} B$, on the other hand, means that other things equal, an increase (decrease) in A brings a decrease (increase) in the value of B.

Variables can be discrete or continuous, binary or multivalued. All that matters is that values are ordered so that it is clear what it means for a variable to increase or decrease its value. Variables can represent physical entities (level of water), abstract entities (prices), utilities (US interests), or whatever.

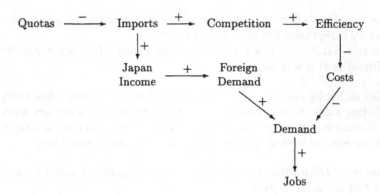

Fig. 2. A Cognitive Map for Friedman's Argument

The cognitive map in Fig. 2 displays some of the knowledge underlying Friedman's argument. It shows that 'quotas' inhibit imports, imports mean competition, competition promotes efficiency, etc.

The standard interpretation of Cognitive Maps is entirely procedural (see [5]). A *path* from a variable A to variable B is *positive* if it has an even number of negative links and *negative* otherwise. The *effect* of A on B is *positive* if all paths between A and B are positive, *negative* if all path between A and B are negative, and *ambiguous* otherwise.

If variable A stands for an action (e.g., setting the level of interest rates) and variable B stands for something we care about (e.g., level of unemployment), a positive effect from A to B may provide the argument to adopt action A, to explain it or even to predict it.

The language of cognitive maps is perhaps too simple and the semantics too ad-hoc, yet cognitive maps have been found useful and have many attractive features.[14] We would like to preserve what's good in cognitive maps and overcome some of their expressive limitations. We can do this by formulating cognitive maps as causal rule systems.[15]

Let us assume that all the variables X have three values: they can go up, they can go down or they can stay where they are. We can represent the three values by three exclusive and exhaustive propositions: x^+, x^- and x^0.[16] Positive links of the form $A \overset{+}{\to} B$ will be mapped into two rules of the form $a^+ \to b^+$ and $a^- \to b^-$, and something similar will be done for negative links. Moreover, since

---

[14] See [24] for a semantics for cognitive maps based on a different qualitative abstraction of Bayesian Network.

[15] We only deal with cognitive maps that are acyclic.

[16] So far we have considered binary variables X associated with the complementary literals x and $\neg x$. The extensions to multivalued variables Y associated with a set of exclusive and exhaustive set of propositional symbols y_1, y_2, ..., is mostly direct, but for reasons of space cannot be detailed here.

all the links really encode the way *perturbations* flow throughout the network, we want to assume that perturbations only occur when there is a reason for them to occur. That is, in the absence of reason for a variable X to change, we want to assume that it remains unchanged. This assumption can be captured by a completion function π such that $\pi(x^0) = 0$ and $\pi(x^+) = \pi(x^-) = 1$. For the resulting encoding we can prove that:[17]

Theorem 9. *Variable A has a positive (negative) effect on variable B iff in the causal rule system that results from adding the link $Z \xrightarrow{+} A$ to the cognitive map, where Z is a new variable, b^+ (b^-) is a consequence of z^+.*

The role of the dummy variable Z here is to explain a^+ away thus eliminating any possible evidential support that may result from observing a^+ without a cause. In the example above, Z would stand for some factor that pushes 'Sales' up without affecting 'Advertising'.

10 Discussion

We haved developed a framework for causal modeling that provides a language for representing knowledge, a semantics based on qualitative probabilities, and a proof-procedure based on the interaction of arguments. The framework ties together existing probabilistic, logical and procedural approaches to causal reasoning like Qualitative Bayesian Networks, logic programs with negation as failure, and Axelrod's cognitive maps.

One goal for the future is to extend cognitive maps and causal rule system to deal with some of the problems that fall into the area of Qualitative Reasoning (e.g., models of how thermostats and bathtubs work, etc.) and use the resulting systems for modeling the types of arguments that are used in the social sciences (economics, political science, etc.)

This requires dealing with both the variable's rate of variation (as in cognitive maps) and their absolute values. We want to infer, for example, that if a cup is placed under an open faucet, eventually the cup will get filled and water will be spilled. This type of problem has been dealt with in the area of Qualitative Reasoning [22], yet the languages and the reasoning methods proposed do not appear to be in close correspondence with the way people reason about these things. An approach that deals with these problems in a more appealing way is Rieger's and Grinberg's [18]. Making sense of the extremely rich but informal language for causal reasoning that they advance (e.g., 10 different types of causal relations, etc.) seems to present a significant challenge.

---

[17] The role of the dummy variable Z in the theorem is to explain a^+ (or a^-) away, thus eliminating any possible evidential support that may result from obtaining a^+ (or a^-) by observation. Recall that in our causal rules, information may flow both in the direction of the rules (predictive reasoning) and against the direction of the rules (evidential reasoning).

References

1. E. Adams. *The Logic of Conditionals*. D. Reiter, Dordrecht, 1975.
2. S. Alvarado. Argument comprehension. In S. Shapiro, editor, *Encyclopedia of Artificial Intelligence (2nd Edition)*, pages 30–52. Wiley, 1992.
3. A. Darwiche. CNETS: A computational environment for generalized causal networks. Technical report, Rockwell International, PARC, 1994.
4. D. Dubois and H. Prade. Epistemic entrenchment and possibilistic logic. *Artificial Intelligence*, 50:223–239, 1991.
5. R. Axelrod (Ed.). *Structure of Decision: The Cognitive Maps of Political Elites*. Princeton University Press, 1976.
6. H. Geffner. *Default Reasoning: Causal and Conditional Theories*. MIT Press, Cambridge, MA, 1992.
7. M. Goldszmidt and A. Darwiche. Action networks. In *Proceedings Spring AAAI Symposium on Decision-Theoretic Planning*, 1994.
8. M. Goldszmidt and J. Pearl. Rank-based systems. In *Proceedings KR'92*, pages 661–672. Morgan Kaufmann, 1992.
9. S. Kraus, D. Lehmann, and M. Magidor. Preferential models and cumulative logics. *Artificial Intelligence*, 44:167–207, 1990.
10. D. Lehmann and M. Magidor. Rational logics and their models. Technical report, Dept. of CS, Hebrew University, Jerusalem 91904, Israel, 1988.
11. J. Lloyd. *Foundations of Logic Programming*. Springer-Verlag, New York, 1984.
12. R. Loui. Defeat among arguments: A system of defeasible inference. *Computational Intelligence*, 3(3):100–107, 1987.
13. D. Makinson. General theory of cumulative inference. In M. Reinfrank *et al.*, editor, *Proceedings Non-Monotonic Reasoning*, pages 1–18, Germany, Springer, 1989.
14. J. Pearl. Embracing causality in default reasoning. *Artificial Intelligence*, 35:259–271, 1988.
15. J. Pearl. *Probabilistic Reasoning in Intelligent Systems*. Morgan Kaufmann, Los Altos, CA., 1988.
16. J. Pollock. Defeasible reasoning. *Cognitive Science*, 11:481–518, 1987.
17. J. Reggia, D. Nau, P. Wang, and Y. Peng. A formal model of abductive inference. *Information Sciences*, 37:227–285, 1985.
18. C. Rieger and M. Grinberg. The declarative representation and procedural simulation of causality. In *Proceedings IJCAI-77*, pages 250–256, 1977.
19. Y. Shoham. *Reasoning about Change: Time and Causation from the Standpoint of Artificial Intelligence*. MIT Press, Cambridge, Mass., 1988.
20. W. Spohn. A general non-probabilistic theory of inductive reasoning. In *Proceedings 4th Workshop on Uncertainty*, pages 315–322, St. Paul, 1988.
21. D. Touretzky, J. Horty, and R. Thomason. A clash of intuitions: The current state of non-monotonic multiple inheritance systems. In *Proceedings of IJCAI-87*, pages 476–482, Milano, Italy, 1987.
22. D. Weld and J. de Kleer, editors. *Readings in Qualitative Reasoning about Physical Systems*. Morgan Kaufmann, Los Altos, CA., 1990.
23. M. Wellman. Fundamental concepts of qualitative probabilistic networks. *Artificial Intelligence*, 44(3):257–303, 1990.
24. M. Wellman. Inference in cognitive maps. *Mathematics and Computers in Simulation*, 36:137–148, 1994.

Goals in Argumentation

Michael A. Gilbert
Department of Philosophy
York University
4700 Keele Street
Downsview, Ontario
CANADA M3J 1P3
gilbert@YorkU.ca
Category: Argumentation Theory

According to the canons of Informal Logic the goal of an argument is to persuade one's opposer of the truth of the proffered claim. The argument, therefore, is always about the claim, and all argumentative activities focus on it. However, arguments occurring between people are more than propositionalizable entities which are Claim-Reason Complexes [CRC] whose structure is locatable. They are also communications occurring between two complex entities with a range of desires, needs and goals. It certainly may be the case that persuading one's opposer of the truth of a claim is *a* goal of a given argumentation, but it will rarely, if ever, be the case that it is the *only* goal of an argumentation.[1]

Within Communication Theory the idea of goals is considered to be inherently manifold. That is, every persuasive interaction has more than one goal. On the one hand there is the obvious goal that is the apparent focus of the encounter, it is what an arguer wants to achieve. But, on the other hand, there are goals relating to the relationship between the arguers as well as goals dealing with the maintenance of the interaction itself. These have been called *primary* and *influence* goals (Dillard, 1990,) *task* or *instrumental* and *face* goals (Tracy & Coupland 1990,) and *instrumental* and *maintenance* goals, (O'keefe 1995.) The underlying conception is that communicative interactions in general and arguments in particular are operating on more than one level, and that there is always a balance between achieving one's immediate objective and maintaining a certain kind of relationship between oneself and one's opposer. I shall use the terms *task goals* to indicate the goals forming the immediate strategic object of the encounter, and *face goals* to indicate the goals concerning the relationship between the participants, including their need to maintain (or terminate) the interaction.

Task goals and face goals differentiate between kinds of goals based on their objectives. A task goal might be quite clear and simple, e.g., gaining an increase on an essay grade, but might also come into conflict with a face goal, e.g., maintaining a good relationship with the professor who assigned the grade. The argument will involve both goals and the arguer will constantly be balancing the needs and dictates of one against the other. Much as when driving a car one has to negotiate between

---

1 Examples of arguments when persuasion is not the goal include devil's advocate arguments, and arguments intentionally undertaken to boost, for example, someone's ego or self-image.

speed, safety, and legal restrictions, conducting an argument must negotiate between the strategic objectives of the encounter and both social and relational restrictions.

Both task and face goals are situation specific. The needs and desires of the proponent determine the task goals, and the relationship between the proponent and opponent determine the face goals. Of course, this is not strictly true. For one thing, task goals might meld over into face goals as when the very object of an argument is to impact in one way or another on the relationship. For another, task goals might change as a result of conflict with face goals during the course of the encounter. As a result, it is important not to think of the two goal classifications as being either independent or invariably separable.

There is one other goal category that will be introduced. This is, to borrow from Dillard (1990) the category of *motives*. "Motives," Dillard writes, "are broad and deep-seated determinants of behaviour" (p. 72). Motives are the sort of goals that determine task and face goals in a broad general way. They delimit, if you will, the sort of goals one considers and acts upon as well as the sorts of actions one might use to obtain the goals. In the following the term 'goal' shall be used to refer to both task and face goals unless further specified, and 'motive' shall be used to refer to the more general goals as indicated above.

It should be clear that goals alone are insufficient to predict or delimit actions. A goal G to achieve an end E can be approached in an infinite number of ways. If Harry wants Jane's car to do an errand he could, for example, just take it. However, stealing the car is not even an option considered, let alone acted upon, because Harry has the motive of remaining honest. In addition, the face goal concerning his relationship with Jane would be violated were he to steal her car.[2]

I have argued elsewhere (Gilbert, 1995a) that understanding an opposer's goals is crucial in what I have called coalescent argumentation. This is an approach to argument that emphasizes locating points of agreement within positions, and builds upon those points to create a coalescent situation incorporating as much as possible of the divergent views. In the effort to bring an argument to a mutually agreeable end with both parties content with the outcome the number of satisfied goals must be maximized on both sides. So, while the very existence of goals, and, specifically goals of different sorts within the same communicative encounter, is descriptive, the utilization of these goals to achieve an harmonious end to an argumentative encounter is prescriptive. The understanding of the role goals play in argumentation, then, becomes important to a deeper understanding of practical reasoning.

Let each individual involved in an argument be considered to have the following *ordered* sets of goals. Note that the sets are ordered by a priority relation which ranks them in importance. Thus, Harry's task goal of doing an errand has a lower priority than Harry's motive of being honest and of offending Jane.

---

2   In some milieus, of course, taking someting without permission is not precluded. Moreoever, it might not even damage or, at least, completely wreck the relationship.

A set of motives, M $= <m_1, ..., m_n>$
A set of task goals, T $= <t_1, ... \ t_n>$
A set of face goals, F $= <f_1, ..., f_n>$
all of which delimit a set of procedures, P $= <p_1, ..., p_n>$

Figure 1 M,G,T,P

'Procedures' are intended to cover arguments and argumentative moves construed broadly (Gilbert, 1995), so they will range from particular arguments offered and the way in which those arguments are presented, to non-linguistic communications intended to persuade or otherwise sway an opposer.

Goals, of course, are very complex things. They both direct and limit actions insofar as one goal might define a strategic objective, but another might restrict the ways in which that objective can be achieved. In addition, it is not uncommon to have goals, even of the same sort, that are in conflict. And tension, if not conflict, is invariably present in goals of different categories. Motives restrict task goals and face goals, and task and face goals limit and constrain each other. To further complicate matters, arguers often have goals that are kept hidden from their opposers and even from themselves, what we might call, following Walton (1989), "dark-side goals". In a negotiation, for example, one might be putting forward one position when the real goal concerns another. Even in a more heuristic argument, it would not be unseemly to present a position as more extreme than it need be in order to provide room for maneuvering.

Goals can be hidden from the person who holds them. We can be unknowingly self-destructive or self-defeating. We can be provocative or antagonistic without realizing that we are trying to evoke a particular reaction. We can think we are doing one thing for one reason only to realize later, with or without help, that we were completely wrong. An individual S's set F for example, might contain a goal f_i which will have a negative face impact. This could result from S's being angry at her opposer, but at the same time not realizing that she in turn wants to anger her opposer.

All of this creates great complexities when considering the question, What is the goal of argumentation? If an argument is viewed as a CRC, then, presumably, the goal is to have one's opposer accept the claim as part of his or her belief set. But, in any particular argumentation, this may not be the actual goal. The proponent might, for example, actually want the opposer to *not* accept the claim, but to move to another related claim. In addition, a proponent might suddenly have another goal, say a face goal, that intervenes to cause a backing off of the original task goals. In Bavelas, et al, (1990) the role of equivocation in a discussion works to avoid conflict between task and face goals. An individual in a socially awkward situation will equivocate rather than offend or lie. One is caught in a snare of conflicting goals. Consequently, it is not possible to state, *simpliciter*, that the goal of an argument is to have the respondent adopt the claim.

The situation is further complicated when we consider that *both* the proponent and respondent have complex goal sets with internal as well as external inconsistencies. Yet, in order to effect an agreement opposer's must have a reasonable idea of what each others' goals are. It is unlikely that two arguers can come to an agreement when they do not understand the terms of the disagreement and the objectives of their opposer. Understanding a position (Gilbert, 1995a) is a more complex and multi-layered endeavour than is usually thought. But, here it is not so much the position, as the goals (including motives) that organize the position that is at issue.

The aim of what I call coalescent argumentation is to bring about an agreement between two arguers based on the conjoining of their positions in as many ways as possible. This means that a full exploration of the positions must be undertaken in order to determine which aspects of a position are crucial, which are peripheral, and which might be held without due consideration. Clearly, the identification of the goals of the partners to the dispute will play a crucial role in bringing about a mutually satisfactory conclusion. Thought of simplistically, the more goals in each position that can be satisfied, the more likely it is that a coalescent termination can occur. This is simplistic, however, because the *number* of goals satisfied will rarely be a major factor. Insofar as goals are ranked in order of importance, the satisfaction of the top several goals may be vastly more important than the satisfaction of numerous lower ranked goals.

Let us consider that each individual S in an argument has a set of goals consisting of each of the sets indicated in Figure 1. Let us call S's set of goals Γs. So, in an argument involving two partners, S and T each will have, respectively a set Γ that can be construed as follows.

$$\Gamma s = \{<m_1, ..., m_n>, <t_1, ..., t_n>, <f_1, ..., f_n>\}$$
$$\Gamma t = \{<m_1, ..., m_m>, <t_1, ..., t_m>, <f_1, ..., f_m>\}$$

Figure 2 Γs

There is a very important question to consider before going further. The first is whether Γs indicates all of S's actual goals, or only those goals that S knows him/herself to have. In other words, what is the extent to which it can be assumed that S is aware of the goals in Γs? The answer is that since the concern here is with actual argumentative practice, Γs must contain only those goals that either S holds in awareness or would agree are held dispositionally. This is important to coalescent argumentation because part of the process is the participants *bringing into awareness* their own and their partners goals. Arguers not infrequently lack complete awareness of the goals they have in an argument. The emphasis is often placed on t_1, the first task goal, and other goals are ignored. In the course of a coalescent argumentation the further members of Γs will be brought out and considered. This not only increases their likelihood of satisfaction, but opens the possibilities of the identification of mutually held goals.

Most arguments have one objective that is considered, by at least the protagonist, to be the main or paramount goal of the interaction. This will often, but not necessarily, be a task goal, and, in particular, the highest ranked task goal. This goal will be called the *apparent strategic goal* [ASG], or, more simply, the strategic goal. Consider the following example.

Example 1

Jim and Richard are arguing about who has done more of the food shopping. Jim insists that he has done far more than Richard. Richard replies that Jim does not mind the chore nearly as much as he, Richard, does. Jim makes one of the following replies:

A] That doesn't matter. We each do things we don't enjoy. It's your turn now.

B] I know that; but what will you do to even things up if I food shop all the time?

C] Well, has it occurred to you that asking me nicely instead of pretending you do as much as I do might work better?

With each of Jim's possible replies we can determine a different motive or goal. The previous argument might be essentially the same, but at this point, Jim's strategic goal leads him in different directions. In [A] the strategic goal is not doing the shopping. It is a task goal, and may very well be t_1 of Γs. In [B] the strategic goal appears to be a face goal with Jim seeking redress for his frequent shopping. Now he does not so much want to avoid shopping as use it as a means of balancing the chores on a broader scale. Finally, in [C], Jim does not want to avoid shopping and does not want any other redress. Rather, he has a motive, m_1, which is to effect an alteration in the way Richard treats him. Now Jim is using the argument to persuade Richard that a motive of his—being up front with people when you want something—should be adopted.

As an argument proceeds one can make judgments concerning the goals of the participants, but caution must be used. It is quite possible that in Example 1 one would guess wrong in cases [B] and [C]. The apparent strategic goal is somewhat hidden. And, yet, it is important to the process of the argument to uncover the goals. If the argument had gone bad and the goals not been properly identified, then numerous opportunities for agreement might have been missed. When it is known that the strategic goal is [B], for example, then avenues of negotiation immediately open up. In the case of [C] the entire footing of the conversation might change, and it might be pursued on a more personal level.

Given the importance of goals in argumentation and how critical is understanding one's own and one's partners goals it is crucial to examine their role carefully. Two main parameters effect the role goals play in a given interaction between S and T. The first is the degree to which S is aware of her *own* goals, Γs, and the second is the degree to which the respondent T is aware of S's goal set Γs.

In the first instance, the requirement is that the arguer have an awareness of her own goals. This statement might seem odd, but it is not. Often an arguer will be aware of her ASG, but unaware of other less obvious goals. The other goals may belong to the set F of face goals, but can also be in T. The former is simple: One may not realize until circumstances force the issue that certain non-task goals play an important role in the argument. Not arousing anger, maintaining a pleasant demeanour, keeping a respondent on one's "good side," can be as or more important than a given task goal.

It is also the case that task goals can be, in the course of an interaction, re-evaluated or adjusted. Sometimes S can perceive her strategic goal as t_1, but subsequently realize that another goal will also satisfy some other need. Consider a very simple example.

Example 2
Susan asks Tom if she can borrow his car. Tom refuses, saying the he needs it himself. Susan argues that she has an important conference 30 miles west of town. Tom says that, as a matter of fact, he is going that way and can drop her off in the morning and pick her up in the afternoon. Susan agrees with thanks.

An analysis of this example shows that Susan went in with the ASG of obtaining the loan of Tom's car. Tom presents Susan with a dispreffered response to her request. At this point, an arguer can do one of two things: First, she can remain focused on her ASG (t_1), or she can open up her set T of task goals to consider other, secondary goals. One way of considering this is to view the encounter as involving a higher goal, perhaps a motive, m_i, of say, fulfilling her work obligations. When the ASG is blocked it might be abandoned, but often the motive that led to it is not. That is, S needs a car to get to the meeting. The ASG is the borrowing of T's car. When that is not possible, the motive that dictated the request to borrow a car then becomes the ASG *if the arguer is sophisticated enough to alter goal strategy.*

Similarly, if we turn to the second of the two aforementioned parameters it becomes obvious that the greater awareness T has of a dispute partner S's set Γs, the greater the likelihood that the argument can come to a jointly satisfactory conclusion. If T is 1) aware of S's goals, and 2) sufficiently sophisticated to be able to consider the goal set beyond the ASG, then there is considerably increased likelihood that T will be able to find a coalescent conclusion. Awareness of an opposer's goals permits a dispute partner to find satisfactory outcomes that might be mutually agreeable, and, hence, coalescent. When, in Example 2 Tom learns that Susan's broader goal is reaching a specific destination, he is able to suggest a means of satisfaction not previously on the table. In other words, there are goals in each arguer's set that can be satisfied by a given outcome.

Another way to express the role of goals in an argument, especially in a coalescent argument, is that the larger the set { $\Gamma s \cap \Gamma t$ }, the greater the likelihood of a mutually agreeable outcome. But it is imperative at this point to recall that a given set Γi, contains only those goals the proponent I is aware of or would agree she/he holds dispositionally. This means that, in the process of argumentation, one very important task is the drawing out of a respondent's goals in order to increase the likelihood that { $\Gamma s \cap \Gamma t$ }will not be empty. When an arguer T increases his awareness of alternates to S's ASG, and, at the same time, opens S up to those same alternates, Γs grows in size. Consequently, the possibility of { $\Gamma s \cap \Gamma t$ }being larger increases as well.

One response to this approach is that it makes all arguments negotiations. Treating all arguments as negotiations ignores the role of truth, correctness, and, in moral arenas, such normative factors as justice, rightness, and so on. But this is just wrong. Both parties to a dispute can hold motives and goals that are very similar in which case there is likely to be greater focus on the heuristic aspects of argumentation. The fact that, more often than not, there are (in particular) face goals that interfere with the purity of inquiry does not mean that, even in those cases, determining the truth of a proposition or the value of a solution is not a highly prioritized motive of both parties. The goal of coalescent argumentation (Gilbert, 1995a) is an agreement based on maximally fulfilling the goals and needs of the arguers involved. As the most common form of argument is not pure inquiry but, at best, eristically tinged inquiry, it does not behoove us to be overly concerned with those few instances, however precious they be, of pure heuristic inquiry.

One important objective for theorists in Argumentation Theory is the development of a formal structure to capture the natural operation of goals as held by one person and as utilized by two persons interacting in a conflictual situation. A number of serious complexities concerning the development of such a formalization must be recognized. First, as discussed above, there is the awareness factor. That is, the extent to which the holder of a goal set Γs is aware of the extent, variability and range of her own goals. Secondly, there are the inconsistencies that always exist between goals held by an individual. These can be, first and foremost, between different *types* of goals, as when face goals limit the kinds of reasons one is willing to present to achieve a strategic goal. But there can also be tension within a given type of goal. One might, for example, have the goal to demonstrate both affection and firmness in an interaction. Or, one might have the goal to both seem clever and bright but not to overshadow one's partner. These situations can create complexities with a set of goals even of the same sort.

A separate sort of difficulty arises in ordering goals. First, the problems just described create ordering difficulties. But, even worse, S might be mistaken about her own goal ordering, it might change en passant, or goals ranked highly by type may well be incommensurable. Many moral dilemmas, for example, involve choices between face and task goals. Questions about how goals can be utilized abound: Can values be assigned to goals within the same category? Or, are there goals within a given set that cannot be ordered? Can values be assigned to goals that permit them to be compared even when of different types? Would such a comparison be useful in determining how to deal with conflict between goals? The formal analysis of goals in argumentation must answer these questions and many others while at the same time not oversimplifying the concept of 'goal' so as to render it sterile.

Coalescent argumentation is argumentation focused on agreement. It builds on the premiss that arguments are complex social activities that involve human egos seeking to satisfy their intellectual, emotional, physical and spiritual needs. In the course of seeking the satisfaction of these needs there is conflict and disagreement. This can be over beliefs, the limited availability of resources, or questions of control and power. By becoming aware of the role goals play in argumentation, arguers can better focus on securing their own needs as well as attempting to satisfy those of their opposers. When the satisfaction of needs is maximized, the opportunity for a mutually agreeable conclusion is maximized as well.

230

Bibliography

Bavelas, Janet Beavin & Black, Alex & Chovil, Nicole & Mullet, Jennifer. "Truth, Lies, and Equivocations. The Effects of Conflicting Goals On Discourse" in Tracy & Coupland, 1990. 135-161.

Craig, R.T., 1986. "Goals In Discourse." In Ellis, D.G., & Donohue, W.A., eds. 1986. *Contemporary Issues In Language and Discourse Processes.* Hillsdale, NJ: Erlbaum.

Dillard, James P. 1990. "The Nature and Substance of Goals in Tactical Communications", in Cody, M.J. and McLaughlin, M.L., eds. *The Psychology of Tactical Communication.* Avon, England: Multilingual Matters Ltd.

Gilbert, Michael A. 1995. "Arguments and Arguers". *Teaching Philosophy.* 18:2:125-138.

Gilbert, Michael A. 1995a. "Coalescent Argumentation." *Argumentation.* 9:5.

O'Keefe, Barbara & McCornack, Steven A. 1987. "Message Design Logic and Message Goal Structure". *Human Communication Research*, 14:68-92.

O'keefe, Barbara.1995. "Influence and Identity In Social Interaction." *Argumentation* 9:5.

O'Keefe, Barbara J., & Shepherd, Gregory J., 1987. "The Pursuit of Multiple Objectives in Face-to-Face Persuasive Interactions: Effects of Construct Differentiation on Message Organization". *Communication Monographs*, 54:396-419.

Tracy, Karen. 1984. "The Effect of Multiple Goals On Conversational Relevance and Topic Shift". *Communication Monographs* 51: 274-287.

Tracy, Karen & Coupland, Nikolas. 1990a. "Multiple Goals in Discourse: An Overview of Issues". in Tracy, Karen & Coupland, Nikolas. 1990.

Tracy, Karen & Coupland, Nikolas. 1990. *Multiple Goals In Discourse.* Avon, England: Multilingual Matters Ltd.,.

Tracy, Karen. 1991. *Undestanding Face to Face Interaction: Issues Linking Goals and Discourse.* Hillsdale, NJ: L. Erlbaum Assoc..

Tracy, Karen. 1991a. "Introduction: Linking Goals with Discourse". in Tracy, Karen. 1991.

Walton, Douglas N. 1989. "Dialogue Theory for Critical Thinking". *Argumentation* 3:2.

An Abductive Proof Procedure
for Conditional Logic Programming *

L. Giordano, A. Martelli, M.L. Sapino

Dipartimento di Informatica, Università di Torino,
Corso Svizzera 185, 10149 Torino, Italy
e-mail: laura,mrt,mlsapino@di.unito.it

Abstract. We develop an abductive extension of the hypothetical language $CondLP^+$ within the argumentation framework. The language supports hypothetical updates together with integrity constraints, and it makes use of a revision mechanism to restore consistency when an update violates some integrity constraint. We introduce a semantics for the language based on the notion of admissible solution, and we define an abductive proof procedure to compute them. The language can be used to perform several types of defeasible reasoning. We discuss how the revision policy of the language can be modified to perform different types of reasoning, like reasoning about actions.

1 Introduction

In [10] a logic programming language $CondLP^+$ has been defined, which is an extension of N_Prolog [8] and provides capabilities for hypothetical and counterfactual reasoning. In $CondLP^+$, a database containing constraints can be hypothetically updated by adding new facts. A revision mechanism allows to deal with the inconsistencies that may arise by removing the information responsible for them.

In this language, as in N_Prolog, hypothetical implications $D \Rightarrow G$ (with G a goal and D a set of hypotheses) may occur both in goals and in clause bodies. For an implication $D \Rightarrow G$ to succeed from a program P, G must succeed from the new program obtained by hypothetically updating P with D. As a difference with N_Prolog, in $CondLP^+$ a program may contain integrity constraints, and, hence, inconsistencies may arise, when updates are performed.

In $CondLP^+$ hypothetical updates are sets of atoms. Moreover, a program consists of a protected part, containing a set of clauses and integrity constraints, and a removable part, consisting of a set of atomic formulas partitioned in different priority classes.

When an inconsistency arises, a *revision* of the program is needed. Revision does not affect the protected part, which is permanent. Instead, it modifies its

---

* This work has been partially supported by ESPRIT Basic Research Project 6471, Medlar II.

removable part, by overruling the atoms which are responsible for the inconsistency and have the least preference. Such atoms are not removed from the database, but they are temporarily made non visible.

Since the removable part of the program may contain several atoms with the same priority, in general, more than one revision of the program can be obtained. Hence it is quite natural to make use of abduction to characterize the alternative solutions.

To capture the non-monotonic behaviour of the language, in [10] an abductive semantics has been presented in the style of Eshghi and Kowalski's abductive semantics for negation as failure [6], (or equivalently, of stable models semantics [11]). However, since the goal directed proof procedure given in [10] is non abductive, no precise correspondence can be established between it and the abductive semantics. What has been shown in [10] is that the procedure is sound at least in the case when an abductive solution exists, and, in particular, the goals proved by the procedure are those true in all the abductive solutions.

Here we define an abductive extension of the proof procedure so that a stronger correspondence between the semantics and the procedure can be established. To this purpose, however, we also modify the abductive semantics, by adopting a weaker one. In particular, we want to give up the totality requirement of stable model semantics (whose consequence is the fact that abductive solutions do not always exist) and we move to a three-valued semantics.

The argumentation framework [5] provides a nice setting in which an abductive semantics and a procedure for computing it can be defined, since it is parametric with respect to both the logic and the abducible symbols adopted. In this paper we develop an abductive extension of $CondLP^+$ within the argumentation framework, by providing an abductive semantics for it, based on the notion of acceptability. In particular, we focus on an approximation of acceptability semantics, namely on admissibility semantics, and we introduce an abductive proof procedure which computes it. Such a procedure is sound with respect to the admissibility semantics and is defined in the style of those proposed in [17] to compute the acceptability semantics. The procedure lacks completeness due to the possibility of infinite loops in the computation.

In [10] it has been shown that $CondLP^+$ can be used to perform several types of defeasible reasoning. In this paper, we will present both the semantics and the proof procedure of the language by making use of a diagnosis example, and in the last section, we will discuss how the revision policy of the language can be modified to perform different types of reasoning, like reasoning about actions.

2   The Operational Semantics

In this section we recall the definition of the syntax of the language $CondLP^+$ and its operational semantics, as originally given in [10]. We will deal only with the propositional case. Let *true* and $\perp$ be distinguished propositions (true and false), and let A denote atomic propositions different from $\perp$. The syntax of the language is the following:

$$G := true \mid A \mid G_1 \wedge G_2 \mid \{A_1, \ldots, A_n\} \Rightarrow G$$
$$D := G \rightarrow A$$
$$I := G \rightarrow \perp.^2$$

In this definition G stands for a goal, D for a clause and I for an integrity constraint. Notice that a goal G may contain nested hypothetical implications, such as $\{a, b\} \Rightarrow (\{c\} \Rightarrow (a \wedge c))$, and that hypothetical implications are allowed in the body of clauses and in constraints, as in the formulas $(\{a, b\} \Rightarrow (\{c\} \Rightarrow a)) \wedge (\{b\} \Rightarrow c) \rightarrow d$ and $(\{a\} \Rightarrow c) \wedge f \rightarrow \perp$.

A *program* P is defined as a set of clauses and integrity constraints Π, and a list of sets of atoms L (the removable part): $P = \Pi \mid L$, where $L = S_1, \ldots, S_n$, each S_i being a set of atoms (facts). While the clauses and constraints in Π cannot be removed, i.e., they are *protected*, the facts in L are *revisable*: each fact $A \in S_i$ in L can be used in a proof unless it is inconsistent with Π together with some facts $B \in S_j$ (with $j > i$) with higher priority. Hence, we assume a total ordering among sets of atoms in the list L, and each S_i is less preferred than S_{i+1}. We assume that atoms in S_n have the highest priority, as Π, and they cannot be removed.

We will now define a goal directed proof procedure for the language. The idea is that, during the computation of a given goal from a program P, the protected part of the program remains fixed, while the list of removable atoms changes, when updates are performed. When an atom from the list L is needed in a proof, it must be verified that it is not inconsistent with the atoms with equal or higher preference (i.e. it must be verified that, assuming that atom, an inconsistency can not be derived from the permanent database and the atoms with equal or higher preference).

The operational derivability of a given goal G from a set of clauses and constraints Π in a context $S_1, \ldots, S_n \mid \{H\}$ (written $\Pi \mid S_1, \ldots, S_n \mid \{H\} \vdash_o G$) is defined by the following proof rules (in which Π is omitted since it does not change during the computation). Each S_i is a set of atomic hypotheses. H represents a temporary hypothesis, which is checked for consistency, and it is virtually part of S_n. However, we keep it separate from S_n since, when entering a new context represented by a further update S_{n+1}, we do not want to assume it anymore.

1. $S_1, \ldots, S_n \mid \{H\} \vdash_o true$;
2. $S_1, \ldots, S_n \mid \{H\} \vdash_o G_1 \wedge G_2$ if $S_1, \ldots, S_n \mid \{H\} \vdash_o G_1$ and
   $S_1, \ldots, S_n \mid \{H\} \vdash_o G_2$;
3. $S_1, \ldots, S_n \mid \{H\} \vdash_o S \Rightarrow G$ if $S_1, \ldots, S_n, S \mid \{\} \vdash_o G$;
4. $S_1, \ldots, S_n \mid \{H\} \vdash_o A$ (including the case $A = \perp$) if
   (i) there is a clause $G \rightarrow A \in \Pi$ such that
       $S_1, \ldots, S_n \mid \{H\} \vdash_o G$, or
   (ii) $A \in S_n$ or $A = H$, or

---

[2] We will regard $\perp$ as a proposition without any special properties, since we will use $\perp$ to express integrity constraints in the program, and we do not want to derive everything from the violation of integrity constraints.

(iii) $A \in S_i$ $(i = 1, \ldots, n-1)$ and **not** $S_i, \ldots, S_n \mid \{A\} \vdash_o \bot$.

By Rule 3, to prove a hypothetical implication $S \Rightarrow G$, the set of atoms S is added to the removable part of the program with the highest priority. By Rule 4, an atomic formula A succeeds either by making use of a clause in the program (item (i)), or when A is the temporary hypothesis (H), or it belongs to the set of atoms on the top (item (ii)), or, finally, when it belongs to one of the other sets S_i in the list (item (iii)). In this last case, the atomic goal A in S_i is regarded as proved if it is not inconsistent with equally or more reliable atoms, i.e., if $\neg A$ cannot be proved in the context $S_i, \ldots, S_n$. In the procedure, this condition is verified by assuming A as the temporary hypothesis H and checking, in that context, the finite failure of $\bot$. Hence, we read **not** $S_i, \ldots, S_n \mid \{A\} \vdash_o \bot$ as "$\bot$ *finitely fails* from $S_i, \ldots, S_n \mid \{A\}$".

We say that a goal G is derivable from a program $P = \Pi \mid S_1, \ldots, S_n$ if $\Pi \mid S_1, \ldots, S_n \mid \{\} \vdash_o G$.

The language is obviously non-monotonic. Let us consider a simple diagnosis example, describing a simple circuit consisting of a battery connected with a bulb, with the constraint that the light cannot be *on* and *off* at the same time.

Example 1. Let Π be the following set of clauses and constraints

    (1) *normal_battery* $\rightarrow$ *voltage*
    (2) *voltage* $\wedge$ *normal_bulb* $\rightarrow$ *light_on*
    (3) *light_on* $\wedge$ *light_off* $\rightarrow \bot$

The propositions *normal_battery* and *normal_bulb* represent normality of components. Let us consider the program $P = \Pi \mid \epsilon$. From this program the query $\{normal\_battery\} \Rightarrow (\{normal\_bulb\} \Rightarrow light\_on)$ succeeds, with the following steps of derivation (for brevity, we will write *n_battery* for *normal_battery* and *n_bulb* for *normal_bulb*).

    $\epsilon \mid \{\} \vdash_o \{n\_battery\} \Rightarrow (\{n\_bulb\} \Rightarrow light\_on)$
    $\{n\_battery\} \mid \{\} \vdash_o \{n\_bulb\} \Rightarrow light\_on$, by rule 3,
    $\{n\_battery\}, \{n\_bulb\} \mid \{\} \vdash_o light\_on$, by rule 3,
    $\{n\_battery\}, \{n\_bulb\} \mid \{\} \vdash_o voltage \wedge n\_bulb$, by rule 4(i),
    $\{n\_battery\}, \{n\_bulb\} \mid \{\} \vdash_o n\_battery \wedge n\_bulb$, by rule 4(i).

By rule 2, the goal $n\_battery \wedge n\_bulb$ succeeds if both of the conjuncts succeed. The second conjunct n_bulb immediately succeeds by rule 4(ii). Let us consider the derivation of the first conjunct $n\_battery$.

    $\{n\_battery\}, \{n\_bulb\} \mid \{\} \vdash_o n\_battery$,
    not $\{n\_battery\}, \{n\_bulb\} \mid \{n\_battery\} \vdash_o \bot$, by rule 4(iii),

which succeeds since $\bot$ fails form $\{n\_battery\}, \{n\_bulb\} \mid \{n\_battery\}$:

    $\{n\_battery\}, \{n\_bulb\} \mid \{n\_battery\} \vdash_o \bot$,
    $\{n\_battery\}, \{n\_bulb\} \mid \{n\_battery\} \vdash_o light\_on \wedge light\_off$, by rule 3.

This derivation finitely fails, since *light_off* cannot be proved. Hence, the derivation of $\perp$ fails, and conversely, the initial goal succeeds.

Note that, in the query above, we have given a higher priority to the hypothesis that the bulb is normal compared to the hypothesis that the battery is normal. Let us now assume that we have the observation that the light is off. Then, the query $\{n\_battery\} \Rightarrow (\{n\_bulb\} \Rightarrow (\{light\_off\} \Rightarrow light\_on))$ fails, for, when *light_off* is added to the database, one of the removable facts *n_battery* and *n_bulb* is deleted, namely *n_battery* which has lower preference. Indeed, the query $\{n\_battery\} \Rightarrow (\{n\_bulb\} \Rightarrow (\{light\_off\} \Rightarrow n\_battery))$ fails, while the query $\{n\_battery\} \Rightarrow (\{n\_bulb\} \Rightarrow (\{light\_off\} \Rightarrow n\_bulb))$ succeeds.

Let us now consider the case when we have the observation that the light is off, but we want to give the same preference to both of the assumptions *n_battery* and *n_bulb*. In such a case, we would expect the query $\{n\_battery, n\_bulb\} \Rightarrow (\{light\_off\} \Rightarrow light\_on))$ to fail, for, when *light_off* is added to the database, one of the removable facts *normal_battery* and *normal_bulb* must be deleted. However, the expected failure is not captured by the operational semantics above, since the given query loops. Indeed, the success of *normal_battery* depends on *normal_bulb* non being proved, and vice versa.

The two hypotheses *normal_battery* and *normal_bulb* appear in the same set in the updatable part of the program P, therefore they have the same preference, and there is no reason for deleting the one instead of the other. The existence of alternative, and equally plausible, solutions for this problem is captured by an abductive semantics, according to which the query $\{n\_battery, n\_bulb\} \Rightarrow (\{light\_off\} \Rightarrow light\_on)$ fails with two different solutions, the one in which the assumption *normal_battery* holds, and the other one in which *normal_bulb* holds. But there is no solution in which both of them hold.

As it should be clear from the example above, an abductive extension of the proof procedure is needed in order to deal with problems having more than one solution. Before doing this, in the next section, we will focus on an abductive semantics for the language, defined within the general setting of the argumentation framework [5]. Then, in section 4, we will present an abductive extension of the proof procedure to compute such a semantics, for the special case of admissibility semantics.

3   An Acceptability Semantics for the Language

To capture the non-monotonic behaviour of the language, in [10] an abductive semantics for it has been presented in the style of Eshghi and Kowalski's abductive semantics for negation as failure [6]. Such semantics relates to the proof procedure in Section 2 as the stable model semantics [11] relates to SLDNF. On the one hand, there is no completeness, since a goal may loop in the proof procedure while being true in the semantics. On the other hand, there is no soundness in general, since a goal may succeed in the procedure even if there is no abductive solution at all. However, in [10] it is shown that the procedure is sound at least in the cases when an abductive solution exists.

In the following, we will define an argumentation-based semantics for our language, which generalizes the semantics proposed in [10]. The notion of argumentation framework has been defined in [5, 12] as a very general framework, and it has been shown to capture many non-monotonic reasoning formalisms, including logic programming. In particular, many of the semantics proposed for logic programming with negation as failure, e.g. stable models, partial stable models, preferred extensions, well-founded models, can be captured within acceptability semantics, which is a particular argumentation semantics [17].

An argumentation framework [5] consists in a *theory* T, a *monotonic logic*, a set of possible *hypotheses* H, and a notion of *attack* between sets of hypotheses. Different instances of this abstract argumentation framework correspond to different non-monotonic formalisms.

In the following, we will define an *acceptability semantics* for our language, by introducing, first, the monotonic logic we use and, then, the proper notion of *attack* between sets of hypotheses. As a first difference with the usual acceptability semantics for logic programming, we will take a modal logic as the chosen monotonic logic, instead of classical logic. In the next section, we will focus on *admissible* sets of hypotheses, and we will provide an abductive extension of the proof procedure in Section 2, in order to compute them.

The monotonic modal logic we are going to propose is based on a modal language L containing: a modal operator $[S]$, for each set of atomic propositions S, and a modal operator $\Box$, which is used to denote those formulas that are permanent. In this modal language, we represent an hypothetical implication $D \Rightarrow G$, where D is a set of atoms, by the modal formula $[D]G$. Moreover, we regard the implication $\rightarrow$ in clauses and constraints as *strict implication*: a clause (or a constraint) $G \rightarrow A$ is regarded as a shorthand for $\Box(G \supset A)$, with the meaning that $G \supset A$ is permanently true.

The modalities $[S]$ are ruled by the axioms of the logic K, while the modality $\Box$ is ruled by the axioms of S4. Moreover, the axiom system for L contains the following *interaction axiom* schema: $\Box F \supset [S]F$, where F stands for an arbitrary formula. In [2] such a modal language has been used to extend logic programming with module constructs.

In our operational semantics, after any sequence of updates ending with the update S, all the atoms in S are true. We can represent this fact by introducing the additional axiom schema $\Box[S]p$, where S ranges over sets of atoms and p ranges over atomic propositions in S^3.

In the acceptability semantics for logic programming [17], as in the abductive interpretation of "negation as failure" [6], *negative* literals are interpreted as abductive hypotheses that can be assumed to hold, provided they are consistent with the program and a canonical set of integrity constraints. Here, on the contrary, we make use of *positive* assumptions of the form $[S_1]\ldots[S_n]p$, with $p \in S_1$, whose meaning is that an atom p in S_1 holds after the sequence

---

of updates $[S_1] \ldots [S_n]$. Assumptions of this kind are needed to model the non-monotonic part of the operational semantics, given by rule 4 (iii). An assumption $[S_1] \ldots [S_n]p$ is blocked whenever the negation of p, $\neg p$ [4], holds after the same sequence of updates (i.e. if p is overridden by atoms with a higher preference).

Actually, in the following we will make use of assumptions (abducibles) of the form $\Box[S_1] \ldots [S_n]p$ (with $p \in S_1$), whose meaning is that p is true after *any* sequence of updates ending with $S_1, \ldots, S_n$.

Let $\models_L$ be the consequence relation in the monotonic modal logic defined above, and let Π be a set of clauses and constraints. We define the notion of attack between sets of abducibles as follows.

Definition 1. Given two sets of *abducibles* Δ_1 and Δ_2, Δ_1 *attacks* Δ_2 if

$$\Pi \cup \Delta_1 \models_L [S_1] \ldots [S_n] \neg p \quad \text{for some} \quad \Box[S_1] \ldots [S_n]p \in \Delta_2.$$

We recall the definition of an acceptable set of hypotheses from [17].

Definition 2. Given two sets of hypotheses Δ_0 and Δ, Δ is *acceptable* to Δ_0 iff for all sets of hypotheses A, if A attacks $\Delta - \Delta_0$, then A is not acceptable to $\Delta \cup \Delta_0$.

A set of hypotheses is *acceptable* if it is acceptable to the empty set of hypotheses. The base case for acceptability is the following: Δ is *acceptable* to Δ_0 if $\Delta \subseteq \Delta_0$.

Note that if Π is itself inconsistent (i.e., $\Pi \models_L \bot$), then the only acceptable set is the empty set.

In [12] it has been shown that many semantics for logic programming correspond to approximations of the full acceptability semantics, by iterating the definition of acceptability above, starting from the base cases. In this paper we will focus on the following approximation at the second iteration, which, in the case of normal programs, captures preferred extensions [4]:

> Δ is *acceptable* iff for all sets of hypotheses A, if A attacks Δ, there is a set of hypotheses D such that D attacks $A - \Delta$, and $D \subseteq \Delta$.

D is called a defence against the attack A. For normal logic programs, Δ is acceptable, according to this last definition of acceptability iff Δ is *admissible* [4], where preferred extensions corresponds to maximal admissible sets.

Given a goal G, we define an *admissible solution* for G in Π to be a set Δ of assumptions that is admissible and, added to Π, entails G.

Example 2. Let us consider the program in Example 1. The goal

$$G_1 = \{n\_battery, n\_bulb\} \Rightarrow (\{light\_off\} \Rightarrow n\_battery)$$

is true in the admissible solution $\Delta_1 = \{\Box[n\_battery, n\_bulb][light\_off]n\_battery\}$, while the goal

$$G_2 = \{n\_battery, n\_bulb\} \Rightarrow (\{light\_off\} \Rightarrow n\_bulb)$$

---

[4] $\neg p$ is to be interpreted as $p \supset \bot$.

is true in the admissible solution $\Delta_2 = \{\Box[n\_battery, n\_bulb][light\_off]n\_bulb\}$.
However, there is no admissible solution Δ containing both the assumption
$\Box[n\_battery, n\_bulb][light\_off]n\_battery$, and the assumption $\Box[n\_battery, n\_bulb]$
$[light\_off]n\_bulb$. In fact, such a Δ would attack itself. This is the reason why the
goal $\{n\_battery, n\_bulb\} \Rightarrow (\{light\_off\} \Rightarrow light\_on)$ has no admissible solution.

4 Abductive procedure

As we have seen from Example 1, the proof procedure in Section 2 fails to return
an answer and enters an infinite loop when there are equally reliable assumptions
which are inconsistent with each other, i.e., in the cases when the program has
alternative abductive solutions. In this section we define an abductive procedure
for our language, which computes admissible solutions for a goal. The general
problem of defining proof procedure for acceptability semantics has been addressed by Toni and Kakas in [17], and we believe that their solutions can be
adapted to the specific case of our conditional language. However, we will not
tackle the general case here, and we will rather focus on the specific case of
admissibility semantics.

The abductive procedure we are going to define, given a program Π and
a goal G, returns an admissible solution, if any, to the given goal, i.e., a set
of abductive hypotheses which, added to the program, entail the goal. Maybe
there is more than one admissible set of hypotheses for the given program, and
maybe the goal does not hold in all of them, but only in someone. For instance,
as we have seen, in the diagnosis example above (Example 2) Δ_1 and Δ_2 are
two alternative admissible sets of hypotheses, which are incompatible with each
other, and while the goal G_1 holds in Δ_1, it does not hold in Δ_2. In this case,
Δ_1 is an admissible solution for G_1, while Δ_2 is not.

The procedure is defined on the line of Eshghi and Kowalski's abductive
procedure for logic programs with negation as failure [6], and of the procedure
to compute admissible sets of hypotheses presented in [17]. It makes use of an
auxiliary set Δ of abductive hypotheses, that is initially empty, and is augmented
during the computation. It interleaves two different phases of computation; the
first phase collects a set of hypotheses which support the success of the goal to
be proved, while the second one assures the failure of all the attacks to such a
set of hypotheses. It may be the case that the second phase requires some new
hypotheses to be added to the current (global) set of hypotheses as defences;
this extension of the set of hypotheses requires to enter again the first phase
of the computation. At the end, if the computation terminates successfully, the
computed (global) set of hypotheses represents an admissible abductive solution
in which the query holds.

A main difference between the abductive problem we are dealing with in this
paper, and the one in terms of which Negation as Failure has been characterized
in [6], concerns the nature of abducibles: in our framework, the abductive hypotheses are positive assumptions of the form $\Box[S_1]\ldots[S_n]p$, with $p \in S_1$, whose

meaning is that the atom $p \in S_1$ can be assumed to hold in any context ending with $S_1, \ldots, S_n$.

The abductive proof procedure is defined in terms of an auxiliary procedure *support*, which carries out the computation for the monotonic part of the language. In particular, given a goal G and a program Π, $support(G, \Pi)$ returns a set $\{\Delta_1, \ldots, \Delta_m\}$ of abductive supports to the goal G in Π, containing at least all the minimal abductive supports. Each support is a set of assumptions that, together with the program Π, monotonically proves the given goal G.

To compute an abductive support Δ_j, $support(G, \Pi)$ performs the monotonic steps of the procedure presented in Section 2. As a difference, *support* makes use of a global set of assumptions and, in step 4.(iii), it does not look for failure, but it adds an assumption to the global assumption set. More precisely, to define $support(G, \Pi)$, the proof procedure in Section 2 is modified as follows:

- Replace step 4.(iii) with the following:
  if $A \in S_i$ ($i = 1, \ldots, n-1$), then $\Box[S_i], \ldots, [S_n]A$ is added to Δ_j
- Add a fifth item:
  $S_1, \ldots, S_n \mid \{H\} \vdash_o \neg A$ if $S_1, \ldots, S_n \mid \{A\} \vdash_o \bot$

The fifth item is motivated by the fact that checking if assumption $\Box[S_1], \ldots, [S_n]A$ can be attacked, requires to verify if $[S_1], \ldots, [S_n]\neg A$ can be proved (see below).

Note that the procedure modified in this way allows an abductive support to be non-deterministically computed. The procedure $support(G, \Pi)$ will collect all the abductive supports computed by that procedure.

Note that the procedure support might be non terminating, for different reasons: it may enter a loop, or it may be the case that the set of supports to the given goal is an infinite set.

Let us move to the definition of the abductive procedure. Since the aspects concerning the dynamic evolution of the program are specifically dealt with by the auxiliary procedure *support*, the abductive procedure does not need any sequence $S_1, \ldots, S_n$ specifying the current context.

The set Δ of abductive hypotheses is initially empty, and is enlarged along the proof of the given goal (this is why the current Δ appears as an argument, and another set Δ', including Δ, is returned as a result from each derivation step). The set returned by the the final step of a success derivation is the admissible solution to the given goal.

The proof rules of the abductive procedure will be of the form:

$$\Delta \vdash_t G \text{ with } \Delta' \text{ (and similarly for } \vdash_f)$$

meaning that the goal G *succeeds* (or *finitely fails*), given the set of assumptions Δ, and its success (finite failure) is supported by the set of hypotheses Δ'.

In the following, we will denote by Δ both a set of abducibles and their conjunction. The atom *true* represents an empty conjunction of hypotheses. When the abductive procedure is called on a goal G, the monotonic procedure *support* computes the sets of hypotheses supporting G in Π. Then, the abductive procedure nondeterministically chooses one of these sets, say Δ_i, and tries to prove that it is an admissible set, or that it can be extended to become an admissible set (rule (S-G)). To do this, it considers each assumption $\Box[S_1]\ldots[S_n]A$ belonging to Δ_i and not yet in Δ: after adding such a hypothesis to the current global set Δ, it starts a failure derivation for all the supports of its complement $[S_1]\ldots[S_n]\neg A$ (rules S-$A(ii)$, F-G). Such supports will be the attacks against $\Box[S_1]\ldots[S_n]A$. All these attacks are considered, to introduce in Δ the defences against them; the process is recursively activated on any defence introduced in Δ.

(S-*true*) $\Delta \vdash_t true$ with Δ

(S-G) $\Delta \vdash_t G$ with Δ' if

         there exists $\Delta_i \in support(G, \Pi)$ such that $\Delta \vdash_t \Delta_i$ with Δ'

(S-$\wedge$) $\Delta \vdash_t (\Delta_1 \wedge \Delta_2)$ with Δ'' if

         $\Delta \vdash_t \Delta_1$ with Δ' and $\Delta' \vdash_t \Delta_2$ with Δ''

(S-$A(i)$) $\Delta \vdash_t \Box[S_1]\ldots[S_n]A$ with Δ if $\Box[S_1]\ldots[S_n]A \in \Delta$

(S-$A(ii)$) $\Delta \vdash_t \Box[S_1]\ldots[S_n]A$ with Δ' if

         $\Delta \cup \{\Box[S_1]\ldots[S_n]A\} \vdash_f [S_1]\ldots[S_n]\neg A$ with Δ'

(F-G) $\Delta \vdash_f G$ with Δ' if

         $support(G, \Pi) = \{\Delta_1, \ldots, \Delta_m\}$ and

         $\Delta \vdash_f \Delta_1$ with Δ_1' and $\Delta_1' \vdash_f \Delta_2$ with Δ_2' and $\ldots$

         and $\Delta_{m-1}' \vdash_f \Delta_m$ with Δ'

(F-$\wedge$) $\Delta \vdash_f (\Delta_1 \wedge \Delta_2)$ with Δ' if $\Delta \vdash_f \Delta_1$ with Δ' or $\Delta \vdash_f \Delta_2$ with Δ'

(F-A) $\Delta \vdash_f \Box[S_1]\ldots[S_n]A$ with Δ' if

         $\Box[S_1]\ldots[S_n]A \notin \Delta$ and $\Delta \vdash_t [S_1]\ldots[S_n]\neg A$ with Δ'

The abductive procedure presents some strong similarities to the one defined in [17] which, given a set of hypotheses Δ, computes an acceptable set of hypotheses Δ' such that $\Delta \subseteq \Delta'$.

We can prove that our abductive procedure is *sound* with respect to the admissibility semantics, i.e., given a program Π and a goal G, if the procedure (starting with $\Delta = \emptyset$) terminates successfully, then the computed global set of hypotheses Δ is an admissible solution for G.

As regards completeness, our procedure is affected by those problems that are common to most procedures computing negation as failure. In particular, since it is strongly based on the auxiliary *support*, which could be non terminating, in general it is not complete. Under the hypothesis that, whenever called along a derivation of a given goal G, the procedure *support* terminates (and thus returns a finite set of hypotheses), the abductive procedure is *complete*.

Example 3. Let us consider the computation of the query

     $G = \{n\_battery, n\_bulb\} \Rightarrow \{light\_off\} \Rightarrow n\_battery$

from program P in Example 1. The proof of this goal succeeds with the set of assumptions $\Delta = \{\Box[n\_battery, n\_bulb][light\_off]n\_battery\}$:

$\emptyset \vdash_t \{n\_battery, n\_bulb\} \Rightarrow (\{light\_off\} \Rightarrow n\_battery)$

$\emptyset \vdash_t \Box[n\_battery, n\_bulb][light\_off]n\_battery$

since $support(G, \Pi) = \{\Delta_1\}$ and $\Delta_1 = \{\Box[n\_battery, n\_bulb][light\_off]n\_battery\}$;
$\Delta_1 \vdash_f [n\_battery, n\_bulb][light\_off]\neg n\_battery$
  by rule (S-A(ii));
$\Delta_1 \vdash_f \Box[n\_battery, n\_bulb][light\_off]n\_bulb$
  since $support(\ [n\_battery, n\_bulb][light\_off]\neg n\_battery, \Pi) = \{\Delta_2\}$ and
  $\Delta_2 = \{\Box[n\_battery, n\_bulb][light\_off]n\_bulb\}$;
$\Delta_1 \vdash_t [n\_battery, n\_bulb][light\_off]\neg n\_bulb$
  by rule (F-A);
$\Delta_1 \vdash_t \Box[n\_battery, n\_bulb][light\_off]n\_battery$
  since $support(\ [n\_battery, n\_bulb][light\_off]\neg n\_bulb,\ \Pi) = \{\Delta_3\}$ and
  $\Delta_3 = \{\Box[n\_battery, n\_bulb][light\_off]n\_battery\}$.

By rule (S-A(ii)), the computation succeeds, since the assumption $\Box[n\_battery, n\_bulb][light\_off]n\_battery$ is already in Δ_1.

To complete the example, let us consider the call of the procedure *support* in one of the intermediate steps of computation. For instance, the computation for *support* $(\ [n\_battery, n\_bulb][light\_off]\neg n\_battery, \Pi\ )$ is the following:

$\epsilon \mid \{\} \vdash_o [n\_battery, n\_bulb][light\_off]\neg n\_battery$
$\{n\_battery, n\_bulb\}, \{light\_off\} \mid \{\} \vdash_o \neg n\_battery$
$\{n\_battery, n\_bulb\}, \{light\_off\} \mid \{n\_battery\} \vdash_o \bot$
$\{n\_battery, n\_bulb\}, \{light\_off\} \mid \{n\_battery\} \vdash_o light\_on \wedge light\_off$
$\{n\_battery, n\_bulb\}, \{light\_off\} \mid \{n\_battery\} \vdash_o light\_on$
$\{n\_battery, n\_bulb\}, \{light\_off\} \mid \{n\_battery\} \vdash_f voltage \wedge n\_bulb$
$\{n\_battery, n\_bulb\}, \{light\_off\} \mid \{n\_battery\} \vdash_f n\_battery \wedge n\_bulb$
$\{n\_battery, n\_bulb\}, \{light\_off\} \mid \{n\_battery\} \vdash_f n\_bulb$

which succeeds with assumption set $\Delta_2 = \{\Box[n\_battery, n\_bulb][light\_off]n\_bulb\}$.

Notice that in the above derivation $\Delta_1 \vdash_f \Box[n\_battery, n\_bulb][light\_off]n\_bulb$. This proves that the assumption $\Box[n\_battery, n\_bulb][light\_off]n\_bulb$ fails, i.e., it cannot be assumed, given the set of assumptions Δ_1. Therefore the goal $light\_off \Rightarrow (n\_battery \wedge n\_bulb)$ fails from P.

By a simple extension of the procedure it would be possible to record also failing assumptions, so as to return them together with the successful ones as a result of the procedure. Not only would this make the procedure more efficient, but also provide more informative results.

5   Top-down vs. Bottom-up Revision Policy

In the previous sections we have used a diagnosis example as our running example, and we have employed the prioritization mechanism to assign different priorities to the assumptions of normal behaviour of components. In [10] it has been shown that the policy adopted in $CondLP^+$ is well suited to perform several types of defeasible reasoning, and, in particular, to represent prioritized normal defaults.

According to the revision policy adopted in $CondLP^+$, rather than removing an inconsistency as soon as it arises, the proof procedure maintains the consistency of logical consequences of the database by making use of the priority among facts. The proof procedure keeps track of the (possibly inconsistent) sequence of updates that have been performed, and it reasons on that to determine the facts which hold in the current situation. Another approach, that has been followed in $CondLP$ [9], is that of removing facts responsible for an inconsistency each time the inconsistency is detected. In particular, $CondLP^+$ resolves the inconsistencies by considering the assumptions from those with the highest preference to those with the lowest preference (top-down), while $CondLP$ follows a bottom-up approach.

This matter has also been discussed in [1], where the problem of combining theories and resolving inconsistencies among them has been studied. As observed in [1], the top-down approach is more informative then the bottom-up one.

The top-down approach we have adopted so far in this paper can be suitably used to perform default reasoning, and in particular, to deal with exceptions. However, there are other applications for which the bottom-up approach is better suited. In fact, the bottom-up approach allows to model more naturally the dynamic change from one state to another one. Let us consider, for instance, the problem of "reasoning about actions". We present an example to show that the bottom-up policy is better suited to deal with the frame problem than the top-down policy.

In the following example we will make use of a first order language, though for the discussion of the predicate case we refer to [10]. We introduce a predicate $holds(G, [A_1, \ldots, A_n])$ which takes a goal G and a sequence of actions $[A_1, \ldots, A_n]$ as arguments, and whose meaning is that G holds after the sequence of actions has been performed. In general, we will formalize a causal law for action A with the following clause:

$$Preconditions \wedge (\{Effects\} \Rightarrow holds(G, [A_1, \ldots, A_n])) \rightarrow holds(G, [A, A_1, \ldots, A_n]),$$

i.e., G holds after a sequence of actions $[A, A_1, \ldots, A_n]$, if the *Preconditions* of A hold in the current state, and the goal G holds after performing the sequence of actions $[A_1, \ldots, A_n]$ in the state obtained by updating the current state with the *Effects* of A. Therefore the atoms in *Effects* will block the persistency of all atoms inconsistent with them, whereas all the other atoms which are true in the current state will also be true in the updated state.

Consider a case when we want to deal with an incompletely specified initial state: the abductive procedure allows to compute those assumptions on the initial state that are needed to explain observations on later states. The following is also an example involving the ramification problem.

Example 4. Assume there are two switches. When both of them are on, the light is on. The action of toggling a switch changes the state of the switch from on to off and vice-versa. Let Π be the following set of clauses and constraints:

$$G \rightarrow holds(G, [])$$
$$off_1 \wedge (on_1 \Rightarrow holds(G, L)) \rightarrow holds(G, [toggle_1 | L])$$

$$off_2 \wedge (on_2 \Rightarrow holds(G, L)) \rightarrow holds(G, [toggle_2|L])$$
$$on_1 \wedge (off_1 \Rightarrow holds(G, L)) \rightarrow holds(G, [toggle_1|L])$$
$$on_2 \wedge (off_2 \Rightarrow holds(G, L)) \rightarrow holds(G, [toggle_2|L])$$
$$on_1 \wedge on_2 \rightarrow light\_on \qquad on_1 \wedge off_1 \rightarrow \bot \qquad on_2 \wedge off_2 \rightarrow \bot.$$

Consider the initial situation in which the state of the two switches is undetermined. We represent the initial situation with an empty set of facts $I = \{\}$. In order to allow assumptions on the initial state, we introduce an (inconsistent) set of assumptions $S = \{off_1, off_2, on_1, on_2\}$. Also, we assume that, after the sequence of actions $toggle_1$, $toggle_2$, the light has been observed to be on. In order to explain this observation, we may ask the goal $G=\{off_1, off_2, on_1, on_2\} \Rightarrow (\{\} \Rightarrow holds(light\_on, [toggle_1, toggle_2]))$. We expect that the goal succeeds by assuming that both $switch_1$ and $switch_2$ are initially off. Actually the top-down abductive procedure succeeds with the set of assumptions: $\Delta = \{\Box[S][I][on_1]off_2, \Box[on_1][on_2]on_1\}$. Note that, the first assumption requires off_2 to be true after the action $toggle_1$, but it does not force off_2 to hold in the initial state. On the contrary, the assumption that on_2 holds in the initial state (i.e. assumption $\Box[S][I]on_2$) is compatible with Δ. This is not the expected behaviour, since we do not want the goal $G'=\{off_1, off_2, on_1, on_2\} \Rightarrow (\{\} \Rightarrow (on_2 \wedge holds(light\_on, [toggle_1, toggle_2])))$ to succeed.

The main problem with the top-down approach, which in some cases may lead to unexpected results, is that an assumption which, at some point, has been rejected can be successively recovered (for instance, off_2 has been rejected in the initial state, in which on_2 holds, but off_2 holds after action $toggle_1$, which should not affect the switch 2).

Both the proof procedure presented for the language and its abductive semantics can be modified to adopt the bottom-up policy, which provides a clearer notion of state change. When we want to adopt a bottom-up policy, we want an assumption A belonging to S_1 to be visible in a context $S_1, \ldots, S_n$, provided it is visible at each intermediate step, i.e., after the update S_2, after S_3, and so on, up to S_n. Of course, this requires a suitable modification of the abductive semantics presented in section 3. In particular, the notion of attack can be modified as follows: given two sets of abducibles Δ_1 and Δ_2, Δ_1 attacks Δ_2 if for some $\Box[S_1]\ldots[S_n]p \in \Delta_2$,

$$\Pi \cup \Delta_1 \models_L [S_1][S_2]\neg p \text{ or } \ldots \text{ or } \Pi \cup \Delta_1 \models_L [S_1]\ldots[S_n]\neg p.$$

Accordingly, the abductive proof procedure in Section 4 can be modified by changing rules (S-A(ii)) and (F-A(i)) as follows.

(S-A(ii)) $\Delta \vdash_t \Box[S_1]\ldots[S_n]A$ with Δ' if
     $\Delta \cup \{\Box[S_1]\ldots[S_n]A\} \vdash_f [S_1][S_2]\neg A$ with Δ' and
     $\ldots$
     and $\Delta \cup \{\Box[S_1]\ldots[S_n]A\} \vdash_f [S_1]\ldots[S_n]\neg A$ with Δ'

(F-A) $\Delta \vdash_f \Box[S_1]\ldots[S_n]A$ with Δ' if $\Box[S_1]\ldots[S_n]A \notin \Delta$ and
     $\Delta \vdash_t [S_1][S_2]\neg A$ with Δ' or

$$\dots$$
$$\text{or } \Delta \vdash_t [S_1] \dots [S_n] \neg A \text{ with } \Delta'$$

Coming back to Example 4 above, in the bottom-up procedure the goal G succeeds computing also the assumption $\Box[S][I] o\!f\!f_2$: for $o\!f\!f_2$ to hold after the action $toggle_1$, it must also hold in the initial state. Hence, goal G' fails.

6 Conclusions and Related Work

In this paper we have proposed an abductive semantics for the hypothetical language $CondLP^+$ within the argumentation framework. Also, we have developed an abductive proof procedure for the language which is sound with respect to the semantics. The language contains a revision mechanism to restore consistency when an hypothetical update violates some integrity constraint, and is capable of performing some counterfactual and defeasible reasoning. We have shown some examples of use of the abductive procedure, in simple cases concerning diagnosis and reasoning about change. In particular, we have discussed how different revision policies may be required to formalize different types of reasoning.

The abductive semantics proposed in this work is different form the one defined in [10] (which has the properties of stable model semantics), and it is a weaker one. In particular, it gives up the totality requirement of stable model semantics (whose consequence is that abductive solutions do not always exist).

Since the logical characterization of $CondLP^+$ is a modal logic, there are strong connections between this language and the one presented in [7], which contains a modal operator $assume[L]$ (where L is a literal) to represent addition and deletion of atomic formulas. However, differently from [7] where a model-based approach to revision is adopted, here we have followed a *formula-based approach*, so that the semantics of an update depends on the syntax of the formulas present in the program (see [18]).

$CondLP^+$ is also strongly related to the work in [3], where the problem of reasoning with multiple sources of information is studied for the case when knowledge sources are (consistent) sets of positive and negative literals, and there is no underlying (protected) theory. Also in [3] a modal language is adopted.

The problem of maintaining consistency in face of contradictory information has been widely studied in belief revision and in database theory (see [18] for a survey). Concerning logical databases, [14] provides a procedure for reasoning over inconsistent data, with the aim of supporting temporal, hypothetical and counter-factual queries. As in [14] in $CondLP^+$ earlier information is superseded by later information, but, as a difference, update operations are explicitly provided inside the language.

The language we have defined is somewhat related to other logic programming languages which deal with explicit negation and constraints and are faced with the problem of removing contradictions when they occur. In particular, Pereira et al. in [15, 16] present a contradiction removal semantics and a proof-procedure for extended logic programs with explicit negation and integrity constraints. Furthermore, Leone and Rullo [13] develop a language in which programs can be structured in partially ordered sets of clauses (with explicit negation). Both of mentioned languages do not support updates.

References

1. C. Baral, S. Kraus, J. Minker, V. S. Subrahmanian. Combining knowledge bases consisting of first-order theories. *J.Automated Reasoning*, vol.8, n.1, pp. 45–71, 1992.
2. M. Baldoni, L.Giordano, and A.Martelli. A multimodal logic to define modules in logic programming. In *Proc. 1993 International Logic Programming Symposium*, pages 473–487, Vancouver, 1993.
3. L. Cholvy. Proving theorems in a multi-source environment. In *Proc. IJCAI-93*, pages 66–71, Chambery, 1993.
4. P. M. Dung. Negations as hypotheses: an abductive foundation for logic programming. In *Proc. ICLP-91 Conference*, pages 3–17, 1991.
5. P. M. Dung. On the acceptability of arguments and its fundamental role in non-monotonic reasoning and logic programming. In *Proc. IJCAI93*, pages 852–857, 1993.
6. K. Eshghi and R. Kowalski. Abduction compared with negation by failure. In *Proc. 6th ICLP*, pages 234–254, Lisbon, 1989.
7. L. Fariñas del Cerro and A. Herzig. An automated modal logic for elementary changes. In P. Smets et al., editor, *Non-standard Logics for Automated Reasoning*. Academic Press, 1988.
8. D. M. Gabbay and N. Reyle. NProlog: An extension of Prolog with hypothetical implications.I. *Journal of Logic Programming*, (4):319–355, 1984.
9. D. Gabbay, L. Giordano, A. Martelli, and N. Olivetti. Conditional logic programming. In *Proc. 11th ICLP*, Santa Margherita Ligure, pages 272–289, 1994.
10. Dov Gabbay, Laura Giordano, Alberto Martelli, and Nicola Olivetti. Hypothetical updates, priority and inconsistency in a logic programming language. In *Proc. LPNMR-95*, LNAI 928, pages 203–216, 1995.
11. M. Gelfond and V. Lifschitz. Logic programs with classical negation. In *Proc. ICLP-90*, pages 579–597, Jerusalem, 1990.
12. A.C. Kakas, P. Mancarella, P.M. Dung. The acceptability semantics for logic programs. In *Proc. 11th ICLP*, Santa Margherita Ligure, pages 504–519, 1994.
13. N. Leone and P. Rullo. Ordered logic programming with sets. *J. of Logic and Computation*, 3(6):621–642, 1993.
14. S. Naqvi and F. Rossi. Reasoning in inconsistent databases. In *Proc. of the 1990 North American Conf. on Logic Programming*, pages 255–272, 1990.
15. L.M. Pereira, J.J. Alferes, J.N. Aparicio Contradiction Removal within the Well Founded Semantics. *Proc. LPNMR-91*, (1991) 105–119.
16. L.M. Pereira, C.V. Damasio, J.J. Alferes Diagnosis and Debugging as Contradiction Removal. *Proc. LPNMR-93*, (1993) 316–330.
17. F. Toni and A. Kakas. Computing the acceptability semantics. In *Proc. LPNMR-95*, pages 401–415, 1995.
18. M. Winslett. *Updating Logical Databases*. Cambridge University Press, 1990.

Commands in Dialogue Logic

Roderic A. GIRLE
Philosophy Department
University of Auckland, Auckland, NEW ZEALAND
e-mail: r.girle@auckland.ac.nz

Keywords: dialogue logic, imperatives, formal models.

Introduction

Dialogue-logic was independently revived this century by the work of Charles Hamblin ([1970]) and Paul Lorenzen ([1960]). This paper follows in the tradition of Hamblin, Mackenzie ([1979, 1984]) and Walton ([1984]).

The motivations for Hamblin's dialogue-logic arose from a set of problems to do with fallacious reasoning. Some logicians have argued that logic has to do only with valid argument. For these logicians the only concern is whether an argument is valid or invalid. Unfortunately, there are some arguments which are valid, but when used in an applied logic context are fallacious. Begging the question arguments are usually taken to be valid but fallacious.

Hamblin argued that there is no satisfactory traditional theory of fallaciousness ([1970]). He proposed that a more general account of argument might provide a framework for a better understanding of fallacious valid arguments. The more general account would be an account of debate and dialogue, called "dialectic" by Hamblin.

Hamblin began work in constructing dialogue-logics in which fallacies such as *begging the question, many questions* and *inconsistency* would be avoided or detected. Most of these systems consist of a set of rules about what can *legally* occur in a (rational debate) dialogue. The systems cope with narrowly defined dialogues in which assertions, questions, reason giving, and the retraction of assertions occur. Dialogue which breaches the rules is *illegal* dialogue. (see Barth and Martens [1982] for a bibliography. We note in particular: Hamblin [1970], Mackenzie [1979, 1984], Walton [1984] and Walton and Krabbe [1995])

We can view these systems from a quite different, and broader, perspective. Dialogue logics can be seen as a sub-species of joint activity systems. Dialogue logics are systems which give necessary conditions for dialogue to be a joint activity of rational (non-fallacious) debate. Breach of the rules indicates a *failure* of the joint activity. The main concern of this paper is to set dialogue-logic in this wider context of *joint activity*.

One of the limitations of the dialogue-logics is that rarely have any provision for *commands* or *instructions*. Even commands common to rational debate, such as the command to clarify a point, are missing. This might partly be due to the difficulty of coping with commands in logic.

Nevertheless, command dialogue occurs in many joint activities. Learning, training, air traffic control, putting plans into effect, and myriads of everyday activities involve a mix of command and debate. The simple case of a parent interacting with a child will show how commands and debate are mixed. The parent says, "Put on a jacket." The child responds, "Why do I need to put on a coat?" The parent answers, "Because it's cold outside."

Such mixtures of command (instruction) and debate can be more or less rational. An instructional dialogue system would assist in analysing such interactions and in developing comprehensible machine human interactions in similar contexts, such as computer assisted learning.

In what follows we will show how dialogue-logic can be extended to broader joint activity systems by the addition of commands. We begin with a brief general introduction to Hamblin style dialogue-logics. In the second main part of the paper we set out a particular Hamblin style logic: **DL3**. In the third part of the paper we extend the logic in the new direction of command dialogue. We set out an instructional dialogue logic, **IDL3**.

The Dialogue Logic: DL3

There are many formal dialogue systems. Despite differences between the systems, they have several things in common.

There are four main elements in most dialogue-logics. The first element is a taxonomy of locutions. The locutions include: statements, responses of various sorts, questions of various kinds, and withdrawals. Locutions are used by the participants in a dialogue to form a sequence of locution events. A dialogue is a *sequence of locution events*. In setting out a dialogue we number locutions to indicate their order in the dialogue. These numbers are somewhat like the numberings of formulas in a proof.

The second element is a set of *commitment stores*, one for each participant in the sequence. Commitment stores are neither deductively closed nor necessarily logically consistent.

The third element is a set of *Commitment Store Rules*. Each participant's commitment store is added to and subtracted from according to what statements, questions, answers and withdrawals are used by participants in the dialogue. For example, if a participant asserts that P, then P is added to everyone's commitment store. If anyone disagrees, then they must explicitly deny P. Such a condition gives expression to the notion that we mostly believe what people say.

The fourth element is a set of *Interaction Rules* to stipulate the legal sequence of locution events. For example, a question of the form "Why do you believe that P?" must be followed either by the reasons from which one is to draw the conclusion that P or a denial that one believes that P.

We set out the rules for a dialogue-logic, **DL3**, which is based on the systems **DL**

(Girle [1993]), **DL2** (Girle [1994]), and **BQD** (Mackenzie [1979, 1984]). For **DL3** there are just two participants, X and Y. In setting out rules below we will use *S* for the speaker and *H* for the hearer. There are nine sorts of locutions allowed: *statements of three kinds, declarations, withdrawals, tf-questions, wh-questions, challenges,* and *resolution demands.*

- The *categorical statements* are statements such as *P, not P, P and Q, P or Q, If P then Q* and *statements of ignorance (I do not know whether or not P)*. The last is abbreviated to ι *P*.
- The *reactive statements* are *grounds (Because P)*, abbreviated to ·.· *P*.
- The *logical statements* are *immediate consequence conditionals* such as: If *P* and *P implies Q*, then *Q*.
- A *term declaration* is the utterance of some *term*, say *t*.
- The *withdrawal* of *P* is of the form *I withdraw P, I do not accept P, not P,* or *I no longer know whether P*. The first and second are abbreviated as − *P*.
- The *tf-questions* are of the form *Is it the case that P?*, abbreviated to *P ?*.
- The *wh-questions* are of the form *What (when, where, who, what, which) is an (the) F ?*. The strict logical form is *(Qx)Fx*, where **Q** is the interrogative quantifier, and for each such formula there will be an associated statement *(∃x)Fx*. (Mackenzie [1987])
- A *challenge* is of the form *Why is it supposed to be that P?*, abbreviated to *Why P?*.
- The *resolution demands* are of the form *Resolve P*.

Each locution event is represented in the formal representation of a dialogue in an ordered triple of a number, an agent and the agent's locution. The number is the number of an event in the dialogue sequence. For example, the statement *P* uttered at the *nth* step in the dialogue by X is represented as *<n, X, P>*. We also allow for *justification sequences*. They are four-tuples consisting of the antecedent of a conditional, the conditional, its consequent, and a challenge of the consequent. For example: *<P, If P then Q, Q, Why Q ?>*

We now set out the rules of **DL3** with comments on their significance and operation. There are seven **Commitment Store Rules**:

(C1)   *Statements*: After an event *<n, S, P>*, where *P* is a statement, unless the preceding event was a challenge, *P* goes into the commitment stores of both participants.
(It is assumed that everyone agrees with statements unless and until they deny them or withdraw them. The inclusion of the full ordered pair is so that there is a record in the commitment store of the historical order of the locutions included.)

(C2)   *Defences*: After the event *<n, S, ·.· P>*, when:
     Why Q? and *Q* are in the speaker's commitment store,
     the *justification sequence* :
     <P, If P then Q, Q, Why Q?>, and

P and *If P then Q* go into the commitment stores of both participants.

The challenge: *Why Q?*

is removed from the commitment stores of both participants.

(If someone gives reasons for a statement *Q*, then the reason, its assumed conditional connection, and exactly what is justified go into the commitment stores of both participants. This allows us to keep track of why statements are in the commitment stores.)

(C3) *Withdrawals*: After any of the following three events:

$<n, S, - P>$ or

$<n, S, not P>$ or

$<n, S, \iota P>$, where *P* is in *S's* commitment store, then

(a) the statement *P* is removed from the speaker's commitment store, and

(b) if the withdrawal was of form *not P* or ιP, then the withdrawal goes into the commitment stores of both participants, and

(c) if *P* is a statement associated with a wh-question, say *(Qx)Fx*, then *P* is removed from the hearer's commitment store and *(Qx)Fx* is removed from the commitment stores of both participants, and

(d) if the withdrawal was preceded by the event $<n-1, H, Why P ?>$, then *Why P ?* is removed from the commitment stores of both participants, and

(e) if $<P, If P then Q, Q, Why Q?>$

is in the speaker's commitment store, then it is removed; and if *Q* is in either participants' store, then it is removed.

(This is a fairly gentle, but not the gentlest, withdrawal rule. Given that commitment stores are neither deductively closed nor essentially consistent, only the obvious justification is withdrawn. A ruthless recursive withdrawal rule is found in Girle [1994]. A far more complex approach is in Walton and Krabbe [1995].)

(C4) *Challenges*: After the event $<n, S, Why P?>$, the challenge, *Why P?*, goes into the commitment stores of both participants.

If *P* is not in the hearer's commitment store then: *P* goes into the hearer's commitment store.

If *P* is in the speaker's commitment store, it is removed.

If the *P* is present in the speaker's commitment store as part of a justification sequence, the justification sequence is removed.

(Although it might seem strange to put *P* into the hearer's commitment store, the hearer can withdraw it or deny it (see *(v)*(a) below and *C3* above). Also, if *P* is in the speaker's commitment store it is withdrawn because, if the speaker has no problem about the statement, the challenge should not have been issued. It should be noted that this is not an altogether unproblematic explanation. The speaker might want to discern whther or not the hearer has reasons for asserting *P* other than the speaker's. We will bypass this point for the moment.)

(C5) *Information*: After the event $<n, S, (Qx)Fx>$, the associated statement, *(∃x)Fx*, goes into the commitment stores all participants, and the wh-question, *(Qx)Fx*, goes into the hearer's commitment store.

(If the speaker asks *"What is the F?"*, then the assumption is that there is at least one *F*.)

(C6) *Reply*: After the declaring of a *term* in an event, say $<n, S, t>$, when the previous event was a wh-question, say $(Qx)Fx$, then Ft goes into the commitment stores of both participants. Ft is known as the *wh-answer to* $(Qx)Fx$.

(C7) *True/false*: After the event $<n, S, P?>$, if the statement P is in the speaker's commitment store it is removed.

There are eight **Interaction Rules**:

(i) *Repstat*:

No statement may occur if it is in the commitment stores of both participants.
(This rule prevents vain repetition and helps stop begging the question. From an everyday rhetorical perspective it is unrealistic, but in the ideal dialogue it is appropriate.)

(ii) *Imcon*:

A conditional whose consequent is an immediate consequence of its antecedent must not be withdrawn.

(iii) *LogChall*:

An immediate consequence conditional must not be withdrawn.
(These rules, *(ii)* and *(iii)*, prevent the withdrawal or challenge of logical principles.)

(iv) *TF-Quest*:

After $<n, S,$ *Is it the case that P?*$>$ the next event must be $<n+1, H, Q>$, where Q is either

(a) a statement that P, or

(b) a statement that *not P*, or

(c) a withdrawal of P, or

(d) a statement of ignorance *(I do not know whether or not P)*.

(This rule must be read in conjunction with *C1* and *C3*.)

(v) *Chall*:

After $<n, S,$ *Why P?*$>$ the next event must be $<n+1, H, Q>$, where Q is either

(a) a withdrawal or denial of P, or

(b) the resolution demand of an immediate consequence conditional whose consequent is P and whose antecedent is a conjunction of the statements to which the challenger is committed, or

(c) a statement of grounds *acceptable* to the challenger.

We require, at this point, a definition of what an *acceptable* statement of grounds is:

A statement of grounds, *Because P*, is *acceptable* to participant S iff either P is not under challenge by S, or if P is under challenge by S then there is a set of statements to each of which S is committed and to none of which is S committed to challenge, and P is an immediate *modus ponens* consequence of the set.

This definition is discussed at length in Mackenzie [1984].

(When the challenge is issued, see *C4*, the person challenged can either (a) deny any adherence to *P*, or (b) throw the challenge back to the challenger by pointing out that the challenger is committed to *P*, or (c) give a reason acceptable to the challenger.)

(vi)   *Resolve*:

The resolution demand in *<n, S, Resolve whether P>* can occur only if either

(a)   *P* is a statement or conjunction of statements which is immediately inconsistent and to which its hearer is committed, or

(b)   *P* is of the form *If Q then R* and *Q* is a conjunction of statements to all of which its hearer is committed, and *R* is an immediate consequence of *Q*, and the previous event was either

<n-1, H, I withdraw P> or *<n-1, H, Why R?>*.

(The rule above opens the way for keeping statements consistent.)

(vii)   *Resolution*:

After the event *<n, S, Resolve whether P>* the next event must be *<n+1, H, Q>*, where *Q* is either

(a)   the withdrawal of one of the conjuncts of *P*, or

(b)   the withdrawal of one of the conjuncts of the antecedent of *P*, or

(c)   a statement of the consequent of *P*.

(viii)   *Enlightenment*:

After the event, *<n, S, (Qx)Fx>*,

(a)   the next event must be *<n+1, H, Q>*, where *Q* is either

(i)   the declartion of some *term*, say *t*, or

(ii)   the withdrawal or denying of the associated statement *(∃x)Fx*, or

(iii)   a statement of ignorance *(I do not know whether or not (∃x)Fx)*.

and

(b)   in the case of the hearer declaring some term, at the earliest subsequent event the asker of the wh-question must state the wh-answer to the wh-question, or its denial, where the earliest subsequent event is the event separated from the term declaration by nothing other than the asker asking further wh-questions and their wh-answers, where the wh-questions contain only predicates and terms from the wh-answers immediately preceding them.

(This rule allows us to introduce the notion that an event or set of events constitute a *discovery block with respect to the wh-question and the participant who asked it.*. A discovery block ends with a statement by the asker of the wh-question, of the wh-answer for the wh-question. In other words, the asker shows that he/she is satisfied.)

It is simpler if we set out the key points in a Rule Operation Table, **TABLE** 1. There are rows for each of the speaker's, **S**, locutions. There are two commitment store columns for the resultant entries to the speaker's, **S**, and hearer's, **H**, commitment stores. We use plus and minus to indicate what is being added to or subtracted from the commitment stores of speaker and hearer. There is a column for any *required* next locution from the hearer.

S LOCUTION at Step n	S STORE	H STORE	H RESPONSE
categorical statement: P	$+P$	$+P$	
reactive statement: $\therefore R$	$+R$ $+If\ R\ then\ S$ $+<\ ...\ Why\ S\ ?>$ $-Why\ S\ ?$	$+R$ $+If\ R\ then\ S$ $+<\ ...\ Why\ S\ ?>$ $-Why\ S\ ?$	
term declaration: t	$+Ft$	$+Ft$	at some later point Ft or $\sim Ft$
withdrawal $-P$	$-P$		
withdrawal $-(\exists x)Fx$	$-(Qx)Fx$ $-(\exists x)Fx$	$-(Qx)Fx$ $-(\exists x)Fx$	
tf-question $P\ ?$	$-P$		one of $P,\ \sim P,\ \iota P$ or $-P$
wh-question $(Qx)Fx$	$+(\exists x)Fx$	$+(Qx)Fx$ $+(\exists x)Fx$	one of $t,\ \sim(\exists x)Fx$ $\iota(\exists x)Fx$ or $-(\exists x)Fx$
challenge $Why\ S\ ?$	$-S$ $+Why\ S\ ?$	$+S$ $+Why\ S\ ?$	one of $acceptable\ \therefore R$ or $\sim S$ or $-S$ or a resolution demand as in (v)
resolution $Resolve\ P$	$+P$ $+Resolve\ P$	$+P$ $+Resolve\ P$	withdraw part P or state part P as in (vii)

TABLE 1

There are three points to note. First, commitment stores contain much more than just categorical statements. They contain relevant portions of the dialogue content. Questions and challenges are important parts of that content.

Second, a participant's commitment store does not have to be logically consistent. Its logical consistency becomes an issue only if the other participants in the dialogue detect *prima face* logical inconsistency and demand that the inconsistency be resolved.

Third, some of the allowed responses are more complex than can be fitted into the box in the table. This is particularly so with respect to the responses to challenges, wh-questions and resolution demands. The response to a challenge must satisfy conditions of "acceptability", as we have seen in *(vi) Chall:* above. The response to a wh-question must be followed by a questioner's acknowledgement or rejection of the answer. The resolution demands require both a set of pre-conditions and a response which can be one of several kinds. There are three critical rules. Two of them, *(viii)* and *(ix)*, are in the Appendix. The other, *(xiii)*, is discussed below.

The operations table shows what constraints the logic imposes on a dialogue. These rules have been used in formal debates in classrooms. They impose a discipline, but can allow utterly inconsequential debates (see Stewart-Zerba and Girle [1993]).

It is worthwhile considering one of the rules set out above. The rule, *(viii)*, allows the introduction of something called a *discovery block* in a dialogue. A similar sort of thing will be introduced later when we discuss commands.

This rule allows us to introduce the notion that an event or set of events constitute a *discovery block with respect to the wh-question and the participant who asked it..* A discovery block with respect to a wh-question and asker ends with a statement by the asker of the wh-question, of the wh-answer for the wh-question. In other words, the asker shows that he/she is satisfied.

A discovery block is not unlike the *scope* of an assumption in an Anderson and Johnstone style natural deduction system. An assumption can be introduced for the sake of a *Conditional Proof*. Until the assumption is discharged and the conditional introduced, the deduction is incomplete. Analogously, when a wh-question is asked a discovery block is started. Until the asker is satisfied, the discovery block remains incomplete. For example:

>   X   *Where is the Capital of Australia ?*
>   [*The discovery block starts.*
>   The associated proposition is:
>   *There is a place, x, and the Capital of Australia is at (or in) x.*]
>   Y   *The Australian Capital Territory, the A.C.T.*
>   [Y makes a declaration of the place.
>   *The Capital of Australia is in the A.C.T.* goes into both stores.]
>   X   *Where is the A.C.T.?*
>   [X is not satisfied, and can ask further wh-questions.
>   The associated proposition is:
>   *There is a place, x, and the A.C.T. is at (or in) x.*]
>   Y   *Southern New South Wales.*
>   [Y makes a further declaration.
>   *The A.C.T. is in southern New South Wales.* goes into both stores.]
>   X   *The Capital of Australia is in the A.C.T. in southern New South Wales.*
>   [X shows satisfaction by accepting the consequence of the declarations.
>   *The discovery block closes.*]

Here is an example sequence in which there are two participants, X and Y. We shall use X+*P* for *P* is added to X's commitment store if not already there, and X−*P* for *P* is taken out of X's commitment store (if it is there).

Example 1.

Locution Event	Store Changes	Commitment Stores
<*1, X, P*>	X+*P*	
	Y+*P*	

(The statement goes into both stores *(C1)*.)

X's Store	*P*
Y's Store	*P*

<*2, Y, Why P ?*> Y+*Why P ?*
Y−*P*
X+*Why P ?*

(This is a challenge to *P*. See rule *(C4)*.)

X's Store	*P Why P ?*
Y's Store	*Why P ?*

<*3, X, ∴ Q*> X+<*Q, If Q then P, P, Why P ?*>
X+*Q*
X+*If Q then P*
X−*Why P ?*
Y+<*Q, If Q then P, P, Why P ?*>
Y+*Q*
Y+*If Q then P*>
Y+*P*>
Y−*Why P ?*

(Grounds are given and added to both stores together with the conditional which links the grounds to what they support *(v)*. At this stage *Q* is not yet under challenge. Since the challenge to *P* has been met, it is removed from both stores.)

X's Store	*P* *Q If Q then P* <*Q, If Q then P, P, Why P ?*>
Y's Store	*P* *Q If Q then P* <*Q, If Q then P, P, Why P ?*>

<4, Y, Why Q ?>   Y+*Why Q ?*
                         Y−*Q*
                         Y−<*Q, If Q then P, P, Why P ?*>
                         X+*Why Q ?*

(This is the challenge to *Q*. *Q* is removed from Y's store, and the justification sequence in which *Q* occurs is also removed. The removal of anything else awaits further challenges or withdrawals.)

X's Store	*P* *Q* *If Q then P* <*Q, If Q then P, P, Why P ?*> *Why Q ?*
Y's Store	*P* *If Q then P* *Why Q ?*

<5, X, − Q>   X−*Q*
                     X−<*Q, If Q then P, P, Why P ?*>
                     X−*Why Q ?*
                     X−*P*
                     Y−*Why Q ?*
                     Y−*P*

(Almost everything in the stores now has to be removed. *If Q then P* does not have to be withdrawn under *(C3)*. But, *P* has to go under *(C3)*, because its proposed justification has been taken away. *P* will have to be restated if X wants to continue to hold to it.)

X's Store	*If Q then P*
Y's Store	*If Q then P*

The commitment stores have not much left in them. Principles of minimal change might have left *P* in the commitment stores of both X and Y at the last step in the dialogue. So, *(C3)* is not a minimal change rule. There are a range of possibilities for withdrawal rules. We have taken the "middle way". Discussion of the range of possibilities will have to wait on another paper.

Instructional Dialogue

There is nothing in **DL3** to deal with commands or instructions. Most dialogue, but especially computer assisted learning dialogue, contains commands, and inputs which are responses to commands. One could certainly also envisage a human-robot dialogue interface in which commands would be an essential element.

It is also clear from the work of rhetoricians that there are instructions which facilitate the dialogue itself. Instructions like "Tell me more" and "Would you repeat that, please", have a very important role in dialogue. These are essentially commands at some sort of meta-level. Since they are not necessarily commands about commands, it might be confusing to call them "meta-commands." So we can call them *hyper-instructions*. We begin by concentrating on lower level commands and instructions.

An instructional dialogue logic will have to contain the additional language elements for commands and rational responses to commands. There will have to be *instructions* of the form *Do* α, where α is an action specification. The syntax we rely on here is to be found in Segerberg [1990]. The form is intended to cope with both "Run away." and "Bring it about that you will have run away from here."

There will have to be command or instruction *acceptances*. In this kind of dialogue, there is no room for anything other than verbal interchange. We have to imagine a situation like the situation where all instructions are sent and received over a telephone. When commands are issued, the recipient has to acknowledge the command. This is not unlike what happens on the bridge of a ship when the helmsman repeats the captain's orders, or says "Aye aye, sir." If one is doing something under instruction over a telephone, then one has to say things like, "Got that", "O.K.", "Go on."

In rational instructional dialogue the instructed person might want to respond in other than a totally obedient way. They might want to refuse an instruction. A *refusal* might trigger a (rational) debate about why the instruction was given or why it is not being obeyed.

The instructed person might want to assert that it's impossible to do what has been commanded. This is not the same as refusing to obey. In response to the instruction to frouse the globulator, the instructed person might say, "I'd love to, but it's impossible, because there is no globulator here." So, we need *impossibility responses*.

There will have to be indications that an instruction has been carried out, when it has been. These can be called *completion* statements. If one is doing something under instruction over a telephone, then one has to say things like, "Done that." or "It's been done." All completion inputs are verbal. But in everyday dialogue they can be of various kinds. A human instructor would, in some contexts, simply *see* whether the instruction is followed or completed.

It will be convenient to group the last three responses under the overall heading of *inputs*. So we will have *acceptance inputs, refusal inputs, impossibility inputs,* and *completion inputs*.

There will have to be questions about how one is to follow out instructions. These are *method questions*. The instructor says, "Frouse the glubulator." The instructed person says, "How do I frouse a glubulator?" Method questions will often start with 'How'.

Some of the syntax we have for argumentative dialogue will have application in instructional dialogue. Challenges (*Why* questions) may well apply. "Why did you order the glubulator to be froused?" would raise the possibility of other things which might accomplish the instructor's goals. On the other hand, such a question might be answered by a set of practical reason premises.

Expanding the language

We can expand the syntax for **DL3** to give an Instructional Dialogue Language: **IDL3**. In **IDL3** there will be four locutions additional to those in **DL3**: *purpose declarations, instructions, method-questions* and *inputs*. Some of the locutions in **DL3** will be extended in the way set out below.

- The *categorical statements* are statements such as *P, not P, P and Q, P or Q, If P then Q* and *statements of ignorance (I do not know whether or not P).* The last is abbreviated to ιP.
  Statements of the form: α *is done* are treated below as *inputs*. But they can be treated as categorical statements, and can occur in conditionals such as *If α is done then β is done.*

- A *purpose declaration* is the utterance of some *aim*, say *In order to do α.* The abbreviation is: *To do α*

- The *instructions* are generally of the form *X orders Y to do α,* This is abbreviated to $X > Y$ *do* α.

- The *withdrawal* of *P* is of the form *I withdraw P, I do not accept P, not P,* or *I no longer know whether P.* Instructions can also be withdrawn: *I withdraw the order for X to do α.* The first and second are abbreviated as $-P$. The last as: $-Y > X$ *do* α

- The *method-questions* are *How is α done ?* The abbreviation is: *How α?*

- The *challenges* are *Why is it supposed to be that P?* or *Why should α be done?* The abbreviations are: *Why P?, Why α?*

- The *resolution demands* are: *Resolve whether P* and *Does X (still) wish to order Y to do α ?.* The abbreviations are: *Resolve P, Resolve X > Y do α*

- The *inputs* are either positive statements of *acceptance (X agrees to do α)* or negative statements of *completion (α is done)* or *refusal (X refuses to do α)* or of *impossibility (It's not possible for X to do α).* (where *X* is the agent agreeing to or refusing to obey the instruction, or finding it impossible to obey).

The rules for **IDL3** follow on from the rules for **DL3**. There are four new **Commitment Store Rules**:

(C8) *Instructions*: After the event *<n, S, S > H do α>.* the instruction goes into the commitment stores of all participants.
(We have to keep track of who gave and who received any instruction.)

(C9) *Instruction Stack*: After the event *<n, S, How α?>* the method question is entered into the commitment stores of all participants.
- (a)   If the next event is *<n+1, H, H > S do β>*, then
        <H > S do β, If β is done then α is done, How α?> and *H > S do β*
        go into the commitment stores of all participants.

(b)   If the next event is $<n+1, H, If \beta$ is done then α is done$>$, then

$<If \beta$ is done then α is done, How α ?$>$

goes into the commitment stores of all participants.

(The conditions *C7* and *C8* are not mutually exclusive, but could both be called into play after a method-question. Note the subtle distinction between the two possible answers to the method question.)

(C10) *Inputs*: After the event $<n, S, I>$, where *I* is any of the four inputs, *I* goes into the commitment stores of all participants. If the input is either a refusal to do α or a statement that it's impossible for *S* to do α or a statement that α is done, then any instruction which contains *S do* α is removed from the commitment stores of all participants.

(See *(xi)* below. If the hearer refuses to obey, or says it's impossible to obey, or says that the instruction is completed, then there is no use going further with that instruction.)

(C11) *Aims*: After the event $<n, S, To\ do\ \beta>$, when the previous event was

$<n-1, H, Why\ do\ \alpha$?$>$, then

$<If\ \alpha$ is done then β is done, To do β, Why do α ?$>$ and

α *is done in order to do* β

go into the commitment stores of all participants.

(If someone asks *"Why do* α?" and the answer is the declaring of an action *aim*, then the assumption is that if α is done then the aimed at action will get done.)

There are five new **Interaction Rules**:

(ix)   *InsReason*:

A challenge of the form *Why do* α? can be made by a participant only if

$X > Y\ do\ \alpha$ is in the participant's commitment store and *Why do* α? is not.

(This rule prevents the challenging of instructions not given or of instructions already challenged and dealt with.)

(x)   *Instr-Chall*:

After the event $<n, S, Why\ do\ \alpha$?$>$ the next event must be $<n+1, H, Q>$, where *Q* is either

(a)   a withdrawal of $X > Y\ do\ \alpha$, or

(b)   a statement of grounds (*Because P*), or

(c)   a declaration of purpose (*In order to do* β).

(This rule should be read together with *C2*, *C4* and *C10*.)

(xi)   *Obey*:

After the event $<n, S, S > H\ do\ \alpha>$ the next event must be $<n+1, H, Q>$, where *Q* is either

(a)   an input of the form *Y agrees to do* α, or

(b)   a method question of the form *How* α?, or

(c)   some *wh-question* concerning an item or set of items mentioned in the instruction, or

(d)   a challenge of the form *Why do* α?, or

(e)   an input of the form α *is done*, or

(f)   an input of the form *Y refuses* α, or

(g)   an input of the form *Y cannot do* α.

(This rule requires verbal indication that either the instruction has been agreed to, or that there is some problem with carrying it out, or that information is required in order to carry it out, or explanation is required, or that it has been carried out, or that it has been declined, or that it cannot be carried out.)

(xii)   Completion:

If at any point in a dialogue after an event $<n, S, S > H \ do \ \alpha>$, there has been neither $<n+i, H, H \ refuses \ \alpha>$ nor $<n+i, Z, H \ cannot \ do \ \alpha>$ nor $<n+i, Z, \alpha \ is \ done>$ (where Z is either S or H), then at some subsequent point in the dialogue there must be either

(a)   the event $<n+i, Z, I>$, where I is an input of the form α *is done*, or

(b)   the event $<n+i, S, W>$, where W is the withdrawal of $S > H \ do \ \alpha$

(This rule requires eventual verbal indication that either the instruction has been carried out, or that it has been withdrawn. This rule introduces the notion that an instructional dialogue can be divided into *instruction completion blocks*. An instruction completion block begins with an instruction and ends with the completion of that instruction. This is analogous to the *discovery block* we discussed earlier.)

(xiii) Method:

After the event $<n, S, How \ is \ \alpha \ done \ ?>$ the next event must be $<n+1, H, Q>$, where Q is either

(a)   an instruction of the form $H > S \ do \ \beta$, where β is necessarily distinct from α, and where β may be complex in the sense set out in Segerberg [1990], or

(b)   an action conditional of the form *If* β *is done then* α *is done*. where β is necessarily distinct from α, and where β may be complex, or

(c)   an input of the form $S \ cannot \ do \ \alpha$

(This rule allows for the respondant to either order something to be done ("You will have to do β"), or state a general sufficient condition, or state that the questioner cannot do α.)

Conclusion

There are several matters which require further work. First, we need to develop further analysis of justification sequences. The conditional which connects the reasons with the questioned statement needs further study. In a deductive logic context this would be seen as an entailment, in an inductive context as a relative probability, in a default reasoning context as a default.

Second, an extended analysis of kinds of withdrawal is needed. The general principle in classical belief revision theory (see Gardenfors [1988]) is that of minimal change. Our rules are stronger than minimal. A more complex approach is given in Walton and Krabbe [1995].

Third, a great deal of work is needed in the study of dialogue *strategy*. The systems set out above give only some of the necessary conditions for rational activity. The rules are like the inference rules of a natural deduction system. We can take those rules and a set of premises, empty or not, and deduce to our heart's content. But mostly we want to get to some specific conclusion by the shortest and most elegant deduction. In dialogue, we want to get to the truth of some statement, or accomplish some worthwhile activity. We need to understand the strategies for getting to the end point.

References

Barth, E.M. and Martens J.L. 1982, *Argumentation Approaches to Theory Formation*, Amsterdam, John Benjamins B.V.

Cohen, P.R. Levesque, H.J. Nunes, J.H.T. and Oviatt, S.L. 1990, "Task-Oriented Dialogue as a Consequence of Joint Activity", *Proceedings of Pacific Rim International Conference on Artificial Intelligence*, Nagoya, Japan, Nov. 14-16, 203-208

Gärdenfors, Peter. 1988, *Knowledge in Flux*, London, MIT Press.

Girle, R.A. 1986, "Dialogue and Discourse", *Proceedings of the Fourth Annual Computer Assisted Learning in Tertiary Education Conference*, Adelaide, South Australia, 123-135

Girle, R.A. 1992, "ELIZA and the Automata", *Proceedings of the Third Annual Conference of AI, Simulation and Planning in High Autonomy Systems*, Perth, Australia, July 8-10, 287-293.

Girle, R.A. 1993, "Dialogue and Entrenchment", *Proceedings of the Sixth Florida Artificial Intelligence Research Symposium*, Ft. Lauderdale, April 18-21, 185-189.

Girle, R.A. 1994, "Knowledge: Organised and Disorganised", *Proceedings of the Seventh Florida Artificial Intelligence Research Symposium*, Pensacola Beach, May 5-7, 198-203.

Hamblin, C.L. 1970, *Fallacies*, London, Methuen.

Lorenzen, P. 1960 "Logik und Agon" *Atti del XII Congresso Internazionale di Filosofia (Venezia, 12-18 Settembre 1958), IV: Logica, linguaggio e communicazione*. Florence: Sansoni, 187-194.

Mackenzie, J.D. 1979, "Question- Begging in Non-Cumulative Systems", *Journal of Philosophical Logic* (8), 117-133

Mackenzie, J.D. 1984, "Begging the Question in Dialogue", *Australasian Journal of Philosophy*, 62(2), 174-181

Mackenzie, J.D. 1987, Private correspondence.

Segerberg, K. 1990, "Validity and Satisfaction in Imperative Logic", *Notre Dame Journal of Formal Logic* 31(2), 203-221

Walton, D.N. 1984, *Logical Dialogue-Games and Fallacies*, Lanham, University Press of America.

Walton, D.N. and E.C.W. Krabbe, 1995, *Commitment in Dialogue*, Albany, State University of New York Press.

Ideal and Real Belief about Belief

Enrico Giunchiglia[1] Fausto Giunchiglia[2,3]

[1] DIST - University of Genoa, Genoa, Italy
[2] DISA - University of Trento, Trento, Italy
[3] IRST, Povo, 38100 Trento, Italy
enrico@dist.unige.it   fausto@irst.itc.it

Abstract. The goal of this paper is to provide a taxonomic characterization of the possible forms of belief about belief. Our analysis is performed in two steps. First, we give a set-theoretic specification of beliefs about beliefs, *i.e.* we characterize them as particular sets satisfying certain closure conditions. Second, we define a set of constructors which generate all and only the objects belonging to these sets. The constructors defined turn out to be inference rules inside a multicontext system. We provide some intuitions about the expressive power and conceptual importance of the multicontext systems defined by proving and discussing some equivalence results with some important modal systems.

1 Introduction

We start from the abstract notion of *reasoner*, where by reasoner we mean anything which is capable of having *beliefs*, *i.e.* facts that it is willing to accept as true. Some examples of reasoners are agents (as discussed in, *e.g.*, [1]), views about agents (as discussed in, *e.g.*, [2]), and frames of mind (as discussed in, *e.g.*, [3]). By *belief about* (a reasoner's) *belief* we mean a belief which states that a reasoner has a certain belief; by *observer*, a reasoner which is capable of having beliefs about beliefs. A reasoner (observer) is said to be *ideal* if its set of beliefs (beliefs about beliefs) satisfies certain idealized properties. Dually, a reasoner (observer) is said to be *real* when it is *incorrect* or *incomplete* —*i.e.* it believes more or less beliefs— with respect to a reasoner (observer) taken as reference.

Our goal in this paper is to provide a taxonomic characterization which uniformly describes the possible forms of ideal and real belief about belief. Our analysis is performed in two steps. First we give a *set-theoretic* specification of belief and belief about belief. The intuition is that a reasoner (an observer) is ideal depending on whether its set of beliefs (beliefs about beliefs) satisfies certain closure conditions. Then, following what is standard practice in software specification, we define a set of constructors which *presents* all and only the beliefs (beliefs about beliefs) of such ideal reasoners (observers). The possible sources of reality can then be characterized by looking at the possible ways in which the identified constructors can be modified to "go wrong".

Following this methodology, the paper is structured as follows. In Section 2 we specify beliefs and beliefs about beliefs as certain sets of formulas constituent the notions of reasoner and observer, respectively. In this Section we also provide the constructors for these sets. These constructors turn out to be inference rules inside a multicontext system. In Section 3 we define the constructors and the multicontext system for ideal belief about belief, MBK⁻. In Section 4 we characterize the possible forms of reality by analyzing how the constructors for ideal belief about belief can be modified to generate incompleteness or incorrectness. The tricky part are the definitions of incompleteness and incorrectness. In Section 5 we compare MBK⁻ and some multicontext systems for real belief with some important modal systems. The goal of this analysis is to provide intuitions about the expressive power and conceptual importance of the multicontext systems defined. Finally, we give some concluding remarks in Section 6.

Three observations. As a first point, it is important to notice that many interesting results can be found in the literature which provide a characterization of examples of ideal and real belief about belief. The notions used are those of *(logical) omniscience* and *non (logical) omniscience* (see *e.g.* [3, 4, 5]). Even though ideality/reality and (non) logical omniscience are closely related (as suggested by the equivalence results in Section 5) these notions present some differences. For instance, as discussed below, reality does not coincide with not ideality, and ideality is a more granular and weaker notion than the notion of omniscience captured by the modal system $\mathcal{K}$. However, the main difference lies in the approach. (Non) omniscience has often (mainly?) been characterized model-theoretically, and, as far as we know, never set-theoretically or proof-theoretically. We believe that the taxonomic analysis provided below would be very hard to do, if not impossible, by using model-theoretic notions. Incompleteness or incorrectness arise because of a failure in the reasoning mechanisms. For instance, a reasoner may not be aware of some inference rules, or not be able to iterate the reasoning process beyond a certain limit. A semantic characterization, instead, is based on the notions of interpretation and model. Interpretations are given inductively following the structure of formulas, and are independent of how theories are defined and used and, therefore, of how the reasoning is done. It is not clear how to provide a uniform map from failures in the reasoning into "appropriate" truth conditions given on the structure of formulas.

As a second observation, the work described in this paper is part of a project which aims at the development of a specification language which will allow us to define formally the way belief about belief is mechanized inside complex reasoning systems. As also briefly discussed in Section 2, all the basic definitions are motivated by and model what is the standard practice in many application areas, *e.g.* natural language understanding, planning and multiagent systems.

Finally, as a third observation, in this paper we treat only the case of non-iterated belief and of a single reasoner having beliefs about another reasoner. This is mainly due to the lack of space. The extension is doable, though not trivial, and can be done along the lines of what already done in some specific cases, see [6, 7]. This will be the topic of a following longer paper.

2   Reasoners and observers

A set-theoretic specification must characterize the relevant sets (in this case the set of beliefs) in terms of well defined set-theoretic constructions performed starting from basic sets whose elements must be obvious to determine. Moreover, for this specification to be meaningful, both the basic sets and the constructions should be defined in a way to provide an extensional characterization of the process of formation of beliefs. Intuitions and motivations about this process come from various implemented applications (see, *e.g.*, [1, 8, 9]). In these applications, reasoners are essentially provided with two capabilities. First, reasoners are provided with *reasoning capabilities*. This is implemented by providing reasoners with a set of facts which constitute their basic knowledge, and with some inference engine which allows them to deduce facts from what they already know. Second, reasoners are provided with *interaction capabilities*. This is implemented by providing reasoners with a set of mechanisms which allow them to compute new facts because of the interaction with other reasoners.

We abstractly represent the reasoning capabilities of a reasoner R as a pair $\langle L, T \rangle$, where L is a set of first order sentences and $T \subseteq L$. L is the *language* and T is the set of *beliefs* or *theorems* of R. (Notationally, in the following, R_i stands for the pair $\langle L_i, T_i \rangle$.) In this paper we restrict ourselves to the propositional case. Formulas are thus propositional combinations of propositional constants or expressions of the form B("A"), where A is a propositional formula. The latter formulas are called *belief sentences*. A in B("A") is called the *argument* of B. To represent abstractly the interaction capabilities of a reasoner we need two reasoners, one having beliefs, the other having beliefs about the first. This is captured by the notion of *belief system*.

Definition 1 Belief System. Let B be a unary predicate symbol. Then a *belief system (for B)* is a pair of reasoners $\langle R_0, R_1 \rangle_{\mathrm{B}}$. The parameter B is *the belief predicate*, R_0 is the *observer* and R_1 is the *observed reasoner* of $\langle R_0, R_1 \rangle_{\mathrm{B}}$.

R_0's beliefs about R_1's beliefs are represented by the set of belief sentences which are part of the beliefs of R_0.

Generating a belief system requires presenting two reasoners plus the fact that a reasoner, *e.g.* the observer, is capable of deducing facts, *e.g.* belief sentences, just because the other reasoner has deduced some other beliefs. We do this by using a special kind of formal system, called *multicontext system (MC system)*. An MC system is a pair $\langle \{C_i\}_{i \in I}, BR \rangle$, where $\{C_i\}_{i \in I}$ is a family of *contexts* and BR is a set of *bridge rules*. A context C is a triple $\langle L, \Omega, \Delta \rangle$, where L is the language, Ω the set of axioms and Δ the set of inference rules of C. C can be intuitively thought as an axiomatic formal system "interacting", via bridge rules, with other contexts. (Notationally, in the following, a context C_i is implicitly defined as $\langle L_i, \Omega_i, \Delta_i \rangle$.) Bridge rules are inference rules with premises and conclusion in different contexts. Thus for instance, the bridge rule

$$\frac{C_1 : A_1}{C_2 : A_2}$$

allows us to derive the formula A_2 in context C_2 just because the formula A_1 has been derived in context C_1. (Notationally, we write $C_i : A$ to mean the formula A in the context C_i.) Derivability in a MC system MS, in symbols $\vdash_{MS}$, is defined in [6]; roughly speaking, it is a generalization of Prawitz' notion of deduction inside a Natural Deduction System [12] [4].

The MC systems which present belief systems are characterized as follows:

Definition 2 MR$^-$. An MC-system $\langle \{C_0, C_1\}, BR \rangle$ is an MR$^-$ system[5] if BR is

$$\frac{C_1 : A}{C_0 : \mathrm{B}(``A")} \; \mathcal{R}^{\mathrm{B}}_{up} \qquad \frac{C_0 : \mathrm{B}(``A")}{C_1 : A} \; \mathcal{R}^{\mathrm{B}}_{dn}$$

and the restrictions include:

$\mathcal{R}^{\mathrm{B}}_{up}$: $C_1 : A$ does not depend on any assumption in C_1.
$\mathcal{R}^{\mathrm{B}}_{dn}$: $C_0 : \mathrm{B}(``A")$ does not depend on any assumption in C_0.

The rule on the left is called reflection up, the one on the right, reflection down. The restrictions on these rules are such that a formula can be reflected up or reflected down only if it does not depend on any assumptions, $i.e.$ if it is a theorem. Reflection up allows C_0 to prove $\mathrm{B}(``A")$ just because $C_1$ has proved $A$, while reflection down has the dual effect, $i.e.$ it allows $C_1$ to prove $A$ just because $C_0$ has proved $\mathrm{B}(``A")$. The intuition is that the reasoner R_0 (whose reasoning capabilities are presented by C_0) believes $\mathrm{B}(``A")$, $i.e.$ it believes that the reasoner $R_1$ (whose reasoning capabilities are presented by $C_1$) believes $A$, if $R_1$ actually believes $A$ and the bridge rules allow $R_0$ to derive $\mathrm{B}(``A")$.

The sense in which an MR$^-$ system presents a belief system is made precise by the following definition:

Definition 3 Belief system presented by an MR$^-$ System. An MR$^-$ System $MS = \langle \{C_0, C_1\}, BR \rangle$ *presents* the belief system $\langle R_0, R_1 \rangle_{\mathrm{B}}$ if $T_i = \{A \mid \vdash_{MS} C_i : A\}$ $(i = 0, 1)$.

Definition 3 says that the beliefs of a reasoner R_i consist of all the theorems proved by the MR$^-$ system, which belong to L_i.

3 Ideal reasoners and observers

Reasoners and observers can be categorized depending on their capability of having beliefs and beliefs about beliefs, respectively. We say that reasoners and observers are *ideal* if they satisfy certain closure properties in the sense made precise by the following definition.

---

[4] Multicontext systems can be thought of as particular Labelled Deductive Systems (LDS)s [10, 11]. In particular, multicontext systems are LDSs where labels are used only to keep track of the context formulas belong to, and where inference rules can be applied only to formulas belonging to the "appropriate" context.

[5] The abbreviations MR and MBK (where the letters "M", "R", "B" and "K" stand for "Multi", "Reflection", "Belief" and the modal logic $\mathcal{K}$, respectively) come from [6]. The minus in the super-script, when used, indicates that the defined system is weaker than the corresponding system in [6].

Definition 4 Ideality. Given a belief system $\langle R_0, R_1 \rangle_{\text{B}}$, we say that:

- R_i $(i = 0, 1)$ is an *Ideal Reasoner* if
  - (i) L_i is closed under the formation rules for propositional languages, and
  - (ii) T_i is closed under tautological consequence.
- R_0 is an *Ideal Observer* if
  - (i) $L_1 = \{A \mid \text{B}(\text{``}A\text{''}) \in L_0\}$, and
  - (ii) $T_1 = \{A \mid \text{B}(\text{``}A\text{''}) \in T_0\}$.

The intuition is that R_0 and R_1 are ideal reasoners if they are able to believe all the (tautological) consequences of what they know. R_0 is an ideal observer if it is able to believe all and only those belief sentences whose argument is a belief of R_1. Notice that no request is made about the specific elements of the sets defining an ideal reasoner or an ideal observer. Thus, for instance, whether a formula belongs to the set of beliefs of an ideal reasoner is left undetermined; we only require that, if this is the case, so must be for all its consequences. This choice has been motivated by the fact that this is how, in practice, ideality is dealt with in the literature. (Thus, for instance, $\mathcal{K}$ is the modal system for omniscience no matter what theoretic axioms are added; adding new axioms will only force us to consider smaller sets of models.)

The closure conditions for ideality can be captured by posing appropriate restrictions on MR^- systems. Let us consider the following definition:

Definition 5. An MR^- system $\langle \{C_0, C_1\}, BR \rangle$ is an MBK^- system if the following conditions are satisfied:

- (i) L_0 and L_1 contain a given set P of propositional letters, the symbol for falsity $\perp$, and are closed under implication[6];
- (ii) Δ_0 (Δ_1) consists of the "Rules for Propositional Logic", *i.e.* of

$$\frac{A_1, \ldots, A_k}{A} \, RPL_k$$

such that $(A_1 \wedge \ldots \wedge A_k) \supset A$ is a tautology and $k \in \{0, 1, 2\}$ [7];
- (iii) $L_1 = \{A \mid \text{B}(\text{``}A\text{''}) \in L_0\}$;
- (iv) the restrictions on the applicability of reflection up/down are only those listed in Definition 2.

Theorem 6. *If $\langle R_0, R_1 \rangle_{\text{B}}$ is the belief system presented by an MBK^- system,*
- (i) *R_0 and R_1 are ideal reasoners, and*
- (ii) *R_0 is an ideal observer.*

The proof is straightforward. It is sufficient to observe that the first two conditions in Definition 5 ensure that each R_i is an ideal reasoner $(i = 0, 1)$; while the last two ensure that R_0 is an ideal observer.

---

[6] We also use standard abbreviations from propositional logic, such as $\neg A$ for $A \supset \perp$, $A \vee B$ for $\neg A \supset B$, $A \wedge B$ for $\neg(\neg A \vee \neg B)$, $\top$ for $\perp \supset \perp$.

[7] Notice that RPL_1 is a derived inference rule of RPL_2, and thus both Δ_0 and Δ_1 are not minimal. However, the case $k = 1$ allows for a more natural formulation of Definition 16 in Section 5.2.

4 Real reasoners and observers

At first, one is tempted to define reality as not ideality, in the same way as not omniscience is usually defined to be absence of omniscience. However we believe that this is not correct. Consider for instance a reasoner which is not aware of a proposition or does not believe a true sentence, but whose formation and inference rules are complete for propositional logic. This reasoner is an ideal reasoner, however we would like to say that it is also a real reasoner. Differently from what is the case for ideality, reality is intrinsically a relative notion which states the absence of certain properties with respect to a specific reference. When talking of a real reasoner (observer) we mean that such a reasoner (observer) believes too little or too much (*i.e.* it is incomplete or incorrect) with respect to another reasoner (observer) taken as reference. This intuition is already informally articulated, even if limited to beliefs and reasoners, in [7]. In particular, in that paper a reasoner is defined real relatively to another reasoner, independently of (what we have called here) the belief system of which it is part. However, the formalization of these ideas is more complex than it might seem, and, as the technical development discussed below shows, the notion of reality informally introduced in [7] is not totally correct. The key observation is that two reasoners or observers cannot be compared independently of the belief system of which they are part[8].

4.1 Realizing MR$^-$ systems

The starting point is to define when a belief system is (in)correct or (in)complete with respect to another (notationally, we write $\langle R_{A_0}, R_{A_1} \rangle_\text{B} \subseteq \langle R_{B_0}, R_{B_1} \rangle_\text{B}$ to mean $R_{A_0} \subseteq R_{B_0}$ and $R_{A_1} \subseteq R_{B_1}$):

Definition 7. A belief system $\langle R_{E_0}, R_{E_1} \rangle_\text{B}$ is *correct* [*complete*] with respect to a belief system $\langle R_{I_0}, R_{I_1} \rangle_\text{B}$ if $\langle R_{E_0}, R_{E_1} \rangle_\text{B} \subseteq \langle R_{I_0}, R_{I_1} \rangle_\text{B}$ [$\langle R_{I_0}, R_{I_1} \rangle_\text{B} \subseteq \langle R_{E_0}, R_{E_1} \rangle_\text{B}$].

The intuition is that, for instance, in a correct belief system, each reasoner maintains a subset of the beliefs of the corresponding reasoner in the reference belief system. If the beliefs of one reasoner (for example of R_{E_0}) are strictly contained in the beliefs of the corresponding reasoner (R_{I_0}) then the belief system is incomplete. We say that a reasoner [observer] is an *incomplete reasoner* [*incomplete observer*] or an *incorrect reasoner* [*incorrect observer*] with the obvious meaning. We say that $\langle R_{E_0}, R_{E_1} \rangle_\text{B}$ is *real* with respect to $\langle R_{I_0}, R_{I_1} \rangle_\text{B}$, to mean that $\langle R_{E_0}, R_{E_1} \rangle_\text{B}$ is incomplete or incorrect with respect to $\langle R_{I_0}, R_{I_1} \rangle_\text{B}$.

The next step is to "propagate" the notions of (in)correctness and (in)completeness from belief systems to MR$^-$ systems. However things are complicated as a comparison between MR$^-$ systems based simply on set inclusion of the components does not work. For instance, it is easy to think of two different

---

[8] The proofs of the theorems in this Section are in Appendix A.

sets of axioms with the same proof-theoretic power. To solve this problem we introduce a new operation $\oplus$ such that, if $C = \langle L, \Omega, \Delta \rangle$ and $C' = \langle L', \Omega', \Delta' \rangle$ are two contexts, then $C \oplus C'$ is the context $\langle L \cup L', \Omega \cup \Omega', \Delta \cup \Delta' \rangle$. This allows us to give the following definition (notationally, in the following, $\mathrm{MS}_E = \langle \{C_{E_0}, C_{E_1}\}, BR_E \rangle$ and $\mathrm{MS}_I = \langle \{C_{I_0}, C_{I_1}\}, BR_I \rangle$ are MR^- systems presenting $\langle R_{E_0}, R_{E_1} \rangle_\mathrm{B}$ and $\langle R_{I_0}, R_{I_1} \rangle_\mathrm{B}$, respectively):

Definition 8. MS_E is a *correct [complete] realization* of MS_I if $\langle \{C_{E_0} \oplus C_{I_0}, C_{E_1} \oplus C_{I_1}\}, BR_E \cup BR_I \rangle$ and MS_I [MS_E] present the same belief system.

MS_E is *equivalent* to MS_I if it is correct and complete with respect to MS_I. We talk of *realization* to emphasize the process by which the constructors of a real belief system [real reasoner, real observer] are defined starting from those of a reference belief system [reasoner, observer]. Consider for instance the notion of correct realization. MS_E is a correct realization of MS_I if adding its proof-theoretic power to that of MS_I results into a system which still has the same proof-theoretic power as MS_I. From the above definition, it trivially follows that MS_E is a correct realization of MS_I if and only if MS_I is a complete realization of MS_E. As trivial examples, the "empty" system, *i.e.* the system with empty languages, is a correct realization of any reference system MS_I. The "absolutely contradictory" system, *i.e.* the system where the theorems of its two contexts are the same as their languages, is complete with respect to any reference system whose two languages stand in a subset relation with its languages. In the following two more complex examples, for any finite set of propositional formulas Γ, MS_E is a correct but incomplete realization of MS_I.

Example 1. Let MS_I be the smallest MBK^- system. Suppose MS_E is defined as MS_I except that Δ_{E_1} consists of the instances of RPL_0 whose conclusion belongs to Γ. Such a system models a situation of an ideal reasoner R_{E_0} ideally interacting with a non ideal reasoner R_{E_1}.

Example 2. Let MS_I be the smallest MBK^- system. Suppose MS_E is defined as MS_I except that $\mathcal{R}_{up}^\mathrm{B}$ has the additional restriction that the premise shall belong to Γ. This system models the situation of an ideal reasoner R_{E_0} whose interaction capabilities are restricted to the subset Γ of the ideal reasoner R_{E_1}'s beliefs.

To save space, from now on, we consider incompleteness only. With some provisos, all the results presented below can be replicated for incorrectness.

The link between MS_E being an incomplete realization of MS_I and the incompleteness of $\langle R_{E_0}, R_{E_1} \rangle_\mathrm{B}$ with respect to $\langle R_{I_0}, R_{I_1} \rangle_\mathrm{B}$ can now be established.

Theorem 9. *If MS_E is a correct realization of MS_I then*
(*i*) $\langle R_{E_0}, R_{E_1} \rangle_\mathrm{B}$ *is correct with respect to* $\langle R_{I_0}, R_{I_1} \rangle_\mathrm{B}$, *and*
(*ii*) $\langle R_{E_0}, R_{E_1} \rangle_\mathrm{B}$ *is incomplete with respect to* $\langle R_{I_0}, R_{I_1} \rangle_\mathrm{B}$ *if and only if MS_E is an incomplete realization of MS_I.*

The proof is a consequence of the fact that $\langle\{C_{E_0}\oplus C_{I_0}, C_{E_1}\oplus C_{I_1}\},\ BR_E\cup$ $BR_I\rangle$ and MS_I present the same belief system. The second item of Theorem 9 states that we have achieved what we wanted, *i.e.* that incompleteness between two MR^- systems corresponds to incompleteness in the belief systems presented. Notice however that this result holds under the hypothesis that MS_E system is a correct realization of MS_I. This hypothesis is necessary in order to guarantee that $\langle R_{E_0}, R_{E_1}\rangle_B$ is correct with respect to $\langle R_{I_0}, R_{I_1}\rangle_B$. In fact, the vice versa of the first item of Theorem 9 does not hold. That is, MS_E can be an incorrect realization of MS_I and $\langle R_{E_0}, R_{E_1}\rangle_B$ be correct with respect to $\langle R_{I_0}, R_{I_1}\rangle_B$. From Definition 8, correct realizations have the property that the result of adding the components of MS_I to MS_E defines an MR^- system which still presents a correct belief system. Intuitively, this property guarantees that correct realizations generate theorems in a way which is consistent with how theorems are generated by the reference MR^- system.

4.2 Realizing contexts and bridge rules

The next step is to find necessary and sufficient conditions for having realizations of MR^- systems. Via Theorem 9 this provides necessary and sufficient conditions on the presented belief systems.

Definition 10. We say that
- C_{E_0} is an *incomplete* realization of C_{I_0} if MS_E is an incomplete realization of $\langle\{C_{E_0}\oplus C_{I_0}, C_{E_1}\}, BR_E\rangle$;
- C_{E_1} is an *incomplete* realization of C_{I_1} if MS_E is an incomplete realization of $\langle\{C_{E_0}, C_{E_1}\oplus C_{I_1}\}, BR_E\rangle$;
- BR_E is an *incomplete* realization of BR_I if MS_E is an incomplete realization of $\langle\{C_{E_0}, C_{E_1}\}, BR_E\cup BR_I\rangle$.

In Example 1, C_{E_0} is a complete realization of C_{I_0}, C_{E_1} is an incomplete realization of C_{I_1} and BR_E is a complete realization of BR_I. In Example 2, C_{E_0} is a complete realization of C_{I_0}, C_{E_1} is a complete realization of C_{I_1} and BR_E is an incomplete realization of BR_I. Notice that in both examples C_{E_0} is a complete realization of C_{I_0} even though R_{E_0} is strictly contained in R_{I_0}. This is exactly what one would expect since the reasoning capabilities of R_{E_0} and R_{I_0} (modeled by C_{E_0} and C_{I_0} respectively) are the same. Section 5.2 provides two substantial examples of context realization and realization of the bridge rules.

Theorem 11. *MS_E is an incomplete realization of MS_I if and only if at least one of the following three conditions is satisfied:*
- *C_{E_0} is an incomplete realization of C_{I_0};*
- *C_{E_1} is an incomplete realization of C_{I_1};*
- *BR_E is an incomplete realization of BR_I.*

4.3 Realizing the components of contexts

The next and final step is to iterate what done in Section 4.2 to the components of contexts. Via Theorem 9 and Theorem 11, this provides necessary and sufficient conditions on the presented belief systems.

(Notationally, in the following, if C_i is a context, P_i and W_i are the set of atomic formulas and the set of construction rules for L_i, respectively. L_i is therefore defined as the smallest set generated from P_i and closed under W_i, in symbols $L_i = Cl(P_i, W_i)$.)

Definition 12. Let A be one of the letters in $\{P, W, \Omega, \Delta\}$. We say that A_{E_i} is an *incomplete realization of* A_{I_i} if MS_E is an incomplete realization of $\langle \{C_0, C_1\}, BR_E \rangle$, where[9]:

$$C_j = \begin{cases} \langle Cl(P_{E_j} \cup P_{I_j}, W_{E_j}), \Omega_{E_j}, \Delta_{E_j} \rangle & \text{if } j = i \text{ and } A = P; \\ \langle Cl(P_{E_j}, W_{E_j} \cup W_{I_j}), \Omega_{E_j}, \Delta_{E_j} \rangle & \text{if } j = i \text{ and } A = W; \\ \langle L_{E_j}, W_{E_j}, \Omega_{E_j} \cup \Omega_{I_j}, \Delta_{E_j} \rangle & \text{if } j = i \text{ and } A = \Omega; \\ \langle L_{E_j}, W_{E_j}, \Omega_{E_j}, \Delta_{E_j} \cup \Delta_{I_j} \rangle & \text{if } j = i \text{ and } A = \Delta; \\ C_{E_j} & \text{otherwise.} \end{cases}$$

In Example 1, Δ_{E_1} is an incomplete realization of Δ_{I_1}. Definition 12 fixes the intuitively correct but formally wrong classification provided in [7]. That paper discusses in detail the intuitions underlying this classification and provides various examples. As already discussed in [7], the various forms of incompleteness (in the signature, formation rules, axioms, inference rules) model very different intuitions. For instance, the incompleteness in the signature models the case in which a reasoner is not aware of some primitive propositions. This is the case, for example, of the Bantu tribesman in [3] who is not aware that personal computer prices are going down. A "more civilized" tribesman might be aware of computers and their prices, but he might not believe that their prices are decreasing. The latter situation is modeled with a reasoner incomplete in the axioms. Incompleteness in the formation rules and/or inference rules are best suited for modeling the limitation of resources that real reasoners have both in constructing sentences and in proving theorems.

Theorem 13. C_{E_i} *is an incomplete realization of* C_{I_i} $(i = 0, 1)$ *if and only if at least one of the following four conditions is satisfied:*

- P_{E_i} *is an incomplete realization of* P_{I_i};
- W_{E_i} *is an incomplete realization of* W_{I_i};
- Ω_{E_i} *is an incomplete realization of* Ω_{I_i};
- Δ_{E_i} *is an incomplete realization of* Δ_{I_i}.

It is important to notice that the classification provided by Definitions 10 and 12 is exhaustive in the sense that it considers all the constructors on MR^-

---

[9] Strictly speaking, $\langle L_{E_j}, W_{E_j}, \Omega_{E_j} \cup \Omega_{I_j}, \Delta_{E_j} \rangle$ is not assured to be a context unless $\Omega_{I_j} \subseteq L_{E_j}$. More carefully, we should write $\langle L_{E_j}, W_{E_j}, \Omega_{E_j} \cup (\Omega_{I_j} \cap L_{E_j}), \Delta_{E_j} \rangle$. Analogously for $\langle L_{E_j}, W_{E_j}, \Omega_{E_j}, \Delta_{E_j} \cup \Delta_{I_j} \rangle$.

systems. This, together with Theorems 11 and 13, achieves the goal set up in Section 1, that is, these definitions provide an exhaustive classification of all the possible forms and sources of reality.

5 A comparison with modal systems

The most common approach to the formalization of beliefs about beliefs is to take a theory, extend it using a modal operator $\mathcal{B}$, and take $\mathcal{B}A$ as representing a belief about belief. In the resulting modal system Σ, beliefs are represented by the formulas A such that A is provable in Σ. Similarly to above, we do not consider nestings of the $\mathcal{B}$ operator. Formally, a *modal system* Σ is a pair $\langle L, T \rangle$ such that

(*i*) L contains the set $P \cup \{\bot\}$ of propositional letters, is closed under implication and the modal operator $\mathcal{B}$; however, the modal operator $\mathcal{B}$ is not allowed to occur inside the scope of another $\mathcal{B}$;

(*ii*) $T \subseteq L$ and T is closed under tautological consequence.

A formula A is a *theorem* of Σ ($\vdash_{\Sigma} A$) if $A \in T$. A is *derivable* from a set Γ of formulas ($\Gamma \vdash_{\Sigma} A$) if $(A_1 \wedge \ldots \wedge A_n) \supset A \in T$ and $\{A_1, \ldots, A_n\} \subseteq \Gamma$. We also say that Σ is $\mathcal{K}_n$-*classical* if for any propositional formula $(A_1 \wedge \ldots \wedge A_n) \supset A \in T$, we have $(\mathcal{B}A_1 \wedge \ldots \wedge \mathcal{B}A_n) \supset \mathcal{B}A \in T$ ($n = 0, 1, 2$). $\mathcal{K}_1$-classical modal systems are said to be *monotone*, $\mathcal{K}_2$-classical modal systems are said to be *regular* and $\{\mathcal{K}_0, \mathcal{K}_2\}$-classical modal systems are said to be *normal* [13].

In the next subsection we consider various modal logics presented in the literature and state equivalence results with (modifications of) MR$^-$ systems[10]. An MC system MS $= \langle \{C_0, C_1\}, BR \rangle$ and a modal system $\Sigma = \langle L, T \rangle$ are said to be *equivalent* if for any formula A in L_0,

$$\vdash_{\text{MS}} C_0 : A \quad \Longleftrightarrow \quad \vdash_{\Sigma} A^+,$$

where A^+ is the modal counterpart of A, *i.e.* it is obtained replacing any monadic atomic formula M("C") with $\mathcal{M}C$ in A.

5.1 Normal modal systems

We start by studying whether the smallest normal modal system $\mathcal{K}$ and the reasoner R_0 presented by the smallest MBK$^-$ system are equivalent. This is motivated by the fact that the observer presented by an MBK$^-$ system is both an ideal reasoner and an ideal observer; in other words, it is saturated with respect to the properties that we have considered (of reasoning, of observing). Analogously, $\mathcal{K}$ is the smallest normal system which is meant to model omniscient agents (see [4]). However they turn out to be not equivalent. In fact, for any set of propositional formulas $\Gamma \cup \{A\}$, we have

$$\Gamma \vdash_{\mathcal{K}} A \Longrightarrow \{\mathcal{B}A : A \in \Gamma\} \vdash_{\mathcal{K}} \mathcal{B}A.$$

---

[10] For lack of space, the proofs of the theorems in this Section are omitted.

This property gives $\mathcal{K}$ a form of ideality with respect to derivations which is much stronger than the form of ideality possessed by R_0, which is only with respect to theorems. One way to fill the gap between $\mathcal{K}$ and R_0 is to add to the set of axioms of the context C_0 the corresponding of the $\mathcal{K}$ axiom $\mathcal{B}A \supset (\mathcal{B}(A \supset B) \supset \mathcal{B}B)$. Namely, for any propositional formulas A and B, $\mathrm{B}(\text{``}A\text{''}) \supset (\mathrm{B}(\text{``}A \supset B\text{''}) \supset \mathrm{B}(\text{``}B\text{''}))$ should be an element of Ω_0. The resulting system can be easily proven equivalent to $\mathcal{K}$. Another way is to drop the restriction on reflection down and take Δ_0 to be the set of inference rules for Classical Natural Deduction Systems.

Definition 14. An MC system $\langle \{C_0, C_1\}, BR \rangle$ is an MBK system if it satisfies all the conditions of Definition 5 except that
(*i*) Δ_0 is the set of Classical Natural Deduction Rules,
(*ii*) $\mathcal{R}^{\mathrm{B}}_{dn}$ has no restrictions.

The effect of unrestricting reflection down is to obtain the desired correspondence between the consequence relations of C_0 and C_1. For any MBK system $MS = \langle \{C_0, C_1\}, BR \rangle$ and any set Γ of formulas in L_1, we have

$$\Gamma \vdash_{\mathrm{MS}} C_1 : A \Longrightarrow \{C_0 : \mathrm{B}(\text{``}B\text{''}) \mid C_1 : B \in \Gamma\} \vdash_{\mathrm{MS}} C_0 : \mathrm{B}(\text{``}A\text{''}).$$

Intuitively, the possibility of applying reflection down to a formula which is not a theorem (*e.g.* an assumption) amounts to imposing, by hypothetical reasoning, a reasoner's new basic belief. This kind of reasoning is very similar to the reasoning informally described in [9, 14] and implemented in many applications. The effect of taking Δ_0 as the set of Classical Natural Deduction Rules is to guarantee that

$$\Gamma, C_0 : A \vdash_{\mathrm{MS}} C_0 : B \Longrightarrow \Gamma \vdash_{\mathrm{MS}} C_0 : A \supset B$$

also when $\Gamma, C_0 : A \vdash_{\mathrm{MS}} C_0 : B$ is proved across languages, *i.e.* by applying (sequences of) reflection down and reflection up.

Theorem 15. Let $\mathcal{K} = \langle L, T \rangle$ be the smallest normal modal system and MBK $= \langle \{C_0, C_1\}, BR \rangle$ the smallest MBK system. $\mathcal{K}$ and MBK are equivalent.

The above Theorem (similar to Theorem 5.1 in [6]) is a corollary of Theorem 18 (see Section 5.2).

5.2 Classical modal systems

In this Section we consider $\mathcal{K}_{p,q}$ modal systems, *i.e.* systems which are $\mathcal{K}_p$-classical and $\mathcal{K}_q$-classical. Consider the smallest $\mathcal{K}_{p,q}$ system. Then,
- if $p = 0$ and $q = 1$ we have Fagin' and Halpern's logic of local reasoning [3, 4];
- if $p = 0$ and $q = 2$ we have the smallest normal modal system $\mathcal{K}$;
- if $p = q = 1$ we have the smallest monotone modal system $\mathcal{M}$ [13];
- if $p = q = 2$ we have the smallest regular modal system $\mathcal{R}$ [13].

Equivalent MC systems can be obtained by starting from an MBK system $\langle\{C_0, C_1\}, BR\rangle$ (presenting the belief system $\langle R_0, R_1\rangle_B$) by weakening the deductive machinery of C_1 or the reflection principles. In the first case R_1 is no longer guaranteed to be an ideal reasoner; in the second R_0 is no longer guaranteed to be an ideal observer. Notice that to obtain the equivalence result we have to start from MBK and not MBK$^-$. In fact the modal systems mentioned above still possess a (limited) form of ideality with respect to derivations. That is, in any $\mathcal{K}_{p,q}$ modal system Σ if $k \in \{p, q\}$ then

$$A_1, \ldots, A_k \vdash_\Sigma A \Longrightarrow \mathcal{B}A_1, \ldots, \mathcal{B}A_k \vdash_\Sigma \mathcal{B}A.$$

Definition 16. An MC system $\langle\{C_0, C_1\}, BR\rangle$ is an MBK$_{p,q}$ system if it satisfies all the conditions of Definition 14 except that RPL_k in Δ_1 has the further restriction that $k \in \{p, q\}$.

Definition 17. An MC system $\langle\{C_0, C_1\}, BR\rangle$ is an MBK$'_{p,q}$ system if it satisfies all the conditions of Definition 14 except that $\mathcal{R}^B_{up}$ has the further restriction that the premise depends on exactly p or q occurrences of formulas in L_0.

Theorem 18. Let $\mathcal{K}_{p,q}$ be the smallest $\mathcal{K}_{p,q}$-classical system and let MBK$_{p,q}$, MBK$'_{p,q}$ be the smallest MBK$_{p,q}$ and MBK$'_{p,q}$ systems respectively. $\mathcal{K}_{p,q}$ is equivalent to MBK$_{p,q}$ and MBK$'_{p,q}$.

Notice that MBK$_{p,q}$ and MBK$'_{p,q}$ present the same observer. However in MBK$_{p,q}$ the source of reality is in one of the contexts, while in MBK$'_{p,q}$ is in the bridge rules. Such a difference is not captured in modal systems, where both beliefs and beliefs about beliefs are modeled in a unique theory.

Finally, it is easy to prove that MBK, as defined in Theorem 15, and MBK$_{0,2}$, as defined in Theorem 18, are equivalent. The two systems differ only for the set of inference rules in C_1. However, these two sets are equivalent: any inference rule in one set is a derived inference rule in the other.

6 Conclusions

In this paper we have provided a taxonomic analysis of ideal and real belief. As far as we know such an analysis has never been done before. Technically the main novelties are:

- Our analysis is based on a set-theoretic (instead of model-theoretic) characterization of belief and belief about belief;
- Reality does not mean not ideality. Ideality characterizes the presence of certain idealized properties; reality the fact that a real reasoner (observer) computes too little or too much with respect to another reasoner (observer);
- We have characterized how each constructor in an MR$^-$ system affects reality.

In a following longer paper we will extend this analysis to nested belief. We will also provide a more extensive set of equivalence results with various important modal systems. So far we have been able to produce MR$^-$ systems equivalent to all the modal systems we know of. This work provides a unified perspective of the various systems discussed in the literature.

References

1. A. Lansky. Localized Event-Based Reasoning for Multiagent Domains. *Computational Intelligence*, 4:319–339, 1988.
2. K. Konolige. *A deduction model of belief and its logics*. PhD thesis, Stanford University CA, 1984.
3. R. Fagin and J.Y. Halpern. Belief, awareness, and limited reasoning. *Artificial Intelligence*, 34:39–76, 1988.
4. R. Fagin, J.Y. Halpern, Y. Moses, and M. Y. Vardi. *Reasoning about knowledge*. MIT press, 1995.
5. H. Levesque. A logic of implicit and explicit belief. In *AAAI-84*, pages 198–202, 1984.
6. F. Giunchiglia and L. Serafini. Multilanguage hierarchical logics (or: how we can do without modal logics). *Artificial Intelligence*, 65:29–70, 1994. Also IRST-Technical Report 9110-07, IRST, Trento, Italy.
7. F. Giunchiglia, L. Serafini, E. Giunchiglia, and M. Frixione. Non-Omniscient Belief as Context-Based Reasoning. In *IJCAI'93*. Also IRST-Technical Report 9206-03, IRST, Trento, Italy.
8. Y. Wilks and J. Biem. Speech acts and multiple environments. In *IJCAI'79*.
9. A. R. Haas. A Syntactic Theory of Belief and Action. *Artificial Intelligence*, 28:245–292, 1986.
10. D. Gabbay. Labeled Deductive Systems. Technical Report CIS-Bericht-90-22, Universität München – Centrum für Informations und Sprachverarbeitung, 1990.
11. D. Gabbay. Labeled Deductive Systems Volume 1 – Foundations. Technical Report MPI-I-94-223, Max-Planck-institut für Informatik, May 1994.
12. D. Prawitz. *Natural Deduction - A proof theoretical study*. Almquist and Wiksell, Stockholm, 1965.
13. B. F. Chellas. *Modal Logic - an Introduction*. Cambridge University Press, 1980.
14. J. Dinsmore. *Partitioned Representations*. Kluwer Academic Publisher, 1991.

A   Proofs of the Theorems in Section 4

In this section C_i is an abbreviation for $C_{E_i} \oplus C_{I_i}$, MS is the MC system $\langle \{C_0, C_1\}, BR_E \cup BR_I \rangle$ and $\langle R_0, R_1 \rangle_B$ is the belief system presented by MS. If MS' and MS'' are two MR$^-$ systems, we also write MS' $\preceq$ MS'', MS' $\asymp$ MS'' and MS' $\prec$ MS'' as abbreviations for

- MS' is a correct realization of MS'',
- MS' is equivalent to MS'' (or, MS' $\preceq$ MS'' and MS'' $\preceq$ MS'),
- MS' is a correct and incomplete realization of MS'' (or, MS' $\preceq$ MS'' and MS' $\not\asymp$ MS''),

respectively.

A.1 Proof of Theorem 9

By hypothesis, we have that MS_E is a correct realization of MS_I, *i.e.*

$$\langle R_{I_0}, R_{I_1} \rangle_{\text{B}} = \langle R_0, R_1 \rangle_{\text{B}}. \tag{1}$$

Considering the first item, we have to prove that

$$\langle R_{E_0}, R_{E_1} \rangle_{\text{B}} \subseteq \langle R_{I_0}, R_{I_1} \rangle_{\text{B}}. \tag{2}$$

Clearly, $\langle R_{E_0}, R_{E_1} \rangle_{\text{B}} \subseteq \langle R_0, R_1 \rangle_{\text{B}}$. The thesis follows from equation (1).

For the second item we prove that $\langle R_{E_0}, R_{E_1} \rangle_{\text{B}}$ is complete with respect to $\langle R_{I_0}, R_{I_1} \rangle_{\text{B}}$ if and only if MS_E is a complete realization of MS_I.

$$\langle R_{E_0}, R_{E_1} \rangle_{\text{B}} \text{ is complete with respect to } \langle R_{I_0}, R_{I_1} \rangle_{\text{B}} \iff \text{(Definition 7)}$$
$$\langle R_{I_0}, R_{I_1} \rangle_{\text{B}} \subseteq \langle R_{E_0}, R_{E_1} \rangle_{\text{B}} \iff \text{(eq. (2))}$$
$$\langle R_{I_0}, R_{I_1} \rangle_{\text{B}} = \langle R_{E_0}, R_{E_1} \rangle_{\text{B}} \iff \text{(eq. (1))}$$
$$\langle R_0, R_1 \rangle_{\text{B}} = \langle R_{E_0}, R_{E_1} \rangle_{\text{B}} \iff \text{(Definition 8)}$$

MS_E is a complete realization of MS_I.

A.2 Proof of Theorem 11

We prove the two directions of the equivalence separately.

From left to right We prove the counter-positive. By hypothesis, we have that

$$MS_E \asymp \langle \{C_{E_0} \oplus C_{I_0}, C_{E_1}\}, BR_E \rangle, \tag{3}$$

and

$$MS_E \asymp \langle \{C_{E_0}, C_{E_1} \oplus C_{I_1}\}, BR_E \rangle, \tag{4}$$

and

$$MS_E \asymp \langle \{C_{E_0}, C_{E_1}\}, BR_E \cup BR_I \rangle. \tag{5}$$

We have to prove that $\langle R_{E_0}, R_{E_1} \rangle_{\text{B}} = \langle R_0, R_1 \rangle_{\text{B}}$.

From equations (3) and (4) we have $L_{I_0} \subseteq L_{E_0}$ and $L_{I_1} \subseteq L_{E_1}$ respectively. Since $L_i = L_{I_i} \cup L_{E_i}$, we also have $L_i = L_{E_i}$.

We now show that $T_i = T_{E_i}$. Let Π be a proof of $C_i : A$ in MS. We show that $C_{E_i} : A$ is provable in MS_E, by induction on the number n of applications of bridge rules in Π.

$\underline{n = 0}$ If $i = 0$ the thesis follows from equation (3). If $i = 1$ the thesis follows from equation (4).

$\underline{n = m + 1}$ For simplicity, we assume $i = 1$. The case for $i = 0$ is analogous. The proof Π has the form

$$\cfrac{\cfrac{\Sigma_1}{\cfrac{C_0 : \text{B}(\text{``}A_1\text{''})}{C_1 : A_1} \mathcal{R}^{\text{B}}_{dn}} \quad \cdots \quad \cfrac{\Sigma_n}{\cfrac{C_0 : \text{B}(\text{``}A_n\text{''})}{C_1 : A_n} \mathcal{R}^{\text{B}}_{dn}}}{\cfrac{\Sigma}{C_1 : A}}$$

where $C_0:\mathrm{B}("A_1"),\ldots,C_0:\mathrm{B}("A_n")$ are the first occurrences of formulas in C_0 met on the thread from $C_1:A$ to the leaves of Π.

Let $k \in \{1,\ldots,n\}$. Given the restrictions on the bridge rules, we have that

$$\frac{\Sigma_k}{C_0:\mathrm{B}("A_k")}$$

is a proof in MS of $C_0:\mathrm{B}("A_k")$. Hence — Σ_k contains less than n applications of bridge rules — $C_{E_0}:\mathrm{B}("A_k")$ is provable in MS_E, and — from equation (5) — there exists a proof Π_k of $C_{E_1}:A_k$ in MS_E. From this it follows that

$$\frac{\Pi_1 \quad \ldots \quad \Pi_n}{\dfrac{\Sigma}{C_1:A}}$$

is a proof of $C_1:A$ in $\langle\{C_{E_0},C_{E_1}\oplus C_{I_1}\}, BR_E\rangle$. The thesis then follows from equation (4).

From right to left We consider only the incompleteness of C_{E_0}. The other two cases are analogous. Suppose the first condition is satisfied. Hence: $\mathrm{MS}_E \prec \langle\{C_{E_0}\oplus C_{I_0},C_{E_1}\}, BR_F\rangle$ and then $\mathrm{MS}_E \prec \langle\{C_{E_0}\oplus C_{I_0},C_{E_1}\oplus C_{I_1}\}, BR_E\cup BR_I\rangle$.

A.3 Proof of Theorem 13

We consider the case $i = 0$ (the case $i = 1$ is analogous).

From left to right Suppose none of the conditions is satisfied. From $\mathrm{MS}_E \asymp \langle\{\langle Cl(P_{E_0}\cup P_{I_0},W_{E_0}),\Omega_{E_0},\Delta_{E_0}\rangle,C_{E_1}\}, BR_E\rangle$ we have $P_{I_0} \subseteq L_{E_0}$. From $\mathrm{MS}_E \asymp \langle\{\langle Cl(P_{E_0},W_{E_0}\cup W_{I_0}),\Omega_{E_0},\Delta_{E_0}\rangle,C_{E_1}\}, BR_E\rangle$ we also have $L_{I_0} \subseteq L_{E_0}$. Given $\mathrm{MS}_E \asymp \langle\{\langle L_{E_0},W_{E_0},\Omega_{E_0}\cup\Omega_{I_0},\Delta_{E_0}\rangle,C_{E_1}\}, BR_E\rangle$ and $\mathrm{MS}_E \asymp \langle\{\langle L_{E_0},W_{E_0},\Omega_{E_0},\Delta_{E_0}\cup\Delta_{I_0}\rangle,C_{E_1}\}, BR_E\rangle$ it is easy to show that for any proof in $\langle\{C_{E_0}\oplus C_{I_0},C_{E_1}\}, BR\rangle$, there exists a proof with the same conclusion in MS_E.

From right to left We consider only the incompleteness of P_{E_0}. The other two cases are analogous. Suppose the first condition is satisfied. Hence,

$$\mathrm{MS}_E \prec \langle\{\langle Cl(P_{E_0}\cup P_{I_0},W_{E_0}),\Omega_{E_0},\Delta_{E_0}\rangle,C_{E_1}\}, BR_E\rangle$$

and then $\mathrm{MS}_E \prec \langle\{C_{E_0}\oplus C_{I_0},C_{E_1}\}, BR_E\rangle$.

Analogical Reasoning of Organic Reactions Based on the Structurized Compound-Reaction Diagram

Hironobu Gotoda[1], Jianghong An[2], and Yuzuru Fujiwara[2]

[1] Research and Development Department, National Center for Science
Information Systems, 3-29-1 Otsuka, Bunkyo-ku, 112 Tokyo, Japan
gotoda@rd.nacsis.ac.jp
[2] Institute of Information Sciences and Electronics, University of Tsukuba
1-1-1 Tennodai, Tsukuba, 305 Ibaraki, Japan
{ajh,fujiwara}@dblab.is.tsukuba.ac.jp

Abstract. As the number of chemical compounds identified increases
as well as the amount of the associated knowledge of their properties
and reactions, the necessity for developing intelligent database or expert
systems is correspondingly expanding. Such systems are often required
to analogically reason new reactions by recognizing the similarities of
compounds and reactions.

This paper presents a novel method to analogically reason new
reactions based on the structural similarities of compounds and reactions, which are represented by a kind of semantic network, termed
the compound-reaction diagram. The diagram integrates the abstraction
hierarchies of compounds and reaction types, and also the parametric
properties of compounds and reaction conditions. An efficient and user-customizable scheme is developed for retrieving and evaluating analogs,
which is the combination of the structure-based reasoning with the statistical, numerical processing of physical and chemical parameters.

Keywords: analogical reasoning, similarity of concepts, conceptual structures, semantic network, organic compounds, organic
reactions.

1 Introduction

Organic synthesis is a field of interdisciplinary research involving chemistry, biology, medicine, and pharmacy. Due to this diversity of the related research communities, various compounds and reactions have been studied from various perspectives, yielding a large amount of knowledge that mutually overlaps. The total number of organic compounds identified so far is already more than 10 million, which is rapidly increasing every year. Storing and reusing that amount of knowledge requires the application of the database technology to this field.

Many database systems have been constructed for this purpose. Most of them can only store and retrieve itemized facts, using for example compound name or reaction type as a search index. However, such systems become almost useless when the desired compounds or reactions are not stored in the *exact* form [8]. Unfortunately, it is either impractical or impossible to store all such information that the database users will request, since the total number of known compounds or reactions are too large. Therefore, the users often have to go through series of complicated procedures of retrieving *relevant* information and *inferring* the properties of the desired compounds or reactions by *analogy*. To facilitate such tedious tasks, it is necessary to develop an intelligent and efficient scheme for processing similarities of compounds and reactions.

This paper presents a new method to analogically reason organic reactions based on the structural similarities of compounds and reactions. The similarities are defined in such a way that simultaneously considers the structural resemblance of compounds and reaction types, and the common parametric properties of compounds and reaction conditions. To facilitate the processing of similarities, the entire information is structured into a kind of semantic network, termed the compound-reaction diagram. The structurization then leads to a novel scheme of analogical reasoning, which is distinct from those based on the probability or fuzzy measures [7].

2 Related Work

Analogical reasoning is an extension of deductive inference. If $A \longrightarrow B$ represents the logical deduction of fact B from fact A, then $A' \longrightarrow B'$ is said to be an *analog* of $A \longrightarrow B$ if A' and B' are respectively similar to A and B. The use of analogs for constructing intelligent inference systems was discussed in [9], and was realized in various expert systems [3]. Many possibilities usually exist in defining the similarity measures, and accordingly, the processing of the resulting analogs varies from system to system. Which measure is the best one should necessarily depend on the application under consideration. In many cases, however, evaluating the appropriateness of similarity measures and the validity of inferred analogs is outside the domain of such systems.

In organic synthesis, similarities can be defined for compounds and for reactions. Each compound has a distinct molecule structure, which decides the physical or chemical properties of the compound, and the reaction conditions under which the compound is manipulated. Therefore, the similarities of compounds or reactions are closely related to those of the underlying molecule structures [4]. Since it is customary to represent a molecule structure by a graph with labels on its vertices, the similarities defined for graphs can serve as natural substitutes for the similarities of compounds. Formula

$$S(A, A') = |A \cap A'|/(|A| + |A'| - |A \cap A'|), \tag{1}$$

expresses one such similarity, where A and A' represent the pair of graphs between which the similarity is evaluated, $|X|$ represents the number of "segments" contained in graph X, and $A \cap A'$ is the common subgraph of A and A'. The "segment" can simply be a vertex, or a connected component that corresponds to a functional group, etc. The resulting similarity depends on how the segment is defined.

There is another practical method to define the similarity of compounds [1]. This is based on the following observation. In synthesizing a new compound from a set of given starting materials, chemists usually try to minimize the number of reaction steps. Thus the shortest path from the starting materials to the target is sought in the diagram that consists of compounds as vertices and reactions as edges. Usually, the reactions are further grouped into several classes, which are called "general reaction types". Each reaction type is a purely geometric transformation of molecule structures. Therefore, any compound can be related to some known compound through a sequence of general reaction types. The length of the sequence then indicates the similarity between the compounds. Notice that the similarity defined in this way does not necessarily reflect the similarity of physical or chemical properties.

It seems that the similarity of reactions is more difficult to formulate than that of compounds. The general reaction types mentioned above give classification of reactions, which can be interpreted as the similarity in some sense. Alternatively, one can employ the parameters of reaction conditions, such as the temperature, atmosphere, or acidity of the solvent, to define the similarity of reactions. Gasteiger et al. derived similarity measures from a set of physical and chemical parameters of compounds involved in particular types of reactions [2]. In their approach, the metric of the space of parameters is derived so that the similarity of reactions is most visible as the clustering of the corresponding points in that space. The distance in the parametric space is then interpreted as the similarity of reactions.

Our approach is similar to the one proposed by Gasteiger et al., but is more structure-oriented in the following respects: (1) the similarities of compounds and reactions are evaluated simultaneously, (2) the reaction types are allowed to form a *hierarchical structure*, and (3) the evaluation of similarities is performed *locally* rather than globally, thus avoiding the traversal of the entire information space every time an reaction is reasoned analogically. These points will be elaborated upon in the following sections.

3   Structurization of the Compound-Reaction Diagram

Let A be the compound that can be transformed to compound B through reaction R. Such a transformation is written as $A \xrightarrow{R} B$. To represent the relationships among a set of transformations, it is natural to construct a labeled graph $G = (V, E)$, termed the *compound-reaction diagram*, whose vertices V correspond to compounds, and edges E to reactions. Each reaction $R \in E$ is a composite entity consisting of *type, pattern, reagent, catalyst, solvent, temperature, atmosphere*, etc., where the type is a character string, such as "oxidation" or "reduction", and the pattern represents the topological change in the molecule (i.e., graph) structure. The reagent, on the other hand, is typically a compound,

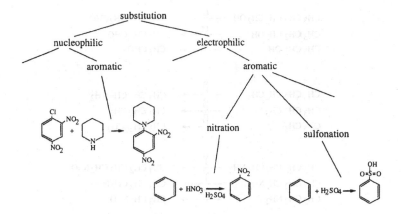

Fig. 1. An abstraction hierarchy of reaction types.

and therefore a pointer to a vertex in the diagram. The other attributes are self-explanatory, and are not further discussed.

There are several reasons to further structurize the compound-reaction diagram. First, the reaction types form a hierarchical structure of generalization and specialization. For instance, the term "electrophilic aromatic substitution" denotes a family of "electrophilic substitution" reactions that apply to aromatic compounds, and "electrophilic substitution" is a subclass of "substitution" reactions. Nitration, sulfonation, and acylation by Friedel-Crafts reaction are examples of electrophilic aromatic substitutions (Fig. 1). Such a hierarchical structure will be useful in analogical reasoning since it restricts the domain in which the validity of particular, generic reaction rules should be evaluated.

The second reason for structurization comes from the algebraic nature of reactions, i.e., when $X \longrightarrow Y$ is a feasible reaction, $ZX \longrightarrow ZY$ is often also a feasible reaction, where ZX represents the addition of the substructure Z to the main structure X. For such kind of additional substructures, $Z = CH_2$ is perhaps the best-known example in organic chemistry. In fact, the series of compounds written as $X(CH_2)_n$ are called *homologous series*, whose chemical properties resemble each other, and the physical properties vary continuously as the number of CH_2 increases or decreases (Fig. 2). Such a resemblance can be a useful clue in analogical reasoning. Similar algebraic nature is also observed for the reagent used in a reaction, i.e., $X \xrightarrow{R} Y$ can often be transformed to $X \xrightarrow{ZR} ZY$ since the reagent R and the left-hand-side compound X participate in the reaction $X \xrightarrow{R} Y$ symmetrically.

Thirdly, there is a general principle of organic chemistry that the carbon skeleton of a compound decides its physical or chemical properties to a certain degree (Fig. 3). Therefore, it might be better to group the set of compounds that have the same carbon skeleton. This observation is further generalized to the following simple principle: the larger the common substructure shared by a pair of compounds becomes, the nearer the pair should be placed to each other in

$$\begin{array}{lll}
CH_3\,CH_2\,CH_2\,CH_2\,OH & \xrightarrow{\ \alpha\ } & CH_3\,CH_2\,CH_2CHO \\
CH_3\,CH_2\,CH_2\,OH & \xrightarrow{\ \alpha\ } & CH_3\,CH_2\,CHO \\
CH_3\,CH_2\,OH & \xrightarrow{\ \alpha\ } & CH_3\,CHO
\end{array}$$

$$\begin{array}{lll}
CH_3\,CH_2\,CH_2\,CH_3 & \xrightarrow{\ \beta\ } & CH_3\,CH_2\,CH=CH_2 \\
CH_3\,CH_2\,CH_3 & \xrightarrow{\ \beta\ } & CH_3\,CH=CH_2 \\
CH_3\,CH_3 & \xrightarrow{\ \beta\ } & CH_3\,CH_2=CH_2
\end{array}$$

$$\begin{array}{lll}
CH_3\,CH_2\,CH_2\,CH_2\,NH_2 & \xrightarrow{\ \gamma\ } & CH_3\,CH_2\,CH_2\,CH_2N=O \\
CH_3\,CH_2\,CH_2\,NH_2 & \xrightarrow{\ \gamma\ } & CH_3\,CH_2\,CH_2N=O \\
CH_3\,CH_2\,NH_2 & \xrightarrow{\ \gamma\ } & CH_3\,CH_2N=O
\end{array}$$

Fig. 2. Examples of homologous series: compounds in the same series exhibit similar properties.

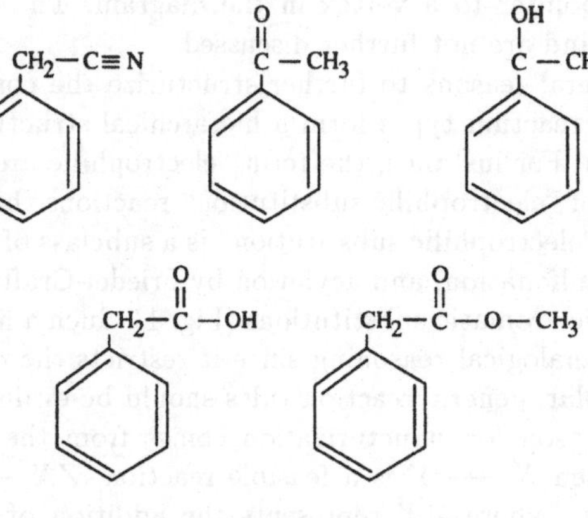

Fig. 3. Compounds that share the same carbon skeleton.

the compound-reaction diagram. Experienced chemists usually do such grouping or placement in mind when planning a synthesis path or estimating the expected properties of a new compound.

These requirements for structurization are met by introducing additional edges to the compound-reaction diagram. To represent the hierarchical structure of reaction types, we span corresponding edges between type values that are associated with the compound-compound (i.e., reaction) edges E in the diagram. To represent the series structure of compounds, we introduce an edge $X \xrightarrow{Z} Y$ between the pair of compounds (X, Y) related to each other by addition or elimination of substructure Z. In practice, functional groups such as hydroxyl and carboxyl groups are chosen as such substructures. Similarly, to represent the sharing of common substructures, edges are added to the diagram among the compounds with labels indicating the substructures involved, e.g., $XZ \xleftrightarrow{Z}$ YZ. Through the addition of these new edges, the compound-reaction diagram becomes similar to the semantic network [6].

For subsequent discussions, we classify the edges in the diagram into three groups: *reaction edges*, *structure edges*, and *hierarchy edges*. The reaction edges are the ones that primarily exist in the diagram and represent real reactions, while the structure edges are the newly introduced ones that corresponds to the series relationships or the sharing of common substructures; these types of edges connect compounds. On the other hand, the hierarchy edges make up the hierarchy of reaction types, and exist only among the reaction edges. In summary, the whole diagram can be expressed as a collection of three graphs

$$G_r = (V_r, E_r) \quad \text{logical structure of compounds}$$
$$G_s = (V_s, E_s) \quad \text{conceptual structure of compounds} \tag{2}$$
$$G_h = (V_h, E_h) \quad \text{conceptual structure of reactions}$$

where E_r, E_s, and E_h represent reaction, structure, and hierarchy edges, respectively. Graph G_r represents the relationships among compounds and reactions, while G_s represents the structural similarities among compounds. Hence,

$$V_r = V_s = \{\text{Set of Compounds}\} . \tag{3}$$

Graph G_h, on the other hand, represents the abstraction hierarchy of reaction types. Since it is a lattice structure constructed on the reaction edges E_r, we have

$$E_r = \{\text{Leaf Node of } V_h\} \subset V_h . \tag{4}$$

Combining (2) with (3) and (4) will yield the structurized compound-reaction diagram, i.e.,

$$\text{Diagram} = \{G_r \cup G_s \cup G_h\}/[V_r = V_s \ \& \ E_r = \text{Leaf}(V_h)] . \tag{5}$$

If each compound is regarded as a concept and each reaction as a logical relationship between a pair of concepts, one can say that G_s represents the *conceptual structure of compounds*, while G_r represents the *logical structure of compounds*. Similarly, the abstraction hierarchy of reactions types G_h can be viewed as the *conceptual structure of reactions*.

The simplest similarity measure of compounds in the structurized diagram is defined to be the length of the shortest path connecting the compounds, i.e.,

$$S_{simple}(\mathbf{cmp}_1, \mathbf{cmp}_2) = d_{G_s}(\mathbf{cmp}_1, \mathbf{cmp}_2) . \tag{6}$$

The path consists of structure edges with certain positive weights assigned. The edge weighting scheme depends on the application domain. For instance, when planning a synthesis path, fewer number of reaction steps and higher yield rates of the intermediate steps are preferred. On the other hand, when estimating the physical properties of compounds, such as the boiling point, that similarity of compounds should further reflect the resemblance of the other physical and chemical properties, i.e.,

$$S_{complex}(\mathbf{cmp}_1, \mathbf{cmp}_2) = f(d_{G_s}(\mathbf{cmp}_1, \mathbf{cmp}_2), d_{E_n}(\mathbf{prop}_1, \mathbf{prop}_2)) \qquad (7)$$

where $\mathbf{prop}_i$ is a real-valued vector in the n-dimensional Euclidean space (termed E^n) consisting of property values of $\mathbf{cmp}_i$, and d_{E^n} is the canonical distance in E^n. Note that function f combines two different metrics d_{G_l} and d_{E_n} to result in an appropriate edge weighting scheme. Finally, when evaluating the feasibility of a certain reaction that has not been well investigated, the reaction and structure edges should be evaluated in concert, i.e., the structural similarity of compounds depends on the type of reaction that the compounds will participate in. More specifically, the similarity of compounds subject to a reaction type $\mathcal{R} \in V_a$ can be written as

$$S_{\mathcal{R}}(\mathbf{cmp}_1, \mathbf{cmp}_2) = f_{\mathcal{R}}(d_{G_s}(\mathbf{cmp}_1, \mathbf{cmp}_2), d_{E_n}(\mathbf{prop}_1, \mathbf{prop}_2)) \qquad (8)$$

and the similarity of reactions as

$$\begin{aligned} S_{\mathcal{R}}(R_1, R_2) = g_{\mathcal{R}}(S_{\mathcal{R}}(\mathbf{cmp}_1^s, \mathbf{cmp}_2^s), S_{\mathcal{R}}(\mathbf{cmp}_1^e, \mathbf{cmp}_2^e), \\ d_{E_m}(\mathbf{cond}_{R1}, \mathbf{cond}_{R2})) \end{aligned} \qquad (9)$$

where each reaction $\mathbf{cmp}_i^s \xrightarrow{Ri} \mathbf{cmp}_i^e$ is of type $\mathcal{R}$, and $\mathbf{cond}_{Ri} \in E^m$ is the vector describing the required reaction condition. Notice that the similarity of reactions is related to that of compounds via function $g_{\mathcal{R}}$. How the similarity measures (8) and (9) can be evaluated in the structurized compound-reaction diagram is our next concern, which will be focused on in the next section.

4 Analogical Reasoning Based on the Structurization

Throughout this section, let us specifically consider the following situation: let A and B be the compounds that belong to the same connected component of the compound-reaction diagram; it is not known whether a feasible reaction R exists that transforms A into B; furthermore, if there is such a reaction, under what condition should the reaction be conducted? Such kind of questions will arise (1) when B is a new compound that has never been synthesized, or (2) when a simpler method for synthesis is sought although there already exists a known path from A to B, that is too complicated to be a practical solution. The typical strategy that chemists employ to solve the above problem is to find a known reaction $A' \longrightarrow B'$ as a reference and evaluate the feasibility of $A \longrightarrow B$ using the similarity relationships $A \sim A'$ and $B \sim B'$ (Fig. 4). This is in fact an example of analogical reasoning.

283

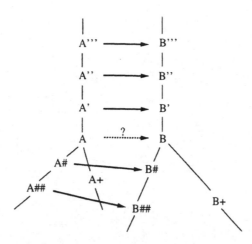

Fig. 4. Simulating the chemists' method: a series of reference reactions (thick arrows) are extracted by traversing the structure edges (thin lines), using which the feasibility of the reaction under consideration (dotted arrow) is evaluated qualitatively.

To simulate the chemists' method of analogical reasoning, let us divide the entire task into two steps: (1) finding the reference reaction, and (2) evaluating the feasibility of the reaction using the similarities of compounds and reactions. For step (1), the crucial issue is to localize the domain in which the reference reactions are searched, because the whole compound-reaction diagram is too immense. The next few paragraphs will details this step. For step (2), each compound is placed in the space of parameters that is isometric to E^n. The key issue here is the reduction of the dimension number n, i.e., the reduction of the number of parameters. Notice that not all parameters are relevant to a specific type of reaction. We apply the principal component analysis frequently used in statistics for this reduction task. Once each compound is located in the parametric space, the similarity of compounds can be measured by the normal distance in E^m ($m \leq n$), and the evaluation of the feasibility of reactions is reduced to a clustering problem in that space. Fig. 5 illustrates this procedure for the compounds and reactions shown in Fig. 4.

Two independent approaches are presented to localize the domain when searching the reference reactions. One is to simply limit the path lengths from A to A' and from B to B'. The other is to restrict the kind of edges to be traversed, i.e., the resulting paths $A \cdots A'$ or $B \cdots B'$ become sequences of edges all of which belong to a predefined class L of edges. For example, suppose that the path to be traversed is restricted to CH_2 addition or elimination edges. Then the path $A \cdots A'$ consists of a homologous series, and so for the path $B \cdots B'$. Suppose also that the CH_2 addition or elimination does not greatly affect the feasibility of the reactions of type R from the A-series of compounds to the B-series. Then evaluating the feasibility of $A \xrightarrow{R} B$ along only these pair of series can be an effective strategy for task reduction. The structure edge class L can

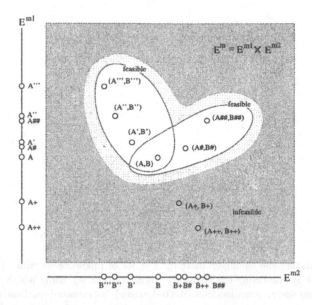

Fig. 5. Evaluating the similarity of compounds and feasibility of reactions in the space of parameters: clustering is performed for pairs of compounds of type $(\mathcal{A}, \mathcal{B})$ in E^m. Projecting these pairs onto the E^{m1} (or E^{m2}) axis induces the similarity relationships among the $\mathcal{A}$ (or $\mathcal{B}$) series of compounds.

be at any level of the abstraction hierarchy, and is subject to the user's control.

In the above description of the second approach, we implicitly assumed that the reactions that transform the A-series of compounds to the B-series were of the same type R. However, this sometimes turns out to be too restrictive to obtain sufficient number of reference reactions comparable to $A \xrightarrow{R} B$. To avoid such situations, the reaction type R can be made controllable by the users.

The procedure of the whole scheme is given below:

1. Select a pair of compounds (A, B), which belong to the same connected component of the compound-reaction diagram.

2. Specify the class L of the structure edges to be traversed.

3. Specify the class R of reaction edges.

4. Specify the maximum path length W.

5. For each pair (L, R), a connected component of the diagram is extracted. The component is in fact a bipartite graph $G_{AB} = (V_A + V_B, E_{LR})$, where any edge that connects a vertex in V_A with a vertex in V_B is of type R, and those that connect vertices in V_A (or V_B) are of type L. Any vertex in V_A (or V_B) can be reached from A (or B) via at most W edges.

6. The vertices of G_{AB} are mapped into E^n, i.e., the space of physical and chemical parameters (e.g, the number of non-hydrogen atoms at the reaction center, bond dissociation energy, partial charges, electronegativity, resonance stabilization of charges, acidity, etc.), which is subsequently reduced to E^m ($m \leq n$) through the principal component analysis. The similarity of compounds is evaluated here by the distance in E^n.

7. Specify the threshold value for clustering.

8. Clustering is performed in E^m using the threshold value. The feasibility of the reaction is evaluated here as the degree of membership to a feasible cluster.

Among the above 8 steps, steps 4 and 7 are optional, and the system can use the default values. Furthermore, steps 2 and 3 can also be skipped. Then, the system will enumerate all possible candidates for the edge type L and the reaction type R, and perform the subsequent steps for all combinations of (L, R). Reaction type R must follow the specific pattern of structural change in $A \longrightarrow B$, and this is in fact the sole constraint on R. On the other hand, L must be chosen as a subset of the structure edges that appear within W distance from either A or B.

Steps 5, 6, and 8 make up the main body of the whole scheme. For steps 6 and 8, various methods have already been reported in the field of statistics or pattern recognition. For step 5, the following simple algorithm, though not very efficient, will work.

```
function retrieve_reference_pairs(A, B, L, R, W)
/* Given a pair of compounds (A, B), retrieve the set of pairs
   that can be reached from (A, B) by traversing at most W edges
   of type L. Each pair (a, b) in the retrieved set should satisfy
   the constraint that a → b exhibits the same structural change
   as observed in reactions of type R. If the pair is actually
   connected together by an R-type reaction edge, the pair is
   said to be feasible; otherwise it is an infeasible pair. */
var pairs Feasible, Infeasible;
var set   Visited;
begin
      Feasible   ← ∅;
      Infeasible ← ∅;
      Visited    ← ∅;
      append_pairs(A, B, L, R, W, Feasible, Infeasible, Visited);
      return(cons(Feasible, Infeasible));
end
```

```
procedure append_pairs(A, B, L, R, W, F, I, V)
var set    N_A, N_B, N_a;
var node  a, b;
begin¹
    if W = 0 or A ∈ V then return;
    V ⇐ A    /* append A to V, i.e., mark A as 'visited' */
    N_A ← get_neighbor(A, L);
    N_B ← get_neighbor(B, L);
    /* N_A (or N_B) is the set of compounds reached from A
       (or B) by traversing one structure edge of type L  */
    foreach a ∈ N_A
        begin²
            N_a ← transform(a, R);
            /* N_a is the set of compounds obtained by applying
               to a the same structural change induced by
               reactions of type R  */
            N_a ← N_a ∩ N_B;
            if N_a ≠ ∅ then
                begin³
                    foreach b ∈ N_a
                    /* If reaction a →R b exists, append (a, b) to
                       the feasible set; otherwise append it to
                       the infeasible set */
                    begin
                        if exist_reaction(a, b) then
                            F ⇐ (a, b);
                        else
                            I ⇐ (a, b);
                        endif
                        append_pairs(a, b, L, R, W - 1, F, I, V);
                    end
                end³
            endif
        end²
end¹
```

5 Implementation

We have implemented a prototype system that automatically constructs the compound-reaction diagram and analogically reasons news reactions based on that structure. The initial data given to the system is a list of reactions (more than 50,000 instances), each of which consists of the molecule structures of the starting and final compounds of the reaction, and the corresponding reaction conditions (reagent, catalyst, solvent, temperature, atmosphere, yield, etc.). This

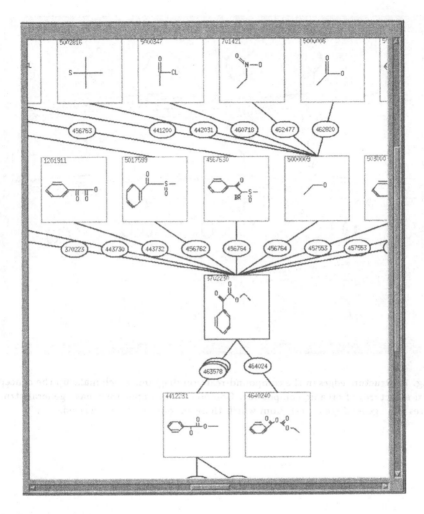

Fig. 6. Reaction edges in the compound-reaction diagram, which make up the logical structure of organic compounds. Numbered ellipses indicate the slots of information associated with the reactions.

immediately results in the initial graph G_r (Fig. 6). Then the system creates structure edges E_s (and equivalently G_s) by comparing every pair of compounds to extract common subgraphs. This is the most time-consuming part in the construction of the compound-reaction diagram. Fig. 7 indicates a part of the graph G_s, where the compounds enclosed by thick rectangles are the ones that exist in the given data, while those enclosed by thin, dotted rectangles are non-existent ones. Synthesizing the dotted compounds is one of the primary objectives of organic chemistry.

To construct the abstraction hierarchy of reaction types, we employed the

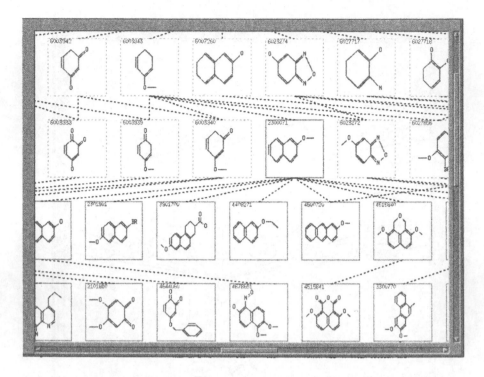

Fig. 7. Structure edges in the compound-reaction diagram, which make up the conceptual structures of organic compounds. Compounds in upper rows have general structures (i.e., general concepts), from which those in lower rows are derived.

method for generating a dynamic thesaurus. The SS-KWIC method was proposed to automatically extract the semantic relationships among technical terms [5], and was found an appropriate tool for our purpose. Once the compound-reaction diagram is structurized in this way, it is possible to analogically reason new reactions. An example of the reasoned reaction is shown in Fig. 8.

Although the result of the reasoning must be ultimately verified by experts, several advantages will follow from using this kind of systems; these include (1) the system can list up all possible reactions without any oversights, which humans are not very good at, and (2) suggestions for reaction conditions can also be presented to the user, which might be inaccurate but could serve as references when conducting experimental verification. At present, our prototype system is being tested by several research groups with some success.

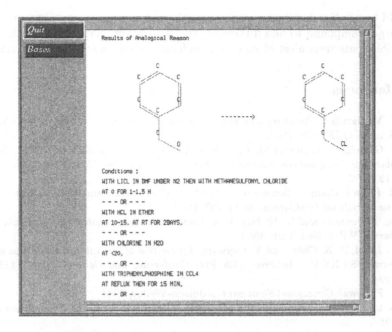

Fig. 8. An example of analogical reasoning, where several candidate reactions are inferred along with the appropriate reaction conditions. These reactions follow the transformation indicated by the arrow.

6 Conclusions

This paper presented a method for analogically reasoning new organic reactions using a set of known reactions based on the structural similarities of compounds and reactions. Toward this end, the knowledge about the known reactions was first structured in the form of a semantic network, called the compound-reaction diagram. The diagram integrated various pieces of information, such as the abstraction hierarchies of compounds and reaction types, and parametric properties of compounds and reaction conditions. An efficient and intelligent scheme was then presented for analogically reasoning new reactions using that structure. This was the combination of a graph theoretical method for extracting bipartite subgraphs and the statistical, numerical methods known as clustering and principal component analysis.

There are many possible directions to extend the current work. First, it is necessary to develop more efficient algorithms for executing the proposed scheme. When more than one (L, R) pairs are to be examined, evaluating the feasibility of a reaction independently for each pair will involve redundant operations, which should be eliminated as much as possible. Secondly, the present scheme can be extended to inferring a sequence of reactions rather than a single reaction. This extension will necessitate the generation of the intermediate compounds between the starting and the target compounds. Finally, abductive inference could be combined with the proposed method of analogical reasoning, i.e., given a target compound to be synthesized, the system automatically finds the most feasible route from a set of available, preferably cheap materials to the target.

References

1. J. M. Barnard. Structure searching methods: old and new. *J. Chem. Inf. Comput. Sci.*, 33(4):532–538, 1993.
2. J. Gasteiger, W. Ihelenfeld, R. Fick, and J. R. Rose. Similarity concepts for the planning of organic reactions and syntheses. *J. Chem. Inf. Comput. Sci.*, 32(6):700–712, 1992.
3. R. P. Hall. Computational approaches to analogical reasoning: A comparative analysis. *Artificial Intelligence*, 39:39–120, 1989.
4. M. A. Johnson and G. M. Maggiora. *Concepts and Applications of Molecular Similarity*. Wiley, New York, 1990.
5. J. J. Lai, H. X. Chen, and Y. Fujiwara. Extraction of semantic relationships among terms—SS-KWIC. In *Proc. 47th FID Conference and Congress*, pages 155–159, 1994.
6. J. F. Sowa. *Conceptual Structures*. Addison-Wesley, 1984.
7. S. L. Tanimoto. *The Elements of Artificial Intelligence*. Computer Science Press, 1987.
8. Z. Q. Wang and Y. Fujiwara. Learning and reasoning in the information-base systems for organic synthesis research. *J. Japan Soc. Information and Knowledge*, 2(1):71–82, 1991.
9. P. H. Winston. Learning and reasoning by analogy. *Comm. ACM*, 23(12):689–703, 1980.

Labelling Ideality and Subideality*

Guido Governatori

CIRFID, University of Bologna
via Galliera 3, 40126 Bologna, Italy
e-mail: governat@cirfid.unibo.it

Abstract. In this paper we suggest ways in which logic and law may usefully relate; and we present an analytic proof system dealing with the Jones Pörn's deontic logic of Ideality and Subideality, which offers some suggestions about how to embed legal systems in label formalism.

1 Introduction

Why should Law need automated proof systems? The answer to this question implies an answer to the following question: Is logic needed in Law? In fact it has been argued that logics are useless for Law, see, for example, (Kelsen 1989). We believe that logic, and deontic logics in particular —but also modal logics— have a role to play in Law; for example if one wants to study what the relationships are among the various degrees of adjudication in Italian Law, one should note that they give rise to a transitive, irreflexive and finite structure, which is the frame of the modal logic of *provability GL*; one of the most important properties of such a logic is that no system, (no court), in this frame, could claim its own correctness without becoming incorrect (Boolos 1993, Smullyan 1988), but the correctness of a lower court can be established by a higher one. This example shows that the study of modal logic can help in finding certain already known properties of legal systems. Moreover, each time we are dealing with the notions of Obligation and Permission, and we are interested in the study of their mutual relationship, we can arrange them into a deontic framework, thus producing a certain kind of deontic logic. Finally a hint for the use of logic in legal reasoning is given, for example in the Italian case, by the law itself; in fact article 192, 1° comma of the "Italian code of criminal procedure" prescribes that the judges state the reasons of their adjudication; moreover several other articles of the same code, state: when evidence is valid, how evidence should be used in order to lead to an adjudication, etc .... On this basis the "Italian code of criminal procedure" can be thought of as a deductive system where its articles act as the inference rules, whereas the articles of the "Italian code of criminal law" are the axioms.

What does a proof system do? A proof system can work in two ways. The first of them consists of producing admissible steps one after the other according

---

* A restricted version of this paper was presented at the XVII IVR World Congress "Challenges to Law at the End of 20th Century", Bologna, June 1995

to the inference rules; in this way each step is guaranteed to be correct, but we are not led to the goal we want to prove. The other one consists of verifying whether a conclusion follows from given premises, i.e. if the adjudication follows logically from the evidence, mainly by refuting the negation of the conclusion.

The system we propose is based on the logic of ideality and subideality developed by Jones and Pörn, and it verifies in the above mentioned logical framework whether a given conclusion follows from given premises. Moreover, due to its basic control structure it can also be used as an analytic direct proof system.

2 Ideal and Subideal Deontic Logic

This logic has been developed in (Jones and Pörn 1985, Jones and Pörn 1986) in order to have a system which permits both factual and deontic detachment; moreover it is possible to define several types of obligation, i.e. ideal obligation and subideal obligation, thus avoiding the drawback that "actual" obligation collapses in the logical necessity of the framework we are using (deontic necessity).

As pointed out by Kelsen

Jurisprudence, by describing the validity of a law system, does not assert what happens regularly, but what ought happen according to a given law system (Kelsen 1989, 458).

Jones and Pörn's (1985, 1986) deontic logic DL has been devised for dealing with ideal as well as sub-ideal situations, i.e. situations which admit some degree of violation of what is ideally the case. Formally it is an extension of standard deontic logic (SDL), which is a normal KD system according to Chellas' (1981) classification, and which incorporates, besides the normal deontic operators O^i and P^i, the deontic operators O^s and P^s. O^i and P^i retain their usual reading. $O^i A$ ($P^i A$), at a world w, mean: A holds in all (some) of w's deontically ideal versions. $O^s A$ ($P^s A$), at a world w, mean: A holds in all (some) of w's sub-ideal versions. DL allows us to define the following notions:

- $N_D A =_{df} (O^i A \wedge O^s A)$ (*Deontic Necessity*)

- $O_T A =_{df} (O^i A \wedge P^s \neg A)$ (*Ought*)

Since DL is a straightforward extension of SDL both O^i, P^i and O^s, P^s behave as normal KD-modalities. Models for DL are thus structures:

$$M = \langle W, R_i, R_s, v \rangle$$

where R_i, $R_s \subseteq W \times W$ are serial (not reflexive) relations on W (intuitive reading: $wR_iv = v$ is an ideal version of w, $wR_sv = v$ is a sub-ideal version of w), subject to the following conditions:

C1: $R_i \cap R_s = \emptyset$

C2: $\{\langle w, w \rangle : w \in W\} \subseteq R_i \cup R_s$

that is to say that there cannot exist ideal worlds that are also sub-ideal, and every world is either ideal or sub-ideal relative to itself (notice that this introduces some form of reflexivity in the model); v is as usual with the following clauses for O^i and O^s respectively

$$\models_w O^i A \Leftrightarrow \forall v \in W : wR_i v, \models_v A$$

$$\models_w O^s A \Leftrightarrow \forall v \in W : wR_s v, \models_v A$$

Remark. Condition **C2** has been dropped in (Jones 1991), so that possible worlds can be both ideal and subideal with respect to themselves; however, when condition **C2** may be parametrized with respect to content matters, see (Epstein 1990), and this may lead to a more satisfactory solution; in fact a section of the "Italian code of criminal procedure" concerns the connected crimes: in a criminal trial pieces of evidence from other trials are examined if and only if they are judged to be relevant (connected) to the subject of the trial; a parking fine, will not usually be considered relevant in an adjudication for murder.

3   The System *KEM*

In this section we shall present *KEM* in its barest outline. We first recall some basic notions. We shall use the letters $X, Y, Z, \ldots$ to denote arbitrary signed formulas (S-formulas), i.e. formulas of the forms SA where $S \in \{T, F\}$. As usual X^C will be used to denote the *conjugate* of X, i.e. the result of changing S to its opposite (with the exception of the following S-formulas[2]: $T\Box A, F\Diamond A, F\Box A$ and $F\Diamond A$ which also have $T\Diamond \neg A, F\Box \neg A, F\Diamond \neg A, T\Box \neg A$ respectively as their conjugates). Two S-formulas X, Z such that $Z = X^C$, will be called *complementary*. As we have already said, *KEM* approach requires us to work with "world" labels. A "world" label is either a constant or a variable "world" symbol or a "structured" sequence of world-symbols we call a "world-path". Intuitively, constant and variable world-symbols stand for worlds and sets of worlds respectively, while a world-path conveys information about access between the worlds in it. We attach labels to S-formulas to yield *labelled signed formulas* (*LS-formulas*), i.e. pairs of the form X, i where X is an S-formula and i is a label. An *LS*-formula SA, i means, intuitively, that A is true (false) at the (last) world (on the path represented by) i. In the course of proof search, labels are manipulated in a way closely related to the semantics of modal operators and "matched" using a (specialized, logic-dependent) unification algorithm. That two world-paths i and k are unifiable means, intuitively, that they virtually represent the same path, i.e. any world which you could arrive at by path i could be reached by path k and vice versa. *LS*-formulas whose labels are unifiable turn out to be true

---

[2] Herein with $\Box$ we mean any modality which acts as $\Box$, i.e. O^i and O^s; and with $\Diamond$ any modality which acts as $\Diamond$, i.e. P^i and P^s.

(false) at the same world(s) relative to the accessibility relation that holds in the appropriate class of models. In particular two LS-formulas X, X^C whose labels are unifiable stand for formulas which are contradictory "in the same world". These ideas are formalized as follows.

3.1 Label Formalism

To treat DL we need three kinds of label world symbols

- Universal $\Phi_W = \{W_1, W_2, \cdots\}$ and $\Phi_w = \{w_1, w_2, \cdots\}$
- Ideal $\Phi_D = \{D_1, D_2, \cdots\}$ and $\Phi_d = \{d_1, d_2, \cdots\}$
- Subideal $\Phi_S = \{S_1, S_2, \cdots\}$ and $\Phi_s = \{s_1, s_2, \cdots\}$

Here the universal world labels denote worlds for which we do not have enough information to specify whether they are ideal or subideal. Let us now define the set of variable world symbols and constant world symbols respectively:

$$\Phi_V = \Phi_W \cup \Phi_D \cup \Phi_S \text{ and }$$
$$\Phi_C = \Phi_w \cup \Phi_d \cup \Phi_s.$$

On this basis the set $\Im$ is now defined as

$$\Im = \bigcup_{1 \leq i} \Im_i \text{ where } \Im_i \text{ is}:$$
$$\Im_1 = \Phi_C \cup \Phi_V;$$
$$\Im_2 = \Im_1 \times \Phi_C;$$
$$\Im_{n+1} = \Im_1 \times \Im_n.$$

In other words a world-label is either (i) an element of the set Φ_C, or (ii) an element of the set Φ_V, or (iii) a path term (k', k) where (iiia) $k' \in \Phi_C \cup \Phi_V$ and (iiib) $k \in \Phi_C$ or $k = (m', m)$ where (m', m) is a label. It is worth noting that such a representation of labels captures the precise meaning of the accessibility relation; in fact, the label $(w_2, (W_1, w_1))$ denotes a path which leads to a world (w_2) accessible from all the worlds accessible from the world denoted by w_1. If labels were sequences of constants and variables, their reading is ambiguous: i.e. what does a label such as $\langle W_1, w_2, w_1 \rangle$ stand for? Does it mean that w_1 sees all the worlds accessible from w_2, or does it have the same meaning as $(W_1, (w_2, w_1))$? The two possible readings of a labels written as a sequence, give rise to different accessibility relations.

A bit of terminology. For any label $i = (k', k)$ we call k' the *head* of i, k the *body* of i, and denote them by $h(i)$ and $b(i)$ respectively. Notice that these notions are recursive: if $b(i)$ denotes the body of i, then $b(b(i))$ will denote the body of $b(i)$, $b(b(b(i)))$ will denote the body of $b(b(i))$; and so on. For example, if i is $(w_4, (W_3, (w_3, (W_2, w_1))))$, then $b(i) = (W_3, (w_3, (W_2, w_1)))$, $b(b(i)) = (w_3, (W_2, w_1))$, $b(b(b(i))) = (W_2, w_1)$, $b(b(b(b(i)))) = w_1$. We call each of $b(i),b(b(i))$, etc., a *segment* of i. Let $s(i)$ denote any segment of i (obviously, by definition every segment $s(i)$ of a label i is a label); then $h(s(i))$ will denote the head of $s(i)$.

For any label i, we define the length of i, $l(i)$, as the number of world-symbols in i, i.e. $l(i) = n \Leftrightarrow i \in \Im_n$.

We shall use $s^n(i)$ to denote the segment of i whose length is n

3.2 Unification Schemes

KEM's label unification scheme involves two kinds of unifications, respectively "high" and "low". "High" unifications are meant to mirror specific accessibility constraints and they are used to build "low" unifications, which account for the full range of conditions governing the appropriate accessibility relation. We then begin by defining the basic notion of "high" unification. First we define a substitution in the usual way as a function

$$\sigma : \Phi_V^0 \longrightarrow \Im^-$$
$$: \Phi_V^i \longrightarrow \Im^i, (1 \leq i \leq n).$$

where $\Im^- = \Im - \Phi_V$. For two labels i, k and a substitution σ, if σ is a unifier of i and k then we shall say that i and k are σ-unifiable. We shall (somewhat unconventionally) use $(i, k)\sigma$ to denote both that i and k are σ-unifiable and the result of their unification.

$$(i,k)\sigma = \begin{cases} \sigma i = \sigma k & l(i) = l(k) = 1 \\ ((h(i), h(k))\sigma, (b(i), b(k))\sigma) \end{cases}$$

In order to get the appropriate unifications we need to define the following substitution acting as σ for universal world symbols:

$$\sigma^\# \Phi_W = \sigma \Phi_W$$

and as follows for ideal an sub-ideal world symbols:

$$\sigma^\# : \Phi_S \to \Phi^s$$
$$: \Phi_D \to \Phi^d$$

where

$$\Phi^d = \{i^r \in \Phi_C : r = d\} \qquad \Phi^s = \{i^r \in \Phi_C : r = s\}$$

Φ^d, Φ^s denote the set of worlds that are respectively an ideal and a subideal version of themselves.

According to the above substitutions we define the

$$(i,k)\sigma^W = (t \times (b(s(i)), b(k))\sigma^J) \iff$$
$$l(k) > 1, \exists s(i) : \forall s'(i)(l(s'(i)) > l(s(i)),$$
$$(h(s'(i)), h(k))\sigma^\# = (h(s(i)), h(k))\sigma = t) \text{ and}$$
$$(b(s(i)), b(k))\sigma^J \text{ or}$$
$$(i,k)\sigma^W = t, \text{ if } l(k) = 1$$

or

$$(i,k)\sigma^W = (t \times (b(i), b(s(k)))\sigma^J) \iff$$
$$l(i) > 1, \exists s(k) : \forall s'(k)(l(s'(k)) > l(s(k)),$$
$$(h(i), h(s'(k)))\sigma^\# = (h(i), h(s(k)))\sigma = t) \text{ and }$$
$$(b(i), b(s(k)))\sigma^J \text{ or }$$
$$(i,k)\sigma^W = t, \text{ if } l(i) = 1$$

where

$$(i,k)\sigma^J = (i,k)\sigma \text{ or } (i,k)\sigma^W.$$

For example the labels $(D_1, (w_2^i, w_1))$ and (w_2^i, w_1) σ^W-unify since $t = w_2^i = (D_1, w_2^i)\sigma^\# = (w_2^i, w_2^i)\sigma$ and obviously $(w_1, w_1)\sigma^J$. A more complex example is given by the labels $(w_2^i, (W_1, (D_1, (S_1, w_1^s))))$ and (D_2, w_1^s) where $((S_1, w_1^s), w_1^s)\sigma^J$ and $t = w_2^i = (D_2, w_2^i)\sigma^\# = (D_2, W_1)\sigma^\# = (D_2, D_1)\sigma$.

We are now able to characterize DL by the notion of σ_{DL}-unification:

$$(i,k)\sigma_{DL} = \begin{cases} (i,k)\sigma \\ (i,k)\sigma^W \end{cases}$$

3.3 Labels, Unifications and Legal Reasoning

What does a possible world denote? We suggest that a possible and plausible answer could be that a possible world of a given type represents an actual fact, and another type of possible world denotes laws. The unifications tell us when two labels are "matchable". If they are, we can compare whatever holds in the worlds they denote; therefore we can decide, analytically, whether a given fact is a violation of a law.

Obviously, according to our philosophical point of view, the formalization of norms will behave in different ways. We believe that our label manipulation could help to examine a few ideas about norms. Let us examine the basic cases of unifications

Case 1. A variable and a constant;

Case 2. Two constants;

Case 3. Two variables.

Roughly, cases 1, 2 and 3 correspond respectively to:

- Distribution axiom $\Box(A \to B) \to (\Box A \to \Box B)$;

- Necessitation rule $\dfrac{A}{\Box A}$;

- Kant's axiom $\Box A \to \Diamond A$.

Combining cases 1, 2 and 3 we can obtain different philosophical positions concerning norms.

Case 1 implies that norms express generic "situations" and we have to detect whether a given "situation" falls into the category of the generic one.

Case 2 implies that each norm expresses a given situation and we have to detect whether a given situation is the same as that of the norm, so each situation should have its specific norm.

Case 3 (idealization) implies the completeness of a normative system in a weak sense. So each instance of a situation should be determined by the norms. If there is a gap in a normative system, this condition states that norms themselves should give tools to fill the gap.

Almost every positive legal system has some mechanism to fulfil the requirement of the last case. For example, in Italian Law, article 12, 2° comma of the "Preleggi" prescribes analogical reasoning.

3.4 Inference Rules

We shall classify our inference rules in two main categories: structural rules and operational rules; the operational rules describe the meaning of the various operators and connectives involved (see (D'Agostino Mondadori 1994) for further explanations), whereas structural rules describe semantic properties holding in the model for the logic we are concerned with. Moreover it is possible to have other non-standard connectives and operators for which we can state their appropriate inference rules using labels, see (D'Agostino Gabbay 1994). The rules for the connectives are stated as follows[3]:

$$\frac{\alpha, j}{\alpha_1, j} \qquad \frac{\alpha, j}{\alpha_2, j}$$

$$\frac{\beta, j}{\beta_2^C, k} \atop \frac{}{\beta_1, (j,k)\sigma_{DL}} \; (j,k)\sigma_{DL} \qquad \frac{\beta, j}{\beta_1^C, k} \atop \frac{}{\beta_2, (j,k)\sigma_{DL}} \; (j,k)\sigma_{DL}$$

For the modal-like operators we have

$$\frac{TN_D A, j}{TA, (W_n, j)} \qquad \frac{\nu_{\{i,s\}} A, j}{\nu_0, (\{D,S\}_n, j)} \; \{D,S\}_n \text{ new}$$

and

$$\frac{FN_D A, i}{FO^i A \wedge O^s A, i} \qquad \frac{\pi_{\{i,s\}}, i}{\pi_0, (\{d,s\}_n, i)} \; \{d,s\}_n \text{ new}$$

The "standard" structural inference rules, respectively the principle of bivalence (PB) and the principle of not contradiction (PNC), are:

$$\frac{}{X, j \qquad X^C, j} \; h(j) \in \Phi_C \qquad \frac{X, j}{X^C, k} \atop \frac{}{\times (j,k)\sigma_{DL}} \; (j,k)\sigma_{DL}$$

---

[3] The following formulation uses a generalized $\alpha, \beta, \nu_{\{i,s\}}, \pi_{\{i,s\}}$ form of Smullyan-Fitting α, β, ν, π unifying notation, see (Fitting 1983).

Here the α-rules are just the familiar linear branch-expansion rules of the tableau method, while the β-rules correspond to such common natural inference patterns as *modus ponens*, *modus tollens*, etc. (i, k, m stand for arbitrary labels). The rules for the modal operators are as usual. "new" in the proviso for the $\nu_{\{i,s\}}$- and $\pi_{\{i,s\}}$-rule means: $\{D, S\}_n$, $\{d, s\}_n$ must not have occurred in any label yet used. Notice that in all inferences via an α-rule the label of the premise carries over unchanged to the conclusion, and in all inferences via a β-rule the labels of the premises must be σ_{DL}-unifiable, so that the conclusion inherits their unification. PB (the "Principle of Bivalence") represents the (LS-version of the) semantic counterpart of the cut rule of the sequent calculus (intuitive meaning: a formula A is either true or false in any *given* world, whence the requirement that i should be restricted). PNC (the "Principle of Non-Contradiction") corresponds to the familiar branch-closure rule of the tableau method, saying that from the occurrence of a pair of LS-formulas X, i, X^C, k such that $(i, k)\sigma_{DL}$ (let us call them σ_{DL}-*complementary*) on a branch we may infer the closure ("×") of the branch. The $(i, k)\sigma_{DL}$ in the "conclusion" of PNC means that the contradiction holds "in the same world".

The peculiar structural inference rules of DL, the rules which represent the conditions of the model, are:

$$\frac{X, (D, j) \\ X, (S, k)}{X, (W_n, (j, k)\sigma_{DL})} \ (j, k)\sigma_{DL}$$

which states that a property holds universally. The main purpose of this rule is to ensure reflexivity with respect to j and k, i.e, each world is either an ideal or a subideal version of itself; in fact a general property of labels and unifications states that

$$(j, k)\sigma_{DL} \Rightarrow ((j, k)\sigma_{DL}, j)\sigma_{DL} \text{ and } ((j, k)\sigma_{DL}, k)\sigma_{DL} .$$

The next rule, RR (Reflexivity Rule) tells us when a world is an ideal or subideal version of itself.

$$\frac{\nu_{\{i,s\}}, j \\ \nu_0^C, k}{\nu_{\{i,s\}}, m^r \\ \nu_0^C, m^r} \ m = (j, k)\sigma_{DL}$$

where

$$i^r = i^s \text{ if } \nu_{\{i,s\}} = TO^i A \,(FP^i A)$$
$$i^r = i^d \text{ if } \nu_{\{i,s\}} = TO^s A \,(FP^s A)$$

and

$$i^x = i : h(i) \in \Phi^x, (x \in \{d, s\})$$

Obviously each $\Phi_X^r \subseteq \Phi_X$. We shall call labels of the form i^x, $(x \in \{d, s\})$ x-reflexive labels.

Besides the usual closure rule (PNC) and the principle of bivalence (PB) we introduce the following rules $LPNC$ and LPB

$$\frac{\begin{array}{c} j \in \Phi^s \\ j \in \Phi^d \end{array}}{\times} \qquad \frac{}{X, j^i \qquad K, j^s}$$

stating, respectively, that no world can be at the same time an ideal and a subideal version of itself and that each worlds is either an ideal or a subideal version of itself.

3.5 Proof search

Let $\Gamma = \{X_1, \ldots, X_m\}$ be a set of S-formulas. Then $\mathcal{T}$ is a *KEM-tree for* Γ if there exists a finite sequence $(\mathcal{T}_1, \mathcal{T}_2, \ldots, \mathcal{T}_n)$ such that (i) $\mathcal{T}_1$ is a 1-branch tree consisting of $\{X_1, i \ldots, X_m, i\}$, where i is an arbitrary constant label; (ii) $\mathcal{T}_n = \mathcal{T}$, and (iii) for each $i < n$, $\mathcal{T}_{i+1}$ results from $\mathcal{T}_i$ by an application of a rule of KEM. A branch τ of a KEM-tree $\mathcal{T}$ of LS-formulas is said to be σ_{DL}-*closed* if it ends with an application of PNC, open otherwise. As usual with tableau methods, a set Γ of formulas is checked for consistency by constructing a KEM-tree for Γ. It is worth noting that each KEM-tree is a (class of) Hintikka's model(s) where the labels denote worlds (i.e. Hintikka's modal sets), and the unifications behave according to the conditions placed on the appropriate accessibility relations. Moreover we say that a formula A is a *KEM-consequence of a set of formulas* Γ if A occurs in all the open branches of a KEM-tree for Γ. We now describe a systematic procedure for KEM. First we define the following notions.

Given a branch τ of a KEM-tree, we shall call an LS-formula X, i *E-analysed in* τ if either (i) X is of type α and both α_1, i and α_2, i occur in τ; or (ii) X is of type β and one of the following conditions is satisfied: (a) if β_1^C, k occurs in τ and $(i, k)\sigma_{DL}$, then also $\beta_2, (i, k)\sigma_{DL}$ occurs in τ, (b) if β_2^C, k occurs in τ and $(i, k)\sigma_{DL}$, then also $\beta_1, (i, k)\sigma_{DL}$ occurs in τ; or (iii) X is of type ν_i and $\nu_0, (m, i)$ occurs in τ for some $m \in \Phi_V$ not previously occurring in τ, or (iv) X is of type π_i and $\pi_0, (m, i)$ occurs in τ for some $m \in \Phi_C$ not previously occurring in τ.

We shall call a branch τ of a KEM-tree *E-completed* if every LS-formula in it is E-analysed and it contains no complementary formulas which are not σ_{DL}-complementary. We shall say a branch τ of a KEM-tree *completed* if it is E-completed and all the LS-formulas of type β in it either are analysed or cannot be analysed. We shall call a KEM-tree *completed* if every branch is completed.

The following procedure starts from the 1-branch, 1-node tree consisting of $\{X_1, i_1, \ldots, X_m, i_m\}$ and applies the rules of KEM until the resulting KEM-tree is either closed or completed.

We shall say that a formula A is a theorem of DL when a closed KEM-tree for FA, w_1 exists.

At each stage of proof search (i) we choose an open non completed branch τ. If τ is not E-completed, then (ii) we apply the 1-premise rules until τ becomes

E-completed. If the resulting branch τ' is neither closed nor completed, then (iii) we apply the 2-premise rules until τ becomes E-completed. If the resulting branch τ' is neither closed nor completed, then (iv) we choose an LS-formula of type β which is not yet analysed in the branch and apply PB so that the resulting LS-formulas are β_1, i' and β_1^C, i' (or, equivalently β_2, i' and β_2^C, i'), where $i = i'$ if i is restricted, otherwise i' is obtained from i by instantiating $h(i)$ to a constant not occurring in i; (v) ("Modal PB") if the branch is not E-completed nor closed, because of complementary formulas which are not σ_{DL}-complementary, then we have to see whether a restricted label unifying with both the labels of the complementary formulas occurs previously in the branch; if such a label exists, or can be built using already existing labels and the unification rules, then the branch is closed; (vi) ("Label PB") if the branch is not E-completed nor closed, because of complementary formulas which are not σ_{DL}-complementary and the heads of their labels, j, k, are respectively in Φ_D and Φ_S, then we have to see whether there exists a restricted non reflexive label, that, when it is i-reflexive, unifies with j and when it is s-reflexive unifies with k; if such a label exists, or can be built using already existing labels and unification rules, then the branch is closed; (vii) we repeat the procedure in each branch generated by PB.

The above procedure is based on a (deterministic) procedure working for *canonical* KEM-trees. A KEM-tree is said to be canonical if it is generated by applying the rules of KEM in the following fixed order: first the α-, $\nu_{\{d,s\}}$- and $\pi_{\{d,s\}}$-rule, then the β-rule and PNC, and finally PB. Two interesting properties of canonical KEM-trees are (i) that a canonical KEM-tree always terminates, since for each formula there are a finite number of subformulas and the number of labels which can occur in the KEM-tree for a formula A (of L) is limited by the number of modal operators belonging to A, and (ii) that for each closed KEM-tree a closed canonical KEM-tree exists. Proofs of termination and completeness for canonical $KEDL$-trees follow by obvious modifications of the proofs given in (Governatori 1995).

Remark. We distinguish between DL-theories, obtained by means of configurations of possible worlds, and DL obtained by means of the above inference rules and unifications. It is worth noting that labels allow us not only to manipulate formulas in deductions but also worlds, which turns out to be very important when dealing with theories. For a similar approach see (Russo 1996).

The following are example proofs of theorems of DL.

1. $F(O^s A \wedge \neg A) \to P^s \neg A$ w_1
2. $T O^s A \wedge \neg A$ w_1
3. $F P^s \neg A$ w_1
4. $T O^s A$ w_1
5. $F A$ w_1
6. $T O^s A$ w_1^s
7. $F A$ w_1^s
8. $T A$ S_1, w_1^s
9. $\times$

The steps leading to the nodes (1)-(5) are straightforward. The nodes (6)-(7) come from the application of the reflexivity rule since the world denoted by w_1 is a sub-ideal version of itself. Closure follows immediately from (7) and (8), which are σ_{DL}-complementary (their labels σ_{DL}-unify because of $(S_1, w_1^s)\sigma^\#$).

1.	$FO^i A \rightarrow (O^s B \rightarrow (\neg A \rightarrow B))$	w_1
2.	$TO^i A$	w_1
3.	$FO^s B \rightarrow (\neg A \rightarrow B)$	w_1
4.	$TO^s B$	w_1
5.	$F\neg A \rightarrow B$	w_1
6.	FA	w_1
7.	FB	w_1
8.		w_1^s
9.		w_1^i
10.		$\times$

Here the steps (8) and (9) are obtained, respectively, from (2), (6) and (4), (7)by RR and the closure follows from an application of $LPNC$.

1.	$F(O^i(A \wedge B) \wedge O^s(C \wedge D)) \rightarrow (A \vee C)$	w_1
2.	$TO^i(A \wedge B) \wedge O^s(C \wedge D)$	w_1
3.	$FA \vee C$	w_1
4.	$TO^i(A \wedge B)$	w_1
5.	$TO^s(C \wedge D)$	w_1
6.	FA	w_1
7.	FC	w_1
8.	$TA \wedge B$	D_1, w_1
9.	$TC \wedge D$	$S_1.w_1$
10.	TA	D_1, w_1
11.	TB	D_1, w_1
12.	TC	$S_1.w_1$
13.	TD	$S_1.w_1$

14. T	w_1^i	15. F	w_1^s
16. $\times$		17. $\times$	

In the left branch, closure follows from $TA, (D_1, w_1)$, FA, w_1 and w_1^i, after we have assumed, through the label version of PB, that w_1 is an ideal version of itself, i.e., w_1^i, we replace, with respect to the left branch, all the occurrences of w_1 with w_1^i thus obtaining D_1, w_1^i and w_1^i which σ_{DL}-unify; on the other hand, in the right branch we have $TC, (S_1, w_1)$, FC, w_1 and w_1^s, and we can repeat the same procedure as for the left side.

4 Final Remarks

Although a satisfactory Logical System for Law is far from being realized, we believe that the approach we have presented may offer a few steps in the right direction. In fact, the label tool we have developed is flexible enough to cope with several types of modal-like notions of obligatoriness at the same time, and to study their mutual relationships through unifications. It often happens that, in a legal system, laws prescribe opposite possibilities for the same fact according to "relevant" pieces of evidence; for example, some legal system could prescribe a murder to be punished unless he/she killed in self-defence. Logically this scenario is contradictory because, in the case of self-defence, both punishment and not punishment are implied; however it is possible to solve this problem as soon as some refinement is assumed; on this point see (Artosi Governatori Sartor 1996). Moreover, as we have already seen, different traits of legal reasoning might involve different kinds of logics (even with different connectives and operators); the resulting overall logic can be embedded in the so called fibred semantics (logic) framework (Gabbay 1994), but the label formalism here presented can be extended, straightforwardly, to deal with it.

The preceding discussion was thus mainly aimed at showing the potential scope of application of the method. In effect, we believe that the method we proposed to determine the ideal/subideal status of world nicely exploits the computational and proof-theoretical advantages offered by the modal theorem proving system KEM. As we have argued elsewhere, this system enjoys most of the features a suitable proof search system for modal (and in general non-classical) logics should have. In contrast with (both clausal and non-clausal) resolution methods, and in general "translation-based" methods (Ohlbach 1991), it works for the full modal language (thus avoiding any preprocessing of the input formulas), and it is flexible enough to be extended to cover any setting having a Kripke-model based semantics (this is clearly shown by our treatment of Jones and Pörn logic DL where the rules specific for such a logic should take care not only of the propositional and modal part but also of the structure of the labels and the relationship between labels and formulas; for example we added another closure rule $\frac{i\in\Phi^i, i\in\Phi^s}{\times}$ which states that no world can be at the same time an ideal and a sub-ideal version of itself; this result is achieved by determining when a deontic word is ideally (sub-ideally) reflexive (i^r) by means of another peculiar inference rule, and finally the principle of bivalence for labels). From this perspective our method is similar to the natural deduction proof method proposed by Russo (1996). Nevertheless, it has several advantages over most tableau/sequent based theorem proving methods: being based on D'Agostino and Mondadori's classical proof system KE, it eliminates the typical redundancy of the standard cut-free methods and, thanks to its label unification scheme, it offers a simple and efficient solution to the permutation problem which notoriously arises at the level of the usual tableau-sequent rules for the modal operators (Fitting 1988). However, unlike e.g. Wallen's (1990) connection method, it uses a natural and easily implementable style of proof construction, and so it appears to provide an adequate basis for combining both efficiency and naturalness.

Acknowledgements

I would like to thank Alberto Artosi and Giovanni Sartor for their helpful suggestions concerning the logical part, and Michele Papa for a useful discussion about the role of evidence in legal systems. Thanks are also due to Charles Hindley for revising the English version.

References

Alberto Artosi, Paola Cattabriga and Guido Governatori. 1994. An automated Approach to Normative Reasoning. In J. Breuker (ed.), *Artificial Normative Reasoning*, Workshop ECAI 1994: 132-145.

Alberto Artosi, Guido Governatori and Giovanni Sartor 1996. Towards a Computational Treatment of Deontic Defeasibility. In M. Brown and J. Carmo (eds). *Deontic Logic, Agency and Normative Systems*. Workshop in Computing, Springer-Verlag, Berlin, 1996: 27–46.

George Boolos 1993. *The Logic of Provability*. Cambridge University Press, Cambridge, 1994.

Brian Chellas 1981. *Modal Logic: An Introduction*. Cambridge University Press, Cambridge, 1981.

Marcello D'Agostino and Marco Mondadori 1994. The Taming of the Cut. *Journal of Logic and Computation* 4, 1994: 285-319.

Marcello D'Agostino and Dov M. Gabbay 1994. A Generalization of Analytic Deduction via Labelled Deductive Systems. Part I: Basic Substructural Logics. *Journal of Automated Reasoning* 13, 1994: 243-281.

Richard D. Epstein 1990. *The semantic Foundations of Logic. Volume 1: Propositional Logics*. Kluwer, Dordrecht, 1990.

Melvin Fitting 1983. *Proof Methods for Modal and Intuitionistic Logic*. Reidel, Dordrecht, 1983.

Melvin Fitting 1988. First Order Modal Tableaux. *Journal of Automated Reasoning* 1988; 4: 191-213.

Dov M. Gabbay 1994. Combining Labelled Deductive Systems. *MEDLAR II*, ESPRIT Basic Research Project 6471, Deliverable DIII.1.2-1P: 285-334.

Guido Governatori 1995. Labelled Tableaux for Multimodal Logics. in P. Baumgatner, R. Hänhle, J. Posegga eds. *Fourth Workshop on Theorem Proving with Analytic Tableaux and Related Methods*, Lecture Notes in Artificial Intelligence, Springer-Verlag, Berlin, 1995: 79-95

Andrew J.I. Jones 1991. On the Logic of Deontic Conditionals. *Ratio Juris* 4, 1991: 355-366.

Andrew J.I. Jones and Ingmar Pörn 1985. Ideality, Sub-Ideality and Deontic Logic. *Synthese* 65, 1985: 275-290.

Andrew J.I. Jones and Ingmar Pörn 1986. "Ought" and "Must". *Synthese* 66, 1986: 89-93.

Hans Kelsen 1989. Sulla logica delle norme (manoscritto). *Materiali per una cultura giuridica* XIX, 1989: 454-468.

Hans J. Ohlbach 1991. Semantics-Based Translation Methods for Modal Logics. *Journal of Logic and Computation* 1, 1991: 691-746.

Alessandra Russo 1996. Generalising Propositional Modal Logic Using Labelled Deductive Systems. In *Proceedings of FroCoS*, Forthcoming.

Raymond Smullyan 1988. *Forever Undecided. A Puzle Guide to Gödel.* Oxford University Press, Oxford, 1988.

Lincoln A. Wallen 1990. *Automated Deduction in Non-Classical Logics.* The MIT Press, Cambridge (Mass.), 1990.

Mind, Morals, and Reasons

by
Marcello Guarini
email: mguarini@julian.uwo.ca
University of Western Ontario
Department of Philosophy
London, ON. Canada
N6A 3K7

Abstract
The purpose of this paper is to show that the appropriate
account of moral reasons will be somewhere between extreme
generalism and extreme particularism. Results from
research in Parallel Distributed Processing will be used to
show how moral knowledge may be thought of as general but
not productively thought of in terms of exceptionless
rules.

Let us consider two views on the nature and role of
principles in moral judgment making. The first view is
this: we arrive at every judgment about a particular case
by deducing it from some exceptionless moral principle.
The second view is quite different: every judgment about a
particular case is absolutely particular, which is to say
that particular judgments are not derived from moral
knowledge having some sort of general form. Let us call
the first view extreme generalism and second view extreme
particularism. I am not claiming that anyone actually
subscribes to these views; rather, I am using them as a
springboard to discuss the nature of moral reasoning.
Extreme Generalism is implausible because some subset
of our moral principles contains ineliminable ceteris
paribus clauses, and a specific judgment about a particular
case cannot be deduced from a principle with a ceteris
paribus clause:

Lying is wrong, ceteris paribus;
Jack lied under circumstance C;
Therefore . . .

Whether we can infer that Jack did anything wrong under C
depends on the contents of the ceteris paribus clause,
which is ineliminable. In light of this, some might be
tempted to say we are not drawing on any sort of general
moral knowledge in making judgments about particular cases.
Such a view is implausible. There are too many possible
cases that we can make judgments on to think that we store
every possible case in our head (together with the relevant

moral judgments) and then simply decide whether the case in question maps on to one of the cases in our heads. In other words, we are not acting like jumped-up, automated look-up tables when we make moral judgments. So far, we have arrived at two results which may appear to conflict:

(1) the ceteris paribus nature of our moral knowledge shows that we cannot be deducing judgments about particular cases from exceptionless rules; and

(2) the cognitive implausibility of treating each case on its own (in other words, without making use of general moral knowledge) suggests that some sort of general moral knowledge is guiding us in making judgments about particular cases.

These claims may appear to conflict to those who make the following assumption:

(3) the only way knowledge can be general is if it is expressed in the form of an exceptionless rule.

The truth of (1) and (2) provide evidence for the falsity of (3). Those who have emphasised the importance of case-by-case reasoning and de-emphasised the role of rules and principles -- call them particularist sympathizers -- have probably been motivated by the truth of (1). Those who have insisted on the importance of rules and principles -- generalist sympathizers -- may have been motivated by the truth of (2). The purpose of this paper will be to begin the project of showing how moral knowledge may be general even if it cannot be cashed in terms of exceptionless rules. If my story is persuasive, it will show that both generalist and particularist sympathizers may be correct.
    The ability to account for persuasion will be an important feature of any good theory of moral reasoning. Pure generalism and pure particularism are suspect not only for the reasons mentioned above, but also as a result of their inability to account for persuasion. How do we persuade a jumped-up look-up table to change the "judgement" it makes? Similarly, how do we persuade a deducer mechanism that it has misapplied some particular rule in some particular case if all it can do is deduce "judgements" about cases from a rule together with the appropriate facts? After sketching out a position somewhere between the extreme versions of generalism and particularism, I will try to show how it might be used to account for persuasion.
    By now, the reader may have had the thought that a position can be very strongly generalist in spirit even if the ceteris paribus clauses of our substantive moral principles are ineliminable. For example, it might be argued that a feature, factor, or property makes the same contribution to the rightness or wrongness of every

situation in which it occurs. In other words, lying (or stealing or promise breaking or whatever) always counts against an action, and it does so with a fixed amount of force. Similarly, other features of actions will always count in favour of specific acts, and they will do so with their (fixed) respective forces. When we evaluate an action in a particular context, we simply add up the constant influences of the various factors. The adherent of such a view can claim that the ceteris paribus clauses of substantive principles are ineliminable because different properties can combine in indefinitely many ways. For example, we cannot say that lying is always wrong because it may be that the only way to keep some promises is to lie; since lying always counts against an act, say, with 30 units of force, and since promise breaking, say, counts against an act with 40 units, then if there are no other relevant factors to consider (i.e. ceteris paribus), we lie in situations where we have a choice between promise breaking and lying. The ceteris paribus clause is not eliminated, but there is something very generalist about this approach: a given feature of an act always counts in the same way (for or against) and with the same force. Jonathan Dancy considers this sophisticated form of generalism -- call it additive generalism -- and rightly rejects it.[1] The fact is, many features of actions do not always count in the same way and with the same force. Consider a married man who has promised to meet a prostitute at a motel for services. Promise keeping often counts in favour of doing an act, but in this case, it seems to count against doing the act. A tough-minded, additive generalist might claim that keeping the promise does count in favour of doing the act; the reason the action is wrong is that it is also an instance of <u>cheating on one's spouse</u>, which greatly outweighs keeping the promise to the prostitute. This move does not aid the generalist cause. In the marriage ceremony, the man in question promised his wife that he would not have sex with other women. Clearly, the promise to remain loyal to his wife outweighs the promise to meet the prostitute for services, which means that not all promises are created equal: keeping some promises is more important than keeping others. If that is true, then additive generalism appears to be false, for it claims that a particular feature of an act always counts with the same amount of force. We could try to formulate more sophisticated versions of additive generalism which focus on the constancy of force of very fine grained features, but I suspect that these more refined versions of generalism also can be hounded to death using counter examples.

---

[1]Jonathan Dancy, <u>Moral Reasons</u> (Oxford: Blackwell, 1993), chapter six.

By now, particularist sympathizers are probably rather pleased that a couple of forms of generalism have been rejected. Generalist sympathizers are probably itching to point out that it is not cognitively possible for us to be glorified look-up tables. Moreover, moral education seems to consist in learning how to treat particular cases and somehow applying what we learn to new cases. That which is learned in treating cases is general in the sense that it can be applied to cases that are different from those we have been taught to deal with. I will show how it might be possible to account for this kind of generality without a significant role for substantive rules or principles.

I will use results from Parallel Distributed Processing (PDP) research to show how both particularist and generalist sympathizers may be correct. One of the interesting results of this research is that artificial (or simulated) neural networks can be "trained" by being exposed to examples to perform specific tasks without being given some rule to follow. Moreover, the network has the ability to apply the "knowledge" it has acquired to new cases. Existing simulated or artificial neural networks are quite different from actual biological neural nets. In many simulated neural nets, the neurons are not physical entities but abstract entities in a complex simulation carried out by a digital computer. There are many other differences in simulated neural nets, but these need not concern us for the purposes of this paper.[2]

To develop an understanding for how computational models of neural nets may shed some light on the particularism-generalism tension, we will look at an example of a neural net (see figure 1)[3] engineered to distinguish between sonar echoes returned from mines and sonar echoes returned by rocks. The input layer of the network has thirteen units (or "neurons"). A given echo is put through a frequency analyzer and is sampled for its relative energy levels at thirteen different frequencies. Each of these thirteen values is entered as an activation level for the respective units of the input or sensory layer. This network contains seven hidden units and two output units. All the input units are connected to all the hidden units, and all the hidden units are connected to the output units. The thirteen input values of a given echo form the input vector for that echo. A well trained

---

[2] For a discussion of the similarities and dissimilarities between artificial neural nets and naturally occurring ones, see Paul Churchland, A Neurocomputational Perspective (Cambridge Mass.: MIT Press, 1989), 163-194. Henceforth cited within the text.

[3] All figures taken from Churchland, 1989. Figures one and three from page 203, and figure two from page 166.

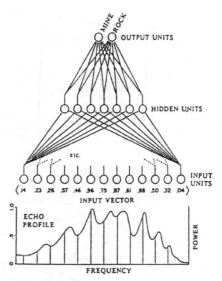

Figure 1

network will produce an output vector at or near <1,0> for a mine echo and an output vector at or near <0,1> for a rock echo. To obtain the vector <1,0>, the output unit designated for mines would have to send out its strongest signal, and the unit designated for rocks would have to send out no signal at all. With random settings on the synaptic connections, the odds are greatly against the production of the correct output vectors. But the network can be "trained up" or "taught" to distinguish mines from rocks by a method that resembles trial and error. A back propagation algorithm is used to correct the responses of the network. The network is "taught" on a set of training examples, which are a collection of vector inputs for which the correct outputs are known. The back propagation algorithm back propagates error, which is to say that when an error message (the recognition that the obtained output vector is not close enough to the correct output vector) is received, it makes minute adjustments in the synaptic weightings in an attempt to bring the obtained output vector closer to the correct output vector. With a sufficiently diverse training set of mine and rock echoes, this network -- under the constant pressure of the back propagation algorithm -- learns to generalize. In fact, it learns to generalize very well. After being trained up with a sample of 100 echoes, it can be given very different echoes which it will recognize nearly as effectively as those in the training set.

The network in figure two has 105 synapses (13x7 + 7x2). A weight-error space of 106 dimensions could be set up to track the gradient descent in that space, where 105 of the dimensions respectively represent the value of the synapses, and the other dimension represents the percentage error (see figure 2). Another kind of mapping can be

Figure 2

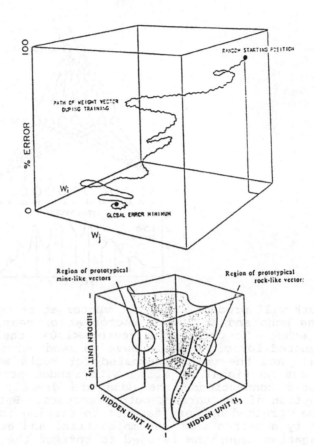

Figure 3

performed. We could represent the activation value of each hidden unit (the weighted sum of the influences reaching a given hidden unit from a lower level on its own axis on a graph. (See figure three for an example of how three of the seven dimensions, or axes, can be graphed.) During the training process, the network searches for a set of synaptic weights that partitions the seven dimensional space such that any mine input produces an activation vector across the hidden units that falls within one large subvolume of this state space and any rock input produces a vector that falls within the complement of that subvolume. Vectors close to the centre of or along a particular path in the rock-vector subvolume represent prototypical rock echoes, which will provide output vectors very close to or at <0,1>. Once the network is trained, the task of recognition takes a fraction of a second.

Are there advantages to representing the knowledge-state of a system in terms of prototypical activation vectors (or prototypes for short)? One advantage is that "prototypes allow us a welcome degree of looseness that is precluded by the strict logic of a universal quantifier"

(Churchland, 1989, 122). Some vectors can be closer to or further from the prototypical vectors, and a geometric account of the relevant similarities and dissimilarities can be given.

> It seems possible that we will also find cognitive significance in surfaces [of partitions], hypersurfaces, and intersections of hypersurfaces, and so forth. What we have opening before us is a "geometrical," as opposed to a narrowly syntactic, conception of cognitive activity. (Churchland, 1989, 108).

All of this could be very useful to us since we are considering the possibility that moral knowledge need not consist in the possession of a set of substantive rules. To make it more clear how it may be useful, let us have a look at a neural net designed to make moral distinctions. Consider the following chart:

	Case Number										
	1	2	3	4	5	6	7	8	9	10	11
Freedom from imposed burden	1	0	0	0	0	0	0	1	0	0	1
Death results	1	1	1	1	1	1	1	1	1	1	0
Doing	1	1	0	0	1	1	1	0	0	1	1
Allowing	0	0	1	1	0	0	0	1	1	0	0
Done in self-defense	0	1	0	0	0	0	0	0	0	0	0
Suffering is relieved	0	0	1	0	0	0	0	0	0	0	1
Protecting lives of many innocents	0	1	0	0	0	0	0	0	0	1	0
Done solely to make money	0	0	0	0	1	0	0	0	1	0	0
Revenge	0	0	0	0	0	1	0	0	0	0	0
Competition is eliminated	0	0	0	0	0	0	1	0	0	0	0
Permissible (1); Impermissible (0)	0	1	1	1	0	0	0	1	0	1	1

The chart describes eleven different moral moral situations. A "1" in a column means the given feature is present in the case; a "0" means that it is not. For example, case one is a situation in which bringing about a death frees someone from an imposed burden -- abortion, say, in the case of rape-induced pregnancy. Case three describes a situation where someone is allowed to die so that the lives of many innocents may be protected. In case nine, someone is allowed to die so that another can profit. You get the idea. The last row in the chart describes the intuition an agent might have about the case: "1" means that the act is morally permissible, and "0" means that it is impermissible. Let us say we had a neural net with ten units in the input layer, one for each feature in our chart. We could convert the eleven cases into eleven vectors. We simply map our top-down description of a case into a vector reading from left to right; for example, <1,1,1,0,0,0,0,0,0,0> represents case one and <1,0,1,0,0,1,0,0,0,0> represents case eleven. Say that we had an output layer with one neuron, which we want to fire strongly (with force 0.9 to 1) if the input vector describes a morally permissible act and weakly (with force 0.1 to 0) if it describes an impermissible act. With only four neurons in the hidden layer, we can successfully train a neural net to make the correct "judgments" about the above cases. I have named this network "Moral Network 1" (or MN1 for short). MN1 can only deal with cases having the ten specified features. Obviously, this is nowhere near enough to be a complete model of human moral cognition, but finding such a model is not the purpose of this paper. Trying to understand how moral knowledge can be, in some sense, general and yet not be usefully described in terms of exceptionless moral principles is the purpose of this paper, and MN1 does help us with that. The examples used to train MN1 are very schematic, and the reader has probably noticed that I am providing fuller descriptions in discussing them than can be found on the chart used to construct the training vectors. Obviously, a much larger neural net would be required to handle the fuller descriptions. For the purpose of exposition, I've limited myself to a small net.

The network is only given descriptions of cases and the results it should achieve. Without being given the equivalent of a substantive moral principle, it finds a distribution of synaptic weights that allows it to make the appropriate moral "judgements." Moreover, MN1 can apply what it has learned to new cases. For example, consider how MN1 responded to the following test cases (not in its training set).

Case 12: <0,1,0,1,1,0,0,0,0,0> as input received a 0.99 as output.
Case 13: <0,1,1,0,0,1,0,0,0,0> as input received a 0.94 as output.
Case 14: <0,1,1,0,0,0,0,0,1,1> as input received a 0.00 as output.

Allowing another to die in the name of self-defense is permissible -- that is what MN1 says about case 12. Killing someone to save the lives of many innocents is acceptable, and killing someone for revenge and to eliminate them as competition is impermissible -- that is how cases 13 and 14 are treated, respectively. These are reasonable responses given the sorts of cases the network was trained on. Just as there were no explicitly represented rules in the rock-mine network, there are no explicitly represented rules in MN1. However, there appears to be some sense in which which the system's "knowledge" is general. After all, it was able to generalize over the examples in its training set and apply what it learned to new cases. What it "learned," however, does not appear to be a set of exceptionless rules. MN1 acquired a set of synaptic weights which allowed it to successfully generalize. That generalization can be understood as a way of partitioning a highly dimensional hidden unit activation vector state space. (Recall the partial graph of the hidden unit activation vector state space for the rock-mine network.) If we wanted to model the respects in which this knowledge was general, we would use algebraic techniques to describe the way the hidden unit activation vector state space was partitioned.

It might be objected that the network really has learned a set of exceptionless rules. After all, the number of inputs in MN1 is finite, and the number of inputs in any neural net is finite. This means that there can only ever be a finite (even if extremely large) number of possible cases that a net can deal with. Provided we are willing to build enough exceptions into our rules, we could claim that, strictly speaking, the ceteris paribus clauses in our substantive moral principles are eliminable. Since, in principle, we could describe everything the network does by using a set of exceptionless rules, perhaps the earlier suggestion that the network is not best characterized as having learned and as applying a set of exceptionless rules is illegitimate.

The only point in the preceding objection which has any merit is the observation that, in principle, ceteris paribus clauses are eliminable. However, even if we concede that point, not very much follows. I will offer two arguments for that claim. The first I will call the argument from cognitive implementation, and the second the argument from methodological practice.

In classical artificial intelligence (AI), rule-based approaches to cognizing are implemented through computer languages that are designed to preserve isomorphism. A high-level language characterizes cognition in terms of following rules, but such a language is inefficient for driving the hardware, so the higher level language is translated into a lower-level language which can efficiently drive the hardware. These languages are designed so that there is an isomorphism between higher and lower level descriptions of cognition. The lower level implements the higher level, and nothing is lost in that implementation. Paul Smolensky has proved that in neural nets having a non-linear activation function, the isomorphism of levels breaks down.[4] In other words, if we

[4]Paul Smolensky, "Neural and Conceptual Interpretation of PDP Models," in J.L. McClelland, D.E. Rumelhart and the PDP Research Group, Parallel Distributed Processing, vol. 2 (Cambridge, Mass.: MIT Press, 1986).

attempted to describe the computation of a non-linear net at the representational level (which is the level of patterns of activation across units), that description would not be isomorphic to a lower level description which characterizes what goes on at the unit-level. The complete, precise, algorithmic account of neural net performance is given at the subrepresentational or unit-level. The upshot of all this is that the classical story does not work for non-linear neural nets. Higher level (or representational rule-level) accounts of such nets are not isomorphic to the more complete lower level accounts. Even if we could describe what a net does in terms of an exceptionless set of representational rules, it does not follow that those rules are <u>implemented</u> by the net since that would require the description of the activation function computing pattern vectors to be isomorphic to the description of the activation function computing unit vectors. Smolensky proved that such an isomorphism does not hold in non-linear nets, and MN1 is a non-linear net. The fact that it might be possible to describe what a neural net does in terms of a set of exceptionless, representational rules does not mean that the net implements those rules. In other words, it is not clear that those rules need play any significant role in cognitive modelling.

The argument from cognitive implementation just given is about the descriptive inadequacy of an exceptionless-rules account of MN1 (or any other non-linear net). My next argument is about the inability of a rules approach to make sense of an important part of moral methodology. In disagreements about moral issues, it is very often the case that whatever substantive moral principle is appealed to by one party of the disagreement is seen as question begging by the other side. In spite of the inability of exceptionless, substantive principles to do any work in such situations, disagreements are often discussed and sometimes resolved by playing off different sorts of cases against one another. The practical ethics literature is filled with case-by-case reasoning. What is interesting about such reasoning is that even if a principle is eventually agreed upon after much case-based reasoning, it is far from obvious that the principle provided any normative guidance. If anything, the status of the principle was in question. But even that is often dubious. After all, if we tried to formulate rules or principles that precisely captured our views on all moral issues, they would be extremely long. If I were to try to capture all my considered intuitions on the permissibility and impermissibility of lying and promise-breaking in a couple of principles, I would need hundreds, perhaps thousands, of words, and maybe even that would not suffice. When principles get that long, it is difficult to believe that a moral disagreement is about the status of some exceptionless principle. Reasoning, as humans do it, is not about deducing a claim from very long and ad hoc principles. So even if it is possible to describe what MN1 is doing in terms of following long principles, it is not clear that such a description would be of much use. To see why, imagine that a much larger and more realistic moral neural net was in someone's head -- Jack; let us also say that the behaviour of this net can be described in terms of following exceptionless (but very, very long) rules. If Jack had a moral disagreement with Jill, would it follow that their disagreement was about the status of some principle? If the alleged principle Jack subscribed to was of encyclopedic length, then I do not think we would characterize the disagreement as being over a principle. Their discussion is about the moral significance of some feature of some act under a particular set of circumstances. Once again, it seems that the in-principle eliminability of ceteris paribus clauses is of little use to those who think that exceptionless rules play an important role in moral reasoning.

Ah, but could MN1 or something like it be adapted to model moral reasoning? Perhaps. Much of the normative ethics literature is filled with the use of contrast cases to decide the significance of a particular features. For example, much has been written on the do-allow distinction. Some claim that killing in certain contexts makes an act impermissible, while allowing someone to die in the same contexts makes an act permissible. One argument that can be used in defense of abortion in cases of rape-induced pregnancies is Judith Thomson's famous violinist exemplar: imagine that you are kidnapped and hooked up to a world famous violinist so that your kidneys may filter his blood; you are free to leave, but if you do not stay hooked up for nine months, the violinist will die.[5] Most people have the intuition that it is acceptable to walk away from violinist even if he dies. Case 8 in our list can be taken to represent the violinist exemplar and the standard response to it. One way to reconcile the acceptability of walking away from the violinist and the unacceptability of abortion in cases of rape-induced pregnancies is to claim that (a) in the abortion case there is killing, (b) in the violinist case we are only allowing someone to die, and (c) there is a morally significant difference between killing and allowing someone to die (or, perhaps more generally, between doing and allowing). To avoid the charge of special pleading, the individual making the preceding argument should be able to come up with other cases in which the killing-letting die distinction is relevant. Contrast cases are often used to show whether a distinction

---

[5] Judith Jarvis Thomson, "A Defense of Abortion," in The Rights and Wrongs of Abortion, eds. M.T. Cohen, T. Nagel, and T. Scanlon (Princeton: Princeton University Press, 1974), 4-5.

is relevant or not. A set of contrast cases consists of two cases which are identical except for the feature meant to be tested. For example, consider the following cases: (A) Jack puts cyanide in Jill's coffee and kills her; (B) there is cyanide in Jill's coffee and Jack knows it, but he did not put it there; nevertheless, Jack allows Jill to drink the coffee, and she dies. There is no difference in how we think of Jack in cases (A) and (B), yet in one case he kills, and in another he allows someone to die. Sometimes, at least, the distinction between killing and allowing someone to die is morally irrelevant. To defend the intuitions outlined above on the abortion and violinist cases, a set of contrast cases (similar to the cases in question) would have to be produced which showed that the killing-letting die distinction does make a difference. If such cases cannot be found, then a change of judgment is in order. MN1 "thinks" that abortion in cases of rape induced pregnancy is impermissible (case 1); however, walking away from the violinist is permissible (case 8). Imagine a neural net that inspects the inputs and outputs of a MN1 and is capable of setting up contrast cases in MN1. Let us say that this secondary net can make us of what it finds out in its inspection of MN1 to make adjustments to MN1 . Such a second-order net should not be difficult to set up. Moreover, it would find that there are no contrast cases that support the moral significance of the do-allow distinction in cases similar to the rape induced abortion and violinist exemplars. The second-order net could then change the "judgement" on the abortion case and run the back propagation algorithm until MN1 is appropriately trained.

The preceding is very crude, but we have to start somewhere. I hope to have shown that it may be possible to account for moral reasoning in a way which is in line with the intuitions of both generalist and particularist sympathizers.

Acknowledgements

I thank Professor Tracy Isaacs for her comments on an earlier version of this paper, and the Social Sciences and Humanities Research Council of Canada for financial support, award 752-94-1127.

Bibliography

Churchland, Paul. A Neurocomputational Perspective: The Nature of Mind and the Structure of Science. Cambridge Mass. : MIT Press, Bradford Book, 1989.

Dancy, Jonathan. Moral Reasons. Oxford: Blackwell, 1993.

317

McClelland, J.L., D.E. Rumelhart. Explorations in
    Parallel Distributed Processing. Cambridge, Mass.:
    MIT Press, Bradford Book, 1988.

Smolensky, Paul. "Neural and Conceptual Interpretation
    of PDP Models," in J.L. McClelland, D.E.
    Rumelhart, and the PDP Research Group. Parallel
    Distributed Processing: Explorations in the
    Microstructure of Cognition, vol. 2. Cambridge
    Mass. : MIT Press, Bradford Book, 1986.

Thomson, Judith Jarvis. "A Defense of Abortion," in The
    Rights and Wrongs of Abortion, eds. M.T. Cohen, T.
    Nagel, and T. Scanlon. Princeton: Princeton
    University Press, 1974.

Aristotle, Whately, and the Taxonomy
of Fallacies

Hans V. Hansen
Brock University

The two most original and important classifications of fallacies in the history of logic are by Aristotle in his *On Sophistical Refutations* and by Richard Whately in his *Elements of Logic*. The present paper proposes to describe the bases of each of these taxonomies; accordingly we must ask ourselves the question, "What principle, or principles, explicit or implicit, underlie and determine the classification of fallacies being presented?"

I. ARISTOTLE

Smith's remark (p. 63) that "Modern textbook treatments of fallacy still follow much of [Aristotle's] classification to a remarkable, if not embarrassing, extent" is a testimony to the importance of Aristotle's work on fallacies. However, in spite of the fact that Aristotle's classification is mentioned by nearly everyone who works in the field, it does not seem to be well understood. Some there are who despair of the project of classifying fallacies altogether; their concerns we shall leave for another time. For now our rationale is that because Aristotle fallacies are themselves in want of clarification, it is important to try to understand his intended way of classifying them. It is anticipated that not only will a study of the taxonomical principles at work help our understanding of the individual fallacies, but also that close attention to what Aristotle says about the fallacies will advance our understanding of his classificatory scheme.

1.1 What is Being Classified?

At the beginning of *On Sophistical Refutations* Aristotle writes that "there exist both reasoning and refutation that is apparent but not real" (165a19-20). What is to be classified are those deceptive kinds of reasonings and refutations that appear to be genuine when they are not. But are *reasoning* and *refutation* two kinds of things or are they the same? This is an important question since if they are not the same, then two kinds of things, sophistical reasonings *and* sophistical refutations are being classified.

Some have implied that there is no difference between sophistical refutations and fallacies, i.e., they think that, for Aristotle, a fallacy is the same as a sophistical refutation. But since sophistical refutations are essentially dialogical (i.e., found essentially in *dialogue*[1]), it might be thought that Aristotle's concept of 'fallacy' is dialogical; or, in other words, that the correct analysis of a fallacy will, necessarily, be in terms of the rules governing a two party argument or discussion. Accordingly, Aristotle's view of fallacies has been thought to have more in common with certain modern dialectical approaches, such and Van Eemeren and Grootendorst's, than with the view of fallacies broadly shared by most logicians, viz. that they are bad arguments (where 'argument' is defined non-socially as a certain kind of an intentional relation between two or more propositions).

However, there are several reasons to think that Aristotle did not equate fallacies outright with sophistical refutations. One reason is that he says that a certain fallacy--form of expression (which is language dependent)--can occur when we are inquiring by ourselves. He explains that the fallacy occurs because we are taking speech as the basis of our inquiry, rather than the things themselves (169b1). But this is a relevant observation for all the language dependent fallacies (165a4-10), and it would be strange if Aristotle thought that language could lead solitary inquirers astray only in the way that this one particular fallacy does. At any rate, Aristotle clearly thinks that some of the fallacies can be committed apart from engagement in dialogue, and this is sufficient to show that, unlike sophistical refutations, fallacies are not essentially dialogical; this in turn is enough to guarantee the conceptual distinction between fallacies and sophistical refutations.

A more general reason for holding that Aristotle distinguished fallacies and sophistical refutations is that he is careful to distinguish deductions and refutations. A refutation is a special kind of deduction, one that has the added feature that it ends with a statement that contradicts the answerer's thesis (165a3, 168a37). Since deductions are not the same as refutations, but a part of them, it will be possible that there are faults in refutations that would not be faults in deductions.

This conceptual distinction between refutations and deductions suggests an hypothesis about the relationship of sophistical refutations and fallacies. It is a relationship that Aristotle never explicitly states, but one that we may reasonably attribute to him; it also simplifies his theory immensely.

(H)   Fallacies are logical errors, and sophistical refutations are sophistical because they *contain* fallacies.

We shall adopt this hypothesis (H) about Aristotle's view for the remainder of this paper.

1.2 What is the Basis of the Classification?
Aristotle's best known classification of fallacies is the one that divides them all into two groups, those dependent on language and those independent of language.

In dictione	*Ex dictione*
ambiguity	accident
amphiboly	*secundum quid*
combination	consequent
division	*ignoratio elenchi*
accent	begging the question
form of expression	non-cause
	many questions

Here the stated operative taxonomical principle is whether or not the item in question 'depends on language'. What can this mean? Van Eemeren, Grootendorst and Kruiger offer a possible explanation.

> Aristotle divides incorrect or false refutations which can be used in dialectical contexts into two groups, distinguishing between Sophistical refutations which are dependent on language (*in dictione*) and Sophistical refutations which are independent of language (*extra dictionem*). ...
> The fallacies which are dependent on language are divided into six types, all connected with ambiguities and shifts of meaning which may occur in ordinary colloquial language. ... The fallacies which are independent of language he divides into seven types, all of which could occur even if colloquial language were perfect.[2]

The basis of the distinction is now made to be ordinary colloquial language rather than language *simpliciter*: fallacies that arise because of ordinary colloquial language are language dependent, and those that occur due to some other factor are language independent.

'Colloquial', however, has a number of senses. It can mean "occurring in spoken rather than in written speech;"[3] or, "being an informal, or ungrammatical, rather than formal use of language". (Notice that 'colloquial language' contrasts with 'formal use of language', not 'formal language'.) At any rate, delimiting the language dependent fallacies to the domain of colloquial language is a narrower interpretation than the text implies. More likely, what Aristotle had in mind would come close to what we think of as *natural language*, and his own explanation of the possibility of language dependent fallacies is that they are due to the fact that things are infinite whereas words and expressions are finite; hence, some bits of language must do double duty (165a13). That some terms must have more than one meaning at once creates the possibility of a shift of meaning and the possibility of the mistake going unnoticed. Aristotle is sure that there are exactly six ways of producing a "false illusion in connection with language" (165b26),[4] and his list contains exactly six items. However, on more than one occasion he sees that some members of the *ex dictione* list of fallacies could also be considered as language dependent, for example, *ignoratio elenchi* (167a35) and many questions (175b39).

If the language dependent fallacies owe their ills to the vagaries of natural language, to what do the *ex dictione* fallacies trace their fallaciousness? Unless we can answer this question we will have no positive characterization of the language independent fallacies. In order to provide an answer some needed concepts must be introduced, viz., (i) the structure of real dialectical refutations, and (ii) Aristotle's concept of 'deduction'.

1.2.1 Dialectical Refutations

Historically, one should understand Aristotle's *On Sophistical Refutations* as an attempt to come to grips with the bad kinds of argumentation practiced by some of the sophists. Examples of such can be found in Plato's *Euthydemus* and some of the very examples in *On Sophistical Refutations* are taken from that dialogue. Broadly speaking, Aristotle's method is to study the variety of sophistical refutations against the background of what he took to be genuine, dialectical refutations and then to highlight the way in which a sophistical refutation failed to meet the standards of a real refutation. A prerequisite, then, to understanding Aristotle's work on fallacies is to be familiar with the general outlines of dialectical argumentation as he conceived of them.

A genuine or real dialectical refutation may be seen as falling into three stages. In the first stage the questioner asks the answerer which of two contradictory theses, or questions, he wants to uphold. By his answer the respondent is committed to a thesis. In the second stage, the interlocutor puts further questions to the respondent to be answered by a Yes or a No. In the third stage, the questioner shows that the answers given to the Yes-No questions imply the contradictory of the answerer's thesis.

The appearance of a real refutation can be gained in one of two different ways that both take place under what we called the 'third stage': the point at which the questioner claims to have deduced the contradictory of the answerer's thesis. (a) If the conclusion deduced does not really follow from the answers to the Yes-No questions then the refutation commits an error in its deductive part; such errors can arise either from a fault with the premises or the subsequent inference based upon them.[5] (b) The other kind of error that can arise is that the conclusion of the (otherwise faultless) deduction fails to contradict the answerer's thesis. The first kind of error we will call *a deductive error*, the second, *an error of (apparent) contradiction*. We introduce this new terminology as an aid to understanding Aristotle's classification of fallacies and sophistical refutations, and remind the reader that it is our distinction, not Aristotle's.

It is tempting to identify the distinction between errors of contradiction and deduction with the one between language dependent and independent fallacies. There is even a reason to do so, for at the end of *On Sophistical Refutations* 7, Aristotle says,

> All the types of fallacy, then, fall under ignorance of what a refutation is, those dependent on language because of the contradiction, which is the proper mark of a refutation, is merely apparent, and the rest because of the definition of *syllogismos* [deduction]. (SR vi 169a19-21)

Let us then suppose, for the moment, that the *in dictione* fallacies are errors of contradiction and that the *ex dictione* fallacies are errors of deduction.

This supposition puts great pressure on the concept of 'fallacy'. Since errors of contradiction are not arguments, they are not fallacious arguments. They are only two propositions that appear to be logically opposed. Hence, a simple identification of language dependent fallacies and errors of contradiction has two unintuitive consequences: (i) the concept of 'fallacy' is stretched beyond its normal theoretical employment, and (ii) for Aristotle, there are no language dependent fallacies of deduction.

But, surely, that was not Aristotle's view as we already know from his comments about the fallacy, form of expression. Moreover, when he introduces the *in dictione* sophistical refutations in chapter 4, he illustrates them as fallacious arguments, not as fallacious refutations, thereby indicating that he considered it possible that they occur as fallacious arguments, and hence that they can also occur in our category of errors of deduction. We must conclude that the language dependent fallacies can occur as both errors of contradiction and deduction and, therefore, obviously, they are not a perfect match with the category of errors of contradiction. For Aristotle, it turns out, "*in dictione* fallacy" is ambiguous; it sometimes denotes an error of contradiction, and sometimes an error of deduction. Moreover, since "sophistical refutation" is defined partly in terms of "fallacy" (according to hypothesis H), this means that the term "*in dictione* sophistical refutation" is also ambiguous with regard to the kind of mistake it contains.

So, some sophistical refutations contain fallacies that are not arguments, but errors of contradiction. These fallacies only occur in sophistical refutations which means that they only occur in dialogical contexts. However, it is important to point out that even in the dialogical context of sophistical refutations an error of contradiction is a *logical mistake*; only in a derivative sense could it be a mistake of 'dialectics'. Moreover, all errors of contradiction can be recast as errors of deduction. This is done by taking the real contradictory of the thesis ('-T') and letting it serve as the conclusion of the deduction in lieu of the proposition giving rise to the error of contradiction ('-T*'). The error of contradiction is now eliminated since T and -T are genuine contradictories, but an error of deduction replaces it since -T is not a consequence of the premises. For example:

STAGE I

Answerer   *Interlocutor*
Thesis (**T**): Bob's money   <-   Is Bob's money in
is in the (river) bank   the bank or not?

STAGE II

Answer *n*   <-   Question *n*
Answer *n*   ->   Premiss *n* of deduction
etc.   etc.
STAGE III

Conclusion 1 (-**T***): Bob's money
is not in the (financial) bank
[follows deductively from answers]

Conclusion 2 (-**T**): Bob's money
is not in the (river) bank
[does *not* follow deductively from answers]

If Conclusion 1 is established then there is an error of contradiction, since Conclusion 1 and the Thesis could both be true. On the other hand, if Conclusion 2 is the one that is claimed to have been established then there is no error of contradiction, but there is an error of deduction. The one kind of error replaces the other kind.[6]

1.2.2 Aristotelian Deduction

We have made some progress in characterizing the *in dictione*, language dependent fallacies, but we are still in want of a positive characterization of the *ex dictione*, language independent fallacies. Identifying them with errors of deduction, it turns out, is exactly the right thing to do; just as errors of contradiction are central to the understanding of language dependent fallacies, so deduction is central to understanding the class of language independent fallacies. However, the operative concept of 'deduction' is narrower than the one we are familiar with. This is highlighted by the fact that Aristotle considers both begging the question and non-cause to be errors of deduction.

Syllogismos is sometimes translated as 'syllogism', sometimes as 'reasoning', sometimes as 'proof', and more recently as 'deduction'. What Aristotle means by *syllogismos* is of the utmost importance. In the *Topics* he introduces the term saying "a *syllogismos* is an argument in which, certain things being laid down, something other than these necessarily comes about through them" (100a25-26), and he repeats this definition in *On Sophistical Refutations* (165a1-2) and *Prior Analytics* (24b19-20). This conception of *syllogismos*, or 'syllogistic deduction' as we shall call it from now on, stipulates two conditions in addition to the requirement that the premises must necessitate the conclusion: (a) the conclusion must be something other than any of the premises, and (b) it must come about *through* the premises.

That the premises must necessitate the conclusion is the condition of syllogistic deduction that is violated by the *ex dictione* fallacies of accident, *secundum quid*, consequent, and *ignoratio elenchi*. The condition that the conclusion must be something other than any of the premises guards against begging the question. That the conclusion must come about *through* the premises is the condition relevant to the analysis of non-cause. This fallacy involves treating non-premises as premises (hence, non-causes as causes). The fallacy of combining two questions into one is that of failing to start the deduction from genuine propositions, a requirement of an Aristotelian deduction (--Aristotle's view is that the answer to 'many questions' is not a proposition).

Aristotle's characterization of the language independent fallacies, therefore, is not just that they have the negative property of not being language dependent; rather they are positively characterized as mistakes of a particular kind of deductive reasoning which specifies certain restrictions on the concepts of 'proposition' and 'deductive consequence'. Moreover, unlike the language dependent fallacies which can function as either errors of contradiction or errors of deduction, the language independent fallacies never function as errors of contradiction; this implies, contrary to popular belief, that there is nothing especially dialogical about any of them.

We should remark on what it is that gives the *ex dictione* fallacies the capacity to deceive. Here there is no one 'semantic fact' that will serve as a general explanation, as there is in the case of the *in dictione* fallacies. We must look to each kind of *ex dictione* fallacy to see how it manages to create an illusion. In the case of begging the question it is because we fail to remember what is different and what is the same (167a39); the fallacy of consequent occurs because people think the relation of consequence is convertible (167b1-2); non-cause happens because the structure of *ad impossibile* reasoning is too hard to follow (157b34-158a2); many questions can arise because people don't see that more than one question is being asked (167b37), etc.

At the beginning of *On Sophistical Refutations* 6 Aristotle says, "it is possible to analyze all the aforesaid modes of fallacy into breaches of the definition of refutation" (168a19), and at the end of the same chapter he writes, "all the kinds of fallacy fall under the heading of ignorance of the nature of refutation" (169a20). Why does Aristotle present us with this alternative classification of fallacies? Since it does not really add anything to the clarification of the nature of the fallacies, and since Aristotle does not seem to express a preference for either classificatory system, we must guess at what his motivation was.

What the discussion of all the fallacies as instances of *ignoratio elenchi* does show is that these thirteen fallacies can all infect a dialogical refutation. Of course, this is not to say that they are dialogical fallacies. However, considering all the fallacies as instances of *ignoratio elenchi* at once reinforces our hypothesis (H), that sophistical refutations contain fallacies, as well as showing that the fallacies Aristotle has identified have genuine application in the analysis of sophistical dialogues.

1.3 Conclusion

When Aristotle introduces the first list, the items under *in dictione*, he refers to them as 'modes' or 'styles' of *elenchus* (refutations) (165b23). At the end of the same chapter he announces that he will next discuss the *ex dictione paralogismoi* (166b21), a term best translated as 'fallacious argument'. Overall, this gives a certain unevenness to any attempted classification since the two categories are not exclusive (nor is there a good reason to think they are exhaustive). Because fallacies are theoretically fundamental to sophistical refutations, the taxonomy is best understood as being about fallacies; but given a broader concept of 'fallacy' it can double as a classification of sophistical refutations. Ironically, the distinction between *in dictione* and *ex dictione* fallacies makes sense only if seen against the backdrop of dialectical refutation and syllogistic deduction, concepts which allow us to distinguish errors of contradiction and deduction.

II. WHATELY

In Whately's view the analysis of fallacies should be based on logical principles and, therefore, it is to be expected that it is logical principles that determine his classification of fallacies. Whately complained that his predecessors had abandoned the principles of logic and "assumed a loose and rhetorical style of writing" (Whately, 155) when they turned to the study of fallacies; accordingly, he promises to give "a scientific analysis of the procedure which takes place in each [fallacy]" (Whately, 153).

2.1 Classification

Whately defined 'fallacy' as "any unsound mode of arguing, which appears to demand our conviction, and to be decisive of the question in hand, when in fairness it is not" (Whately, 153).[7] The genus here is modes of arguing that *appear* to demand our conviction and to be decisive, the species is arguments that really are neither convincing nor decisive. The genus, then, is shared with Aristotle, the point of difference will be found in the way that the species is divided, and this turns out to be Whately's fundamental principle of classifying fallacies.

Whately's main distinction is between logical fallacies and non-logical fallacies. The former are fallacies because they are invalid arguments, the latter are valid arguments which qualify as fallacies nonetheless. The non-logical fallacies he also calls 'material fallacies', indicating thereby that the source of fallacy lies in the *matter* of the argument rather than in the form.

Each of these two broad classes is again divided one more time. The logical fallacies contain the subclasses, *purely logical fallacies* and *semi-logical fallacies*. The former consist of some explicit violation of the rules of the valid syllogism, e.g., undistributed middle, illicit process. Members of the subclass of semi-logical fallacies, are reducible to purely formal fallacies once some ambiguity of the middle term (e.g., equivocation, division or composition) has been unmasked.

The non-logical, or material, fallacies also fall into one or another of two subclasses, according to whether the (non-logical) fault lies with the material of the argument that is a premiss, or the material of the argument that is the conclusion. In *petitio principii* and false cause, Whately thinks the fault is in one of the premises. If the fallacy is traced to a problem with the conclusion, however, Whately will subsume it under the sub-heading *ignoratio elenchi*, and it is here that the various *ad* arguments, if fallacious, are placed. (e.g., *ad hominem, ad populum, ad verecundiam* and, possibly, *ad ignorantiam*).

WHATELY'S CLASSIFICATION OF FALLACIES[8]

Fallacies			
Logical		Non-logical	
Purely Logical	Semi-logical	Premiss wrongly assumed	*ignoratio elenchi*
undistributed middle	division	*petitio principii*	*ad hominem*
illicit process	composition	false cause	*ad verecundiam*
	equivocation		*ad populum*
	accident		appeal to the passions

Whately complained bitterly about Aristotle's classification of fallacies, and the use logicians had made of it.

> The division of Fallacies into those in the words (IN DICTIONE,) and those in the matter (EXTRA DICTIONEM) has not been, by any writers hitherto, grounded on any distinct principle: at least, not on any that they have themselves adhered to. (Whately, 158)

But Whately was not complaining only about Aristotle, he is also critical of the main text of his day, Aldrich's *Compendium*.

> [A]fter having distinguished Fallacies into those in the *expression*, and those in the *matter* ("*in dictione*," and "*extra dictionem*,") he [Aldrich] observes of one or two of *these last*, that they are not properly called *Fallacies* as not being *syllogisms faulty in form*; ... as if anyone, that was such, could be "*Fallacia extra dictionem*." (Whately, 155n)[9]

What we can gather from this is that Aldrich recast Aristotle's distinction as one between form and matter, and perhaps that he thought of these as exclusive and exhaustive categories for classifying fallacies. Having made this distinction he also seems to have noticed that some of the *material* fallacies were not faulty syllogisms. A passage in Hamblin (p. 49) suggests that these were non-cause, begging the question and many questions.[10] Accordingly, Aldrich doubted that some of Aristotle's fallacies were really fallacies.

What is interesting, and important, about Whately is that he agreed with Aldrich that non-cause and begging the question were not faulty by form; however, he also preserved the Aristotelian tradition by considering them to be fallacies; hence the category of material fallacies evolves into the category of *non-logical* fallacies. The emergence of the category of non-logical fallacies is aided by the fact that Aristotle and Whately meant different things by 'syllogistic reasoning (deduction)', as we can see by comparing their treatments of three of the fallacies, *petitio principii*, non-cause, and *ignoratio elenchi*.

2.2 Begging the Question

For Aristotle, 'syllogism' (or 'deduction') places two restrictions on deduction over and above *following necessarily*, namely (i) the conclusion must not be identical to any of the premises, and (ii) the conclusion must be based on the premises. The first of these conditions makes begging the question a syllogistic error. Although Whately's definition of 'argument' is very similar to Aristotle's definition of 'syllogism',[11] Whately does not find begging the question to be a logical fallacy. Instead he subsumes it under the chief heading of non-logical (material) fallacies, and the sub-heading of 'Premiss unduly assumed', and he characterizes the fallacy as being that "the premiss either appears manifestly to be the same as the conclusion, or it is actually proved from the conclusion, or is such as would be naturally and properly so proved" (Whately, 200). But it is not possible to mark precisely the distinction between the Fallacy in question and fair argument; since that may be correct and fair reasoning to one person, which would be, to another, "begging the question;" inasmuch as to one, the conclusion might be more evident than the premiss, and to the other, the reverse. (Whately, 163)

Like some of the other non-logical fallacies, the fallaciousness of begging the question depends on the *use* of the argument. In other words, some tokens of an argument may be fallacious and others not.

2.3 Non-Cause

Aristotle's explanation of the fallacy of non-cause (in *On Sophistical Refutations*) is that it is an argument in which it looks like the conclusion is based on the premises present, but in fact one of the premises is idle. Thus, for Aristotle, this fallacy runs afoul of the third condition of a syllogism, that the conclusion be based on the premises given, and not just on *some* of them. Unlike the non-cause fallacy introduced in Aristotle's *Rhetoric*, the one in *On Sophistical Refutations* is an Aristotelian deductive error.

The Port-Royal logicians (Arnauld and Nicole) treated non-cause as a fallacy of physical causation. If they thought they were developing the Aristotelian tradition then they overlooked the distinction between a premiss as a logical cause and as a physical cause. Whately remarked that, "the logical writers ... were confounding together *cause* and *reason*; the sequence of *conclusion* from *premises* being perpetually mistaken for that of *effect* from physical *cause*" (Whately, 206). However, although there is a mistake in thinking that the premises of an argument are physical causes of the conclusion, according to Whately, this is not the fallacy of false cause. He locates the fault in "arguing from that which is *not* a sufficient cause as if it *were* so" (Whately, 203). Consequently, this fallacy will be found in a premiss that states a false causal connection. But since the argument might well be valid, the fallacy will be a non-logical, material fallacy rather than a logical one. What Whately has done is to substitute for the non-cause fallacy of *On Sophistical Refutations* a different fallacy, and then make a point of saying that it doesn't fit Aristotle's classification.

2.4 Ignoratio Elenchi

On Aristotle's view *ignoratio elenchi* is, in its first incarnation, a deductive error not covered by the other *ex dictione* fallacies, i.e., the failure to prove the conclusion in the "same respect and relation and manner and time" (167a26). In its second guise *ignoratio elenchi* is any failure of a refutation identical to one or another of the thirteen fallacies, either errors of contradiction or deduction. Either version of *ignoratio elenchi*, by Aristotle's lights, contains a logical mistake.

However, once Whately has established the category of non-logical, material fallacies this opens the door for a different interpretation of *ignoratio elenchi*. Whately treats *ignoratio elenchi* as a sub-heading of non-logical fallacies in which "the Conclusion is not the one required, but irrelevant" (Whately, 162). He does consider treating *ignoratio elenchi* as a logical fallacy since 'elenchus' means 'proof of the *contradictory*'; and since it is logic that defines what 'contradiction' is, *ignoratio elenchi* should qualify as a logical fallacy. But he rejects this for the following reason, among others:

> it seems an artificial and circuitous way of speaking, to suppose in all cases an *opponent* and a *contradiction*; the simple statement of the matter being this,--I am required, by the circumstance of the case, (no matter why) to prove a certain conclusion; I prove, not that, but one which is likely to be mistaken for it; in this lies the Fallacy. (Whately, 162)

Whately, like the modern generation of logicians he inspired, saw the dialogical trappings of the Aristotelian tradition as unnecessary.

John Locke and Isaac Watts had considered *ad* arguments as a distinct kind, or sort, of argument; but they had not thought of them as fallacies. However, with *ignoratio elenchi* in place as a sub-heading of non-logical fallacies, Whately was in the position to make a significant innovation. He was the first to include the *ad* arguments in a classification of fallacies, although he warned that "we should by no means universally call [them] Fallacies" (Whately, 215). The *ad hominem* argument, for example, even though it is valid, will be fallacious only if it is used unfairly.[12]

III. CONCLUSION

Aristotle's classification of fallacies into two kinds, *in dictione* and *ex dictione*, involves two criteria: the first criterion is that of language dependency, the second involves the special character of Aristotelian deduction. However, all the thirteen fallacies of *On Sophistical Refutations* are 'logical fallacies' in the sense that, at bottom, they contain a logical mistake. How well Whately understood Aristotle's attempted distinction is not clear. However, his avowed intention to present a classification of fallacies based on purely logical principles has the interesting result that it leads to the creation of the category of non-logical fallacies. So, like Aristotle, Whately classification also involves two criteria: the first criterion identifies logical fallacies, and invalidity is a necessary condition of membership in this class; the second criterion identifies the non-logical fallacies and it holds that valid arguments can be *used* fallaciously. Whately's basic distinction, then, divides logical and what we may call 'pragmatic fallacies'.

There is an alternative way in which we can view the difference between Aristotle's and Whately's classification. Aristotle's classification presupposes the background of correct dialectical argumentation; yet all his fallacies are types of logical mistakes. Remove the dialogical context, as Whately does, and some of Aristotle's fallacies become non-logical, pragmatic fallacies.[13]

REFERENCES

Aristotle, *On Sophistical Refutations*.

Aristotle, *Topics*.

Carney, J. D., and R. K. Scheer (1980) *Fundamentals of Logic*. New York: Macmillan.

Eemeren, Frans H. van, Rob Grootendorst (1992) *Argumentation, Communication and Fallacies*. Hillsdale, NJ: Erlbaum.

Eemeren, Frans H. van, Rob Grootendorst and Tjark Kruiger (1987) *Handbook of Argumentation Theory*. Dordrecht: Foris.

Hamblin, Charles L. (1970) *Fallacies*. London: Methuen.

Hansen, Hans V. (1993) "Sophistical refutations as fallacious deductions," Ontario Philosophical Society, Univ. of Waterloo.

Hansen, Hans V. (forthcoming) "Whately on the *ad hominem*: A liberal exegesis," *Philosophy and Rhetoric*.

Smith, Robin (1995) "Logic," in *The Cambridge Companion to Aristotle*, J. Barnes, editor. Cambridge: Cambridge University Press. pp. 27-65.

Whately, Richard (1844) *Elements of Logic*, 8th ed. London: B. Fellowes. Originally published 1826.

NOTES

1. I use 'dialogical' to mean 'dependent on dialogue' or 'occurring only in dialogue'. 'Dialectical' I reserve (with Aristotle) for a fallacy-free kind of argumentation whose premises are *endoxa* (accepted by either the many or the wise).

2. Van Eemeren, Grootendorst and Kruiger, p. 79. Notice the easy assimilation of fallacies and sophistical refutations.

3. Yet Aristotle finds it difficult to give examples of some of the fallacies in spoken speech, e.g., the fallacy of accent (166b1-2).

4. Aristotle thinks this can be proved both inductively and deductively (165b28) although he never produces the proofs.

5. Smith writes:

> Sophistical arguments only *appear* to be good dialectical arguments: either they appear to be deductions but are not, or they rest on premises which appear to be acceptable but are not. (Smith 1995: 63)

According to this there are two sources of error for sophistical refutations: either a deductive error, or an error of accepting a premise one ought not to have accepted. Smith is right that a refutation can fail to be dialectical because its premises are not dialectical (as Aristotle suggests at 165b8); but this fault, as far as I can tell, never gets catalogued among Aristotle's thirteen fallacies. The possible exception to this is the fallacy of combining two questions into one (many questions). This fallacy I subsume under deductive fallacies, since Aristotle thinks that no deduction can be based on the answers to many questions. So, even though Aristotle gives a general characterization of contentious reasoning as falling into one of the two categories Smith reports, Aristotle's actual discussion belies this distinction. Note, by the way, that Smith completely ignores the role of *in dictione* fallacies as errors of contradiction.

6. One is tempted to think that any *in dictione* error of contradiction (amphiboly, say) shares something with its fallacious deductive counterpart, viz., the exploitation of the same grammatical ambiguity. This may have been Aristotle's view but that it is a general truth about errors of contradiction and deduction is in want of demonstration.

7. In the Index of the Principal Technical Terms, Whately defines 'fallacy' as follows: "Any argument, or apparent argument, which professes to be decisive of the matter at issue, while in reality it is not" (*Elements of Logic*, 396).

8. This is a simplified version of the chart that appears in Whately's *Logic*, p. 164. There are a number of fallacies enumerated by Whately that we shall not consider, e.g., the fallacy of objections, the fallacy of shifting ground.

9. From this we see that the idea of a material fallacy can be dated back at least as far as Aldrich book of 1691.

10. Carney and Scheer (p. 30) adopt a variation on Aldrich: they read Aristotle's classification as being between fallacies of ambiguity and material fallacies but by material fallacy they mean the kind of fallacy that can only be solved by familiarity with the subject matter.

11.   An argument is an expression in which *"from something laid down and granted as true (i.e., the Premises) something else (i.e. the Conclusion) beyond this must be admitted to be true, as following necessarily [resulting] from the other* [" (sic)]. (Whately, 79)

We may discern in this passage the same three conditions on argument that Aristotle placed on syllogism.

12. See Hansen (forthcoming) for a detailed examination of when the *ad hominem*, by Whately's lights, is a fallacy.

13. I am grateful to Larry Powers (Wayne State University) and John Woods (University of Lethbridge) for continuing dialogue on the nature of Aristotle's fallacies.

Nonmonotonic Reasoning
with Multiple Belief Sets

Joeri Engelfriet[1], Heinrich Herre[2], Jan Treur[1]

[1] Free University Amsterdam,
Department of Mathematics and Computer Science,
De Boelelaan 1081a, 1081 HV Amsterdam, The Netherlands
e-mail: {joeri, treur}@cs.vu.nl
[2] University of Leipzig,
Department of Computer Science,
Augustplatz 10-11, 04109 Leipzig, Germany,
e-mail: herre@informatik.uni-leipzig.de

Abstract. In the present paper we introduce nonmonotonic belief set operators and selection operators to formalize and to analyze multiple belief sets in an abstract setting. We define and investigate formal properties of belief set operators as absorption, congruence, supradeductivity and weak belief monotony. Furthermore, it is shown that for each belief set operator satisfying strong belief cumulativity there exists a largest monotonic logic underlying it, thus generalizing a result for nonmonotonic inference operations. Finally, we study abstract properties of selective inference operations connected to belief set operators and which are used to choose one of the possible views.

keywords : nonmonotonic inference, knowledge representation, belief sets.

1 Introduction

In a broad sense, reasoning can be viewed as an activity where an agent, given some initial information (or set of beliefs) X, performs some manipulation to this information and arrives at a new state with different information. So a (partial) view on a situation (in the domain the agent is reasoning about) is transformed to another partial view. In general the mechanism may be non-deterministic in the sense that multiple possible views on the world can result from the reasoning process. In the current paper we present an approach to formalize and to analyze structural aspects of reasoning of an agent with multiple belief sets.

If we want to formalize reasoning in this way, we must describe the input-output behavior of the agent's reasoning process. We propose to use belief set operators for this purpose. A belief set operator is a function B which assigns to a set of beliefs (information) X, given in some language L, a family of belief sets $B(X)$, described in the same language.

Different modes of reasoning give rise to different kinds of belief set operators. If we consider exhaustive propositional reasoning, a set of propositional beliefs

X is mapped to the set $Cn(X)$ of propositional consequences of X, which is unique; so in this case there is only one belief set: $B(X) = \{Cn(X)\}$. However, if we look at nonmonotonic logics such as Autoepistemic Logic or Default Logic, an initial set of beliefs X may have none or more than one possible expansion (or extension).

In these two cases, the reasoning is conservative: the resulting belief sets extend the set of initial beliefs. But there are also modes of reasoning in which beliefs are retracted. This is the case in, for instance, contraction in belief revision, in which the contraction of a belief from a belief set is not uniquely determined. Also, when the set of initial beliefs is contradictory, and we want to remove the contradiction, one can select a consistent subset; this again can be done in more than one way.

Even though we have argued that in general a reasoning process may have multiple possible outcomes, an agent which has to act in a situation must commit itself somehow to one set of conclusions by using the information in the possible belief sets. In nonmonotonic logics, two different approaches to this problem are well-known: the credulous approach, where the agent believes anything from any possible extension (thus taking the union of the possible belief sets), and the sceptical approach, in which it only believes those facts which appear in all of the possible belief sets (taking their intersection).

A third approach is based on the situation where the agent has additional (control) knowledge allowing it to choose one of the possible belief sets as the "preferred" one. (Many nonmonotonic formalisms such as Autoepistemic Logic, Default Logic and Logic Programming have a prioritized or stratified variant.) As the different belief sets are usually based on different assumptions, and may even be mutually contradictory, we feel the credulous approach is not very realistic. Looking at belief revision in the AGM framework ([AGM85]), when we retract a sentence φ from a belief set K, the maximally consistent subsets of K which do not contain φ (denoted $K \perp \varphi$), in a sense play the role of the possible belief sets. Contraction with φ is always the result of intersecting a number of these belief sets. Special cases of contraction are full meet contraction, in which all elements of $K \perp \varphi$ are intersected (analogously to sceptical inference), and maxi-choice contraction, in which just one element of $K \perp \varphi$ is selected (analogously to prioritized nonmonotonic logics).

In the current paper, in Section 2 some basic background notions are introduced. In Section 3 the notion of belief set operator is introduced, some illustrative examples are described (default logic, belief revision) and a number of more specific properties are discussed. In Section 4 results are obtained on the semantics of a belief set operator in terms of the semantical notion belief state operator. Moreover, results are obtained on the existence of a greatest underlying (monotonic) deductive system. In Section 5 the notion of selection operator is introduced, formalizing an agent's commitment to one of its belief sets. Selection functions applied to the results of a belief set operator provide a set of (selective) inference operations. Such a set of inference operations can be viewed as an alternative formalization of multiple belief sets. The formal relationship between sets of inference operations and belief set operators is established. Properties of selection operators are related to properties of the belief set operator and the inference operations resulting after selection. In Section 6 conclusions are drawn and perspectives on further research are sketched.

2 Background and Preliminaries

Let L be a nonempty language whose elements are denoted by ϕ, ψ, χ; $\mathcal{P}(X)$ denotes the power set of the set X. An operation $C : \mathcal{P}(L) \to \mathcal{P}(L)$ is called an *inference operation*, and the pair (L, C) is said to be an *inference system*. The operation C represents the notion of logical inference. An inference system (L, C_L) is a *closure system* and C_L a closure operation if it satisfies the following conditions: $X \subseteq C_L(X)$ (*inclusion*), $C_L(C_L(X)) = C_L(X)$ (*idempotence*), $X \subseteq Y \Rightarrow C_L(X) \subseteq C_L(Y)$ (*monotony*). An inference operation C_L satisfies *compactness* if $\phi \in C_L(X)$ implies the existence of a finite subset $Y \subseteq X$ such that $\phi \in C_L(Y)$. A closure system (L, C_L) is a *deductive system* if C_L satisfies compactness; then C_L is said to be a *deductive (inference) operation*. A set $X \subseteq L$ is closed under C_L if $C_L(X) = X$. $\mathcal{L}_0 = (L_0, Cn)$ denotes the inference system based on classical propositional logic.

A semantics for a closure system (L, C_L) can be defined by a *logical system*. A *logical system* $(L, M, \models)$ is determined by a language L, a set (or class) M whose elements are called *worlds* and a *relation of satisfaction* $\models \subseteq M \times L$ between worlds and formulas. Given a logical system $(L, M, \models)$, we introduce the following notions. Let $X \subseteq L$, $Mod^\models(X) = \{m : m \in M \text{ and } m \models X\}$, where $m \models X$ if for every $\phi \in X : m \models \phi$. Let $K \subseteq M$, then $Th^\models(K) = \{\phi : \phi \in L \text{ and } K \models \phi\}$, where $K \models \phi$ if for all $m \in K : m \models \phi$. $C^\models(X) = \{\phi : Mod^\models(X) \subseteq Mod^\models(\phi)\}$, $X \models \phi$ if $\phi \in C^\models(X)$. Obviously, $(L, C^\models)$ is a closure system and if $C^\models(X) = X$ then $Th^\models(Mod^\models(X)) = X$. $(L, M, \models)$ is said to be compact if the closure operation $C^\models$ is compact. The inference system (L, C_L) is *correct* (*complete*) with respect to the logical system $(L, M, \models)$ if $C_L(X) \subseteq C^\models(X)$ $(C_L(X) = C^\models(X))$. In case of completeness we say also that $(L, M, \models)$ *represents* (or *is adequate for*) (L, C_L).

The study of the general properties of inference operations $C : \mathcal{P}(L) \to \mathcal{P}(L)$ that do not satisfy monotony is well-established (see e.g. [Ma94]). A condition on inference operations is said to be *pure* if it concerns the operation alone without regard to its interrelations to a deductive system (L, C_L) representing a monotonic and compact logic. The most important pure conditions are the following: $X \subseteq Y \subseteq C(X) \Rightarrow C(Y) \subseteq C(X)$ (*cut*), $X \subseteq Y \subseteq C(X) \Rightarrow C(X) \subseteq C(Y)$ (*cautious monotony*), $X \subseteq Y \subseteq C(X) \Rightarrow C(X) = C(Y)$ (*cumulativity*). Some impure conditions are: $C(X) \cap C(Y) \subseteq C(Cn(X) \cap Cn(Y))$ (*distributivity*), $Cn(X) \neq L \Rightarrow C(X) \neq L$ (*consistency preservation*).

C is said to be *supraclassical* if it extends the consequence operation Cn of classical logic, i.e. $Cn(X) \subseteq C(X)$ for all $X \subseteq L$. If we assume this condition for an arbitrary deductive systems (L, C_L) then we get the more general condition of *supradeductivity*: $C_L(X) \subseteq C(X)$. A system $\mathcal{IF} = (L, C_L, C)$ is said to be an *inference frame* if L is a language, C_L is a deductive inference operation on L, and $C_L(X) \subseteq C(X)$ (supradeductivity) is fulfilled. The operation C satisfies *left absorption* if $C_L(C(X)) = C(X)$; and C satisfies *congruence* or *right absorption* if $C_L(X) = C_L(Y) \Rightarrow C(X) = C(Y)$. C satisfies *full absorption* if C satisfies left absorption and congruence. An inference frame $\mathcal{DF} = (L, C_L, C)$ is said to be a *deductive inference frame* if it satisfies full absorption. In this case C is said to be *logical over* C_L, and (L, C_L) is a *deductive basis* for C.

The semantics of a deductive frame can be described by introducing a model operator based on a logical system ([DH94]). $\mathcal{SF} = (L, M, \models, \Phi)$ is a *semantical frame* if $(L, M, \models)$ is a logical system and $\Phi : \mathcal{P}(L) \to \mathcal{P}(M)$ is a functor (called *model operator*) such that $\Phi(X) \subseteq Mod^\models(X)$. Let $C_\Phi(X) = Th^\models(\Phi(X))$. The operator Φ is said to be C_L-*invariant* if $(\forall X \subseteq L)(\Phi(X) = \Phi(C_L(X)))$. The inference operation C_Φ satisfies extension and left absorption, and hence $(L, C^\models, C_\Phi)$ is an inference frame associated to $\mathcal{SF}$ and denoted by $IF(\mathcal{SF})$. An inference frame $\mathcal{I} = (L, C_L, C)$ is said to be *complete* for a semantical frame $(L, M, \models, \Phi)$ if (L, C_L) is complete with respect to $(L, M, \models)$ and $C = C_\Phi$. Representation theorems for classes of inference frames can be proved by using semantical frames based on the *Lindenbaum-Tarski* construction of maximal consistent sets. We collect some elementary results that can be formulated and proved within this framework ([DH94]).

Proposition 1 *Let $\mathcal{F} = (L, C_L, C)$ be an inference frame satisfying left absorption. Then there exists a semantical frame $\mathcal{SF} = (L, M, \models, \Phi)$ such that $\mathcal{F}$ is complete with respect to $\mathcal{SF}$, i.e. $C_L = C^\models$ and $C = C_\Phi$.*

Left absorption does not imply congruence. We get an adequateness result for deductive inference frames by using invariant semantical frames, [Li91], [DH94].

Proposition 2

1. *Let $\mathcal{F} = (L, C_L, C)$ be a deductive inference frame. Then there exists a semantical frame $\mathcal{S} = (L, M, \models, \Phi)$ such that Φ is an invariant model operator and $\mathcal{S}$ represents $\mathcal{F}$.*
2. *If Φ is an invariant model operator for the logical system $(L, M, \models)$ then $(L, C^\models, C_\Phi)$ is a deductive inference frame.*

3 Belief Set Frames

Usually, there can be many alternative sets of beliefs that can be justified on the base of a set X of given knowledge. A set of such alternative belief sets will be called a *belief set family*. In this section we adapt and generalize the framework of deductive and semantical frames to the case of belief set operators.

Definition 1 *A belief set operator B is a function that assigns a belief set family to each set of initial facts: $B : \mathcal{P}(L) \to \mathcal{P}(\mathcal{P}(L))$.*

1. *B satisfies inclusion if $(\forall X)(\forall T \in B(X))(X \subseteq T)$.*

2. B satisfies non-inclusiveness if $(\forall X)(\forall U \in B(X))(\forall V \in B(X))$ $(U \subseteq V \Rightarrow U = V)$.
3. The kernel $K_B : \mathcal{P}(L) \to \mathcal{P}(L)$ of B is defined by $K_B(X) = \bigcap B(X)$.

We give two examples of belief set operators.

Example 1: Default Logic

Let D be a set of defaults. For $X \subseteq L$, and $\Delta = (X, D)$ let $\mathcal{E}(\Delta)$ denote the set of (Reiter) extensions of the default theory Δ. The belief set operator B_D can be defined as follows: $B_D(X) = \mathcal{E}(\Delta)$. The kernel of B_D gives the sceptical conclusions of a default theory.

Example 2: Belief Revision

Let φ be a sentence. For a deductively closed belief set K, define
$K \perp \varphi = \{T \mid T \subseteq K \backslash \{\varphi\}, T = Cn(T)$ and T is maximal with respect to these properties $\}$. These maximal subsets can play the role of possible belief sets resulting from the contraction of φ from K. Define a belief set operator $B_{-\varphi}$ by $B_{-\varphi}(X) = Cn(X) \perp \varphi$. The kernel of this operator yields a special contraction function, called a *full meet contraction function*.

A belief set operator B has one of the properties monotony, inclusion, idempotence if its kernel K_B satisfies these properties. Refined versions of these properties have to take into consideration the structure of the belief sets. Whereas for belief sets we have a natural notion of degree of information (a belief set T contains more information than a belief set S if $S \subseteq T$), we also need a more refined notion for belief set families:

Definition 2 *Let $\mathcal{A}, \mathcal{B}$ be belief set families. We say $\mathcal{B}$ contains more information than $\mathcal{A}$, denoted $\mathcal{A} \preceq \mathcal{B}$, if $(\forall T \in \mathcal{B})(\exists S \in \mathcal{A})(S \subseteq T)$. We write $\mathcal{A} \equiv \mathcal{B}$ if $\mathcal{A} \preceq \mathcal{B}$ and $\mathcal{B} \preceq \mathcal{A}$.*

If one of the arguments in the above definition is a singleton belief set family, we will often omit the parentheses and write $X \preceq \mathcal{A}$ instead of $\{X\} \preceq \mathcal{A}$. Thus, we can also write $X \preceq Y$ instead of $X \subseteq Y$. The intuition behind this definition is that a belief set family $\mathcal{B}$ has more information than $\mathcal{A}$ if any of the alternatives of $\mathcal{B}$ extends some alternative of $\mathcal{A}$. Obviously, the condition $\mathcal{A} \preceq \mathcal{B}$ implies $\bigcap \mathcal{A} \subseteq \bigcap \mathcal{B}$. We introduce the following formal properties of belief set operators capturing essential features of a rational agent.

Definition 3 *Let B be a belief set operator.*

1. *B satisfies belief monotony if $(\forall X \forall Y)(X \preceq Y \Rightarrow B(X) \preceq B(Y))$.*
2. *B satisfies weak belief monotony if*
   $(\forall XY)(X \preceq Y \preceq B(X) \Rightarrow B(X) \preceq B(Y))$.
3. *B satisfies belief transitivity if*
   $(\forall XYT)(T \in B(X)$ and $X \subseteq Y \subseteq T \Rightarrow K_B(Y) \subseteq T)$.[3]

---

[3] This property is called in [Vo93] *cumulative transitivity*.

4. B satisfies belief cut if $(\forall XY)(X \preceq Y \preceq B(X) \Rightarrow B(Y) \preceq B(X))$.
5. B satisfies belief cumulativity if it satisfies weak belief monotony and belief cut.
6. B satisfies strong belief cumulativity if it satisfies belief cumulativity and belief transitivity.

In [Vo93] a belief set operator B satisfying inclusion is said to be cumulative if it satisfies belief transitivity and the following condition that we call in the present paper local belief monotony: $(\forall TXS)(T \in B(X)$ and $X \subseteq S \subseteq T \Rightarrow B(S) \subseteq B(X))$. A weaker form of this notion is defined by the following condition: $(\forall XYS)(T \in B(X)$ and $X \subseteq S \subseteq T \Rightarrow (\forall S_1 \in B(S))(\exists T_1 \in B(X))(T_1 \subseteq S_1))$. All these properties are generalizations of the notion of cautious monotony for inference operations to the case of belief set operators. Similar notions can be formulated for the conditions of cut and cumulativity. There is not yet a complete analysis of these properties and their interrelations. Obviously, the following holds:

Proposition 3 *Let B be a belief set operator satisfying inclusion.*

1. *If B is belief monotonic then K_B is monotonic.*
2. *If B satisfies belief transitivity or belief cut then K_B satisfies cut.*
3. *If B satisfies weak belief monotony then K_B satisfies cautious monotony.*

We now connect a belief state system with a compact logic which can be considered as a deductive basis. Many non-classical forms of reasoning are built "on top of" a monotonic logic (L, C_L).

Definition 4

1. *A system $\mathcal{BF} = (L, C_L, B)$ is said to be a belief set frame if the following conditions are satisfied:*
   (a) L is a language and C_L is a deductive inference operation on L.
   (b) B is a belief set operator on L satisfying non-inclusiveness.
   (c) K_B satisfies $C_L(X) \subseteq K_B(X)$ (supradeductivity).
2. *B satisfies belief left absorption iff $C_L(T) = T$ for every $T \in B(X)$, and B satisfies belief congruence of C_L-invariance iff $C_L(X) = C_L(Y)$ implies $B(X) = B(Y)$. B satisfies full absorption iff B satisfies left belief absorption and congruence.*
3. *A belief set frame $\mathcal{DF} = (L, C_L, B)$ is said to be a deductive belief set frame if it satisfies full absorption. In this case the system (L, C_L) is called a deductive basis for B.*

Proposition 4 *Let $\mathcal{BF} = (L, C_L, B)$ be a belief set frame satisfying strong belief cumulativity. Then $\mathcal{BF}$ satisfies belief left absorption and belief congruence, i.e. $\mathcal{BF}$ is a deductive belief set frame.*

Proof: 1. From belief transitivity follows that for every $T \in B(X)$ the condition $K_B(T) \subseteq T$ is satisfied, hence $K_B(T) = T$. By supradeductivity we get $C_L(T) \subseteq K_B(T)$, thus $C_L(T) = T$.

2. Assume $C_L(X) = C_L(Y)$. Since $K_B : \mathcal{P}(L) \to \mathcal{P}(L)$ is cumulative it follows that (L, C_L, K_B) is a deductive frame, hence $K_B(X) = K_B(Y)$. It is sufficient to prove $B(X) = B(K_B(X))$, because this condition implies $B(X) = B(Y)$.

Let $S \in B(X)$, by belief cut there is an extension $T \in B(K_B(X))$ such that $T \subseteq S$. By belief monotony there exists a $S_1 \in B(X)$ satisfying $S_1 \subseteq S$. Because the sets in $B(X)$ are pairwise non-inclusive we get $S = S_1$, which implies $T = S$, hence $S \in B(K_B(X))$.

Let $T \in B(K_B(X))$; by belief monotony there is a $S \in B(X)$ such that $S \subseteq T$. By the previous proved condition this implies $S \in B(K_B(X))$, hence by non-inclusiveness of B we get $T = S$. $\square$

Further important impure properties of inference frames can be generalized to belief set frames.

Definition 5 *Let* (L, C_L, B) *be a belief set frame.*

1. *B satisfies belief distribution if*
   $(\forall XYS)((S \in B(C_L(X) \bigcap C_L(Y)) \Rightarrow (S \in B(X) \text{ or } S \in B(Y)))).$
2. *B satisfies belief consistency preservation if*
   $(\forall X)(C_L(X) \neq L \Rightarrow B(X) \neq \{L\}).$

The following proposition is obvious.

Proposition 5

1. *If B satisfies belief distribution then K_B satisfies distribution.*
2. *If B satisfies belief consistency preservation then K_B satisfies consistency preservation.*

The semantics of a belief set is a set of models. Since there can be many belief sets we have to take into consideration functors associating to sets of assumptions sets of sets of models. Such functors are called *belief state operators*.

Definition 6

1. *A belief state operator Γ is a function $\Gamma : \mathcal{P}(L) \to \mathcal{P}(\mathcal{P}(Mod))$*
2. *The tuple $(L, M, \models, \Gamma)$ is said to be a belief state frame.*
3. *Γ satisfies non-inclusiveness if $\forall KJ \in \Gamma(X) : J \subseteq K \Rightarrow K = J$.*
4. *Γ satisfies inclusion if $(\forall X)(\forall K \in \Gamma(X))(K \subseteq Mod(X))$.*
5. *Γ satisfies left absorption, or L-invariance, if $\Gamma(X) = \Gamma(C_L(X))$ for all $X \subseteq L$.*

For a given belief state operator Γ the following belief set operator B_Γ can be introduced $B_\Gamma(X) = \{Th(K) : K \in \Gamma(X)\}$. The notion of a belief state operator is a generalization of the notion of a model operator. The following examples summarize some types of belief set operators investigated in the literature.

Example 1: Default Logic (continued)
Remember that we associated a belief set operator B_D with a set of defaults D. Then (L, Cn, B_D) is a deductive belief set frame. Then also the following belief state operator can be defined for $X \subseteq L$: $\Gamma_D(X) = \{Mod(E) : E \in \mathcal{E}(\Delta)\}$.

Example 2: Poole Systems
Let $\Sigma = (D, E)$, $D \cup E \subseteq L$; the elements of D are called defaults, the elements of E are said to be constraints. A set $\delta \subseteq D$ is a basis for $X \subseteq L$ if the set $X \cup \delta \cup E$ is consistent and δ is maximal with this property. Let $Cons_\Sigma(X) = \{\delta : \delta \subseteq D$ and δ is a basis for $X\}$. Then define $B_\Sigma(X) = \{Cn(X \cup \delta) : \delta \in Cons_\Sigma(X)\}$. Obviously, B_Σ is a belief set operator. A belief state operator Γ_Σ providing a semantics for B_Σ can be introduced by $\Gamma_\Sigma(X) = \{Mod(T) : T \in B_\Sigma(X)\}$. Obviously, Γ_Σ is Cn-invariant.

Example 3: Generalized Belief Revision
Let $A \subseteq L$ be an arbitrary fixed consistent deductively closed set and $X \subseteq L$ an arbitrary set. Define $Cons(A, X) = \{Y : Y \subseteq A, Y \cup X$ is consistent and Y is maximal with this property $\}$. Let $B(X) = \{Cn(Y \cup X) : Y \in Cons(A, X)\}$. If $A \cup X$ is consistent then $B(X) = \{Cn(A \cup X)\}$. If $A \cup X$ is inconsistent then $B(X)$ contains all complete extensions of X. This can be shown using a generalization of results in [Gr88]. To get belief set operators derived from A, subsets from $Cons(A, X)$ have to be selected. Let $S : \mathcal{P}(L) \to \mathcal{P}(\mathcal{P}(L))$ satisfying $S(X) \subseteq Cons(A, X)$ such that $S(X) \neq \emptyset$ if $Cons(A, X) \neq \emptyset$. Then the following belief set operators B_S can be introduced: $B_S(X) = \{Cn(Y \cup X) : Y \in S(X)\}$. Again, we may introduce a belief state operator Γ_S for B_S by defining $\Gamma_S(X) = \{Mod(T) : T \in B_S(X)\}$.

4 Representation Theorems

The methods described in Section 2 can be generalized to the case of belief set operators and belief set frames. In particular, there is a canonical method to introduce a semantics for a given belief set frame.

Proposition 6 *Let $\mathcal{F} = (L, C_L, B)$ be a belief set frame satisfying belief left absorption. Then there exists a belief state frame $S\mathcal{F} = (L, M, \models, \Gamma)$ such that $\mathcal{L} = (L, C_L)$ is complete with respect to $(L, M, \models)$ and $B = B_\Gamma$. If $\mathcal{F}$ is a deductive belief set frame then $S\mathcal{F}$ can be taken to be $\mathcal{L}$ -invariant.*

The question arises whether a belief set operator B can be extended to a deductive belief set frame (L, C_L, B). Of course, there is the following trivial solution: $C_L(X) = X$, which cannot be considered as adequate. It is reasonable to assume that the desired logic for B should be as close as possible to K_B; i.e. C_L should be maximal below K_B with respect to the following partial ordering between inference operations $C_1, C_2 : C_1 \leq C_2 \Leftrightarrow (\forall X \subseteq L)(C_1(X) \subseteq C_2(X))$.

Proposition 7 *Let B be a belief set operator on L satisfying strong belief cumulativity. Then there exists a deductive system (L, C_L) such that following conditions are satisfied:*

1. (L, C_L, B) is a deductive belief set frame.
2. If (L, C_1, B) is a deductive belief set frame then $C_1 \leq C_L$, i.e. C_L is the greatest deductive system for (L, B).

Proof: Since B is strong cumulative the inference system (L, K_B) is cumulative. By the main result in [Di94] there exists a largest deductive operation $C_L \leq K_B$ such that (L, C_L, K_B) is a deductive inference frame. Since $\mathcal{BF} = (L, C_L, B)$ is a strong cumulative belief set frame it follows by Proposition 4 that $\mathcal{BF}$ is a deductive belief set frame. $\mathcal{BF}$ satisfies the desired properties. $\square$

The semantical approach presented here can be summarized as follows. We start with a belief set operator B on a language L; in the next step we construct a belief set frame (L, C_L, B) such that the compact logic (L, C_L) satisfies additional properties, e.g. maximality. Then for (L, C_L, B) we may introduce the standard semantics indicated in Proposition 6.

Finally, we consider connections between deductive frames and deductive belief set frames. Obviously, deductive frames (L, C_L, C) can be considered as a special case of belief set frames by taking $B_C(X) = \{C(X)\}$. On the other hand, for every deductive belief set frame (L, C_L, B) there exists exactly one deductive frame defined by the intersection K_B of all extensions. The converse is not true: for a given deductive inference frame there can be many deductive belief set frames with the same intersection. Belief set frames can be understood as approximations of deductive inference frames and a deductive inference frame can be interpreted as a condensed representation of a family of deductive belief set frames. To make this precise let $\mathcal{F} = (L, C_L, C)$ be a deductive inference frame and let $\Omega(\mathcal{F}) = \{B : (L, C_L, B)$ is a consistency preserving deductive belief set frame such that $C = K_B\}$. The binary relation $\preceq$ between belief set operators in $\Omega(\mathcal{F})$ is defined as follows: $B_1 \preceq B_2$ if $(\forall X)(B_1(X) \preceq B_2(X))$, and $B_1 \equiv B_2$ iff $B_1 \preceq B_2$ and $B_2 \preceq B_1$. Obviously, $\Omega^*(\mathcal{F}) = (\Omega(\mathcal{F})/ \equiv, \preceq)$ is a partial ordering. Let $Max(X) = \{S : S$ is a maximal consistent extension of $X\}$; $B \in \Omega(\mathcal{F})$ is said to be a *maximization operator* iff $(\forall X \subseteq L)(B(X) \subseteq Max(C(X)))$.

Proposition 8 *Let $\mathcal{F} = (L_0, Cn, C)$ be a deductive inference frame over classical logic (L_0, Cn). The system $\Omega^*(\mathcal{F})$ is a partial ordering with a least element and it has a least maximization operator.*

Let $\mathcal{F} = (L_0, Cn, C)$; the least element B_{min} of $\Omega(\mathcal{F})$ is defined by $B_{min}(X) = \{C(X)\}$, and the least maximization operator is determined by $B_{max}(X) = Max(C(X))$. These operators have the following congruence property. A belief set operator $B \in \Omega(\mathcal{F})$ satisfies *C-congruence* iff $(\forall XY \subseteq L)(C(X) = C(Y) \Rightarrow B(X) = B(Y))$. The following observation is obvious.

Proposition 9 *Let $\mathcal{F} = (L_0, Cn, C)$ be a cumulative deductive inference frame. Then every C-congruent belief set operator in $\Omega(\mathcal{F})$ satisfies belief cumulativity, i.e. weak belief monotony and belief cut.*

Remark. Concerning the structure of $\Omega(\mathcal{F})$ there is the following question. Let P be a property on belief set frames, and $\mathcal{F}$ is a cumulative deductive inference frame. Does there exist an element in $\Omega(\mathcal{F})$ which is maximal with respect to the property P? Examples of such properties are distributivity or strong belief cumulativity.

5 Selection Operators

In the previous sections we concentrated on the multiple belief set view. The kernel of a belief set operator represents the most certain inferences the agent can make. But there is also another way in which the agent can handle the multiple views, and that is by selecting one (or a subset) of the possible views and focusing on this view. In the area of design, given some requirements a designing agent may have multiple (partial) descriptions of objects that do not contradict the requirements. It may have one of these descriptions (views) in focus, which it will try to complete. Here the selection indicates which view is in focus. On the other hand, for many nonmonotonic formalisms in which a theory can have multiple extensions (or expansions), a prioritized or stratified version exists, in which control knowledge (such as a preference ordering on the nonmonotonic rules) is used to designate one of the extensions as the most preferred one ([Br94], [ABW88], [Ko88], [TT92]). This focusing mechanism can be studied abstractly through *selective inference operations* for a given belief set operator which choose one of the sets of beliefs.

Definition 7 *Let B be a belief set operator. A selective inference operation for B is an inference operation C such that $\forall X \subseteq L : C(X) \in B(X)$.*

To structure the connections between belief set operators and selective inference operations, we give the following definition:

Definition 8

1. *Let a belief set operator B be given. The family of selective inference operations for B, denoted by $\mathcal{C}_B$ is defined by*
   $\mathcal{C}_B = \{C \mid C$ *is a selective inference operation for $B\}$.*
2. *Let $\mathcal{C}$ be a family of inference operations. Define the belief set operator $B_{\mathcal{C}}$ by $B_{\mathcal{C}}(X) = \{C(X) \mid C \in \mathcal{C}\}$.*

It is easy to see that for a belief set operator B, the set $\mathcal{C}_B$ is non-empty as soon as $B(X) \neq \emptyset$ for all X.

A selective inference operation for a belief set operator will in general be more informative than the associated kernel: $K_B(X) \subseteq C(X)$. But even if the belief set operator is well-behaved, a selective inference operation can be badly behaved. The question arises whether a well-behaved selective inference operation always exists. That is, given a belief set operator B, the question is whether there exists a $C \in \mathcal{C}_B$ with certain nice properties. This is a very hard question. Sufficient conditions can be found, for instance for monotony: $\forall Y \exists T \in B(Y) \forall X \subseteq Y \forall S \in B(X) : S \subseteq T$. But this condition implies that $B(X)$ is a singleton for all X. Necessary conditions are easier to find, but quite trivial. For a belief set operator B and a a selective inference operation C for B we have the following. If C satisfies cut then $(\forall X)(\exists S \in B(X))(\forall Y)(X \subseteq Y \subseteq S \Rightarrow ((\exists T \in B(Y))(T \subseteq S))$ (*1). If C satisfies cautious monotony then $(\forall X)(\exists S \in B(X))(\forall Y)(X \subseteq Y \subseteq S \Rightarrow ((\exists T \in B(Y))(S \subseteq T))$ (*2). If C satisfies cumulativity then $(\forall X)(\exists S \in B(X))(\forall Y)(X \subseteq Y \subseteq S \Rightarrow ((\exists T \in B(Y))(S = T))$ (*3). If C satisfies monotony then $(\forall XY)(X \subseteq Y \Rightarrow ((\exists S \in B(X))(\exists T \in B(Y))(S \subseteq T))$.

The preceding paragraph pertains to the situation when a belief set operator B is given, and we want to study C_B. Questions about the second item in Definition 9 are easier to answer. We will say a family C of inference operations satisfies one of the properties of cut, cautious monotony, cumulativity and monotony if all of the inference operations in C satisfy this property. Then we have:

Proposition 10 *Let C be a family of inference operations.*

1. *If C satisfies monotony then B_C satisfies belief monotony.*
2. *if C satisfies cautious monotony then B_C satisfies weak belief monotony.*
3. *if C satisfies cut, then B_C satisfies both belief transitivity and belief cut.*

One way of defining selective inference operations for a given belief set operator is through *selection operators*. Given a set of views, such a selection operator selects one (or some) of them:

Definition 9 *A selection operator is a function $s : \mathcal{P}(\mathcal{P}(L)) \rightarrow \mathcal{P}(\mathcal{P}(L))$ if for $A \subseteq \mathcal{P}(L) : s(A) \subseteq A$. s is single-valued iff for all non-empty $A \subseteq \mathcal{P}(L)$: $card(s(A)) = 1$. A single-valued selection function s can be understood as a choice function $s : \mathcal{P}(\mathcal{P}(L)) \rightarrow \mathcal{P}(L)$ satisfying $s(A) \in A$ for all non-empty A.*

Using selection operators we can generate inference operations:

Definition 10 *Let a belief set operator B and a selection operator s be given. We define the inference operation C_s by: $C_s(X) = \bigcap s(B(X))$.*

Single-valued selection operators generate *selective* inference operations. A first observation about when a selective inference operation can be generated by a single-valued selection operator:

Proposition 11 *Let a selective inference operation C for a belief set operator B be given. Then $C = C_s$ for some single-valued selection operator iff $(\forall X \forall Y)(B(X) = B(Y) \Rightarrow C(X) = C(Y))$.*

We can study properties of selection operators and the relation with properties of belief state operators and selective inference operations. Although a full treatment is beyond the scope of this paper, we will give an example.

Definition 11 *A selection operator s satisfies selection monotony if for all belief set families A, B we have $A \preceq B \Rightarrow s(A) \preceq s(B)$.*

Then we have the following:

Proposition 12

1. *Let a belief set operator B and a selection function s be given. If B satisfies belief monotony and s satisfies selection monotony then C_s satisfies monotony.*
2. *Let a selection function s be given. If for any belief set operator B which satisfies belief monotony, C_s satisfies monotony, then s satisfies selection monotony.*

Remark: The considerations in the sections 4 and 5 reflect certain aspects of knowledge dynamics [Pop77]. Let X_0 be a deductively closed set representing the knowledge at a certain time point. X_0 can be extended by a combined application of a belief set operator B_0 whose kernel is X_0 and a generalized selection operator s_0. The new knowledge stage X_1 is defined by $X_1 = \bigcap s_0(B_0(X_0))$. The forming of belief sets for a knowledge base can be understood as theory formation or hypothesis building; after new observations are performed those belief sets are left out which contradicts the observations.

6 Conclusions and Future Research

In research on nonmonotonic reasoning often an ambivalent or negative attitude is taken towards the phenomenon of multiple (belief) extensions. Of course, from the classical viewpoint it may be considered disturbing when a reasoning process may have alternative sets of outcomes, often mutually inconsistent. Many approaches try to avoid the issue by adding additional control knowledge to decide which extension is intended, thus obtaining a parameterization of the possible sets of outcomes of the reasoning by the chosen control knowledge: for each control knowledge base a unique outcome (e.g, [Br94], [TT92]). Another approach to avoid the multiple extension issue is to concentrate on the intersection of them: the sceptical approach. A number of results have been developed on nonmonotonic inference operations that are a useful formalization of this approach (e.g, [KLM90]). However, in the sceptical approach the remaining conclusions may be very limited, insufficient for an agent to act under incomplete information in a dynamic environment.

In the current paper we address the multiple extension issue in an explicit manner by introducing nonmonotonic (multiple) belief set operators and their semantical counterpart: belief state operators. Many properties and results on nonmonotonic inference operations (and model operators) turn out to be generalizable to this notion.

Introducing alternative belief sets that can serve as the outcomes of a nonmonotonic reasoning reasoning process, the question becomes how to formalize the process of committing to one belief set. To this end in the current paper selection operators are introduced that formalize this process. The specification of a selection operator expresses the strategic (control) knowledge used by an agent to choose between the different alternatives.

Agents often construct belief sets to which they commit in a step by step manner, using some kind of inference rules. Specification of such a nonmonotonic reasoning process is easier to obtain if the reasoning patterns leading to the outcomes are specified instead of (only) the outcomes of the reasoning. In related and future research the notion of a trace for a nonmonotonic reasoning process is taken as a point of attention. We are studying the notion of controlled forcing to describe reasoning patterns to multiple sets of outcomes. This enables us for each belief set operation to obtain a set of (multiple) reasoning traces that generate them. Furthermore, we are addressing a formalization of reasoning traces based on temporal logic (following the line of [ET93, ET94]). The dualism between multiple outcomes and multiple reasoning traces of a nonmonotonic reasoning process is also studied in the context of default logic, leading to a representation theory: for which set of outcomes a default theory can be found with these outcomes (see [MTT96]). A preliminary abstract on this perspective, including some of the semantical notions involved can be found in [EHT95].

7 Acknowledgments

Parts of the research reported here has been supported by the ESPRIT III Basic Research project 6156 DRUMS II on Defeasible Reasoning and Uncertainty Management Systems. Stimulating discussions about the subject have taken place with Jens Dietrich.

References

[ABW88]   Apt, K.R., Blair, H.A., Walker, A.: Towards a Theory of Declarative Knowledge, in: Minker, J. (ed), Foundations of Deductive Databases and Logic Programming, Morgan Kaufmann, 1988, pp. 89–142

[AGM85]   Alchourrón, C.E., Gärdenfors, P., Makinson, D.: On the Logic of Theory Change: Partial Meet Contraction and Revision Functions; *Journal of Symbolic Logic* 50, 510–530 (1985)

[Be89]   Besnard, P.: An Introduction to Default Logic; Berlin, Springer-Verlag, 1989

[Br91]   Brewka, G.: Nonmonotonic Reasoning: Logical Foundations of Commonsense, Cambridge University Press, 1991

[Br94]   Brewka, G.: Adding Priorities and Specificity to Default Logic; in: C. MacNish, D. Pearce, L.M. Pereira (eds.) Logics in Artificial Intelligence, JELIA'94, Springer Verlag, 1994

[Di94]   Dietrich, J.: Deductive Bases of Nonmonotonic Inference Operations: *NTZ-Report*,7/94, University of Leipzig, 1994

[DH94]   Dietrich, J., H. Herre: Outline of Nonmonotonic Model Theory, *NTZ-Report*, 5/94, University of Leipzig, 1994

[EHT95]   Engelfriet, J., H. Herre, J. Treur: Nonmonotonic Belief State Frames and Reasoning Frames, in: C. Froidevaux, J. Kohlas (eds.), Proc. ECSQARU'95, Lecture Notes in AI, vol. 946, Springer Verlag, 1995, pp. 189-196

[ET93]   Engelfriet, J., J. Treur: A Temporal Model Theory for Default Logic, in:
          M. Clarke, R. Kruse, S. Moral (eds), Proc. ECSQARU'93, Lecture Notes in
          Computer Science, vol. 747, Springer-Verlag, 1993, pp. 91-96

[ET94]   Engelfriet, J., J. Treur: Temporal Theories of Reasoning. In: C. MacNish,
          D. Pearce, L.M. Pereira (eds.) *Logics in Artificial Intelligence*, Proceedings
          of the 4th European Workshop on Logics in Artificial Intelligence, JELIA'94,
          Lecture Notes in AI, vol. 838, Springer Verlag, pp. 279-299; Also in: *Journal
          of Applied Non-Classical Logics*, vol. 5 (2), 1995, pp. 239-261

[Et87]   Etherington, D.W.: A semantics for Default Logic, Proc. IJCAI-87, pp. 495-
          498; see also in: D.W. Etherington, Reasoning with Incomplete Information,
          Morgan Kaufmann, 1988

[Ga85]   Gabbay, D.: Theoretical Foundations for Non-monotonic Reasoning in ex-
          pert systems; in Apt, K. (ed.): *Logic and Models of Concurrent Systems*,
          Springer, Berlin,1985

[Gr88]   Grove, A.: Two modelings for theory change; *Journal of Philosophical Logic*
          17, 157-170 (1988)

[He94]   Herre, H.: Compactness Properties of nonmonotonic Inference Operations,
          In: C. MacNish, D. Pearce, L.M.Pereira (eds.) *Logics in Artificial Intel-
          ligence*, Proceedings of the 4th European Workshop on Logics in Artifi-
          cial Intelligence, JELIA'94, Lecture Notes in AI, vol. 838, Springer Verlag,
          pp. 19-33; Also in: *Journal of Applied Non-Classical Logics*, vol. 5, 1995,
          pp. 121-136 (Special Issue with selected papers from JELIA'94)

[Ko88]   Konolidge, K.: Hierarchic Autoepistemic Theories for Nonmonotonic Rea-
          soning, in: Proceedings AAAI'88, Minneapolis, 1988

[Li91]   Lindström, S.: A semantic approach to nonmonotonic reasoning: inference
          operations and choice; Dept. of Philosophy, Uppsala University, Preprint,
          1991

[Ma94]   Makinson, D.: General Patterns in Nonmonotonic Reasoning; in:
          D.M. Gabbay, C.J. Hogger, J.A. Robinson (eds.), Handbook of Logic in
          Artificial Intelligence and Logic Programming, Vol.3, Oxford Science Pub-
          lications, 1994

[MT93]   Marek, V., M. Truszczynski: Nonmonotonic Logic, Springer-Verlag, 1993

[MTT94]   Marek, V., J. Treur, M. Truszczynski: Representation Theory for Default
          Logic, Proc. Symposium on AI and Mathematics, 1996

[Po88]   Poole, D.: A logical Framework for Default Reasoning;Artificial Intelligence,
          36: 27-47 (1988)

[Pop77]   Popper, K.: The Logic of Scientific Discovery; Hutchinson, London, 1977,
          9th edition.

[Re80]   Reiter, R.: A logic for default reasoning; *A.I.* , vol. 13 , 81-132 , 1980

[Sh88]   Shoham, Y.: Reasoning about Change; MIT-Press, Cambridge/USA, 1988

[Ta56]   Tarski, A.: Logic, Semantics, Metamathematics. Papers from 1923–1938.
          Clarendon Press, Oxford, 1956

[TT92]   Tan, Y.H., J. Treur: Constructive Default Logic and the Control of Defeasi-
          ble Reasoning; in: B. Neumann (ed.), Proc. ECAI'92, Wiley and Sons, 1992,
          pp. 299-303

[Vo93]   Voorbraak, F.: Preference-based semantics for nonmonotonic logics,in:
          Bajcsy, R.(ed), Proc. IJCAI-93, Morgan Kaufmann, 1993, pp. 584-589

SEdit – Graphically Validating Technical Systems

Gerd Große, Christoph S. Herrmann, Enno Sandner*

Technische Hochschule Darmstadt, FB Informatik, FG Intellektik
Alexanderstr. 10, 64283 Darmstadt, Germany
*Corresponding author's email: enno@intellektik.informatik.th-darmstadt.de

Abstract. In order to minimize the cost of rapid prototyping, SEdit offers the possibility to validate a technical system before its actual realization. For this purpose, the system has to be described logically by a set of axioms (system specification). By means of deduction, it can be proven that the system will show the desired behaviour. For the sake of user acceptance and understandability, SEdit is a graphical, interactive tool that may be regarded as a shell for easily invoking logical proofs. The applicability has been demonstrated at the 1995 MEDLAR review meeting [5].

1 Introduction

The development of complete and correct specifications of technical systems is a very tedious task. Usually, it starts from the first very shallow idea of the desired system behavior, then iterative improvements lead to an exact description of the system in terms of (sub) components and their relationships. As part of the MUSE[1] project, we have developed a prototypical environment for specifying technical systems. The name, SEdit, is an abbreviation for "State Event Logic Editor" referring to the State Event Logic (SEL), mentioned later. The user is enabled to describe the system as a structure of components where each component is represented by a set of logical axioms. One then has the opportunity to improve the formalization quickly by running a simulation tool which interprets the axioms and starting from a given situation creates a tree of possible future behaviors. Once one assumes that the specification seems to behave properly one can run a theorem prover for checking more sophisticated requirements such as liveness or safety.

2 A guided tour

To introduce how the different components of SEdit work together, we will demonstrate a typical validation session. First, a specification with at least one

---

[1] Multimedia System Development – a project of the German Research Foundation "Deutsche Forschungsgemeinschaft" (DFG)

module has to be created. A specification may consist of several modules. A module has an unique name, an optional documentation and a set of State Event Logic formulae. State Event Logic [2] is a modal logic, where the universe becomes a set of pairs in which one component is a state and the other is one of the events following the state. The connection between two subsequent pairs is expressed by an accessibility relation. This extension permits an elegant treatment of causality and simultaneity. The module may export predicates and *use* other modules. These features are similar to the ones supported by new PROLOG systems. After the initial module is selected the user can edit the set of formulae. We have restricted the possible formulae to simply "boxed" implications with the premise and conclusion in disjunctive normal form. We do not see a need for additional operators like "eventually" or "until" in the given problem-domain. At any time the user can use the simulation tool to compare the so far described specification with his intention. The user is asked to enter an initial formula consisting of atomic state and event formulae and the simulation tool computes the tree of possible future worlds. In case the system is deterministic, i.e., no rule contains a disjunction in its conclusion, the result is one linear ordering otherwise the alternatives are represented by different worlds. The user can move through the worlds, add new events and see how the modifications affect future worlds. Sometimes it is reasonable to use a more concrete representation of the worlds. The 3D representation tool uses a so-called graphic specification to visualize a given world. The tool maps the world to a graphical state consisting of several graphical elements. In other words, the tool maps the abstract representation of the world given by the SEL-formula to a more concrete one. The graphic specification consists of rules which translate the SEL-conjunction to a set of well-known graphic-predicates. The 3D-representation tool visualizes these predicates. By moving through different worlds, the graphical state is updated. Some of the graphical elements can be used to interact with the simulation tool. For example pressing a button may result in adding a new event to the visualized world. The association between the graphical element and the SEL-event is also part of graphic specification. By using this tool in an iterative way, a user of the environment is enabled to produce a sophisticated specification in a very short time. One gets a very good feeling for the behavior of the system in different situations and knows that the specification is a formalization of the concept one had in mind. Once the user has enough confidence in the specification, it is time to prove some requirements. To perform a proof, the user adds some extra knowledge and a query. The system tries to show whenever the formula described by the specification and the additional knowledge is valid the query most also hold. The resulting implication is translated by an adaption of Ohlbach/Nonnegart's approach [4] to a first-order logic formula and passed to the KoMeT [1] prover.

Figure 1 shows a part of the directed acyclic graph (DAG) of the *Car Painting Scenario* [6] that we modeled for the MEDLAR demonstration [5]. Two cars (car(a), car(b)) with white color (color(a, white)) can be moved from their storage room (sr) to different painting rooms (pr1, pr2) by robots r1 or 2

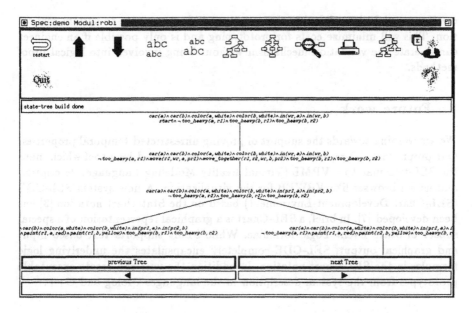

Fig. 1. Car Painting Scenario

(move(r1, wr, a, pr1)) where they will be painted (paint(r1, b, yellow)). Depending on which robot carries out which of the possible actions, different states occur—resulting in a split of the DAG. Since the cars end up with the same color irrespective of which robot carried them, the edges of the DAG will join after their splitting.

3 Implementation

The graphic user interface (GUI) has been implemented in C++ and Motif. All additional work like the deductive logic computing or 3D-scenario representation is delegated to external programs. The systems supports two ways of interprocess communications: via a file and by remote procedure call. The communication uses the client-server model where the GUI always acts as client and an auxilary program as server. The resulting encapsulation of the system components makes it easy to exchange or add components without affecting the whole system.

4 Discussion

By offering a graphical tool for the purpose of validating a specified technical system with logical deduction, we allow the user to apply logical mechanisms without having theoretical background on logics. The MUSE project cooperates with the German Ministry of Transport[2] (MOT) since the goal is to get a MOT

---

[2] Technischer Überwachungsverein (TÜV)

certificate solely by validating the system specification prior to system realization. This will minimize costs for prototyping and is only possible if an engineer can operate the validation mechanism without being involved into logical proof methods.

5 Future work

We are working towards the support of proving unrestricted temporal properties. Our propriatary 3D-representation tool should be replaced by a tool which maps the SEL-formulae to a VRML (Virtual Reality Modeling Language) description and uses a browser like VRWEB for the visualization. A new system SEL-CDE (SEL-Chart Development-Enviroment) based on the Statechart notation [3] has been developed [7]. In brief, a SEL-Chart is a graphical representation of a special subset of State Event Logic formulae. While SEdit supports text-based input and graphical output, SEL-CDE completely encapsulates the underlying logic formulas in the SEL-Chart notation. In addition the user can create executable prototypes from the system-description in the languages Prolog and C++.

References

1. W. Bibel, S. Brüning, U. Egly, and T. Rath. KoMeT. In A. Bundy, editor, *Proceedings of the 12th Conference on Automated Deduction (CADE)*, pages 783–787. Springer LNAI 814, 1994.
2. G. Große and H. Khalil. State Event Logic. *MEDLAR Special Issue of the Bulletin of the Interest Group in Pure and Applied Logics (IGPL)*, 1995. to appear.
3. D. Harel. Statecharts: a visual formalism for complex systems. *Science of Computer programming*, 8:231–274, 1987.
4. H. J. Ohlbach. Semantics based translation methods for modal logics. *Journal of Logic and Computation*, 1(5):691–746, 1991.
5. J. Pfalzgraf, U. Sigmund, V. Sofronie, and K. Stokkermans. Towards a cooperating robot demonstrator. In J. Cunningham and J. Pitt, editors, *Periodic Progress Report 3 – Mechanizing Deduction in the Logics of Practical Reasoning (MEDLAR II, ESPRIT Basic Research Project 6471)*, pages 181–197, 1995.
6. J. Pfalzgraph, U. Sigmund, and K. Stokkermans. Modeling cooperating agent scenarios by deductive planning methods and logical fiberings. In *Proceedings of AISMC-2*. Springer LNCS, 1995. in press.
7. E. Sandner. SEL-Charts, Master's thesis, TH Darmstadt, FG Intellektik, 1995.

The Need for a Dialectical Tier in Arguments

DR. RALPH H. JOHNSON

University of Windsor
Windsor Ontario Canada
N9B 394

Abstract. In this century, the dominant approach in logic to the study of argument has been *formal*. An argument has been understood as discourse that displays a certain form. The systematic study of such forms is undertaken by formal, deductive logic (hereafter: FDL). In the last 25 years, logicians have been pursuing alternative approaches to argument. One such approach has been characterized as *informal* rather than formal. Closely related is an approach that has been characterized as *pragmatic* rather than syntactic or semantic. The purpose of this research paper is to show how these alternative approaches can shed light on the basic question: How are we to understand (and represent) arguments?

1 Introduction

In this century, the dominant approach in logic to the study of argument has been *formal*. An argument has been understood as discourse that displays a certain form. The systematic study of such forms is undertaken by formal, deductive logic (hereafter: FDL). In the last 25 years, logicians have been pursuing alternative approaches to argument. One such approach has been characterized as *informal* rather than formal (Johnson and Blair (1980, 1985). Closely related is an approach that has been characterized as *pragmatic* rather than syntactic or semantic (Barth, 1987; Walton, 1990).

The purpose of this research paper is to show how these alternative approaches can shed light on the basic question: How are we to understand (and represent) arguments? The importance of this question for those interested in the task of representing and modelling reasoning is obvious. For argument is among the most important forms of reasoning. The shift that has taken place in logicians' understanding of argumentation has important repercussions.

I begin by sketching briefly the approach to argument taken in FDL, then discuss some limitations of that approach. That will pave the way for a discussion of an alternative to the formal approach. I then consider objections to this way of defining argument and conclude with some brief remarks about how to represent argument so understood.

2   The traditional approach to argument

A. FDL's conception of argument

In the late 19th century, a revolution occurred in both logic and the foundations of mathematics. The architects of this revolution were Frege (1879), Russell and Whitehead (1910-1913) and Wittgenstein (1922). In this period, logic becomes mathematicized, inextricably linked with questions about the nature of and foundation of mathematics. What happens to argument in the new regime? It virtually disappears[1].

When argument reappears in the textbook tradition that develops out of mathematical logic, it is now conceived of as a group of propositions in which some (the premises) support the other (the conclusion). Whether that support is adequate becomes the question of whether that argument instantiates a valid logical form. A logical form is valid if it is impossible for the premises to be true and the conclusion false. A good argument (sound) is a valid argument with true premises. That, in essence, is the FDL story.

Thus, in Copi's *Introduction to Logic*, one of the oldest introductory logic textbooks in North America, you will find the following definition:

An argument, in the logician's sense, is any group of propositions of which one is claimed to follow from the others which are regarded as providing evidence for the truth of that one. (1961: 7)

(In later editions, "evidence" is replaced by "support or grounds" (1986:6).) A few lines later, Copi adds: "An argument is not a mere collection of propositions, but has a structure" 1986:7).

This traditional view was welcome to those interested in formal representation, because certain portions of validity theory are formalizable. Hence the promise exists, not only that arguments can be represented formally, but also that a decision theory for determining validity can be found. Prospects for formalization of the theory of argument looked good.

But there are problems with the traditional view, and slowly these come into view. The Gödel result took most of the wind out of the logicist's sails, but there were developments closer to home that need to be considered.

B. Problems with the traditional view

For one thing, the formal concept of validity does not seem to capture the kind of relationship we expect between premises and conclusion. The following turns out to be a valid argument:

(S1)   If the moon is not made of green cheese, then arithmetic is complete.
       The moon is not made of green cheese. So, arithmetic is complete.

---

[1] For example, in Quine's *Introduction to Mathematical Logic* (1944), the concern for argumentation as a naturally occurring process has disappeared. The term "argument" occurs precisely once, and here in its mathematical sense.

To many this example seems specious and certainly far removed from any-thing one would expect to find in the realm of practical reasoning. The problem here involves the representation of "if, then" by the horseshoe (⊃), the symbol for the truth-functor sometimes known as material implication.

The unhappiness of this result led some logicians to pursue a better opera-tor. In this quest, Lewis developed the systems of strict implication, and these systems are responsible for the development of modal logic (Lewis and Lang-ford, 1935). But strict implication has its shortcomings as a rendering of the conditional which lies at the heart of logic and argument[2].

Later, Belnap and Anderson attempt to develop relevance logic (1975). By all accounts, it, too, fails to capture formally the required sense of "if-then." Thus, the attempt to render formally the entailment relationship (the soul of FDL's idea of an argument) turns out to be a *feu de follet*.

The shortcomings of the traditional view are now well-known within the AI community. Thus Gabbay (1995:309):

> The traditional modelling of an argument, as assumptions leading to a conclusion, is too narrow. There is a need to see the argument within a changing and evolving context, involving not only deduction but also other mechanisms such as abduction, action updates and consistency maintenance.

Yet many in the AI community, apparently unaware of this shift, remain under the influence of the traditional view. Thus, Hendricks, Elvang-Gorannson and Pedersen (1995: 361):

> Arguments are constructed using natural deduction over a propositional database. Arguments are pairs (δ, p) where δ is the set of facts in the database from which the conclusion of the argument, p is derived.

One prominent alternative development is the view that conceives argument as part of the *practice* of argumentation, and sees argumentation as *process*, as dialogue[3]. But I shall not track this line of development, as I am concerned in this paper with the argument as the *product*, or the distillate, of the process of arguing (Johnson, 1994).

Recently I have been speaking of the limitations of the traditional view which, let it be added, was not all wrong. It focused on one important aspect of argument–its core structure. Yet when the following text is presented as an argument because it satisfies the structural definition, the notion of argument appears to be on the verge of evaporating.

---

[2] See Kneale and Kneale (1970:548 ff) for a fuller account of the rise of modal logic and the problems associated with strict implication.

[3] For examples, see Barth and Krabbe (1982), van Eemeren and Grootendorst (1984), Walton (1984) and, most recently, Walton and Krabbe (1995). For an approach to argumentation that looks more closely at the process of argumentation, see Farley and Freeman (1995).

(S2) The sky is blue. Grass is green.

Therefore, it is not the case that tigers are carnivorous[4].

C. Causes of the limitations of the traditional view

The causes of these limitations appear to be three.

The first important limitation is that the definition was formal (syntactical) rather than functional. It did not take into account the wider context, the purpose and goals that brought this structure into being. If it did, the specimen above would not pass muster. Another reason is that this formalistic understanding of argument won't serve to separate argument from other forms of reasoning. I may give reasons in support of a claim to influence a person's belief and yet not be making an argument. This can happen when, for example, I am giving instructions, or giving an explanation or giving advice–none of which is the same as arguing. So the traditional view needs to be supplemented by an approach which takes into account the purposes of argument.

A second reason for its failure is that it did not give adequate attention to the dialectical character of argumentation. A purely formal account abstracts from purpose, function and context. But arguments are not just pieces of discourse without location or context. Rather they take place against a certain background, within a context of competing points of view, within, one might say, dialectical space[5]. It is difficult to imagine any context for specimen (S2) above.

A third limitation is the way in which the core has been understood, namely as consisting of a set of premises plus an *inference* from them to the conclusion. The inference is typically represented as a bridge or a link. But now it becomes problematic how to render this notion of a bridge or a link in a non-metaphorical way. We have seen that the attempt to capture formally the notion of entailment failed, so that the tightest kind of link (deductive) cannot be modelled formally.

Problems multiply when the realization is added that many arguments are not deductive. One might broaden the notion of induction to cover the rest of the territory, but that seems to be stretch the category beyond the breaking point. The same status will have to be assigned to the following specimens:

(S3) 99% of ravens are black.

That bird is a raven.

Probably, that bird is black.

(S4) Tweety is a bird.

Probably, Tweety can fly.

(S5) Such and such a candidate is ahead in the polls.

The pollster is reputable and the poll was well done.

Probably, such and such a candidate will win.

(S6) You ought to take your son to the movies because you promised to so, it is a good movie, and you have nothing better to do this afternoon.

---

[4] See Lambert and Ulrich (1980). See also Govier (1987:182).

[5] In (1987), E. M. Barth presents and develops the notion of a dialectical field which is closely related to what I have here dubbed the dialectical tier.

The difference between (S3) and (S6) seems glaring. The idea of gathering all of them under the label "inductive" is not promising, if we seek to model real arguments.

Another possibility is to look for some third type of link. This quest has led some, Govier (1987) and Hitchcock (1983) to name but two, to embrace what has been called, following Wellman (1971) a "conductive inference." This is one in which the premises furnish good reasons for the conclusion. At this point, the concept of inference has been mightily compromised[6].

Our discussion of the limitations of the traditional view has brought forth a number of points that I now seek to incorporate into a revised view.

3   A pragmatic approach to argument

A. Introduction

As the limitations of FDL came into view, alternative approaches to the understanding of argument were developed. The one I favour has been called a pragmatic approach, for it takes seriously the idea that an argument emerges within the *practice* of argumentation. The place to begin our search for a better understanding of argument is here.

B. Argumentation as teleological

Practices exists to serve our purposes. The pragmatic approach encourages us to ask: What is the purpose of the practice of argumentation? The best answer seems to be: There are many purposes. Argumentation may be used to persuade, to reinforce, to inquire, to castigate, and so on. PreŒeminent is the function of persuading someone (I shall call this someone 'The Other') of the truth of something (I shall call this something 'The Thesis') by reasoning, i.e., by producing a set of reasons whose function is to lead The Other rationally to accept The Thesis.

From the pragmatic point of view, then, an arguØment is discourse aimed at rational persuasion. By "rational per-suasion," I mean that the Arguer wishes to persuade the Other to accept the conclusion on the basis of the reasons and considerations cited, and those alone. The Arguer agrees to foreswear all other methods that might be used to achieve this: force, flattery, trickery, and so forth.

Contrast argumentation with practice of majority rule, another reputable method of achieving closure on an issue. It says that the verdict of the majority shall stand as the verdict of the whole. This is an efficient method and *eo ipso* a rational one. But the idea behind the practice of argumentation, as I understand it, is that we reach closure not by counting noses, but rather by aiming at

---

[6] Perhaps we should not think of arguments as inferences at all. Nor yet as containing inferences; nor yet as inviting inferences. Perhaps we should distinguish between inference and argument. I pursue this thought in my forthcoming book: *Manifest Rationality*.

consensus in which all parties agree that the force of the better reasoning, and that alone, has determined the outcome.Θ

C. Argument: the illative core

As we saw earlier, the traditional (formal) approach to defining an argument was structural in character. It conceived argument as a form of discourse with a certain structure, premises leading to a conclusion. This view was not altogether wrong. Because rational persuasion is the *telos* (at least as for purposes of this discussion, participants in the practice recognize that any claim made must be supported by reasons, or evidence of some sort.

Hence, in the first instance an argument appears as a premise–conclusion structure: reasons are produced to justify a target proposition, which is the conclusion. This first-tier has been called the *illative core* of the argument[7]. But there is more to the story.

D. The dialectical tier

Does the illative core suffice? My answer is "No," because although the arguer has given some reasons or evidence for his conclusion, in the dialectical situation, that is not enough. (It might be, if we were to abstract from the context, but that we may not do.) Because the arguer's purpose is rational persuasion, a second tier is required as well. Why?

We have seen that the practice of argumentation presupposes a background of controversy. The first tier (the illative core) is meant to initiate the process of converting the Other(s), winning them over to the arguer's thesis. *But they will not easily be won over, nor should they be, if they are rational.* For the participants know that there are objections to the Arguer's premises and criticisms of the Arguer's position.

Indeed, the Arguer must know this herself and so it is typical that to attempt to defuse such objections within the course of the argument[8]. If she does not deal with the objections and criticisms, then to that degree her argument is not going to satisfy the dictates of rationality. More, to that very degree her argument falls short of what is rationally required in terms of structure (never mind the content; i.e., the adequacy of the Arguer's response to those objections.) For those at whom it is directed, "those who know" about The Issue, will naturally wish to know how the Arguer proposes to handle these objections and criticisms, this dialectical material.

Hence, if the Arguer really wishes to persuade the Other rationally, the Arguer is obligated take account of these objections, these opposing points of view, these criticisms. For to ignore them, not to mention them, or to suppress them— these cannot the moves of someone engaged in rational persuasion. An argument, then, must have a *dialectical tier* in which objections and criticisms are dealt

---

[7] I owe this phrase to my colleague, Prof. J. A. Blair (1995:191).

[8] The fallacy of *straw man* occurs when this attempt goes awry.

with. Whether the argument is a good one is at least in part a function of how well the Arguer discharges these dialectical obligations.

Accordingly, I propose the following as a better way to conceptualize argument:

> An argument is a type of discourse or text, the distillate of the practice of argumentation, in which the Arguer seeks to persuade the Other(s) of the truth of a thesis rationally by producing the grounds or evidence that justify it.

In addition to that illative core, an argument also has a dialectical tier in which the Arguer seeks to dispose of his dialectical obligations.

Under this alternative approach, two important features have emerged. We have seen that argumentation as a practice is both *teleological* (governed by the purpose of rational persuasion) and *dialectical* (inherently directed to the Other). I want next to say a bit more about this most recently emerged feature.

4 Argumentation as dialectical

The force of this allŒtooŒfamiliar characterization has not been fully appreciated. The root meaning of "dialectical" is dialogue – a *logos* (reasoned speech) that takes place *between* two (or more) people. That requires more than just speech between two parties, because as we all know, such talking may be nothing more than a monologue conducted in the presence of another. Genuine dialogue requires not merely the presence of the Other, or speech between the two, but the real possibility that the logos of the Other will influence one's own logos. An exchange is dialectical when, as a result of the intervention of the Other, one's own "logos" (discourse/reasoning/thinking) is affected in some way.

Specifically, the Arguer "agrees" to let the feedback from the Other affect the product. The Arguer consents to take criticism and to take it seriously. Indeed, she not only agrees to take it, when it comes, as it typically does. The Arguer will often make a point of soliciting criticism. In this sense, argumentation is a (perhaps even *the*), dialectical process par *excellence*. If (as is likely) the Arguer now modifies that argument as a result of the intervenØtion by the Other, the result will likely be an improved product–a better argument. IntervenØtion of the Other is thereby seen to lead to the improvement of the product. It has become a better argument, a "more rational" product. Criticism (the dialectical moment) results in enhanced rationality.

Thus do these first two features of argumentaØtion–its being rational and dialectical–reinforce each other.

5 Matters dialectical

A. Introduction

In the course of this paper, I have accumulated some dialectical obligations of my own. First, *the question* will asked: How many objections must be dealt with in the dialectical tier, and how does the Arguer determine which ones to deal with? Second, there is *the objection* that my conception of argument is much too idiosyncratic. Third, there is *the problem*: how would argument so understood be represented?

B. How many objections?

An important theoretical issue emerges: what precisely is the arguer's responsibility in this matter of handling objections? It is clear that the arguer is under an obligation to discharge his/her dialectical obligations; i.e., to answer objections, anticipate criticisms, and so on. Clearly the Arguer cannot be expected to deal with *all* of the possible and or actual objections, else an argument becomes an incredibly prolix affair. So how does one specify which ones? I do not know the answer to that question, and must here be content simply to raise it and acknowledge that an adequate theory of argument must answer it.

C. Conception too narrow

One objection is that my approach to defining an argument is much too narrow. There are three prongs to this objection.
(O1) It would result in withholding the term "argument" from ever so many specimens which we would ordinarily call an argument.
(O2) The objection continues: since a defense of something is external to that which it defends, it seems wrong to make the dialectical tier part of the argument itself[9].
(O3) Why not, it will be asked, simply say that arguments typically require what you have called the dialectical tier? Then call that whole, the illative core plus the dialectical tier, something else, like: the case[10].

As to (O1): It is true that in practice, many arguments consist of the only the first tier. That is only to be expected, since often those engaged in the practice are casual rather than dedicated participants. What I mean by that is they have argue on occasion; it is not their life-blood. If, on the other hand, we look to the best practice, to the practice of those who have the most at stake and are most heavily invested in this process–philoØsophers and logicians and others who have a vested interest in this practice–we will find that their arguØments almost always take account of the standard objections. Their arguments typically include a dialectical tier. They realize that to establish their own positions in a rationally compelling way, they will have to do so by facing off against their dialectical opponents.

As to (O2): That a reasoned defense is *external* to that which it defends is not at all evident to me. Allen says that the premise is external to the conclusion it defends. Yet it is internal to the argument of which it is part; as the conclusion is. And what I am proposing is that we extend the logic that leads us to see how

---

[9] Thanks to Derek Allen for this way of putting the objection.
[10] Thanks to Derek Allen for this way of putting the objection.

it is that the premise is internal. For that same logic shows that the handling of objections is likewise *internal*, provided we focus on the demands of rationality.

As to (O3): I acknowledge that this is an alternative way of approaching the matter. What leads me to my proposal is the fear that the notion of argument was so impoverished by FDL that it needs considerable beefing up if it is to remain the forceful intellectual practice it has been.

D. Representation

How might an argument as conceived here be modelled and represented? *In Logical Self-Defense* (1993/1994), Blair and I have made a proposal for this using the strategy known as tree diagramming.

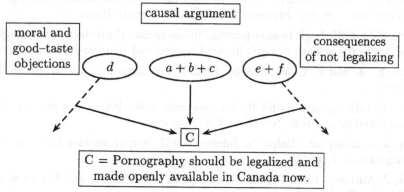

Note that we have given a code name to each of the parts of the argument. The code name will vary from one argument to the next. And we have put the letters of the paragraphs from the synopsis in the circle to indicate that "branch" of the argument. Beginning at the left of the diagram, we have placed what we have coded as the moral and goodŒtaste objections (comprising paragraph **d**). And notice that we have shown the objection with a broken arrow leading away from the conclusion (because that is the logical direction of an objection of this sort) and then we show a straight arrow taking that objection back to the conclusion, indicating that the arguer has attempted to counter that objection. (Whether successfully at not is another matter.) In the middle we have what we have called the causal argument, comprising paragraphs **a**, **b** and **c**, and leading directly to the conclusion. Then at the right we have a branch of the argument, we have a branch of argument called "consequences of not legalizing." Again we shown this as an objection.

But I make no claim here for the adequacy of this approach, but present it as one way of representing the dialectical tier.

6 Conclusion

I have argued here that the traditional formal approach to understanding argument must give way to more pragmatic approach which sees argument as a

358

product of the practice of argumentation. It has the structure it has because of its telos. and that structure is more complex because more situated than the traditional account supposes.

Now, however, the task of representing argument has become more, not less, difficult. The representation must include what I have called the dialectical tier. Thus, as we shift away from simplistic formal conceptions of argument and toward more realistic pragmatic ones, the challenges for AI become greater still.

REFERENCES

Anderson, Alan Ross and Nuel D. Belnap, Jr. 1975. Entailment: the logic of relevance and necessity. Princeton: Princeton University Press.

Barth, E. M. 1987. Logic to some purpose: theses against the deductive-nomological paradigm in the science of logic. In van Eemeren and Grootendorst (1987).

Barth, E. M. and E. C. W. Krabbe. 1982. From axiom to dialogue. Berlin: De Gruyter.

Blair, J. Anthony and Ralph H. Johnson, eds. 1980. Informal logic: The first international symposium. Inverness, CA: Edgepress.

Blair, J. Anthony and Ralph H. Johnson. 1987. Argumentation as dialectical. Argumentation 1, 41-56.

Blair, J. Anthony. 1995. Premiss adequacy. In van Eemeren and Grootendorst (1995), Vol. II, 191-202.

Copi, Irving. 1961. Introduction to Logic. New York: Macmillan. (6th edition, 1986)

van Eemeren, Frans and Rob Grootendorst, Charles A. Willard and J. Anthony Blair, eds. 1987. Argumentation: Across the Lines of Discipline. Dordrecht: Foris Publications.

van Eemeren, Frans and Rob Grootendorst, Charles A. Willard and J. Anthony Blair, eds. 1995. Proceedings of the Third International Conference on Argumentation. 5 Volumes. Amsterdam: Sicsat.

Farley, Arthur M. and Katlneen Freeman. 1995. Burden of proof in a computational model of argumentation. In van Eemeren and Grootendorst at al (1995).

Frege, Gottlob. 1879. Begriffschrift.

Gabbay, Dov. 1995. Labelled deductive systems and the informal fallacies: a preliminary analysis. In Van Eemeren and Grootendorst (1995).

Govier, Trudy. 1985. A practical study of argument. Belmont, CA: Wadsworth. (3rd edition, 1993)

Govier, Trudy. 1987. Problems in argument analysis and evaluation. Dordrecht: Foris.

Habermas, Júrgen. 1984. The Theory of communicative practice. Vol. I: Reason and the rationalization of society. Tr. by Thomas McCarthy. Boston: Beacon Press.

Hintikka, Jaakko. 1989. The role of logic in argument. The Monist, 72: 3-24.

Hitchcock, David. 1983. Critical thinking: a guide to evaluating information. Toronto: Methuen.

Johnson, Ralph H. and J. Anthony Blair. 1980. The recent development of informal logic. In Blair and Johnson (1980).

Johnson, Ralph H. and J. Anthony Blair 1985. Informal logic: The past five years. American Philosophical Quarterly, 22: 181-196.

Johnson, Ralph H. and J. Anthony Blair. 1993. Logical Self-Defense. Toronto: McGraw-Hill Ryerson. (3rd edition.)

Johnson, Ralph H. 1994. Argument: a pragmatic perspective. Inquiry, 13: 3-9.

Kahane, Howard. 1971. Logic and contemporary rhetoric: The use of reasoning in everyday life. Belmont, CA: Wadsworth. (6th edition)

Kneale, William and Martha Kneale. 1970. The development of logic. Oxford: The Clarendon Press.

Lambert, Karel and William Ulrich. 1980. The nature of argument. New York: Macmillan.

Lewis, C. I. and C. H. Langford. 1932. Symbolic logic. New York: Dover.

Quine, Willard van Orman. 1944. Introduction to mathematical logic. New York: Harper.

Siegel, Harvey. 1988. Educating reason: Rationality, critical thinking and education. New York: Routledge.

Toulmin, Stephen E. 1958. The uses of argument. Cambridge: Cambridge University Press.

Toulmin, Stephen, Richard Reike and Allan Janik. 1984. An introduction to reasoning. New York: Macmillan. (2nd edition)

Walton, Douglas. 1983. Topical relevance in argumentation. Amsterdam: John Benjamins.

Walton, Douglas. 1984. Logical dialogue games and fallacies. Lanham, MD: University Press of America.

Walton, Douglas. 1985. Arguer's position: A pragmatic study of 'ad hominem' attack, criticism, refutation and fallacy. Westport, CT: Greenwood Press.

Walton, Douglas. 1987. Informal fallacies: Towards a theory of argument criticisms. Amsterdam: John Benjamins.

Walton, Douglas. 1989. Informal logic: A handbook for critical argumentation. Cambridge: Cambridge University Press.

Walton, Douglas. 1990. What is Reasoning? What is an Argument? Journal of Philosophy

Walton, Douglas N. and Erik C. W. Krabbe. 1995. Commitment in dialogue: Basic concepts of interpersonal reasoning. Ithaca: SUNY Press.

Wellman, Carl. 1971. Challenge and response: Justification in Ethics. Carbondale, IL: University of Illinois Press.

Whitehead, Alfred North and Bertrand Russell. Principia Mathematica. 1910-13. Cambridge: Cambridge University Press.

Willard, Charles. 1983 Argumentation and the social grounds of knowledge. Tuscaloosa: The University of Alabama Press.

Willard, Charles. 1989. A theory of argumentation. Tuscaloosa: The University of Alabama Press.

Wittgenstein, Ludwig. 1921. Tractatus logico-philosophicus. London.

Two Kinds of
Non-Monotonic Analogical Inference*

Manfred Kerber[1] and Erica Melis[2]

[1] The University of Birmingham, School of Computer Science
Birmingham, B15 2TT, England
e-mail: M.Kerber@cs.bham.ac.uk
WWW: http://www.cs.bham.ac.uk/~mmk

[2] Universität des Saarlandes, FB Informatik
D-66041 Saarbrücken, Germany
e-mail: melis@cs.uni-sb.de
WWW: http://jswww.cs.uni-sb.de/~melis

Abstract. This paper addresses two modi of analogical reasoning. The first modus is based on the explicit representation of the justification for the analogical inference. The second modus is based on the representation of typical instances by concept structures. The two kinds of analogical inferences rely on different forms of relevance knowledge that cause non-monotonicity. While the uncertainty and non-monotonicity of analogical inferences is not questioned, a semantic characterization of analogical reasoning has not been given yet. We introduce a minimal model semantics for analogical inference with typical instances.

1 Introduction

Analogical reasoning is a process whereby similarities between a source and a target are used to infer the probable existence of further similarities. Thus, under certain conditions, an analogical inference can be employed to provide a description of an aspect of the target, if the description of the same aspect is known for the source.

What are the conditions that justify an analogical inference? There are several answers to this question, but most of them agree in characterizing some relevance knowledge as a justification for analogical inferences. Goebel [12], for instance, emphasizes some *relevant similarity* as the knowledge required for an analogical inference. Gentner [9] prefers by her systematicity criterion a mapping of predicates which are connected by a higher order relation explicitly to be known for analogical inferences. The higher order relations she examined, namely *cause* and *implies*, actually represent relevance knowledge of the form "aspect A_1 is relevant for aspect A_2".

The maybe best-recognized proponent of justification is Russell, who defined the so-called determinations as the crucial relevance knowledge. His total determinations represent a strict form of connections between two aspects, expressing "the values of a predicate A_1 determine the values of a predicate A_2".

---

* This work was supported by the Deutsche Forschungsgemeinschaft, SFB 314 (D2)

Let us look at an example: Knowing that the car car_S was produced in 1990, that its make is Honda-civic, and that its price is 13,000.00 DM you would infer analogically that a car car_T which was produced in 1990 and the make of which is Honda-civic would probably cost about 13,000.00 DM. The relevance knowledge that justifies this particular analogical inference is the determination of the price of a car by its make and year of construction. But what happens if you gain the additional knowledge that car_T is rusty and was involved in many severe accidents? Then you would no longer infer the price of 13,000.00 DM for car_T. This example provides a clue for the non-monotonicity of inferences called connection-based analogical inferences.

For understanding the non-monotonicity of another kind of analogical reasoning, remember the classical Tweety story of non-monotonic reasoning: In this story the non-monotonicity has a lot to do with the break down of an inheritance in the concept "bird". In (semantic net and frame) representations that use prototypes, such as the TypicalElephant of [4], certain analogical inferences provide new information by copying facts from the prototype to the individual. Even though known as "inheritance", actually this reasoning is another common kind of analogical reasoning, which we shall call typical-example-based. As the Tweety story shows, it is non-monotonic as well.

Analogical inference is commonly considered to be non-deductive, hypothetical, tentative, and non-monotonic. These features of analogical reasoning have been addressed explicitly in [12, 13, 24, 19] and implicitly presupposed in many approaches to reasoning by analogy. Other questions have usually been the center of attention of analogy research though. We discuss the additional justifying knowledge needed for inferring a further similarity from a similarity of a source and a target for two kinds of analogical reasoning. We propose how to represent this knowledge and present a semantic characterization for the analogical inference with typical instances that is actually designed for non-monotonic logics.

This paper is organized as follows: We first recall the basic ingredients of connection-based analogical reasoning. Then we introduce analogical reasoning based on typical instances, which makes use of an exemplary knowledge representation. We present the inference schemas for the two forms of analogical reasoning and discuss a semantic that is appropriate for the analogical inference with typical instances and which captures its non-monotonicity. At last we propose a hybrid framework integrating these two forms of analogical reasoning using a hybrid knowledge representation[3].

2 Connection-Based Analogical Reasoning

An important concept in analogical reasoning is the so-called aspect. In many approaches to computational analogy, aspects are represented just by predicates or fixed formulae. In order to capture a wider range of analogical reasoning and to include inter-domain analogies, we define an *aspect A* as a partial function,

---

[3] Thereby we add to the debate about logic-based versus non-propositional knowledge representation in artificial intelligence, which dates back to McCarthy's Advice Taker [18] and Sloman's analogical representations [26].

mapping the individuals (instances c of a concept C) to non-tautological formulae with at most one free variable x such that $A\langle c\rangle[x/c]$ is **true**[4].

Informally speaking, if c is an instance and A an aspect then $A\langle c\rangle$ is a formula describing A of c. For example: The value of $safety\langle c\rangle$ of a car c might be $airbag(x) \wedge antiblock(x) \wedge max\_speed(x) < 100$. It is assumed that $airbag(c) \wedge antiblock(c) \wedge max\_speed(c) < 100$ is **true**. For a bicycle b, $safety\langle b\rangle$ might be $frame\_diameter(x, 3) \wedge age(x) < 10$.

The transfer of the value of an aspect A_2 from a source case s to a target case t based on the similarity of s and t with respect to another aspect A_1 is the *standard form of justified reasoning by analogy*, investigated, e.g. , in [7]. This kind of analogical inference requires for its justification some relevance knowledge expressing that A_1 is relevant for A_2. The relevance knowledge "A_1 is relevant for A_2", written as $[A_1, A_2]$, is usually represented explicitly and propositionally, as determinations [7], as schemata [11], as connections [19], or as similarity transforms [6]. In modeling human analogical reasoning, the relevance knowledge must allow for exceptions and uncertainty rather than being a logical implication or a total determination. In the following we use the most general notion, connection, that does not require a specific representation and may have exceptions.

A *connection* is a pair of aspects $[A_1, A_2]$. An example of such connections is $[population, cars]$ expressing "if two cities have the same number of inhabitants, then probably the same number of cars is registered in the cities". This connection is used in an analogical inference that yields a value of the aspect *cars* for the target instance *Rome*. This inference takes as inputs the similarity of *Rome* and another city, say *Madrid*, with respect to the number of inhabitants and the connection. It infers the correspondence of *Rome* and *Madrid* with respect to the aspect *cars*. Using the additional information of the actual value of $cars\langle Madrid\rangle$ the value of $cars\langle Rome\rangle$ can be inferred.

Utilizing connections, the connection-based analogical inference can formally be described by the schema

$$(\mathbf{AR}) \qquad \frac{A_1\langle t\rangle = A_1\langle s\rangle, [A_1, A_2]}{A_2\langle t\rangle := A_2\langle s\rangle}$$

A connection-based analogical inference is confirmed only if the resulting target aspect $A_2\langle t\rangle$ does not contradict the knowledge inferable for the target by deduction. *The uncertainty of the connection and the consistency constraint are responsible for the non-monotonicity of the connection-based analogical inference.*

But, what happens, if such explicit connections are not available? The kind of analogical inference presented next allows for justified analogical reasoning that is not necessarily based on explicit connections.[5]

---

[4] Note that $A\langle c\rangle$ is a formula with a free variable x. $A\langle c\rangle[x/c]$ denotes the formula $A\langle c\rangle$ in which the free variable x is substituted by c.

[5] Although the field of analogical reasoning is concerned with reasoning based on examples, surprisingly the importance of reasoning by *typical* instances, as for example, investigated by Rosch [21] and Lakoff [17], has not been elaborated. Only an attempt of Winston [30] was influenced by statistic prototypicality.

3 Analogical Reasoning Based on Typical Instances

Russell [24] tries to interpret analogical inference based on typical instances as a connection-based inference with "belonging to the same class" as the aspect A_2 of a connection $[A_1, A_2]$. However, psychological investigations provided evidence that typical instances are the only, or at least the preferred sources for analogical reasoning corresponding to inheritance among instances of a concept. Consequently, connection-based analogical inference does not cover analogical inferences which can have only a typical instance as the source rather than an arbitrary instance.

The typicality of instances of concepts is a phenomenon investigated in empirical psychology [17, 20, 23]. Reproducible typicality ratings that distinguish typical instances have been found. Some experimental methods [17] for the extraction of this typicality rating are the direct rating of representativity, the examination of the reaction time to decide whether an instance belongs to a category, the test of the reproduction of instances, and the use of instances in generalizations and in analogical reasoning. Hence you find the notation *structured concept* in the psychological literature that refers to a set of instances with a typicality relation which we denote[6] by $\sqsubseteq_C$. The notion "concept structure" denotes the structure of one single concept rather than the relationship between different concepts like in KL-ONE. The typicality relations $\sqsubseteq_C$ compare for each concept C the degrees of typicality (for example, a hammer is commonly considered a more typical tool than a compass saw) of similar instances. Within such concept structures we define typical instances:

DEFINITION: An instance $c \in C$ is called a *typical instance* if it is maximal (that is, there is no $c' \neq c$ with $c \sqsubseteq_C c'$), written as typex(c).

For example, a hammer is commonly considered a typical tool and a violin a typical musical instrument. An example of a concept structure is given in figure 1.

There are two kinds of aspects of typical instances: aspects that are important for the typicality of the instance (e.g. the size of a city) and aspects that are accidental (e.g. the number of research institutes of a city). Of course, justified analogical reasoning transfers only relevant information. This motivates the following definition that describes a relevance different from that in the previous section:

DEFINITION: An aspect A is called *relevant* for an instance c (written as *relevant*(A, c)) iff $A\langle c \rangle [x/c]$ is true and for all instances c' with $c' \sqsubseteq_C c$ holds $A\langle c \rangle [x/c']$ is either true or undefined.

Analogical inferences based on typical instances can formally be described by the schema

---

[6] The subscript specifies the concept.

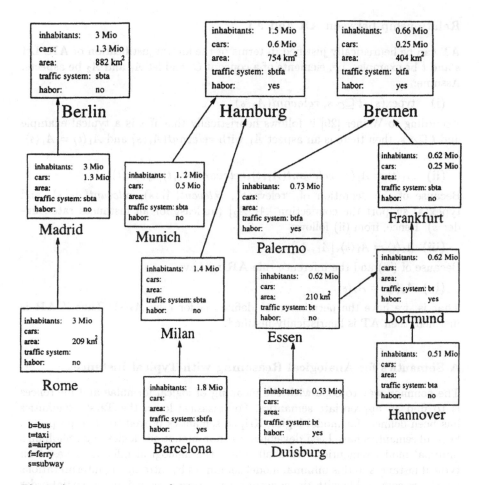

Fig. 1. A concept structure of cities

$$(\mathbf{AT}) \qquad \frac{typex(s), t \sqsubseteq_C s, relevant(A_2, s)}{A_2\langle t\rangle := A_2\langle s\rangle}$$

Again, an analogical inference based on typical instances is confirmed only if the result does not contradict the knowledge inferable for the target by deduction. *This consistency constraint and the uncertainty of the knowledge about the relevance of an aspect for a typical instance of a concept are responsible for the non-monotonicity of analogical inference based on typical instances.*

Relationship between AR and AT

AT can be heuristically justified in terms of the known justification of **AR**: Let s and t be examples, i.e. elements of a concept C, and let A_1 and A_2 be aspects. Assume:

(i) typex(s), $t \sqsubseteq_C s$, relevant(A_2, s).

According to Weiner [29] it follows heuristically that if s is a typical example and $t \sqsubseteq_C s$, then there is an aspect A_1 with relevant(A_1, s) and $A_1\langle t \rangle = A_1\langle s \rangle$. Hence, we have

(ii) $A_1\langle t \rangle = A_1\langle s \rangle$, relevant($A_1, t$), relevant($A_2, t$), typex($t$).

Because of the definition of "relevant", relevant(A_1, s), relevant(A_2, s), and typex(s) support the connection $[A_1, A_2]$ (at least for all instances rated under s). Hence, from (ii) follows

(iii) $A_1\langle t \rangle = A_1\langle s \rangle, [A_1, A_2]$.

Because of (iii) and the inference rule **AR** we have

(iv) $A_2\langle t \rangle := A_2\langle s \rangle$.

That is, we have the permission to define A_2 for t by $A_2\langle s \rangle$. Thus if **AR** is justified, then **AT** is heuristically justified.

A Semantics for Analogical Reasoning with Typical Instances

The common way to cope with the meaning of logical formulae and inferences is to find an appropriate semantics. For classical logic, the Tarski semantics has been defined, for modal logics, Kripke semantics proved to be appropriate. Several semantics have been developed for non-monotonic logics, e.g., Shoham's minimal model semantics. We shall relate the analogical inferences based on typical instances to this minimal model semantics by introducing interpretations which are compatible with the concept structure and by defining a partial order on these interpretations.

Let $\partial\mathcal{I}$ be a partial interpretation of formulae, that assigns to each pair (*instance, formula with one variable*) one of the values true, false, unknown. It can be thought of as the result of an inspection of the instances. In the case of our city example

$$\partial\mathcal{I}(Rome, no\_of\_inhabitants(x) \doteq 3million) = \texttt{true}$$

since a corresponding entry can be found in the concept structure. In contrast $\partial\mathcal{I}(Rome, no\_of\_cars(x) \doteq 1million) = \texttt{undef}$, since there is no entry for *cars* in the instance of *Rome*.

More precisely, for the following we assume a sorted (first-order) logic $\mathcal{L}$, where each sort can be viewed as a concept like *car* or *city*. We denote the sorts by lowercase Greek letters such as κ or μ. $\mathcal{E}$ is a set of sets $\{\mathcal{E}_\kappa\}_\kappa$, where each $\mathcal{E}_\kappa$ is called the *set of examples of sort* κ. The $\mathcal{E}_\kappa$ are such that their structure corresponds to the sort structure of $\mathcal{L}$, that is, if $\mu \sqsubseteq \kappa$ (i.e. μ is *subsort* of κ) then $\mathcal{E}_\mu \subseteq \mathcal{E}_\kappa$. $\mathcal{E}$ forms the *frame* (the collection of universes) for the partial interpretation of the terms. $\partial\mathcal{I}$ is a fixed partial interpretation function (corresponding to three-valued strong Kleene logic $\mathcal{L}^K$ [16, 28]) in the frame $\mathcal{E}$. Each term t of sort κ is either interpreted by an example in $\mathcal{E}_\kappa$ or by the bottom element $\perp$. Formulae may be evaluated by $\partial\mathcal{I}$ to true, to false, or to undef. Furthermore, we assume that for every element $e^i_\kappa \in \mathcal{E}_\kappa$ there exists a constant c^i_κ of sort κ with $\partial\mathcal{I}(c^i_\kappa) = e^i_\kappa \in \mathcal{E}_\kappa$ and that there are only finitely many examples in $\mathcal{E}$.

The partial interpretation of composed formulae is defined as usual, based on the propositional connectives as defined by the following truth tables:

$\neg$		$\vee$	false	undef	true
false	true	false	false	undef	true
undef	undef	undef	undef	undef	true
true	false	true	true	true	true

In order to fix the semantics of the universal quantifier, assignments ξ for the interpretation of the variables into the frame are necessary. If ξ is an arbitrary assignment, $\xi[x \leftarrow a]$ denotes the assignment equal to ξ for all variables except for x and $\xi[x \leftarrow a](x) = a$.

$$\partial\mathcal{I}_\xi(\forall x_\kappa \varphi) := \begin{cases} \text{true} & \text{if } \partial\mathcal{I}_{\xi[x \leftarrow a]}(\varphi) = \text{true for all } a \in \mathcal{E}_\kappa \\ \text{false} & \text{if } \partial\mathcal{I}_{\xi[x \leftarrow a]}(\varphi) = \text{false for one } a \in \mathcal{E}_\kappa \\ \text{undef} & \text{else} \end{cases}$$

The semantics of $\wedge$, $\Rightarrow$, $\Leftrightarrow$, and $\exists$ can then be defined in the usual manner.

Note that these definitions do not assume a concrete representation of the examples—the only requirement is that we get an answer to certain questions, thus fixing the interpretation function. In other words, $\partial\mathcal{I}$ has to be effectively computable for all ground formulae (i.e., variable free formulae) and, consequently, for all formulae, since we assume the number of examples in $\mathcal{E}$ to be finite.

As usual, we give an (extended) set theoretic semantics for a formula set. Our semantics is such that it is compatible with the examples. An interpretation of a knowledge base Δ is defined as an extension of the partial interpretation, given by $\langle\mathcal{E}, \partial\mathcal{I}\rangle$.

DEFINITION ($\langle\mathcal{E}, \partial\mathcal{I}\rangle$-INTERPRETATION): Let $\mathcal{E} = \{\mathcal{E}_\kappa\}_\kappa$ be a given set of example sets and let $\partial\mathcal{I}$ be a partial interpretation function in $\mathcal{E}$. An interpretation $\langle\{\mathcal{D}_\kappa\}_\kappa, \mathcal{I}\rangle$ is called an $\langle\mathcal{E}, \partial\mathcal{I}\rangle$-*interpretation* iff

- there are injective mappings $\text{inj}_\kappa : \mathcal{E}_\kappa \hookrightarrow \mathcal{D}_\kappa$ with $\mathcal{I}(c_\kappa) = \text{inj}_\kappa(\partial\mathcal{I}(c_\kappa))$ for all constant symbols c_κ with $\partial\mathcal{I}(c_\kappa) \neq \perp$. (When the sort is not important, we omit the index κ and simply write inj.)
- for all terms t and arbitrary ground instances $\sigma(t)$ with $\partial\mathcal{I}(\sigma(t)) \neq \perp$ holds $\mathcal{I}(\sigma(t)) = \text{inj}(\partial\mathcal{I}(\sigma(t)))$.
- for all formulae φ and all ground instances σ of φ holds, if $\partial\mathcal{I}(\sigma(\varphi)) \neq \text{undef}$ then $\mathcal{I}(\sigma(\varphi)) = \partial\mathcal{I}(\sigma(\varphi))$

If $\mathcal{I}_\xi(\varphi) = \mathbf{true}$ for all assignments ξ then $\mathcal{I}$ is called an $\langle \mathcal{E}, \partial \mathcal{I} \rangle$-*model* of φ. If φ has no $\langle \mathcal{E}, \partial \mathcal{I} \rangle$-model, it is said to be $\langle \mathcal{E}, \partial \mathcal{I} \rangle$-*unsatisfiable*. Γ $\langle \mathcal{E}, \partial \mathcal{I} \rangle$-*entails* the formula φ iff each $\langle \mathcal{E}, \partial \mathcal{I} \rangle$-model of Γ is an $\langle \mathcal{E}, \partial \mathcal{I} \rangle$-model of φ, too (i.e. $\Gamma \models_{\langle \mathcal{E}, \partial \mathcal{I} \rangle} \varphi$).

So far all possible interpretations $\mathcal{I}$ have to be considered for modeling formulae. This is sufficient for capturing *deductive* reasoning as shown in [15]. However, in order to model any form of non-monotonic reasoning (in particular analogical reasoning), each conclusion drawn must potentially be withdrawn. In order to describe non-monotonicity semantically, Shoham [25] has introduced an order on the interpretations and weakened the notions of satisfiability, of consequence etc. His key idea is to only consider distinguished *minimal* models for the satisfiability and consequence relations. This approach is well-suited for analogical reasoning which is based on preferred instances—the typical instances of a concept.

Let us recall the notions of minimal model and preferential entailment in Shoham's approach:

DEFINITION: Assuming a partial order $<$ on interpretations, an interpretation $\mathcal{I}$ *preferentially satisfies* a formula φ (written as $\mathcal{I} \models_< \varphi$) if $\mathcal{I} \models \varphi$ and if there is no other interpretation $\mathcal{I}'$ such that $\mathcal{I}' < \mathcal{I}$ and $\mathcal{I}' \models \varphi$, that is, $\mathcal{I}$ is a minimal model of φ.

This definition attains a general semantic characterization of non-monotonic inferences. The only point to be specified is that of a partial order $<$ on the interpretation $\mathcal{I}$. Now we want to give a semantic characterization for the inference scheme **AT** of section 3. For the case of analogical reasoning with typical instances we define an interpretation to be smaller then another if it agrees in more individuals with the typical instances. Put formally,

DEFINITION: $\mathcal{I}' \leq \mathcal{I}$ if and only if[7,8]
1. for all instances a with $\partial \mathcal{I}(a) = \perp$ and b is typical instance with $\partial \mathcal{I}(a) \sqsubseteq \partial \mathcal{I}(b)$ and $\partial \mathcal{I}(b) \neq \perp$ holds if $\mathcal{I}'(a) = \mathsf{inj}(\partial \mathcal{I}(b))$ then $\mathcal{I}(a) = \mathsf{inj}(\partial \mathcal{I}(b))$ and

---

[7] Nota bene: While the optimal models in Shoham's approach are *minimal*, the typical instances are *maximal* with respect to the typicality relation.

[8] In an alternative definition, the information is not taken from the typical instances, but from the instances immediately above the instance in consideration. In this case the definition looks as follows:

$\mathcal{I}' \leq \mathcal{I}$ if and only if

1. for all instances a with $\partial \mathcal{I}(a) = \perp$ and b is a least instance with $\partial \mathcal{I}(a) \sqsubseteq \partial \mathcal{I}(b)$ and $\partial \mathcal{I}(b) \neq \perp$ holds if $\mathcal{I}'(a) = \mathsf{inj}(\partial \mathcal{I}(b))$ then $\mathcal{I}(a) = \mathsf{inj}(\partial \mathcal{I}(b))$ and

2. for all predicates P with $\partial \mathcal{I}(P(a_1, \ldots, a_n)) = \mathbf{undef}$, for all i, the b_i are least instances such that $\partial \mathcal{I}(a_i) \sqsubseteq \partial \mathcal{I}(b_i)$ and $\partial \mathcal{I}(P(b_1, \ldots, b_n)) \neq \mathbf{undef}$ holds if $\mathcal{I}'(P(a_1, \ldots, a_n)) = \mathsf{inj}(\partial \mathcal{I}(P(b_1, \ldots, b_n)))$ then $\mathcal{I}(P(a_1, \ldots, a_n)) = \mathsf{inj}(\partial \mathcal{I}(P(b_1, \ldots, b_n)))$.

3. an analogous relation holds for functions f (replace $\mathbf{undef}$ by $\perp$.)

2. for all predicates P with $\partial\mathcal{I}(P(a_1,\ldots,a_n)) = \mathtt{undef}$; for all i, the b_i are typical instances such that $\partial\mathcal{I}(a_i) \sqsubseteq \partial\mathcal{I}(b_i)$ and $\partial\mathcal{I}(P(b_1,\ldots,b_n)) \neq \mathtt{undef}$ and P is relevant for the b_i then holds:
if $\mathcal{I}'(P(a_1,\ldots,a_n)) = \mathsf{inj}(\partial\mathcal{I}(P(b_1,\ldots,b_n)))$ then
$\mathcal{I}(P(a_1,\ldots,a_n)) = \mathsf{inj}(\partial\mathcal{I}(P(b_1,\ldots,b_n)))$.

3. an analogous relation holds for functions f (replace $\mathtt{undef}$ by $\perp$.)

That is, in minimal models all *relevant* information that is not fixed by the knowledge base is transferred from typical instances.

Here is an example of how the semantics works: By definition, for a preferred model $\mathcal{I}$ with the typical instance *Berlin* and the instance *Rome* the information about *public_transportation* is transferred from *Berlin* to *Rome* if no additional information is known. Concretely, since $\partial\mathcal{I}(Rome, no\_of\_cars(x)) = \perp$, in a minimal model $\mathcal{I}^1_{\min}$ holds

$$\mathcal{I}^1_{\min}(Rome, no\_of\_cars(x)) = \mathsf{inj}(\partial\mathcal{I}(Berlin, no\_of\_cars(x))) = 1.3 \text{ Mio}$$

Of course this is only the case if $\mathcal{I}^1$ is a model of the first formula at all, that is, if there is no information that contradicts to the assumption that there are 1.3 Mio cars in *Rome*. It is easy to see that this semantics is non-monotonic: For instance, if the information that in *Rome* 2 Mio cars are one the roads, that is, $no\_of\_cars(Rome) \doteq 2$ Mio, is added to the knowledge base then $\mathcal{I}^1$ is no longer a model at all, and in particular no minimal model. Hence, the analogical conclusion cannot be inferred any longer.

The non-monotonicity can be concretely seen in the following form. Let $\varphi = no\_of\_cars(Rome) \doteq 1.3$ Mio and $\psi = no\_of\_cars(Rome) \doteq 2$ Mio, then we have:

$$\emptyset \models^{\langle\mathcal{E},\partial\mathcal{I}\rangle}_< \varphi \quad \text{but}$$
$$\emptyset \cup \{\psi\} \not\models^{\langle\mathcal{E},\partial\mathcal{I}\rangle}_< \varphi$$

4 Hybrid Framework

In order to *computationally* realize both kinds of analogy, a hybrid framework is needed, that provides propositional *and* exemplary knowledge representations and procedures to extract information from the non-propositional representation (see, e.g., [27, 22]). Our framework consists of three parts: a hybrid knowledge base, a reasoner, and procedures which deliver information from the knowledge base to the reasoner [14]. The knowledge base itself has two parts: a collection of propositions and non-propositional representations of concepts. The reasoner consists of inference methods that operate on the propositional part of the knowledge base and of methods that use information contained in the conceptual part of the knowledge base. Figure 2 displays this framework.

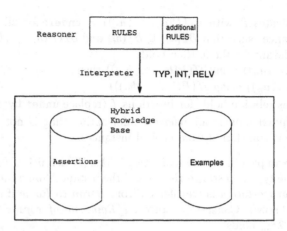

Fig. 2. System structure

The Propositional Subsystem

The propositional subsystem consists, as usual, of a set Γ of (sorted first order) formulae. Aspects, as mentioned above, can be defined in this subsystem. The propositionally represented connections of aspects—as far as they are available—belong to this subsystem as well.

The Conceptual Subsystem

We extend the knowledge base by a *conceptual* part consisting of *concept structures* which are non-propositional representations of concepts. We use directed acyclic graphs consisting of a set of instances and the typicality relation $\sqsubseteq_C$ as concept structures.

The instances themselves might be represented by neural nets, maps, diagrams or some other means including symbolic representations. The particular type of these representations is of no concern for the rest of the paper. We consider concepts C to be represented by a set of instances with a partial order $\sqsubseteq_c$. The elements of these concept structures represent concept instances. A concept structure is displayed as a directed acyclic graph as in figure 1. In this city example, the instances are not directly represented as maps, but we deal with another concept representation which is similar to that employed by Barwise and Etchemendy [3]. It encodes instances as tables. Nevertheless, the conceptual part of the knowledge base is non-propositional because of the concept structure.

The Reasoner

In order to integrate the conceptual subsystem into the framework, its informational content has to be accessible. The semantics corresponds exactly to the

371

semantics given in section 3. Several inspection procedures work on the conceptual subsystem and provide the information that is needed by the rules of the reasoner:

- A TYP-procedure provides access to the structural content of the concept structures in that it finds a typical instance s with $t \sqsubseteq_C s$ for an instance t of a concept C. For example, TYP yields for the instance *Rome* the typical instance *Berlin* by looking up the concept structure *city*.[9]
- An interpreting partial procedure, called ASP, computes the values of aspects A for the *typical* instances c out of the representation of the instances. If there is a value it is required to be a formula $A\langle c \rangle$ with $\mathcal{I}(A\langle c \rangle) = \text{true}$. For example, ASP yields a value of the aspect *public_transportation* for the typical instance *Berlin*: $subway(x) \wedge bus(x) \wedge taxi(x) \wedge airport(x)$ by looking up the representation of *Berlin*.
- A RELV-procedure provides true/false-information about the relevancy of aspects for the typical instances that is encoded in the concept representation. An aspect A is *relevant* for an instance c iff $\mathcal{I}(A\langle c \rangle[x/c]) = \text{true}$ and for all instances $c' \sqsubseteq_C c$ holds $\mathcal{I}(A\langle c \rangle[x/c']) \in \{\text{true}, \text{undefined}\}$. This notion of relevancy corresponds to a kind of modal operator. We assume that every typical examples has at least one relevant aspect.

AR which finally provides information about a target case t takes as input:

- The similarity of a source case s and the target case t expressed by the equivalence w.r.t. an aspect A_1. For example, let be $s = Madrid$, $t = Rome$, and $A_1 = population$. The input is $population\langle Rome \rangle = population\langle Madrid \rangle$.
- A connection $[A_1, A_2]$ which belongs to the propositional part of the knowledge base. Such a connection is $[population, cars]$ which means that, if the populations of two cities agree, then probably the number of cars registered in these cities agrees too.

AR yields $A_2\langle t \rangle = A_2\langle s \rangle$. For the example input above: $cars\langle Rome \rangle = cars\langle Madrid \rangle$. Using the additional information about the actual value of $A_2\langle s \rangle$, which is explicitly given by the propositional subsystem, we can infer the value of $A_2\langle t \rangle$.

At a glance: Let the following information be given

[population, cars]	
$population\langle Madrid \rangle =$ $no\_of\_inhabitants(x, 3million)$	$population\langle Rome \rangle =$ $no\_of\_inhabitants(x, 3million)$
$cars\langle Madrid \rangle =$ $no\_of\_cars(x, 1million)$	$cars\langle Rome \rangle=$?

[9] This example shows that typical examples are not independent of the culture background. For Italians in general, *Rome* would be the typical example and not *Berlin*.

AR infers by analogy $cars\langle Rome \rangle = cars\langle Madrid \rangle$.

Hence it is possible to infer $cars\langle Rome \rangle = no\_of\_cars(x, 1 million)$

The **AT** rule of the reasoner takes as inputs:

- a typical instance s with $t \sqsubseteq_C s$ which is computed by the TYP-procedure, and
- information about the relevancy of A_2 for this s. This information is either extracted by the RELV-procedure or explicitly represented in the propositional part of the knowledge base as, e.g., suggested by Gentner [10].

The **AT** rule then infers $A_2\langle t \rangle = A_2\langle s \rangle$. Using the additional information about the actual value of $A_2\langle s \rangle$, which is computed by the ASP-procedure, $A_2\langle t \rangle$ can be inferred.

Let us look at an example where the individual concept structure of the concept *city* is given. We want to compute $A_2\langle t \rangle = public\_transportation\langle Rome \rangle$ by analogy to a typical city. If there is no explicit connection for *public_transportation*, then **AR** cannot be applied and we proceed as follows:

- The TYP-procedure computes the typical instance *Berlin* as a typical city that is rated over *Rome*.
- RELV tests whether *public_transportation* is a relevant aspect for *Berlin*. Provided that the result is **true**, **AT** yields
  $public\_transportation\langle Berlin \rangle = public\_transportation\langle Rome \rangle$.
- With the additional information
  $public\_transportation\langle Berlin \rangle = (subway(x) \wedge bus(x) \wedge taxi(x) \wedge airport(x))$
  which is provided by the ASP-procedure or explicitly given in the propositional subsystem, we infer $public\_transportation\langle Rome \rangle = (subway(x) \wedge bus(x) \wedge taxi(x) \wedge airport(x))$ by analogy.

At a glance: let the following information be given

$relevant(public\_transportation, Berlin)$	
typex(*Berlin*) and $Rome \sqsubseteq_{city} Berlin$.	
$public\_transportation\langle Berlin \rangle =$	$public\_transportation\langle Rome \rangle =$
$(subway(x) \wedge bus(x) \wedge$	?
$taxi(x) \wedge airport(x))$	

AT infers by analogy

$$public\_transportation\langle Berlin \rangle = public\_transportation\langle Rome \rangle.$$

Hence it is possible to infer

$$public\_transportation\langle Rome \rangle = (subway(x) \wedge bus(x) \wedge taxi(x) \wedge airport(x)).$$

5 Conclusion

In this paper we discussed two different kinds of analogical reasoning, connection-based analogical reasoning and analogical reasoning based on typical instances. The non-monotonicity of the connection-based analogical inferences stems from the justifying connections that represent the relevance of an aspect for another aspect. On the other hand, the non-monotonicity of the analogical inference based on typical instances is caused by the knowledge about the relevance of an aspect for a typical instance of a concept. For the latter we presented an appropriate semantics that is a special case of Shoham's minimal model semantics.

The paper dealt with two different analogical inference schemas and with their computational realization in a hybrid framework that is equipped with different kinds of knowledge representation. Related work with typical instances has been done for the machine learning systems PROTOS and COBWEB [2, 8] that also use a representations of concepts with typical instances.

References

1. J. Arima and K. Sato. Non-monotonic reasoning. In K. Furukawa and F. Mizoguch, editors, *Knowledge Programming*, pages 189–214. Kyritso Shuppan Co. Ltd., 1988. in Japanese.
2. E.R. Bareiss, B.W. Porter, and C.C. Wier. Protos: An exemplar-based learning apprentice. *International Journal of Man-Machine Studies*, 29:549–561, 1988.
3. J. Barwise and J. Etchemendy. *Hyperproof.* CSLI Lecture Notes. University of Chicago press, Chicago, 1993.
4. R.J. Brachman and J.G. Schmolze. An overview of the KL-ONE knowledge representation system. *Cognitive Science*, 9:171–216, 1985.
5. G. Brewka. Belief revision in a framework for default reasoning. In Gerhard Brewka and Ulrich Junker, editors, *Aspects of Non-Monotonic Reasoning*, pages 2–18. GMD, St. Augustin, Germany, TASSO Report No.1, March 1990.
6. A.M. Collins and R. Michalski. The logic of plausible reasoning: A core theory. *Cognitive Science*, 13(1):1–50, 1989.
7. T.R. Davis and S.J. Russell. A logical approach to reasoning by analogy. In *Proceedings of the Tenth International Joint Conference on Artificial Intelligence*, pages 264–270, Milan Italy, 1987. Morgan Kaufmann.
8. D.H. Fisher. Conceptual clustering, learning from examples, and inference. In *Proceedings of the 4th International Machine Learning Workshop*, pages 38–49, 1987.
9. D. Gentner. Structure mapping: A theoretical framework for analogy. *Cognitive Science*, 7(2):155–170, 1983.
10. D. Gentner. The mechanisms of analogical learning. In S. Vosniadou and A. Ortony, editors, *Similarity and Analogical Reasoning*, pages 199–241. Cambridge University Press, 1989.
11. M.L. Gick and K.J. Holyoak. Schema induction and analogical transfer. *Cognitive Psychology*, 15(1):1–38, 1983.
12. R. Goebel. A sketch of analogy as reasoning with equality hypotheses. In K.P. Jantke, editor, *Analogical and Inductive Inference*, volume 397 of *Lecture Notes on Computer Science*, pages 243–253. Springer, 1989.

13. R. Greiner. Learning by understanding analogies. *Artificial Intelligence*, 35:81–125, 1988.
14. M. Kerber, E. Melis, and J. Siekmann. Analogical reasoning based on typical instances. in Proceedings of the IJCAI-Workshop on Principles of hybrid reasoning and representation, July 1993. Chambéry, France.
15. M. Kerber, E. Melis, and J. Siekmann. Reasoning with assertions *and* examples. AAAI Spring Symposium on AI and Creativity, März 1993. Stanford, California, USA.
16. S. C. Kleene. *Introduction to Metamathematics*. Van Nostrand, Amsterdam, The Netherlands, 1952.
17. G. Lakoff. *Women, Fire and Dangerous Things*. The University of Chicago Press, London, 1987.
18. J. McCarthy. Programs with common sense. In Marvin Minsky, editor, *Semantic Information Processing*, pages 403–418. MIT Press, Cambridge, Massachusetts, 1968.
19. E. Melis. Study of modes of analogical reasoning. TASSO-report 5, Gesellschaft für Mathematik und Datenverarbeitung, Birlinghoven, Germany, 1990.
20. C. Mervis and E. Rosch. Categorization of Natural Objects. *Annual Review of Psychology*, 32:89–115, 1981.
21. C.B. Mervis and E. Rosch. Family resemblance: Studies in the internal structure of categories. *Cognitive Psychology*, 7:573–605, 1975.
22. K.L. Myers and K. Konolige. Reasoning with analogical representations. In *Proceedings of the Third International Conference on Knowledge Representation KR'92*, pages 189–200, 1992.
23. L. Rips. Inductive judgements about natural categories. *Journal of Verbal Learning and Verbal Behavior*, 14:665–681, 1975.
24. S.J. Russell. *The Use of Knowledge in Analogy and Induction*. Pitman, London, 1989.
25. Y. Shoham. *Reasoning about Change - Time and Causation from the Standpoint of Artificial Intelligence*. MIT Press, Cambridge, Massachusetts, 1988.
26. A. Sloman. Interactions between philosophy and artificial intelligence: The role of intuition and non-logical reasoning in intelligence. *Artificial Intelligence*, 2:209–225, 1971.
27. A. Sloman. Afterthoughts on analogical representation. In *Proceedings of Theoretical Issues in Natural Language Processing*, 1975.
28. A. Urquhart. Many-valued logic. In D. Gabbay and F. Guenthner, editors, *Handbook of Philosophical Logic*, chapter III.2, pages 71–116. D.Reidel Publishing Company, Dordrecht, The Netherlands, 1986. Volume III: Alternatives to Classical Logic.
29. E.J. Weiner. A knowledge representation approach to understanding metaphors. *Computational Linguistics*, 10(1):1–14, 1984.
30. P. Winston. Learning by creating and justifying transfer frames. *Artificial Intelligence*, 10(2):147–172, 1978.

The Normative Reconstruction of Analogy Argumentation in Judicial Decisions: a Pragma-Dialectical Perspective

Harm Kloosterhuis
Erasmus University Rotterdam
Faculty of Law
The Netherlands

Abstract

In this paper I try to describe, analyse and explain the elements of analogy-argumentation advanced in the justification of legal decisions, to explore the criteria for the assessment of this type of argumentation, to relate that to the general theory of law and to do all that in the framework of the pragma-dialectical approach to argumentation.

1   Analogy Argumentation as an Interpretative Argumentation Scheme

In the Dutch newspaper 'de Volkskrant' dated Saturday 29th May 1993 ex-journalist Herman Wigbold argues that the mildness of attitudes in society as a whole is abating, except amongst the judicial powers. Wigbold is offended by half-baked sentences, by the endlessly extended rights of the accused and by the fact that mistakes made by the judiciary lead to acquittal. Wigbold gives the following example of the latter. In Amsterdam in 1991 the 29 year-old M.C. was arrested, suspected of committing at least nine cases of rape and sexual assault. There was no doubt about his guilt. However, the man was acquitted because a sample of saliva was taken for a DNA-test without his consent. The test proved the guilt of the accused, but the judge acquitted him because his personal inviolability had been breached.

According to Wigbold a different decision would have been defensible. Because, he argues, anybody found in his car having drunk an excess of alcohol can be forced to take a bloodtest and there is no fundamental difference between taking a blood sample or taking a sample of saliva. In his opinion, the judge could have referred to this argument and nobody would have been surprised.

In legal terms Wigbolds complaint amounts to the following: he is surprised that in the opinion of the judge, the regulation regarding the taking of blood samples could not be applied *analogically* to the taking of a saliva sample for a DNA-test. Opting for this solution would amount to applying the interpretative argumentation scheme that is based on an analogy relationship. The problem of interpretation arising from this *analogy argumentation* is resolved by the judge by analogically applying an existing statutory rule that is meant for another (albeit comparable) case to an unregulated matter. The judge is hereby filling a gap in the legal system by *constructing* a new legal norm.

When a judge uses analogy argumentation in solving a problem of interpretation, the elements of this argumentation should, either explicitly or implicitly, be part of the argumentation that is put forward to justify the decision taken. It must then be established that relevant points of the non-regulated matter correspond with those in cases that are regulated. For this purpose, different types of interpretative arguments can be brought forward, each involving one or more specific methods of interpretation. In addition, the principle of equality will be - in most cases implicitly - a starting point of the argumentation.

Anybody who wants to pass judgment about the *acceptability* of the analogy argumentation in a legal decision, must solve a number of problems of analysis before he can arrive at an evaluation of this argumentation. First, he must *identify* the analogy argumentation as such. This means that he must establish *if* at a certain point in the legal decision analogy argumentation is evident and what function this argumentation fulfils in the legal decision. Second, he must *interpret* the analogy argumentation, which means that he must determine which judgment in the argumentation is defended as standpoint and which judgments are brought forward as explicit or implicit arguments in defence of the standpoint. Third, he must *analyse* the analogy argumentation, which means that he must examine what bearing the arguments have on each other and on the standpoint. In practice, the identification, interpretation and analysis of analogy argumentation seems to pose problems with respect to the rational reconstruction of argumentation in legal judgments.[1] It is not always clear if analogy argumentation is applied in legal judgments and which function this analogy argumentation fulfils in the motivation of the decision. Often, there is also a problem in determining which judgments must be interpreted as standpoint and argumentation and how the structure of the argumentation must be analysed.

In this paper I will indicate how the problems arising from the rational reconstruction of analogy argumentation can be tackled in a systematic manner. First, I will briefly discuss the connection between the questions asked during the reconstruction of analogy argumentation and the criteria for an evaluation of this argumentation. Second, assuming a choice has been made for pragma-dialectical norms for the evaluation of analogy argumentation, I will describe a method for analysing legal analogy argumentation.

---

[1]   A rational reconstruction of the argumentation in a legal judgment is a reformulation in view of a testing of the rationality of the judgment. See MacCormick and Summers (1991).

2   The Acceptability of Analogy Argumentation

Giving a rational reconstruction of analogy argumentation means reconstructing the argumentation in such a way that its acceptability can be evaluated. The way in which the argumentation is reconstructed is therefore indissolubly related to the standards that are used to assess this argumentation. Generally speaking, two approaches can be discerned in the literature on legal analogy argumentation.

In the *first* approach to analogy argumentation the reconstruction of the argumentation is mainly directed towards the substantive question of the acceptability of the premisses used in the decision. The acceptability of the premiss in which the comparison is expressed, is paramount in this analysis. In the *second* approach to analogy argumentation the reconstruction is not only aimed at establishing the acceptability of the premisses but also at answering the *formal* question of whether the derivation in analogy argumentation is logically valid. Authors such as Alexy (1983), Klug (1982) and Tammelo (1969) consider analogy argumentation to be a specifically legal form of reasoning that can be reconstructed as a logically valid argument.

What these two approaches have in common is that the analysis of the analogy argumentation is exclusively aimed at the question as to whether the analogy argumentation has been applied in the right way. Besides the substantive question of the acceptability of the premisses and the formal question of the validity of the derivation in the analogy reasoning, these analyses do not take account of a preliminary - one could say procedural - standard of judgment: the question of whether in a concrete case the analogy argumentation may be used by a judge to solve a certain interpretation problem.[2][3]

This incompletion in the reconstruction of analogy argumentation can be remedied by using the standards for the judgment of argumentation schemes as formulated in the pragma-dialectical argumentation theory. According to these standards, the acceptability of analogy reasoning must be tested by using the following two criteria:[4]

---

[2]   See Wróblewski (1974) for the differences between normal inference rules and systematic inference rules, such as, analogy reasoning. Wróblewski assumes that the use of normal inference rules is not bound by legal standards, whereas, the use of systematic inference rules is.

[3]   In the more practically orientated literature relating to legal argumentation also little attention is given to this procedural aspect. Henket and van den Hoven, for example, indicate in their book *Juridische vaardigheden in argumentatief verband* (1990), that the following two issues should be answered when raising the question whether a regulation may be analogically applied: what is the tenor of the regulation? and in what way does the non-regulated case differ from the regulated case and what are their similarities? They note that establishing what the tenor of a regulation is, is a question of interpretation, but they do not give any precise indications about the considerations this interpretation should be based on.

[4]   See Van Eemeren, Grootendorst and Kruiger (1986) and Van Eemeren and Grootendorst (1992).

(1)   Is the analogy argumentation an admissible argumentation scheme?
(2)   Has the analogy argumentation been correctly applied?

When evaluating an argument, first it must be determined, whether the argumentation scheme has been *well chosen*. This means that it must be assessed whether the chosen scheme belongs to those argumentation schemes that are in principle allowed to be used in a given discussion context to defend a certain standpoint. Only if this question can be answered positively, can it be judged whether the chosen analogy argumentation has been *correctly applied* - the second criterion of evaluation. This means, among other things, that it must be determined whether the comparative criterion used is valid and whether the cases compared using this comparative criterion are indeed comparable.[5]

Next, I will indicate the consequences of choosing these two general standards for the judgment of analogy argumentation for the analysis of legal analogy argumentation. I will attempt, hereby, to relate the insights gained from the pragma-dialectical argumentation theory to those from legal theory, so that these can be integrated into a method for analysing legal analogy argumentation.

3   Guidelines for the Analysis of Analogy Argumentation

In the analysis of legal analogy argumentation that is aimed at an evaluation of the argumentation using the two aforementioned general standards of judgment, six stages of analysis can be distinguished.[6]

The *first stage* of the analysis consists in determining the precise *content* of the legal norm, establishing to which *field of law* it must be accounted and to which *type of legal norm* it belongs.

In legal theory it is generally accepted that the field of law to which a legal norm belongs, can determine whether or not this norm can be applied analogically. An example which catches the eye most is the prohibtion of analogy in criminal law: extending the meaning of norms of criminal law on the grounds of analogy contradicts the nature of criminal law.[7] Another example of a field of law in which limitations exist to analogically applying legal rules is tax law. It is assumed that analogical application is only admissible if it is advantageous to those liable for tax. Finally, in civil law it is assumed, in principle, that the possibility does exist to analogically

---

5   See Kloosterhuis (1991) for an elaboration of this second standard of judgment.

6   These stages in analysing analogy argumentation form a further elaboration of the directives for the analysis and judgment of argumentation. See Van Eemeren, Grootendorst and Kruiger (1983) and Van Eemeren, Grootendorst and Kruiger (1986).

7   See Henket (1991) for an elaboration of the prohibtion of analogy in Criminal Law.

applying legal rules (there are, as will be shown later, in civil law also limitations to the possibility of applying rules analogically).[8]

Besides having knowledge of the field of law to which a norm belongs it is also of importance to know to which *type* of legal norm the analogically applied norm belongs. In legal theory it is pointed out that it depends on the type norm whether analogical application is admissible. Aarnio (1987:106), for example, assumes that in this respect a distinction should be made between substantive and procedural norms. In his opinion, the principle of legal certainty prevails in the interpretation of procedural standards. Another distinction which is important in this respect, is the distinction between duty imposing norms, permissive norms, power conferring norms and norms of adjudication. Peczenik (1989:396) indicates, for example, that duty imposing norms must be restrictively interpreted and, may only in exceptional cases be analogically applied.

Furthermore, for an adequate analysis of analogy argumentation it is of importance to reconstruct the analogically applied legal norm as a conditional norm-sentence. A starting-point for the *second stage* in the analysis is that every legal norm can be reconstructed as a *conditional norm* that consists of a conditional combination of the descriptions of certain types of legal facts (the application condition or norm-condition) and their legal consequence. As a basis for our reconstruction of this conditional norm-sentence we use the following *model* of a conditional norm-sentence, in which all the components of a norm are made explicit:[9]

> *if* ... (description of legal facts (1), *then* ... (determination of legal consequence: norm-subject (2), norm-object (3), deontic modality (4), indication of time and place (5).

> *model of conditional norm-sentence*

The norm-condition or condition of application (1), which indicates under which conditions the determination of legal consequence is applied, is formulated in the description of the legal facts. In the norm-subject component (2) a description is given of the person whose action is being regulated. The norm-object or a description of the act (3) is a description of the action being regulated. The deontic modality (4) indicates the character of the norm: is the action described in the norm prohibited, obligated or permitted. The indication of time and place (5) shows us when and where the action is regulated. Finally, the sixth norm-component: the conditional connection, which makes explicit the nature of the connection between the description of the legal facts and the determination of legal consequence.

---

[8] See MacCormick and Summers (1991) for a summary of the possibilities and limitations of analogy argumentation in the different fields of law.

[9] See Brouwer (1990) for the components of conditional legal norm and the problems in reconstructing thereof.

This model of the conditional norm-sentence functions as a heuristic tool in the reconstruction of the conditional norm-sentence: different components of the legal norm can be reconstructed with the aid of this model; any implicit components can be made explicit.

The reconstruction of the conditional norm-sentence is of twofold importance for a systematic analysis of analogy argumentation.

First, the question of the content of the analogical application can be formulated much more precisely if there is knowledge of the conditional norm-sentence: the question is no longer solely which legal norm should be applied analogically, but also, which *component* of the legal norm should be applied analogically (this will be elaborated upon later in the third stage of analysis).

Second, the reconstruction of the conditional norm-sentence clearly identifies the *nature* of the conditional connection between the description of the legal facts and the determination of the legal consequence. There are three types of connections. The description of the legal facts may already contain *sufficient* conditions for the determination of the legal consequence; the conditional connection can then be characterized as an implication. The description of the legal facts may also have formulated *necessary* conditions for the determination of the legal consequence; the conditional connection can then be characterized as a replication. Finally, both *necessary* and *sufficient* conditions may have been formulated in the description of the legal facts for the determination of the legal consequence; the conditional connection can then be characterized as an equivalence.

Insight into the nature of the conditional connection is of importance with regard to the question whether analogy argumentation is an applicable argumentation scheme (judgment standard no.1). An analogical application which refers to the description of the legal facts is only possible in legal theory if sufficient conditions have been specified in the description of the legal facts. An analogical application which refers to the description of the legal facts is not possible when this description only mentions necessary or necessary and sufficient conditions.[10]

When the conditional norm-sentence is reconstructed, it can be determined precisely during the *third stage* of analysis which component of the legal norm is at stake in the analogical application. We can hereby indicate the following alternatives.

First, the analogical application can refer to one or more norm-conditions from the description of the legal facts. As previously stated, the nature of the conditional connection between the description of the legal facts and the determination of the legal consequence determines the extent to which the analogical application is admissible in these cases.

Second, analogical application can also refer to one or to more components in the determination of legal consequence. It is also important in judging whether analogy argumentation is an admissible argumentation scheme, to know to which element of the determination of legal consequence the analogical application applies.

---

[10]   See Klug (1982) and Peczenik (1989).

There are, for example, limitations to the analogical application of definitions of time and place.[11] Moreover, the judge has more freedom to interpret vague standards than relatively concrete ones. It can be assumed that the latter has consequences for the permissibility of analogy argumentation.

On the grounds of the outcome of the first three stages of analysis it is possible during the *fourth stage* to precisely determine the result of the analogical application: the legal standard constructed by the judge to solve the legal question. For a precise analysis of the analogy argumentation this constructed legal standard should be formulated as a conditional norm-sentence: the comparison between the existing legal norm and the constructed legal norm can then be optimally assessed.

In the *fifth stage of analysis* both the explicit and implicit arguments that are put forward to defend the analogical application are identified. As one would expect on the grounds of the constitutional legal requirements, a legal norm constructed by the judge will have to be well motivated in the legal decision: the plausibility that the new rule fits in with the legal system must be established.

This argumentation can be very complex and not always so easy to analyse. It is not often clear which elements form the argumentation, how these elements are connected and what their legal status is. The solution to this problem of analysis is important when answering the question as to whether the analogy argumentation has been applied correctly: only when one has insight into the different elements of the argumentation, can the question whether this argumentation has been correctly applied be put forward.

First of all, the arguments must be identified. This identification is sometimes complicated by the fact that the arguments are not always made explicit: in these cases the implicit arguments must be identified and explicitized. To defend the analogical application, the relevant comparisons between the non-regulated case and the regulated cases are made by means of *interpretative* arguments that have a systematic or teleological character.[12]

Only when all the explicit and implicit arguments brought forward to defend the analogical application have been identified, can the structure of the argumentation be analysed. This happens in the *sixth* and final stage of the analysis. In most cases the argumentation brought forward to justify the analogical application will have a complex structure, consisting of arguments in which legal rules, legal principles and interpretative judgments are expressed. During this final stage of analysis, the relationship between the different arguments, and between the arguments and the standpoint, must be examined.[13]

---

[11]   See Peczenik (1989) and others.

[12]   See Alexy (1983) among others.

[13]   For a comparison of the interpretation of argumentation in legal pleas and the analysis of the structure of argumentation, see Van Eemeren et al. (1991).

The analysis of analogy argumentation using the six aforementioned stages of analysis results in an analytical overview, in which is spelled out which constructed legal norm the judge is defending, which existing legal norm is being analogically applied, which analogy relationship is being assumed and, further, with which arguments the analogical application is being justified. This analytical overview forms the starting-point for an evaluation of the argumentation, in which the two aforementioned standards of judgment take a central position: was the judge allowed to use analogy argumentation and if so, did the judge apply this argumentation correctly?

4   Conclusion

In this paper I have indicated how a number of problems in the reconstruction of legal analogy argumentation can be tackled by making use of insights gained from both argumentation theory and legal theory. The analysis procedure with its different stages not only enables the analyst to reconstruct legal analogy argumentation in a systematic way, but also in an *efficient* manner. For example, if it appears during the first stage of analysis that the analogical application of a rule is not allowed, then a further judgment and therefore, a further reconstruction of the analogy argumentation is no longer necessary.

This is the case, for example, in Wigbolds argumentation that was reproduced in the beginning of this paper. In short, Wigbolds analogy argumentation boils down to the following. The judge should have permitted the taking of a saliva sample against the will of the accused, in order to make a DNA-test possible. His argument runs as follows: anybody found in his car having drunk an excess of alcohol can be forced to take a bloodtest and there is no fundamental difference between taking a blood sample or taking a sample of saliva for a DNA-test. Wigbold here anticipates criticism resulting from an application of the second standard of judgment: are we talking of comparable cases here? But he neglects to consider the *first and preliminary judgment issue* in his analysis: the question whether in this specific case the judge is allowed to solve a problem of interpretation by using analogy argumentation. Or, to put it more concretely: is the judge allowed to analogically apply a rule that violates the rights of the accused to cases for which the rule was not intended? Everybody who is acquainted with the rules of criminal law knows how problematic this is - and contrary to what is claimed by Wigbold, the sentence in this respect could not have been different.

Bibliography

Aarnio, A.
1987   *The Rational as Reasonable. A Treatise on Legal Justification*. Dordrecht: Reidel.

Alexy, R.
1983   *Theorie der juristischen Argumentation. Die Theorie des rationalen Diskurses als Theorie der juristischen Begründung*. Frankfurt: Suhrkamp.

Brouwer, P.W.
1990   Samenhang in recht. Een analytische studie. Groningen: Wolters-
                   Noordhoff.
Eemeren, F.H. et al.
1991   Argumenteren voor juristen. Het analyseren en schrijven van juridi-
                   sche betogen en beleidsteksten. Groningen: Wolters Noordhoff.
                   (Tweede, herziene druk. 1e druk 1987.)
Eemeren, F.H. van en E.T. Feteris (eds.)
1991   Juridische argumentatie in analyse. Opstellen over argumentatie en
                   recht. Groningen: Wolters Noordhoff.
Eemeren, F.H. van en R. Grootendorst
1992   Argumentation, Communication and Fallacies. Hillsdale: Erlbaum.
Eemeren, F.H. van, R. Grootendorst en T. Kruiger.
1983   Het analyseren van een betoog. Argumentatieleer 1. Groningen:
                   Wolters Noordhoff.
Eemeren, F.H. van, R. Grootendorst en T. Kruiger
1986   Drogredenen. Argumentatieleer 2. Groningen: Wolters Noordhoff.
Henket, M.
1991   'Het analogieverbod in het strafrecht'. In: F.H. van Eemeren en
                   E.T. Feteris (red.), Juridische argumentatie in analyse. Opstellen
                   over argumentatie en recht. Groningen: Wolters Noordhoff.
Henket, M. M. en P.J van den Hoven
1990   Juridische vaardigheden in argumentatief verband. Groningen: Wol-
                   ters Noordhoff.
Kloosterhuis, H.
1991   'Argumentatie bij analogische toepassing van rechtsregels'. In: F.H.
                   van Eemeren en E.T. Feteris (red.), Juridische argumentatie in
                   analyse. Opstellen over argumentatie en recht. Groningen: Wolters
                   Noordhoff.
Klug, U.
1982   Juristische Logik, Berlin/Heidelberg/New York: Springer-Verlag.
MacCormick, D. N., R. S. Summers (eds.)
1991   Interpreting Statutes, A Comparative Study. Aldershot: Dartmouth.
Peczenik, A.
1989   On Law and Reason. Dordrecht: Reidel.
Tammelo, I.
1969   Outlines of modern legal logic. Wiesbaden: Franz Steiner Verlag.
Wróblewski, J.
1974   'Legal Syllogism and Rationality of Judicial Decision'. Rechtstheo-
                   rie, 1974, p. 33 - 46.

Formal Reasoning about Modules, Reuse and their Correctness*

Christoph Kreitz,[1] Kung-Kiu Lau[2] and Mario Ornaghi[3]

[1] Fachgebiet Intellektik, Fachbereich Informatik
Technische Hochschule Darmstadt
Alexanderstr. 10, D-64283 Darmstadt, Germany
kreitz@intellektik.informatik.th-darmstadt.de
[2] Department of Computer Science
University of Manchester, Manchester M13 9PL, UK
kung-kiu@cs.man.ac.uk
[3] Dipartimento di Scienze dell'Informazione
Universita' degli studi di Milano, Via Comelico 39/41, Milano, Italy
ornaghi@hermes. mc.dsi.unimi.it

Abstract. We present a formalisation of modules that are *correct*, and (*correctly*) *reusable* in the sense that composition of modules preserves both correctness and reusability. We also introduce a calculus for formally reasoning about the construction of such modules.

1 Introduction

Modular programming has been around for a long time, and has more recently evolved into object-oriented programming (e.g. [11]). Various forms of *modules* and *objects* can be found in a variety of modern programming languages. They are important because they facilitate structured design as well as code *reuse*. However, for formal program development, i.e. developing programs that are *formally correct* wrt their (formal) specifications, current modular and object-oriented programming languages lack a suitable formal semantics in our view, even though some of them do have type-system based rules for program composition (see e.g. [3, 13]).

In this paper, we define modules as first-order theories (with isoinitial semantics) that contain logic programs. We express reuse in model-theoretic terms. Our formalisation provides a logical basis for formal reasoning about modules, reuse, and their *correctness*; it is therefore suitable for formal program development in an object-oriented fashion. We shall use the logic programming paradigm for simplicity, but our approach is very general, and should be applicable to any programming paradigm with suitable logical semantics.

The paper is organised as follows. In Section 2, we introduce the background theory together with our notation and terminology. In Section 3, we briefly define our notion of a module, its correctness and reusability. Then in Section 4, which is the bulk of the paper, we show rules to construct modules in such a way that correctness and reusability are both preserved. The rules are based on the notion of dependency type, and refine and expand our first results in earlier sections.

---

* This work was done during the second author's visit to TH Darmstadt, supported by the European Union HCM project LPST, contract no. 93/414. He wishes to thank Prof Wolfgang Bibel for his invitation and hospitality.

2 Background

2.1 Many-sorted First-order Languages and Interpretations

We will denote a many-sorted *signature* by $\Sigma = \langle S, F, R \rangle$, where S is a set of *sort* symbols, F is a set of *function* declarations and R is a set of *relation* declarations. Signatures will be introduced as shown in the following example.

Example 1. The signature of stacks is defined thus:

SORTS: *Elem, Stacks*;

FUNCTIONS: *empty* : $\rightarrow$ *Stacks*;
            push : (*Elem, Stacks*) $\rightarrow$ *Stacks*;

RELATIONS: = : (*Elem, Elem*);
            = : (*Stacks, Stacks*);

Each function declaration has the form $f : a \rightarrow s$, where f is the declared function symbol, $a = (s_{i_1}, \ldots, s_{i_n})$ is its arity, and s is its sort, and each relation declaration has the form $r : a$, where r is the declared relation symbol, and $a = (s_{k_1}, \ldots, s_{k_m})$ is its arity. Constants will be functions with empty arity, i.e. $c : \rightarrow s$. We allow *overloaded* relation symbols, i.e. symbols that have many declarations, like overloaded identity $=: (Elem, Elem)$, $=: (Stacks, Stacks)$ in the example.

A Σ-*interpretation* $\mathcal{I}$ interprets the symbols of the signature as follows:

- Every sort symbol s is interpreted as a domain $s^{\mathcal{I}}$.
- Every function declaration $f : a \rightarrow s$ is interpreted as a function $f^{\mathcal{I}} : a^{\mathcal{I}} \rightarrow s^{\mathcal{I}}$, where, if $a = (s_1, \ldots, s_n)$, then $a^{\mathcal{I}}$ is $s_1^{\mathcal{I}} \times \cdots \times s_n^{\mathcal{I}}$.
- Every relation declaration $r : a$ is interpreted as a relation $r^{\mathcal{I}} \subseteq a^{\mathcal{I}}$.

The first-order language L_{Σ} generated by a signature Σ (and by some set of sorted variables) is defined in the usual way, as are terms and formulas. We will write $\tau : s$ to indicate that the sort of τ is s. Formulas of L_{Σ} will also be called Σ-formulas.

An *assignment* **a** over an interpretation $\mathcal{I}$ is a map that associates with every variable $x : s$ an element $\mathbf{a}(x)$ of $s^{\mathcal{I}}$. In a given $\mathcal{I}$ and **a**, we have that:

- Every term $\tau : s$ denotes an element of $s^{\mathcal{I}}$, that we will call the value of τ and denote by $val_{\mathcal{I}}(\tau, \mathbf{a})$. Since the value of a ground term τ does not depend on **a** it will be denoted by $val_{\mathcal{I}}(\tau)$.
- Every formula evaluates to *true* or *false*. We will write $\mathcal{I} \models_{\mathbf{a}} F$ to indicate that the formula F evaluates to *true* (in $(\mathcal{I}, \mathbf{a})$). If F is closed, we will simply write $\mathcal{I} \models F$.

A Σ-homomorphism $h : \mathcal{I}_1 \rightarrow \mathcal{I}_2$, for Σ-interpretations $\mathcal{I}_1$ and $\mathcal{I}_2$, is an S-indexed family of maps $h_s : s^{\mathcal{I}_1} \rightarrow s^{\mathcal{I}_2}$ that preserve functions and relations, i.e. $h(f^{\mathcal{I}_1}(\alpha)) = f^{\mathcal{I}_2}(h(\alpha))$[4] and $\alpha \in r^{\mathcal{I}_1} \Rightarrow h(\alpha) \in r^{\mathcal{I}_2}$. Starting from (many-sorted) homomorphisms, isomorphism and isomorphic embeddings can be defined in the usual way. To indicate that two Σ-interpretations $\mathcal{I}_1$ and $\mathcal{I}_2$ are isomorphic, we will write $\mathcal{I}_1 \sim \mathcal{I}_2$.

---

[4] We omit the subscript s in h_s. Moreover, if $\alpha = \alpha_1, \ldots, \alpha_n$, then $h(\alpha_1, \ldots, \alpha_n)$ means $h(\alpha_1), \ldots, h(\alpha_n)$.

2.2 Theories and Logic Programs

A Σ-theory T is a set of axioms. As usual, a model $\mathcal{M}$ of a theory T is an interpretation such that $\mathcal{I} \models T$. We say that $\mathcal{M}$ is *reachable* iff, for every sort s and $\alpha \in s^{\mathcal{M}}$, there is a ground term τ of sort s such that $\alpha = val_{\mathcal{M}}(\tau)$. $\mathcal{M}$ is an *isoinitial model* of T iff, for every other model $\mathcal{N}$ of T, there is a unique isomorphic embedding $\mu : \mathcal{M} \to \mathcal{N}$.

For theories with reachable models, the following very useful theorem holds:

Theorem 1. *Let T be a theory that has a reachable model $\mathcal{M}$. Then $\mathcal{M}$ is an isoinitial model of T iff, for every closed atomic formula A:*

$$T \vdash A \quad or \quad T \vdash \neg A \tag{1}$$

A theory T is *atomically complete* if it satisfies (1). For such a theory any two reachable models are isomorphic, since they are both isoinitial.[5]

We will often use theories related to logic programs (see e.g. [10] for a general introduction to logic programming). We say that a predicate symbol is *defined* by a logic program P, if it occurs in the head of at least one clause of P. If a predicate is not defined by P, it is said to be *open* (in P).

We associate with a program P the theory $T(P) = free(P) \cup Ocomp(P)$, where $free(P)$ is the set of freeness axioms for the constant and function symbols of P; and $Ocomp(P)$, the *open completion* of P, is the set of completed definitions of the predicates defined by P.[6]

2.3 Signature and Theory Morphisms

First-order theories are an institution [6]. Here we briefly introduce relevant properties of an institution, and some terminology.

A signature $\Sigma = \langle S, F, R \rangle$ is a *subsignature* of $\Delta = \langle S', F', R' \rangle$, written $\Sigma \preceq \Delta$, iff $S \subseteq S'$, $F \subseteq F'$ and $R \subseteq R'$.

Let Σ be a signature and ρ be a function that maps the symbols of Σ into another set of symbols. Then ρ *generates* $\rho(\Sigma) = \langle \rho(S), \rho(F), \rho(R) \rangle$, where $\rho(S)$ is the image of S, $\rho(F)$ contains the declarations $\rho(f) : \rho(a) \to \rho(s)$ such that $f : a \to s \in F$, and $\rho(R)$ the declarations $\rho(r) : \rho(a)$ such that $r : a \in R$.

If ρ is a map from the symbols of Σ to those of another signature Δ, and if $\rho(\Sigma) \preceq \Delta$, then we say that ρ is a *signature morphism* from Σ to Δ. If it is an injective map, then ρ is called a *signature extension*. If bijective, it is called a *renaming*.

Let $\rho : \Sigma \to \Delta$ be a signature morphism. The interpretations $int(\Delta)$ of Δ are related to the interpretations $int(\Sigma)$ of Σ by the *reduct* operation $|\rho : int(\Delta) \to int(\Sigma)$ defined thus:

$$s^{\mathcal{I}|\rho} = \rho(s)^{\mathcal{I}}$$
$$f^{\mathcal{I}|\rho} : a \to s = \rho(f)^{\mathcal{I}} : \rho(a) \to \rho(s)$$
$$r^{\mathcal{I}|\rho} : a = \rho(r)^{\mathcal{I}} : \rho(a)$$

$\mathcal{I}|\rho$ will be called the ρ-*reduct* of $\mathcal{I}$.

---

[5] [1] gives a proof of Theorem 1. Isoinitial semantics is closely related to initial semantics used in algebraic ADTs [14] (see also [7, 12]).

[6] We can also use induction principles in $T(P)$.

If $\rho(\sigma) = \sigma$ for every symbol σ, i.e. $\Sigma \preceq \Delta$, we get the usual notion of reduct to a subsignature (see e.g. [5]). In this case, the reduct will also be denoted by $\mathcal{I}|\Sigma$.

It is immediate to extend a morphism $\rho : \Sigma \to \Delta$ to a map $\rho : L_\Sigma \to L_\Delta$ and the following satisfaction condition (see [6]) can be easily proved:

Theorem 2. *For every closed Σ-formula F and every Δ-interpretation $\mathcal{I}$:*

$$\mathcal{I} \models \rho(F) \Leftrightarrow \mathcal{I}|\rho \models F \tag{2}$$

Finally, given a signature morphism $\rho : \Sigma \to \Delta$ and a theory $T \subseteq L_\Sigma$, we denote the ρ-image of T by $\rho(T)$.

3 Modules

Our notion of a module is that it consists of a (first-order) theory $\mathcal{F}$ which axiomatises a problem domain, together with logic programs $prog_1$, $prog_2$, ..., that can be specified and derived in $\mathcal{F}$. It can be visualised as follows:

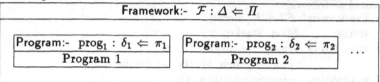

We call $\mathcal{F}$ a *framework*. It contains axiomatisations of abstract data types (ADTs), *any* relevant axioms for reasoning about these ADTs (e.g. induction axioms), as well as other axioms for reasoning about the domain (e.g. axioms in a library framework stating that each book has a title, a list of authors, and so on). In particular $\mathcal{F}$ axiomatises function and relation symbols Δ.

$\mathcal{F}$ may be *open*, i.e. it may have *parameters* Π. In this case, $\mathcal{F}$ may axiomatise the symbols in Δ in terms of Π, and we write $\mathcal{F} : \Delta \Leftarrow \Pi$, where $\Delta \Leftarrow \Pi$ is called $\mathcal{F}$'s *dependency type*.

Similarly, a program $prog_i$ in $\mathcal{F}$ may be open, and computes relations in δ in terms of its parameters π, where δ are the defined relations of P and π are its open relations (see Section 2). In this case, we write $prog_i : \delta \Leftarrow \pi$, and $\delta \Leftarrow \pi$ is also called $prog_i$'s dependency type.

A module M therefore contains two kinds of parameters: *framework parameters*, viz. Π, and *program parameters*, viz. π. This means that the issues of *reuse* and *correctness* occur at two levels. M can be reused with different instantiations of both Π and π, i.e. M can be composed with other modules that (partially or completely) define Π and π in different ways.

The correctness of a module M refers to the consistency of its framework $\mathcal{F}$, as well as the correctness of each program $prog_i$ wrt its specification in $\mathcal{F}$. Moreover, when M is composed with another module, the composition must preserve both the consistency of $\mathcal{F}$ and the correctness of all the $prog_i$'s. By contrast, current object-oriented programming lacks a formalisation of reuse and correctness.

Our notion of a module is in spirit similar to that of a *class* in object-oriented programming, which is an ADT implementation together with procedures for the ADT. Indeed we can also have an *object* by endowing a module with an internal state.

3.1 Frameworks

A framework is a first-order theory that axiomatises an ADT, or a problem domain, or both. A framework is *closed* iff it axiomatises an isoinitial model (that is unique up to isomorphism). Otherwise, it is *open*. An open framework axiomatises a class of isoinitial models.

Example 2. The following is an example of a *closed* framework:

Framework $\mathcal{K}_{a,b}$;
SORTS: $Elab$;
FUNCTIONS: $a, b :\to Elab$;
D-AXIOMS: $\neg a = b$;

The interpretation $\mathcal{I}$ such that $Elab^{\mathcal{I}} = \{a, b\}$ is the isoinitial model of $\mathcal{K}_{a,b}$.

Example 3. As an example of an *open* framework, the theory of stacks can be axiomatised as follows.

Framework $STACK(Elem)$;
SORTS: $Elem, Stacks$;
FUNCTIONS: $empty : \to Stacks$;
           $push : (Elem, Stacks) \to Stacks$;
D-AXIOMS: $\neg empty = push(x, S)$;
           $push(x_1, S_1) = push(x_2, S_2) \to x_1 = x_2 \wedge S_1 = S_2$;
P-AXIOMS: $\exists x, y \, (\neg x = y)$;

where we have omitted the universal closure of the axioms, upper-case variables are of sort *Stacks*, and lower-case variables are of sort *Elem*.

We have a parameter *Elem* and two kinds of axioms. The *p*-axioms, that state some assumptions on the parameter *Elem* (in this case we require that *Elem* contains at least two distinct elements). The *d*-axioms completely characterise the non-parametric symbols in terms of the parameter. This means that via *Elem* the framework *STACK* axiomatises a class of isoinitial models. That is, whenever we choose a particular set of elements (fixed by a closed framework), we get an axiomatisation of stacks with the chosen elements. For example, if we instantiate *Elem* by the sort *Elab* of $\mathcal{K}_{a,b}$, we get a closed (reachable and atomically complete) framework that has the isoinitial model $\mathcal{I}$ such that $Elem^{\mathcal{I}}$ is $\{a, b\}$, $Stacks^{\mathcal{I}}$ is the set of stacks with elements from $\{a, b\}$, and $push^{\mathcal{I}}$ is the usual push operation on stacks.

A framework also contains a list of *theorems*. For example, in $STACK(Elem)$ we can prove $\forall S \exists x, y (\neg push(x, S) = push(y, S))$, and add it as a theorem to the framework.

In the sequel, a framework will be indicated by $\mathcal{F}(\Pi) = \langle \Sigma, \mathcal{T}, \mathcal{TH}, \Pi \rangle$, where Π is the set of parameters, Σ the signature, $\mathcal{T}$ the theory (a set of axioms), and $\mathcal{TH}$ the current set of proved theorems. We will write $\mathcal{F}$ if no confusion arises.

In a closed framework, Π is empty. In an open one, any sort symbol s or relation declaration $r : a$ can be used as a parameter. Identity $=: (s, s)$ is a parameter iff s is a parameter too. The set of *p*-axioms is simply the restriction $\mathcal{T}|\Pi$, namely the set of axioms that contain *only* symbols of Π. The other axioms are the *d*-axioms. Note that in closed frameworks there are no *p*-axioms.

3.2 Programs

In a module M, the logic programs $prog_i$ are not just ordinary programs. They are open programs that must be correct in all possible instances of M. We call this kind of correctness *steadfastness*, and a model-theoretic formalisation can be found in [9]. Here we give a brief explanation so as to make clear what kind of programs we can find in a module.

If $\mathcal{F}$ in M is a closed framework, then a closed program $P : \delta$ (that is, a program without open relations) is correct in M iff its minimum Herbrand model is isomorphic to the intended model of $\mathcal{F}$ restricted to the signature of P. Steadfastness of an open program can be defined as follows:

Definition 3. A program $P : \delta \Leftarrow \pi$ is *steadfast* in a closed framework $\mathcal{C}$ iff, for every closed program $Q : \delta_1$ that is correct in $\mathcal{C}$, if $\pi \subseteq \delta_1$ and $\delta_1 \cap \delta = \emptyset$ then $P \cup Q : \delta \cup \delta_1$ is correct in $\mathcal{C}$.

A program $P : \delta \Leftarrow \pi$ is *steadfast* in an open framework $\mathcal{F}$ iff it is steadfast in every closed instance $\mathcal{F} \cup \mathcal{C}$, where $\mathcal{C}$ is a closed framework.

This means that $P : \delta \Leftarrow \pi$ is (closed and) correct wrt to its specification whenever $\mathcal{F} : \Delta \Leftarrow \Pi$ becomes closed, and correct closed programs for computing π are supplied. The programs for computing the framework parameters in π can be supplied only after the instantiation $\mathcal{F} \cup \mathcal{C}$ by a closed framework $\mathcal{C}$, while those for the rest of π can be supplied either before or after such an instantiation.

Example 4. In $\mathcal{LIST} : \Delta \Leftarrow Elem, \lhd$, the framework of lists which axiomatises the usual list relations Δ, e.g. $ord, perm, \ldots$, with parametric element type $Elem$ and ordering relation $\lhd$, sorting can be defined in the usual way:[7]

$$sort(x, y) \leftrightarrow perm(x, y) \wedge ord(y) \tag{3}$$

From (3) we can synthesise the following open program for $sort$:[8]

Framework:- $\mathcal{LIST} : \Delta \Leftarrow Elem, \lhd$

Program:- merge sort : $sort \Leftarrow merge, split$

$sort(nil, nil) \leftarrow$
$sort(x.nil, x.nil) \leftarrow$
$sort(x.y.A, W) \leftarrow split(x.y.A, I, J) \wedge sort(I, U) \wedge sort(J, V) \wedge$
$\qquad merge(U, V, W)$

$$(4)$$

The meaning of the program parameters *split* and *merge* can be partially specified. For example, to avoid an overdetermined specification, $merge(U, V, W)$ may be required to give an ordered W if U and V are ordered, whilst its meaning may be left open for unordered U and V. Such partially specified relations are called *internal parameters*. They are used only in program derivation, but not as parameters of the framework. By a suitable partial characterisation of *split* and *merge*, we can prove that the program (4) is steadfast wrt the specification (3), that is, it is correct for every interpretation of the internal parameters that

---

[7] This definition is adequate, see Section 4.2 later.

[8] The synthesis of steadfast programs is discussed in [8].

satisfies their partial definitions, as well as in every closed instance of $\mathcal{LIST}$. Thus it can be composed with different programs for *merge* and *split* to yield different sorting algorithms based on merging, e.g. insertion sort, and with every program for deciding the framework parameter $\lhd$, in any possible instance of the parameters *Elem* and $\lhd$.

Thus a (steadfast) program prog$_i$ is not only correct wrt to its specification within its parent module M, but it can also be reused correctly in different instances of M, i.e. in different instances of the framework $\mathcal{F}$ in M.

Finally, it is important to note that the correctness of prog$_i$ in M is defined in model-theoretic terms ([9]), by comparing the (minimum Herbrand) model of prog$_i$ with the (isoinitial) model of $\mathcal{F}$.

4 Module Construction

To construct a module M, we need to construct the framework $\mathcal{F}$ and the programs prog$_i$ that it contains. We want the programs developed in M to be *correct* and (*correctly*) *reusable*, that is to remain correct in larger modules built up by composition with M. To achieve this, we must construct $\mathcal{F}$ and prog$_i$ in special ways: prog$_i$ must be correct in M, and the framework $\mathcal{F}$ must be *adequate*, i.e. it must behave in a sound way with respect to composition.

Note that we have two levels of rules. The first level contains the rules for program synthesis, that build up programs that are formally correct wrt a framework, where correctness has a precise formal definition. The second level contains rules to build up frameworks, in order to build a formalisation that intuitively meets some informal problem specification. For this second level, we can only have informal correctness: the intended model that we incrementally construct intuitively agrees with the problem domain and the specification.

In the preceding section, we have briefly discussed the construction of correct and reusable prog$_i$. In the rest of the paper, we will concentrate on the construction of such a framework $\mathcal{F}$.

4.1 Operations on Frameworks

First we introduce basic operations on frameworks that will be employed in framework construction.

Renaming Let $\mathcal{F} = \langle \Sigma, \mathcal{T}, \mathcal{TH}, \Pi \rangle$ be a framework. Let ρ be a bijective map from the symbols of Σ to another set of symbols. We define $\rho(\mathcal{F}) = \langle \rho(\Sigma), \rho(\mathcal{T}), \rho(\mathcal{TH}), \rho(\Pi) \rangle$. We will introduce renaming by the syntax shown in the following example:

> **Framework** $\mathcal{LIST}_0(Elem)$;
> RENAMES $\mathcal{STACK}_0(Elem)$;
> RENAMING: *Stacks* $\mapsto$ *List*;
> *empty* $\mapsto$ *nil*;
> *push* $\mapsto$ ·;

Composition Let $\mathcal{F}(\Pi) = \langle \Sigma, \mathcal{T}, \mathcal{TH}, \Pi \rangle$ be an open framework. We say that a sort symbol is used in the parameters Π iff it belongs to Π or if it occurs in some declaration of Π, whilst a declaration is used iff it belongs to Π.

The composition of $\mathcal{F}(\Pi)$ with $\mathcal{G}(\Pi_1) = \langle \Sigma_1, T_1, T\mathcal{H}_1, \Pi_1 \rangle$ is performed through a Π-renaming ρ defined thus: $\rho(\Sigma)$ is a Π-*renaming for* Σ and Σ_1 if for every σ used in Π, $\rho(\sigma)$ belongs to Σ_1, whilst the sort symbols and declarations not used in Π are mapped into names that do not occur in Σ_1. For a Π-renaming ρ, we can build ρ-amalgamation of the signatures

$$\rho(\Sigma) + \Sigma_1 = \langle \rho(S) \cup S_1, \rho(F) \cup F_1, \rho(R) \cup R_1 \rangle$$

where $\Sigma = \langle S, R, F \rangle$ and $\Sigma_1 = \langle S_1, R_1, F_1 \rangle$.

Now, let ρ be a Π-renaming. The composition operation $\mathcal{F}[\rho, \mathcal{F}_1]$ is defined if the following proof obligation is satisfied:

$$\rho(T|\Pi) \subseteq T_1 \cup T\mathcal{H}_1 \tag{5}$$

That is, we require that the p-axioms become theorems or axioms of $\mathcal{F}_1$.
If $\mathcal{F}[\rho, \mathcal{F}_1]$ is defined, the resulting composite $\mathcal{F}[\rho, \mathcal{F}_1](\Pi_c) = \langle \Sigma_c, T_c, T\mathcal{H}_c, \Pi_c \rangle$ is defined as follows:

- $\Pi_c = \Pi_1$, i.e. the parameters of the composite are the (possible) ones of $\mathcal{F}_1$.
- $\Sigma_c = \rho(\Sigma) \cup \Sigma_1$, i.e. the new signature enriches Σ_1 by $\rho(\Sigma)$.
- $T_c = T_1 \cup (\rho(T) - T\mathcal{H}_1)$, i.e. we consider the union of the axioms, and we eliminate possible axioms of $\rho(T)$ that are theorems of $\mathcal{F}_1$.[9]
- $T\mathcal{H}_c = T\mathcal{H}_1 \cup (\rho(T\mathcal{H}) - T_1)$

If ρ is the identity, we indicate the corresponding composition by $\mathcal{F}[\mathcal{F}_1]$.

Example 5. $STACK(Elem)$ can be composed with $\mathcal{K}_{a,b}$ by the renaming $Elem \mapsto Elab$ ($\rho(\sigma) = \sigma$ for the other symbols). We will use the syntax:

Framework $STACK_{a,b}$;
COMPOSES: $STACK(Elem)[\mathcal{K}_{a,b}]$
WITH: $Elem \mapsto Elab$;
OBLIGATION: $\exists x, y (\neg x = y)$;

The composition is defined. Indeed, the proof obligation $\exists x, y (\neg x = y)$ immediately follows from the axiom $\neg a = b$ of $\mathcal{K}_{a,b}$.

Composition is associative, i.e.

Theorem 4. *If* $(\mathcal{F}_0[\rho_1, \mathcal{F}_1])[\rho_2, \mathcal{F}_2]$ *is defined, then* $\mathcal{F}_0[\rho_2\rho_1, \mathcal{F}_1[\rho_2, \mathcal{F}_2]]$ *is defined and*

$$(\mathcal{F}_0[\rho_1, \mathcal{F}_1])[\rho_2, \mathcal{F}_2] = \mathcal{F}_0[\rho_2\rho_1, \mathcal{F}_1[\rho_2, \mathcal{F}_2]]$$

where $\rho_2\rho_1$ *is the composition of the two renamings.*

Proof. The proof follows from the definition of composition. □

Moreover, it is related to framework extension.

Definition 5. Let $\mathcal{F}(\Pi) = \langle \Sigma, T, T\mathcal{H}, \Pi \rangle$ and $\mathcal{G}(\Pi') = \langle \Sigma', T', T\mathcal{H}', \Pi' \rangle$ be two frameworks, and $\rho : \Sigma \to \Sigma'$ be a signature extension. $\mathcal{G}(\Pi')$ is a ρ-*extension* of $\mathcal{F}$, written $\mathcal{F}(\Pi) \preceq_\rho \mathcal{G}(\Pi')$,[10] iff

- $\rho(T) \subseteq T' \cup T\mathcal{H}'$;
- for every σ not used in Π, $\rho(\sigma)$ does not occur in Π'.

We can easily prove the following theorem:

---

[9] This does not mean that axioms are independent, but simply that they remain disjoint from theorems, to avoid redundancies.

[10] If ρ is the identity, we write $\mathcal{F}(\Pi) \preceq \mathcal{G}(\Pi')$.

Theorem 6. *Let* $\mathcal{F}(\Pi)$ *and* $\mathcal{F}_1(\Pi_1)$ *be two frameworks. If* $\mathcal{F}[\rho, \mathcal{F}_1]$ *is defined,* *then* $\mathcal{F}(\Pi) \preceq_\rho \mathcal{F}[\rho, \mathcal{F}_1](\Pi_1)$ *and* $\mathcal{F}_1(\Pi_1) \preceq \mathcal{F}[\rho, \mathcal{F}_1](\Pi_1)$.

These theorems will be used to introduce sound *extension rules* as compositions of a particular kind.

Theory morphism, as defined in [6], can be seen as ρ-compositions, where ρ is any signature morphism (i. e. it may be non-injective). Moreover, our definition of composition essentially coincides with that of [6]. However, the theory of institutions is not useful for our purposes, since it does not deal with the problem of preserving consistency.[11]

4.2 Constructing Adequate Frameworks

As we mentioned in Section 3, correctness of a framework $\mathcal{F}$ in a module M refers to $\mathcal{F}$'s consistency. Thus to ensure that M is correct and (correctly) reusable, we need to have rules for framework construction that preserve consistency. Moreover, as we indicated in Section 3.2, in order to keep steadfastness meaningful for programs, we also have to preserve the isoinitial model of $\mathcal{F}$. To achieve both of these targets, we introduce the notion of *adequate frameworks*, and use framework composition as the main operation to construct frameworks in such a way that adequacy is preserved. We shall first define adequacy, and then give a set of simple adequate frameworks that can be used as basic units to build composite adequate frameworks incrementally.

Definition 7. A closed framework is *adequate* iff it has a reachable isoinitial model. (In the sequel, we will simply say 'closed framework' instead of 'adequate closed framework'.)

Definition 8. $\mathcal{G}$ is an *adequate closed ρ-extension* of a closed framework $\mathcal{C}$, written $\mathcal{C} \sqsubseteq_\rho \mathcal{G}$,[12] iff $\mathcal{G}$ has a reachable isoinitial model $\mathcal{I}$,[13] i.e. it is closed, and the reduct $\mathcal{I}|\rho$ is an isoinitial model of $\mathcal{C}$.

Definition 9. An open framework $\mathcal{F}(\Pi)$ is *adequate* iff, for every closed framework $\mathcal{C}$, if $\mathcal{F}[\rho, \mathcal{C}]$ is defined, then $\mathcal{C} \sqsubseteq \mathcal{F}[\rho, \mathcal{C}]$.

Definition 10. $\mathcal{G}(\Pi')$ is an *adequate ρ-extension* of an adequate open framework $\mathcal{F}(\Pi)$, written $\mathcal{F}(\Pi) \sqsubseteq_\rho \mathcal{G}(\Pi')$, iff it is adequate, $\mathcal{F}(\Pi) \preceq \mathcal{G}(\Pi')$ and, for every closed framework $\mathcal{C}$, if $\mathcal{G}[\rho_1, \mathcal{C}]$ is defined, then $\mathcal{F}[\rho_1\rho, \mathcal{C}]$ is defined and $\mathcal{F}[\rho_1\rho, \mathcal{C}] \sqsubseteq \mathcal{G}[\rho_1, \mathcal{C}]$.

It is easy to see that renaming does not change any feature of an adequate framework. Furthermore, we have the following theorem about framework composition:

Theorem 11. *If* $\mathcal{F}(\Pi)$ *and* $\mathcal{G}(\Pi')$ *are adequate, and* $\mathcal{F}[\rho, \mathcal{G}]$ *is defined, then* $\mathcal{F}[\rho, \mathcal{G}]$ *is adequate.*

Proof. Let $\mathcal{F}[\rho, \mathcal{G}][\rho_1, \mathcal{C}]$ be defined. Since $\mathcal{F}$ and $\mathcal{G}$ are adequate, $\mathcal{C} \sqsubseteq \mathcal{G}[\rho_1, \mathcal{C}] \sqsubseteq \mathcal{F}[\rho_1\rho, \mathcal{G}[\rho_1, \mathcal{C}]] = \mathcal{F}[\rho, \mathcal{G}][\rho_1, \mathcal{C}]$. Thus $\mathcal{F}[\rho, \mathcal{G}](\Pi')$ is adequate. □

---

[11] Simple forms of axioms, like equational or strict Horn axioms, are guaranteed to be consistent.

[12] If ρ is the identity, we write $\sqsubseteq$ instead of $\sqsubseteq_\rho$.

[13] Note that if $\mathcal{C} \sqsubseteq_\rho \mathcal{G}$, then $\mathcal{G}$ is consistent by definition, since it has a model.

Moreover we can prove that, if $\mathcal{F}(\Pi)$ and $\mathcal{G}(\Pi')$ are adequate, $\mathcal{F}[\rho, \mathcal{G}]$ does not introduce new constant or function symbols with sorts in $\mathcal{G}$, and $\mathcal{F}[\rho, \mathcal{G}]$ is defined, then $\mathcal{G}(\Pi') \sqsubseteq \mathcal{F}[\rho, \mathcal{G}](\Pi')$. This and Theorem 11 show that framework composition behaves in a sound way. It provides the basis for our strategy to start with small, simple, but adequate closed and open frameworks, and then use composition to construct larger, adequate frameworks incrementally.

The first example of a basic framework is $free(K)$, where K is a set of constant and function symbols. K may be open or closed, depending on whether it contains open sort symbols or not. A sort s is open (wrt K) iff no declaration of the form $f : a \rightarrow s$ belongs to K. We have the following theorem on the adequacy of K.

Theorem 12. *If K is closed, then $free(K)$ is an adequate closed framework, and the structure freely generated by K is an isoinitial model of $free(K)$.*

If K contains open sorts $s_1, \ldots, s_n$, then $free(K)(s_1, \ldots, s_n)$ is an adequate open framework.

Proof. If K is closed, we get our result by Theorem 1, since $free(K)$ is atomically complete and the structure freely generated by K is a reachable model of it.

If K is open, then let $t(u), t'(v)$ be terms with variables u, v of open sorts, $free(K)[\mathcal{C}]$ be a closed instance, and $t(\alpha) = t'(\beta)$ be ground. Since $\mathcal{C} \vdash \alpha = \beta$ or $\mathcal{C} \vdash \neg \alpha = \beta$, we can prove that $free(K)[\mathcal{C}] \vdash t(\alpha) = t'(\beta)$, or $free(K)[\mathcal{C}] \vdash \neg t(\alpha) = t'(\beta)$. Moreover $free(K)[\mathcal{C}]$ is reachable. Then, by Theorem 1, it is closed. $\square$

The second kind of frameworks are explicit definitions.

Theorem 13. *Let Σ be a signature containing a relation declaration $r : (a, s)$, a function declaration $f : a \rightarrow s$, and the sort symbols of a, s. Then*

$$\mathcal{D}_{r(x, f(x))} = \langle \Sigma, \{\forall x\, r(x, f(x)), \forall x \exists! y\, r(x, y)\}, \emptyset, \{r : (a, s), a, s\}\rangle$$

is an adequate open framework.

Let Σ be the signature of a definition axiom $r(x) \leftrightarrow F(x)$, where $F(x)$ is a quantifier-free formula that does not contain r, and let Π be the signature of $F(x)$ (i.e. Π does not contain r). Then

$$\mathcal{D}_{r(x) \leftrightarrow F(x)}(\Pi) = \langle \Sigma, \{\forall x\, (r(x) \leftrightarrow F(x))\}, \emptyset, \Pi\rangle$$

is an adequate open framework.

Proof. Suppose $\mathcal{D}_{r(x, f(x))}[\mathcal{C}]$ is defined, where $\mathcal{C}$ is closed. Then $\mathcal{C}$ proves the p-axiom $\forall x \exists! y\, r(x, y)$. Hence the explicit definition $\forall x\, r(x, f(x))$ determines an expansion $\mathcal{I}_f$ of the isoinitial model $\mathcal{I}$ of $\mathcal{C}$. Since $r(x, f(x))$ is atomic, we can prove that $\mathcal{I}_f$ is an isoinitial model of $\mathcal{D}_{r(x, f(x))}[\mathcal{C}]$.

Suppose $\mathcal{D}_{r(x) \leftrightarrow F(x)}[\mathcal{C}]$ is defined. Since $\mathcal{D}_{r(x) \leftrightarrow F(x)}[\mathcal{C}]$ contains only the new axiom $\forall x(r(x) \leftrightarrow F(x))$, it is reachable and, since $F(x)$ is quantifier-free, it is atomically complete. Hence it has an isoinitial model $\mathcal{I}$. The reduct of $\mathcal{I}$ to the language of $\mathcal{C}$ is reachable, hence it is an isoinitial model of $\mathcal{C}$. $\square$

The third basic framework is the theory $T(P)$ of a program P. Note that P is open if it contains open sorts. For closed programs we have the following theorem.

Theorem 14. *If P is a definite closed program, and it terminates for every ground atomic goal $\leftarrow A$, then the theory $T(P)$ is an adequate closed framework, and the minimum model of P is an isoinitial model of $T(P)$.*

Proof. The minimum model of P is reachable and, by termination and completeness of finite failure, $\mathcal{T}(P)$ is atomically complete. By Theorem 1 we get our result. □

Example 6. Consider the following framework $\mathcal{PA}_0$:

Framework $\mathcal{PA}_0$;
SORTS: Nat;
FUNCTIONS: $0 : \to Nat$;
            $s : (Nat) \to Nat$;
D-AXIOMS: $free(0, s : Nat)$;
            $sum(x, y, z) \leftrightarrow (y = 0 \wedge x = z)\vee$
                    $\exists i, v \, (y = s(i) \wedge z = s(v) \wedge sum(x, i, v))$;

By Theorem 14, $\mathcal{PA}_0$ is an adequate closed framework, since it is the theory $\mathcal{T}(\mathbf{sum})$ of the usual (terminating) program $\mathbf{sum}$ for computing sum.

Now we can prove $\forall x, y \, \exists! z \, sum(x, y, z)$ (by induction) in $\mathcal{PA}_0$. So the composition $\mathcal{PA}_1 = \mathcal{D}_{sum(x,y,x+y)}[\mathcal{PA}_0]$ is defined, and by Theorem 11, it is an adequate closed framework that extends $\mathcal{PA}_0$ by the (computable) function $+$.

Now consider the following framework $\mathcal{LIST}_0$:

Framework $\mathcal{LIST}_0(Elem)$;
SORTS: $List, Elem$;
FUNCTIONS: $nil : \to List$;
           $. : (Elem, List) \to List$;
D-AXIOMS: $free(nil, .)$;

By Theorem 12, $\mathcal{LIST}_0$ is an adequate open framework. Then by Theorem 11, the composition $\mathcal{LIST}_0[\mathcal{PA}_1]$ is an adequate closed framework.

We could also associate open frameworks with open programs. However, here the situation is more complex. In particular, even non-adequate open frameworks could be used in a sound way for framework composition if suitable extra conditions are satisfied by the composition operation. To treat these cases, we will define adequacy with respect to a *dependency type*. This will yield a calculus to reason about framework composition, that extends our results so far.

4.3 Rules for Dependency Types

By Theorem 1, an equivalent definition of adequacy of an open framework $\mathcal{F}(\Pi)$ is the following: for every framework $\mathcal{C}$, if $\mathcal{C}$ is reachable and *atomically complete*, then $\mathcal{F}[\mathcal{C}]$ is reachable and *atomically complete*. We want to generalise this notion of adequacy: for every framework $\mathcal{C}$, if $\mathcal{C}$ is reachable and satisfies some property possibly stronger than atomic completeness, then $\mathcal{F}[\mathcal{C}]$ is reachable and satisfies some *other* property stronger than atomic completeness. To this end, we introduce the following notion of *constructive evaluation*.

Definition 15. Let I be a set of closed atomic or negated Σ-formulas,[14] and F be a closed formula. Then I *(constructively) evaluates* F in Σ, written $ev(I, F)$, iff
_____

[14] A negated formula is any formula of the form $\neg H$, with H possibly non-atomic.

- F is atomic or negated, and $F \in I$.
- $F = A \wedge B$ and: $ev(I, A)$ and $ev(I, B)$.
- $F = A \vee B$ and: $ev(I, A)$ or $ev(I, B)$.
- $F = \exists x A(x)$ and: $ev(I, A(t))$ for some ground term t of the appropriate sort.
- $F = \forall x A(x)$ and: $ev(I, A(t))$ for every ground term t of the appropriate sort.

A (possibly closed) framework $\mathcal{F} = \langle \Sigma, T, \mathcal{TH}, \Pi \rangle$ *evaluates* F iff F is evaluated by the set of closed atomic or negated formulas classically provable by T. To indicate that $\mathcal{F}$ evaluates F, we will write $\mathcal{F} : F$.

For example, to say that C is atomically complete, we can write $C : \forall x \, (r(x) \vee \neg r(x))$ for every relation symbol r of the signature.

We will abbreviate $\forall x \, (r(x) \vee \neg r(x))$ by $\mathbf{dec}(r : a)$, where a is the arity of r, and, if Π is a set of relation declarations, $\mathbf{dec}(\Pi)$ will contain $\mathbf{dec}(r : a)$ for every declaration $r : a \in \Pi$ and $\mathbf{dec}(=: (s, s))$ for every sort symbol used in Π.

Now adequacy can be treated by dependency types as follows.

Definition 16. We say that $\mathcal{F}(\Pi) = \langle \Sigma, T, \mathcal{TH}, \Pi \rangle$ *preserves reachability* iff, for every framework $\mathcal{G}$ with at least one reachable model, $\mathcal{F}[\mathcal{G}]$ (when defined) has a reachable model.

Definition 17. We say that $\mathcal{F}(\Pi)$ has *evaluation type* $A \Rightarrow B$, written $\mathcal{F} : A \Rightarrow B$, iff, for every reachable $\mathcal{G}$ such that $\mathcal{F} \preceq \mathcal{G}$, we have $\mathcal{G} : A$ entails $\mathcal{G} : B$.

Theorem 18. *Let $\mathcal{F}(\Pi)$ be a parametric framework such that, for every sort s used in Π, it does not contain constant or function symbols of sort s. If $\mathcal{F}(\Pi)$ preserves reachability, and has dependency type $\mathbf{dec}(\Pi) \Rightarrow \mathbf{dec}(\Delta)$, where Δ contains the relation symbols not in Π, then $\mathcal{F}(\Pi)$ is adequate.*

Proof. Let C be closed and $\mathcal{F}[C]$ be defined. Since $C : \mathbf{dec}(\Pi)$ and for the sorts used in Π no new constant or function symbol is added, $\mathcal{F}[C] : \mathbf{dec}(\Pi)$. We get $\mathcal{F}[C] : \mathbf{dec}(\Delta)$ and hence $\mathcal{F}[C]$ is atomically complete. Since $\mathcal{F}[C]$ is reachable, it has an isoinitial model $\mathcal{I}$. Since $\mathcal{I}|\Sigma_C$ is a reachable model of C (no new constant or function symbols of sorts in C are added), it is an isoinitial model of C. $\square$

The idea here is to extend Theorem 11 to a more general kind of dependency types, in order to get a richer calculus for module composition. Here we consider the beginnings of such a calculus, which contains rules for reasoning about dependency types and for using them in framework composition.

Rules for Dependency Types

These rules work for any kind of framework.

$$\frac{\mathcal{F}_1(\Pi_1) : A \Rightarrow B, \quad \mathcal{F}_2(\Pi_2) : \rho(B) \Rightarrow C}{\mathcal{F}_1[\rho, \mathcal{F}_2](\Pi_2) : \rho(A) \Rightarrow C} \tag{6}$$

$$\frac{\mathcal{F}_1(\Pi_1) : B \Rightarrow C, \quad \mathcal{F}_2(\Pi_2) : A \Rightarrow \rho(B)}{\mathcal{F}_1[\rho, \mathcal{F}_2](\Pi_2) : A \Rightarrow \rho(C)} \tag{7}$$

$$\frac{\mathcal{F}(\Pi) : A \Rightarrow B, \quad \mathcal{F}(\Pi) : B \Rightarrow C}{\mathcal{F}(\Pi) : A \Rightarrow C} \tag{8}$$

Extension-by-composition Rules. In the following rules $\mathcal{F}(\Pi)$ is adequate, whilst $\mathcal{E}(\Pi')$, which we call an *extension framework*, may be non-adequate. Δ are the relation symbols of $\mathcal{E}$ not in Π'. We require that $\mathcal{E}(\Pi')$ preserves reachability, the (possible) function or constant symbols of sort used in Π' are parameters, and in $\mathcal{E}[\rho, \mathcal{F}](\Pi)$ and $\mathcal{F}[\mathcal{E}](\Pi)$ there are no constant or function symbols of sort in Π. Soundness follows from the previous rules, from the fact that reachability is preserved by $\mathcal{E}$ and from Theorem 18.

$$\frac{\mathcal{E}(\Pi') : B \Rightarrow \mathbf{dec}(\Delta), \quad \mathcal{F}(\Pi) : \mathbf{dec}(\Pi) \Rightarrow \rho(B)}{\mathcal{F}(\Pi) \sqsubseteq \mathcal{E}[\rho, \mathcal{F}](\Pi)} \tag{9}$$

$$\frac{\mathcal{F}(\Pi) : \mathbf{dec}(\Pi) \Rightarrow B, \quad \mathcal{E}(\Pi) : B \Rightarrow \mathbf{dec}(\Delta), \quad \mathcal{T}_{\mathcal{F}} | \Pi \sqsubseteq \mathcal{T}_{\mathcal{E}} | \Pi}{\mathcal{F}(\Pi) \sqsubseteq \mathcal{F}[\mathcal{E}](\Pi)} \tag{10}$$

We can show that if $\mathcal{F}(\Pi)$ and $\mathcal{G}(\Pi')$ preserve reachability, then $\mathcal{F}[\mathcal{G}](\Pi')$ (if defined) preserves reachability too. Therefore we can build extension frameworks by reasoning with their dependency types, starting from a few kinds of basic extension frameworks.

As an example, the theory $\mathcal{T}(P)$ of an open program $P : \delta \Leftarrow \pi$ preserves reachability and, by reasoning about termination properties of P, we can prove $\mathcal{T}(P) : A \Rightarrow B$ for suitable A and B.

Example 7. Let us consider the open program $\mathbf{iter}(iter \Leftarrow unit, op)$:

$$\begin{aligned} iter(X, 0, U) &\leftarrow unit(U) \\ iter(X, s(I), Y) &\leftarrow iter(X, I, W), op(X, W, Y) \end{aligned} \tag{11}$$

We associate **iter** with the extension framework:

Framework $\mathcal{ITER}(unit, op, 0, s, Nat)$;

SORTS: Nat;

FUNCTIONS: $0 :\rightarrow Nat$; $s : (Nat) \rightarrow Nat$;

RELATIONS: $unit \; : (Nat)$;

                $iter, op : (Nat, Nat, Nat)$;

D-AXIOMS: $\mathbf{CD}(iter, iter)$

P-AXIOMS: $free(0, s : Nat)$; $construct(0, s : Nat)$;

where $\mathbf{CD}(iter, iter)$ is the completed definition of $iter$ in the program $\mathbf{iter}$ and $construct(0, s : Nat)$ requires that, if a composition introduces operations on Nat different from $0, s$, then for every ground term $t : Nat$ there is a numeral n (of the form $0, s(0), \ldots$) such that $n = t$ is provable. The framework $\mathcal{ITER}$ is not adequate. However we can prove that it has the following dependency types:

$$\exists! x \, unit(x) \wedge \forall x, y \exists! z \, op(x, y, z) \Rightarrow \forall x, y \exists! z \, iter(x, y, z)$$
$$\forall x, y \, (x = y \vee \neg x = y) \wedge \exists! x \, unit(x) \wedge \forall x, y \exists! z \, op(x, y, z) \Rightarrow$$
$$\forall x, y, z \, (iter(x, y, z) \vee \neg iter(x, y, z))$$

where the second type follows from the first since $\emptyset : \forall x, y (x = y \vee \neg x = y) \wedge \forall x, y \exists! z \, iter(x, y, z) \Rightarrow \forall x, y, z \, (iter(x, y, z) \vee \neg iter(x, y, z))$.

We can use this framework to introduce product in $\mathcal{PA}_1$ (see Example 6). First we introduce in $\mathcal{PA}_1$ the explicit definition $zero(X) \leftrightarrow X = 0$ (we get an adequate expansion by Theorem 13). Since $\mathcal{PA}_1 : \forall x, y (x = y \vee \neg x = y) \wedge \exists! x \, zero(x) \wedge \forall x, y \exists! z \, sum(x, y, z)$, by Rule (9) we get $\mathcal{PA}_1 \sqsubseteq \mathcal{ITER}[\rho, \mathcal{PA}_1]$, where ρ maps $unit$ to $zero$, op to sum and $iter$ to $prod$.

By rule (7) we also get that $\mathcal{ITER}[\rho, \mathcal{PA}_1] : \forall x, y \exists! z \, prod(x, y, z)$, and we could now use $\mathcal{ITER}$ to obtain exponentiation, and so on.

Moreover, we can extend the notion of a steadfast program to frameworks with dependency types. In this way we get a way of composing programs within a framework whilst preserving their steadfastness (i.e. correctness) properties. Furthermore, these properties will also be preserved by framework composition. Thus we get both *correctness* within a framework and *reusability* with respect to module composition. We will not discuss this issue any further here. Rather, we will focus on a calculus of dependency types (see Section 4.4 later).

4.4 Towards a Calculus of Dependency Types

To conclude the paper, we introduce a preliminary, incomplete calculus for proving dependency types. It allows us, for example, to prove $\forall u, v\, (u = v \vee \neg u = v) \wedge \forall x \exists! y\, r(x, y) \Rightarrow \forall x, y\, (r(x, y) \vee \neg r(x, y))$, where $\exists! u\, H(u)$ is an abbreviation of $\exists u\, H(u) \wedge \neg \exists a, b\, (H(a) \wedge H(b) \wedge \neg a = b)$.

The proof system works on sequents of the form $\Gamma \vdash_{ev} F$, where Γ is a sequence of formulas and F is a formula. For simplicity, we consider the one-sorted case, but the extension to the many-sorted case should be obvious. Free variables are allowed in Γ and F.

A dependency type $A \Rightarrow B$ is *valid*[15] iff, for every signature Δ containing those of A and B, every ground instance $A^* \Rightarrow B^*$ using terms from Δ, and every Δ-theory T, if $T : A^*$, then $T : B^*$. We say that $A_1, \ldots, A_n \vdash_{ev} F$ is *evaluation-valid* iff $A_1 \wedge \ldots \wedge A_n \Rightarrow F$ is a valid dependency type. We can show that, if $\Gamma, A \vdash_{ev} B$ is evaluation-valid and Γ is a set of atomic or negated closed formulas provable in a framework $\mathcal{F}(\Pi)$, then $\mathcal{F}(\Pi) : A \Rightarrow B$. This shows how we can use the calculus below for our purposes.

Now we can introduce our proof system, which is sound with respect to evaluation-validity.

(i) The axiom $A \vdash_{ev} A$, and the usual structural rules of sequent calculus.
(ii) The left and right rules for $\wedge$, $\vee$, $\exists$, $\forall$, but not those for negation and implication.
(iii) If the formulas in Γ and A are atomic or negated, and $\Gamma \vdash A$ in classical sequent calculus, then we have the axiom $\Gamma \vdash_{ev} A$.
(iv) The cut rule.

Note that (iii) introduces a recursively enumerable set of axioms, and so the set of provable sequents still remains recursively enumerable. Also, due to (iii), cut cannot be eliminated. However, it can be restricted to atomic or negated formulas. The axiom $A \vdash_{ev} A$ is trivially sound wrt evaluation-validity. The soundness of the axioms introduced by (iii) follows from the fact that a theory evaluates an atomic or closed formula iff it proves it. The inference rules ((ii) and (iv)) can be shown to be sound, by proving that if the premises of a rule are evaluation-valid, then the consequence is evaluation-valid (we omit the proof for brevity).

This system is a subsystem of the sequent calculus for classical logic. It is not a subsystem of intuitionistic logic. Conversely, intuitionistic logic is not a subsystem of our system. Finally, our system is not a complete system for proving valid dependency types.

---

[15] With A and B possibly containing free variables.

5   Conclusion

We have presented a formalisation of a module that is correct, and correctly reusable in the sense that module composition preserves both correctness and reusability. We have also given a calculus for constructing modules that behave soundly with respect to composition. Our notion of a module is very general, and it should be equally applicable to any modular or object-oriented programming paradigm.

To extend our work, we intend to continue our study of dependency types, and expand the preliminary calculus we have introduced here. For instance, we could introduce $\Rightarrow$ in the formulas of a sequent $\Gamma \vdash_{ev} F$, and study how the rules for implication behave. At present the implication $A \rightarrow B$ is read as an abbreviation of $\neg(A \wedge \neg B)$ and we do not have rules for it.

Moreover, we could use stronger requirements in the definition of evaluation set, that is we could further require that all the universally quantified formulas that are involved belong to I. A calculus for this kind of strong evaluation would give a logic that is intermediate between intuitionistic logic and classical logic. However, strong principles like the Gregorzyck principle become unsound with respect to strong evaluation. Thus it is not clear if the stronger calculus is an advantage. The best choice may well be to use a mixture of different calculi.

References

1. A. Bertoni, G. Mauri and P. Miglioli. On the power of model theory in specifying abstract data types and in capturing their recursiveness. *Fundamenta Informaticae* VI(2):127–170, 1983.
2. A. Brogi, P. Mancarella, D. Pedreschi and F. Turini. Modular logic programming. *ACM TOPLAS* 16(4):1361-1398, 1994.
3. K.M. Bruce. A paradigmatic object-oriented programming language: Design, static typing and semantics. *J. Functional Programming* 4(2):127–206, 1994.
4. M. Bugliesi, E. Lamma and P. Mello. Modularity in logic programming. *J. Logic Programming* 19,20:443–502, 1994. Special issue: Ten years of logic programming.
5. C.C. Chang and H.J. Keisler. *Model Theory*. North-Holland, 1973.
6. J.A. Goguen and R.M. Burstall. Institutions: Abstract model theory for specification and programming. *J. ACM* 39(1):95–146, 1992.
7. K.K. Lau and M. Ornaghi. On specification frameworks and deductive synthesis of logic programs. In L. Fribourg and F. Turini, editors, *Proc. LOPSTR 94 and META 94*, *LNCS* 883, pages 104–121, Springer-Verlag, 1994.
8. K.K. Lau, M. Ornaghi and S.-Å. Tärnlund. The halting problem for deductive synthesis of logic programs. In P. van Hentenryck, editor, *Proc. 11th Int. Conf. on Logic Programming*, pages 665–683, MIT Press, 1994.
9. K.K. Lau, M. Ornaghi and S.-Å. Tärnlund. Steadfast logic programs. Submitted
10. J.W. Lloyd. *Foundations of Logic Programming*, Springer-Verlag, 1987.
11. B. Meyer. *Eiffel the Language*. Prentice Hall, 1992.
12. P. Miglioli, U. Moscato and M. Ornaghi. Abstract parametric classes and abstract data types defined by classical and constructive logical methods. *J. Symb. Comp.* 18:41-81, 1994.
13. J. Palsberg and M.I. Schwartzbach. *Object-Oriented Type Systems*. Wiley, 1994.
14. M. Wirsing. Algebraic specification. In J. Van Leeuwen, editor, *Handbook of Theoretical Computer Science*, pages 675–788. Elsevier, 1990.

A Tableau Calculus for First-Order Branching Time Logic

Wolfgang May[1] and Peter H. Schmitt[2]

[1] Institut für Informatik, Universität Freiburg, Germany,
may@informatik.uni-freiburg.de,
[2] Institut für Logik, Komplexität und Deduktionssysteme,
Universität Karlsruhe, Germany, pschmitt@ira.uka.de

Abstract. Tableau-based proof systems have been designed for many
logics extending classical first-order logic. This paper proposes a sound
tableau calculus for temporal logics of the first-order CTL-family. Until
now, a tableau calculus has only been presented for the propositional
version of CTL. The calculus considered operates with prefixed formulas
and may be regarded as an instance of a labelled deductive system. The
prefixes allow an explicit partial description of states and paths of a
potential Kripke counter model in the tableau. It is possible in particular
to represent path segments of finite but arbitrary length which are needed
to process reachability formulas. Furthermore, we show that by using
prefixed formulas and explicit representation of paths it becomes possible
to express and process fairness properties without having to resort to full
CTL*. The approach is suitable for use in interactive proof-systems.

1 Introduction

Interactive proof-systems for verification of processes are gaining increasing interest. A very popular approach is to use temporal logic. Following from the
observation that most of the specification can be expressed in first-order CTL,
an extension of existing proof-systems to temporal logic of the CTL-family seems
adequate. This paper presents an intuitive, straightforward extension of the first-
order tableau calculus to first-order CTL with additional fairness requirements
well-suited for use in an interactive proof-system. The main ideas are

- explicit representation of the "geographical" structure of a fictive model
  by way of naming of states and paths,
- encoding of this information in a special type of formulas,
- abstraction of path segments of unknown, but finite length in order to
  process and represent eventualities.

The paper is structured as follows: In section 2 the temporal logic CTL and the
notion of a Kripke-structure are reviewed. In section 3 the tableau semantics
is presented. Section 4 gives the tableau rules and casts a short glance on correctness and completeness: A complete calculus for first-order CTL cannot be
achieved. In section 5 fairness requirements are analyzed and included into the

---

[1] Most of this work has been done while the first author was student at Universität
Karlsruhe. At present his work at Universität Freiburg is supported by grant no.
GRK 184/1-96 of the Deutsche Forschungsgemeinschaft.

calculus and some further extensions are pointed out. Section 6 completes the work with some concluding remarks.

2 The Temporal Logic CTL

The base of first-order CTL is a language of first-order predicate logic, including the symbols "(" and ")", the boolean connectives $\neg$, $\wedge$, $\vee$, $\rightarrow$, the quantifiers $\forall$, $\exists$, and an infinite set of variables $\mathsf{Var} := \{x_1, x_2, \ldots\}$. A particular language is given by its *signature* Σ consisting of function symbols and predicate symbols with fixed arities $\mathrm{ord}(f)$ resp. $\mathrm{ord}(p)$. Terms, first-order formulas, and the notions of bound and free variables are defined in the usual way, $\mathrm{free}(\mathcal{F})$ denoting the set of variables occurring free in a set $\mathcal{F}$ of formulas.

A *substitution* (over a signature Σ) is a mapping $\sigma : \mathsf{Var} \rightarrow \mathsf{Term}_\Sigma$ where $\sigma(x) \neq x$ for only finitely many $x \in \mathsf{Var}$. $\sigma : \sigma(x) = t$ is written as $[x \leftarrow t]$. Substitutions are extended to terms and formulas as usual.

A *first-order interpretation* $\mathbf{I} = (I, \mathbf{U})$ over a signature Σ consists of a non-empty set $\mathbf{U}$ (*universe*) and a mapping I which maps every function symbol $f \in \Sigma$ to a function $I(f) : \mathbf{U}^{\mathrm{ord}(f)} \rightarrow \mathbf{U}$ and every predicate symbol $p \in \Sigma$ to a relation $I(p) \subseteq \mathbf{U}^{\mathrm{ord}(p)}$.

A *variable assignment* is a mapping $\chi : \mathsf{Var} \rightarrow \mathbf{U}$. For a variable assignment χ, a variable x, and $d \in \mathbf{U}$, the *modified* variable assignment χ_x^d is identical with χ except that it assigns the element $d \in \mathbf{U}$ to the variable x. Let Ξ denote the set of variable assignments.

The notion of an interpretation is extended to an *evaluation* $\mathbf{I} : \mathsf{Term}_\Sigma \times \Xi \rightarrow \mathbf{U}$:

$$\mathbf{I}(x, \chi) := \chi(x) \quad \text{for } x \in \mathsf{Var},$$
$$\mathbf{I}(f(t_1, \ldots, t_n), \chi) := (I(f))(\mathbf{I}(t_1, \chi), \ldots, \mathbf{I}(t_n, \chi))$$
$$\text{for } f \in \Sigma, \, \mathrm{ord}(f) = n \text{ and } t_1, \ldots, t_n \in \mathsf{Term}_\Sigma.$$

To indicate the truth of a formula F in an interpretation $\mathbf{I}$ under a variable assignment χ, the standard notation $\models_{\mathrm{FO}}$ (or simply $\models$) is used: Let s, t be terms, p a predicate symbol, $\mathrm{ord}(p) = n$, $t_1, \ldots, t_n$ terms, x a variable, A and B formulas. Then

$$
\begin{aligned}
(\mathbf{I}, \chi) &\models \mathsf{true} \quad , \\
(\mathbf{I}, \chi) &\models p(t_1, \ldots, t_n) &:&\Leftrightarrow \quad (\mathbf{I}(t_1, \chi), \ldots, \mathbf{I}(t_n, \chi)) \in I(p) \quad , \\
(\mathbf{I}, \chi) &\models \neg A &:&\Leftrightarrow \quad \text{not } (\mathbf{I}, \chi) \models A \quad , \\
(\mathbf{I}, \chi) &\models A \vee B &:&\Leftrightarrow \quad (\mathbf{I}, \chi) \models A \text{ or } (\mathbf{I}, \chi) \models B \quad , \\
(\mathbf{I}, \chi) &\models \exists x : A &:&\Leftrightarrow \quad \text{there is a } d \in \mathbf{U} \text{ with } (\mathbf{I}, \chi_x^d) \models A \quad .
\end{aligned}
$$

The symbols $A \wedge B := \neg(\neg A \vee \neg B)$, $A \rightarrow B := \neg A \vee B$ and $\forall x : F := \neg \exists x : \neg F$ are defined as usual.

Definition 1. A first-order *Kripke-structure* over a signature Σ is a triple $\mathbf{K} = (\mathbf{G}, \mathbf{R}, \mathbf{M})$ where $\mathbf{G}$ is a set of states, $\mathbf{R} \subseteq \mathbf{G} \times \mathbf{G}$ an *accessibility relation*, and for every $g \in \mathbf{G}$, $\mathbf{M}(g) = (M(g), \mathbf{U}(g))$ is a first-order interpretation of Σ with universe $\mathbf{U}(g)$. $\mathbf{G}$ and $\mathbf{R}$ are called the *frame* of $\mathbf{K}$.

In this paper, only Kripke-structures with constant universe (i.e. $\mathbf{U}(g) = \mathbf{U}(g')$ for all $g, g' \in \mathbf{G}$) are considered. The notion of a *variable assignment* is then defined as in the first-order case.

Definition 2. For a Kripke-structure $\mathbf{K}$ over a signature Σ the state-independent portion $\Sigma^c \subseteq \Sigma$ consists of all function symbols $f : (M(g))(f) = (M(g'))(f)$ for all $g, g' \in \mathbf{G}$ and all predicate symbols $p : (M(g))(p) = (M(g'))(p)$ for all $g, g' \in \mathbf{G}$. This induces a state-independent evaluation $\mathbf{K}(t)$ for $t \in \mathrm{Term}_{\Sigma^c}$.

Definition 3. A *path* p in a Kripke-structure $\mathbf{K} = (\mathbf{G}, \mathbf{R}, \mathbf{M})$ is a sequence $p = (g_0, g_1, g_2, \ldots)$, $g_i \in \mathbf{G}$ with $\mathbf{R}(g_i, g_{i+1})$ holding for all i. It induces a mapping $p : \mathbb{N} \to \mathbf{G}$ with $p(i) = g_i$. Let $p|_i := (g_i, g_{i+1}, \ldots)$.

The family CTL of temporal logics of branching time used in this paper is defined in [BMP81], [CE81], and [EH83] in its propositional version. It uses the unary modal operators $\square$ ("always"), $\Diamond$ ("sometimes"), $\circ$ ("nexttime"), the binary modal operator until, and two path-quantifiers $\mathbf{A}$ and $\mathbf{E}$. For this paper, only a short review of CTL – the basic logic of this family – is given. Two classes of formulas are distinguished: *state formulas* holding in states, and *path formulas* holding on paths:

Definition 4. The syntax of CTL-formulas is given as follows:
- (S0) Every first-order formula is a CTL-state formula.
- (S1) With F and G CTL-state formulas, $\neg F$, $F \wedge G$, and $F \vee G$ are CTL-state formulas.
- (P1) With F and G CTL-state formulas, $\circ F$, $\square F$, $\Diamond F$, and $(F$ until $G)$ are CTL-path formulas.
- (P2) With P a CTL-path formula, $\neg P$ is a CTL-path formula.
- (S2) With P a CTL-path formula, $\mathbf{A}P$ and $\mathbf{E}P$ are CTL-state formulas.
- (SQ) With F a CTL-state formula and x a variable, $\forall x : F$ and $\exists x : F$ are CTL-state formulas.
- (F)   Every CTL-state formula is a CTL-formula.

The definition shows that in CTL every modality (modal operators and negated modal operators) is immediately preceded by a path-quantifier. CTL* is obtained by weakening this requirement [EH83].

Definition 5. The truth of formulas, $\models_{\mathrm{CTL}}$ (or simply $\models$), in a first-order Kripke-structure $\mathbf{K} = (\mathbf{G}, \mathbf{R}, \mathbf{M})$ is defined separately for state- and path-formulas:
Let $g \in \mathbf{G}$ be a state, $p = (g_0, g_1, \ldots)$ a path in $\mathbf{K}$, A an atomic formula, F and G CTL-state formulas, P a CTL-path formula, and χ a variable assignment:

- (S0) $(g, \chi) \models A$ $\quad$:$\Leftrightarrow$ $(M(g), \chi) \models_{\mathrm{FO}} A$.
- (S1a) $(g, \chi) \models \neg F$ $\quad$:$\Leftrightarrow$ not $(g, \chi) \models F$.
- (S1b) $(g, \chi) \models F \vee G$ $\quad$:$\Leftrightarrow$ $(g, \chi) \models F$ or $(g, \chi) \models G$.
- (P1a) $(p, \chi) \models \circ F$ $\quad$:$\Leftrightarrow$ $(g_1, \chi) \models F$.
- (P1b) $(p, \chi) \models F$ until G $\quad$:$\Leftrightarrow$ there is an $i \geq 0$ such that $(g_i, \chi) \models G$ and for all $j : 0 \leq j < i$ $(g_j, \chi) \models F$ holds.
- (P2) $(p, \chi) \models \neg P$ $\quad$:$\Leftrightarrow$ not $(p, \chi) \models P$.
- (S2) $(g, \chi) \models \mathbf{E}P$ $\quad$:$\Leftrightarrow$ there is a path $p = (g_0, g_1, \ldots)$ in $\mathbf{K}$ and an i such that $g_i = g$ and $(p|_i, \chi) \models P$.
- (SQ) $(g, \chi) \models \exists x : F$ $\quad$:$\Leftrightarrow$ there is a $d \in \mathbf{U}(g)$ with $(g, \chi_x^d) \models F$.

The other symbols are defined as $\Diamond F :=$ true until F, $\Box F := \neg \Diamond \neg F$, and $AP := \neg E \neg P$. A state formula is *valid in a Kripke-structure* $\mathbf{K} = (\mathbf{G}, \mathbf{R}, \mathbf{M})$ iff it is valid in all states $g \in \mathbf{G}$. A formula is *valid* iff it is valid in all Kripke-structures.

Fairness:
Different kinds of fairness requirements are distinguished [LPS81]. In this paper only the strongest (and most important) type is considered:
Compassion (strong Fairness):
Every action which is enabled infinitely often in the future will be carried out eventually. [La80], [EC80] and [EH83] state that strong fairness cannot be expressed in CTL. The CTL* expression is given as follows:

$$\text{CTL}^*\text{: } A((\Box\Diamond(\text{action enabled})) \to \Diamond(\text{action is carried out})) \ .$$

2.1 Related Work

In [CES86] and [EL85], a model checking procedure for propositional CTL is presented. The inclusion of fairness requirements is done by extensions to the algorithm.

In [BMP81], [EH82], and [Wol85], a tableau semantics and -calculus for propositional CTL is presented. The paths of the tableau represent paths in a fictive model. Cycles in the tableau are allowed. After termination, which is guaranteed, eventuality formulas have to be postprocessed. In case of a non-closable tableau where no inconsistency is found by postprocessing, the whole tableau represents a model of the initial formula. An extension to CTL* or at least to fairness requirements does not exist.

Both methods cannot be extended to first-order variants because the finite number of possible different states is the central point in their concept.

Facing these problems, it seems necessary to make basic changes in the processing of eventualities: it has to be possible to abstract from finitely many states in-between. In turn, it also seems desirable to have a 1:1-correspondence of branches of the tableau to Kripke-structures.

3 A Tableau Semantics for Branching Time

To achieve a strict distinction between the two graph structures "Kripke-structure" and "tableau", the terms "path" and "state" will be used for Kripke-structures whereas the terms "branch" and "node" will be used for tableaux.

Like in traditional tableau proving, for a proof of the validity of a formula F, the inconsistency of the formula $\neg F$ is proven. It is systematically tried to construct a model for $\neg F$, with the intention to show the impossibility of that attempt. So the situation from first-order theorem proving to find a model for a given set of formulas occurs multiply: Every state is such a first-order interpretation. For this purpose the well-known first-order tableau calculus will be embedded in the temporal tableau calculus which is constructed. Moreover, from these first-order interpretations a branching time temporal Kripke-structure has to be built.

Therefore it is necessary to describe many individual states as well as the relations between them in the tableau. The latter include the ordering of states on a path together with the connections between different paths.

Thus three kinds of entities have to be described: Elements of the universe inside states, states, and paths. In the chosen semantics these will be explicitly named when their existence is stated by a formula:

- Elements of the universe: as in the first-order tableau calculus a new constant resp. function symbol is introduced by a δ-rule when an $\exists$-quantor is processed.
- States: states are named when required by an existence formula (type $\Diamond F$ or $\circ F$). In the chosen semantics a newly named state has to be positioned on an existing path, retaining the linear ordering of all states on this path.
- Paths: paths are named when required by an existence formula of the kind EP. A newly introduced path is assumed to branch off in the state where its existence is claimed.

In general, between two known states there can be many other still unknown states. These can be named when needed. Thus, a straightforward dissolving of eventualities at any time is possible.

To allow the naming of states at any position of the model, the descriptions of paths contain, apart from the (partial) ordering of known states, additional information about formulas which have to be true in still unknown states on the segments in-between. These are used when new states are explicitly named.

As a conceptional extension of first-order tableaux, every branch of the tableau (resp. the set of formulas on it) corresponds to a complete Kripke-structure.

3.1 Representation

Starting with a formula F over a signature Σ, it is systematically attempted to create a Kripke-structure satisfying F. As every branch of the tableau represents a complete Kripke-structure, apart from the first-order portion, information about the frame of the Kripke-structure has to be coded in tableau nodes. For distinguishing and naming of states, a tableau calculus based on the free variable tableau calculus from [Ree87],[Fit90] augmented with *prefixes* is used for the first-order portion: A state formula F, assumed to be true in a certain state, occurs in the tableau as *prefixed formula* $\gamma : F$. The paths described in the tableau are named by *path descriptors*. For these, *path information formulas* contain the information about the prefixes situated on this path.

Thus the signature Σ_T used in the tableau is partitioned into Σ_L (first-order part) and Σ_F (frame part).

In a first step, Σ is augmented with a countable infinite set of n-ary (skolem) function symbols for every $n \in \mathbb{N}$ and a countable infinite set of variables X_i.

Σ_F consists of a set $\hat{\Gamma}$ of *prefix symbols* and a set $\hat{\Lambda}$ of *path symbols*, each containing a countable infinite set of n-ary prefix- resp. path symbols for every

$n \in \mathbb{N}$. The construction of prefixes and path descriptors from these corresponds to the use of skolem functions in the first-order tableau calculus. Here the prefix- and path symbols take the role of the skolem functions. With this, the free variables resulting from invocations of the γ-rule have to be considered. Thus prefixes γ and path descriptors λ are terms consisting of a prefix symbol $\hat{\gamma}$ resp. a path symbol $\hat{\lambda}$ of an arity n and an n-tuple of terms as arguments. Additionally, there is a 0-ary symbol $\hat{\infty}$ that is not a prefix symbol, but is used in a similar way.

Definition 6. Let $\hat{\Gamma}$ be the set of prefix symbols, $\hat{\Lambda}$ the set of path symbols, Σ^c the state-independent portion of Σ. Then the following sets Σ_L, Γ, and Λ are simultaneously recursively enumerable:

$\Sigma_L := \Sigma \cup \{f : f \text{ an } n\text{-ary skolem function symbol}\} \cup \{f_\gamma : f \in \Sigma \setminus \Sigma^c, \gamma \in \Gamma\}$,

with $\text{ord}(f_\gamma) = \text{ord}(f)$ and Skolem functions and all f_γ interpreted state-independently, thus

$\Sigma_L^c = \Sigma^c \cup \{f : f \text{ an } n\text{-ary skolem function symbol}\} \cup \{f_\gamma : f \in \Sigma \setminus \Sigma^c, \gamma \in \Gamma\}$.

$\Gamma := \{\hat{\gamma}(t_1, \ldots, t_n) : \hat{\gamma} \in \hat{\Gamma} \text{ an } n\text{-ary prefix symbol}, t_1, \ldots, t_n \in \text{Term}_{\Sigma_L^c}\}$

is the set of *prefixes*, and

$\Lambda := \{\hat{\lambda}(t_1, \ldots, t_n) : \hat{\lambda} \in \hat{\Lambda} \text{ an } n\text{-ary path symbol}, t_1, \ldots, t_n \in \text{Term}_{\Sigma_L^c}\}$

is the set of *path descriptors*, and $\Sigma_F := \hat{\Gamma} \cup \hat{\Lambda}$.

In both sets $\Gamma, \Lambda \subset \text{Term}_{\Sigma_T}$ of terms it is precisely the leading function symbol which is a prefix- resp. path symbol taken from Σ_F and all argument terms are in $\text{Term}_{\Sigma_L^c}$. Those are interpreted state-independently by $\mathbf{K}$.

An interpretation of Σ_T – describing a Kripke-structure – is accordingly partitioned: The interpretation of Σ_L is taken over by a suitable set $\{\mathbf{M}(g) : g \in \mathbf{G}\}$ of first-order interpretations. Complementary to this, an "interpretation" of the prefix- and path symbols in Σ_F is defined. The corresponding evaluations map prefixes and path descriptors to the entities described by them:

Definition 7. A P&P-interpretation ("prefixes and paths interpretation") of the sets $\hat{\Gamma}$ and $\hat{\Lambda}$ to a Kripke-structure $(\mathbf{G}, \mathbf{R}, \mathbf{M})$ with a constant universe $\mathbf{U}$ and a set $\mathbf{P}(\mathbf{K})$ of paths is a triple $\Omega = (\phi, \pi, \psi)$ where

$\phi : \hat{\Lambda} \to \mathbf{U}^n \to \mathbf{P}(\mathbf{K})$ maps every n-ary $\hat{\lambda} \in \hat{\Lambda}$ to a function $\phi(\hat{\lambda}) : \mathbf{U}^n \to \mathbf{P}(\mathbf{K})$
resp. $\phi(\hat{\lambda}) : \mathbf{U}^n \to \mathbb{N} \to \mathbf{G}$,

$\pi : \hat{\Lambda} \times (\hat{\Gamma} \cup \{\hat{\infty}\}) \to (\mathbf{U}^n \times \mathbf{U}^m) \to \mathbb{N} \cup \{\infty\}$ is an (in general not total) mapping of pairs of n-ary $\hat{\lambda} \in \hat{\Lambda}$ and m-ary $\hat{\gamma} \in \hat{\Gamma}$ to functions $\pi(\hat{\lambda}, \hat{\gamma})$: $\mathbf{U}^n \times \mathbf{U}^m \to \mathbb{N} \cup \{\infty\}$ with $\pi(\lambda, \gamma) = \infty \Leftrightarrow \gamma = \hat{\infty}$, and

$\psi : \hat{\Gamma} \to \mathbf{U}^n \to \mathbf{G}$ maps every n-ary $\hat{\gamma} \in \hat{\Gamma}$ to a function $\psi(\hat{\gamma}) : \mathbf{U}^n \to \mathbf{G}$.

Ω is organized similarly to a first-order interpretation $\mathbf{I} = (I, \mathbf{U})$ if the corresponding mappings, "universes", and induced evaluations are considered:

$$\Phi = (\phi, \mathbf{P}(\mathbf{K})) \quad , \quad \Pi = (\pi, \mathbb{N} \cup \{\infty\}) \quad , \quad \Psi = (\psi, \mathbf{G})$$

Based on ϕ, π, and ψ, the evaluations

$\Phi : \Lambda \times \Xi \to \mathbf{P(K)}$ of path descriptors,

$\Pi : \Lambda \times (\Gamma \cup \{\hat{\infty}\}) \times \Xi \to \mathbb{N} \cup \{\infty\}$ of pairs of path descriptors and prefixes,

and $\Psi : \Gamma \times \Xi \to \mathbf{G}$ of prefixes

are defined as follows: Let $\lambda = \hat{\lambda}(t_1, \ldots, t_n) \in \Lambda$ and $\gamma = \hat{\gamma}(s_1, \ldots, s_m) \in \Gamma$, thus $t_i, s_i \in \mathrm{Term}_{\Sigma_L^c}$. Then

$$\Phi(\lambda, \chi) := (\phi(\hat{\lambda}))(\mathbf{K}(t_1, \chi), \ldots, \mathbf{K}(t_n, \chi)) \quad ,$$
$$\Pi(\lambda, \gamma, \chi) := (\pi(\hat{\lambda}, \hat{\gamma}))(\mathbf{K}(t_1, \chi), \ldots, \mathbf{K}(t_n, \chi), \mathbf{K}(s_1, \chi), \ldots, \mathbf{K}(s_m, \chi)) \quad ,$$
$$\Psi(\gamma, \chi) := (\psi(\hat{\gamma}))(\mathbf{K}(s_1, \chi), \ldots, \mathbf{K}(s_m, \chi)) \quad .$$

Finally, the interpretation of the derived function symbols f_γ is defined state-independent for all $g \in \mathbf{G}$ as

$$(\mathbf{M}(g))(f_\gamma(t_1, \ldots, t_n), \chi) := (\mathbf{M}(\Psi(\gamma, \chi)))(f(t_1, \ldots, t_n), \chi) \quad .$$

Tableau formulas:

On this foundation the syntax used in the tableaux can be worked out: Let $\mathcal{L}$ be the language of state formulas (CTL or CTL*).

The frame of the Kripke-structure is encoded in *path information formulas* of the form $\lambda : [\gamma_0, L_0, \gamma_1, L_1, \ldots, \gamma_n, L_n, \hat{\infty}]$ with $\lambda \in \Lambda$, $\gamma_i \in \Gamma$ and $L_i \in \mathcal{L} \cup \{\circ\}$. Logical formulas occur in the tableau as *prefixed formulas* of the form $\gamma : F$ with $\gamma \in \Gamma$ and the same branch of the tableau containing a path information formula $\lambda : [\ldots, \gamma, \ldots]$ and $F \in \mathcal{L}$ being a state formula.

Following the explicit naming of paths in the calculus, the formulas used internally to the tableau have a more detailed syntax than ordinary CTL/CTL*-formulas. A syntactic facility to use path descriptors in logical formulas is added: To state the validity of a path formula P on the suffix of a path p (described by a path descriptor λ) beginning in a fixed state g (described by a prefix γ) on that path, the symbol λ can syntactically take the role of a path quantifier. In this role, λ is a *path selector*. This results in the following syntax for node formulas in all tableaux tracing this concept:

Definition 8.   (TA)   Every atomic formula is a $\mathcal{TK}$-state formula.

(TS1)   With F und G $\mathcal{TK}$-state formulas, $\neg F$, $F \wedge G$, $F \vee G$ and $F \to G$ are $\mathcal{TK}$-state formulas.

(TSQ)   With F a $\mathcal{TK}$-state formula and x a variable, $\forall x : F$ and $\exists x : F$ are $\mathcal{TK}$-state formulas.

(TP1)   With F and G $\mathcal{TK}$-state formulas, $\circ F$, $\Box F$, $\Diamond F$, and $(F \text{ until } G)$ are $\mathcal{TK}$-path formulas.

(TP2)   With P a $\mathcal{TK}$-path formula, $\neg P$ is a $\mathcal{TK}$-path formula.

(TS2)   With P a $\mathcal{TK}$-path formula, $\mathsf{A}P$ and $\mathsf{E}P$ are $\mathcal{TK}$-state formulas.

(TC1)   Every $\mathcal{TK}$-state formula is a $\mathcal{TK}$-pre-node formula.

(TC2)   With P a $\mathcal{TK}$-path formula and $\lambda \in \Lambda$, λP is a $\mathcal{TK}$-pre-node formula.

(TK1)   Every path information formula is a $\mathcal{TK}$-node formula.

(TK2)   With F a $\mathcal{TK}$-pre-node formula and $\gamma \in \Gamma$ a prefix, $\gamma : F$ is a $\mathcal{TK}$-prefixed formula.

(TN)   All $\mathcal{TK}$-prefixed formulas are $\mathcal{TK}$-node formulas.

Semantics:

Definition 9. For a P&P-interpretation $\Omega = (\phi, \pi, \psi)$, a path information formula $I = \lambda : [\gamma_0, L_0, \gamma_1, L_1, \ldots, \gamma_n, L_n, \infty]$ is *consistent with* Ω for a variable assignment χ, if every $\hat{\gamma}$ occurs in I at most once, and for all i

$$\Pi(\lambda, \gamma_0, \chi) = 0 \quad , \quad \Pi(\lambda, \gamma_i, \chi) < \Pi(\lambda, \gamma_{i+1}, \chi) \quad ,$$
$$\text{and} \quad \Psi(\gamma_i, \chi) = \Phi(\lambda, \chi, \Pi(\lambda, \gamma_i, \chi)) \quad .$$

This means that the path $\Phi(\lambda, \chi) = (g_0, g_1, \ldots)$ of $\mathbf{K}$ begins in state $g_0 = \Psi(\gamma_0, \chi)$ and passes through the other known states $g_{\Pi(\lambda, \gamma_1, \chi)} = \Psi(\gamma_1, \chi), \ldots,$ $g_{\Pi(\lambda, \gamma_n, \chi)} = \Psi(\gamma_n, \chi)$ in the specified order.

Definition 10. The relation $\models$ of a Kripke-structure $\mathbf{K} = (\mathbf{G}, \mathbf{R}, \mathbf{M})$ with a set $\mathbf{P(K)}$ of paths, a P&P-interpretation Ω, a set $\mathcal{F}$ of formulas and a variable assignment χ to free($\mathcal{F}$) is defined as follows, based on the truth of formulas in Kripke-structures, $\models_{\text{CTL}}$ resp. $\models_{\text{CTL}^*}$,

1a. for every prefixed formula $\gamma : F$, F not containing a path selector:

$$(\mathbf{K}, \Omega, \chi) \models \gamma : F \quad :\Leftrightarrow \quad (\Psi(\gamma, \chi), \chi) \models_{\text{CTL}} F \quad ,$$

i.e. in the state corresponding to the prefix γ under variable assignment χ, the (state) formula F holds.

1b. for every prefixed formula $\gamma : F$, F containing a (leading) path selector:

$$(\mathbf{K}, \Omega, \chi) \models \gamma : \lambda P \quad :\Leftrightarrow \quad (\Phi(\lambda, \chi)|_{\Pi(\lambda, \gamma, \chi)}, \chi) \models P \quad ,$$

i.e. on the suffix of the path $\Phi(\lambda, \chi)$ beginning in the $\Pi(\lambda, \gamma, \chi)$th state (which is $\Psi(\gamma, \chi)$ by concistency), the path formula P holds.

2. for all path information formulas $I = \lambda : [\gamma_0, L_0, \gamma_1, L_1, \ldots, \gamma_n, L_n, \infty]$:

$$(\mathbf{K}, \Omega, \chi) \models \lambda : [\gamma_0, L_0, \gamma_1, L_1, \ldots, \gamma_n, L_n, \infty]$$

iff I is consistent with Ω for the variable assignment χ, and for all $0 \leq i \leq n$: $L_i = \circ \Rightarrow \Pi(\lambda, \gamma_{i+1}, \chi) = \Pi(\lambda, \gamma_i, \chi) + 1$,
$L_i \neq \circ \Rightarrow$ for all j with $\Pi(\lambda, \gamma_i, \chi) < j < \Pi(\lambda, \gamma_{i+1}, \chi)$:
$(\Phi(\lambda, \chi, j), \chi) \models L_i$,

i.e. if $L_i = \circ$, then $\Pi(\lambda, \gamma_i, \chi)$ and $\Pi(\lambda, \gamma_{i+1}, \chi)$ are immediately succeeding indices, else for all (finitely, but arbitrary many) states g_j situated between $\Phi(\lambda, \chi, \Pi(\lambda, \gamma_i, \chi))$ and $\Phi(\lambda, \chi, \Pi(\lambda, \gamma_{i+1}, \chi))$ on path $\Phi(\lambda, \chi)$ the relation $(g_j, \chi) \models L_i$ holds.

For a set $\mathcal{F}$ of path information formulas and prefixed formulas, its truth in a Kripke-structure $\mathbf{K} = (\mathbf{G}, \mathbf{R}, \mathbf{M})$ with a set $\mathbf{P(K)}$ of paths under a variable assignment χ to free($\mathcal{F}$) is defined as follows:

$$(\mathbf{K}, \chi) \Vvdash \mathcal{F} :\Leftrightarrow \text{ there is a P&P-interpretation } \Omega = (\phi, \pi, \psi)$$
$$\text{such that } (\mathbf{K}, \Omega, \chi) \models \mathcal{F} \text{ holds.}$$

Since a branch of a tableau is a set of formulas like this, $\Vvdash$ is a relation on Kripke-structures and branches.

The construction of Kripke-structures and consistent P&P-interpretations to a given set of formulas plays an important role in the proof of correctness.

4 The Tableau Calculus $\mathcal{TK}$

For proving the validity of a formula F, the inconsistency of $\neg F$ is proven: it is shown that there is no Kripke-structure $\mathbf{K} = (\mathbf{G}, \mathbf{R}, \mathbf{M})$ with any state $g_0 \in \mathbf{G}$ where F does not hold.

Thus the initialization of the tableau is $\boxed{\widehat{0} : \neg F}$.

The tableau calculus is based on the well-known first-order tableau calculus, consisting of α-, β-, γ- and δ-rules and the atomic closure rule [Ree87, Fit90].

Let F and G be $\mathcal{TK}$-state formulas, A an atomic formula. In the sequel, $F[t/x]$ denotes the formula F with all occurences of x replaced by t.

$$
\begin{array}{ll}
\alpha : \dfrac{\gamma : F \wedge G}{\begin{array}{c}\gamma : F \\ \gamma : G\end{array}} \quad \dfrac{\gamma : \neg(F \vee G)}{\begin{array}{c}\gamma : \neg F \\ \gamma : \neg G\end{array}} & \beta : \dfrac{\gamma : F \vee G}{\gamma : F \mid \gamma : G} \quad \dfrac{\gamma : \neg(F \wedge G)}{\gamma : \neg F \mid \gamma : \neg G}
\end{array}
$$

$$
\gamma : \dfrac{\gamma : \forall x : F}{\gamma : F[X/x]} \quad \dfrac{\gamma : \neg \exists x : F}{\gamma : \neg F[X/x]} \qquad \text{with } X \text{ a new variable.}
$$

$$
\delta : \dfrac{\gamma : \exists x : F}{\gamma : F[f(\mathsf{free}(T))/x]} \qquad \dfrac{\gamma : \neg \forall x : F}{\gamma : \neg F[f(\mathsf{free}(T))/x]} \qquad \begin{array}{l}\text{with } f \text{ a new function} \\ \text{symbol and } T \text{ the current} \\ \text{branch.}\end{array}
$$

Atomic closure rule:
For a substitution σ and a prefix γ, σ_γ is the γ-*localization*, i.e. $\sigma_\gamma(X) = \sigma(X)$ where every function symbol $f \in \Sigma \backslash \Sigma^c$ is replaced by its localized symbol f_γ. So the substitutes in σ_γ contain only function symbols which are interpreted state-independently.

$$
\dfrac{\begin{array}{c}\gamma : A \\ \gamma : A' \\ \sigma(A) = \neg\sigma(A')\end{array}}{\bot}
$$

apply σ_γ to the whole tableau.

For dissolving modalities, the information about the frame of the Kripke-structure has to be considered. It is encoded in the path information formulas. In one step a prefixed formula is dissolved "along" a path information formula, inducing the following form of tableau rules:

$$
\dfrac{\begin{array}{c}\text{prefixed formula} \\ \text{path information formula}\end{array}}{\begin{array}{c}\text{prefixed formulas} \\ \text{path information formulas}\end{array}}
$$

where the premise takes the latest path information formula on the current branch for the path symbol to be considered. The connection between the prefixed formula being dissolved and the path information is established by the prefix and, if exists, the leading path selector of the prefixed formula. For dissolving prefixed formulas, a path quantifier resp. -selector is broken up together with the subsequent modal operator:

- For dissolving a formula of the form $\gamma : \mathsf{E}P$, a path satisfying P is named and the path formula is bound to that path:

$$
\dfrac{\begin{array}{c}\gamma : \mathsf{E}P \\ \lambda : [\widehat{0}, \ldots, \gamma, \ldots]\end{array}}{\begin{array}{c}\hat{\kappa}(\mathsf{free}(T)) : [\widehat{0}, \ldots, \gamma, \mathsf{true}, \widehat{\infty}] \\ \gamma : \hat{\kappa}(\mathsf{free}(T))P\end{array}}
$$

- Formulas of the form $\gamma : \mathsf{A}P$ are dissolved once for every path information formula on this branch containing the prefix γ.
- Formulas of the form $\gamma : \lambda P$ are dissolved along the path information formula for λ.

In the latter cases, the claim that the state described by the current prefix satisfies some formula is decomposed in some less complex claims:

- Which formulas hold in the current state?
- Which state should be regarded as the "next relevant state" on the path?
- Which formulas hold in this next relevant state?
- Which formulas hold in all states in-between?

Special Properties of CTL:

For CTL some *propagation theorems* can be stated [May95] which simplify the dissolving of universally path-quantified formulas along branching paths: The validity of a formula $F = \mathsf{A}P$ can be decomposed into the validity of a formula G in the current state and the validity of a formula Q on *all* outgoing paths, concerning only proper successor states. Especially, for parallel paths, only one of them has to be considered.

According to this, for CTL, the rule for $\gamma : \mathsf{E}P$ can be modified: $\triangleright$

Additionally the dissolving of universally path-quantified formulas is divided in two parts. The

$$\frac{\gamma : \mathsf{E}P}{\hat{\kappa}(\mathsf{free}(T)) : [\gamma, \mathsf{true}, \tilde{\infty}]}$$
$$\gamma : \hat{\kappa}(\mathsf{free}(T))P$$

syntax of tableau formulas is enriched with the syntactic element (A), meaning "on all paths, concerning only proper successor states", which can replace the leading A of a state formula, leading to the following enlargement to Def. 8:

(TC3)   With P a $\mathcal{TK}$-path formula, $(\mathsf{A})P$ is a $\mathcal{TK}$-pre-node formula.

The above-mentioned decomposition is formalized as

$$(\mathbf{K}, \Omega, \chi) \Vdash \gamma : \mathsf{A}P \Leftrightarrow \bigvee((\mathbf{K}, \Omega, \chi) \Vdash \gamma : G_i \text{ and } (\mathbf{K}, \Omega, \chi) \Vdash \gamma : (\mathsf{A})Q_i) \quad ,$$

where $\bigvee$ counts over some possible decompositions (G_i, Q_i).

There is the following survey over the basic types of state formulas extending Def. 10:

(TS2a) $(\mathbf{K}, \Omega, \chi) \Vdash \gamma : \mathsf{A}P$:$\Leftrightarrow$ For all paths $p = (g_0, g_1, \ldots)$ in $\mathbf{K}$ and all n with $g_n = \Psi(\gamma, \chi)$ $(p|_n, \chi) \models P$ holds.

(TS2b) $(\mathbf{K}, \Omega, \chi) \Vdash \gamma : \mathsf{E}P$:$\Leftrightarrow$ there is a path $p(\chi) = (g_0, g_1, \ldots)$ in $\mathbf{K}$ and an $n(\chi)$ so that $g_{n(\chi)} = \Psi(\gamma, \chi)$ and $(p(\chi)|_{n(\chi)}, \chi) \models P$ holds.

(TC2) $(\mathbf{K}, \Omega, \chi) \Vdash \gamma : \lambda P$:$\Leftrightarrow$ $(\Phi(\lambda, \chi)|_{\Pi(\lambda, \gamma, \chi)}, \chi) \models P$.

(TC3) $(\mathbf{K}, \Omega, \chi) \Vdash \gamma : (\mathsf{A})Q_i$:$\Leftrightarrow$ For all paths $p = (g_0, g_1, \ldots)$ in $\mathbf{K}$ and all n with $g_n = \Psi(\gamma, \chi)$, $(\Psi(\gamma, \chi)) \models G_i$ implies that $(p|_n, \chi) \models P$ holds.

Because of this decomposition, each formula of the form $\gamma : \mathsf{A}P$ is dissolved exactly once, resulting in pairs of formulas $\gamma : G_i$ (for the current state) and $\gamma : (\mathsf{A})Q_i$ (describing a property of all outgoing paths). Thus, for CTL, formulas of the form $\gamma : (A)P$ are dissolved once for every path information formula on the same branch containing the prefix γ.

The tableau rules for CTL for formulas which are universally path-quantified or explicitly bound to named paths are as follows. In the sequel, T denotes the current branch of the tableau, $\hat{\gamma}$ is a new prefix symbol and $\hat{\kappa}$ is a new path symbol. P is a path formula, F is a state formula.

$$
\begin{array}{l}
\dfrac{\alpha : A\square F}{\begin{array}{l}\alpha : F\\ \alpha : (A)\square F\end{array}}
\qquad
\dfrac{\begin{array}{c}\alpha : (A)\square F\\ \lambda : [\ldots, \alpha, L, \beta, \ldots],\ L \neq \circ\end{array}}{\begin{array}{c}\lambda : [\ldots, \alpha, L \wedge F \wedge (A)\square F, \beta, \ldots]\\ \beta : A\square F\quad \text{if } \beta \neq \hat{\infty}\end{array}}
\qquad
\dfrac{\begin{array}{c}\alpha : (A)\square F\\ \lambda : [\ldots, \alpha, \circ, \beta, \ldots]\end{array}}{\beta : A\square F}
\end{array}
$$

$$
\dfrac{\begin{array}{c}\alpha : \lambda\square F,\\ \lambda : [\ldots, \alpha, L, \beta, \ldots],\ L \neq \circ\end{array}}{\begin{array}{c}\lambda : [\ldots, \alpha, L \wedge F, \beta, \ldots]\\ \alpha : F\\ \beta : \lambda\square F\quad \text{if } \beta \neq \hat{\infty}\end{array}}
\qquad
\dfrac{\begin{array}{c}\alpha : \lambda\square F\\ \lambda : [\ldots, \alpha, \circ, \beta, \ldots]\end{array}}{\begin{array}{c}\alpha : F\\ \beta : \lambda\square F\end{array}}
$$

$$
\dfrac{\alpha : A\Diamond F}{\begin{array}{c|c}\alpha : F & \alpha : \neg F\\ & \alpha : (A)\Diamond F\end{array}}
\qquad
\dfrac{\begin{array}{c}\alpha : (A)\Diamond F\\ \lambda : [\ldots, \alpha, \circ, \beta, \ldots]\end{array}}{\beta : A\Diamond F}
\qquad
\dfrac{\begin{array}{c}\alpha : \lambda\Diamond F\\ \lambda : [\ldots, \alpha, \circ, \beta, \ldots]\end{array}}{\begin{array}{c|c}\alpha : F & \alpha : \neg F\\ & \beta : \lambda\Diamond F\end{array}}
$$

$$
\dfrac{\begin{array}{c}\alpha : (A)\Diamond F\\ \lambda : [\ldots, \alpha, L, \beta, \ldots],\ L \neq \circ\end{array}}{\begin{array}{c|c}\begin{array}{c}\lambda : [\ldots, \alpha, L \wedge \neg F \wedge (A)\Diamond F,\\ \hat{\gamma}(\text{free}(T)), L, \beta, \ldots]\\ \hat{\gamma}(\text{free}(T)) : L\\ \hat{\gamma}(\text{free}(T)) : F\end{array} & \begin{array}{c}\text{if } \beta \neq \hat{\infty}:\\ \lambda : [\ldots, \alpha, L \wedge \neg F \wedge (A)\Diamond F, \beta, \ldots]\\ \beta : A\Diamond F\end{array}\end{array}}
$$

$$
\dfrac{\begin{array}{c}\alpha : \lambda\Diamond F\\ \lambda : [\ldots, \alpha, L, \beta, \ldots],\ L \neq \circ\end{array}}{\begin{array}{c|c|c}\alpha : F & \begin{array}{c}\lambda : [\ldots, \alpha, L \wedge \neg F,\\ \hat{\gamma}(\text{free}(T)), L, \beta, \ldots]\\ \alpha : \neg F\\ \hat{\gamma}(\text{free}(T)) : L\\ \hat{\gamma}(\text{free}(T)) : F\end{array} & \begin{array}{c}\text{if } \beta \neq \hat{\infty}:\\ \lambda : [\ldots, \alpha, L \wedge \neg F, \beta, \ldots]\\ \alpha : \neg F\\ \beta : \lambda\Diamond F\end{array}\end{array}}
$$

The rules for $\neg\square F = \Diamond\neg F$ and $\neg\Diamond F = \square\neg F$ are analogous.

$$
\dfrac{\begin{array}{c}\alpha : A\circ F\\ \lambda : [\ldots, \alpha, L, \beta, \ldots],\ L \neq \circ\end{array}}{\begin{array}{c|c}\begin{array}{c}\lambda : [\ldots, \alpha, \circ, \hat{\gamma}(\text{free}(T)), L, \beta, \ldots]\\ \hat{\gamma}(\text{free}(T)) : L\\ \hat{\gamma}(\text{free}(T)) : F\end{array} & \begin{array}{c}\text{if } \beta \neq \hat{\infty}:\\ \lambda : [\ldots, \alpha, \circ, \beta, \ldots]\\ \beta : F\end{array}\end{array}}
\qquad
\dfrac{\begin{array}{c}\alpha : A\circ F\\ \lambda : [\ldots, \alpha, \circ, \beta, \ldots]\end{array}}{\beta : F}
$$

$$\frac{\alpha : A(F \text{ until } G)}{\alpha : G \quad \begin{array}{|c} \alpha : F \\ \alpha : \neg G \\ \alpha : (A)(F \text{ until } G) \end{array}} \qquad \frac{\alpha : A\neg(F \text{ until } G)}{\alpha : \neg G \quad \begin{array}{|c} \alpha : F \\ \alpha : \neg F \\ \alpha : (A)\neg(F \text{ until } G) \end{array}}$$

$$\frac{\alpha : (A)(F \text{ until } G)}{\lambda : [\ldots, \alpha, L, \beta, \ldots], \ L \neq \circ}$$

$\lambda : [\ldots, \alpha, L \wedge F \wedge \neg G \wedge (A)(F \text{ until } G),$ $\hat{\gamma}(\text{free}(T)), L, \beta, \ldots]$ $\hat{\gamma}(\text{free}(T)) : L$ $\hat{\gamma}(\text{free}(T)) : G$	if $\beta \neq \hat{\infty}$: $\lambda : [\ldots, \alpha, L \wedge F \wedge \neg G$ $\wedge (A)(F \text{ until } G), \beta, \ldots]$ $\beta : A(F \text{ until } G)$

$$\frac{\alpha : (A)(F \text{ until } G)}{\lambda : [\ldots, \alpha, \circ, \beta, \ldots]} \qquad \frac{\alpha : (A)\neg(F \text{ until } G)}{\lambda : [\ldots, \alpha, \circ, \beta, \ldots]}$$
$$\frac{}{\beta : A(F \text{ until } G)} \qquad \frac{}{\beta : A\neg(F \text{ until } G)}$$

$$\frac{\alpha : (A)\neg(F \text{ until } G)}{\lambda : [\ldots, \alpha, L, \beta, \ldots], \ L \neq \circ}$$

$\lambda : [\ldots, \alpha, L \wedge F \wedge \neg G$ $\wedge (A)\neg(F \text{ until } G),$ $\hat{\gamma}(\text{free}(T)), L, \beta, \ldots]$ $\hat{\gamma}(\text{free}(T)) : L$ $\hat{\gamma}(\text{free}(T)) : \neg F$ $\hat{\gamma}(\text{free}(T)) : \neg G$	if $\beta \neq \hat{\infty}$: $\lambda : [\ldots, \alpha, L \wedge F \wedge \neg G$ $\wedge (A)\neg(F \text{ until } G), \beta, \ldots]$ $\beta : A\neg(F \text{ until } G)$ if $\beta = \hat{\infty}$: $\lambda : [\ldots, \alpha, L \wedge F \wedge \neg G$ $\wedge (A)\neg(F \text{ until } G), \hat{\infty}]$

The rules for $\lambda \circ F$, λF until G, and $\lambda \neg(F \text{ until } G)$ are analogous.
Two rules are provided to instantiate states on path segments:

$$\frac{\lambda : [\ldots, \alpha, L, \beta, \ldots], \ L \neq \circ}{\beta \neq \hat{\infty}}$$
$$\frac{}{\lambda : [\ldots, \alpha, L, \hat{\gamma}(\text{free}(T)), \circ, \beta, \ldots] \ \big| \ \lambda : [\ldots, \alpha, \circ, \beta, \ldots]}{\hat{\gamma}(\text{free}(T)) : L}$$

$$\frac{\lambda : [\ldots, \alpha, L, \hat{\infty}], \ L \neq \text{true}}{\lambda : [\ldots, \alpha, L, \hat{\gamma}(\text{free}(T)), L, \hat{\infty}],}$$
$$\hat{\gamma}(\text{free}(T)) : L$$

Definition 11. A tableau $\mathcal{T}$ is *satisfiable* if there is a Kripke-structure $\mathbf{K} = (\mathbf{G}, \mathbf{R}, \mathbf{M})$ and a P&P-interpretation $\Omega = (\phi, \pi, \psi)$ of $\hat{\Lambda}$ and $\hat{\Gamma}$ such that for every variable assignment χ of free$(\mathcal{T})$ there is a branch T in $\mathcal{T}$ with $(\mathbf{K}, \Omega, \chi) \models T$. A branch T in $\mathcal{T}$ is *closed* if it contains the formula $\bot$. A tableau $\mathcal{T}$ is *closed*, if every branch T in $\mathcal{T}$ is closed.

Theorem 12 (Substitution Lemma). *Let* $\mathbf{K}$ *be a Kripke-structure over a signature* Σ, Σ^c *the state-independent portion of* Σ, $\Omega = (\phi, \pi, \psi)$ *a P&P-interpretation,* χ *a variable assignment,* X *a free variable,* $g \in \mathbf{G}$, $s \in \mathrm{Term}_{\Sigma^c}$, $t \in \mathrm{Term}_\Sigma$, $\gamma \in \Gamma$ *a prefix,* $\lambda \in \Lambda$ *a path descriptor,* F *a* $\mathcal{TK}$-*state formula,* I *a path information formula, and* $a := \mathbf{K}(s, \chi) \in \mathbf{U}(\mathbf{K})$. *Then*

$$\mathbf{K}([X \leftarrow s]t, \chi) = \mathbf{K}(t, \chi_X^a) \qquad , \Phi([X \leftarrow s]\lambda, \chi) = \Phi(\lambda, \chi_X^a) \quad ,$$
$$\Pi([X \leftarrow s]\lambda, \gamma, \chi) = \Pi(\lambda, \gamma, \chi_X^a) \qquad , \Psi([X \leftarrow s]\gamma, \chi) = \Psi(\gamma, \chi_X^a) \quad ,$$
$$(g, \chi) \models [X \leftarrow s]F \Leftrightarrow (g, \chi_X^a) \models F \; ,$$
$$(\mathbf{K}, \Omega, \chi) \models [X \leftarrow s](\gamma : F) \Leftrightarrow (\mathbf{K}, \Omega, \chi_X^a) \models (\gamma : F) \quad ,$$
$$(\mathbf{K}, \Omega, \chi) \models [X \leftarrow s]I \Leftrightarrow (\mathbf{K}, \Omega, \chi_X^a) \models I \qquad and$$
$$[X \leftarrow s]I \; consistent \; with \; \Omega \; for \; \chi \Leftrightarrow I \; consistent \; with \; \Omega \; for \; \chi_X^a.$$

The proof is done separately for terms and formulas by structural induction. This shows the necessity of the substitutes s of σ_γ in the atomic closure rule being interpreted state-independently (i.e. $s \in \mathrm{Term}_{\Sigma^c}$): Then s has a well-defined global interpretation $a := \mathbf{K}(s, \chi) \in \mathbf{U}(\mathbf{K})$ needed for the modification of χ.

Theorem 13 (Correctness of $\mathcal{TK}$).

(a)   If a tableau $\mathcal{T}$ is satisfiable and $\mathcal{T}'$ is created from $\mathcal{T}$ by an application of any of the rules mentioned above, then $\mathcal{T}'$ is also satisfiable.

(b)   If there is any closed tableau for $\mathcal{F}$, then $\mathcal{F}$ is unsatisfiable.

The proof of (a) is done by case-splitting separately for each of the rules. By assumption, there is a Kripke-structure $\mathbf{K}$ and a P&P-Interpretation $\Omega = (\phi, \pi, \psi)$ such that for every variable assignment χ there is a branch T_χ in $\mathcal{T}$ with $(\mathbf{K}, \Omega, \chi) \models T_\chi$. In all cases apart from the atomic closure rule, $\mathbf{K}$ and Ω are extended such that they witness the satisfiability of $\mathcal{T}'$. In case of the atomic closure rule the Substitution Lemma guarantees the existence of a branch for every variable assignment to $\mathrm{free}(\mathcal{T}')$. (b) follows directly from (a).

It is well known that the set of first-order tautologies of CTL and even of less expressive systems is not recursively enumerable, see, e.g. [GHR94, Theorem 4.6.1, p. 130]. In [May95] the following is shown:

Theorem 14. *a) First-order CTL is not compact.*
b) Any calculus for first-order CTL cannot be complete.

The calculus is complete modulo inductive properties. For such cases induction rules for temporal properties and well-founded data-structures have to be included. In this setting the notion of completeness has to be relativized to that any proof done in a mathematical way can be completely redone formally.

The calculus is even incomplete for propositional CTL because it cannot use its finite-state-property, so the induction problem remains. For PCTL, the methods mentioned in section 2.1 are complete and efficient. As mentioned there, propositional CTL and first-order CTL require completely different, even contrary, concepts. By introducing abstraction, the presented calculus shows a new concept designed for first-order CTL, accepting not to be optimal for propositional CTL.

5   Fairness and Other Extensions

Fairness is not expressible in CTL. It requires the class of path formulas called "reactivity" [MP92] which is expressible in CTL*. In linear time temporal logic, fairness is expressed as $(\Box\Diamond(\text{action enabled})) \rightarrow \Diamond(\text{action is carried out})$.

A formula P of linear temporal logic can be bound to a path as λP. Complex formulas of linear temporal logic can be processed on single paths by some extensions to the calculus:

- Obvious rules for $\lambda : P \wedge Q$ resp. $\lambda : P \vee Q$.
- All tableau rules copy the leading path selector of the premise in front of the consequent if otherwise the consequent would start with a modal operator not preceded by a path quantifier/selector.

The observation that fairness is a property of a path which is decided "near infinity" makes it tractable in the presented calculus (and intractable in the calculus presented in [BMP81]):

Definition 15. A formula P of linear time temporal logic is of *type* ω iff for every Kripke-structure $\mathbf{K} = (\mathbf{G}, \mathbf{R}, \mathbf{M})$, every path $p \in \mathbf{P}(\mathbf{K})$, every variable assignment χ, and all $n \in \mathbb{N}$

$$(\text{for all } i < n : (p\,|_i, \chi) \models P) \Leftrightarrow p\,|_n \models P \ .$$

This establishes the tableau rule   ▷

Since λP is a linear time formula bound to a single path it can be processed by the calculus on this path.

$$\frac{\gamma_i : AP, \quad P \text{ of type } \omega}{\lambda : [\gamma_0, L_0, \gamma_1, L_1, \dots, \gamma_n, L_n, \hat\infty]}$$
$$\gamma_n : \lambda P$$
$$\text{for all } j > i: \ \gamma_j : AP$$

Theorem 16. *For first-order formulas F and G, $\Box(\Diamond\Box\neg F \vee \Diamond G)$ is of type ω. Fairness is expressible by a formula of type ω.*

The following extensions are pointed out in [May95]:
The handling of state-independent interpreted atomic formulas can be improved. In the pure form, such formulas can only be propagated by frame-axioms which have to be included into the specification and the set of input formulas.
Based on the idea of binding complex formulas of linear time temporal logic to paths the calculus can be used to process CTL*-formulas with only little changes.

6   Conclusion

The presented tableau semantics and -calculus shows new perspectives for formal reasoning in first-order CTL, enabling a formal verification of processes with first-order specifications. Due to the embedding of first-order tableaux all recent techniques such as universal formulas, free variables, liberalized δ-rule, and equality-handling can be made full use of. Because of the complexity, pure computational as well as intellectual, which results in a very large search space including many occurrences of inductions, interactive proving seems appropriate. This also reflects the point of view that these inductions are part of the specification, and thus are to be proven on one side, and can be exploited on the other.

References

[BMP81]   M. Ben-Ari, Z. Manna, A. Pnueli: *The Temporal Logic of Branching Time*. Proc. of the 8th ACM Symp. on Principles of Programming Languages, 1981.

[CE81]   E. M. Clarke, E. A. Emerson: *Design and Synthesis of Synchronization Skeletons using Branching Time Temporal Logic*. Proc. of the IBM Workshop on Logics of Programs, Springer LNCS 131, 1981.

[CES86]   E. M. Clarke, E. A. Emerson, A. P. Sistla: *Automatic Verification of Finite-State Concurrent Systems Using Temporal Logic Specifications*. ACM Tr. on Programming Languages and Systems, Vol. 8, No. 2, 1986.

[EC80]   E. A. Emerson, E. M. Clarke: *Characterizing Properties of Parallel Programs as Fixpoints*. Proc. of the 7th Int. Coll. on Automata, Languages and Programming. Springer LNCS 85, 1980.

[EH82]   E. A. Emerson, J. Y. Halpern: *Decision Procedures and Expressiveness in the Temporal Logic of Branching Time*. Proc. of the 14th ACM Symp. on Computing, 1982.

[EH83]   E. A. Emerson, J. Y. Halpern: *"Sometimes" and "not never" revisited: On Branching Time versus Linear Time in Temporal Logic*. Proc. of the 10th ACM Symp. on Principles of Programming Languages, 1983.

[EL85]   E. A. Emerson, C.-L. Lei: *Modalities for Model Checking: Branching Time Strikes Back*. Proc. of the 12th ACM Symp. on Principles of Programming Languages, 1985.

[Fit90]   M. Fitting: *First Order Logic and Automated Theorem Proving*. Springer, New York, 1990.

[GHR94]   D. M. Gabbay, I. Hodkinson, M. Reynolds: *Temporal Logic. Mathematical Foundations and Computational Aspects. Vol. 1.* Clarendon Press, Oxford Logic Guides No. 28, 1994.

[La80]   L. Lamport: *"Sometimes" is Sometimes "Not Never"*. Proc. of the 7th ACM Symp. on Principles of Programming Languages, 1980.

[LPS81]   D. Lehmann, A. Pnueli, J. Stavi: *Impartiality, Justice, and Fairness: The Ethics of Concurrent Termination*. Proc. of the 8th Int. Coll. on Automata, Languages and Programming. Springer LNCS 115, 1981.

[May95]   W. May: *Protokollverifikation in Temporallogik: Evolving Algebras und ein Tableaukalkül*. Diplomarbeit, Universität Karlsruhe, 1995.

[MP92]   Z. Manna, A. Pnueli: *The Temporal Logic of Reactive and Concurrent Systems: Specification*. Springer, 1992.

[Ree87]   S. V. Reeves: *Semantic Tableaux as a Framework for Automated Theorem-Proving*. Dept. of Comp. Sc. and Statistics, Queen Mary College, Univ. of London.

[Wol85]   P. Wolper: *The Tableau Method for Temporal Logic*. Logique et Analyse, 110-111, vol. 28, 1985.

Possible World Semantics for Analogous Reasoning

J.-J.Ch. Meyer[1] and J.C. van Leeuwen[2]

[1] Utrecht University, Dept of Comp. Sc., P.O. Box 80089, NL-3508 TB Utrecht
(email: jj@cs.ruu.nl)
[2] Eindhoven University of Technology, Den Dolech 2, NL-5600 MB Eindhoven

Abstract. Analogous reasoning is a form of reasoning that is often used in daily life situations. It is also a form of reasoning that appears in certain AI applications such as learning and knowledge acquisition. Mostly a kind of quantitative notion of (dis)similarity is employed. In this paper we present a modal model for a qualitative notion of similarity and thus we obtain a basis for qualitative analogous reasoning. Topic areas: formal models for reasoning, nonmon. practical reasoning mechanisms.

1 Introduction

Analogies are ubiquitous in common-sense situations. Often we reason by analogy to predict the outcome of a situation at hand on the basis of a *similar* case we have encountered in the past. In fact, reasoning by analogy is the very hart of learning by intelligent agents. By matching new cases to familiar ones we may extend our knowledge by transposing (assuming or investigating) what we know to hold in the old case with what it would correspond to in the new case. Analogies are used in a wide spectrum of cases ranging from poetry where metaphors are employed to explain things in other (more familiar or expressible) terms via science where models of familiar notions may guide the exploration of new concepts to even such a rigourous discipline as mathematics, where in proofs often a phrase is used such as "we have proven the case in detail for such and such; the case so and so is analogous".

Not surprisingly, also in AI reasoning by analogy has an important role. For instance, it is employed for automatic (machine) learning and for classifying newly obtained knowledge into the scheme of knowledge so far in knowledge acquisition. In the literature we find many classifications and variants of reasoning by analogy (e.g. transformational and derivational analogy, bottom up analogy (Evans 68, Winston 80), top down analogy (Burstein 86)) as well as a number of programs that are available (cf. Hall 89) and perform some form of analogical reasoning in a concrete context (e.g. CARL in the context of learning assignment statements in BASIC (Burstein 86, 88), a full treatment of which is beyond the scope of this paper. We will focus on the following dichotomy proposed by Indurkhya. (Actually, Indurkhya also distinguishes a third form, proportional analogy, which is of a slightly different nature and which we shall ignore here. In (van Leeuwen 95) it is indicated how also this form of analogous reasoning can be fitted into our model framework.)

Indurkhya distinguishes the following types of analogy: *analogy by rendition* and *predictive analogy* (Indurkhya 89).

Analogy by rendition (translation) views a situation (target) or object as if it were another (source). Elements are mapped and translated. It is a way of interpreting a target situation in the light of a known source situation in order to gain more or different information about the target or to get a better understanding of it. This is accomplished by projecting 'framework' and terminology from source to target. The source domain may be an artificial one, a model in which one may focus on the relevant level of abstraction. It is thus closely related to the use of models in problem solving and metaphors. Poets use this kind of analogy all the time and designers obtain creative ideas through analogy by rendition.

Predictive analogy is also based on rendition, but goes further than just stating renditions or similarities between two domains that are both known completely: when a rendition is possible between two domains predictions are made about more similarities by considering relations on the already mapped elements. In this context rendition must agree on the ontology of elements in both domains. The emphasis here is on making an *inference* (prediction) about the target domain on the basis of what is known about a usually more familiar source domain. This form of analogical reasoning is mostly considered in AI applications.

However, the distinction between analogy by rendition and predictive analogy is not entirely clear-cut, since it depends on what is known exactly about the (source and particularly the) target domain whether a conclusion is a prediction or rather a mere rendition. Below we shall consider rendition on the basic elements of the logic at hand (atomic propositions in the propositional case), while we view prediction as the result following from this rendition (translation) to more complex formulas. Of course, if both domains are completely known, these 'predictions' are then not much more than simple translations / renditions. Later in the paper we will see how we can extend this idea to a first-order language (and logic), where we can express more refined notions of rendition and prediction. Also the idea that some predictions are not completely certain and are some more or less 'educated guesses' will be discussed and treated formally in the paper.

In this paper we shall give a semantical treatment of the above two forms of reasoning, put into a possible world framework. Proofs of propositions and theorems are omitted here; those of sections 2 and 3 can be found in (van Leeuwen 95).

2 A Propositional Modal Logic of Analogy

2.1 Defining the concept of similarity

In this section we will try to identify some key concepts of analogical reasoning within the context of propositional logic. The most important of these concepts is that of *similarity*. It is important to note that we will develop a *semantical* theory of these concepts. We consider this semantics-based treatment of analogical reasoning one of the main contributions of our paper, which is lacking in most approaches in the literature (a notable exception is (Thiele 86)). Furthermore, we embed our theory into a possible world semantics. For this modal approach we were influenced by (Morgan 79), who was in his turn inspired by early work by C.S. Peirce. As compared to the work of Morgan, we make the model much more explicit, both with respect to the modal aspect and the aspect of the rendition mapping.

From a practical stand-point propositional logic is clearly too 'poor' as to expressive power to enable one to represent 'real-life' examples of analogical reasoning: viewing propositional logic as predicate logic without function, relation, and constant symbols, obviously there is little room for similarity between models left, because of the sheer simplicity of these models. However, we believe that the simple setting of propositional logic enables one to concentrate on some important issues without having the to deal with the complexity of richer logics. The results of attempting to describe analogical reasoning in this setting will also serve as a natural basis for our further development of the theory in the sequel of this paper.

In commonsense use, analogical reasoning manifests itself between two domains of knowledge, which may be represented by formal theories. The very essence of analogical reasoning indicates that these two domains between which it takes place should display some form of correspondence or *similarity*. When looked upon semantically this means that the models of these theories should show some form of similarity as well. In our simple propositional setting the only way to express this similarity is to consider these on the level of propositional atoms. We will do this in the most simple way conceivable: as a (*similarity*) mapping $\mathbf{T}$ from propositional atoms from the one domain (which we shall call the *source domain*) to the other (the *target domain*), representing that the propositional atom $\mathbf{T}(p)$ in the target domain is similar to the propositional atom p in the source domain (as far as the context of reasoning at hand is concerned). In fact, this function $\mathbf{T}$ may be viewed as the (formal counterpart of the) translation mapping (*rendition*) in the rendition type of analogy mentioned in the introduction. Of course, one might also consider more general similarity mappings $\mathbf{T}$ from *formulas* in $\mathcal{L}(\mathcal{P}_1)$ to *formulas* in $\mathcal{L}(\mathcal{P}_2)$ as primitive, but in our view this yields a rather non-compositional theory in the sense that it is then not clear at all how to determine the similarity mapping for complex formulas of which the function $\mathbf{T}$ is not given.

To make a start with our formal treatment, we assume two sets of propositional atoms $\mathcal{P}_1$ and $\mathcal{P}_2$, which, for convenience, we assume to be disjoint. On the basis of a set $\mathcal{P}$ of propositional atoms we construct a propositional language $\mathcal{L}(\mathcal{P})$ as usual: the smallest set containing $\mathcal{P}$ and closed under the propositional connectives $\neg$, $\wedge$ and $\vee$ (and other connectives which may be introduced as abbreviations in terms of these, such as $\rightarrow$). We furthermore use $\top$ and $\bot$ to denote the constants for 'truth' and 'falsehood', respectively. The symbols φ and ψ are used as metavariables for formulas in a propositional language. Unless stated otherwise, we shall use the language $\mathcal{L}(\mathcal{P}_1)$ for the description of the source domain and $\mathcal{L}(\mathcal{P}_2)$ for that of the target domain. As usual in propositional logic we describe the semantics of formulas by means of *valuations*. We use *tt* for the truth value true and *ff* for the truth value false.

A propositional model over the set $\mathcal{P}$ of propositional atoms is a valuation function v with $v : \mathcal{L}(\mathcal{P}) \to \{tt, ff\}$ (induced by a function $v : \mathcal{P} \to \{tt, ff\}$). In this propositional context we may call $\mathcal{P}$ the *signature* of model v. The class of valuations (propositional models) over $\mathcal{P}$ is denoted $VAL(\mathcal{P})$.

It may not be necessary or even desirable to map all of the propositions because some elements may be irrelevant. Therefore we choose $\mathbf{T}$ to be a *partial* function. We will further assume this function to be injective, since this will enable us to speak about its inverse later on. We do not assume surjectivity of the function $\mathbf{T}$, since it might well be that the target domain has some elements (viz. propositional atoms) that do not correspond (have no counterpart) in the source domain. First we define the concept of *rendition* as discussed above, which is similarity on the smallest elements (atomic propositions):

Assuming an injective partial function $\mathbf{T} : \mathcal{P}_1 \to \mathcal{P}_2$ with $dom(\mathbf{T}) \neq \emptyset$, the *similarity mapping* $R_\mathbf{T}$ from $VAL(\mathcal{P}_1)$ to models $VAL(\mathcal{P}_2)$ induced by $\mathbf{T}$ is given by:

$$\forall\, v_1 \in VAL(\mathcal{P}_1), v_2 \in VAL(\mathcal{P}_2)\ \forall\, p \in \mathcal{P}_1 \cap dom(\mathbf{T}) : R_\mathbf{T}(v_1)(\mathbf{T}(\mathcal{P}_1)) = v_1(\mathcal{P}_1)$$

We can lift $\mathbf{T}$ to formulas which will create a base for the predictive part of analogy. We just define:

- $\mathbf{T}(\varphi \wedge \psi) = \mathbf{T}(\varphi) \wedge \mathbf{T}(\psi)$
- $\mathbf{T}(\varphi \vee \psi) = \mathbf{T}(\varphi) \vee \mathbf{T}(\psi)$
- $\mathbf{T}(\varphi \to \psi) = \mathbf{T}(\varphi) \to \mathbf{T}(\psi)$
- $\mathbf{T}(\neg\varphi) = \neg\mathbf{T}(\varphi)$

Proposition 1. *The above map* $\mathbf{T} : L(\mathcal{P}_1) \to L(\mathcal{P}_2)$ *induced by a similarity mapping* $\mathbf{T} : \mathcal{P}_1 \to \mathcal{P}_2$ *satisfies the property: for all* $v \in VAL(\mathcal{P}_1)$ *we have:*

$$\forall\, \varphi \in L(\mathcal{P}_1) \cap dom(\mathbf{T}) : v \models \varphi \iff R_\mathbf{T}(v) \models \mathbf{T}(\varphi)$$

$\Diamond$

2.2 Kripke models and modalities for analogous reasoning

To build a modal logic of analogical reasoning based on similarity mappings, we start with a notion of Kripke model tailored for this purpose. We assume $\mathcal{P} \neq \emptyset$ to be the universe of propositional atoms. For convenience we consider partial valuations v over $\mathcal{P}$, which means that the function $v : \mathcal{P} \to \{tt, ff\})$ (and thus also the function $v : \mathcal{L}(\mathcal{P}) \to \{tt, ff\})$ is partial and may not be defined for all arguments. The set of partial valuations over $\mathcal{P}$ is denoted VAL. To simplify notation we will use these valuations directly as our set of worlds. Thus our set of worlds are exactly the valuations over $\mathcal{P}$. The use of *partial* valuations enables us to effectively vary the domain of propositional atoms defined in a world without having to bother with distinct sets of propositional atoms per world. For a valuation v we denote its domain (i.e. the set of propositional variables on which v is defined) by $dom(v)$.

For the accessibility relations we use relations induced by similarity mappings $VAL \to VAL$: given a set $\{\mathbf{T}_i | i = 1, ..., n\}$ of partial functions $\mathcal{P} \to \mathcal{P}$, we define $R_{\mathbf{T}_i} \subseteq VAL \times VAL$ (overloading notation slightly) as:

$$R_{\mathbf{T}_i}(v_1, v_2) \Leftrightarrow R_{\mathbf{T}_i}(v_1) = v_2$$

Note that since we have stipulated that $R_{\mathbf{T}_i}$ is a partial injective function we can define its inverse function $R_{\mathbf{T}_i}^{-1}$ as:

$$R_{\mathbf{T}_i}^{-1}(v_1, v_2) \Leftrightarrow R_{\mathbf{T}_i}(v_2, v_1)$$

Naturally, $R_{\mathbf{T}_i}^{-1}$ is the accessibility relation associated with the inverse $\mathbf{T}_i^{-1}$ of the similarity mapping $\mathbf{T}_i$, thus $R_{\mathbf{T}_i}^{-1} = R_{\mathbf{T}_i^{-1}}$.

Our notion of Kripke model now comes down to the following. A (simplified) Kripke model is an ordered tuple $\mathcal{M} = (VAL, \{R_{\mathbf{T}_i} | i = 1, ..., n\})$.

On the basis of this notion of a Kripke model, we introduce a modal language which we can evaluate with these models. This modal language consists of the propositional language over the propositional atoms $\mathcal{P}$ together with a clause for modal operators $\Box_{\mathbf{T}_i}$. The latter are interpreted on a Kripke model $\mathcal{M} = (VAL, \{R_{\mathbf{T}_i} | i = 1, ..., n\})$ as follows:

$$\mathcal{M}, v \models \Box_{\mathbf{T}_i} \varphi \Leftrightarrow \text{ for all } w \text{ with } R_{\mathbf{T}_i}(v, w) : \mathcal{M}, w \models \varphi.$$

We can also introduce further modalities derived from these $\Box_{\mathbf{T}_i}$, viz. $\Box_{\mathbf{T}_i}^{-1}$, $\Box_i$, $\Box_i^*$, $\Box$ and $\Box^*$, based on the (derived) relations $R_{\mathbf{T}_i}^{-1}$, $R_i = R_{\mathbf{T}_i} \cup R_{\mathbf{T}_i}^{-1}$, R_i^*, the transitive, reflexive closure of the relation R_i, $R = \bigcup_{i=1}^{n} R_i$, and R^*, the transitive, reflexive closure of the relation R, respectively. So, for instance,

$$\mathcal{M}, v \models \Box_i^* \varphi \Leftrightarrow \text{ for all } w \text{ with } R_i^*(v, w) : \mathcal{M}, w \models \varphi,$$

and similarly for the other operators. Also we may use the duals $\Diamond_{\mathbf{T}_i}$, $\Diamond_{\mathbf{T}_i}^{-1}$, $\Diamond_i$, $\Diamond_i^*$, $\Diamond$ and $\Diamond^*$ of these operators, defined as usual.

These operators enable us to express properties of analogies transcending just the relation between a source and a target domain. For instance, note that the relation R_i^* yields the equivalence class (*analogy class*) associated with similarity mapping $\mathbf{T}_i$, i.e. all models that are 'analogous with respect to $\mathbf{T}_i$'. Thus, the modality $\Box_i^*$ states something about what is common between domains that are related on the basis of $\mathbf{T}_i$. The modality $\Box^*$ states even more general properties, viz. those common to all domains that are related with respect to *some* similarity mapping. (Perhaps these properties are even too general to be useful, but this may depend on the context.)

As usual we define validity of a formula φ in a model $\mathcal{M}$, denoted $\mathcal{M} \models \varphi$, by $\mathcal{M}, v \models \varphi$ for all valuations v in $\mathcal{M}$, and validity of φ, denoted $\models \varphi$, by $\mathcal{M} \models \varphi$ for all models $\mathcal{M}$.

Since this gives us a normal modal logic in the sense of (Chellas 80), the modal operators above satisfy the K-axiom:

$$\models \Box_{\mathbf{T}_i}\varphi \wedge \Box_{\mathbf{T}_i}(\varphi \to \psi) \to \Box_{\mathbf{T}_i}\psi$$

(If in every i-similar world both φ and $\varphi \to \psi$ holds, then in all those worlds ψ holds.)

and the necessitation rule:

$$\models \varphi \Rightarrow \models \Box_{\mathbf{T}_i}\varphi$$

Moreover, we can now directly put Proposition 1 in modal terms:

Theorem 2.
$$\models \varphi \leftrightarrow \Box_{\mathbf{T}_i}\mathbf{T}_i(\varphi)$$

$\Diamond$

This theorem states precisely how rendition can be obtained by considering a similar world.

The other modal operators satisfy the following validities (Here $\Box^*_{(i)}$ stands for a modal operator in the set $\{\Box^*, \Box^*_i\}$):

- $\models \Box_i\varphi \leftrightarrow \Box_{\mathbf{T}_i}\varphi \wedge \Box_{\mathbf{T}_i}^{-1}\varphi$
- $\models \varphi \to \Box_{\mathbf{T}_i}\Diamond_{\mathbf{T}_i}^{-1}\varphi$
- $\models \varphi \to \Box_{\mathbf{T}_i}^{-1}\Diamond_{\mathbf{T}_i}\varphi$
- $\models \Box^*_{(i)}\varphi \to \varphi$
- $\models \Box^*_{(i)}\varphi \to \Box_{(i)}\Box^*_{(i)}\varphi$
- $\models \Box^*_{(i)}(\varphi \to \Box_{(i)}\varphi) \to (\varphi \to \Box^*_{(i)}\varphi)$
- $\models \Box\varphi \leftrightarrow (\Box_1\varphi \wedge \ldots \wedge \Box_n\varphi)$

This follows directly from the definition of the relations that are associated with these operators.

Moreover, since always R_i and R are symmetrical and R^* is an equivalence relation, independent of the properties of the relation $R_{\mathbf{T}_i}$ itself, we also immediately have the following validities:

1. $\models \varphi \to \Box_i\Diamond_i\varphi$
2. $\models \varphi \to \Box\Diamond\varphi$
3. $\models \varphi \to \Box^*_{(i)}\Diamond^*_{(i)}\varphi$
4. $\models \Box^*_{(i)}\varphi \to \Box^*_{(i)}\Box^*_{(i)}\varphi$
5. $\models \Diamond^*_{(i)}\varphi \to \Box^*_{(i)}\Diamond^*_{(i)}\varphi$

2.3 Analogical inferences

We have seen above how the concept of analogy by rendition can be formalized by means of our similarity mappings. We now show how we can also formulate true analogical reasoning in the sense of making inferences in our setting.

Let $\mathcal{R}$ be an inference: $\mathcal{R} = \varphi_1 \vdash \varphi_2 \vdash ... \vdash \varphi_m$. Now we would like to make an 'analogous' reasoning in another domain: $\mathcal{R}' = \varphi_1' \vdash \varphi_2' \vdash ... \vdash \varphi_m'$.

By using our similarity mappings and associated modal operators we can now do this in a formal way. Let us say that the similarity mapping involved is $\mathbf{T}$. Then we would expect that the 'source' inference $\mathcal{R} := \varphi_1 \vdash \varphi_2 \vdash ... \vdash \varphi_m$ could be transformed into: $\mathcal{R}' = \mathbf{T}(\varphi_1) \vdash \mathbf{T}(\varphi_2) \vdash ... \vdash \mathbf{T}(\varphi_m)$. That this is indeed the case is justified by the following derived rule:

$$\frac{\varphi \to \psi}{\Box_{\mathbf{T}}\mathbf{T}(\varphi) \to \Box_{\mathbf{T}}\mathbf{T}(\psi)}$$

Via this rule we can now reason as follows: suppose $\varphi \vdash \psi$. Then by the deduction theorem of classical logic we obtain $\vdash \varphi \to \psi$, and so by the above rule, $\vdash \Box_{\mathbf{T}}\mathbf{T}(\varphi) \to \Box_{\mathbf{T}}\mathbf{T}(\psi)$, and hence $\Box_{\mathbf{T}}\mathbf{T}(\varphi) \vdash \Box_{\mathbf{T}}\mathbf{T}(\psi)$. This is indeed the formal statement of the 'translated' inference above. Note, moreover, that our modal framework exactly pin-points the worlds where this analogous reasoning takes place. If one did not have this, one would be forced to use a 'semantically polluted' inference rule (so not really an inference rule at all) such as

$$\frac{v \models \varphi \to \psi}{w \models \mathbf{T}(\varphi) \to \mathbf{T}(\psi)}$$

where $w = R_{\mathbf{T}}(v)$.

3 Extension to First-Order Logic

If we want to extend our propositional logic to a first-order one, we need to enrich the structure of our models, and consequently redefine our notion of similarity between models. Here we build on work by (Thiele 86) but again provide for a possible world semantics. The way we consider similarity mappings in the first-order case is also reminding of work done on so-called *interpretability logics* in a completely different context, viz. in metamathematics for the proof of consistency and undecidability of mathematical theories (cf. e.g. (Tarski, Mostowski & Robinson 53)).

3.1 Similarity in a first-order setting

First of all, we extend our propositional language to a first-order one: we assume a set VAR of variables, a set $FUNC$ of function symbols and a set $PRED$ of predicate symbols. As usual, function and predicate symbols have an arity associated with them determining the number of arguments. (0-ary function symbols are called constants; 0-ary predicates are called atomic propositions.) We call a pair $\Sigma = (FUNC, PRED)$ a *signature*. A signature $\Sigma_1 = (FUNC_1, PRED_1)$ is a *subsignature* of $\Sigma_2 = (FUNC_2, PRED_2)$ if $FUNC_1 \subseteq FUNC_2$ and $PRED_1 \subseteq PRED_2$. We can also speak about the intersection $\Sigma_1 \sqcap \Sigma_2$ of two signatures Σ_1 and Σ_2 in the obvious way (just take the intersections of the sets of function symbols and of the sets of predicate symbols).

The set $TERM(VAR, FUNC, PRED)$, or just abbreviated $TERM$, is the minimal set containing VAR and closed under the construction $g(t_1, ..., t_n)$ for function symbols $g \in FUNC$ and $t_i \in TERM$ $(i = 1, ..., n)$ where the arity of g is n. The set $AT(VAR, FUNC, PRED)$, or AT, of atomic formulas is given as the smallest set closed under the constructions

- $P(t_1, ..., t_n)$ for function symbols $P \in PRED$ and $t_i \in TERM$ $(i = 1, ..., n)$ where the arity of P is n, and
- $t_1 = t_2$ for $t_1, t_2 \in TERM$.

The set $\mathcal{L}(VAR, FUNC, PRED)$, usually abbreviated $\mathcal{L}$, of first-order formulas is the minimal set closed under the classical connectives and the construction $\forall x\ \varphi$ and $\exists x\ \varphi$ with $x \in VAR$ and $\varphi \in \mathcal{L}$, and containing the set AT of atomic formulas. We denote the set of free variables of a formula φ by $FV(\varphi)$.

As usual, a language $\mathcal{L}(VAR, FUNC, PRED)$ over signature $\Sigma = (FUNC, PRED)$ is interpreted on Σ-structures of the form $w = (\mathcal{A}, \Sigma, \Phi, \Pi)$, where $\mathcal{A}$ is a domain of values for the interpretation of the variables and constants, Φ is a function such that, for all $g \in FUNC$, $\Phi : g \mapsto (\mathcal{A}^n \rightarrow_{part} \mathcal{A})$ where n is the arity of g, and Π is a function such that, for all $P \in PRED$, $\Pi : P \mapsto (\mathcal{A}^n \rightarrow_{tot} \{tt, ff\})$ where n is the arity of P. (Here $X \rightarrow_{part} Y$ and $X \rightarrow_{tot} Y$ stand for the classes of partial and total functions from X to Y, respectively.) When convenient we may also denote $\Pi(P)$ as a subset of $\mathcal{A}^n$. We denote the class of Σ-structures as $STRUCT(\Sigma)$, while the class of Σ-structures with fixed domain $\mathcal{A}$ is denoted $STRUCT(\mathcal{A}, \Sigma)$. We omit the standard clauses for the interpretation of the language on these structures, since they can be found in any textbook on logic (such as e.g. (van Dalen 89)).

Given a structure $w = (\mathcal{A}, \Sigma, \Phi, \Pi)$, with $\Sigma = (FUNC, PRED)$. Let $\mathcal{A}_0$ be another domain and let $\Sigma_0 = (FUNC_0, PRED_0)$ be another signature. We define the substructure $w \downarrow \mathcal{A}_0, \Sigma_0$ of w as the structure $(\mathcal{A} \cap \mathcal{A}_0, \Sigma \sqcap \Sigma_0, \Phi \downarrow \mathcal{A}_0, \Sigma_0, \Pi \downarrow \mathcal{A}_0, \Sigma_0)$, where $\Phi \downarrow \mathcal{A}_0, \Sigma_0$ is a function interpreting only the function symbols in $FUNC \cap FUNC_0$ in the domain (and range) $\mathcal{A} \cap \mathcal{A}_0$, and similarly for $\Pi \downarrow \mathcal{A}_0, \Sigma_0$.

In this more refined set-up we can also be more precise about the similarity mapping. Instead of just stipulating a mapping from atomic formulas to atomic formulas, we now define a mapping between signatures $\Sigma_1 = (FUNC_1, PRED_1)$ and $\Sigma_2 = (FUNC_2, PRED_2)$ where we assume $FUNC_1, FUNC_2 \subseteq FUNC$ and $PRED_1, PRED_2 \subseteq PRED$. In fact, we shall define a similarity mapping based on a signature *isomorphism*:

A pair (T^1, T^2) is a signature isomorphism from $\Sigma_1 = (FUNC_1, PRED_1)$ to $\Sigma_2 = (FUNC_2, PRED_2)$ if

1. T^1 is a bijective, arity-preserving mapping from $FUNC_1$ to $FUNC_2$.
2. T^2 is a bijective, arity-preserving mapping from $PRED_1$ to $PRED_2$.

We can extend a signature isomorphism to a function T on $TERM$ and AT as follows:

Let $T^1 : FUNC \to FUNC'$ and $T^2 : PRED \to PRED'$ be arbitrary functions, then we define $T : AT(VAR, FUNC, PRED) \to AT(VAR, FUNC', PRED')$ on atoms as follows:

- $T(x_i) = x_i$ for $x_i \in VAR$
- $T(f) = T^1(f)$ for $f \in FUNC$
- $T(P) = T^2(P)$ for $P \in PRED$
- $T(f(t_1, ..., t_n)) = T(f) (T(t_1), ..., T(t_n))$ if f has arity n
- $[T(t_1 = t_2)] = [T(t_1) = T(t_2)]$
- $T(P(t_1, ..., t_i)) = T(P) (T(t_1), ..., T(t_n))$ if P has arity n

An isomorphism between signatures together with a bijection between the domains of the structures based on these signatures induces a similarity mapping between these structures .

Assuming a signature isomorphism (T^1, T^2) from $\Sigma = (FUNC, PRED)$ to $\Sigma' = (FUNC', PRED')$, and a bijective function T^0 from the domain $\mathcal{A}$ to the domain $\mathcal{A}'$, the *similarity mapping* $R_{\mathbf{T}}$ from $STRUCT(\mathcal{A}, \Sigma)$ to structures $STRUCT(\mathcal{A}', \Sigma')$ induced by $\mathbf{T} = (T^0, T^1, T^2)$ is given by:

for all $w = (\mathcal{A}, \Sigma, \Phi, \Pi) \in STRUCT(\mathcal{A}, \Sigma), R_{\mathbf{T}}(w) = (\mathcal{A}', \Sigma', \Phi', \Pi') \in STRUCT(\mathcal{A}', \Sigma')$ satisfying

1. $\Phi(f)(a_1, ..., a_n)$ exists $\Leftrightarrow$ $\Phi'(T^1(f))(T^0(a_1), ..., T^0(a_n))$ exists, and
   $T^0(\Phi(f)(a_1, ..., a_n)) = \Phi'(T^1(f))((T^0(a_1), ..., T^0(a_n))$
   for all $f \in FUNC$ and $(a_1, ..., a_n) \in \mathcal{A}^n$
   Note: for constants this means $T^0(\Phi(c)) = \Phi'(T^1(c))$
2. $\Pi(P)(a_1, ..., a_n) = \Pi'(T^2(P))(T^0(a_1), ..., T^0(a_n))$
   for all $P \in PRED$ and $(a_1, ..., a_n) \in \mathcal{A}^n$

Let $\bar{t}^w[a_1/x_1, ..., a_n/x_n]$ be the interpretation of the term $t(x_1, ..., x_n)$ in structure w under the valuation $x_1 \mapsto a_1, ... x_n \mapsto a_n$. (This can be defined inductively, which we omit here.)

Lemma 3. *Let $w \in STRUCT(\mathcal{A}, \Sigma)$ and $w' = R_{\mathbf{T}}(w)$ for $\mathbf{T} = (T^0, T^1, T^2)$. For any $a_1, ... a_n \in \mathcal{A}$ and any term t:*

$$T^0(\bar{t}^w[a_1/x_1, ..., a_n/x_n]) = \overline{T(t)}^{w'}[T^0(a_1)/x_1, ..., T^0(a_n)/x_n]$$

$\diamond$

In the sequel of this paper we employ the notation $w \models \varphi\, [a_1/x_1, ..., a_n/x_n]$ (for $a \in \mathcal{A}$) meaning that φ is true in w under the valuation $x_1 \mapsto a_1, ... x_n \mapsto a_n$.

The translation lemma above for terms gives rise to a first-order version of rendition of atomis formulas as put in the following proposition.

Proposition 4. *Let $\Sigma = (FUNC, PRED)$ and $\Sigma' = (FUNC', PRED')$. Suppose $w \in STRUCT(\mathcal{A}, \Sigma)$, and $w' = R_{\mathbf{T}}(w) \in STRUCT(\mathcal{A}', \Sigma')$ for $\mathbf{T} = (T^0, T^1, T^2)$. Let T be the function $T : AT(VAR, FUNC, PRED) \to AT(VAR, FUNC', PRED')$ from the above definition, in terms of T^1 and T^2.
Then for any $a_1, ..., a_n \in \mathcal{A}$ and $P(x_1, ..., x_n) \in AT(VAR, FUNC, PRED)$:*
$$w \models P(x_1, ..., x_n)\, [a_1/x_1, ..., a_n/x_n] \iff$$
$$w' \models T(P)(x_1, ..., x_n)\, [T^0(a_1)/x_1, ..., T^0(a_n)/x_n] \quad \Diamond$$

T can be lifted to formulas $\varphi \in \mathcal{L}(VAR, FUNC, PRED)$ (providing a base for prediction again) as follows:

- $T(\forall x_1 \varphi(x_1, ..., x_n)) = \forall x_1\, T(\varphi)(x_1, ..., x_n)$
- $T(\exists x_1 \varphi)(x_1, ..., x_n)) = \exists x_1\, T(\varphi)(x_1, ..., x_n)$
- $T(\varphi \vee \psi) = T(\varphi) \vee T(\psi)$
- $T(\varphi \wedge \psi) = T(\varphi) \wedge T(\psi)$
- $T(\varphi \to \psi) = T(\varphi) \to T(\psi)$
- $T(\neg\varphi) = \neg T(\varphi)$

Proposition 5. *Let $\Sigma = (FUNC, PRED)$ and $\Sigma' = (FUNC', PRED')$. Suppose $w \in STRUCT(\mathcal{A}, \Sigma)$, and $w' = R_{\mathbf{T}}(w) \in STRUCT(\mathcal{A}', \Sigma')$ for $\mathbf{T} = (T^0, T^1, T^2)$. Let T be the function $T : AT(VAR, FUNC, PRED) \to AT(VAR, FUNC', PRED')$ from the above definition, in terms of T^1 and T^2.
then for any formula $\varphi \in \mathcal{L}(VAR, FUNC, PRED)$ with $FV(\varphi) = \{x_1, ..., x_n\}$ and any $a_1, ..., a_n \in \mathcal{A}$*

$$w \models \varphi\, [a_1/x_1, ..., a_n/x_n] \iff w' \models T(\varphi)\, [T^0(a_1)/x_1, ..., T^0(a_n)/x_n]$$

In particular, if $FV(\varphi) = \emptyset$ then

$$w \models \varphi \iff w' \models T(\varphi)$$

$\Diamond$

3.2 First-order modal logic of analogy

We now enhance our Kripke models to cater for our first-order language. Essentially, worlds are changed from simple valuations on propositional atoms to first-order structures containing a signature $(FUNC, PRED)$ and a domain of interpretation for the variables and constants.

For the accessibility relations we use relations induced by similarity mappings $STRUCT(\mathcal{A}, \Sigma)$ to structures $STRUCT(\mathcal{A}', \Sigma')$: given a set $\{\mathbf{T}_i | i = 1, ..., n\}$

of the form $\mathbf{T}_i = (T_i^0, T_i^1, T_i^2)$ as above, we define $R_{\mathbf{T}_i} \subseteq STRUCT(\Sigma) \times STRUCT(\Sigma')$ as:

$$R_{\mathbf{T}_i}(w_1)(w_2) \Leftrightarrow R_{\mathbf{T}_i}(w_1) = w_2.$$

Note that since we have stipulated that $R_{\mathbf{T}_i}$ is a triple of bijective functions we can again define its inverse $R_{\mathbf{T}_i}^{-1}$ as:

$$R_{\mathbf{T}_i}^{-1}(w_1)(w_2) \Leftrightarrow R_{\mathbf{T}_i}(w_2)(w_1).$$

Our notion of Kripke model now becomes the following. We assume a universal domain $\mathcal{A}$ and signature $\Sigma = (FUNC, PRED)$, and consider worlds as first-order structures of type $w = (\mathcal{A}', \Sigma', \Phi', \Pi')$, where $\mathcal{A}' \subseteq \mathcal{A}$, Σ' is a subsignature of Σ, and Φ and Π are interpretation functions of the function and predicate symbols, respectively, of the signature Σ' in the domain $\mathcal{A}'$, as explained above. The class of all Σ'-structures for all Σ' that are subsignatures of Σ is denoted by $STRUCT$. The worlds in our first-order Kripke models will be just a subset of $STRUCT$. Note that this more liberal than in the propositional setting where we took as universe of worlds just *all* valuation functions. We think that in the first-order setting this liberality is more realistic, since we are very unlikely to consider all possible Σ'-structures. On the other hand, in the propositional setting we also considered *partial* valuations in order to cope with similarities between worlds with different relevant propositions.

In the first-order setting that we propose, we do this a little different. For convenience, we will work with similarity mappings between structures which are signature isomorphic and for which there is a bijection between the domains, but we will allow that the actual structures in the Kripke models have a richer structure (but irrelevant as to the similarity under consideration). Technically, we allow for this by looking at *relevant* subdomains and subsignatures. So, for example, we might consider a similarity mapping between two structures $w_1 = (\mathcal{A}_1, \Sigma_1, \Phi_1, \Pi_1)$ and $w_2 = (\mathcal{A}_2, \Sigma_2, \Phi_2, \Pi_2)$ with respect to the restriction to the domain/signature pairs $\sigma_0 = (\mathcal{A}_0, \Sigma_0)$ and $\sigma'_0 = (\mathcal{A}'_0, \Sigma'_0)$ which means that actually only there is a similarity mapping from the substructure $w_1 \downarrow \mathcal{A}_0, \Sigma_0$ to the substructure $w_1 \downarrow \mathcal{A}'_0, \Sigma'_0$ of w_2. We denote such a mapping as $\mathbf{T}(\sigma_0, \sigma'_0)$.

A (first-order) Kripke model is an ordered tuple $\mathcal{M} = (STRUCT', \{R_{\mathbf{T}_i(\sigma_i, \sigma'_i)} | i = 1, ..., n\})$, where $STRUCT' \subseteq STRUCT$ and $\sigma_i = (\mathcal{A}_i, \Sigma_i)$ with $\mathcal{A}_i \subseteq \mathcal{A}$ and Σ_i is a subsignature of Σ, and similarly for σ'_i.

Here the relations $R_{\mathbf{T}_i(\sigma_i, \sigma'_i)}$ give the similarity mappings that are considered. Note that in view of the above, with each such mapping it is specified with respect to which restricting signature one should take the similarity.

On the basis of this Kripke model, we can interpret modal operators of type $\Box_{\mathbf{T}_i(\sigma_i, \sigma'_i)}$ and derived modalities $\Box_{\mathbf{T}_i(\sigma_i, \sigma'_i)}^{-1}$, $\Box_i$, $\Box$ and $\Box^*$, as in the propositional case: for instance,

$$\mathcal{M}, w \models \Box_{\mathbf{T}_i(\sigma_i, \sigma'_i)}\varphi \Leftrightarrow \text{ for all } w' \text{ with } R_{\mathbf{T}_i(\sigma_i, \sigma'_i)}(w, w') : \mathcal{M}, w' \models \varphi.$$

$\Box_{\mathbf{T}_i(\sigma_i, \sigma'_i)}^{-1}$, $\Box_i$, $\Box$ and $\Box^*$ are based on the (derived) relations $R_{\mathbf{T}_i(\sigma_i, \sigma'_i)}^{-1}$, $R_i = R_{\mathbf{T}_i(\sigma_i, \sigma'_i)} \cup R_{\mathbf{T}_i(\sigma_i, \sigma'_i)}^{-1}$, $R = \bigcup_{i=1}^{n} R_i$, and R^*, the transitive, reflexive closure of the relation R, respectively. So, for instance,

$$\mathcal{M}, v \models \Box^*\varphi \Leftrightarrow \; for \; all \; w \; with \; R^*(v, w) \; : \; \mathcal{M}, w \models \varphi,$$

and similarly for the other operators. Also we may use the duals $\Diamond_{\mathbf{T}_i(\sigma_i, \sigma'_i)}$, $\Diamond^{-1}_{\mathbf{T}_i(\sigma_i, \sigma'_i)}$, $\Diamond_i$, $\Diamond$ and $\Diamond^*$ of these operators, defined as usual.

As usual we define validity of a formula φ in a model $\mathcal{M}$, denoted $\mathcal{M} \models \varphi$, by $\mathcal{M}, w \models \varphi$ for all worlds w in $\mathcal{M}$, and validity of φ, denoted $\models \varphi$, by $\mathcal{M} \models \varphi$ for all models $\mathcal{M}$.

We can now put Proposition 3 in modal terms if we abuse our language slightly:

Theorem 6.

$$\models \; \varphi \leftrightarrow \Box_{\mathbf{T}_i} \mathbf{T}_i(\varphi)[T^0(x_1)/x_1, ..., T^0(x_n)/x_n]$$

$\Diamond$

Note that this formula contains the substitution of a variable x by the formula $T^0(x)$, not to be confused with $T(x)$, which is just equal to x itself. The interpretation of this nonstandard formula is as follows: in a world w of a model $\mathcal{M}$, if the variable x has the value a, $T^0(x)$ denotes the element $T^0(a)$.

In the full paper we present the example of the solar system as similar to the atom in some detail, where e.g. the sun and some of its associated predicates (such as 'Large') is mapped onto the nucleus of the atom and predicates associated with this and likewise for planets and electrons, respectively.

4 Defeasible Analogical Reasoning

In the previous sections we have treated reasoning by analogy in such a way that the conclusions of such reasoning are certain and inescapable. This is obviously an over-idealization and not true in general. If one knows exactly to what degree two similar structures are similar, then drawing certain conclusions seems to be a correct procedure. However, in practice this is almost never the case. We have uncertainties about the target domain which is the very reason why we try to compare it with a familiar source domain and draw some conclusions from it which are meant to be tentative and might be defeated by the discovery of new information about the target domain. In this case it seems even to be very undesirable to be able only 'hard' conclusions, since this implies that either we cannot infer interesting but uncertain things about the target domain (and this is what we are aiming for), or when we derive something which is contradictory with later observations we are faced with a hard inconsistency.

For instance, consider the solar system - Atom example. Here there is a similarity between the Sun and the nucleus of the atom, and we might transfer what we know about the Sun to the nucleus to a certain degree. This we might want to specify completely in the sense that we specify exactly which predicates are considered in the similarity mapping. For instance, the predicate 'is the centre of the (sun) system' can be mapped onto the predicate 'is the centre of the (atom) system'. On the other hand, the predicate 'is yellow' for the sun cannot be mapped so easily to something similar in the domain of the atom. In general, however, we do not know exactly which predicates 'carry over' and which do not.

For this reason we introduce the notion of *defeasibility* for analogical reasoning. The upshot of this is that conclusions concerning certain predicates are 'likely', but not quite certain, and can be defeated by new information about the target domain (such as direct observations, although we will not specify the source of information here). Some argue that this likelihood has to do with (high) probabilities. This might well be the case in particular contexts but as in studies of default reasoning we are more inclined to look at this from a more qualitative perspective and view 'rendition rules' in analogical reasoning as *default rules*, i.e. rules that under normal circumstances can be applied, but allow exceptions if one has the disposal of additional information.

So, in a sense, what we will do, is combining our modal framework of analogical reasoning with defeasible reasoning, and, in particular, some form of default reasoning. The way we will do this is inspired by earlier work on default and counterfactual reasoning we have done (Meyer & Van der Hoek 93, 95).

Before engaging into the formal details we will give the general idea. Suppose we have two domains that we think are similar in certain respects. And suppose we have information φ about the source domain that is likely to translate to the target domain (so, technically we are then speaking about a predicate that is in the domain of our similarity mapping, but which must be considered defeasible). Then we can use our framework to derive its translation $T(\varphi)$ for the target domain. Now, we first check whether this assertion $T(\varphi)$ is *consistent* with what we know already about the target domain. If this is not the case, we forget about the conclusion, If it is, we draw the tentative conclusion that $T(\varphi)$ holds in the target domain. In the formalism we will use, it will be indicated or marked as tentative by means of a modal operator, which basically states that the formula is true in a selected or *preferred* subset of the worlds that describe the (knowledge about the) target domain. Naturally, this allows for the possibility that when new information about the target domain is obtained, this will lead us to a (set of) world(s) that are not within this preferred subset of worlds, which means that the tentatively believed conclusion about the target domain will not be true after all.

Formally we go about as follows:

We extend our modal language with some new operators. Firstly, we introduce operators $\Box'_{T_i}$. These operators will be used to point to a (preferred) subset of the set of worlds that pertain to the target domain with respect to the similarity mapping T_i. For further convenience, we also introduce modal operators $[S]$, where S is a subset of the set VAL of all worlds. This will facilitate speaking about the source and target domains, since these operators give us so-to-speak a direct pointer to the worlds pertaining to these, whereas the modalities associated with the similarity mappings do this in a rather indirect manner. Finally, we also consider the duals $\Diamond'_{\mathbf{T}_i}$ and $< S > \varphi$ defined as usual.

We enrich our Kripke structures with some extra elements:

An (enriched) Kripke model is an ordered tuple $\mathcal{M} = (VAL, \{R_{\mathbf{T}_i} | i = 1, ..., n\}, \{R'_{\mathbf{T}_i} | i = 1, ..., n\})$, where, for all i and j, $R'_{\mathbf{T}_i} \subseteq R_{\mathbf{T}_i}$.

The additional operators are interpreted on these enriched Kripke models as follows:

$$\mathcal{M}, v \models \Box'_{\mathbf{T}_i} \varphi \Leftrightarrow \quad for\ all\ w\ with\ R'_{\mathbf{T}_i}(v, w) : \quad \mathcal{M}, w \models \varphi$$

and

$$\mathcal{M}, v \models [S]\varphi \Leftrightarrow \quad for\ all\ w \in S : \quad \mathcal{M}, w \models \varphi$$

One now immediately obtains the following validities:

1. $\models \Box_{\mathbf{T}_i} \varphi \to \Box'_{\mathbf{T}_i} \varphi$
2. $\models [S_1]\varphi \to [S_2]\varphi$ for $S_1 \supseteq S_2$
3. $\models < S_1 > \top \to ([S_1][S_2]\varphi \leftrightarrow [S_2]\varphi)$
4. $\models < S_1 > \top \to ([S_1] < S_2 > \varphi \leftrightarrow < S_2 > \varphi)$

Using the expressibility of our modal language we can now give defeasible versions of analogical derivation rules:

Instead of the (strict) rule

$$\varphi \to \Box_{\mathbf{T}_i}(\mathbf{T}_i(\varphi))$$

which was a validity in the previous sections, where we based the modality $\Box_{\mathbf{T}_i}$ on a strict rendition function $\mathbf{T}_i$, we now relax this and use the defeasible rule:

$$\varphi \wedge \Diamond_{\mathbf{T}_i}(\mathbf{T}_i(\varphi)) \to \Box'_{\mathbf{T}_i}(\mathbf{T}_i(\varphi)) \tag{1}$$

which expresses formally what we described informally above: if the $\mathbf{T}_i$-translation of the source domain information φ is compatible with the information about the target domain determined by the similarity mapping $\mathbf{T}_i$, then we prefer a set of worlds within the target domain satisfying this translated information $\mathbf{T}_i(\varphi)$.

In the example of the solar system - atom we might, for instance, when we are not completely sure about this, try to render the predicate (light) emitting in the solar system to a predicate (radiation) emitting (added to the signature) in the atom system to be used as a kind of hypothesis. For this rendition one now can use the above *defeasible* rule

$$Emitting(Sun) \wedge \Diamond_{\mathbf{T}_i}(\mathbf{T}_i(Emitting(Sun))) \to \Box'_{\mathbf{T}_i}(\mathbf{T}_i(Emitting(Sun)))$$

yielding the rule:

$$Emitting(Sun) \wedge \Diamond_{\mathbf{T}_i}(Emitting(Nucleus)) \to \Box'_{\mathbf{T}_i}(Emitting(Nucleus))$$

428

This means that *unless there is information in the target domain that it is not the case that Emitting(Nucleus) is true*, it is assumed that it is so.

Note that the target domain is generally dependent on the similarity mapping $\mathbf{T}_i$ and the world where you are (in or rather pertaining to the source domain), since in general we allow the same similarity mapping $\mathbf{T}_i$ to point to different target domains from different source domains.

This is the reason why we do *not* have 'S5-like' validities such as

- $\Diamond_{\mathbf{T}_i}(\top) \rightarrow (\Box_{\mathbf{T}_i}\Box'_{\mathbf{T}_i}(\varphi) \leftrightarrow \Box'_{\mathbf{T}_i}(\varphi))$
- $\Diamond'_{\mathbf{T}_i}(\top) \rightarrow (\Box'_{\mathbf{T}_i}\Box_{\mathbf{T}_i}(\varphi) \leftrightarrow \Box_{\mathbf{T}_i}(\varphi))$

To ease working with these defeasible rules and reasoning about a particular source and target domain, we may use the modalities $[S]\varphi$, which are 'S5-like'. The way to use these in practice is the following: identify your (fixed) source domain and the set S of worlds associated with this. Determine the target domain(s) S_i with respect to the similarity mapping(s) $\mathbf{T}_i$ under consideration. If this can be done, formula (1) can be written in a more 'rigid' or direct way:

$$[S]\varphi \wedge <S_i> (\mathbf{T}_i(\varphi)) \rightarrow [S'_i](\mathbf{T}_i(\varphi)) \text{ for some } S'_i \subseteq S_i \qquad (2)$$

When reasoning with this representation we have the disposal of the validities 1.-4. mentioned above.

5 Conclusion

In this paper we have given a semantical approach to analogical reasoning based on a possible worlds framework. Starting out from a very simple and rather naive notion of similarity and associated models and modal operators expressing such similarity in a purely propositional setting we have extended our approach to a first-order language and logic, and ended with the incorporation of the notion of defeasibility in analogical inferences. Of course, the semantics proposed does not yet capture the full complexity of all forms of reasoning by analogy. To begin with, only a qualitative form of analogical reasoning is treated, where it is abstracted away from all possible kinds of similarity measures. This issue should bear a relation with the defeasible nature of this form of reasoning, which can also be viewed both qualitatively and quantitatively. This will be subject of further study.

Acknowledgements. The authors wish to thank Wiebe van der Hoek, Dirk van Dalen, Peter van Emde Boas, Johan van Benthem, Rineke Verbrugge and Dick de Jongh for discussions on the topic of this work. Also the partial support of ESPRIT III BRA No 6156 DRUMS (Defeasible Reasoning and Uncertainty Management Systems) is gratefully acknowledged.

429

References

1. (Burstein 86)
   Burstein M.H. Incremental Analogical Reasoning. *Machine Learning, an Artificial Approach* Vol. II. Michalski R. S., Carbonel J.G., Mitchel T.M. (eds.), Morgan Kaufman, 1986.
2. (Burstein 88)
   Burstein M.H. Combining Analogies in Mental Models. *Analogical Reasoning.* Helman D.H. (ed.), Kluwer Academic Publishers, Boston, 1988.
3. (Chellas 80)
   Chellas B.F. *Modal Logic: An Introduction.* Cambridge University Press, Cambridge/London 1980.
4. (van Dalen 89)
   van Dalen D. *Logic and Structure.* Springer-Verlag, second edition 1989.
5. (Evans 68)
   Evans T.G. A Program for the Solution of Geometric Analogy Intelligence Test Questions. *Semantic Information Processing.* Minsky M.L. (ed.), MIT Press, Cambridge 1968.
6. (Hall 89)
   Hall R.P. Computational Approaches to Analogical Reasoning. A Comparative Analysis. *AI Journal* 39(1) 1989, pp. 39-120.
7. (Indurkhya 89)
   Indurkhya B. Modes of Analogy. *International Workshop on AII.* Jantke K.P. (ed.), Lecture Notes in Artificial Intelligence, Springer Verlag 1989, pp. 217-229.
8. (van Leeuwen 95)
   J.C. van Leeuwen. Analogical Reasoning. A Semantical Approach, Master's Thesis, Utrecht University, 1995.
9. (Meyer & Van der Hoek 93)
   Meyer J.-J. Ch. & van der Hoek W. Counterfactual Reasoning by (Means of) Defaults. *Annals of Mathematics and Artificial Intelligence* 9 (III-IV), 1993, pp. 345-360.
10. (Meyer & Van der Hoek 95)
    Meyer J.-J. Ch. & van der Hoek W. A Default Logic Based on Epistemic States. *Fundamenta Informaticae* 23(1), 1995, pp. 33-65.
11. (Morgan 79)
    Morgan C.G. Modality, Analogy and Ideal Experiments according to C.S. Peirce. *Synthese* 41, 1979 (1), pp. 65-83.
12. (Peirce)
    Peirce C.S. *Collected Papers of Charles Saunders Peirce Vols. I - VIII.* ed. by Ch. Hartsborne, P. Wriss & A.W. Burks, Harvard University Press, Cambridge (1933-1958).
13. (Tarski, Mostowski & Robinson 53)
    Tarski A., Mostowski A. & Robinson R. *Undecidable Theories.* North-Holland 1953.
14. (Thiele 86)
    Thiele H. A Model Theoretic Approach to Analogy. *International Workshop on AII.* Jantke K.P. (ed.), Lecture Notes in Computer Science, Springer Verlag 1986, pp 196-208.
15. (Winston 68)
    Winston P. Learning and Reasoning by Analogy. *Communications of the ACM* 23(12), 1968, pp 683-703.

Using Temporary Integrity Constraints to Optimize Databases

Danilo Montesi[1], Chiara Renso[2] and Franco Turini[2]

[1] School of Information Systems, University of East Anglia
Norwich NR4 7TJ, UK
dm@sys.uea.ac.uk
[2] Dipartimento di Informatica, Università di Pisa
Corso Italia, 40, 56125 Pisa, Italy
renso/turini@di.unipi.it

Abstract. Integrity constraints are usually assumed to be permanent properties that must be satisfied by any database state. However, there are many situations requiring also temporary constraints, that is, constraints that must hold only for a single database state. In this paper we propose a schema to define both permanent and temporary constraints, that supports efficient constraint checking and semantic query optimization. The proposed schema associates integrity constraints to rules and queries, and it uses methods, that were originally defined for permanent constraints only, to perform constraint checking and semantic query optimization.

1 Introduction

Integrity constraints play an important role in databases and artificial intelligence [8, 4]. In databases they are used to enforce properties of relations, like, for example, that "the age of a person is within the interval 0 − 150". While we can consider this as a permanent constraint, there are other cases where integrity constraints are temporary, that is, they are required to hold only on a single database state [9]. Similarly, in the design of intelligent agents there is often the need for constraints that hold only for a specific session [6]. Unfortunately, temporary and permanent integrity constraints very often are not considered within a homogeneous schema, that allows one to take advantage of both of them. For instance, in databases permanent constraints do not only enforce specific conditions, but they are also important for performing query optimization [5].

The contribution of this paper is to show that it is possible to have both temporary and permanent constraints within the same schema. Furthermore, we extend methods already developed for permanent constraints to the checking and semantic query optimization of temporary constraints. This is achieved through an extended version of Datalog that allows us to have constraints in rules and queries [7]. The resulting $Datalog^{IC}$ language allows us to express knowledge evolution within a logical and well know framework, that is the CLP(X)

schema [3]. Then we show that temporary constraint checking can take advantage of permanent constraints and that query evaluation can take advantage of temporary constraints. The paper is organized as follows. Section 2 sketches the basics of $Datalog^{IC}$. Section 3 provides an algorithm that compiles permanent constraints into the rules of the database, and discusses its soundness. Section 4 discusses the constraints and semantic query optimization in the presence of temporary and permanent constraints. Finally, section 5 summarizes the results.

2 $Datalog^{IC}$

Here we briefly recall the rule-based language named $Datalog^{IC}$. The extensional database (EDB) is a set of extensional ground atoms, while the intensional one is a set of rules of the form

$$H \leftarrow IC_1, \ldots, IC_n, B_1, \ldots, B_m$$

where $B_1, \ldots, B_m$ (as in Datalog) is the query part, that cannot be empty, H is an intensional atom and $IC_1, \ldots, IC_n$ is the integrity constraint part (possibly empty). The two parts do not share variables. Each IC_i is a denial of the form $\leftarrow L_1, \ldots, L_m$ where the L_i are atoms which use only extensional relations or order atoms, i.e., atoms using comparison operators. The intuitive meaning of the above rule is that "if $B_1, \ldots, B_m$ is true, and $IC_1, \ldots, IC_n$ are not violated (in EDB), then H is true". A query is a rule with no head of the form ? $IC_1, \ldots, IC_n, B_1, \ldots, B_m$. A $Datalog^{IC}$ database DB consists of the extensional database and of the intensional database. Note that the $Datalog^{IC}$ becomes Datalog whenever the constraint part is empty. The following example illustrates the main features of the language. We represent a timetable for trains. In particular, we consider trains connecting Rome and Pisa. The predicate `train` states that there is a direct train between two cities and it contains information about the type of the train and the departure and arrival time. The predicate `trip` defines, in a recursive way, all the possible trips between cities.

TIMETABLE

`train(intercity,rome,pisa,0930,1210)`	*EDB1*
`train(TGV,rome,florence,1015,1215)`	*EDB2*
`train(expr,florence,pisa,1230,1320)`	*EDB3*

```
trip(StartC, ArrivalC, StartT, ArrivalT) ←                    IDB1
              train(X, StartC, ArrivalC, StartT, ArrivalT)
trip(StartC, ArrivalC, StartT, ArrivalT) ←                    IDB2
              trip(TempC, ArrivalC, TempT, ArrivalT),
              train(X, StartC, TempC, StartT, TempT)
```

IC_p

```
← train(TGV, X, Y , T1, T2),
      train(intercity, X, Y, T3, T4), (T2 - T1) > (T3 - T4)          IC1

Q₁
← ← train(TGV, rome, Y, T1, _ ), T1 > 2300, T1 < 0600,               IC2
      trip(rome,pisa, X, Y),
```

The constraint $IC1$ states that a train "TGV" is faster than an "intercity". For the sake of simplicity, we indicate with the symbol "$-$" a more complex operation on time which should take into account hours and days. Suppose that, due to technical reasons, temporarily no "TGV" can leave from the Railway Station of Rome during the night. Then, we may associate a temporary constraint representing this state to each query. Notice however that a temporary constraint is independent from the specific query. It just represent a temporary illegal state of the database. This is the case of the query Q_1, in which we ask for a trip from Rome to Pisa with the temporary constraint ($IC2$) that the trains classified "TGV" cannot leave from Rome during the night. We have that the temporary constraint associated to the query Q_1 satisfies EDB, i.e. $EDB \models IC2$. So, the query can be evaluated.

It is worth noting that temporary constraints behave as checks that should be performed before the evaluation of the query. They are independent from the query and the evaluation of the query can start only when they have been checked.

3 From Datalog and IC to Datalog^{IC}

The main feature of the language $Datalog^{IC}$ is the presence of integrity constraints inside the body of rules and queries. Using integrity constraints in rules allows one to model permanent constraints, while adding integrity constraints to goals allows one to model temporary constraints. This is due to the fact that the intensional database is not assumed to change, while the queries do change.

The process of associating integrity constraints to rules is, in general, a hard task for the programmer. We propose an algorithm that implements this process automatically. Furthermore, the algorithm performs a kind of optimization on Datalog^{IC} since it removes "useless" rules. A rule is useless when its body can never be satisfied. Since we assume that the permanent constraints are satisfied in the current database state whenever a constraint is "included" in the body of a rule, the body will never be true, so the rule can be deleted. The notion of inclusion between clauses is formally defined as *subsumption* [1].

Definition 1. A clause C *subsumes* a clause D if there is a substitution σ such that Cσ is a subclause of D

The idea is to apply the subsumption algorithm to an integrity constraint and the body of a rule. When a constraint subsumes the body of a rule, the head will never be derived, and the rule can be deleted from the database. Let DB be a Datalog program and IC_p be the set of permanent constraints. Given DB and IC_p, the algorithm returns a Datalog^{IC} program equivalent to the original one.

Algorithm ADDCONSTR
Input: *An IDB Datalog program and IC_p constraints*
Output: *An IDB' Datalog^{IC} program*
begin
Var *TempC, R, C;*
    forall *rules $R_j \in IDB$ of the form $H \leftarrow B_1, \ldots, B_n$* **do**
    forall *constraints C_i in IC_p of the form $\leftarrow A_1, \ldots, A_k$* **do**
        begin
            if SUBSUMES(C_i, R_j) **then**
            begin
                $R'_j := \emptyset;$
                exit;
            end
            else
            begin
                $TempC := TempC \wedge C_i;$
                $R'_j := H \leftarrow TempC, B_1, \ldots, B_n;$
            end
        end
    $IDB' := \bigcup_j R'_j$
end.

The algorithm ADDCONSTR adds permanent constraints to a rule body whenever this one is not invalidated from the constraints themselves. We denote with SUBSUMES the subsumption algorithm presented in [1]. It is worth noting that our algorithm performs $n * k$ times the subsumption algorithm if n is the number of rules and k the number of constraints. However, this process of attaching permanent constraints is done at compile time. The resulting rules have the form

$$H \leftarrow (\leftarrow A_1, \ldots, A_k) \wedge \cdots \wedge (\leftarrow A'_1, \ldots, A'_h), B_1, \ldots, B_n.$$

As a shorthand, we denotes the above rule as $H \leftarrow IC_1 \wedge IC_2 \wedge \cdots \wedge IC_r, B_1, \ldots, B_n$. The transformation obtained through ADDCONSTR produces a database that, even if written in a different language, is equivalent to the former one. In the following definition we formalize this notion of equivalence. In the following we rely on the operational semantics described in [7] for Datalog^{IC}. Since the operational semantics coincides with SLD in case of logic programs without constraints, we can prove the following theorem, that states the equivalence between the resulting Datalog^{IC} program and the original Datalog program with constraints IC_p.

Theorem 2. *Let DB_D and DB'_{DIC} be programs, written in Datalog and Datalog^{IC} respectively, and IC_p a set of constraints. If DB'_{DIC} is obtained from DB_D and IC_p by applying algorithm* ADDCONSTR, *then DB_D is semantically equivalent to DB'_{DIC}.*

4 Optimizations

Integrity constraint checking takes into account the hypothesis that the constraints are satisfied in the previous state of the database. Therefore efficient constraint checking procedures verify the constraints by considering only the "difference" between the previous and the current states of the database [2].

Semantic query optimization is a technique that, given a database DB and a set of integrity constraints IC, transforms DB and IC into a new database DB' such that the evaluation of a query in DB' is more efficient than in DB. In other words, the transformation considers (fragments of) the constraints, that are supposed to be satisfied in the current database state, to improve the efficiency of the query evaluation process [5]. This method can be applied to Datalog^{IC} as far as permanent constraints are concerned. Due to the dynamic nature of temporary constraints, they cannot be taken into account during the transformation, since the hypothesis that they are satisfied in the current database state does not hold.

However, there are cases in which temporary constraints can be used both for efficient constraint checking and semantic query optimization. We use the constraint checking and semantic query optimization techniques already developed for permanent constraints. The semantic query optimization technique allows one to compile fragments of permanent constraints into rules. Then the algorithm ADDCONSTR adds extra (redundant) constraints that will be used to check if the temporary constraints are included in the permanent ones. If this is the case, the temporary constraints are deleted, since they are logical consequences of the permanent ones, that are (by definition) assumed to be satisfied. Thus the temporary constraints that are included in the permanent one are expected to be satisfied. If the temporary constraints are inconsistent with the permanent ones, then a policy for handling inconsistency has to be defined. For instance, the temporary constraints can be deleted assuming that they are weaker than the permanent ones. If the temporary constraints pass the check, then the query can be evaluated. Notice that now we can assume that all the constraints (both temporary and permanent) are satisfied. So we can perform the semantic query optimization based, this time, on all the constraints. The permanent constraints were compiled into the rules to take advantage of the traditional semantic query optimization, and the temporary ones are propagated through the constraints defined in the query and in the rules of Datalog^{IC}. In particular, when the temporary constraints are subsumed by the query, then the query can be deleted because it will never be satisfied.

5 Conclusion

We have seen that permanent and temporary integrity constraints can be defined in a uniform schema within the CLP(X) approach, and that we can extend the techniques already known for permanent constraint checking and semantic query optimization This approach is promising for further investigation in the field of textual databases.

References

1. C.L. Chang and R.C.T. Lee Symbolic Logic and Mechanical Theorem Proving, Academic Press, Inc., 1973
2. H. Decker. Integrity Enforcement in Deductive Databases. In *Proc. Int'l Conf. on Expert Database Systems*, pages 271–285, 1986.
3. J. Jaffar and M. Maher. Constraint Logic Programming: a survey. *Journal of Logic Programming*, 19:20, 503–581, 1994.
4. R. Kowalski Using Meta-logic to reconcile Reactive with Rational Agents Meta-Logics and Logic Programming, K. Apt and F. Turini editors, MIT Press, 1995.
5. A. Y. Levy and Y. Sagiv. Semantic Query Optimization in Datalog Programs. In *Proc. of the ACM Symposium on Principles of Database Systems*. ACM, New York, USA, 1995.
6. P. Maes. Designing Autonomous Agents. MIT, 1991.
7. D. Montesi and F. Turini. Knowledge Evolution in Deductive Databases. *Proc. International Symposium on Knowledge Retrieval, Use and Storage for Efficiency*, Santa Cruz, pp. 59–63, 1995.
8. F. Sadri and R. Kowalski. Integrity Checking in Deductive Databases. In P. Hammersley, editor, *Proc. Thirteenth Int'l Conf. on Very Large Data Bases*, pages 61–69, 1987.
9. J. D. Ullman. *Database and Knowledge-Base Systems*. Computer Science Press, 1989.

Graded Inheritance Nets for Knowledge Representation

Ingrid Neumann

Institut für Logik, Komplexität
und Deduktionssysteme
Universität Karlsruhe
76128 Karlsruhe, Germany
e-mail: ingrid@ira.uka.de

Abstract. Inheritance nets are used for the representation of hierarchical knowledge. If we allow the representation of positive and negative information, conflicts can occur. They are resolved according to the *deduction strategy* used (cp. [T 86], [HTT 87], [THT 87]) that is based on two fundamental principles: 1. A link is better than a compound path. 2. More specific information is better than less specific information.
We doubt these principles and give an alternative approach instead. It is more general in that it allows to label the links in an inheritance net. The decision which path to believe depends on a partial order given on the labels of the net.
We will show that we can model preemption as defined by [T 86] and get the set of sceptical valid paths as the intersection of all possible sets of credulous valid paths.

1 Introduction

Inheritance nets as proposed by Touretzky and others (cp. [T 86], [HTT 87], [THT 87]) are used for the representation of hierarchical information. They consist of nodes, that represent sets, and links for relations between these sets. Links represent *explicit* knowledge, concatenations of links yield paths, that is *deduced* knowledge. Conflicts may occurr if we allow the representation of positive and negative information. A suitable deduction strategy then has to decide which paths to believe. The set of valid paths is traditionally determined by the choice of 3 parameters:

  i. chaining strategy: the decision, how the validity of a compound paths relies on the validity of its subpaths (generally, forward or backward chaining).
 ii. preemption: resolves conflicts between "comparable" paths according to the following two strategies:
    (a) "Believe the link: As direct links represent explicit information whereas paths represent *deduced* information a link traditionally overrides a path contradicting it.
    (b) "Specificity criterion": The most specific path wins.

iii. mode of reasoning: decides between contradictory paths that are incomparable with respect to the specificity criterion. We can choose either

(a) credulous reasoning (choose one of the contradictory paths as valid), or

(b) sceptical reasoning (believe none of them).

In traditional inheritance nets ([T 86], [HTT 87], [THT 87]) the decisions on
these parameters are interlaced. We get complicated fix point constructions of
the set of paths that can be believed. There exist lots of different definitions, but
none of them is generally accepted. The most subtle differences lie in the exact
definition of the preemption strategy. If the nets become more complicated it is
hard to see which paths are valid and which are not. The intuition about the
paths that should be valid seems also to depend on the labels that are written
on the nodes.

We adopt a different approach. We regard inheritance nets as constructed
iteratively. Then, it is essential to get a feeling for the set of valid paths according
to the set of links the inheritance net consists of. For that reason, we are looking
for a straightforward definition of the set of valid paths.

We doubt the general principles underlying the definitions of preemption because we see the construction of an inheritance net for knowledge representation
as an iterative process. We cannot always be sure that more specific concepts lie
below more general concepts and further refinements, that is replacing a direct
link by a compound path, should not have an effect on the previous conclusions
of the net. We will illustrate our criticism of the concept "believe the link" that
also underlies the specificity criterion by way of the following example.

Example 1. i. This famous example from the literature (cp. [SL 93]) contains
the sentences "Royal elephants are elephants. Elephants are gray. Royal elephants are not gray." and is represented below.

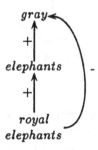

In this net, the path (royal elephants, elephants, gray)$^{+}$ will be preempted
by the direct link (royal elephants, gray)$^{-}$.

ii. Assume now, that we learn "Royal elephants are white animals" and we know
"White animals are not gray". Then, we can refine the link (royal elephants, gray)$^{-}$
and get the following representation:

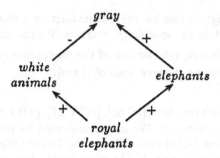

This net is also known as the "Nixon Diamond". There are two contradictory paths from node "royal elephants" to node "gray" and neither is more specific.

iii. Sometimes, it is argued that negative links can be read in both directions. If we change the direction of the negative link in our example, we have again a preemption situation:

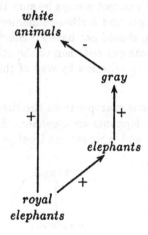

Here, the direct link (royal elephants, white animals)$^+$ preempts the path (royal elephants, elephants, gray, white animals)$^-$. On the other hand, it seems strange that the concept "gray" should be more specific than the concept "white animals".

iv. We could try to restore a convincing hierarchy and just use the colours as concepts. We change the concept "white animals" to the concept "white". Then, it would seem natural to put both colour-concepts on the same level and connect them with a bidirectional link:

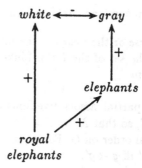

Then, we get the conclusion that royal elephants are white. But we cannot decide whether they are gray or not, because there are two contradictory paths that are incomparable with respect to the specificity criterion.

One idea to cope with the situation might be to use mixed inheritance. Then, it would be allowed to use strict and defeasible links and preemption has to take into account whether the contradictory paths are composed of strict and/or defeasible links. But then, the preemption concept becomes even more complicated.

For that reason, we adopt a different strategy. We give a more general definition of inheritance nets in that we allow to *label* the links in the net. The labels can denote, for example,

- the reliability of the information represented by the link (strict or defeasible, or a probability),
- the kind of the relation between the concepts connected by the link (with traditional inheritance nets positive or negative links, resp., in knowledge representation the relations used like is-a, subset, is-element, has-part, etc.),
- the *relative* reliability of the information, if there exists a partial order on the labels.

So, we are able to represent *more* information in the net. The valid paths are then chosen according to a partial order on the labels. If we extend traditional inheritance nets by this means, we can do with a much simpler deduction strategy.

The paper is organized as follows: We start with the definition of graded inheritance nets, followed by a section on traditional inheritance nets as defined in [T 86]. Afterwards, we show how to model traditional inheritance nets with graded inheritance nets and compare the results. We will conclude with a summary.

2 Graded Inheritance Nets

In this section, we will give the definition of graded inheritance nets. We allow links with arbitrary labels and compute the label of the paths according to a *path labelling function*.

Definition 1. A *graded inheritance net* is a quintupel $(U, L, M, pl, \prec)$ where

- U is a set, the *universe* or the *nodes* of the inheritance net,
- $L \subseteq U \times U \times M$ is the set of the *links*. Links are written $(x, y)^m$ or drawn as arrows with label m.
- M is a set of *labels*,
- $pl : M \times M \mapsto M$ is a partial, associative function, the *path labelling function*,
- ex. a set of *grades*, G, so that $M := G \times \{+, -\}$, and
- $\prec \subseteq G \times G$ is a partial order on G. For $m, m' \in M$, $m = (g, s)$, $m' = (g', s')$, we also write $m \prec m'$ iff $g \prec g'$.

An inheritance net $\mathcal{N}$ is a directed graph if there is only one link between every ordered pair of nodes, otherwise it becomes a multigraph. Traditional inheritance nets assume a graph but we do not restrict ourselves to them because it will not be important. We could imagine a situation where a positive and a negative link between two concepts exist simultaneously or different links that represent different strengths of arguments.

Links can be put together in order to build *paths*. The path labelling function pl computes the label of a path from the labels of its subpaths. It may be partial and thus defines *which* links are allowed to be put together (and which are not). We demand associativity in order to allow arbitrary concatenation of paths and links and do not restrict ourselves to a specific kind of concatenation (e.g. forward, backward concatenation).

Definition 2. Let $\mathcal{N} := (U, L, M, pl, \prec)$ be a graded inheritance net, n a natural number, $x_1, \ldots, x_n \in U$, $(x_1, \ldots, x_n)$ a finite sequence of nodes. Then $\nu = (x_1, \ldots, x_n)^m$ is a path in $\mathcal{N}$, if

- $n = 2$ and $\nu \in L$ or
- $n > 2$ and ex. $i \in \{2, \ldots, n-1\}$, $m', m'' \in M$:
  - $(x_1, \ldots, x_i)^{m'}$ is a path in $\mathcal{N}$,
  - $(x_i, \ldots, x_n)^{m''}$ is a path in $\mathcal{N}$ and
  - $pl(m', m'') = m$.

$paths(\mathcal{N})$ is the set of all paths in $\mathcal{N}$.

Definition 3. To concatenate paths, we define the function $\bullet$ by pl

$$(x_1, \ldots, x_i)^{m'} \circ (x_i, \ldots, x_n)^{m''} := (x_1, \ldots, x_i, \ldots, x_n)^{pl(m', m'')}.$$

Example 2. Let $M := \{+, -\}$. Then

$$pl(+, +) = +, \; pl(+, -) = -$$

is the path labelling function that allows to construct all paths of an inheritance net without any grades on them.

Definition 4. We define the following notation:

$$\sim s = \begin{cases} - & ,if \quad s = + \\ + & ,if \quad s = - \end{cases}$$

Now, we will define conflicting paths.

Definition 5. Let $\mathcal{N} = (U, L, M, pl, \prec)$ be a graded inheritance net, $\nu = (x_1, \ldots, x_n)^{(g,s)}$ and $\nu' = (y_1, \ldots, y_m)^{(g',s')}$ paths in $\mathcal{N}$, Φ a set of paths.

- ν and ν' are called *contradictory* iff
  - $x_1 = y_1$, $x_n = y_m$, and
  - $s = \sim s'$.

  We also say ν *contradicts* ν' or ν *is contradicted by* ν'.
- ν is called *contradicted in* Φ iff Φ contains a path ν' that contradicts ν.

If we can build conflicting paths we want to decide which to believe and which not, that is which paths are *valid*. This is the point, where the relation $\prec$ is used.

If we want to define valid paths, we have different alternatives. At least, we want to believe all paths that are not contradicted. If we have contradictory paths, we want to believe the best one (according to $\prec$). We do not need a fixpoint-construction any longer, because we believe the best paths from all paths that can be build in a net, that is from $paths(\mathcal{N})$. In a second step, we can restrict this set of "believed paths" by demanding that a path can only be believed if (some of) its subpaths are believed as well. That is the point, where forward or backward concatenation can be regarded. Compared to traditional inheritance nets and their definition of valid paths we will in general get a subset of their valid paths.

With graded inheritance nets, a difference between sceptical and credulous reasoning will occur, if the $\prec$-relation is not defined on all contradictory labels.

As we do not longer rely on the notion of a more specific link or path we have no problems with acyclic nets.

Definition sceptical valid paths *Let Φ be a set of paths.*
$\nu = (x_1, \ldots, x_n)^m \in VSpaths(\Phi)$ iff

- $\nu \in \Phi$ *and*
- *for every contradictory path $\nu' \in \Phi$ with label m': $m' \prec m$.*

If $\Phi = paths(\mathcal{N})$ for a graded inheritance net $\mathcal{N}$, we also write $VSpaths(\mathcal{N})$ instead of $VSpaths(paths(\mathcal{N}))$.

Definition credulous valid paths *Let Φ a set of paths.*
$\nu = (x_1, \ldots, x_n)^m \in VCpaths(\Phi)$ iff

- $\nu \in \Phi$ *and*
- *for every contradictory path $\nu' \in \Phi$ with label m': $m \not\prec m'$.*

If $\Phi = paths(\mathcal{N})$ for a graded inheritance net $\mathcal{N}$, we also write $VCpaths(\mathcal{N})$ instead of $VCpaths(paths(\mathcal{N}))$.

Example 3. $\mathcal{N} = (\{re, e, g, wa\}, L, \{+, -\}, pl, \emptyset)$ with $L = \{(re, e)^+, (e, g)^+, (re, wa)^+, (wa, g)^-\}$ and $pl(+, s) = s$ for $s \in \{+, -\}$ represents the example "Royal elephants are elephants. Elephants are gray. Royal elephants are white animals. White animals are not gray." from the beginning. $VSpaths(\mathcal{N}) = L$, $VCpaths(\mathcal{N}) = L \cup \{(re, wa, g)^-, (re, e, g)^+\}$.

Note, that $VCpaths(\Phi)$ might still contain contradictory paths, but only those with incomparable labels. A *credulous path set* of an inheritance net is a maximal subset of $VCpaths(\mathcal{N})$ that does not contain contradictory paths.

Definition 6. Let Φ be a set of paths. $\Pi \in Cred(\Phi)$ iff Π is a maximal subset of Φ that does not contain contradictory paths.

Example 4. Let $\mathcal{N}$ be the graded inheritance net defined in example 3. Then $Cred(VCpaths(\mathcal{N})) = \{L \cup \{(re, wa, g)^-\}, L \cup \{(re, e, g)^+\}$.

$Cred(VCpaths(\mathcal{N}))$ is the set of all credulous path sets of the inheritance net $\mathcal{N}$. If we assume $\prec$ to be acyclic (that is $m \not\prec m$ for all $m \in M$), the set of all sceptical valid paths of a graded inheritance net equals the intersection of its credulous path sets.

Theorem 7. Let $\mathcal{N}$ be a graded inheritance net, $\prec$ acyclic. Then

$$VSpaths(\mathcal{N}) = \bigcap Cred(VCpaths(\mathcal{N})).$$

Proof. Let $\sigma := (x_1, \ldots, x_n)^m \in paths(\mathcal{N})$.

i. If σ is not contradicted in $paths(\mathcal{N})$, then $\sigma \in VSpaths(\mathcal{N})$ and $\sigma \in \Pi$ for every $\Pi \in Cred(VCpaths(\mathcal{N}))$.
ii. If exists a path $\nu \in paths(\mathcal{N})$ with label m' that contradicts σ and $m \prec m'$, $\sigma \notin VSpaths(\mathcal{N})$ and $\sigma \notin \Pi$ for all $\Pi \in Cred(VCpaths(\mathcal{N}))$.
iii. If exists a contradictory path ν with label m', $m' \prec m$ and all other paths contradicting σ have a label s with $m \not\prec s$, then ($\sigma \in VSpaths(\mathcal{N})$ iff for all σ' with label s' that contradict σ: $s' \prec m$) and ($\sigma \in VCpaths(\mathcal{N})$ iff for all σ' with label s' that contradict σ: $m \not\prec s'$). As $\prec$ is acyclic we know that $m \not\prec m'$. If exists a contradictory path ν with incomparable label, than both paths σ and ν belong to $VCpaths(\mathcal{N})$, but then both do not belong to $\bigcap Cred(VCpaths(\mathcal{N}))$ (see next case).
iv. Assume, that for all paths $\sigma' \in paths(\mathcal{N})$ with label m' that contradict σ, $m \not\prec m'$ and $m' \not\prec m$. Then σ, $\sigma' \notin VSpaths(\mathcal{N})$. On the other hand, σ, $\sigma' \in VCpaths(\mathcal{N})$. Then, (for all $\Pi \in Cred(VCpaths(\mathcal{N}))$: $\sigma \notin \Pi$ or $\sigma' \notin \Pi$) and (exist Π, $\Pi' \in Cred(VCpaths(\mathcal{N}))$: $\sigma \in \Pi$,, $\sigma' \in \Pi'$). So, σ, $\sigma' \notin \bigcap Cred(VCpaths(\mathcal{N}))$.

Example 5. An example often discussed in the literature (cp. [THT 87]) is the "double diamond":
$\mathcal{N} = (\{a, b, c, d, e, f\}, L, \{+, -\}, pl, \emptyset)$ with
$L := \{(a, b)^+, (a, c)^+, (c, d)^-, (b, d)^+, (c, e)^+, (e, f)^-, (d, f)^+\}$ and
$pl(+, s) = s$ for $s \in \{+, -\}$.

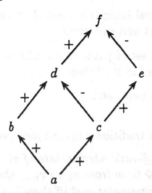

$VSpaths(\mathcal{N}) = L \cup \{(a, c, e)^+\}$ does not contain a path from a to d or from a to f. We would get the same set with the representation as suggested in section 4, a solution that up to now no direct sceptical approach could reach.

In general, arguments are constructed by the concatenation of valid subarguments. Usually, forward or backward chaining is demanded. We perform this selection in an independant step and give the definition for forward chaining.

Definition forward chaining *Let $\mathcal{N} = (U, L, M, pl, \prec)$ be a graded inheritance net, Φ a set of paths.*
$\nu = (x_1, \ldots, x_{n-1}, x_n)^{(g,s)} \in fwpaths(\Phi)$ *iff*

- $\nu \in \Phi$,
- *ex.* $(g', +) \in M$: $(x_1, \ldots, x_{n-1})^{(g', +)} \in fwpaths(\Phi)$,
- *ex.* $(g'', s) \in M$: $(x_{n-1}, x_n)^{(g'', s)} \in fwpaths(\Phi)$, *and*
- $pl((g', +), (g'', s)) = (g, s)$.

Example 6. In example 3, $fwpaths(VCpaths(\mathcal{N})) = VCpaths(\mathcal{N})$.

3   Traditional Inheritance Nets

We will now give the definition of Touretzky's proposal ([T 86]).

Definition 8. A *traditional inheritance net* is a pair (K, L), where

- K is a set of nodes and
- $L \subseteq K \times K \times \{+, -\}$ a set of links. + or - is called the *polarity* of a link. Links with a + (-) are called positive (negative).

Definition 9. Let $T = (K, L)$ be a traditional inheritance net. A *path* is a sequence of n links (x_i, x_{i+1}, s_i) for $i = 1, \ldots, n-1$, $n \geq 1$, where $s_1, \ldots, s_{n-2}$ are positive. The *polarity of a path* is the polarity of the final link. We will draw paths as $x_1 \to \ldots \overset{s}{\to} x_n$. $paths(T)$ is the set of all paths that are built from links of T.

Definition 10. A traditional inheritance net T is called *acyclic* iff it contains no path with the same start and end point.

Definition 11. Let Φ be a set of paths of a traditional inheritance net and σ a path. Then σ *is contradicted in* Φ iff there is a path in Φ with

- the same start and end point and
- opposite polarity.

Definition 12. Let T be a traditional inheritance net and Φ a set of paths.

- A node x is called an *off-path intermediate of* $x_1 \to \ldots, \overset{s}{\to} x_n$ in Φ iff there is a positive path ν in Φ from from x_1 to x_{n-1} that contains x.
- $\sigma = x_1 \to \ldots, \overset{s}{\to} x_n$ is *preempted in* Φ iff there is a node x that is an off-path intermediate of σ and a link $x \overset{s}{\to} x_n$ with the opposite polarity of σ.
- σ is an *upward concatenation of paths in* Φ iff the last link of σ is in Φ and all but the last link of σ is also in Φ.

Definition 13. Let T be a traditional inheritance net, Φ a set of paths and σ a path in T. σ is *inheritable in* Φ iff

i. σ is an upward concatenation of paths in Φ,
ii. σ is not contradicted in Φ and
iii. σ is not preempted in Φ.

If a net contains contradictory paths and neither is preempted we can either choose to believe one of those paths (credulous reasoning) or none of them (sceptical reasoning). [T 86] defines a credulous approach.

Definition 14. Let $T = (K, L)$ be a traditional inheritance net. Φ is a *credulous grounded extension of* T iff

- $L \subseteq \Phi$ and
- for all σ: $\sigma \in \Phi \setminus L$ iff σ is inheritable in Φ.

Example "Credulous Grounded Extensions" *If we define the traditional inheritance net $T = (\{re, e, g, wa\}, \{(re, e)^+, (e, g)^+, (re, wa)^+, (wa, g)^-\}$ for example 3, the credulous grounded extensions are the same than the credulous path sets. If we extend example 3 in that we come to know about a dirty royal elephant named Clyde, we will see a difference. We want to represent "Royal elephants are elephants. Elephants are gray. Royal elephants are white animals. White animals are not gray. Clyde is a royal elephant. Clyde is not a white animal."*
    We define
$$T = (\{Cl, re, e, g, wa\}, \{(Cl, re)^+, (Cl, wa)^-, (re, e)^+, (e, g)^+, (re, wa)^+, (wa, g)^-\},$$
$pl(+, s) = s$ *for* $s \in \{+, -\}$.

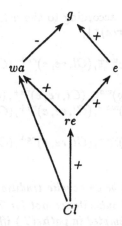

Then, there is only one credulous grounded extension, namely $L \cup \{(Cl, wa)^-, ((Cl, e)^+, (Cl, e, g)^+\}$. This is due to the fact, that the link $(Cl, wa)^-$ preempts the path $(Cl, re, wa)^+$. So, $(Cl, re, wa, g)^-$ is not a concatenation of paths of the extension.

4 Traditional Inheritance Nets as Graded Inheritance Nets

We will represent traditional inheritance nets as graded inheritance nets.

Definition 15. Let $T = (K, L)$ be a traditional inheritance net. We call $\mathcal{N}(T) = (K, L, M, pl, \prec)$ the *graded inheritance net for T* iff

- $M = \{(x, s)/x \in K, \ s \in \{+, -\}\}$,
- $pl((x, +), (y, s)) = (y, s)$ and
- $x \prec y$ iff there is a positive path from y to x in $paths(\mathcal{N}(T))$.

Example 7. The graded inheritance net for T from example "Credulous Grounded Extensions" is:

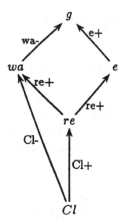

We will build sets of paths according to the definitions above:

$VSpaths(\mathcal{N}(T)) = L \cup \{(Cl, ree)^+\}$,

$VCpaths(\mathcal{N}(T)) =$

$\quad L \cup \{(re, e, g)^{e+}, (re, wa, g)^{wa+}, (Cl, re, e)^{re+}, (Cl, re, e, g)^{e+}, (Cl, re, wa, g)^{wa+}\}$,

$Cred(VCpaths(\mathcal{N}(T)) =$

$\quad \{L \cup \{(re, e, g)^{e+}, (re, wa, g)^{wa-}, (Cl, re, e)^{re+}, (Cl, re, e, g)^{e+},$

$\quad L \cup \{(re, e, g)^{e+}, (re, wa, g)^{wa-}, (Cl, re, e)^{re+}, (Cl, re, wa, g)^{wa-}\}\}$,

$fwpaths(VCpaths(\mathcal{N}(T))) =$

$\quad L \cup \{(re, e, g)^{e+}, (re, wa, g)^{wa-}, (Cl, re, e)^{re+}, (Cl, re, e, g)^{e+}\}$.

Theorem 16. *Let $T = (K, L)$ be an acyclic traditional inheritance net, $\mathcal{N}(T) = (K, L_{\mathcal{N}}, M_{\mathcal{N}}, pl, \prec)$ the graded inheritance net for T, $\sigma = x_1 \to \ldots \to x_{n-1} \overset{s}{\to} x_n \in paths(T)$. Then σ is preempted in $paths(T)$ iff $(x_1, \ldots, x_{n-1}, x_n)^{(x_{n-1}, s)} \notin VCpaths(\mathcal{N}(T))$.*

Proof. $\sigma = x_1 \to \ldots \to x_{n-1} \overset{s}{\to} x_n$ is preempted in $paths(T)$

iff

ex. off-path intermediate $x \in K$ and link $(x, x_n, \sim s) \in L$

iff

ex. $y_1, \ldots, y_m \in K : y_1 = x_1$, $y_m = x_{n-1}$, $y_1 \to \ldots \overset{+}{\to} y_m \in paths(T)$, exists

$i \in \{1, \ldots, m\} : x = y_i$ and $(x, x_n, \sim s) \in L$

iff

ex. $y_1, \ldots, y_m \in K : y_1 = x_1$, $y_m = x_{n-1}$, $(y_1, \ldots, y_m)^{y_{m-1}, +} \in paths(\mathcal{N}(T))$,

exists $i \in \{1, \ldots, m\} : x = y_i$ and $(x, x_n)^{y_i, \sim s} \in L_{\mathcal{N}}$

iff

ex. $y_1, \ldots, y_m \in K : (y_1, \ldots, y_i, x_n)^{y_i, \sim s} \in paths(\mathcal{N}(T))$,

$(y_i, \ldots, x_{n-1})^{y_{m-1}, +} \in paths(\mathcal{N}(T))$ because $y_m = x_{n-1}$,

iff

exists paths $(x_1, \ldots, x_n)^{y_i, \sim s}$ and $(y_1, \ldots, y_i, x_n)^{y_i, \sim s} \in paths(\mathcal{N}(T))$,

$x_{n-1} \prec y_i$

iff

$(x_1, \ldots, x_{n-1}, x_n)^{(x_{n-1}, s)} \notin VCpaths(\mathcal{N}(T))$.

Now, it would be interesting to find a relationship between the credulous grounded extensions of traditional inheritance nets and the credulous path sets of graded inheritance nets. The problem here is the fix point construction of extensions. A path might be preempted by the set of all paths of an inheritance net, but not necessarily by an extension (that often is a proper subset of the set of paths). The following example will demonstrate the differences.

Example 8. Let $T := (\{a, b, c, d, e\}, L)$ with

$L := \{(a, b, +), (b, c, +), (a, c, -), (a, d, +), (c, d, +), (c, e, -), (d, e, +)\})$.

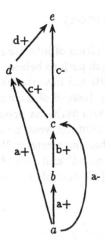

This inheritance net contains only one credulous extension. It contains the path $a \to d \xrightarrow{+} e$ because the contradictory path $a \to b \to c \xrightarrow{-} e$ is not valid as its subpath $a \to b \xrightarrow{+} c$ is preempted by the link $(a, c, -)$. On the other hand, $\sigma := (a, b, c, e)^{c,-} \in VCpaths(\mathcal{N}(T))$. $(a, d, e)^{d,+} \notin VCpaths(\mathcal{N}(T))$ because it is preempted by σ.

Even though, $VCpaths(\mathcal{N}(T))$ contains only one credulous path set, this is not a convincing set of valid paths. It is rather unnatural, that $(a, b, c, e)^{c,-}$ should be believed if $(a, b, c)^{b,+}$ is not believed. We can get rid of this path if we build $fwpaths(VCpaths(\mathcal{N}(T)))$, that are the credulous paths sets that *obey forward chaining*. Then, we get a proper subset of the credulous extension of T that seems natural. Because it is rather difficult to decide only because of the structure of the net whether $(a, d, e)^+$ should be believed or not. The decision depends on the strength of the links of the net that can vary according to the application. We prefer to be more cautious and not to decide on the validity of path $(a, d, e)^+$ because the better path $(a, b, d, e)^-$ might be valid. We prefer to use less sophistication in the deduction mechanism and to alter inheritance nets that yield undesired results by adding additional links or by changing the grades on the links.

With traditional inheritance nets, preemption and forward chaining are computed in parallel whereas with graded inheritance nets it is separated. If we only regard credulous paths sets that obey forward chaining we get subsets of credulous extensions of traditional inheritance nets.

Corollary 17. *Let T be a traditional inheritance net, $\mathcal{N}(T)$ the graded inheritance net for T. For every grounded extension Φ of T there exists a $\Pi \in Cred(fwpaths(VCpaths(\mathcal{N}(T))))$ with $\Pi \subseteq \Phi$.*

Proof. Induction on the length of the paths in $paths(T)$.

5 Summary (and Outlook)

We presented a generalized definition of inheritance nets that allows to label the links of the net and decides which path to believe according to a partial order on the labels. Preemption in traditional inheritance nets as defined by [T 86] can be modelled with our approach. Instead of a complicated fix point construction we get a simpler definition of valid paths, as preemption depends on the partial order of the labels and the decision between credulous and sceptical reasoning is separated from that of the chaining strategy. We get the right set of sceptical valid paths for the "double diamond" and proved that the set of sceptical valid paths equals the intersection of all possible sets of credulous valid paths.

References

[H 94]   Horty, John F.: Some Direct Theories of Nonmonotonic Inheritance, in: Gab-
         bay, Dov M.: Hogger, C. J.; Robinson, J. A. (ed.): Handbook of Logic in Artifi-
         cial Intelligence and Logic Programming, Volume 3, Clarendon Press, Oxford,
         1994.

[HTT 87] Horty, J. F.; Thomason, R. H.; Touretzky, D.S.: A skeptical theory of inher-
         itance in nonmonotonic semantic nets. Proceedings of AAAI-87, 1987.

[SL 93] Selman, Bart; Levesque, Hector J.: The complexity of path-based defeasible
         inheritance, Artificial Intelligence 62 (1993), pp. 303-339.

[T 86]   Touretzky, D.S.: The Mathematics of Inheritance Systems, Morgan Kaufmann,
         Los Altos, CA, 1986.

[THT 87] Touretzky, D.S.: Horty, J. F.; Thomason, R. H.: A Clash of Intuitions: The
         Current State of Nonmonotonic Multiple Inheritance Systems, IJCAI-87, pp
         476-482, 1987.

Defining Normative Systems for Qualitative Argumentation

Simon Parsons[1][2]

[1] Advanced Computation Laboratory, Imperial Cancer Research Fund,
P.O. Box 123, Lincoln's Inn Fields, London WC2A 3PX, United Kingdom.
[2] Department of Electronic Engineering, Queen Mary and Westfield College,
Mile End Road, London E1 4NS, United Kingdom.
S.Parsons@qmw.ac.uk

Abstract. Inspired by two different approaches to providing a qualitative method for reasoning under uncertainty—qualitative probabilistic networks and systems of argumentation—this paper attempts to combine the advantages of both by defining systems of argumentation that have a probabilistic semantics.

1 Introduction

In the last few years there have been a number of attempts to build systems for reasoning under uncertainty that are of a qualitative nature—that is they use qualitative rather than numerical values, dealing with concepts such as increases in belief and the relative magnitude of values. In particular, two types of qualitative system have become well established, namely qualitative probabilistic networks (QPNs) [4, 18], and systems of argumentation [8, 11, 12]. While the former are built as an abstraction of probabilistic networks where the links between nodes are only modelled in terms of the qualitative influence of the parents on the children, and therefore have an underlying probabilistic semantics, some of the latter lack such a sound foundation. Instead they offer a greater degree of resolution, allowing more precise deductions to be made.

In this paper we present several normative systems of argumentation. These are systems of argumentation which have a probabilistic semantics, and are thus normative in that they behave according to the norms of probability theory. Such systems aim to extend both QPNs in the sense of reducing the degree of abstraction of the former, and argumentation in the sense of providing it with a probabilistic semantics whilst using only qualitative or semi-qualitative information[3]. Of course this extension might not always be desired, but may be useful at times to ensure that a given system reasons within probabilistic norms. The systems are built upon the framework introduced by Fox, Krause and their colleagues [8, 11], and we begin by introducing this framework.

---

[3] If we don't have any commitment to qualitative information, we can use ordinary probabilities as suggested by Krause *et al.* [11].

2 Introducing systems of argumentation

In classical logic, an argument is a sequence of inferences leading to a conclusion. If the argument is correct, then the conclusion is true. Consider the simple database Δ_1 which expresses some very familiar information in a Prolog-like notation in which variables are capitalised and ground terms and predicate names start with small letters.

$$f1 : human(socrates). \qquad \Delta_1$$
$$r1 : human(X) \rightarrow mortal(X).$$

The argument $\Delta_1 \vdash mortal(socrates)$ may be correctly made from this database because $mortal(socrates)$ follows from Δ_1 given the usual logical axioms and rules of inference. Thus a correct argument simply yields a conclusion which in this case could be paraphrased '$mortal(socrates)$ is true in the context of $f1$ and $r1$'. In the system of argumentation proposed by Fox and colleagues [11] this traditional form of reasoning is extended to allow arguments to indicate support and doubt in propositions, as well as proving them, by assigning labels to arguments which denote the confidence that the arguments warrant in their conclusions. This form of argumentation may be summarised by the following schema:

$$\text{Database} \vdash_{ACR} (\text{Sentence, Grounds, Sign})$$

where $\vdash_{ACR}$ is a suitable consequence relation. Informally, Grounds (G) are the facts and rules used to infer Sentence (St), and Sign (Sg) is a number or a symbol drawn from a dictionary of possible numbers or symbols which indicate the confidence warranted in the conclusion.

To formalise this kind of reasoning we start with a language, and we will take $\mathcal{L}$, a set of propositions, including $\perp$, the contradiction. We also have a set of connectives $\{\rightarrow, \neg\}^4$, and the following set of rules for building the well formed formulae of the language:

- If $l \in \mathcal{L}$ then l is a well formed formula (*wff*).
- If $l \in \mathcal{L}$ then $\neg l$ is a *wff*.
- If $l, m \in \mathcal{L}$ then $l \rightarrow m$, $l \rightarrow \neg m$, $\neg l \rightarrow m$ and $\neg l \rightarrow \neg m$ are *wffs*.
- Nothing else is a *wff*.

The members of $\mathcal{W}$, the set of all *wffs* that may be defined using $\mathcal{L}$, may then be used to build up a database Δ where every item $d \in \Delta$ is a triple $(i : l : s)$ in which i is a token uniquely identifying the database item (for convenience we will use the letter 'i' as an anonymous identifier), l is a wff, and s is a sign. With this formal system, we can take a database and use the argument consequence relation $\vdash_{ACR}$ given in Figure 1 to build arguments for propositions in $\mathcal{L}$ that we are interested in.

---

[4] Note that both the set of connectives and the rules for building *wffs* are more restrictive than for other similar systems of argumentation [11]. A normative system which does not suffer from these limitations is discussed in [13].

$$\text{Ax} \frac{\qquad\qquad\qquad}{\Delta \vdash_{ACR} (St, \{i\}, Sg)} \quad (i : St : Sg) \in \Delta$$

$$\to\text{-E} \frac{\Delta \vdash_{ACR} (St, G, Sg) \qquad \Delta \vdash_{ACR} (St \to St', G', Sg')}{\Delta \vdash_{ACR} (St', G \cup G', \text{comb}(Sg, Sg'))}$$

$$\to\text{-I} \frac{\Delta \cup (St, \emptyset, Sg) \vdash_{ACR} (St', G, Sg')}{\Delta \vdash_{ACR} (St \to St', G, \text{comb}'(Sg, Sg'))}$$

Fig. 1. Argumentation Consequence Relation

Typically we will be able to build several arguments for a given proposition, and so, to find out something about the overall validity of the proposition, we will *flatten* the different arguments to get a single sign.

Together $\mathcal{L}$, the rules for building the formulae, the connectives, and $\vdash_{ACR}$ define a formal system of argumentation, which, for want of a name we will call $\mathcal{SA}$. In fact, $\mathcal{SA}$ is really the basis of a family of systems of argumentation, because one can define a number of variants of $\mathcal{SA}$ by using different dictionaries of signs. Each dictionary will have its own combination functions comb and comb$'$, and its own means of flattening arguments, and the meanings of the signs, the flattening function, and the combination function delineate the semantics of the system of argumentation. Thus $\mathcal{SA}$ gives us a general framework for expressing logical facts which can incorporate different models of uncertainty by varying the signs and their associated combination and flattening functions as well as a means of representing default information and of handling inconsistent information [15].

3 A first normative system

One commonly used system of argumentation within the framework of $\mathcal{SA}$ is one in which the dictionary consists of three symbols, $+$, $-$ and 0, which represent the notion of an increase, a decrease and no change in belief respectively. When a proposition is labelled with $+$, it is taken to represent the fact that there is an increase in belief in the proposition, while labelling the rule:

$$human \to mortal$$

with a $+$ is taken to represent the fact that showing that there is an increase in belief in the proposition "*human*" causes an increase in belief in the proposition "*mortal*". The combination function comb for this system of argumentation is $\otimes$ of Table 1, while comb$'$ is its inverse $\otimes^{-1}$, also given in Table 1—blank spaces mark impossible combinations. As with all combinator papers in this paper, the first argument of the function described by the table is drawn from the first column, and the second argument is drawn from the first row. The flattening

$\otimes$	$+$	0	$-$	$?$
$+$	$+$	0	$-$	$?$
0	0	0	0	0
$-$	$-$	0	$+$	$?$
$?$	$?$	0	$?$	$?$

$\otimes^{-1}$	$+$	0	$-$	$?$
$+$	$+$	0	$-$	$?$
0		$?$		
$-$	$-$	0	$+$	$?$
$?$		0		$?$

Table 1. The functions $\otimes$ and $\otimes^{-1}$

function is one that implements a form of improper linear model with uniform weights and no constant term [3]. This counts the number of $+$ and $-$ weighted arguments for a proposition, takes the sign that occurs most often and makes that the sign of the proposition, thus taking the sum of all the arguments while giving each argument equal weight. We will refer to the system of argumentation which uses this dictionary and set of functions along with the argument building capabilities of $\mathcal{SA}$ as $\mathcal{NA}_1$. The question that faces us here is how $\mathcal{NA}_1$ may be given a probabilistic semantics. Now, the use of $+$ and $-$ to represent changes in belief suggests a link between this system of argumentation and QPNs [18] since the latter make use of a similar notion. Indeed, it turns out that we can modify the notion of a probabilistic influence in a QPN to give our database facts and rules a probabilistic interpretation. In particular we take triples $(i : l : +)$, where $l \in \mathcal{W}$ and l does not include the connective $\rightarrow$, to denote the fact that $\Pr(l)$ is known to increase, and similar triples $(i : l : -)$, to denote the fact that $\Pr(l)$ is known to decrease. Triples $(i : l : 0)$, clearly denote the fact that $\Pr(l)$ is known to neither increase nor decrease. With this interpretation facts correspond to the nodes in a QPN, and as in QPNs we deal with changes in their probability.

Database rules can similarly be given a probabilistic interpretation by making the triple $(i : n \rightarrow m : +)$, where m and n are members of $\mathcal{W}$ which do not include the connective $\rightarrow$, denote the fact that:

$$\Pr(m \mid n, X) \geq \Pr(m \mid \neg n, X)$$

for any $X \in \{x, \neg x\}$ for which there is a triple $(i : x \rightarrow m : s)$ or $(i : \neg x \rightarrow m : s)$ (where s is any sign), while the triple $(i : n \rightarrow m : -)$ denotes the fact that:

$$\Pr(m \mid n, X) \leq \Pr(m \mid \neg n, X)$$

again for any X for which there is a triple $(i : x \rightarrow m : s)$ or $(i : \neg x \rightarrow m : s)$. We do not make use of triples such as $(i : n \rightarrow m : 0)$ since such rules have no useful effect. As a result a rule $(i : n \rightarrow m : +)$ means that there is a probability distribution over the propositions m and n such that an increase in the probability of n makes m more likely to be true, and a rule $(i : n \rightarrow m : -)$ means that there is a probability distribution over the propositions m and n such that an increase in the probability of n makes m less likely to be true. With this interpretation, rules correspond to qualitative influences in QPNs. It should be noted that the effect of declaring that there is a rule $(i : n \rightarrow m : +)$

is to create considerable constraints on the probability distribution over m and n to the extent that the effect of other rules relating m and n are determined absolutely. That is, a necessary consequence of $(i : n \to m : +)$ is that we have constraints on the conditional probability distributions across the propositions in the database equivalent to the rules $(i : \neg n \to m : -)$, $(i : n \to \neg m : -)$ and $(i : \neg n \to \neg m : +)$, and similar restrictions are imposed by rules like $(i : n \to m : -)$.

With this interpretation of rules and facts, the combination function $\otimes$ has a natural probabilistic interpretation as the function by which changes in probability are combined with probabilistic influences. Indeed $\otimes$ is the function used to combine the two in QPNs, and $\otimes^{-1}$ remains as defined above. The flattening function also has an obvious probabilistic interpretation in terms of calculating the overall change in probability of a proposition. However, in order for the improper linear model to make sense probabilistically, it is necessary to apply a restriction to the sizes of changes in probability represented by $(i : l : +)$ and $(i : l : -)$. In particular, it requires that all arguments, irrespective of how many steps they contain, have the same strength. This clearly places a great restriction on the number of probabilistic models that can be captured by this system.

In the kind of minimal logic we have taken as the base language for our system, any negated formula $\neg l$ is taken as shorthand for a formula $l \to \bot$. Thus in our system we should replace any formula $(i : \neg l : s)$ with $(i : l \to \bot : s)$, and any formula $(i : \neg l \to m : s)$ with $(i : (l \to \bot) \to m : s)$ before constructing any arguments. However, our probabilistic semantics give us an alternative way of replacing negated propositions, since $(i : \neg l : +) \equiv (i : l : -)$ and $(i : \neg l \to m : +) \equiv (i : l \to m : -)$, which does not involve introducing the contradiction (from which $x \to \bot$ follows where x is any proposition), and this is the method we prefer.

As an example of the kind of reasoning that can be performed in $\mathcal{N}\mathcal{A}_1$, consider the following simple database Δ_2 of propositional rules and facts. What these rules say is that there are three events that may have an effect on whether or not I lose my job—I have a high research output, I am ill, and I am a good tutor to my students. The first and third have a negative influence on the probability of me losing my job, while the other has a positive influence. The database facts say that there is evidence which leads to an increased probability of me being ill and having a high research output, and a decreased probability of me being a good tutor:

$$f1 : high\_research\_output : +. \qquad\qquad \Delta_2$$
$$f2 : ill : +.$$
$$f3 : good\_tutor : -.$$
$$r1 : high\_research\_output \to lose\_job : -.$$
$$r2 : ill \to lose\_job : +.$$
$$r3 : good\_tutor \to lose\_job : -.$$

From Δ_2 we can build the arguments:

$$\Delta_2 \vdash_{ACR} (lose\_job, (f1, r1), -).$$
$$\Delta_2 \vdash_{ACR} (lose\_job, (f2, r2), +).$$
$$\Delta_2 \vdash_{ACR} (lose\_job, (f3, r3), +).$$

And the improper linear model will flatten them to come up with the overall conclusion that there is an increase in the probability of me losing my job after the facts of my situation are known.

4 A second normative system

As stated above, $\mathcal{NA}_1$ is highly restrictive because its flattening function requires all arguments to have the same strength. To relax this restriction we clearly need a new flattening function. One suitable function is that used by QPNs for combining the effect of several influences on one variable. This function is $\oplus$ as specified in Table 2. The use of this function to define a new system of argumentation $\mathcal{NA}_2$ is straightforward after the dictionary of signs is extended to become $\{+, -, 0, ?\}$ where labelling a fact with ? indicates that the change in probability of that fact is unknown, and a rule $(i : n \to m : ?)$ denotes that the relationship between $\Pr(m \mid n, X)$ and $\Pr(m \mid \neg n, X)$ is unknown, so that if the probability of n increases it is not possible to say how the probability of m will change.

With this interpretation, there is a direct correspondence between a database of formulae drawn from $\mathcal{W}$ and a qualitative probabilistic network, and it is quite easy to see that any conclusion drawn by $\mathcal{NA}_2$ from a database would also be drawn by the corresponding QPN. The fact that qualitative multiplication distributes over addition ensures that the fact that argumentation builds separate arguments for the same proposition and then flattens them does not mean that it gives a different answer to the equivalent QPN.

To illustrate the difference between $\mathcal{NA}_1$ and $\mathcal{NA}_2$, consider what $\mathcal{NA}_2$ would conclude from Δ_2. Firstly it would build the same arguments as $\mathcal{NA}_1$:

$$\Delta_2 \vdash_{ACR} (lose\_job, (f1, r1), -).$$
$$\Delta_2 \vdash_{ACR} (lose\_job, (f2, r2), +).$$
$$\Delta_2 \vdash_{ACR} (lose\_job, (f3, r3), +).$$

$\oplus$	+	0	−	?
+	+	+	?	?
0	+	0	−	?
−	?	−	−	?
?	?	?	?	?

Table 2. The function $\oplus$

But this time the flattening function would conclude that the overall change in belief in the proposition *lose_job* was ?, indicating that it cannot be accurately identified. This is, of course, probabilistically correct—without information on the relative effects of the various causes of a loss of job, the way in which its probability will change cannot be predicted.

5   A more subtle normative system

Now, in the kind of applications for which $\mathcal{SA}$ was developed [7, 9], it is necessary to represent information of the form "X is known to be true", and "If X is true then Y is true"—information that we might term categorical. It is therefore interesting to investigate if $\mathcal{NA}_2$ can be extended to cover categorical relationships. To do so we first extend the dictionary of signs to be $\{++, +, -, --\}$ as suggested in [8, 11], where $++$ and $--$ are labels for categorical information. It then turns out that we can give $++$ and $--$ a probabilistic semantics, giving a system of argumentation $\mathcal{NA}_3$ which is $\mathcal{NA}_2$ extended by allowing triples such as $(i : l : ++)$ and $(i : l : --)$ and rules such as $(i : n \rightarrow m : ++)$ and $(i : n \rightarrow m : --)$.

The meaning of $(i : l : ++)$, where l is a *wff* which does not contain $\rightarrow$, is that the probability of l becomes 1, and $(i : l : --)$ means that the probability of l decreases to 0, and to make this clear, we write $(i : l : \bar{\uparrow})$ for $(i : l : ++)$, and $(i : l : \downarrow)$ for $(i : l : --)$. The meaning of the rules is slightly more complicated. We want a rule $(i : n \rightarrow m : ++)$, where neither m or n contain $\rightarrow$, to denote a constraint on the probability distribution across m and n such that if $\Pr(n)$ becomes 1, so does $\Pr(m)$. This requires that:

$$\Pr(m \,|\, n, X) = 1$$

for all $X \in \{x, \neg x\}$ such that the database contains $(i : x \rightarrow m : s)$ or $(i : \neg x \rightarrow m : s)$. [13]. Similarly, a probabilistic interpretation of a rule $(i : n \rightarrow m : --)$ requires that:

$$\Pr(m \,|\, n, X) = 0$$

for all $X \in \{x, \neg x\}$ such that the database contains $(i : x \rightarrow m : s)$ or $(i : \neg x \rightarrow m : s)$. Considering the constraints on the conditional probabilities imposed by $++$ and $--$ rules, a further pair of rules are suggested. These are a rule $(i : n \rightarrow m : -+)$ which requires that:

$$\Pr(m \,|\, \neg n, X) = 1$$

for all $X \in \{x, \neg x\}$ such that the database contains $(i : x \rightarrow m : s)$ or $(i : \neg x \rightarrow m : s)$ (s now being able to take any value in the set $\{++, +-, +, -, -+, --\}$), and a rule $(i : n \rightarrow m : +-)$ which requires that:

$$\Pr(m \,|\, \neg n, X) = 0$$

for all $X \in \{x, \neg x\}$ such that the database contains $(i : x \rightarrow m : s)$ or $(i : \neg x \rightarrow m : s)$. Once again, the introduction of such rules imposes restrictions on other

⊗*	++	+-	+	0	-	-+	--	?
T̄	T̄	+	+	0	-	-	↓	?
+	+	+	+	0	-	-	-	?
0	0	0	0	0	0	0	0	0
-	-	-	-	0	+	+	+	?
↓	-	↓	-	0	+	T̄	+	?
?	?	?	?	0	?	?	?	?

⊗*⁻¹	T̄	+	0	-	↓	?
T̄	++	+	0	-	--	?
+		+	0	-		
0			0			
-		-	0	+		
↓	-+	-	0	+	+-	?
?			0			

Table 3. Variants of ⊗ and ⊗⁻¹

rules involving the same propositions so that $(i : n \rightarrow m : ++)$ implies that there must be restrictions equivalent to the rules $(i : \neg n \rightarrow m : --)$, $(i : n \rightarrow \neg m : --)$ and $(i : \neg n \rightarrow \neg m : ++)$, and similar restrictions are imposed by the other rules. As before, having introduced new qualitative values and ensured that they have a probabilistic meaning, we have to give a suitably probabilistic means of combining them if we want the whole system to be normative. It is reasonably clear that suitable functions comb and comb′ are those variants of ⊗ and ⊗⁻¹ given in Table 3 [13]. Note the asymmetry in the tables. Once again, the base logic compels us to replace negated literals before constructing any arguments, and again we do this by replacing facts and rules. It is clear that facts and rules with signs $+$ and $-$ are handled as before, and that $(i : \neg l : \overline{\top}) \equiv (i : l : \downarrow)$. Categorical rules are also handled by the appropriate substitution using, for instance, the equivalencies $(i : l \rightarrow \neg m : ++) \equiv (i : l \rightarrow m : --)$, $(i : \neg l \rightarrow m : ++) \equiv (i : l \rightarrow m : -+)$ and $(i : \neg l \rightarrow \neg m : ++) \equiv (i : l \rightarrow m : +-)$ [13].

The correct way to flatten normative arguments, some of which are categorical, is a little complex. The problem is that the very strong constraint that a rule $(i : n \rightarrow m : ++)$ puts on the distribution over m and n greatly restricts the values of other rules whose consequent is m. In fact, if we have $(i : n \rightarrow m : s)$, $s \in \{++, -+\}$ then for any other $(i : x \rightarrow m : s')$, $s' \in \{++, +, -, -+\}$ and if we have $(i : n \rightarrow m : s)$, $s \in \{+-, --\}$ then for any other $(i : x \rightarrow m : s')$, $s' \in \{+-, +, -, --\}$ [13, 14]. This means that we have a revised flattening op-

⋔*	T̄	+	0	-	↓	?
T̄	T̄	T̄	T̄	T̄	U	T̄
+	T̄	+	+	?	↓	?
0	T̄	+	0	-	↓	?
-	T̄	?	-	-	↓	?
↓	U	↓	↓	↓	↓	↓
?	T̄	?	?	?	↓	?

Table 4. A new flattening function

erator $\oplus_*$ as given in Table 4 where the symbol U indicates that the result is not defined. U may also be taken to indicate that if this is the result of flattening, then the database on which its deduction is based violates the laws of probability. Equipping $\mathcal{N}\mathcal{A}_3$ with these extensions ensures that it is normative in the sense that all its conclusions will either be in accordance with probability theory or indicate that there has been a violation of the theory.

To see how the system incorporates categorical knowledge, consider the following variation on our example, which includes the categorical rule that being found to embezzle funds from my organisation would lead to me losing my job:

$$f1 : embezzle\_funds : \overline{\uparrow}. \qquad\qquad \Delta_3$$
$$f2 : good\_research\_output : \overline{\uparrow}.$$
$$r1 : embezzle\_funds \rightarrow lose\_job : ++ .$$
$$r2 : good\_research\_output \rightarrow lose\_job : - .$$

From this using $\mathcal{N}\mathcal{A}_3$ we can build the arguments:

$$\Delta_3 \vdash_{ACR} (lose\_job, (f1, r1), \overline{\uparrow}).$$
$$\Delta_3 \vdash_{ACR} (lose\_job, (f2, r2), -).$$

which will flatten to tell us that I will definitely lose my job since the categorical positive effect of embezzling outweighs the negative effect of not being ill.

6 Using order of magnitude information

As the example of Δ_3 demonstrated, $\mathcal{N}\mathcal{A}_3$ extends the kind of representation and reasoning provided by QPNs by allowing the explicit handling of categorical information. This is not the only extension that is possible. Another is to use some form of order of magnitude reasoning. This would make it possible to say, for instance, that because $\Pr(a)$ increases much more than $\Pr(b)$, and $\Pr(a)$ influences $\Pr(c)$ much more strongly than $\Pr(b)$ influences $\Pr(d)$, it is clear that $\Pr(c)$ will undergo a much larger change in value than $\Pr(d)$. A particularly appropriate system for performing this kind of reasoning, known as ROM[K], is provided by Dague [1]. ROM[K] works by manipulating expressions about the relative size of two quantities Q_1 and Q_2. There are four possible ways of expressing this relation: Q_1 is *negligible with respect to* Q_2, $Q_1 \ll Q_2$, Q_1 is *distant from* Q_2, $Q_1 \not\sim Q_2$, Q_1 is *comparable to* Q_2, $Q_1 \sim Q_2$, and Q_1 is *close to* Q_2, $Q_1 \approx Q_2$. Once the relation between pairs of quantities is specified, it is possible to deduce new relations by applying the axioms and properties of ROM[K], some of which are reproduced in Figure 2.

We can use ROM[K] to define a system of argumentation $\mathcal{N}\mathcal{A}_4$ which extends $\mathcal{N}\mathcal{A}_2$ with relative order of magnitude reasoning about the size of the changes in probability with which the system deals. As usual, we need to define combination and flattening functions, though here they differ from those of other systems in that they are comparative and additional to those used by $\mathcal{N}\mathcal{A}_2$. Once the argument is established as being $+$ or $-$ using the function $\otimes$ from $\mathcal{N}\mathcal{A}_2$, this

(A1)	$A \approx A$	(A9)	$A \sim 1 \rightarrow [A] = [+]$
(A2)	$A \approx B \rightarrow B \approx A$	(A10)	$A \ll B \leftrightarrow B \approx (B + A)$
(A3)	$A \approx B, B \approx C \rightarrow A \approx C$	(A11)	$A \ll B, B \sim C \rightarrow A \ll C$
(A4)	$A \sim B \rightarrow B \sim A$	(A12)	$A \approx B, [C] = [A] \rightarrow (A + C) \approx (B + C)$
(A5)	$A \sim B, B \sim C \rightarrow A \sim C$	(A13)	$A \sim B, [C] = [A] \rightarrow (A + C) \sim (B + C)$
(A6)	$A \approx B \rightarrow A \sim B$	(A14)	$A \sim (A + A)$
(A7)	$A \approx B \rightarrow C.A \approx C.B$	(A15)	$A \not\approx B \leftrightarrow (A - B) \sim A$ or $(B - A) \sim B$
(A8)	$A \sim B \rightarrow C.A \sim C.B$		

(P3)	$A \ll B \rightarrow C.A \ll C.B$	(P26)	$A \sim B \rightarrow B \sim A$
(P35)	$A \not\approx B \rightarrow C.A \not\approx C.B$	(P38)	$A \not\approx B, C \approx A, D \approx B \rightarrow C \not\approx D$

Fig. 2. Some of the axioms and properties of ROM[K]

new combination function comb$_*$ gives the relation between the changes based on the strength of the influences that cause the change while comb$'_*$ may be used to identify the relation between the influences of two rules based upon the relation between the changes in probability of their antecedents and the change in probability of their consequents. Similarly, the new flattening function identifies the greatest influence on a given hypothesis allowing a ? caused by two conflicting arguments to resolved into a $+$ or a $-$. The combination function comb$_*$ is defined in Table 5—if the change in $\Pr(a)$ stands in relation rel$_1$ to the change in $\Pr(b)$ (where rel$_1$ is one of the relations of ROM[K]) and the strength of the influence of $\Pr(a)$ on $\Pr(c)$ stands in relation rel$_2$ to the strength of the influence of $\Pr(b)$ on $\Pr(d)$ (rel$_2$ also being one of the relations of ROM[K]), then the relation rel$_3$ between the changes in $\Pr(c)$ and $\Pr(d)$ is given by the combinator table. Note that Table 5 only covers the cases in which the change in $\Pr(a)$ is less than or equal to that in $\Pr(b)$ and the strength of the influence between $\Pr(a)$ and $\Pr(c)$ is less than or equal to that of $\Pr(b)$ on $\Pr(d)$. Obvious permutations of the table will cover the other cases. Also note that the letter V indicates that rel$_3$ may not be determined from the particular values of rel$_1$ and rel$_2$ because to make any prediction would be to step outside the bounds of probability.

Table 5 also defines comb$'_*$, which as before is the inverse of comb$_*$ with rel$_1$ and rel$_3$ determining rel$_2$. Note that for some combinations of input, the output is ambiguous.

		rel$_2$						rel$_3$			
	comb$_*$	$\approx$	$\sim$	$\not\approx$	$\ll$		comb$'_*$	$\approx$	$\sim$	$\not\approx$	$\ll$
	$\approx$	$\approx$	$\sim$	$\not\approx$	$\ll$		$\approx$	$\approx$	$\sim$	$\not\approx$	$\ll$
rel$_1$	$\sim$	$\sim$	$\sim$	V	$\ll$	rel$_1$	$\approx$	$\approx, \sim$			$\ll$
	$\not\approx$	$\not\approx$	V	V	$\ll$		$\not\approx$		$\approx$		
	$\ll$	$\ll$	$\ll$	$\ll$	$\ll$		$\ll$				$\approx, \sim, \not\approx, \ll$

Table 5. Combining ROM[K] relations.

For the flattening function, if the change in $\Pr(a)$ stands in relation rel_4 to the change in $\Pr(b)$ (where rel_4 is one of the relations of ROM[K]) and the strength of the influence of $\Pr(a)$ on $\Pr(c)$ stands in relation rel_5 to the strength of the influence of $\Pr(b)$ on $\Pr(c)$ (rel_5 also being one of the relations of ROM[K]), the sign of the change in $\Pr(c)$ is given in Table 6 (where $[\Delta \Pr(b)]$ indicates the sign of the change in $\Pr(b)$). Note that Table 6 only covers the cases in which the change in $\Pr(a)$ is less than or equal to that in $\Pr(b)$ and the strength of the influence between $\Pr(a)$ and $\Pr(c)$ is less than or equal to that of $\Pr(b)$ on $\Pr(c)$. Obvious permutations of the table will cover the other cases.

As an example of the kind of reasoning that may be performed in $\mathcal{N}A_4$, consider the following variant of our running example.

$$f1 : ill : +. \qquad\qquad \Delta_4$$
$$f2 : embezzle\_funds : -.$$
$$r1 : ill \rightarrow lose\_job : +.$$
$$r2 : ill \rightarrow hospital : +.$$
$$r3 : embezzle\_funds \rightarrow lose\_job : +.$$

In addition, consider we know that the relationship between the strengths of $r1$ and $r2$ is $\ll$, while the changes in probability implied by $f1$ and $f2$ stand in relation $\sim$. From the database we can build the arguments:

$$\Delta_4 \vdash_{ACR} (lose\_job, (f1, r1), -).$$
$$\Delta_4 \vdash_{ACR} (hospital, (f1, r2), +).$$
$$\Delta_4 \vdash_{ACR} (lose\_job, (f2, r2), +).$$

using the combination function from $\mathcal{N}A_2$. Considering the first two arguments, the **comb**$_*$ may then be used to establish which has stronger support. Since both arguments are based upon the same fact, rel_1 is '$\approx$', so that we can conclude that the relation rel_3 between the changes in probability of 'hospital' and $lose\_job$ must be '$\ll$' so that the increase in belief that I will lose my job is much smaller than the increase in belief that I will go to hospital. Similarly, flattening the arguments for $lose\_job$ with the old flattening function will give ?, while the new flattening function will establish that the probability of $lose\_job$ will increase as can be seen by looking at the intersection of $\not\approx$ and $\approx$ in Table 6.

		$\approx$	$\sim$	$\not\approx$	$\ll$
				rel_5	
	$\approx$	?	?	$[\Delta \Pr(b)]$	$[\Delta \Pr(b)]$
rel_4	$\sim$	?	?	?	$[\Delta \Pr(b)]$
	$\not\approx$	$[\Delta \Pr(b)]$	?	?	$[\Delta \Pr(b)]$
	$\ll$	$[\Delta \Pr(b)]$	$[\Delta \Pr(b)]$	$[\Delta \Pr(b)]$	$[\Delta \Pr(b)]$

Table 6. How to flatten arguments in ROM[K]

7 Using numerical information

Further precision may be obtained by incorporating numerical information about the size of changes in probability and the strengths of influences. Inspired by Dubois *et al.* [6], we build a new system of argumentation $\mathcal{N}\mathcal{A}_5$ with the same base language as the other systems, but which has a dictionary which includes a set of "linguistic"[5] labels, each of which is an identifier for an interval probability, and may be used to give the strength of rules. A suitable set is:

Strongly Positive $\geq$ Weakly Positive $\geq$ Zero $\geq$ Weakly Negative $\geq$ Strongly Negative
   (SP)   (WP)   (Z)   (WN)   (SN)

$$(1, \alpha] \quad \geq \quad [\alpha, 0) \quad \geq 0 \geq \quad (0, -\alpha] \quad \geq \quad [-\alpha, 1)$$

though we could take any set of intervals we desire—a larger set will give us a finer degree of resolution but be more tedious to use as an example. Note that the open intervals explicitly do not allow the modelling of categorical influences (if these are required we can simply add additional labels at either end of the scale). The dictionary also includes a second set of labels which quantify changes in probability:

Complete Positive $\geq$ Big Positive $\geq$ Medium Positive $\geq$ Little Positive $\geq$ Zero
     (CP)   (BP)   (MP)   (LP)   (Z)

$$1 \quad \geq \quad (1, 1 - \beta] \geq \quad [1 - \beta, \beta] \quad \geq \quad [\beta, 0) \quad \geq 0$$

The definition of the changes Little Negative (LN), Medium Negative (MN), Big Negative (BN) and Complete Negative (CN) are symmetrical, and again we could use a different set if desired. Like the other systems of argumentation, $\mathcal{N}\mathcal{A}_5$ uses the argument consequence relation $\vdash_{ACR}$ to build arguments for hypotheses, and so in order to be able to determine the strength of arguments we must define combination functions comb and comb$'$ which say how to combine the "linguistic" labels. To do so we must first choose suitable values of α and β, and on the grounds that we would like our intervals to be evenly sized, we choose $\beta \approx 0.33$ and $\alpha \approx 0.5$. This then gives us the combination functions of Table 7 where [MP, LP] stands for the interval whose upper limit is the upper limit of MP and whose lower limit is the lower limit of LP. Results of combining with negative influences and changes can be obtained by symmetry

To combine several arguments for one proposition we need a suitable flattening function, and this is provided by interval addition. Furthermore, if we are to use the precision of the system we need a way to compare intervals in order to identify which arguments have the greatest support. This may be done using $\leq_{int}$ where $[a, b] \leq_{int} [c, d]$ iff $a \leq c$ and $b \leq d$ [5]. To illustrate the use of $\mathcal{N}\mathcal{A}_5$

---

[5] The scare quotes denoting that no claim is being made that the probability intervals with which we deal are in any way related to interpretations of natural language—we are just adopting Dubois *et al.*'s terminology.

consider the database:

$$f1 : embezzle\_funds : CP. \qquad \Delta_5$$
$$f2 : ill : BP.$$
$$r1 : embezzle\_funds \rightarrow lose\_job : SP.$$
$$r2 : ill \rightarrow lose\_job : WP.$$
$$r3 : ill \rightarrow hospital : SP.$$

From this we can build the arguments:

$$\Delta_5 \vdash_{ACR} (lose\_job, (f1, r1), [BP, MP]).$$
$$\Delta_5 \vdash_{ACR} (lose\_job, (f2, r2), [MP, LP]).$$
$$\Delta_5 \vdash_{ACR} (hospital, (f2, r3), [BP, MP]).$$

The two arguments for *lose_job* may be flattened to give the overall value of [CP, MP] and using $\leq_{int}$ we learn that the increase in probability of *hospital* is less than or equal to that of *lose_job*.

8 Discussion

This paper began with the claim that it would present a number of normative systems of argumentation, taking this to mean that they have a probabilistic semantics, and that they would thus be an improvement on non-normative systems of argumentation for those cases in which such norms are desirable. Furthermore, the claim was made that these systems would also be an improvement on qualitative systems for reasoning with probability such as QPNs since they would allow more precise predictions to be made. In the event five different systems, $\mathcal{NA}_1$–$\mathcal{NA}_5$, which meet these objectives to varying degrees, have been presented.

$\mathcal{NA}_1$, uses a probabilistic notion of qualitative influences between variables to give meaning to logical rules. The fact that $\mathcal{NA}_1$ has a strict probabilistic semantics means that it is an extension of non-normative systems of argumentation. However, the restrictions on the meaning of the rules imposed by the improper linear model mean that $\mathcal{NA}_1$ is not an extension of QPNs. $\mathcal{NA}_2$ is a system of argumentation which is roughly equivalent to QPNs. Thus $\mathcal{NA}_2$ whilst an extension of non-normative systems of argumentation on which it is based, is not an extension of QPNs.

The problem of extending QPNs was addressed by $\mathcal{NA}_3$. The extension takes the form of allowing the representation of categorical influences between variables. Giving these a qualitative representation and a probabilistic meaning

comb	CP	BP	MP	LP	Z
SP	[BP, MP]	[BP, MP]	[MP, LP]	LP	Z
WP	[MP, LP]	[MP, LP]	[MP, LP]	LP	Z
Z	Z	Z	Z	Z	Z

comb'	[BP, MP]	[MP, LP]	LP	Z
SP	[CP, BP]	MP	LP	Z
WP		[CP, MP]	LP	Z
Z				Z

Table 7. Combining "linguistic" labels

makes $\mathcal{N}\mathcal{A}_3$ a system which is both normative and can represent and reason with a wider range of information than is possible in a QPN whilst retaining the latter's qualitative nature. Thus it meets overall objectives of the paper. Two further extensions were introduced in the form of $\mathcal{N}\mathcal{A}_4$ and $\mathcal{N}\mathcal{A}_5$ which use order of magnitude and interval information respectively.

9 Relation to other work

There are a number of connections with the work of other authors. The close relation between qualitative approaches to probabilistic reasoning in networks and probabilistic systems based on logic was suggested by Wellman [17] while the idea of a database of influences which is equivalent to a probabilistic network has been discussed by, among others, Poole [16] and Wong [19]. The attempt to give an essentially logical system a probabilistic semantics makes our efforts similar to Goldszmidt's work on normative systems for defeasible reasoning [10]. This clearly has some similarities with our work, but differs in its intent. Goldszmidt aims to build defeasible systems whose behaviour is justified by their probabilistic semantics while we are intent on a more general system. The use of a probabilistic semantics is not our only goal—we are just interested in being able to provide a normative system when one is required, with the choice of alternative combination and flattening functions allowing a broad range of possible systems to be adopted. In addition, our work has strong connections with that of Darwiche [2], this time differing in the way it is approached. His aim was "...to relax the commitment to numbers while retaining the desirable features of probability theory", which is rather different to the aim of the work described here. We started from the opposite position, taking a completely abstract model of reasoning and seeing how it could be instantiated to behave in a probabilistic way if so desired (which often it won't be since probability theory often imposes overly strict constraints for the kind of reasoning that argumentation was designed to provide), and the fact that we did so suggests that the work presented here and that in [2] are to some extent complementary.

Acknowledgements

This work was partially supported by Esprit Basic Research Action 6156 DRUMS II (Defeasible Reasoning and Uncertainty Management Systems). Many thanks to Paul Krause, John Fox, Kathy Laskey and Jack Breese for making me think more clearly about some of the issues discussed in this paper.

References

1. Dague, P. (1993) Symbolic reasoning with relative orders of magnitude, *Proceedings of the 13th International Joint Conference on Artificial Intelligence*, Chambery, France.

463

2. Darwiche, A. (1993) A symbolic generalization of probability theory, Ph.D. Thesis, Stanford.
3. Dawes, R. (1979) The robust beauty of improper linear models, *American Psychologist*, **34**, 571–582.
4. Druzdzel, M. J. and Henrion, M. (1993) Efficient reasoning in qualitative probabilistic networks, *Proceedings of the 11th National Conference on Artificial Intelligence*, Washington.
5. Dubois, D. and Prade, H. (1979) Fuzzy real algebra: some results, *Fuzzy sets and systems*, **2**, 327–348.
6. Dubois, D., Prade, H., Godo, L., and Lopez de Mantaras, R. (1992) A symbolic approach to reasoning with linguistic quantifiers, *Proceedings of the 8th Conference on Uncertainty in Artificial Intelligence*, Stanford.
7. Fox, J. (1990) Automating assistance for safety critical decisions, *Philosophical Transactions of the Royal Society*, B, **327**, 555–567.
8. Fox, J., Krause, P. and Ambler, S. (1992) Arguments, contradictions and practical reasoning, *Proceedings of the 10th European Conference on Artificial Intelligence*, Vienna.
9. Glowinski, A., O'Neil, M., and Fox, J. (1987) Design of a generic information system and its application to primary care, *Proceedings of AIME Conference*, Marseille.
10. Goldszmidt, M. (1992) Qualitative probabilities: a normative framework for commonsense reasoning, Ph.D. Thesis, UCLA.
11. Krause, P., Ambler, S., Elvang-Gøransson, M., and Fox, J. (1995) A logic of argumentation for reasoning under uncertainty, *Computational Intelligence*, 11, 113–131.
12. Loui, R. P. (1987) Defeat among arguments: a system of defeasible inference, *Computational Intelligence*, **3**, 100–106.
13. Parsons, S. (1996) Normative argumentation and qualitative probability, Technical Report 317, Advanced Computation Laboratory, Imperial Cancer Research Fund.
14. Parsons, S. (1995) Refining reasoning in qualitative probabilistic networks, *Proceedings of the 11th Conference on Uncertainty in Artificial Intelligence*, Montreal.
15. Parsons, S. and Fox, J. (1994) A general approach to managing imperfect information in deductive databases, *Proceedings of the Workshop on Uncertainty in Databases and Deductive Systems*, Ithaca, NY.
16. Poole, D. (1991) Representing Bayesian networks within probabilistic horn abduction, in *Proceedings of the 7th Conference on Uncertainty in Artificial Intelligence*, Los Angeles, CA.
17. Wellman, M. P. (1994) Some varieties of qualitative probability, *Proceedings of the 5th International Conference on Information Processing and the Management of Uncertainty*, Paris.
18. Wellman, M. P. (1990) *Formulation of tradeoffs in planning under uncertainty*, Pitman, London.
19. Wong, S. K. M., Xiang, Y., and Nie, X. (1994) Representation of Bayesian networks as relational databases, *Proceedings of the 5th International Conference on Information Processing and the Management of Uncertainty*, Paris.

Complex Argumentation in Judicial Decisions.
Analysing Conflicting Arguments.

José Plug
Erasmus University Rotterdam
Faculty of Law
The Netherlands

1 Introduction

Judicial decisions are often justified by means of complex argumentation, that is to say by more than only one argument. It is only to be expected that studies on the justification of judicial decisions pay attention to the complexity of argumentation. Both in legal theory as in studies on legal practice different types of complex argumentation are distinguished. When analysing argumentation in judicial decisions it is, however, often quite difficult to decide how the arguments are structured because the distinctions between the various types are not very clearly characterised. This problem may find its cause in the fact that research in these fields is not in the first place focused on analysing the structure of argumentation but rather on questions as to what kind of arguments can be deployed in a certain situation. It is nevertheless very important to get a clear insight into in the way arguments are structured, since vagueness as to the relations between arguments may have serious consequences for the evaluation of the acceptability of the decision.

In legal theory, several authors choose a dialogical point of view to study legal reasoning. They look upon legal argumentation as part of a discussion in which a decision is being defended against criticism. Summers and MacCormick (1991), for instance, emphasize that criticism by means of counter-argumentation may play an important role in the justification of a decision.

Recent studies in argumentation theory by, amongst others, Snoeck Henkemans (1992) indicate that the analysis and evaluation of argumentation should be studied in a dialogical context. This dialogical approach to argumentation not only provides an insight into how arguments and counter-arguments result in complex argumentation, it also provides clues for reconstructing the structure of the argumentation. This common, dialogical, ground may be a fruitful starting-point to gain more insight into the various types of complex argumentation that are distinguished in legal theory.

In this paper I intend to make clear how counter-argumentation can be both part of the structure of argumentation and of the justification of the judge's standpoint. The basic assumption is that an adequate reconstruction of complex argumentation needs to deal with the way complex argumentation evolves in a discussion as well as with the function of various argumentation structures that can be used in the defence of a standpoint.

First I will give a short account of the factors that contribute to the complexity of argumentation in judicial decisions and of the various types of complex argumentation that are distinguished in legal theory. Then, using insights developed in pragma-dialectic argumentation theory, I will indicate what clues may help to establish the structure of arguments.[1] Some examples will illustrate how to reconstruct counter-argumentation concerning the facts that are part of the justification of a decision. Finally I will try to analyse types of counter-argumentation proposed in legal theory to justify decisions concerning the interpretation of statutory rules.

2 Complexity of legal argumentation

In the Netherlands the requirements as to what elements should constitute a judicial decision are laid down by law. These requirements, however, are not very specific. A variety of decisions, all construed in a different manner is the result. Nevertheless, the basic outline usually shows a common pattern. This pattern, moreover, appears to correspond with the global structure of judicial decisions in other countries.[2] Although the decisions show many similarities, a great variety is apparent as to the complexity and elaborateness of the motivation.

According to Summer and Taruffo (1991:492) the application of complex justification forms, finds an explanation in cultural and institutional factors. They assume, moreover, that the application of complex justification forms in judicial decisions are indicative of a shift from a rather authoritative towards a more dialogical approach. From this follows that the decision is looked upon as a reaction to the arguments put forward by the parties. In the Netherlands this shift has been observed by, for instance, Franken (1982:457):

> To sum up, we may conclude, that the civil judge should do more than simply act the referee, making sure the parties observe the rules of fair play. He participates in the trial and endeavours to reach a verdict effective in the relationship between the parties. He is no longer attempting to 'find justice' in order to enforce it but he is the one to make a choice between alternatives.

These observations are in keeping with those of Van Eemeren et.al. (1991) and Feteris (1992) that the judicial process may be regarded as a reasonable discussion in which case the judge is not merely a guardian of procedural matters but also the one to judge the contributions to the discussion. The decision is regarded as part of the discussion which is brought up to solve a dispute. Feteris (1992:51) specifies the judge's role in this context as follows:

---

[1]   This theory is developed by Van Eemeren and Grootendorst (1984, 1992).

[2]   See the comparative analysis by Summers and Taruffo (1991: 490). This study has been conducted in Argentina, Germany, Finland, France, Italy, Poland, Great Britain and the United States of America.

Since the parties do not themselves solve their dispute in consultation, but rather have the judge, as a neutral third party, decide on the eventual outcome, the parties should be granted insight into the grounds the judge has taken into consideration in reaching his verdict. In dialectic terms, stating the considerations underlying his decision amounts to stating the factors which were instrumental in his assessment of the acceptability of the propositional content as well as the justificatory potential of the argumentation of the party asking for a decision.

Dutch jurisprudence prescribes that the judge, in principle, need not go into each and every proposition and argument put forward by the legal parties in support of their standpoints. In case of essential propositions, however, or when the decision is of singular importance to the legal parties, there is always the possibility of the decision being quashed, because certain propositions have not been considered (among others Veegens 1989:234-236). In cases such as these the Supreme Court may rule as follows:

Because of the far-reaching nature of the decision under consideration, high requirements should be met as to the motivation. [...] The Court should have weighed the interests of both parties. The motivation should reflect this these considerations. (Supreme Court 17 September 1982, NJ 1983, 46)

Differences in elaborateness and complexity of motivation may thus depend on the question in how far and in what way the judge, in his motivation, responds to the argumentation of the legal parties or anticipates possible objections.

The complexity of the argumentation may also follow from the complexity of the casus. Legal theory distinguishes between 'clear' cases on the one hand, and 'hard' cases on the other.[3] In clear cases the decision follows from the legal rule as well as from the qualification of the facts. Argumentation in these cases is complex when more than just one fact will have to be qualified or when a number of legal rules apply. In clear cases there is no difference of opinion as to the interpretation of the relevant legal rule. Nor do opinions differ regarding the qualification of the facts.

In hard cases differences of opinion as to interpretation and qualification do occur. In cases like these, it is the judge who takes an *interpretative decision* as to the facts and the relevant legal rule.[4] In the case of an interpretative decision, argumentation will be put forward (justification of the second order), in which an appeal can be made to methods of interpretation, precedents or dogmatics. This argumentation, in its turn, may be justified by making an appeal to ends, values

---

[3]   See, for example, Aarnio (1987:1), Alexy (1989:8) and MacCormick (1978: 100).

[4]   Aarnio (1991:129) distinguishes four types of hard cases in which interpretative decisions have to be taken: 'a provision is ambiguous or vague; there is no provision that would apply to the case (a gap); two provisions with different content apply to the same case (a conflict); or the same issue is dealt with in the same way in several different provisions (overregulation).'

and principles acknowledged in the legal community. This type of justification has been described as 'the non-judicial justification which [...] in the end is founded on the judge's image of society' (Franken et al. 1995:145, 160).

Argumentation produced in hard cases, is often of the complex kind. MacCormick and Summers (1991:525) distinguish two types of complex argumentation: the 'Cumulative-arguments Form' and the 'Conflict-settling Forms'. By the 'Cumulative-arguments Form' they mean both arguments which are mutually independent and arguments which are interdependent, stressing the fact that in the case of interdependent arguments the force of the whole may be greater than the sum of the parts. 'Conflict-settling Forms' include argumentation in which arguments for mutually conflicting interpretations are balanced. The arguments which are put forward can be weighed in a number of ways:

(1)   The other argument proves to be, on close analysis, unavailable inasmuch as the very conditions required for it to exist simply are not present.
(2)   The other argument is deprived of all or most of its prima facie force by the prevailing argument.
(3)   The other argument is mandatorily subordinated pursuant to a general rule or maxim of priority.
(4)   The other argument is simply outweighed.

MacCormick and Summers (1991:528) consider the whole topic of conflict-settling, the weighing mode in particular, to be of considerable importance but state that it 'requires much further work'. Their aim is merely to indicate the general problem. On the basis of their description of these four types of argumentation, I will try to shed some light on the question if and how these types of complex argumentation can be recognized and could be reconstructed.

3 Clues for reconstructing complex argumentation

In the pragma-dialectic studies on argumentation, Van Eemeren et al. (1991) and Snoeck Henkemans (1992) examine the problems one may encounter when reconstructing complex argumentation. They distinguish multiple argumentation and coordinatively compound argumentation. They propose a dialogical approach to argumentation. That is to say, both the way complex argumentation is generated in a discussion and the functions these various argumentation structures fulfil in defending a standpoint, characterize these types of argumentation.

In multiple argumentation, each argument in itself constitutes sufficient support for the standpoint. Should one of the arguments in multiple argumentation be contested successfully, the standpoint is still sufficiently supported by the remaining arguments. Coordinatively compound argumentation consists of a number of argumentations, horizontally linked and which in conjunction provide sufficient support for the standpoint. In such cases, a successful attack on one of the arguments can be detrimental to the argumentation as a whole. Snoeck

Henkemans (1992:96) points out that, within coordinatively compound argumentation, a further distinction can be made between cumulative and complementary argumentation. Cumulative argumentation involves adding a new argument of a different type to the first argument. In complementary argumentation the new argument is presented in order to remove doubts as to the importance of the first argument. For the purpose of evaluating both forms of coordination she emphasises however, that it is only the conjunction of arguments that can constitute an adequate defence of the standpoint.

To be able to determine in what way arguments are structured, the point of departure should always be the verbal presentation of a text.[5] A text may sometimes contain verbal directions as to the way arguments are related - so-called indicators like *'Apart from this consideration*, it is also true that...' or *'In connection with* this it should be considered...'. If the structure of the arguments cannot be traced by means of explicit indicators, it may still be possible to find clues as to the manner in which the arguments are linked. The phrasing of the statutory rule to which the judge refers, may constitute a clue as to how the structure of the judge's argumentation should be reconstructed (see Van Eemeren et al. 1991:68 et seq. and Plug 1994).

In the reasons given in the decision, the judge will often react to the arguments presented by the parties in support of their standpoint. The very structure of these arguments may also provide clues for the reconstruction of the argumentation (see Plug 1994). When reconstructing the argumentation structure of a judicial decision, it is often necessary to view these various clues in conjunction. Even if an explicit indicator is used, the other clues can be used to decide what interpretation should be preferred (see Plug 1995).

In the following paragraph I will demonstrate that references to the dialogical character of argumentation can also constitute part of the argumentation structure of the motivation. In doing so I will link the critical reactions of both actual and eventual opponents on the one hand to the argumentation structure of the motivation of the decision on the other.

4 Reconstructing counter-argumentation concerning the facts

In *Analysing complex argumentation* (1992:129 et seq.) Snoeck Henkemans indicates how refutations of counter-arguments provide useful information about the way in which a standpoint is defended. Although this approach has not been drawn up primarily for legal argumentation, it may be very helpful when analysing the argumentation in judicial decisions. As was shown before, it is, after all, only

---

[5]   Van Eemeren and Grootendorst (1992:22): 'In principle, it must always be assumed that the verbal presentation of his communication reflects the intentions of the writer. In practice, of course, there may well be something wrong with this presentation.'

to be expected that the judge, in his argumentation, refers to the critical opponent's reaction, i.e. reactions to arguments of legal parties or anticipations to criticism on appeal. With regard to references to a counter-argumentation, three possible situations may arise.

(1)   The judge mentions an argument against his standpoint and then refutes this counter-argument.

(2)   The judge mentions an argument against one of his arguments and then refutes this counter-argument.

(3)   The judge acknowledges a counter-argument, but considers his standpoint to be defensible.

The counter-arguments are different in each of these instances. In the first situation the judge mentions an argument used to directly criticize his standpoint. This criticism does not so much concern the argumentation with which the judge justifies his standpoint, but is rather meant as a support for the opposite standpoint. In the next example the counter-argument arguing against the standpoint is refuted. The example deals with a neighbours' quarrel over a row of conifers.

> (...) Defendant argues that the conifers have been planted to reduce draught in his house, but this argument is absolutely unsound, since most of the window posts are closed and the window that does open, is located on a point higher than the tops of the conifers and has not been fitted with any antidraught facilities. (...)
> Whereas the defendant has no considerable interest in these conifers, removal is of significant concern to the claimant since they block his view and take away the light. (...)
> (2981. Country court Enschede 6 October 1988)

The judge defends the standpoint that the claimant's interest in the removal of the conifers is greater than the defendant's interest in leaving them untouched. In support of this standpoint he argues that the conifers block the view and take away the light. In his preceding remarks the judge mentions the defendant's argument in support of the opposite standpoint: he does have a considerable interest in the conifers since they reduce draught in his house. The judge refutes this argument referring to the fact that most of the window posts are closed and the opening window, which has no antidraught facilities whatsoever, is located higher that the tops of the conifers.

So the judge's argumentation consists of a pro-argument and the refutation of a counter-argument which, in conjunction, form sufficient support for his standpoint. Subsequently, the argumentation with which the judge refutes the counter-argument is coordinate as well. The argumentation structure can be represented as follows.

Figure 1

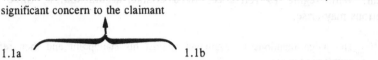

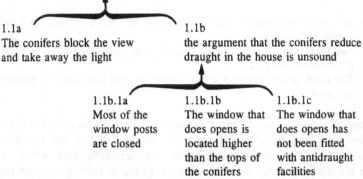

1.

Whereas the defendant has no considerable interest in these conifers, removal is of significant concern to the claimant

1.1a
The conifers block the view and take away the light

1.1b
the argument that the conifers reduce draught in the house is unsound

1.1b.1a
Most of the window posts are closed

1.1b.1b
The window that does opens is located higher than the tops of the conifers

1.1b.1c
The window that does opens has not been fitted with antidraught facilities

In second situation the judge mentions an argument which indirectly criticizes the standpoint. The criticism is aimed at the argumentation with which the judge justifies his standpoint. The counter-argument against his own argumentation may hold that the argumentation would be *untrue* or *unacceptable*. The counter-argument may also be targeted at the *sufficiency* or the *relevance* of the argumentation. Subsequently, the judge may refute the criticism expressed in the counter-argument by showing that the objection is *untrue* or *unacceptable*, *insufficient* or *irrelevant*. An example of a situation like this one is about a dispute over the question whether an employer is allowed to fire a female employee who has been employed by him over a longer period than her male colleague.

> (...) However, since we learned that Ms Been is not at all unable or unwilling to carry large orders and to deliver these three floors up, and such over a longer period of time, we can't help but feeling that, in this respect, Buitelaar's use for Groenendijk's firm is hardly greater than the contributions Ms Been can offer. Groenendijk's fear that customers would be astonished or even annoyed if a woman would be lugging objects for them, seems to us not to be well-founded at this juncture in which emancipation is gaining more and more ground. (...)
> (2349. Country court Rotterdam, 12 June 1985)

The judge is of the opinion that both employees are of more or less equal use to the firm. By means of justification he puts forward that the employee Ms Been did not seem unable or unwilling to carry large orders. Subsequently the judge anticipates his argument not to be sufficient. True, Ms Been is neither unable nor unwilling to carry orders, but customers may be astonished or annoyed. The judge shows that this objection is unacceptable. In this case too there is a coordinate relation between the judge's pro-argument and the refutation of the counter-argument. The structure of this argumentation is shown in figure 2.

Figure 2

1.
Buitelaar's use for Groenendijk's firm is hardly greater
than the contributions Ms Been can offer

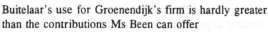

1.1a
Ms Been is not unable or unwilling
to carry large orders and to deliver
these three flours up, and such over
a long period of time

1.1b
Groenendijk's fear that customers
would be astonished or even annoyed
if a woman would be lugging objects
for them seems not well founded

1.1b.1
At this junction emancipation is
gaining more and more ground.

In the third situation too, the judge mentions an argument which is used to, indirectly, criticize the standpoint. The difference with the second case is that the judge does not refute the criticism, but rather holds that the criticism does not counter balance the argumentation in favour of his standpoint. The judge does acknowledge the counter-argument, but maintains that his standpoint is defensible. In a case like this he does not criticise the propositional content of the objection. The justification potential, however, is considered not to be sufficient because the objection is irrelevant or insufficient. This situation would arise if the judge of the previous example acknowledged that customers would indeed be astonished and annoyed but refuted the claim that this counter-argument is more important than the fact that Ms Been is able and willing to carry and deliver large orders. The judge could argue his case, for instance, by pointing out that the reactions of the customers have not resulted in any loss of customers. Since the judge attempts to refute an objection concerning the sufficiency of his first argument, the main arguments form a coordinatively compound argumentation. The argumentation structure is schematized in figure 3.

Figure 3

1.
Buitelaar's use for Groenendijk's firm is hardly greater
than the contributions Ms Been can offer

1.1a
Ms Been is not unable or unwilling
to carry large orders and to deliver
these three flours up, and such over
a long period of time

1.1b
That Ms Been is not unable or unwilling
to deliver large orders is of more
importance than the fact that customers
are astonished and annoyed.

1.1b.1
The reactions of the customers have not
resulted in any loss of costumers

These examples show how the judge's reaction to the argumentation of the legal parties may be part of the motivation of his decision. An analysis of the different types of counter-arguments and the different ways in which they, in their turn, can be refuted, can offer an insight into the argumentation structure of the motivation. With a reference to appeal cases, this will not only make clear where to aim argumentative criticism but also what consequences to expect from such action as far as the acceptability of the decision is concerned. The decisions in these examples are all concerned with the facts presented by the legal parties. In the following paragraph I will work out how counter-arguments can form part of the argumentation for the benefit of interpretative decisions as to the legal rule.

5 Reconstructing argumentation concerning the legal rule

5.1 Cumulative-arguments Form

An interpretative decision concerning the statutory rule has to be taken if, for example, the connection between the conditions stated in the rule is ambiguous. In the next example vagueness is caused by the ambiguous conjunction 'and', used to express the relation between the conditions. The conjunction 'and' may express either alternative conditions or cumulative conditions. This ambiguity gave rise to a difference of opinion as to the relations between the conditions in the statutory rule. Section 12 Royal Decree 1913 reads:

> No person is allowed to trespass upon the grounds outside the public highways and footpaths, carrying, among other things, trap-nets or dragnets, the wings of which have a length of more than 10 metres, [...], *and* a width of more than 10 metres, measured   according to the lengths of the bars, *and* having meshes measuring less than three by three centimetres.

It was considered a proven fact that 'the defendant has trespassed the grounds, outside the public highways and footpaths, carrying more than one trap-net or dragnet having meshes measuring less than three by three centimetres.' The court, however, judged that this fact did not result in a violation of section 12. The defendant would only be guilty if, on top of these facts, the nets' wings had not met the requirements of the article. In cassation, the public prosecutor argued that the conditions could just as well have been interpreted in an alternative way. The Supreme Court put the public prosecutor in the right and quashed the verdict of the district court. It judged that the conditions in this statute should be interpreted as being alternative and gave the following arguments.

> (...) since the purpose of the injunction, catching too many birds at a time or of birds that are too small, can only be reached, when it is prohibited to use nets other than those which meet every one of the three requirements in section 12 of the Royal Decree;

that this intention can be found in the text of this section in which "and" can evidently be regarded in a disjunctive fashion. (...)
(Supreme Court 3 November 1924, W 11289)

In legal theory different arguments are distinguished in order to establish the interpretation of a legal rule.[6] In this example the Supreme Court refers to a teleological argument (the purpose of the rule) and a linguistic argument (the text). This complex argumentation of the Supreme Court can, in Summers and MacCormick's terms, be characterized as a cumulative-arguments Form and the argumentation structure could be schematized as follows.

Figure 4

1.
The conditions in the rule (r) should be interpreted as being alternative (a).

1.1a
Interpretation (a) is in keeping with the purpose of the rule

1.1b
This intention can be found in the text of this section in which 'and' can evidently be regarded in a disjunctive fashion

These arguments are most likely to be interpreted as coordinate compound, because the second argument seems to be put forward in case the first argument is criticised because its is insufficient. Taking this example as a starting point, I will try to establish in what way the argumentation could be structured if a judge would justify his decision by means of conflict-settling forms of argumentation.

5.2 Conflict-settling forms of argumentation

Conflict-settling forms of argumentation, which I mentioned earlier, presuppose conflicting arguments for rival interpretations of the statutory rule to be cumulated on each side, discussed and the conflict resolved in the opinion. In settling the conflict, the court may have deployed one or more of several distinct modes of resolution. From a pragma-dialectic point of view, these modes of resolution could be seen as different ways in which the court responds to criticism. The court can be seen as a protagonist who explicitly defends his standpoint against counter-arguments. I will try to find out whether the analysis of refutations of counter-arguments in decisions concerning the facts can be of any help for the analysis of the four modes of resolution in decisions concerning the statutory rules as distinguished by MacCormick and Summers.

---

[6] Linguistic arguments, historical arguments, systematic arguments, and teleological arguments are the best known.

The first conflict-settling form they mention occurs when two or more arguments come into conflict, and one argument may prevail because the other argument is inapplicable and therefore rebutted. They give the following description of this type:

(1)   an argument is *rebutted* if it can be shown that, despite first appearances the relevant interpretative conditions do not exist.

If the court would have used this mode of resolution to justify the interpretative decision that the conditions in Section 12 of the Birds Protection Act are alternative, the argumentation could read as follows:

> (...) since the purpose of the injunction, catching too many birds at a time or of birds that are too small, can only be reached, when it is prohibited to use nets other than those which meet every one of the three requirements in section 12 of the Royal Decree;
> that, moreover, ordinary language does know many examples of situations in which 'and' is indeed used to express alternative conditions.

To justify an alternative interpretation of the conditions in the legal rule the court deploys a teleological argument. The court subsequently refutes the counter-argument in which is stated that in ordinary language the conditions should be interpreted as being cumulative. This counter-argument is refuted by demonstrating that the propositional content of the counter-argument is untrue (not acceptable); in ordinary language too, 'and' appears to be used for expressing alternative conditions. Because the refutation of the counter-argument is by itself not sufficient support for the conclusion, the main argumentation is coordinatively compound. The argumentation structure of this example may be schematized as follows.

Figure 5

1.
The conditions in the rule (r) should be interpreted as being alternative (a) (and not as being cumulative).

1.1a
Interpretation (a) is in keeping with the purpose of the rule

1.1b
That the conditions of rule (r) because of the ordinary meaning of 'and' should be interpreted as being cumulative is not correct

1.1b.1
In ordinary language too, 'and' appears to be used for expressing alternative conditions

The second form of conflict-settling argumentation mentioned by MacCormick and Summers, is called cancellation. Their description of this type of complex argumentation reads as follows.

(2)   an argument is *cancelled* if, despite its interpretative conditions being fulfilled, another applicable argument wholly nullifies its justificatory force.

If the court had used this mode of resolution to justify that the conditions in Section 12 of the Birds Protection Act are indeed alternative, the argumentation could look like this:

> (...) since the purpose of the injunction, catching too many birds at a time or of birds that are too small, can only be reached, when it is prohibited to use nets other than those which meet every one of the three requirements in section 12 of the Royal Decree:
> that, now it appears a specific meaning should be attached to 'and' if the purpose of the injunction be met, the fact that 'and' in ordinary language is only used in a cumulative sense, is not considered of any importance.

In defence of the standpoint that the conditions in the legal rule should be interpreted in an alternative way, the court advances a teleological argument. Apart from this teleological argument, the court anticipates the objection that the fact that 'and' in ordinary language is used in a cumulative way only, might be just as relevant or even more relevant than the teleological argument. Subsequently the court refutes the counter-argument by arguing that in legal rules only know alternative 'and'. In this case the court does not attack the propositional content of the counter-argument, but refutes its justificatory force by considering it to be irrelevant. The argumentation structure may be schematized as follows.

Figure 6

1.
The conditions in the rule (r) should be interpreted as being alternative (a) (and not as being cumulative).

1.1a
Interpretation (a) is in keeping with the purpose of the rule

1.1b
That the conditions of rule (r) because of the ordinary meaning of 'and' should be interpreted as being cumulative is irrelevant

1.1b.1
A specific meaning should be attached to 'and' to meet the purpose of the statute

The third form of conflict-settling argumentation mentioned by MacCormick and Summers, is overriding and priorities:

(3)   an argument can be *overridden* when some other argument takes priority over it under a priority-rule established within the system.

Dutch legal doctrine assumes no order of ranking in methods of interpretation. Nor is there a rule which prescribes a judge to resort to precedents or dogmatics. If this were the case, the court could have presented the following argumentation to justify their point that the conditions in Section 12 of the Birds Protection Act are alternative.

> (...) since the purpose of the injunction, catching too many birds at a time or of birds that are too small, can only be reached, when it is prohibited to use nets other than those which meet every one of the three requirements in section 12 of the Royal Decree;
>
> that the correct observation that 'and' in ordinary language is used only in a cumulative sense cannot be of overriding importance, since the purpose of the rule always takes precedence over the everyday meaning of the terms.

In this case too, the court justifies an alternative interpretation of the conditions in the legal rule by means of a teleological argument. The counter-argument is not refuted because of a defective propositional content, but rather on the grounds of a priority rule stating which argument should prevail (in this case that the argumentative power of the teleological argument takes precedence over the argumentative power of the argument which refers to ordinary language). This example resembles the previous case. The previous case implicitly illustrates that a specific meaning is considered to be of more importance than an ordinary meaning. In this example the priority-rule has been made explicit. The argumentation structure here may be schematized as follows.

Figure 7

1.
The conditions in the rule (r) should be interpreted as being alternative (a) (and not as being cumulative).

1.1a
Interpretation (a) is in keeping with the purpose of the rule

1.1b
The correct observation that 'and', in ordinary language, is used only in a cumulative sense cannot be of overriding importance

1.1b.1
The purpose of the law always takes precedence over the everyday meaning of the terms.

The fourth and last form of conflict-settling argumentation mentioned by MacCormick and Summers deals with outweighing and is described thus:

(4)   an argument can be *outweighed* when, although its interpretative conditions are satisfied, its force is not cancelled, and it is not overridden by virtue of a priority-rule, there is nevertheless a counter argument leading to a different interpretation which counts as a weightier argument in the prevailing circumstances.

If the court had used this mode of resolution to justify the interpretative decision claiming that the conditions in Section 12 of the Birds Protection Act are in fact cumulative, the argumentation could be as follows.

(...) since the purpose of the injunction, catching too many birds at a time or of birds that are too small, can only be reached, when it is prohibited to use nets other than those which meet every one of the three requirements in section 12 of the Royal Decree:
that this consideration should here outweigh the fact that 'and' in ordinary language is used only in a cumulative sense, because recent ornithologist surveys show that the bird population is under threat.

The court first defends its standpoint, again, by means of a teleological argument. Then it goes on to argue that, on the basis of the meaning of 'and' in common usage, it is possible to interpret the conditions cumulative, but that the teleological argument prevails because birdlife is under threat and birds should be protected. In other words, he acknowledges the counter-argument, but considers it to be outweighed. The argumentation structure of this example may be schematized as follows.

Figure 8

1.
The conditions in the rule (r) should be interpreted as being alternative (a) (and not as being cumulative).

1.1a
Interpretation (a) is in keeping with the purpose of the rule

1.1b
The interpretation that is in keeping with the purpose of the rule outweigh the fact that 'and' in ordinary language is used only in a cumulative sense

1.1b.1
Recent ornithologist surveys show that the bird population is under threat

The counter-arguments in all the these cases are not so much attacks on the court's arguments, but are rather objections to the positions taken up by the court. In the first case, in which 'the other argument is rebutted', the propositional content of the counter-argument is not accepted by the court. In the next three cases, the propositional content is accepted, but its refutatory potential is not. The court, in these cases, acknowledges the counter-argument as a possible objection to its standpoint, but it considers the counter-argument to be irrelevant or insufficient. The objections to the standpoint are overcome in different ways. In the second case, where 'the other argument is cancelled', the court implicitly makes an appeal to a hypothetical priority-rule. Since this rule does not seem to be established within the system, this implicit argument could be criticised. In the third case, in which the other argument is overridden, the court does make an appeal to an existing rule embedded in the system. In the last case, the other argument is outweighed. Here, the argumentation of the court is found on, what Franken (see par. 2) calls, 'the judge's image of society'.

By way of these examples it may be concluded that the conflict-settling forms of argumentation are relevant to the contents of the argumentation, rather than to its specific structure and can, therefore, be considered as a sub-category of the cumulative-arguments form.

Bibliography

Aarnio, A.
1987   *The Rational as Reasonable. A Treatise on Legal Justification.* Dordrecht: Reidel.
Alexy, R.
1989   *A theory of legal Argumentation. The theory of Rational Discourse as Theory of Legal Justification.* Oxford: Clarendon press.

Eemeren, F.H. van & R. Grootendorst
1984   *Speech Acts in Argumentative Discussions. A Theoretical Model for the Analysis of Discussions Directed towards Solving Conflicts of Opinion.* Berlin/New York: Foris/De Gruyter.
Eemeren, F.H. van (et al.)
1991   *Argumenteren voor juristen. Het analyseren en schrijven van juridische teksten en beleidsteksten.* Groningen: Wolters-Noordhoff.
Eemeren, F.H. van & R. Grootendorst
1992   *Argumentation, Communication and Fallacies.* Hillsdale: Erlbaum.
Feteris, E.T.
1992   'Discussies in an juridisch proces: an pragma-dialectisch perspectief'. *Tijdschrift voor taalbeheersing, 14*, pp. 44-54.
Franken, H.
1982   'Van rechtspleging tot rechtsbescherming'. *RM Themis, 5/6*, pp. 450-461.
Franken, H. (et al.)
1995   *Inleiden tot de rechtswetenschap.* Arnhem: Gouda Quint, 7th. ed.
MacCormick, D.N.
1978   *Legal reasoning and legal theory.* Oxford: Oxford University Press.

MacCormick, D.N. en R.S. Summers,
1991   'Interpretation and Justification'. In: MacCormick, D.N. & R.S.
                Summers, (eds.) *Interpreting Statutes. A comparative study.*
                Dartmouth: Aldershot, pp. 511-545.

Plug, H.J.
1994   'Reconstructing complex argumentation in judicial decisions'. In:
                F.H. van Eemeren & R. Grootendorst (eds.), *Studies in Pragma-*
                Dialectics, Amsterdam: Sic Sat, pp. 246-254, Ch. 22.

Plug, H.J.
1995   'The rational reconstruction of additional considerations in
                judicial decisions'. In: F.H. van Eemeren & R. Grootendorst et
                al. (eds.), *Special Fields and Cases. Proc. of the Third ISSA*
                Conference on Argumentation, Amsterdam: Sic Sat pp. 61-73.

Snoeck Henkemans, A.F.,
1992   *Analysing complex argumentation. The reconstruction of multiple*
                and coordinatively compound argumentation in a critical
                discussion. Sicsat, Amsterdam, 'Dissertation Amsterdam'.

Summers, R.S. en M. Taruffo,
1991   'Interpretation and Comparative Analysis'. In: MacCormick,
                D.N. & R.S. Summers, (eds.) *Interpreting Statutes. A*
                comparative study. Dartmouth: Aldershot, pp. 461-510.

Veegens, D.J.
1989   *Cassatie in burgerlijke zaken.* (3rd. ed. revised by E. Korthals
                Altes & H.A. Groen). Zwolle: Tjeenk Willink.

Combining Partitions and Modal Logic for User Modeling

Wolfgang Pohl*

University of Essen, Dept. of Mathematics and Computer Science;
and German National Research Center for Information Technology (GMD),
Human-Computer Interaction Research Group (FIT.MMK),
Schloß Birlinghoven, D-53754 St. Augustin
email: Wolfgang.Pohl@gmd.de

Abstract. User modeling means acquiring, representing and managing assumptions about users of software systems. Inference processes as part of the representation component can be used to extend the base of assumptions, thus supporting the acquisition task. In the user modeling shell system BGP-MS, a partition approach to user model representation and reasoning has been taken. It offers the possibility to maintain different types of assumptions about the user in a partition hierarchy; reasoning about the assumptions of one type is possible. In addition, useful inferences are already built into the partition mechanism of BGP-MS. However, negative assumptions about what users do not know or want, or relationships between assumption types cannot be maintained in BGP-MS. Since such user modeling knowledge can be formulated with modal logic, a method was developed that combines techniques of reasoning with modal logic and the partition mechanism. Algorithms for translating modal expressions into first-order logic are employed. Thus, new representation and reasoning facilities were implemented into BGP-MS. One of the main qualities of the combination method is that the advantages of the partition approach are preserved.

1 Introduction

Software systems that are supposed to behave in a flexible way according to the individual needs of their users can take advantage of user modeling techniques in order to accomplish this task. In general, user modeling means acquiring, representing and managing assumptions about system users. If not only explicitly observed assumptions shall be maintained, inference processes as part of the representation component can be used to extend the base of assumptions, thus supporting the acquisition task.

One characteristic aspect of user modeling is that different types of assumptions about the user exist. First, assumptions may concern user beliefs or user

---

* This research has been supported by the German Science Foundation (DFG) under Grant No. Ko 1044/4. Thanks to Alfred Kobsa and Jörg Schreck for valuable contributions.

goals. Second, also a user's beliefs about system knowledge or goals may be of interest. Third, it can be important to know which beliefs are shared by system and user or even mutually believed.

The user modeling shell system BGP-MS (Kobsa and Pohl 1995) can assist interactive software systems in adapting to users, based on assumptions about the user that belong to such different types. In order to distinguish between assumption types, BGP-MS uses a partition mechanism, with partitions representing assumption types. Basic relations between different types like inheritance can be modeled by forming partition hierarchies. But if there is a need to model more complex relationships involving one or more assumption types, the partition mechanism cannot help. Since typed assumptions correspond to simple modal formulas with nested belief and goal operators, complex relations between types can also be expressed in modal logic. In order to use the resulting formulas in reasoning processes, an integration with partition contents becomes necessary.

This paper presents the mechanisms that enable BGP-MS to represent knowledge about complex relations between assumption types and to use this knowledge for drawing inferences about the user. In the next section, the partition mechanism will be described, and the facilities for processing assumptions of one type will be presented. The third section gives an overview over techniques of automated reasoning with modal logic, that could be used for BGP-MS. Section 4 then shows, how such techniques have been integrated with the partition mechanism. Section 5 discusses further issues of current and future work. Section 6 concludes the paper with a summary.

2 User Model Representation with Partitions

2.1 The Use of Partition Hierarchies in BGP-MS

The partition mechanism KN-PART (Scherer 1990; Fink and Herrmann 1993), which forms the outer knowledge representation layer in BGP-MS, allows one to collect different types of assumptions about the user as well as the system's domain knowledge in separate partitions. A partition approach for belief and goal modeling has already been used by Cohen (1978), Kobsa (1985), Ballim and Wilks (1991), and others. In KN-PART, this approach was enhanced by subordination relations that allow for the definition of partition hierarchies with partition inheritance. If two or more partitions $P_1, \ldots, P_n$ have a common superordinate partition P_s, then KN-PART will automatically *propagate* the common contents of $P_1, \ldots, P_n$ to P_s (unless the application developer prohibits this in the definition of P_s), and each of the $P_1, \ldots, P_n$ will *inherit* the contents of P_s at the time of retrieval. These properties of the partition mechanism will lateron be referred to as *propagation* and *inheritance*.

KN-PART allows the definition of arbitrary partition hierarchies (i.e., directed acyclic graphs). Leaf partitions in a hierarchy (with all their directly or indirectly inherited contents) are called *views*. A non-leaf partition is called *vista*, if the fact that it may also inherit contents from superpartitions shall be stressed (cf. (Hendrix 1979)). For the purposes of this paper, only the term 'view' will be used for both leaf and non-leaf partitions, if inheritance is important and unless the distinction is relevant. By propagation and inheritance, it is guaranteed that contents of more than one view will be stored in as few partitions of a given hierarchy as possible.

Hierarchical relationships do not extend the expressive power of partitions. However, they offer a convenient means for collecting the commonalities of two or more views and, when used for the modeling of beliefs and goals, can efficiently implement certain frequently-used inferences. Views seem to fulfill Levesque's (1986) criteria for a vivid representation (see (Kobsa 1992)). In BGP-MS they are employed for collecting the domain knowledge of BGP-MS, the different types of assumptions that are contained in the individual user model, and the user stereotypes that are known to BGP-MS.

Representation of the Domain Knowledge of BGP-MS. Domain knowledge which BGP-MS possibly needs for carrying out its user modeling tasks must be stored in a separate partition, which is usually named 'SB' for 'System Believes'. For applications where BGP-MS does not need domain knowledge, SB need not be present (or may be present but empty). If SB does contain domain knowledge, the application can enter new expressions into SB as well as query SB contents. Hence the application developer may choose to represent some or all domain knowledge of the application in BGP-MS rather than in the application, or duplicate some of the domain knowledge of the application into BGP-MS.

Representation of the Individual User Model. The individual user model in BGP-MS contains all current assumptions and information about the user. The types of assumptions that should be distinguished depend largely on the user modeling needs of the application system. Frequently employed types include the following:

- the "privately" held assumptions of the system about the user's beliefs about the application domain (named SBUB for System Believes User Believes), including those user beliefs which the system does not share (i.e., the user's misconceptions),
- the mutual beliefs of the system and the user about the user's beliefs about the application domain (SBMBUB), including those user beliefs which the system does not share (i.e., the mutually known user misconceptions),
- the mutual beliefs of the system and the user about the application domain (SBMB),
- the "privately" held assumptions of the system about the user's goals with respect to the application domain (named SBUW for System Believes User Wants),
- system assumptions about "objective facts" concerning the user (which are part of SB).

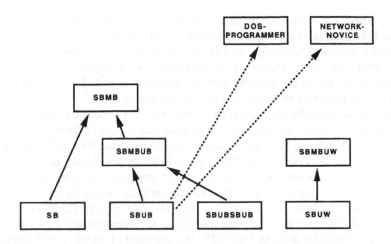

Fig. 1. Domain knowledge, individual user model and stereotypes represented in a partition hierarchy

In BGP-MS, each of these assumption types is collected in a separate partition. Fig. 1 shows an example of a simple partition hierarchy that contains all the types listed above. A solid arrow from one partition to another means that the first partition inherits from the latter.The user model developer may select the partitions and the nesting of assumptions that are appropriate in the application domain. Experience with natural-language dialog systems give rise to the assumption that the maximum depth of embedding is quite limited in cooperative dialog situations (Taylor and Carletta 1994).

The names of BGP-MS assumption types and their associated partitions (with the exception of stereotype partitions, see below) must be sequences of actor/modality pairs. Actors may be S (for System), U (for User), and M (for Mutual), and modalities may be B (for Believes) and W (for Wants). So, partition contents can be seen as qualified by a nesting of modalities, with each modality being further specified by an actor. Note that partition names start with SB (with the same exception as above). This is to stress that all user model contents are only assumptions which are not claimed to objectively hold in the world.

Representation of Stereotypes. In the stereotype approach to user model acquisition, the application developer must identify subgroups within the expected user population whose members are very likely to possess certain homogeneous application-relevant characteristics. The characteristics of the identified user groups must be formalized in an appropriate representation system. The collection of all represented characteristics of a user subgroup is called a stereotype for this subgroup. For each stereotype, the application developer must also identify a small number of key characteristics which allow a user modeling system to identify the user as belonging to the corresponding user subgroup, or as not belonging to this subgroup. Knowledge about these key characteristics is encoded in so-called activation and retraction conditions for each stereotype.

In BGP-MS, each stereotype is represented by a separate partition. If a stereotype applies to the current user, its partition is linked by an inheritance link to an appropriate partition of the individual user model, which thereby automatically inherits the contents of the stereotype. Fig. 1 contains two stereotype partitions that collect the presumable domain knowledge of DOS programmers and network novices. They can become linked to partition SBUB if their activation conditions have been met (this is symbolized by the dotted lines).

In contrast to other partitions, stereotype partitions do currently not represent specific assumption types. Stereotype contents belong to the type SBUB and are dynamically added or removed from the view SBUB. However, it is certainly possible to introduce different modalities for stereotypes. Particularly goal stereotypes could be used in BGP-MS, the contents of which would belong to the assumption type SBUW.

2.2 Partition and View Contents

Representation within Partitions. As we have shown in the previous section, partitions may serve as storages for the different types of assumptions about the user, allowing the developer to separate different parts of the user model. Within partitions, assumption contents can be represented in a hybrid formalism that consists of two components. The formalism embodied in the conceptual knowledge representation system SB-ONE (Kobsa 1990, 1991) can be used to formulate assumptions concerning the concepts and terminology of the application domain. SB-ONE fits loosely into the KL-ONE paradigm (Brachman 1978; Brachman and Schmolze 1985);a detailed discussion of this system is beyond the scope of this paper. The second means for representing assumption contents within partitions is first-order predicate calculus (FOPC). It was introduced into BGP-MS to allow more assumptions to be made that cannot be expressed in SB-ONE, notably assumptions that include disjunction, negation, quantification and implication. Examples include the user's rule-like knowledge about the application domain.

Inferences within Views. The representation formalisms available in BGP-MS offer a variety of inference possibilities, which are carried out by two main inference processes operating on the contents of a given view. The representation system for conceptual knowledge, SB-ONE, offers numerous built-in inferences, like computing the transitive closure of the 'is-a' relation.

The resolution-based theorem prover OTTER (McCune 1994) is employed for inferences in FOPC. Since OTTER is designed for batch processing, the program had to be extended and modified in several ways in order to become usable for on-line processing of logical expressions. Mainly, the interface of OTTER was changed from file processing to processing messages via the inter-process

communication management system KN-IPCMS. Messages to KN-OTTER, as
the modified system is called within BGP-MS, may contain OTTER flag and
parameter settings, clause or formula lists, or special KN-OTTER commands.
These commands can cause KN-OTTER to return diverse results as well as
traces of the OTTER proof process (see Fig. 2).

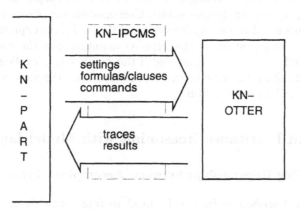

Fig. 2. Using KN-OTTER for logical reasoning

Since the expressive power of SB-ONE is considerably weaker than that of
first-order predicate calculus, OTTER was chosen to be the central processor
for inferences that take place within views. The exclusive use of SB-ONE mechanisms is limited to situations in which only terminological knowledge is concerned. Otherwise, KN-OTTER will be employed, but is not restricted to process
logical knowledge only. In order to allow hybrid processing of logical formulas
and SB-ONE constructs within views, KN-OTTER was also enabled to make
use of terminological knowledge, if needed or probably helpful for completing
a proof (Zimmermann 1994). For example, if, while performing a view-internal
reasoning task, OTTER is trying to find resolution partners for the unit clause
'apple_lw(lw_plus)', it can ask for subsuming concepts of the SB-ONE concept
'apple_lw' and then generate new unit clauses, e.g. 'laser_printer(lw_plus)'. Similar strategies are applied in order to integrate further SB-ONE constructs into
the logical reasoning process. KN-OTTER requests for terminological knowledge
are also handled via KN-IPCMS message processing.

2.3 The Application of Reasoning in BGP-MS

The representation and reasoning mechanisms described above are employed for
mainly the following user modeling tasks:

- Processing observations that were made by the user-modeling application
  and reported to BGP-MS: Observations are transformed into user model

contents. Their assumption type is determined, and an entry into the corresponding partition is made either as an SB-ONE construct, if the assumption is of a terminological nature, or as an FOPC expression otherwise.

- Checking user model consistency: When a new assumption content is entered into a partition P, the consistency of the views inheriting from P is checked. For this view-internal consistency checking, either built-in mechanisms of SB-ONE or logical reasoning with KN-OTTER are employed.
- Answering queries to the user model: User-modeling applications can specify an expression and an assumption type and then ask if the expression is among the assumption contents of this type as maintained in the user model. To answer such a query, it is examined if the expression is explicitly contained in the corresponding view or can be inferred from the view contents using SB-ONE and/or KN-OTTER.

3 Beyond Partitions: Reasoning with Modal Logic

3.1 Modeling Relationships between Assumption Types

The facilities described so far can be used to represent different types of assumptions with views, the assumption contents being terminological structures or, exceeding this possibility, formulas of first-order logic. The possible assumption types are of a positive nature only, i.e. only assumptions about what the user *does* believe or want are allowed, but not assumptions about what the user *does not* believe or want. Although in many cases these mechanisms will satisfy the representational and inferential needs of a user modeling system, we wanted to additionally enable the application developer to formulate statements expressing relationships between different assumption types (i.e., between different views) as well as negative assumptions.

Up to now, user model contents of BGP-MS can be characterized by a view V that represents an assumption type and a logical expression α that represents an assumption content.[2] The notation for such *view-internal expressions* will be $V : \alpha$. In Sect. 2.1, it has been noted that view names V are sequences of actor/modality pairs which represent nestings of indexed modalities. So, view names are closely related to sequences of indexed belief and goal operators of modal logic. Now, it is a simple, but basic insight that view-internal expressions $V : \alpha$ can be reformulated with simple formulas of a modal logic with indexed belief and goal operators. E.g., the view-internal expression 'SBUW:printed(userdoc)' (meaning that the logical expression 'printed(userdoc)' is contained in view SBUW) can be expressed with the modal formula

$$B_S W_U \text{printed(userdoc)}$$

---

[2] In the following, it is assumed that for SB-ONE constructs their FOPC equivalents are used as α. Other methods for the integration of SB-ONE constructs into view-external reasoning are still to be investigated.

This formula is of the form $\mathcal{M}\alpha$, where $\mathcal{M}$ is a sequence of modal operators corresponding to the view, and α is the content expression. For each view V, there exists a corresponding modal operator sequence $\mathcal{M}(V)$.

Based on this correspondence, there is a straightforward way to represent relationships between views as well as negative assumptions, namely by using more complex expressions of the underlying modal logic. For instance, the following formula states that the expression 'printed-on(userdoc,lw-plus)' is not mutually believed:

$$B_S \neg B_M \text{ printed-on(userdoc, lw-plus)} \tag{1}$$

The complete version of the modal logic, which includes the standard logical connectives, can be used to express relationships between specific contents of different views. For example,

$$\forall doc, p \left[B_S W_U \text{ printed-on}(doc, p) \rightarrow B_S B_U \text{ printable-on}(doc, p) \right] \tag{2}$$

represents a rule for inferring an assumption about a user belief from an assumption about a user goal: if the user is believed to want a document to be printed on some printer, then he is also assumed to believe that this is possible.

General relationships between views may be expressed by modal formula schemes, which (for our purposes) result from replacing FOPC parts in modal expressions by formula variables. For example,

$$(B_S W_U \Phi \wedge B_S B_U (\Phi \rightarrow \Psi)) \rightarrow B_S W_U \Psi \tag{3}$$

(with Φ and Ψ being formula variables) represents the rule that users want any implication of their immediate goals if they know the implication relation.

Furtheron, modal formulas like (1), (2) and (3) that cannot be represented within views will be called *view-external expressions*.

3.2 Reasoning with Modal Formulas

Functional Translation for Modal Formulas. Modal logic inferences can be performed using a first-order logic reasoner. To achieve this, a procedure is needed that translates modal expressions into first-order predicate calculus while preserving their original semantics. The so-called *functional translation* (Ohlbach 1991) is a method based on the standard possible-worlds semantics for modal logic that transforms sequences of modal operators into complex logical terms. The transformation uses *context access functions* that map worlds to accessible worlds, and consecutively applies one function variable for each modal operator to the initial world. Thereby the generated terms represent (generic) paths through the possible-worlds structure. Basically, modal formulas are translated as follows: Operator sequences are replaced by adding the corresponding *world terms* as additional arguments to all atoms within their scope in the formula. [3]

---

[3] For the purposes of BGP-MS, a simplified version of functional translation is employed. This is mainly based on a constant domain assumption and a more technical assumption about the seriality of the possible-worlds structure.

To illustrate this approach, we will first show the translation of a modal formula that corresponds to a view-internal expression:

$$B_S W_U \text{ printed(userdoc)}$$

is translated into

$$\forall f, g \text{ printed}(g_{W_U}(f_{B_S}(w_0)), \text{userdoc}) \tag{4}$$

where f and g are variables that represent context access functions and w_0 denotes the initial world. So, $g_{W_U}(f_{B_S}(w_0))$ is the complete world term referring to worlds that are accessible from w_0 through the B_S and W_U operators. These operators are parametrized versions of the standard modal operator $\Box$, namely $\Box(B, S)$ and $\Box(W, U)$. In the transformation of these operators, the context access function variables have to be applied to the same parameters. Moreover, the application of functions to the initial world or complex world terms has to be explicitly formulated such that the first-order nature of the quantified context access function variables is conserved. So, world terms that can be used by a first-order reasoner like OTTER are quite complex and do not read very well. Therefore, in the above translation (4) and in the following examples, context access function variables are left in a second order position and indexed with their corresponding operators, allowing world terms to be written in an abbreviated but legible way.

Negations within the modal operator sequence are simply retained, so that the modal-negated formula (1) translates to

$$\forall f \, \neg \forall g \text{ printed-on}(g_{B_M}(f_{B_S}(w_0)), \text{userdoc}, \text{lw}+) \tag{5}$$

The following formula is the translation of the view-external expression (2). Note that the two world paths in this formula are different.

$$\forall f, g, t, u \, \forall doc, p \tag{6}$$
$$[\text{printed-on}(g_{W_U}(f_{B_S}(w_0)), doc, p) \rightarrow \text{printable-on}(u_{B_U}(t_{B_S}(w_0)), doc, p)]$$

SCAN for Modal Formula Schemes. The functional translation even helps in integrating modal schemes into first-order inference procedures. Modal schemes are second-order by nature because they contain implicitly (universally) quantified formula variables. First-order reasoners therefore cannot be applied directly. However, there is a method to avoid this problem. The SCAN algorithm (Gabbay and Ohlbach 1992) eliminates second-order quantifiers and transforms modal schemes into a set of first-order equalities employing functional translation.[4] These equalities between world terms represent the relationships between views that were originally expressed by the modal scheme. They can be processed by

---

[4] Not all modal formula schemes correspond to first-order expressions (cf. (v. Benthem 1984)). It still has to be investigated, which class of interesting general relationships between views can safely be handled by SCAN.

equation handling mechanisms like paramodulation, which is a built-in proof method of OTTER.

For example, from the modal scheme (3) which relates user beliefs and goals, the following conjunction of two equations would be generated:

$$\forall w \, \forall q, r \, \exists f, g, t, u \tag{7}$$
$$\left[g_{W_U}(f_{B_S}(w)) = u_{B_U}(t_{B_S}(w)) \; \wedge \; u_{B_U}(t_{B_S}(w)) = r_{W_U}(q_{B_S}(w)) \right]$$

In these equations, the initial world constant w_0 is replaced by the universally quantified variable w. In a proof, w may be substituted with arbitrary world terms, which reflects the fact that formula variables in modal schemes may be instantiated by arbitrary modal formulas.

4 Combining Modal Reasoning with Partitions

4.1 Modal Reasoning without Partitions?

The techniques for translating modal formulas and formula schemes into first-order predicate logic could be used to realize user model representation and reasoning in BGP-MS, including negative assumptions and relationships between different assumption types.

The first and most straightforward way would be to maintain BGP-MS user models simply as one big set of translated modal expressions. Since OTTER with its built-in mechanisms for equality handling (mainly paramodulation) is able to process all translation results, i.e. also the equations generated by SCAN, KN-OTTER could on principle be employed to perform the reasoning tasks of BGP-MS. However, such an approach would make assumption-type-specific, i.e. view-internal reasoning very inefficient. The reason is that all translations would have to be parsed in order to identify the subset that represents the assumptions of a certain type. However, note that such a procedure exists, i.e. all assumptions of a given type have an identical translation pattern.

A second possibility is to store the translated assumptions of each type separately. View-internal reasoning would be more simple, but still the built-in inferences of partition hierarchies like inheritance and propagation could not be taken advantage of. Because these features have proven useful for user modeling, they shall be preserved. Therefore, it becomes necessary to look for a solution that uses the partition mechanism and integrates the use of translation techniques for modal reasoning smoothly.

It is obvious that view-external expressions will be translated using the standard procedures and kept aside of the partition hierarchy. The more problematic issue is: How to deal with view-internal expressions? A first approach is to store view-internal expressions both in the partition hierarchy and, in translated form, together with the translations of view-external expressions. For view-external reasoning, the resulting set of translated formulas could be used.

But, this approach seems to be a waste of space. Not only are the translations of view-internal expressions redundant; moreover, more than one translation would have to be stored for contents of non-leaf partitions. An expression α that is contained in a non-leaf partition will be inherited to views $V_1, \ldots, V_n$. Thus, it represents the view-internal expressions $V_1 : \alpha, \ldots, V_n : \alpha$, which correspond to the modal formulas $\mathcal{M}(V_1)\alpha, \ldots, \mathcal{M}(V_n)\alpha$. So, n formulas would have to be translated for one partition content expression. Also, the set of translations would have to be updated not only because of new entries into the user model, but also in case of upward propagation of partition contents. So, this solution is space-intensive and causes additional maintenance problems.

4.2 Modal Reasoning with Partitions!

In this section, an alternative approach will be described that avoids the disadvantages discussed above. It is based on the simple fact that all expressions α in a view V are of the same modal logical form $\mathcal{M}(V)\alpha$. This implies that also their functional translations follow an identical pattern. Thus, a very efficient way of translating view-internal expressions can be developed. This allows dynamical translation on demand (i.e., when translations are needed for view-external reasoning), which relieves modal mechanisms from taking care of partition inferences.

An example of a view-internal translation has already been presented (see Formula 4): The view-internal expression 'SBUW:printed(userdoc)' corresponds to the modal formula

$$B_S W_U \text{printed(userdoc)}$$

which is translated into

$$\forall f, g \; \text{printed}(g_{W_U}(f_{B_S}(w_0)), \text{userdoc})$$

In general, all expressions of the form

$$B_S W_U \alpha$$

will be translated into

$$\forall f, g \; \alpha[g_{W_U}(f_{B_S}(w_0))] \,,$$

where $\alpha[t]$ means adding a first argument term t to all atoms in α. It is easy to see that both the quantifier list $\forall f, g$ and the world term $g_{W_U}(f_{B_S}(w_0))$ can be generated directly from the modal operator sequence $B_S W_U$. Of course, also this can be generalized: From a modal operator sequence $\mathcal{M}$, a world term $wt(\mathcal{M})$ as well as a quantifier list $Q(\mathcal{M})$ can be generated. These are the prerequisites for the main statement of this section:

Generating the functional translation for a view-internal expression $V : \alpha$ means translating $\mathcal{M}(V)\alpha$ and therefore results in
$Q(\mathcal{M}(V)) \, \alpha[wt(\mathcal{M}(V))]$.

In BGP-MS, the world term $wt(\mathcal{M}(V))$ and the quantifier list $Q(\mathcal{M}(V))$ are computed once and stored along with each view V (i.e., with the corresponding partition) in the hierarchy. Translating a view-internal expression $V : \alpha$ then consists of the following steps:

1. From α and $wt(\mathcal{M}(V))$, generate $\alpha[wt(\mathcal{M}(V))]$.
2. Put $Q(\mathcal{M}(V))$ in front of the result.

The first step can be simplified further: Note that for the view-internal expressions $V_1 : \alpha, \ldots, V_n : \alpha$, where all of $V_1, \ldots, V_n$ correspond to or inherit from the same partition P, the content expression α is stored only in P. Along with α we can store its "translation frame" $\alpha[]$, which is the view-independent part of $\alpha[wt(\mathcal{M}(V))]$, as sequence of formula substrings. An example: For $\alpha =$ 'printed(userdoc)' we obtain $\alpha[] = ($"printed(" ",userdoc)"$)$. Since $\alpha[]$ is view-independent, it needs to be computed only once and stored at no more places in the partition hierarchy than α itself. Now, generating $\alpha[wt(\mathcal{M}(V))]$ simply means inserting $wt(\mathcal{M}(V))$ between each two adjacent substrings of $\alpha[]$. Thus, we have a very efficient method of generating the translation

$$T_V(\alpha) := Q(\mathcal{M}(V)) \, \alpha[wt(\mathcal{M}(V))]$$

of an expression α contained in view V. The translation of a complete view is the set

$$\{T_V(\alpha) \mid \alpha \text{ in } V\}$$

Reasoning that is to involve both view-internal and view-external user model contents can be done using the permanently stored functional translations of the view-external expressions together with the translations of all views of the current partition hierarchy. KN-OTTER can be applied to process the resulting formula set. A crucial property of the presented approach to view-external reasoning is that the mechanisms of view-internal reasoning remain unchanged.

4.3 Extended User Modeling Capabilities of BGP-MS

Since the partition mechanism has been preserved, the features mentioned in Sect. 2.3 are still available. Based on the techniques described above, now the following additional facilities can be used:

- Processing view-external expressions and modal formula schemes: Both kinds of expressions can be entered into BGP-MS by the developer of a user-modeling application to specify inferential relations between different assumption types. Only negative assumptions may also be entered as observations at application run time.
- Checking the consistency of the complete user model: Given the translations of both view-external and view-internal user model contents, KN-OTTER can be used to detect inconsistencies of the whole user model. Such inconsistencies will most probably be caused by conflicts between several assumptions of different types due to one or more view-external formulas relating these types.
- Answering queries to the user model: Still, queries must refer to one assumption type. Again, the only exception are negative assumptions, which may also be queried. The latter kind of queries will be handled with view-external

reasoning, while queries related to an assumption type can alternatively be handled by view-internal reasoning or, if specified by the application, by view-external reasoning, potentially involving the whole user model.

5 Further Issues

5.1 Implementation

The implementation of view-external reasoning in BGP-MS has been designed and realized by Schreck and Simon (Schreck 1995; Simon 1995). They implemented the general mechanism of functional translation, the SCAN algorithm (using KN-OTTER for the resolution steps of SCAN), and the integration of the translation methods into the partition mechanism KN-PART. Especially Schreck (1995) argues in detail that the on-demand generation of view-internal translations is advantageous to alternative solutions, if an efficient and optimized generation method like the one presented in Section 4.2 is used.

5.2 Limiting Resource Usage

View-external reasoning based on translated modal logic expressions can be quite costly. Therefore, it has been examined if the usage of resources can be limited. In the current implementation, it has been exploited that view-external expressions may set up relationships between views. For instance, the view-external expression (2) relates the views SBUB and SBUW. If this were the only view-external expression, only the contents of SBUB and SBUW would have to be considered in answering a query to SBUB with view-external reasoning. BGP-MS analyzes a given set of view-external expressions and tries to form a relation R over the set of views. For a query concerning a view V, only views V' with $R^*(V, V')$ have to be considered for view-external reasoning (R^* being the transitive closure of R). By assigning a positive integer n to a variable, the set of considered views may further be limited to those views V' with $R^n(V, V')$.

The latter limitation is motivated by the thought that the reasoning of the user who is being modeled will probably be more limited than that of an automatical reasoner. Therefore, the level of implicitness of assumptions about the user can be restricted. The mechanism described above is a first approach to limit the (view-external) reasoning processes in BGP-MS. In future work, possible limitations of the proof level and length of OTTER will be investigated in order to make logical reasoning produce or verify assumptions with a controlled level of implicitness. Such methods will also allow limitations of the resources used by logical (view-internal or view-external) reasoning if this is necessary for the user modeling tasks of BGP-MS.

6 Summary

In this paper, techniques for combining the partition approach with the modal logic approach (Allgayer et al. 1992; Hustadt 1995) to user modeling have been discussed, which have been applied to and implemented in the user modeling shell system BGP-MS. The partition mechanism of BGP-MS and its use for user model representation has been described. It has been shown that the limitations of partition hierarchies can be overcome by the application of modal logic to user modeling. However, since the representation and reasoning techniques of the partition mechanism are worth being preserved, and since pure modal reasoning may be quite inefficient, a method has been developed that leaves the existing view-internal reasoning techniques unchanged, but allows the easy integration of view contents into view-external, modal reasoning. Future work will be concerned with resource-limited reasoning and with the evaluation of the developed method in application scenarios. A first study (Eisenmann and Mast 1995) has shown that view-external reasoning could be beneficial in systems supporting coordination in computer-supported cooperative work. These ideas must be further elaborated.

References

J. Allgayer, H. J. Ohlbach, and C. Reddig: Modelling Agents with Logic. In *Proc. of the Third International Workshop on User Modeling*, pages 22–34, Dagstuhl, Germany, 1992.

A. Ballim and Y. Wilks: Beliefs, Stereotypes and Dynamic Agent Modeling. *User Modeling and User-Adapted Interaction*, 1(1):33–65, 1991.

J. v. Benthem: Correspondence Theory. In D. Gabbay and F. Guenthner, editors, *Handbook of Philosophical Logic*, volume II, pages 167–247. D. Reidel Publishing Company, Dordrecht, 1984.

R. J. Brachman and J. G. Schmolze: An Overview of the KL-ONE Knowledge Representation System. *Cognitive Science*, 9(2):171–216, 1985.

R. J. Brachman: A Structural Paradigm for Representing Knowledge. Technical Report 3605, Bolt, Beranek, and Newman Inc., Cambridge, MA, 1978.

P. R. Cohen: On Knowing What to Say: Planning Speech Acts. Technical Report 118, Department of Computer Science, University of Toronto, Canada, 1978.

M. Eisenmann and R. Mast: Entwurf von Benutzermodellen am Beispiel von adaptiven Print- und Email-Tools. Project report, WG Knowledge-Based Information Systems, Department of Information Science, University of Konstanz, Germany, 1995.

J. Fink and M. Herrmann: KN-PART: Ein Verwaltungssystem zur Benutzermodellierung mit prädikatenlogischer Wissensrepräsentation. WIS Memo 5, WG Knowledge-Based Information Systems, Department of Information Science, University of Konstanz, Germany, 1993.

D. Gabbay and H. J. Ohlbach: Quantifier Elimination in Second-Order Predicate Logic. In B. Nebel, C. Rich, and W. Swartout, editors, *Principles of Knowledge Representation and Reasoning: Proc. of the Third International Conference (KR'92)*, pages 425–435. Morgan Kaufmann, San Mateo, CA, 1992.

G. Hendrix: Encoding Knowledge in Partitioned Networks. In N. V. Findler, editor, *Associative Networks, Representation and Use Of Knowledge by Computers*. Academic Press, New York, 1979.

494

U. Hustadt: Introducing Epistemic Operators into a Description Logic. In A. Laux and H. Wansing, editors, *Knowledge and Belief in Philosophy and Artificial Intelligence*, Logica Nova, pages 65–86. Akademie Verlag, Berlin, 1995.

A. Kobsa and W. Pohl: The User Modeling Shell System BGP-MS. *User Modeling and User-Adapted Interaction*, 4(2):59–106, 1995.

A. Kobsa: Benutzermodellierung in Dialogsystemen. Springer-Verlag, Berlin, Heidelberg, 1985.

A. Kobsa: Modeling The User's Conceptual Knowledge in BGP-MS, a User Modeling Shell System. *Computational Intelligence*, 6:193–208, 1990.

A. Kobsa: Utilizing Knowledge: The Components of the SB-ONE Knowledge Representation Workbench. In J. Sowa, editor, *Principles of Semantic Networks: Exploration in the Representation of Knowledge*, pages 457–486. Morgan Kaufmann, San Mateo, CA, 1991.

A. Kobsa: Towards Inferences in BGP-MS: Combining Modal Logic and Partition Hierarchies for User Modeling. In *Proc. of the Third International Workshop on User Modeling*, pages 35–41, Dagstuhl, Germany, 1992.

H. J. Levesque: Making Believers out of Computers. *Artificial Intelligence*, 30:81–108, 1986.

W. W. McCune: OTTER 3.0 Reference Manual and Guide. Technical Report ANL-94/6, Argonne National Laboratory, Mathematics and Computer Science Division, Argonne, IL, 1994.

H. J. Ohlbach: Semantics-Based Translation Methods for Modal Logics. *Journal of Logic and Computation*, 1(5):691–746, 1991.

J. Scherer: SB-PART: Ein Partitionsverwaltungssystem für die Wissensrepräsentationssprache SB-ONE. Memo 48, Project XTRA, Department of Computer Science, University of Saarbrücken, Germany, 1990.

J. Schreck: Konzeption der Erweiterung von partitionenorientierter Wissensrepräsentation um modallogische Inferenzen: Diplomarbeit, AG Wissensbasierte Informationssysteme, Informationswissenschaft, Universität Konstanz, 1995.

R. Simon: Realisierung der Erweiterung von partitionenorientierter Wissensrepräsentation um modallogische Inferenzen: Diplomarbeit, AG Wissensbasierte Informationssysteme, Informationswissenschaft, Universität Konstanz, 1995.

J. A. Taylor and J. C. Carletta: Limiting Nested Beliefs in Cooperative Dialogue: In *Proc. of the 16th Annual Conference of the Cognitive Science Society*, pages 858–863, Atlanta, GA, 1994.

J. Zimmermann: Hybride Wissensrepräsentation in BGP-MS: Integration der Wissensverarbeitung von SB-ONE und OTTER. WIS Memo 12, WG Knowledge-Based Information Systems, Department of Information Science, University of Konstanz, Germany, 1994.

Reason in a Changing World

John L. Pollock
Department of Philosophy
University of Arizona
Tucson, Arizona 85721
(e-mail: pollock@arizona.edu)

A rational agent (artificial or otherwise) residing in a complex changing environment must gather information perceptually, update that information as the world changes, and combine that information with causal information to reason about the changing world. Using the system of defeasible reasoning that is incorporated into the OSCAR architecture for rational agents (Pollock 1995 and 1995a), a set of reason-schemas will be proposed for enabling an agent to perform some of the requisite reasoning. Along the way, solutions will be proposed for the Frame Problem and the Yale Shooting Problem. The principles and reasoning described below have all been implemented in OSCAR.

1. Reasoning from Percepts

In a complex changing environment, an agent cannot be equipped from its inception with all the information it needs. It must be capable of gathering new information by sensing its surroundings. This is perception, in a generic sense. Perception is a process that begins with the stimulation of sensors, and ends with beliefs about the agent's immediate surroundings. In perceptual reasoning, the basic inference is from a percept to a conclusion about the world. This enables us to assign propositional contents to percepts. The content of a percept will be taken to be the same as the content of the belief for which the percept provides a reason. But of course, the percept is not the same thing as the belief. Having the percept consists of having a perceptual experience that is *as if* the belief were true. Given this understanding of the content of percepts, we can, as a first approximation, formulate the reasoning from percepts to beliefs as follows (Pollock, 1987):

(1)   Having a percept with the content P is a prima facie reason for the agent to believe P.

In this principle, the variable 'P' ranges over the possible contents of percepts. That range will depend upon the perceptual apparatus of the agent in question.

This formulation of the reasoning captures the obvious but important point that perceptual reasoning must be defeasible—appearances can be deceptive. However, that this formulation is not entirely adequate becomes apparent when we consider temporal updating. The world changes, and accordingly percepts produced at different times can support inferences to conflicting conclusions. We can diagram this roughly as follows, where t_1 is a later time than t_0, '- - - - - - ▶' symbolizes defeasible inference, and '⋯⋯⋯' symbolizes defeat relations:

The reasoning seems to produce a case of collective defeat. But it shouldn't. The initial percept supports the belief that P holds *at t_0*, and the second percept supports the belief that P does not hold *at t_1*. These conclusions do not conflict. We can hold both beliefs simply by acknowledging that the world has changed.

We can accommodate this by building temporal reference into the belief produced by perception, and giving the percept a date. This allows us to reformulate the above prima facie reason as follows:

PERCEPTION
Having a percept at time t with the content P is a prima facie reason for the agent to believe P-at-t.

with the result that the apparent conflict has gone away.

It will be convenient to build the time of the percept into the formula representing the content of the percept. Accordingly, I will adopt the convention of saying that a percept at time t with content P is a percept of P-at-t.

2. Perceptual Reliability

When giving an account of a species of defeasible reasoning, it is as important to characterize the defeaters for the defeasible reasons as it is to state the reasons themselves. This paper assumes the theory of defeasible reasons and reasoning implemented in OSCAR and described in detail in Pollock 1995 and 1995a. One of the central doctrines of that theory is that there are just two kinds of defeaters—*rebutting defeaters* and *undercutting defeaters*. Any reason for denying P-at-t is automatically a rebutting defeater for *PERCEPTION*. An undercutting defeater for an inference from a belief in P to a belief in Q attacks the connection between P and Q rather than merely denying the conclusion. An undercutting defeater is a reason for the formula $(P \otimes Q)$ (read "It is false that P would not be true unless Q were true", or abbreviated as "P does not guarantee Q"). It is desirable to represent undercutting defeaters for *PERCEPTION* similarly, but we cannot use the content of the percept as the first formula, because that would give us $(P \otimes P)$, which is logically false. In order to have a formula to represent the content of the undercutting defeater, I will write it as $((\text{It appears to me that P}) \otimes P)$. The only obvious undercutting defeater is a reliability defeater, which is of a general sort applicable to all defeasible reasons:

PERCEPTUAL-RELIABILITY
Where r is the strength of *PERCEPTION* , and s < r, "R-at-t, and the probability is less than or equal to s of P's being true given R and that I have a percept with content P" is a conclusive reason for $((\text{It appears to me that P-at-t}) \otimes \text{P-at-t})$.

A perplexing problem remains for this account. *PERCEPTUAL-RELIABILITY* requires us to know that R is true at the time of the percept. We will typically know this only by inferring it from the fact that R was true earlier. The nature of this inference is the topic of the next section.

3. Temporal Projection

The reason-schema *PERCEPTION* enables an agent to draw conclusions about its current surroundings on the basis of its current percepts. However, that is of little use unless the agent can also draw conclusions about its current surroundings on the basis of earlier (at least fairly recent) percepts. For instance, imagine a robot whose task is to visually check the readings of two meters and then press one of two buttons depending upon which reading is higher. This should not be a hard task, but if we assume that the robot can only look at one meter at a time, it will not be able to acquire the requisite information about the meters using only the reason-schema *PERCEPTION*. The robot can look at one meter and draw a conclusion about its value, but then when the robot turns to read the other meter, it no longer has a percept of the first and so is no longer in a position to hold a justified belief about what that meter reads *now*. In order to perform this task, the robot needs some basis for believing that the first meter still reads what it read a moment ago. In other words, the robot must have some basis for regarding the meter reading as a *stable property*—one that tends not to change quickly over time.

It is natural to suppose that a rational agent, endowed with the ability to learn by induction, can discover inductively that some properties (like the meter readings) are stable to varying degrees, and can bring that knowledge to bear on tasks like the meter-reading task. However, I argued long ago (Pollock 1974) that such inductive learning is not epistemically possible—it presupposes the very stability that is the object of the learning. The argument for this somewhat surprising conclusion is as follows. To say that a property is stable is to say that objects possessing it tend to retain it. To confirm this inductively, an agent would have to re-examine the same object at different times and determine whether the property has changed. The difficulty is that in order to do this, the agent must be able to reidentify the object as the same object at different times. Although this is a complex matter, it seems clear that the agent make essential use of the perceptible properties of objects in reidentifying them. If all perceptible properties fluctuated wildly, we would be unable to reidentify anything. If objects tended to exchange their perceptible properties abruptly and unpredictably, we would be unable to tell which object was which.[1] The upshot of this is that it is epistemically impossible to investigate the stability of perceptible properties inductively without presupposing that most of them tend to be stable. If we make that general supposition, then we can use induction to refine it by discovering that some perceptible properties are more stable than others, that particular properties tend to be unstable under specifiable circumstances, etc. But our conceptual framework must include a general presumption of stability for perceptible properties before any of this refinement can take place. In other words, the built-in epistemic arsenal of a rational agent must include reason-schemas of the following sort for at least some choices of P:

TEMPORAL-PROJECTION
If $t_0 < t_1$ then believing P-at-t_0 is a prima facie reason for the agent to believe P-at-t_1.

---

[1] A more detailed presentation of this argument can be found in Pollock 1974.

There are interesting interactions between *TEMPORAL-PROJECTION* and *PERCEPTION*. Suppose an agent has a percept of P at time t_0, and a percept of ~P at a later time t_1. What an agent *should* conclude under these circumstances is that the world has changed between t_0 and t_1, and although P was true at t_0, it is no longer true at t_1 and hence no longer true at a later time t_2. However, there is no way to get this conclusion out of the principles thus far endorsed. From the conclusion ~P-at-t_1, the agent can defeasibly infer ~P-at-t_2, as we would expect, but from the conclusion P-at-t_0 the agent can similarly infer P-at-t_2. Once more, we have a case of collective defeat. To avoid this, there must be an independent source of defeat for the latter inference. My proposal is that the only way to get P-at-t_1 defeated but ~P-at-t_1 undefeated is to have P-at-t_1 defeated by the percept of ~P-at-t_1:

(2)   If $t_0 < t_1 \leq t_2$, a percept of ~P-at-t_1 is an undercutting defeater for the inference from P-at-t_0 to P-at-t_2.

This has the desired consequence that ~P-at-t_1 is undefeated but P-at-t_2 is defeated, and hence ~P-at-t_2 is undefeated.

This principle must be supplemented by another. Suppose the inference from the *percept* of ~P-at-t_1 to the *conclusion* ~P-at-t_1 were defeated. Then perception does not provide us with an undefeated reason for revising our assessment of the truth value of P, and so the original inference from P-at-t_0 to P-at-t_1 should not be defeated. This observation can be accommodated by taking the undercutting defeater to be defeasible (as indicated in the inference-graph), and taking any undercutting defeater for the inference from the percept of ~P-at-t_1 to ~P-at-t_1 to be an undercutting defeater for the defeater as well:

(3)   ((It appears to me that ~P-at-t_1) $\otimes$ ~P-at-t_1) is an undercutting defeater for the inference from the percept of ~P-at-t_1 to (P-at-t_0 $\otimes$ P-at-t_2).

Principles (2) and (3) handle the cases for which they were designed, but they must be extended to handle the full range of perceptual updating. For example, although principles (2) and (3) handle a case in which an object first looks red to me and then later does not look red to me, they do not handle a case in which the object first looks red to me and then later looks blue to me. In the latter case, instead of having a percept of ~P-at-t_1, I have a percept of Q-at-t_1, where Q is incompatible with P. The incompatibility need not be logical incompatibility. For instance, I might look out the window and read a thermometer, from which I infer that the temperature is around 70°. Several hours later I look out at a different thermometer and conclude that the temperature has risen to around 80°. The logic of this reasoning is as follows. We begin by having a percept of the first thermometer, from which we infer by *PERCEPTION* that it reads "70", and from that together with contingent generalizations about thermometers we infer that the outside air temperature is around 70°. Later, we have a percept of the second thermometer, from which we infer that it reads "80", and from that together with the generalizations about thermometers we infer that the outside air temperature is around 80°. The latter entails that the air temperature is not around 70°, from which we can infer, using our generalizations about thermometers, that the first thermometer does not read "70". Thus we have a reason for believing that if the second thermometer reads "80" then the first does not read "70", and that together with the fact that the second reading is based upon a later percept is taken to defeat the temporal projection that the first thermometer still reads "70". This suggests that (2) and (3) should be generalized as follows:

TEMPORAL-UPDATE

If $t_0 < t_1 \le t$, the set consisting of a percept of Q-at-t_1 together with a belief in (Q-at-t $\supset$ ~P-at-t) is a defeasible undercutting defeater for the inference from P-at-t_0 to P-at-t (i.e., a defeasible reason for (P-at-$t_0 \otimes$ P-at-t)).

DEFEAT-FOR-TEMPORAL-UPDATE

((It appears to me that Q-at-t_1) $\otimes$ Q-at-t_1) is an undercutting defeater for the inference to (P-at-$t_0 \otimes$ P-at-t) from the set consisting of a percept of Q-at-t_1 together with a belief in (Q-at-t $\supset$ ~P-at-t) (i.e., a reason for (((It appears to me that Q-at-t_1) & (Q-at-t $\supset$ ~P-at-t)) $\otimes$ (P-at-$t_0 \otimes$ P-at-t))).

Now let us return to the problem noted above for *PERCEPTUAL-RELIABILITY*. This is that we will typically know R-at-t_1 only be inferring it from R-at-t_0 for some t_0 < t_1 (by *TEMPORAL-PROJECTION*). To facilitate this, I propose combining *PERCEPTUAL-RELIABILITY* and *TEMPORAL-PROJECTION* into a single reason-schema:

PERCEPTUAL-RELIABILITY+

"$t_0 \le t_1$, R-at-t_0, and the probability is less than r of P's being true given R and that I have a percept with content P at t_1" is an undercutting defeater for *PERCEPTION* as a reason of strength $\ge$ r.

This subsumes the original *PERCEPTUAL-RELIABILITY*, so I will rename it "*PERCEPTUAL-RELIABILITY*" and dispense with the earlier principle.

4. Temporal Indexicals

An agent that did all of its temporal reasoning using the reason-schemas described above would be led into crippling computational complexities. Every time it wanted to reuse a belief about its surroundings, it would have to reinfer it for the present time. Inference takes time, so by the time it had reinferred the belief, other beliefs with which the agent might want to combine this belief in further inference would themselves no longer be current. To get around this difficulty, the agent would have to make inferences about some time in the near future rather than the present, inferring a number of properties of that time, and then combine those properties to make a further inference about that time, and finally project that new property into the future for use in further inference. This would not be computationally impossible, but it would make life difficult for an agent that had to reason in this way.

Human beings achieve the same result in a more efficient way by employing the temporal indexical "now". Rather than repeatedly reinferring a property as time advances, they infer it once as holding *now*, and that single belief is retained until it becomes defeated. The mental representation (i.e., formula) believed remains unchanged as time advances, but the content of the belief changes continuously in the sense that at each instant it is a belief about *that instant*. This has the effect of continuously updating the agent's beliefs in accordance with the prima facie reasons of section three, but no actual reasoning need occur.

Percepts are always percepts of the agent's present situation, so we can regard them as providing defeasible reasons for inferences about the present:

INDEXICAL-PERCEPTION
Having a percept at time t with the content P is a prima facie reason for the agent to believe P-now.

INDEXICAL-PERCEPTUAL-RELIABILITY
If $t_0 \leq t_1$, then "R-at-t_0, and the probability is less than r of P's being true given R and that I have a percept with content P at t" is a defeasible reason for ((It appears to me that P-at-t) $\otimes$ P-now).

The conclusion P-now is automatically projected into the future just by retaining it, so *INDEXICAL-PERCEPTION* can be viewed as combining *PERCEPTION* and *TEMPORAL-PROJECTION* into a single reason-scheme. The undercutting defeater for temporal projection can then be brought to bear as follows:

PERCEPTION-UPDATE
If $t_0 < t_1 \leq$ now, the set consisting of a percept of Q-at-t_1 together with a belief in (Q-now $\supset$ ~P-now) is a defeasible undercutting defeater for the inference from a percept of P-at-t_0 to P-now (i.e., a defeasible reason for ((It appears to me that P-at-t_0) $\otimes$ P-now)).

DEFEAT-FOR-PERCEPTION-UPDATE
((It appears to me that Q-at-t_1) $\otimes$ Q-now) is an undercutting defeater for the inference to ((It appears to me that P-at-t_0) $\otimes$ P-now)) from the set consisting of a percept of Q-at-t_1 together with a belief in (Q-now $\supset$ ~P-now).

These principles can be illustrated by the following problem. First, Fred looks red to me. Later, I am informed by Merrill that I am then wearing blue-tinted glasses. Later still, Fred looks blue to me. All along, I know that the probability is not high of Fred being blue given that Fred looks blue to me, but I am wearing blue tinted glasses. What should I conclude about the color of Fred? By *INDEXICAL-PERCEPTION* I first infer that Fred is red. Later I infer that Fred is blue, and by *PERCEPTION-UPDATE*, the earlier conclusion that Fred is red is defeated. By I also infer that I am wearing blue-tinted glasses, which by *INDEXICAL-PERCEPTUAL-RELIABILITY* defeats the conclusion that Fred is blue and reinstates the conclusion that Fred is red. OSCAR performs this reasoning as follows:

```
# 1
((The probability of (The color of Fred is blue) given ((I have a percept with content (The color of Fred
is blue)) & I am wearing blue tinted glasses)) <= 0.8)
given
# 2
(Merrill is a reliable informant)
given
                    # 1
              interest: ((The color of Fred is ^@v0) now)
              This is of ultimate interest
||||||||||||||||||||||||||||||||||||||||||||||||||||||||||||||||||||||||||||||||||||||||||||||||||||||||||||||||||||||||
It appears to me that ((the color of fred is red) (at 1))
||||||||||||||||||||||||||||||||||||||||||||||||||||||||||||||||||||||||||||||||||||||||||||||||||||||||||||||||||||||||
# 3
It appears to me that ((The color of Fred is red) (at 1))
# 4
((The color of Fred is red) now)
Inferred by:
        support-link #3 from { 3 } by *INDEXICAL-PERCEPTION* defeaters: { 15 }
This discharges interest 1
```

2
interest: ((it appears to me that (The color of Fred is red) (at 1)) ⊗ ((The color of Fred is red) now))
Of interest as a defeater for support-link 3 for node 4
===
Justified belief in ((The color of Fred is red) now)
answers #<Query #1: ((? x)((The color of Fred is x) now)>
===
||||||||||||||||||||||||||||||||||||||||||||||||||||||||||||||||||||||||||||||||||||||||||||||||||||||||||||
It appears to me that ((merrill reports that i am wearing blue tinted glasses) (at 20))
||||||||||||||||||||||||||||||||||||||||||||||||||||||||||||||||||||||||||||||||||||||||||||||||||||||||||||
6
It appears to me that ((Merrill reports that I am wearing blue tinted glasses) (at 20))
8
((Merrill reports that I am wearing blue tinted glasses) (at 20))
Inferred by:
        support-link #6 from { 6 } by *perception*
9
(I am wearing blue tinted glasses (at 20))
Inferred by:
        support-link #7 from { 2 , 8 } by *reliable-informant*
||||||||||||||||||||||||||||||||||||||||||||||||||||||||||||||||||||||||||||||||||||||||||||||||||||||||||||
It appears to me that ((the color of fred is blue) (at 30))
||||||||||||||||||||||||||||||||||||||||||||||||||||||||||||||||||||||||||||||||||||||||||||||||||||||||||||
10
It appears to me that ((The color of Fred is blue) (at 30))
11
((The color of Fred is blue) now)   DEFEATED
Inferred by:
        support-link #8 from { 10 } by *INDEXICAL-PERCEPTION* defeaters: { 17 }
This discharges interest 1
                # 12
                interest: ((it appears to me that (The color of Fred is blue) (at 30)) ⊗ ((The color
of Fred is blue) now))
                Of interest as a defeater for support-link 8 for node 11
        ===
        Justified belief in ((The color of Fred is blue) now)
        answers #<Query #1: ((? x)((The color of Fred is x) now)>
        ===
                # 16
                interest: (30 <= now)
                For interest 2 by *perception-update-from-competing-percept*
13
(30 <= now)
Inferred by:
        support-link #10 from { } by is-past-or-present
This node is inferred by discharging interest #16
                # 17
                interest: (1 < 30)
                For interest 2 by *perception-update-from-competing-percept*
14
(1 < 30)
Inferred by:
        support-link #11 from { } by strict-arithmetical-inequality
This node is inferred by discharging interest #17
15
((it appears to me that (The color of Fred is red) (at 1)) ⊗ ((The color of Fred is red) now))
Inferred by:
        support-link #12 from { 14 , 13 , 10 } by *perception-update-from-competing-percept*
defeaters: { 19 }
 defeatees: { link 3 for node 4 }
 This node is inferred by discharging interest #2
                # 18
                interest: (((1 < 30) & ((30 <= now) & (it appears to me that (The color of Fred is
                blue) (at 30)))) ⊗ ((it appears to me that (The color of Fred is red) (at 1)) ⊗
                ((The color of Fred is red) now)))
                Of interest as a defeater for support-link 12 for node 15

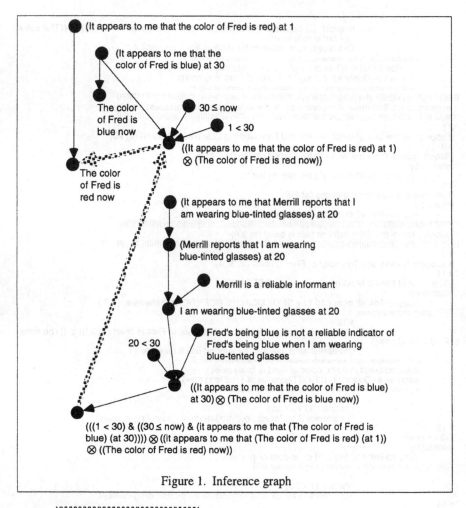

Figure 1. Inference graph

vvvvvvvvvvvvvvvvvvvvvvvvvvvvvvv
#<Node 4> has become defeated.
vvvvvvvvvvvvvvvvvvvvvvvvvvvvvvv
==
Lowering the undefeated-degree-of-support of ((The color of Fred is red) now)
retracts the previous answer to #<Query #1: ((? x)((The color of Fred is x) now)>
==

```
                        # 20
                        interest: (20 < 30)
                        For interest 12 by *indexical-perceptual-reliability*
# 16
(20 < 30)
Inferred by:
        support-link #13 from { } by strict-arithmetical-inequality
This node is inferred by discharging interest #20
# 17
((it appears to me that (The color of Fred is blue) (at 30)) ⊗ ((The color of Fred is blue) now))
Inferred by:
        support-link #14 from { 16 , 9 , 1 } by *indexical-perceptual-reliability*
defeatees: { link 8 for node 11 }
```

This node is inferred by discharging interest #12
vvvvvvvvvvvvvvvvvvvvvvvvvvvvvvv
#<Node 11> has become defeated.
vvvvvvvvvvvvvvvvvvvvvvvvvvvvvv

==
Lowering the undefeated-degree-of-support of ((The color of Fred is blue) now)
retracts the previous answer to #<Query #1: ((? x)((The color of Fred is x) now)>
==
19
(((1 < 30) & ((30 <= now) & (it appears to me that (The color of Fred is blue) (at 30)))) ⊗ ((it appears to
me that (The color of Fred is red) (at 1)) ⊗ ((The color of Fred is red) now)))
Inferred by:
    support-link #16 from { 17 } by *defeat-for-perception-update-from-competing-percept*
defeatees: { link 12 for node 15 }
This node is inferred by discharging interest #18
vvvvvvvvvvvvvvvvvvvvvvvvvvvvvvv
The undefeated-degree-of-support of #<Node 4> has increased to 0.9
vvvvvvvvvvvvvvvvvvvvvvvvvvvvvvv
#<Node 15> has become defeated.
vvvvvvvvvvvvvvvvvvvvvvvvvvvvvv

==
Justified belief in ((The color of Fred is red) now)
answers #<Query #1: ((? x)((The color of Fred is x) now)>
==

This reasoning diagramed in figure 1, where "fuzzy" arrows indicate defeat relations:

5. The Frame Problem

Reasoning about what will change if an action is performed or some other change occurs often presupposes knowing what will not change. Early attempts to model such reasoning deductively proceeded by adopting a large number of "frame axioms", which were axioms to the effect that if something occurs then something else will not change. For instance, in a blocks world one of the frame axioms might be "If a block is moved, its color will not change". It soon became apparent that complicated situations required more frame axioms than axioms about change, and most of the system resources were being occupied by proofs that various properties did not change. In a realistically complicated situation, this became unmanageable. What became known as the *Frame Problem* is the problem of reorganizing reasoning about change so that reasoning about non-changes can be done efficiently (McCarthy and Hayes, 1969; Janlert, 1987).

AI hackers, as Hayes [1987] calls them, avoided this problem by adopting the "sleeping dog strategy" (Haugeland 1987). Starting with STRIPS, actual planning systems maintained databases of what was true in a situation, and with each possible action they stored lists of what changes those actions would produce. For planning systems intended to operate only in narrowly circumscribed situations, this approach is effective, although for general-purpose planning it quickly becomes unwieldy. In the attempt to provide a more general theory that justifies this approach as a special case, several authors (Sandewall 1972, McCarthy) proposed reasoning about change defeasibly and adopting some sort of defeasible inference scheme to the effect that it is reasonable to believe that something doesn't change unless you are forced to conclude otherwise. But to make the idea work, one needs both a precise framework for defeasible reasoning and a precise formulation of the requisite defeasible inference schemes. That proved to be a difficult problem.

The stability principles defended above can be regarded as a precise formulation of the defeasible inference schemes sought. For the reasons given in the preceding discussion of temporal indexicals, I do not think that these principles by themselves constitute a solution to the Frame Problem. However, I propose that when they are coupled with the additional principles regarding temporal indexicals, an adequate solution to the Frame Problem is forthcoming. These principles have all been implemented in OSCAR.

6. Reasoning about Change

The frame problem concerns efficient reasoning about non-change. This only arises because we must often know what does not change in order to reason about what changes. Steve Hanks and Drew McDermott [1986] seem to have been the first to observe that even with defeasible principles of non-change, a reasoner will often be unable to determine what changes and what does not. (This could be viewed as an argument that the frame problem has not yet been solved, but it more convenient to regard this as a separate problem.) They illustrated this with what has become known as "the Yale shooting problem". The general form of the problem is this. Suppose we have a causal law to the effect that if P is true at a time t and action A is performed at that time, then Q will be true shortly thereafter. (More generally, A could be anything that becomes true at a certain time. What is significant about actions is that they are changes.) Suppose we know that P is true now, and Q false. What should we conclude about the results of performing action A in the immediate future? Hanks and McDermott illustrate this by taking P to be "The gun is loaded, in working condition, and pointed at Jones", Q to be "Jones is dead", and A to be the action of pulling the trigger. We suppose (simplistically) that there is a causal law dictating that if the trigger is pulled on a loaded gun that is in working condition and pointed at someone, that person will shortly be dead. Under these circumstances, it seems clear that we should conclude that Jones will be dead shortly after the trigger is pulled.

The difficulty is that all we can deduce from what we are given is that when A is performed either P will no longer be true or Q will be true shortly thereafter. Intuitively, we want to conclude (at least defeasibly) that P will remain true at the time A is performed and Q will therefore become true shortly thereafter. But none of our current machinery enables us to distinguish between P and Q. Because P is now true and Q is now false, we have a defeasible reason for believing that P will still be true when A is performed, and we have a defeasible reason for believing that Q will still be false shortly thereafter. We know that one of these defeasible conclusions will be false, but we have no basis for choosing between them, so this becomes a case of collective defeat. That, however, is the intuitively wrong answer.

When we reason about causal mechanisms, we think of the world as "unfolding" temporally, and changes only occur when they are forced to occur by what has already happened. In our example, when A is performed, nothing has yet happened to force a change in P, so we conclude defeasibly that P remains true. But given the truth of P, we can then deduce that at a slightly later time, Q will become true. Thus when causal mechanisms force there to be a change, we conclude defeasibly that the change occurs in the later states rather than the earlier states. This seems to be part of what we mean by describing something as a causal mechanism. Causal mechanisms are systems that force changes, where "force" is to be understood in terms of temporal unfolding.[2]

When reasoning about such a causal system, part of the force of describing it as

---

[2] This intuition is reminiscent of Shoham's (1987) "logic of chronological ignorance", although unlike Shoham, I propose to capture the intuition without modifying the structure of the system of defeasible reasoning.

causal must be that the defeasible presumption against the effect occurring is somehow removed. Thus, although we normally expect Jones to remain alive, we do not expect this any longer when he is shot. To remove a defeasible presumption is to defeat it. This suggests that there is some kind of general "causal" defeater for the temporal projection principles adumbrated above. The problem is to state this defeater precisely.

As a first approximation we might try:

(4)   If ϵ and δ are greater than 0, then "A-at-t & P-at-(t–ϵ) & (A&P causes Q)" is a defeasible undercutting defeater for the defeasible inference from ~Q-at-(t–ϵ*) to ~Q-at-(t+δ).

Suppose the inference from P-at-(t–ϵ) to P-at-t were defeated by observing P-at-(t–μ) where $\mu < \epsilon$. For instance, suppose we discover that the gun was unloaded before the trigger was pulled. Then it would be reasonable to expect Jones to remain alive, despite the fact that P-at-(t–ϵ), A-at-t, and (A&P causes Q) remain undefeated. The only way to obtain this result within this framework is to take the inference to ~Q-at-t $\otimes$ ~Q-at-(t+δ) to itself be defeasible. Any undercutting defeater for the inference from P-at-(t–ϵ) to P-at-t must also be an undercutting defeater for the inference to ~Q-at-t $\otimes$ ~Q-at-(t+δ). This can all be accomplished as follows:

(5) (P-at-(t–ϵ) $\otimes$ P-at-t) is an undercutting defeater for (4).

This account of causal reasoning requires causation to be temporally asymmetric. That is, "A&P causes Q" means, in part, that if A&P becomes true then Q will *shortly* become true. This precludes *simultaneous* causation, in which Q is caused to be true at t by A&P being true at t. This may seem problematic, on the grounds that simultaneous causation occurs throughout the real world. For instance, colliding billiard balls in classical physics might seem to illustrate simultaneous causation. However, this is a mistake. If two billiard balls collide at time t with velocity vectors pointing towards each other, they do not also have velocity vectors pointing away from each other at the very same time. Instead, this illustrates what I have elsewhere (1984) called *instantaneous* causation. Instantaneous causation requires that if A&P becomes true at t, then for some $\delta > 0$, for every ϵ such that $0 < \epsilon \le \delta$, Q will true at t+ϵ.[3] I believe that instantaneous causation is all that is required for describing the real world.

I have followed AI-convention here in talking about describing causal change in terms of causation. However, that introduces unnecessary complexities. For example, it is generally assumed in the philosophical literature on causation, that if P causes Q then Q would not have been true if P were not true. This has the consequence that in cases in which there are two independent factors each of which would be sufficient by itself to cause the same effect, if both occur then neither cause it. These are cases of causal overdetermination. A familiar example of causal overdetermination occurs when two assailants shoot a common victim at the same time. Either shot would be fatal. The result is that neither shot is such that if it had not occurred then the victim would not have died, and hence, it is generally maintained, neither shot caused the death of the victim. However, this kind of failure of causation ought to be irrelevant to the kind of causal reasoning under discussion in connection with change. Principle (12) ought to apply to cases of causal overdetermination as well as to genuine cases of causation.

---

[3] I assume that time has the structure of the reals, although that assumption is not required for the implementation.

I take it (and have argued in my [1984]) that all varieties of causation (including causal overdetermination) arise from the instantiation of "causal laws". These are what I have dubbed *nomic generalizations*, and have discussed at length in my 1990. Nomic generalizations are symbolized as "P ⇒> Q", where P and Q are formulas and '⇒>' is a variable-binding operator, binding all free occurrences of variables in P and Q. An informal gloss on "P ⇒ Q" is "Any physically-possible P would be a Q". For example, the law that electrons are negatively charged could be written "(x is an electron) ⇒> (x is negatively charged)". The free occurrences of 'x' are bound by '⇒>'. I propose that we replace "(A&P causes Q)" in (12) by "(A&P ⇒> Q will shortly be true)", where the latter typically results from instantiating more general laws. More precisely, let us define "A when P is causally sufficient for Q after an interval ε" to mean

$$(\forall t)[(A\text{-at-}t \ \& \ P\text{-at-}t) \Rightarrow> (\exists\gamma)(\forall\delta)(\epsilon < \delta \leq \gamma \supset Q\text{-at-}(t+\delta))].$$

Instantaneous causation is causal sufficiency with an interval 0.

My proposal is to replace 'causes' by 'causal sufficiency' in (12). Modifying it to take account of the interval over which the causation occurs:

CAUSAL-UNDERCUTTER
"A-at-t & P-at-t_0 & t_0 < t & t_1 < t & (A when P is causally sufficient for ~Q after an interval ε) & (t+ε) < t_2" is a defeasible undercutting defeater for the defeasible inference from Q-at-t_1 to Q-at-t_2.

For causal reasoning, we also want to use the causal connection to support inferences to conditionals. This is more complicated than it might initially seem. The difficulty is that, for example,

the gun is fired when the gun is loaded is causally sufficient for ~(Jones is alive) after an interval 20

does not imply that if the gun is fired at t and the gun is loaded at t then Jones is dead at t+20. Recall the discussion of instantaneous causation. All that is implied is that Jones is dead over some interval open on the left and with t+20 as the lower bound. We can conclude that *there is* at time > t+20 at which Jones is dead, but it does not follow as a matter of logic that Jones is dead at any particular time because, at least as far as this causal law is concerned, Jones could become alive again after becoming dead. To infer that Jones is dead at a particular time after t+20, we must combine the causal sufficiency with temporal projection. This yields the following principle:

CAUSAL-IMPLICATION
If Q is temporally projectible, then "(A when P is causally sufficient for Q after an interval ε) & A-at-t & P-at-t & ((t+ε) < t*)" is a defeasible reason for "Q-at-t*".

With the use of these principles, it is concluded that Jones is not alive at 50:

```
# 1
(the gun is loaded at 0)
given
# 2
((Jones is alive) at 20)
given
```

\# 3
(the gun is fired at 30)
given
\# 4
(the gun is fired when the gun is loaded is causally sufficient for ~(Jones is alive) after an interval 10)
given

> \# 1
> interest: (~(Jones is alive) at 50)
> This is of ultimate interest
> \# 3
> interest: ((Jones is alive) at 50)
> This is of ultimate interest
> \# 4
> interest: (20 < 50)
> For interest 3 by *temporal-projection*

\# 5
(20 < 50)
Inferred by:
        support-link #5 from { } by strict-arithmetical-inequality
This discharges interest 4
\# 6
((Jones is alive) at 50)   DEFEATED
Inferred by:
        support-link #6 from { 2 , 5 } by *temporal-projection* defeaters: { 14 , 11 }
This node is inferred by discharging interest #3

> \# 5
> interest: ((((Jones is alive) at 20) & (20 < 50)) @ ((Jones is alive) at 50))
> Of interest as a defeater for support-link 6 for node 6

==
Justified belief in ((Jones is alive) at 50)
answers #<Query 1: ((Jones is alive) at 50)>
==

> \# 7
> interest: (the gun is loaded at 30)
> For interest 1 by *causal-implication*
> For interest 5 by *causal-undercutter*
> \# 8
> interest: (0 < 30)
> For interest 7 by *temporal-projection*

\# 7
(0 < 30)
Inferred by:
        support-link #7 from { } by strict-arithmetical-inequality
This discharges interest 8
\# 8
(the gun is loaded at 30)
Inferred by:
        support-link #8 from { 1 , 7 } by *temporal-projection*
This node is inferred by discharging interest #7

> \# 11
> interest: ((30 + 10) < 50)
> For interest 1 by *causal-implication*
> \# 12
> interest: (20 <= 30)
> For interest 5 by *causal-undercutter*

\# 9
((30 + 10) < 50)
Inferred by:
        support-link #9 from { } by strict-arithmetical-inequality
This discharges interests (11 16)
\# 10
(~(Jones is alive) at 50)
Inferred by:
        support-link #10 from { 4 , 3 , 8 , 9 } by *causal-implication* defeaters: { 12 }
This node is inferred by discharging interest #1

==
Justified belief in (~(Jones is alive) at 50)
answers #<Query 2: (~(Jones is alive) at 50)>
==

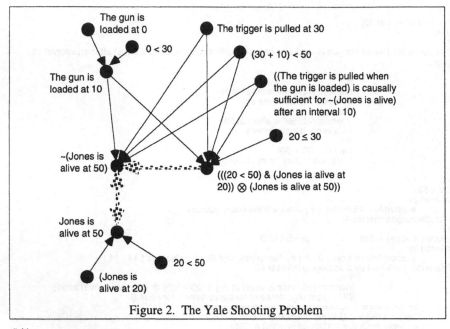

Figure 2. The Yale Shooting Problem

11
~((Jones is alive) at 50)
Inferred by:
        support-link #11 from { 10 } by neg-at-elimination
defeatees: { link 6 for node 6 }
        vvvvvvvvvvvvvvvvvvvvvvvvvvvvvv
        #<Node 6> has become defeated.
        vvvvvvvvvvvvvvvvvvvvvvvvvvvvvv

        ===
        Lowering the undefeated-degree-of-support of ((Jones is alive) at 50)
        retracts the previous answer to #<Query 1: ((Jones is alive) at 50)>
        ===
12
~(~(Jones is alive) at 50)
Inferred by:
        support-link #12 from { 6 } by inversion from contradictory nodes 11 and 6
defeatees: { link 10 for node 10 }
        vvvvvvvvvvvvvvvvvvvvvvvvvvvvvv
        #<Node 12> has become defeated.
        vvvvvvvvvvvvvvvvvvvvvvvvvvvvvv
        #<Node 10> has become defeated.
        vvvvvvvvvvvvvvvvvvvvvvvvvvvvvv
        #<Node 11> has become defeated.
        vvvvvvvvvvvvvvvvvvvvvvvvvvvvvv

        ===
        Lowering the undefeated-degree-of-support of (~(Jones is alive) at 50)
        retracts the previous answer to #<Query 2: (~(Jones is alive) at 50)>
        ===
13
(20 <= 30)
Inferred by:
        support-link #13 from { } by arithmetical-inequality
This discharges interest 12
                        # 16
                        interest: ((30 + 10) < 50)
                        For interest 5 by *causal-undercutter*

```
# 14
((((Jones is alive) at 20) & (20 < 50)) @ ((Jones is alive) at 50))
Inferred by:
        support-link #14 from { 3 , 4 , 8 , 13 , 9 } by *causal-undercutter*
defeatees: { link 6 for node 6 }
This node is inferred by discharging interest #5
        vvvvvvvvvvvvvvvvvvvvvvvvvvvvvv
        The undefeated-degree-of-support of #<Node 11> has increased to 1.0
        vvvvvvvvvvvvvvvvvvvvvvvvvvvvvvvv
        The undefeated-degree-of-support of #<Node 10> has increased to 1.0
        vvvvvvvvvvvvvvvvvvvvvvvvvvvvvv
        ==========================================
        Justified belief in (~(Jones is alive) at 50)
        answers #<Query 2: (~(Jones is alive) at 50)>
        ==========================================
```

This reasoning is diagramed in figure 2.

This, as far as I know, is the first *implemented* solution to the Yale Shooting Problem. It is noteworthy how simple it is to implement such principles in OSCAR, making OSCAR a potent tool for epistemological analysis.

References

Hanks, Steve, and McDermott, Drew

1986   "Default reasoning, nonmonotonic logics, and the frame problem", AAAI-86.

Haugland, John

1987   "An overview of the frame problem", in Z. Pylyshyn (ed.) *The Robot's Dilemma,,* MIT Press.

Hayes, Patrick

1987   "What the frame problem is and isn't", in Z. Pylyshyn (ed.) *The Robot's Dilemma,,* MIT Press.

Janlert, Lars-Erik

1987   "Modeling change—the frame problem", in Z. Pylyshyn (ed.) *The Robot's Dilemma,,* MIT Press.

McCarthy, John, and Hayes, Patrick

1969   "Some philosophical problems from the standpoint of artificial intelligence". In B. Metzer & D. Michie (eds.), *Machine Intelligence 4*. Edinburgh: Edinburgh University Press.

Pollock, John

1974   *Knowledge and Justification*, Princeton University Press.

1984   *The Foundations of Philosophical Semantics*, Princeton University Press.

1987   *Contemporary Theories of Knowledge*, Rowman and Littlefield.

1990   *Nomic Probability and the Foundations of Induction*, Oxford University Press.

1995   *Cognitive Carpentry*, MIT Press.

1995a   *The OSCAR Manual*, available online from http://info-center.ccit.arizona.edu/~oscar.

Sandewall, Erik

1972   "An approach to the frame problem and its implementation". In B. Metzer & D. Michie (eds.), *Machine Intelligence 7*. Edinburgh: Edinburgh University Press.

Shoham, Yoav

1987   *Reasoning about Change*, MIT Press.

A System for Defeasible Argumentation, with Defeasible Priorities

Henry Prakken*[1] and Giovanni Sartor[2]

[1] Computer/Law Institute, Free University, De Boelelaan 1105 Amsterdam
email: henry@rechten.vu.nl
[2] CIRFID, University of Bologna, Via Galliera 3, 40121, Bologna
IDG-CNR, Via Panchiatichi 56/16, Firenze
email sartor@cirfid.unibo.it

Abstract. Inspired by legal reasoning, this paper presents an argument–based system for defeasible reasoning, with a logic–programming–like language, and based on Dung's argumentation–theoretic approach to the semantics of logic programming. The language of the system has both weak and explicit negation, and conflicts between arguments are decided with the help of priorities on the rules. These priorities are not fixed, but are themselves defeasibly derived as conclusions within the system.

1 Introduction

This paper presents an argument–based system for defeasible reasoning, with a logic–programming–like language. Argument–based systems analyze defeasible reasoning in terms of the interactions between arguments for alternative conclusions. Defeasibility arises from the fact that arguments can be defeated by stronger counterarguments.

Argumentation has proved to be a fruitful paradigm for formalising defeasible reasoning (cf. [13, 17, 18]). Not only does the notion of an argument naturally point at possible proof theories, but also do notions like argument, counterargument, attack and defeat have natural counterparts in the way people think, which makes argument–based systems transparent in applications. Especially in legal reasoning these notions are prevalent, which explains why several argument–based systems have been applied to that domain ([14, 10, 8, 16]).

Also the present system is inspired by the legal domain. In particular, we want to capture the following features of legal reasoning (but also of some other domains, such as bureaucracies). The first is that in law the criteria for comparing arguments are themselves part of the domain theory. For instance, in Italy, in town planning regulations we can find a priority rule stating that rules on the protection of artistic buildings prevail over rules concerning town planning.

---

* Henry Prakken was supported by a research fellowship of the Royal Netherlands Academy of Arts and Sciences, and by Esprit WG 8319 'Modelage'. The authors wish to thank Mark Ryan for his comments on an earlier version of this paper.

Apart from varying from domain to domain, priority rules can also be debatable, in the same way as 'ordinary' domain information can be. For instance, if an artistic–buildings rule is of an earlier date than a conflicting town planning rule, the just–mentioned conflict rule is in conflict with the temporal principle that the later rule has priority over the earlier rule. Other conflict rules may apply to this conflict, and this makes that also reasoning about priorities is defeasible.

The second feature is that in law specificity is not the overriding standard for comparing arguments. In most legal systems it is subordinate to the hierarchical criterion (e.g. 'the constitution has priority over statutes') and to the temporal criterion. This means that systems like [7], making specificity the overriding standard for comparison, are for our purposes inadequate.

Finally, we want to model the fact that legal reasoning combines the use of priorities to choose between conflicting rules with the use of assumptions, or 'weak' negation, within a rule to make it inapplicable in certain circumstances. An example of such a rule is section 3:32–(1) of the Dutch Civil Code, which declares every person to have the capacity to perform juridical acts, "unless the law provides otherwise". Accordingly, our language will have both explicit and weak negation, which yields two different ways of attacking an argument: by stating an argument with a contradictory conclusion, or by stating an argument with a conclusion providing an 'unless' clause of the other argument.

It is not our aim to present a general theory of defeasible argumentation. Rather, we will analyse these phenomena within a logic–programming–like setting, in particular Dung's [6] argument–based approach to the semantics of extended logic programming. With the choice for a logic–programming like language we hope to increase the prospects for implementation, while our choice for Dung's approach is motivated by his emphasis on argumentation.

We will present our system in two phases. In the first phase the priorities are still externally given and fixed (section 2) and in the second phase they are derived within the system itself (section 3). After that, the system is compared with related research (section 4). This paper is a revised version of [16], with more emphasis on formal aspects and less attention to legal applications. For an extensive discussion of examples and applications the reader is referred to [16].

2 The Formal System I: Fixed Priorities

As most systems of defeasible argumentation, our system contains the following elements. To start with, it has an underlying formal language and, based on the language, a notion of an argument. Then it has a definition of when an argument is in conflict with, or attacked by other arguments, a way of comparing conflicting arguments and, most importantly, a definition of the ultimate status of an argument, in terms of three classes: arguments with which a dispute can be 'won', respectively, 'lost' and arguments which leave the dispute undecided.

2.1 The Language

The object language of our system is of familiar logic–programming style: it contains a twoplace one–direction connective that forms rules out of literals. The language has two kinds of negation, weak and classical negation. An atomic first-order formula is a *positive* literal; a positive literal preceded by $\neg$ is a *negative* literal; a positive or negative literal is a *weak* literal if preceded by $\sim$; otherwise it is a *strong* literal. For any atom $P(x)$ we say that $P(x)$ and $\neg P(x)$ are the complement of each other; in the metalanguage $\overline{L}$ denotes the complement of L.

Now a *rule* is an expression of the form

$$r : L_0 \wedge \ldots \wedge L_j \wedge \sim L_k \wedge \ldots \wedge \sim L_m \Rightarrow L_n$$

where r is the name of the rule and each L_i $(0 \leq i \leq k)$ is a strong literal. The conjunction at the left of the arrow is the *antecedent* and the literal at the right of the arrow is the *consequent* of the rule. As usual, a rule with variables is a scheme standing for all its ground instances.

The input information of our system does not only contain rules, but also priorities. We call the input an *ordered theory*, which is a pair $(T, <)$, where T is a set of rules, and $<$ is a noncircular ordering on T. That $r < r'$ means that r' is preferred over r.

The rules are intended to express defeasible information. They can be defeasible in two ways: a rule can contain unwarranted assumptions, and it can still be overridden by stronger rules with a contradictory consequent, even if it does not contain assumptions. Note, finally, that the antecedent of a rule can be empty; such rules can be used to express facts. However, being rules, also the facts are subject to defeat.

2.2 Arguments

The basic notion of a system for defeasible argumentation is that of an argument. In general, the idea is that an argument for a certain proposition is a proof of that proposition in the logic of the underlying language. In our system, the simple language gives rise to a simple notion of an argument, viz. as a sequence of rules that can be chained together, and that is 'grounded' in the facts. This is captured by a slight variant of Dung's [6] notion of a 'defeasible proof'.

Definition 1 An *argument* is a finite sequence $[r_n, \ldots, r_m]$ of ground instances of rules such that

1. for every $i, n \leq i \leq m$, for every positive or negative literal L in the antecedent of r_i there is a $j < i$ such that L is the consequent of r_j; and
2. for no r_i its consequent is also the consequent of some r_j $(j < i)$.

An argument A is *based on* the ordered theory $(T, <)$ iff all rules of A are ground instances of rules in T.[3]

---

[3] We will often leave the phrase 'based on $(T, <)$' implicit.

For any ordered theory Γ we will denote the set of all arguments on the basis of Γ with Args_Γ. Likewise, for any set of rules T, Args_T stands for the set of all arguments that consist of only ground instances of rules in T.

We will also use the following notions. For any argument A, an argument A' is a *(proper) subargument* of A iff A' is a (proper) subsequence of A. A literal L is a *conclusion* of A iff it is the consequent of some rule in A. And L is an *assumption* of A iff $\sim \overline{L}$ occurs in some rule in A. The following example illustrates these notions: for the argument $A = [r_1: \sim \neg a \Rightarrow b, r_2: b \wedge \sim d \Rightarrow c]$ we have that A's conclusions are $\{b, c\}$, its subarguments are $\{[\,], [r_1], [r_1, r_2]\}$, and its assumptions are $\{a, \neg d\}$.

2.3 Relations between Arguments

So far the notions have been fairly standard; now we present the adversarial aspects of our system, starting with the ways in which an argument can attack, i.e. be a counterargument of another argument. This definition does not yet evaluate arguments; it only tells us which arguments are in conflict. As noted in the introduction, the two different kinds of negation give rise to two ways of attacking an argument. We formalise this as follows.

Definition 2. An argument A_1 *attacks*, or, is a *counterargument of* an argument A_2 iff some conclusion of A_1 is the complement of some conclusion or assumption of A_2. If one argument attacks another, they are *in conflict* with each other.

Note that in order to attack an argument, a counterargument can point its attack at that argument itself, but also at one of its proper subarguments, thereby indirectly attacking the entire argument. In fact, since every argument is a subargument of itself, the definition implies that A_1 attacks A_2 iff a subargument of A_1 attacks a subargument of A_2. So, for instance, if we have

$$r_1: \Rightarrow a \quad r_2: a \Rightarrow b$$
$$r_3: \Rightarrow \neg b \quad r_4: \neg b \Rightarrow \neg a$$

we have that not only $[r_1, r_2]$ and $[r_3]$ attack each other, and $[r_3, r_4]$ and $[r_1]$ attack each other, but also that $[r_1, r_2]$ and $[r_3, r_4]$ attack each other.

The concept of 'attack/counterargument' is very important, since obviously, any system for defeasible argumentation should say that if two arguments are in conflict with each other, they cannot be accepted together as justified. We will prove that our system fulfils this requirement. To that end, we also need to define when an argument is coherent, and when a set of arguments is conflict–free.

Definition 3. An argument is *coherent* iff it does not attack itself.

Two examples of incoherent arguments are $[r_1: \Rightarrow a, r_2: a \Rightarrow \neg a]$ and $[r_1: \sim a \Rightarrow b, r_2: b \Rightarrow a]$.

Definition 4. A set *Args* of arguments is *conflict–free* iff no argument in *Args* attacks an argument in *Args*.

Now that we know which arguments are in conflict with each other, the next step is to compare conflicting arguments. To this end we define, in two steps, a binary relation of defeat among arguments. It is important to realise that this comparison does not yet determine with which arguments a dispute can be won; it only tells us something about the relation between two individual arguments (and their subarguments). Note also that the argument relations and properties defined below are relative to an implicitly assumed ordered theory.

To capture the different 'force' of the two ways of attack, we define 'defeat' in terms of two other evaluative notions, depending on whether an attack is on a conclusion or on an assumption of another argument. Only in the first case the priorities will be used. The definition of defeat will state how to use these two notions in combination.

Definition 5. Let A_1 and A_2 be two arguments. Then

- A_1 *rebuts* A_2 iff for some pair of rules r_1 in A_1 and r_2 in A_2:
  - r_1 and r_2 have complementary consequents; and
  - $r_1 \not< r_2$.
- A_1 undercuts A_2 iff some conclusion of A_1 is the complement of some assumption of A_2.

This is almost the definition of attack; the only difference is that if arguments have complementary conclusions, they are compared with the help of priorities. Thus, since undercutting will imply defeat, an attack on an assumption always succeeds. This is motivated by our reading of $\sim A$ simply as the assumption that there is no evidence of A.

The way this definition uses the priorities is that it compares arguments with contradictory conclusions on the priority relation between the rules with these conclusions as the consequent.

As with attack, also rebutting and undercutting an argument can be 'direct', or 'indirect', by providing the complement of a conclusion, or assumption, of a proper subargument. Note also that if A rebuts or undercuts B, then A attacks B. Obviously, it does not hold that if A undercuts B, then B undercuts A. However, it neither holds that if A undercuts B, then B does not undercut A. A counterexample is $A = [r_1 : \sim b \Rightarrow a], B = [r_2 : \sim a \Rightarrow b]$. It is also not the case that if A rebuts B, then B rebuts A. Just assume we have $A = [r_1 : \Rightarrow a]$, $B = [r_2 : \Rightarrow \neg a]$, and $r_1 < r_2$. Then B rebuts A but not the other way around. However, rebutting involving a $<$ relation between conflicting rules is not always one–sided. To see this, assume in the above example illustrating attack, that $r_1 < r_4$ and $r_3 < r_2$; then $[r_1, r_2]$ and $[r_3, r_4]$ rebut each other.

Finally, our definition of defeat states how in evaluating arguments the notions of undercutting and rebutting attack are combined. As a boarder case it regards any incoherent argument as defeated by the empty argument. Apart from this, the definition is based on two ideas, of which the first is inherited from the definition of rebutting and undercutting: defeat of an argument can be direct, or indirect, by defeating one of its proper subarguments; this captures that an argument cannot be stronger than its weakest link. The other idea is that our reading of $\sim A$ not only makes attacks on assumptions always succeed, but also makes an attack on an assumption stronger than rebutting: if one argument undercuts the other, and the other does not undercut but only rebut the first, the second does not defeat the first.

Definition 6. Let A_1 and A_2 be two arguments. Then A_1 *defeats* A_2 iff A_1 is empty and A_2 is incoherent, or else if

- A_1 undercuts A_2; or
- A_1 rebuts A_2 and A_2 does not undercut A_1.

We say that A_1 *strictly defeats* A_2 iff A_1 defeats A_2 and A_2 does not defeat A_1.

Corollary 7. *If A_1 attacks A_2, then A_1 defeats A_2 or A_2 defeats A_1. And if A_1 defeats A_2, then A_1 attacks A_2.*

The following example illustrates that undercutting an argument is stronger than rebutting it. Consider $r_1 : \sim \neg Innocent(OJ) \Rightarrow Innocent(OJ)$ and $r_2 : \Rightarrow \neg Innocent(OJ)$ and assume that $<= \emptyset$. Then, although $[r_1]$ rebuts $[r_2]$, $[r_1]$ does not defeat $[r_2]$, since $[r_2]$ undercuts $[r_1]$. So $[r_2]$ strictly defeats $[r_1]$.

Our notion of defeat is a weak one, since for rebutting an argument no $>$ relation is needed between the relevant rules, but only a $\not<$ relation. So, rather than having to be 'really better' than the argument that is to be defeated, a defeating argument only has to be not inferior. This is since we want our notion of 'justified arguments', to be defined next, to really capture only those arguments that, given the premises, are beyond any doubt or challenge. And doubt can already be cast on an argument by providing a counterargument that is at least not inferior to it. For the same reason we also need the notion of strict defeat, which, being asymmetric, captures the idea of 'really being better than'. This notion will be used to ensure that a counterargument fails to cast doubt if it is inferior to at least one counterargument that is itself protected from defeat.

2.4 The Status of Arguments

Since defeating arguments can themselves be defeated by other arguments, comparing just pairs of arguments is not sufficient; what is also needed is a definition that determines the status of arguments on the basis of all ways in which they interact. In particular, the definition should allow for reinstatement of defeated arguments, if their defeater is itself (strictly) defeated by another argument. This definition, then, is the central element of our system. It takes as input the set of all possible arguments and their mutual relations of defeat, and produces as output a division of arguments into three classes: arguments with which a dispute can be 'won', respectively, 'lost' and arguments which leave the dispute undecided. As remarked above, the winning arguments should be only those arguments that, given the premises, are beyond any doubt: the only way to cast doubt on these arguments is by providing new premises, giving rise to new defeating counterarguments. Accordingly, we want our set of justified arguments to be unique, and to be conflict–free.

The intuitive idea is that the set of *justified* arguments is constructed step–by–step (note that everything here is relative to an implicitly assumed input theory). We first collect into $JustArgs_1$ all arguments which are directly justified, by their own strength: those are the ones which are not defeated by any counterargument. Then we add all arguments that are justified indirectly, i.e. with the help of arguments in $JustArgs_1$. More exactly, each argument that has defeating counterarguments, is still added to $JustArgs_1$ if all those counterarguments are strictly defeated by an argument already in $JustArgs_1$. The resulting set is $JustArgs_2$. We repeat this step until we obtain a set $JustArgs_n$ to which no new argument can be added, which, then, is the set of all justified arguments. Then we can define the losing, or 'overruled' arguments as those that are attacked by a justified argument and, finally, the undeciding or 'defensible' arguments as all the arguments that are neither justified nor overruled.

We formalise this with a definition based on Dung's [6] grounded (sceptical) semantics of extended logic programs. It makes use of a variant of Dung's notion of the acceptability of an argument with respect to a set of arguments.

Definition 8. An argument A is *acceptable* with respect to a set $Args$ of arguments iff each argument defeating A is strictly defeated by an argument in $Args$.

Next we use this notion in defining the set of justified arguments.

Definition 9. Let Γ be an ordered theory. Then we define the following sequence of subsets of $Args_\Gamma$.

- $F_\Gamma^0 = \emptyset$
- $F_\Gamma^{i+1} = \{A \in Args_\Gamma \mid A \text{ is acceptable with respect to } F_\Gamma^i\}$.

Then the set $JustArgs_\Gamma$ of arguments that are justified on the basis of Γ is $\cup_{i=0}^\infty (F_\Gamma^i)$.

In terms of the set of justified arguments we define the overruled and defensible arguments. We also define the corresponding notions for conclusions.

Definition 10. For any ordered theory Γ and argument A we say that on the basis of Γ:

1. A is *overruled* iff A is attacked by an argument in $JustArgs_\Gamma$;
2. A is *defensible* iff A is neither justified nor overruled.

And for any literal L, we say that L is a *justified conclusion* iff it is a conclusion of a justified argument, L is a *defensible conclusion* iff it is not justified and it is a conclusion of some defensible argument, and L is an *overruled conclusion* iff it is not justified or defensible, and a conclusion of an overruled argument.

In the following subsection this definition will be illustrated. Now we state some formal properties. Firstly, by definition the set of justified arguments is unique, as we wanted. Furthermore, Proposition 11 says that the set of justified arguments can be constructed step–by–step by just adding new arguments; at no step arguments are deleted. And Proposition 12 says that the result of the construction is conflict–free.

Proposition 11. *Let* $JustArgs_\Gamma = \cup_{i=0}^{\infty}(F_\Gamma^i)$. *Then for all* i, $F_\Gamma^i \subseteq F_\Gamma^{i+1}$.

Proposition 12. *For each ordered theory* Γ, $JustArgs_\Gamma$ *is conflict-free.*

The reader will have noticed that Definition 10 does not explicitly require that all proper subarguments of an argument are justified, as, e.g. [11, 18, 14] do. Instead, as the following proposition states, this requirement is implicit in our definitions, as also in e.g. [13, 17, 7].

Proposition 13. *If an argument is justified on the basis of* Γ, *all its subarguments are justified on the basis of* Γ.

2.5 Illustration of the Definitions

The first example illustrates the step–by–step construction of $JustArgs$.[4]

$$r_0: \Rightarrow a \qquad r_1: a \Rightarrow b$$
$$r_2: \sim b \Rightarrow c$$
$$r_3: \Rightarrow \neg a$$

where $r_0 < r_3$. First we identify the relations of defeat. The argument $A_1 = [r_0, r_1]$ defeats the argument $A_2 = [r_2]$, since it undercuts it. Furthermore, the argument $A_3 = [r_3]$ defeats A_1, by rebutting its proper subargument $A_0 = [r_0]$ and thereby also rebutting A_1 itself.

With these relations we can construct the set of justified arguments as follows. A_3 is not defeated by any argument, since its only counterargument A_1 is too weak. So $F^1 = \{A_3\}$ How now about A_1? We cannot add it, since it is defeated by A_3, which in turn is not strictly defeated by any argument in F^1. Also A_0 cannot be added, for the same reasons. How then about A_2? Although it is defeated by its only counterargument A_1, we can still add it, since A_1 is in turn strictly defeated by an argument in F^1, viz. A_3. Thus A_3 reinstates A_2 and we have that $F^2 = \{A_3, A_2\}$. Repeating this process adds no new relevant arguments, so we can stop here: $F^3 = F^2$. Thus we obtain that A_2 and A_3 are justified, while A_0 and A_1 are overruled. Note that the weakest link principle is respected: A_1 is overruled since its subargument A_0 is overruled.

Next we illustrate the relation between undercutting and rebutting attack.

---

[4] In the rest of this paper we will leave the empty argument, and combinations of independently justified arguments, implicit.

$$r_1 : \Rightarrow Has\_Porsche \qquad r_2 : Has\_Porsche \Rightarrow Rich$$
$$r_3 : \sim Rich \Rightarrow \neg Has\_Porsche$$

Assume that $r_1 < r_3$. Definition 6 ignores this ordering; $B = [r_3]$ does not defeat $A = [r_1, r_2]$ since A undercuts B. This causes A to be justified, as well as well as its subargument $A' = [r_1]$, although that is defeated by B. A' is reinstated since A strictly defeats its counterargument B.

Finally we illustrate that our definitions respect the 'step–by–step' nature of argumentation: conflicts about conclusions or assumptions earlier in the chains are dealt with before 'later' conflicts.

$$r_1 : \Rightarrow a \qquad r_2 : a \Rightarrow b \qquad r_3 : b \Rightarrow c$$
$$r_4 : \Rightarrow \neg a \qquad r_5 : \neg a \Rightarrow d \qquad r_6 : d \Rightarrow \neg c$$

Assume that $r_4 < r_1$ and $r_3 < r_6$. Then the arguments $[r_1 - r_3]$ and $[r_4 - r_6]$ defeat each other. However, $[r_1]$ (and also $[r_1, r_2]$), strictly defeats $[r_4 - r_6]$ and has no other counterarguments; so $F^1 = \{[r_1], [r_1, r_2]\}$. But then $[r_4 - r_6]$, although defeating $[r_1 - r_3]$, is not in F^2, while $[r_1 - r_3]$ is in F^2, reinstated by $[r_1]$.

3 The Formal System II: Defeasible Priorities

3.1 Definitions

So far we have simply assumed that there is an input ordering on the rules. Now we will study the situation where this ordering is derived from the premises. To this end we assume that our language contains a distinguished twoplace predicate symbol $\prec$, with which information on the priorities can be stated in the object language. This makes the ordering component of an ordered theory redundant, so an *ordered theory* is from now on just a set of rules. Next we change the definition of rebutting and undercutting arguments, to make sure that arguments which together would make the ordering circular cannot be accepted simultaneously.

Definition 14. Let A_1 and A_2 be two arguments. Then

1. A_1 *rebuts* A_2 iff
   - for some pair of rules r_1 in A_1 and r_2 in A_2:
     - r_1 and r_2 have complementary consequents; and
     - $r_1 \not\prec r_2$.
   - or for some sequence of rules $r_1, \ldots, r_n$ in A_1 and some rule r_m in A_2:
     - the consequent of r_m is $x \prec y$ and the consequents of $r_1, \ldots, r_n$ are a chain $y \prec z, \ldots, z' \prec x$; and
     - for all r_i $(1 \leq i \leq n) : r_i \not\prec r_m$.
2. A_1 *undercuts* A_2 iff
   - some conclusion of A_1 is the complement of some assumption of A_2; or
   - A_2 contains an assumption $x \prec y$ and A_1 has a chain of conclusions $y \prec z, \ldots, z' \prec x$.

Note that in priority conflicts between arguments more than one rule of an argument may directly contribute to a conclusion, viz. when the chain of priority conclusions has more than one element. In that case it is natural to require for rebutting that not just the last of those rules, but all of them are not less than the relevant rule of the other argument. To illustrate this with an example, assume we have $r_4: \Rightarrow r_1 \prec r_2$, $r_5: \Rightarrow r_2 \prec r_3$ and $r_6: \Rightarrow r_3 \prec r_1$, and assume that $< = \{r_4 < r_5, r_4 < r_6\}$. With Definition 14 we have that $[r_4]$ is strictly defeated by $[r_5, r_6]$, which itself is not defeated by any argument, so that $r_2 \prec r_3$ and $r_3 \prec r_1$ are justified conclusions, while $r_1 \prec r_2$ is an overruled conclusion. By contrast, if only the last rule of the sequence were taken into account, we would have that $[r_4, r_5]$ defeats $[r_6]$, and all the conclusions would be defensible, which seems a less natural outcome.

We now define the consequences of the extended framework with a revision of the fixpoint definition. The idea is that now with the set of justified arguments also the ordering is 'constructed' step–by–step. The resulting definition is inspired by an earlier version of [4].

Note that we now have to make the dependence of the various notions on a rule ordering explicit, since this ordering now varies with the conclusions that have been drawn so far. To this end we first define the following notation. For any set $Args$ of arguments

$$<_{Args} = \{r < r' \mid r \prec r' \text{ is a conclusion of some } A \in Args\}$$

Then we say for any set $Args$ of arguments and any set T of rules that A (strictly) $Args$–defeats B on the basis of T iff according to Definition 6 A (strictly) defeats B on the basis of $(T, <_{Args})$.[5] Occasionally, we will also use the analogous notion $Args$–rebuts.

Corollary 15. *If A_1 attacks A_2, then for any conflict-free set $Args$ of arguments, A_1 $Args$–defeats A_2 or A_2 $Args$–defeats A_1; and if A_1 $Args$–defeats A_2, then A_1 attacks A_2.*

Next we incorporate the new notation in the definition of acceptability.

Definition 16. An argument A is *acceptable* with respect to a set $Args$ of arguments iff all arguments $Args$–defeating A are strictly $Args$–defeated by some argument in $Args$.

Thus acceptability of an argument with respect to a set $Args$ now depends on the priority conclusions of the arguments in $Args$. With this change, the definition of justified arguments can stay the same (recall that Γ is now a set of rules).

Definition 17. Let Γ be an ordered theory. Then we define a sequence of subsets of $Args_\Gamma$ as follows.

---

[5] Below we will leave the phrase 'on the basis of T' implicit. Note also that now $<_{Args}$ can for arbitrary $Args$ be circular (although not for $JustArgs$: see Corollary 20).

$-\ G_\Gamma^0 = \emptyset$

$-\ G_\Gamma^{i+1} = \{A \in Args_\Gamma \mid A \text{ is acceptable with respect to } G_\Gamma^i\}.$

Then the set $JustArgs_\Gamma$ of arguments that are justified on the basis of Γ is $\cup_{i=0}^\infty(G_\Gamma^i)$. Overruled and defensible arguments are defined as above.

Also for this definition it holds that each new step only adds arguments.

Proposition 18. *Let* $JustArgs_\Gamma = \cup_{i=0}^\infty(G_\Gamma^i)$. *Then for all* i, $G_\Gamma^i \subseteq G_\Gamma^{i+1}$.

Note that this time each step in the construction $JustArgs$ also implicitly extends the ordering on rules; this is because G uses the notion of acceptability, which in turn uses the notion of $Args$–defeat, and the point is that $Args$, and thus also $<_{Args}$, changes with each new step. So what happens is that at each new step the defeat relations are redetermined on the basis of the set of justified arguments that has been constructed thus far.

Finally, we can again prove that the set of justified arguments is conflict–free.

Proposition 19. *For each ordered theory* Γ, $\cup_{i=0}^\infty(G_\Gamma^i)$ *is conflict-free.*

Corollary 20. *For any set of justified arguments,* $<_{JustArgs}$ *is noncircular.*

3.2 Illustrations

We will illustrate the new definitions with an example in which the abovementioned Italian priority rule on building regulations conflicts with the temporality principle that the later rule has priority over the earlier one. They state contradicting priorities between a town planning rule saying that if a building needs restructuring, its exterior may be modified, and an earlier, and conflicting, artistic–buildings rule saying that if a building is on the list of protected buildings, its exterior may not be modified. In the example we will use a common method for naming rules. Every rule with terms $t_1, \ldots, t_n$ is named with a function expression $r(t_1, \ldots, t_n)$, where r is the informal name of the rule.

Note that rule r_9 states that rule r_3 is later than the Lex Posterior principle T, which implies that r_3 prevails over T, according to T itself. The application of a priority rule to itself (or better, to one of its instances), is an interesting peculiarity of this example.

$r_1(x)$:	x is protected $\Rightarrow \neg\ x$'s exterior may be modified
$r_2(x)$:	x needs restructuring $\Rightarrow x$'s exterior may be modified
$r_3(y,x)$:	x is a rule about the protection of artistic buildings $\wedge$
	y is a town planning rule $\Rightarrow y \prec x$
$T(x,y)$:	x is earlier than $y \Rightarrow x \prec y$
$r_4(r_1(x))$:	$\Rightarrow r_1(x)$ is about the protection of artistic buildings
$r_5(r_2(x))$:	$\Rightarrow r_2(x)$ is a town planning rule
$r_6(r_1(x), r_2(y))$:	$\Rightarrow r_2(x)$ is later than $r_1(y)$
$r_7(Villa_0)$:	$\Rightarrow Villa_0$ is protected
$r_8(Villa_0)$:	$\Rightarrow Villa_0$ needs restructuring
$r_9(r_3(x,y), T(x,y))$:	$\Rightarrow r_3(x,y)$ is later than $T(x,y)$

To maintain readability, we will below only give the function symbol part of the rule names, except when we want to stress that a rule is applied to other rules. We also use the following abbreviations: the fact instance $r_6(r_2, r_1)$, stating that r_2 is later than is r_1, is abridged as r_6'; the rule instance $r_3(r_2, r_1)$, giving precedence to artistical protection rule r_1, is written as r_3', and the rule instance $T(r_1, r_2)$, giving precedence to the later rule r_2, is shortened to T'.

First we collect all arguments which have no G^0– defeating counterarguments (for readability, we do not list subarguments).

$$G^1 = \{[r_4], [r_5], [r_6'], [r_7], [r_8], [r_6(r_3', T'), T(T', r_3')]\}$$

The last argument in G^1 concludes that, as a criterion for solving the conflict between r_1 and r_2, r_3 is better than T itself, i.e., that r_3' is better than T':

$$T(T', r_3'): r_3' \text{ is later than } T' \Rightarrow T' \prec r_3'$$

Hence $<^1 = \{T' < r_3'\}$. With this ordering relation we can solve the conflict between the argument $[r_6', T']$, saying that rule r_2 prevails over r_1, according to temporality, and the argument $[r_4, r_5, r_3']$, saying that rule r_1 prevails over rule r_2, according to r_3: the latter argument now strictly G^1–defeats the first, so $G^2 = G_1 \cup \{[r_4, r_5, r_3']\}$ and $<^2 = \{T' < r_3', r_2 < r_1\}$. With this priority relation added, we can finally solve the conflict between the arguments $[r_7, r_1]$ and $[r_8, r_2]$ in favour of the first: $G^3 = G_2 \cup \{[r_7, r_1]\}$. Having considered all arguments, we can stop here: our set of justified arguments is G^3, which means that the exterior of $Villa_0$ may not be modified.

4 Comparison with Related Work

With respect to related research, we here focus on logic– programming–like systems for defeasible reasoning. Related research in Artificial Intelligence and Law is discussed in [16]. [6] has developed various argumentation based semantics of extended logic programming, based on more abstract research in [2] and [5]. As already said above, we have adapted Dung's version of [12]'s well–founded semantics of extended logic programs. However, there are some differences. Firstly, Dung does not consider reasoning with or about priorities. Moreover, since Dung only reformulates the various semantics of logic progams, in the case of well–founded semantics he inherits some of its features that we find less attractive. In particular, Dung defines arguments as being the set of assumptions of what we define as an argument. For instance, in his terms the program $r_1 : \Rightarrow a, r_2 : \Rightarrow b$, $r_3 : \Rightarrow \neg a$, has only one argument, viz. [ ], which defeats itself: so $F^1 = F^0 = \emptyset$. We, by contrast, have that $F^1 = \{[r_3]\}$ and $F^2 = F^1$, which seems more natural.

Also [1] deviate from well–founded semantics, for similar reasons. Although their system is not argument–based, it has two kinds of negations and prioritized rules, but not reasoning about priorities. It would be interesting to study the relation between [1] and the special case of our framework with fixed priorities.

In [4] Brewka has defined an extension of [12]'s well–founded semantics to programs with defeasible priorities. Since Brewka defines a conservative extension of this semantics, he inherits its features. Another difference with our system is that Brewka reduces rebutting to undercutting defeat, by adding to every rule with consequent L a literal $\sim \overline{L}$ to its antecedent. Although this makes the definitions simpler, it also gives up some expressive power. In particular, the difference between rules like 'If A then B' and 'If A then B, unless the contrary can be shown' cannot be expressed.

Two other relevant logic–programming–like systems also need to be mentioned: 'ordered logic' (see e.g. [9]) and Nute's [11] 'defeasible logic', which both have prioritized rules but no undercutting arguments or reasoning about priorities. Nute also has a second category of so–called 'strict rules', of which the use in an argument cannot cause the argument to be defeated. Currently we are investigating how strict rules can be included in our theory with avoiding some features of Nute's treatment that we find less attractive.

Finally, an alternative approach to reasoning about priorities, suitable for extension– or model–based nonmonotonic logics, was developed by [3] for prioritised default logic, and generalised and slightly revised by [15] for any extension– or model–based system. Roughly, the method checks whether what an extension says about the priorities corresponds to the ordering on which the extension is based. Space limitations prevent a detailed comparison here.

5 Conclusion

Inspired by legal reasoning, and in a logic–programming like setting, this paper has studied the combination of the following features of defeasible argumentation: reasoning with two kinds of negation, reasoning with priorities, and reasoning about priorities. As already indicated, one aim of future research is to extend the language with strict rules, in order to make the expression possible of properties of relations, such as transitivity, and of incompatibility of predicates, like 'bachelors are not married'. We also plan to develop a proof theory, in terms of 'dialectical proof trees'.

6 Proofs

Proposition 11 *Let $JustArgs_\Gamma = \cup_{i=0}^{\infty}(F_\Gamma^i)$. Then for all i, $F_\Gamma^i \subseteq F_\Gamma^{i+1}$.*

Proof. The proof is by induction on the definition of $JustArgs_\Gamma$. Clearly $F^0 = \emptyset \subseteq F^1$. Assume next that $F^{i-1} \subseteq F^i$ and consider any $A \in F^i$ and $B \in Args_\Gamma$ such that B defeats A. Then some $C \in F^{i-1}$ strictly defeats B, and since also $C \in F^i$, A is acceptable with respect to F^i and so $A \in F^{i+1}$.

Proposition 12 *For each ordered theory Γ, $\cup_{i=0}^{\infty}(F_\Gamma^i)$ is conflict–free.*

Proof. We first prove with induction that each set F^i is conflict–free. Clearly $F^0 = \emptyset$ is conflict–free. Consider next any F^i that is conflict–free, and assume that $A, A' \in F^{i+1}$ and A attacks A'. Then by Corollary 7 A defeats A' or A' defeats A. Assume without loss of generality that A defeats A'. Then, since

$A' \in F^{i+1}$, some $B \in F^i$ defeats A. But then, since $A \in F^{i+1}$, some $C \in F^i$ defeats that B. But then F^i is not conflict–free. Contradiction, so also F^{i+1} is conflict–free.

Now with Proposition 11 it follows that also $\cup_{i=0}^{\infty}(F_{\Gamma}^i)$ is conflict–free.

Proposition 13 *If an argument is justified on the basis of Γ, all its subarguments are justified on the basis of Γ.*

Proof. Suppose $A \in JustArgs_{\Gamma}$, A' is a subargument of A and B defeats A'. Observe first that for some $F^i \subseteq JustArgs_{\Gamma}$ A is acceptable wrt F^i. Then two cases have to be considered. Firstly, if B does not defeat A, this is because A undercuts B. But then A strictly defeats B. Secondly, if B defeats A, then some $C \in F^i$ strictly defeats B. In both cases A' is acceptable with respect to F^i and so $A' \in F^{i+1}$ and by Proposition 11 $A' \in JustArgs_{\Gamma}$.

Proposition 18 *Let $JustArgs_{\Gamma} = \cup_{i=0}^{\infty}(G_{\Gamma}^i)$. Then for all i, $G_{\Gamma}^i \subseteq G_{\Gamma}^{i+1}$.*

Proof. It sufficies to prove the following properties of $Args$–defeat.

Lemma 21. *For any two conflict–free sets of arguments S and S' such that $S \subseteq S'$, and any two arguments A and B we have that*

1. *If A S'–defeats B, then A S–defeats B.*
2. *If A strictly S–defeats B, then A strictly S'–defeats B.*

Proof. Observe first that undercutting is independent of S and S'. Then (1) follows since if no argument in S' has a conclusion $r < r'$, also no argument in S has that conclusion. For (2) observe first that if S contains an argument A with conclusion $r < r'$, also S' contains A and by conflict–freeness S' does not contain an argument for $r' < r$; so for any B and C, if B strictly S–defeats C, then B S'–defeats C. Moreover, by contraposition of (1) C does not S'–defeat B if C does not S–defeat B, so C strictly S'–defeats B.

With this lemma the proof of Proposition 11 can easily be adapted to the present case.

Proposition 19 *For each ordered theory Γ, $\cup_{i=0}^{\infty}(G_{\Gamma}^i)$ is conflict–free.*

Proof. Observe first that for any conflict–free G^{i-1}, $A \in G^{i-1}$ and $B \in Args_{\Gamma}$ attackking A, by Corollary 15 A G^{i-1}–defeats B or B G^{i-1}–defeats A. Then the proof can be completed in the same way as for Proposition 12.

Corollary 20 *For any set of justified arguments, $<_{JustArgs}$ is noncircular.*

Proof. Assume for contradiction that some conclusions of arguments in $JustArgs$ form a sequence $r_1 \prec r_2, \ldots, r_m \prec r_n, r_n \prec r_1$. Then $JustArgs$ contains an argument A_1 with conclusion $r_n \prec r_1$, and an argument A_2 containing a sequence of conclusions $r_1 \prec r_2, \ldots, r_m \prec r_n$ (A_2 is the combination of the arguments for all atoms in the sequence). Then A_1 and A_2 $JustArgs$–rebut each other and by Lemma 21(1) they also G^i–rebut each other for any $G^i \subseteq JustArgs$. Then by Corollary 15 we have that A_1 and A_2 attack each other, which by Proposition 19 implies a contradiction.

References

1. A. Analyti and S. Pramanik, Reliable semantics for extended logic programs with rule prioritization. *Journal of Logic and Computation* 5 (1995), 303–324.
2. A. Bondarenko, F. Toni, R.A. Kowalski, An assumption–based framework for non-monotonic reasoning. *Proceedings of the Second International Workshop on Logic Programming and Nonmonotonic Reasoning*, MIT Press, 1993, 171–189.
3. G. Brewka, Reasoning about priorities in default logic. *Proceedings AAAI-94*, 247–260.
4. G. Brewka, What does a defeasible rule base with explicit prioritiy information entail? *Proceedings of the second Dutch/German Workshop on Nonmonotonic Reasoning*, Utrecht 1995, 25–32.
5. P.M. Dung, On the acceptability of arguments and its fundamental role in non-monotonic reasoning, logic programming, and n–person games. *Artificial Intelligence* 77 (1995), 321–357.
6. P.M. Dung, An argumentation semantics for logic programming with explicit negation. *Proceedings of the Tenth Logic Programming Conference*, MIT Press 1993, 616–630.
7. H. Geffner and J. Pearl, Conditional entailment: bridging two approaches to default reasoning. *Artificial Intelligence* 53 (1992), 209–244.
8. T.F. Gordon, The pleadings game: an exercise in computational dialectics. *Artificial Intelligence and Law*, Vol. 2, No. 4, 1994, 239–292.
9. E. Laenens and D. Vermeir, A fixed points semantics for ordered logic. *Journal of Logic and Computation* Vol. 1 No. 2, 1990, 159–185.
10. R.P. Loui, J. Norman, J. Olson, A. Merrill, A design for reasoning with policies, precedents, and rationales. *Proceedings of the Fourth International Conference on Artificial Intelligence and Law*, ACM Press, 1993, 202–211.
11. D. Nute, Defeasible logic. In D. Gabbay (ed.) *Handbook of Logic and Artificial Intelligence*, Vol. 3. Oxford University Press, 1994, 353–395.
12. L.M. Pereira and J.J. Alferes, Well–founded semantics for logic programs with explicit negation. *Proceedings ECAI-92*.
13. J.L. Pollock, Defeasible reasoning. *Cognitive Science* 11 (1987), 481–518.
14. H. Prakken, An argumentation framework in default logic. *Annals of Mathematics and Artificial Intelligence*, 9 (1993) 91–131.
15. H. Prakken, A semantic view on reasoning about priorities (extended abstract) *Proceedings of the Second Dutch/German Workshop on Nonmonotonic Reasoning*, Utrecht 1995, 152–159.
16. H. Prakken and G. Sartor, On the relation between legal language and legal argument: assumptions, applicability and dynamic priorities. *Proceedings of the Fifth International Conference on Artificial Intelligence and Law*. ACM Press 1995, 1–9.
17. G.R. Simari and R.P. Loui, A mathematical treatment of defeasible argumentation and its implementation. *Artificial Intelligence* 53 (1992), 125–157.
18. G. Vreeswijk, *Studies in defeasible argumentation*. Doctoral dissertation Free University Amsterdam, 1993.

Modal Logic for Modelling Actions and Agents

Helmut Prendinger

Int'l Forschungszentrum, Mönchsberg 2a, A-5020 Salzburg, Austria,
Email: Helmut.Prendinger@mh.sbg.ac.at
Fax: 0043–662–8044-214

Abstract. We present a modal logic approach to reasoning about actions and agents where propositional dynamic logic is used to describe plans and reasoning about belief is performed in a modal system as well. Eventually, the modal logic of belief is *embedded* within dynamic logic such that we may reason about changes of the agent's belief state as a result of executing actions. As a theory of actions, the proposed extension of standard dynamic logic has two distinguished features: it is *monotonic* and gives reasonable solutions to various 'frame problems' even when *multiagent* domains are considered. Moreover, *updates* are established as actions of a special kind within the same framework. Formal properties of the resulting system of belief and action can be proved by a methodology recently introduced by Finger and Gabbay.

Keywords: theories of actions, theories of agents, combining reasoning formalisms, formal models for reasoning.

1 Introduction

Agents that are designed to operate in real-world environments need to reason about actions and plans, and keep track of changes which are either caused by themselves or other agents present in the domain. In this paper, we propose a *modal logic* approach to describe actions and agents. Actually, we present a theory of actions and identify the mental state of the agent with a set of *beliefs*. More elaborate frameworks of mental categories can be found elsewhere [2, 25].

Actions are specified within *propositional dynamic logic* (*PDL*), the modal logic of actions and computer programs [20]. The reading of a formula $[\alpha]A$ is "after every terminating execution of α, A is true." *PDL* is especially apt to describe *compound* actions such as sequential composition of actions α and β, written $\alpha; \beta$, choice between α and β, written $\alpha + \beta$, and iteration of α, written α^*. Moreover, test, written $A?$, and 'doing nothing' (stationary waiting), denoted by λ, are considered as actions. We should stress that unlike most work on reasoning about action and change [13], we prefer a *monotonic* solution to various 'frame problems' [18, 8]. As pointed out in [23], the (global) minimization strategy employed by most nonmonotonic logics is inappropriate when the domains under consideration are populated with multiple agents. In multi-agent domains other agents may bring about changes *independently* from the agent's actions;

so deviations from the 'normal' course of events need not be minimal. Dynamic logic permits natural solutions to the combinatorial (frame) problem (or extended prediction problem) and problems of ramification and overcommitment [23]. This paper extends the range of applicability of the monotonic approach to reasoning about action to the *update* problem. We show that the minimization strategy pursued by Katsuno and Mendelzon [15] is not adequate for update actions and provide a simple solution within the dynamic logic setting.

So far, relatively little formal research has been concerned with logics of action *and* belief (or knowledge), although the agent's beliefs (or knowledge) play an important role in planning [19]. Moore [19] observes that there is a mutual dependence between knowledge and action: (i) the agent needs certain knowledge in order to take reasonable actions, and (ii) the execution of actions may modify the agent's knowledge. Rather than just combining modalities, we consider the case where a logic of belief is *embedded* within dynamic logic, thereby following the methodology of Finger and Gabbay [6] (for the general case of combining different modalities we refer the reader to [5]). Their 'externally' applied (*global*) logic is a temporal logic **T**, the underlying (or *local*) logic any completely axiomatizable logic **L**. In this paper, we add a *dynamic* dimension to a modal logic of belief. This is advantageous since in dynamic logic actions are explicit in the language; on the other hand, temporal logic considers a single, fixed plan. The approach most similar to ours is Moore's [19] theory of knowledge and action where a modal logic of knowledge is combined with a possible worlds version of situation calculus [18]. Although some simple forms of compound actions may be expressed in this framework, no proof theory is provided for the action part of the theory.

The main contribution of our paper is the establishment of dynamic action logic as a general framework to reason about action and plans that covers certain aspects of temporal logic [10], process logic [14], and even deontic logic [3]. Formally, we may thereby rely on well known proof techniques from modal logic [10]. More recently developed methodologies [6, 5] also allow us to combine different modalities. We explore the case where a dynamic modality is externally applied to an epistemic modality.

The rest of the paper is organized as follows. Section 2 gives an introduction to (standard) *PDL*. In Section 3, we adapt *PDL* for a uniform treatment of multiagent domains and extend the logic by various action constructs. The resulting logic, *propositional dynamic action logic* (*PDAL*), lends itself to elegant formulations and solutions of some well-known frame problems. Furthermore, we show that update actions can be integrated within the same framework. Then (in Section 4) our conception of update actions is compared to the famous approach of Katsuno and Mendelzon [15]. Section 5 explains how the logic of belief *B* can be embedded within dynamic logic. Finally, in Section 6, we discuss related work and hint at topics for further research.

2 Standard Propositional Dynamic Logic

Basics. We postulate as given a set $S = \{s, t, ...\}$ of possible total states of the world. A *proposition* is identified with the set of states in which it is true; an *action* α is a binary relation R_α on S, that is, a set of ordered pairs $\langle s, t \rangle$ of states where s is the initial state of some execution of α and t is the final state. Since a final state need not be uniquely determined the modelling of actions is *semantically non-deterministic*. There are two reasons for non-determinism: on the one hand, an action can always be performed in different ways, on the other, multiple agents may interfere and thus contribute to state-changes as well. This notion of non-determinism must not be confused with "non-determinism with respect to control-flow" or *procedural* non-determinism. For dynamic logic provides no machinery to give priority of executing one action over executing the other, choice is non-deterministic in this sense. In case of iteration α^*, an action α is performed some non-deterministically chosen finite number $n \geq 0$ of times.

Of course, also the *specification* of actions can be non-deterministic. In the case of flipping a coin, the result is inherently unpredictable although the range of possible outcomes is definite, we have $< \texttt{flip} > \text{HEADS} \wedge < \texttt{flip} > \neg \text{HEADS}$. Definite action specifications are constrasted in [1] to *indefinite* (or indeterminate) specifications of actions where one is ignorant of possible outcomes. Indefinite specifications are encoded by formulas like $< \alpha > (A \vee B)$ and are weaker than their definite (possibly non-deterministic) counterparts.

We start with the description of the *standard* view of propositional dynamic logic (see [10]).

Language. Let $\mathcal{P}_0 = \{p_1, p_2, ..., p, q, r, ...\}$ be a denumerably infinite set of propositional variables and $\mathcal{A}_0 = \{a_1, a_2, ..., a, b, c, ...\}$ be a denumerably infinite set of action variables. We use A, A_1, B,... to denote arbitrary formulas and α, α_1, β,... to denote arbitrary terms.

$$A ::= p \mid \neg A \mid A_1 \vee A_2 \mid [\alpha]A$$
$$\alpha ::= a \mid \alpha_1; \alpha_2 \mid \alpha_1 + \alpha_2 \mid \alpha^* \mid A?$$

where $p \in \mathcal{P}_0$, $a \in \mathcal{A}_0$, $A \in \mathcal{L}(PDL)$ and $\alpha \in \mathcal{A}$ (the set of action terms). We assume the usual definitions of $\wedge$, $\supset$, $\top$ (*verum*), $\perp$ (*falsum*), and so on. For example, $< \alpha > A \overset{\text{def}}{=} \neg[\alpha]\neg A$. In particular, we define an action constant λ called *stationary waiting* by $\lambda \overset{\text{def}}{=} \top?$.

Semantics. By a *frame* $\mathcal{F}$ we mean a structure $\mathcal{F} = \langle S, \{R_a : a \in \mathcal{A}_0\} \rangle$ such that S is a nonempty set (of world-states), $\{R_a : a \in \mathcal{A}_0\}$ is a set of binary relations, where $R_a \subseteq S \times S$ for each action variable $a \in \mathcal{A}_0$. A *model* $\mathcal{M}$ based on the frame $\mathcal{F} = \langle S, \{R_a : a \in \mathcal{A}_0\} \rangle$ is a structure $\mathcal{M} = \langle S, \{R_a : a \in \mathcal{A}_0\}, v \rangle$ where v is a function $v : \mathcal{P}_0 \to \text{Pow}(S)$ ($\text{Pow}(S)$ is the powerset of S).

A *standard* model $\mathcal{M} = \langle S, \{R_\alpha : \alpha \in \mathcal{A}\}, v \rangle$ is uniquely determined by the model $\langle S, \{R_a : a \in \mathcal{A}_0\}, v \rangle$ through the following conditions that inductively define R_α for compound action terms $\alpha \in \mathcal{A}$: $R_{\alpha;\beta} = R_\alpha \circ R_\beta$; $R_{\alpha+\beta} = R_\alpha \cup R_\beta$;

$R_{\alpha^*} = (R_\alpha)^*$ (the reflexive and transitive closure of R_α); and $R_{A?} = \{\langle s, s \rangle : \mathcal{M}, s \models A\}$. From this and the definition of λ it follows that $R_\lambda = \{\langle s, s \rangle : s \in S\}$.

Truth of a formula at a state s in a model $\mathcal{M}$ is inductively defined as follows:

- $\mathcal{M}, s \models p$ iff $s \in v(p)$ (for $p \in \mathcal{P}_0$).
- $\mathcal{M}, s \models \neg A$ iff $\mathcal{M}, s \not\models A$.
- $\mathcal{M}, s \models A \vee B$ iff $\mathcal{M}, s \models A$ or $\mathcal{M}, s \models B$.
- $\mathcal{M}, s \models [\alpha]A$ iff $\forall t (\langle s, t \rangle \in R_\alpha \Rightarrow \mathcal{M}, t \models A)$.

Logics and axiomatization. *PDL* is the smallest subset $\mathbf{L} \subseteq \mathcal{L}(PDL)$ that contains all instances of the following *axiom* schemes:

Taut	all classical propositional tautologies
K	$[\alpha](A \supset B) \supset ([\alpha]A \supset [\alpha]B)$
Comp	$[\alpha; \beta]A \equiv [\alpha][\beta]A$
Union	$[\alpha + \beta]A \equiv ([\alpha]A \wedge [\beta]A)$
Mix	$[\alpha^*]A \supset (A \wedge [\alpha][\alpha^*]A)$
Ind	$[\alpha^*](A \supset [\alpha]A) \supset (A \supset [\alpha^*]A)$
sTest	$[A?]B \equiv (A \supset B)$

and which is closed under the following *rules of inference*:

MP	from $\vdash A$ and $\vdash A \supset B$ infer $\vdash B$
NEC	from $\vdash A$ infer $\vdash [\alpha]A$

Correctness and completeness. A formula A is said to be *valid in* a model $\mathcal{M}$ (written $\mathcal{M} \models A$) iff A is true at all states s in $\mathcal{M}$, and A is *valid with respect to* a class of models $\mathbf{M}$ iff A is valid in all models $\mathcal{M} \in \mathbf{M}$. For convenience, we will use the notion of the *truth-set* of a formula A, $\|A\|^{\mathcal{M}}$, which is defined as $\{s \in S : \mathcal{M}, s \models A\}$.

A logic $\mathbf{L}$ is *correct* with respect to a class of models $\mathbf{M}$ iff all theorems of $\mathbf{L}$ are valid in all models in $\mathbf{M}$. A logic $\mathbf{L}$ is *(weakly) complete* with respect to a class of models $\mathbf{M}$ iff all formulae which are valid in $\mathbf{M}$ are theorems of $\mathbf{L}$. Finally, a logic $\mathbf{L}$ is *(weakly) determined* (characterised) by a class of models $\mathbf{M}$ iff $\mathbf{L}$ is correct and (weakly) complete with respect to $\mathbf{M}$.

Utilizing a canonical model construction and the filtration technique *PDL* can be shown to be (weakly) characterised by the class of all standard models [24, 10]; furthermore, *PDL* has the *finite model property* and is decidable (note that *PDL* is finitely axiomatizable). It is now a well known fact that the very nature of iteration prohibits *PDL* to have the compactness property.

3 Reasoning About Action in Dynamic Logic

A full-fledged theory of actions has to provide some solution to problems adherent to reasoning about action and change. The approach advocated in [23] meets this demand by adding various action constructs to *PDL*.

First of all, λ (stationary waiting) and A? (stationary test) are complemented with *non-stationary* versions, ω and τA, respectively. Thus multi-agent domains (where other agents may change states) are uniformly handled for initial and final states of the relations R_ω and $R_{\tau A}$ need not be identical. Note that actions of other agents are only considered in the semantics but are not explicit in the language. For several purposes, a stronger notion of test, called *safe* nonstationary test, is needed. The operation $\tau^\circ A$ always terminates if applied to a state in which A holds. Second, ν (the 'any' action) is included in order to express that something is true at every state in a (generated) model. Finally, two notions of preservation are introduced. Formulas of format $\mathbf{tpres}(\alpha, A)$ (terminal preservation) are intended to mean that if a fact A is true when the execution of an action α is initiated it it also true upon termination of α. If the fact A has to be true chronologically (at all the intermediate states of the performance of α) then formulas of format $\mathbf{cpres}(\alpha, A)$ (chronological preservation) are used. Observe that formulas of the latter sort are compact representations of the *ongoing* behavior of a compound action or plan.

We extend the framework of [23] by introducing *update actions*, written $\diamond A$, with intended meaning "the action consisting of updating with A." The logic resulting from incorporating all these actions and formulas to *PDL* will be called *propositional dynamic action logic (PDAL)*.

Language. The vocabulary of $\mathcal{L}(PDL)$ is enriched by the designated action variable $\omega \in \mathcal{A}_0$, the action constant ν, and the (unary) operators τ, τ° and $\diamond$; moreover the operators $\mathbf{tpres}$ and $\mathbf{cpres}$ are added to the vocabulary of $\mathcal{L}(PDL)$.

$$A ::= p \mid \neg A \mid A_1 \vee A_2 \mid [\alpha]A \mid \mathbf{tpres}(\alpha, A) \mid \mathbf{cpres}(\alpha, A)$$
$$\alpha ::= a \mid \nu \mid \alpha_1; \alpha_2 \mid \alpha_1 + \alpha_2 \mid \alpha^* \mid A? \mid \tau A \mid \tau^\circ A \mid \diamond A$$

Here, $p \in \mathcal{P}_0$, $A \in \mathcal{L}(PDAL)$, and $\alpha \in \mathcal{A}$.

Semantics. An *agent frame* $\mathcal{F}$ is a structure $\mathcal{F} = \langle S, \{R_\alpha : \alpha \in \mathcal{A}_{at}\}, T, T^\bullet, U, \mu \rangle$ such that S is a nonempty set of states, $\{R_\alpha : \alpha \in \mathcal{A}_{at}\}$ is the set of relations corresponding to atomic action terms, where the set $\mathcal{A}_{at}$ of atomic action terms is defined as $\mathcal{A}_0 \cup \{\nu\}$. The functions T ($T^\bullet$) and U mirror (safe) non-stationary test and update on the level of frames, respectively. T is a function $T : \text{Pow}(S) \to \text{Pow}(S \times S)$ such that for all $X \in \text{Pow}(S)$, $TX \subseteq \{\langle s, t \rangle : s \in X, t \in X\}$; $T^\bullet$ is like T except that it meets the further condition that for each $s \in X$, $T^\bullet X$ contains a pair $\langle s, t \rangle$ (for $t \in S$). U is a function $U : \text{Pow}(S) \to \text{Pow}(S \times S)$ such that for all $X, Y \in \text{Pow}(S)$, $UX \subseteq \{\langle s, t \rangle : t \in X\}$ and $U(X \cup Y) = UX \cup UY$.

A *standard agent frame* is an agent frame $\mathcal{F}$ where R_ν is *universal*. Hence $\mathcal{F}$ is R_ν-generated and $R_\alpha \subseteq R_\nu$ for all $\alpha \in \mathcal{A}$. Note that the restriction to R_ν-generated frames does not produce new theorems. μ is called the *smoothness* function. It is an *definitional extension* of standard agent frames, a function $\mu : \{R_\alpha : \alpha \in \mathcal{A}_{at}\} \times \text{Pow}(S) \to \text{Pow}(S)$ such that for all $s \in S$, $X \in \text{Pow}(S)$: $s \in \mu(R_\alpha, X)$ iff $s \in X \Rightarrow (\{t : \langle s, t \rangle \in R_\alpha\} \subseteq X)$.

A *standard agent model* $\mathcal{M}$ based on a standard agent frame $\mathcal{F}$ is a structure $\mathcal{M} = \langle S, \{R_\alpha : \alpha \in \mathcal{A}\}, T, T^\bullet, U, \mu_t, \mu_c, v \rangle$ where all conditions for standard models are satisfied, $R_{\tau A} = T\|A\|^{\mathcal{M}}$, and $R_{\tau^\circ A} = T^\bullet \|A\|^{\mathcal{M}}$, and $R_{\diamond A} = U\|A\|^{\mathcal{M}}$.

Loosely speaking, T ($T^\bullet$, U) assign to the proposition denoted by A the action consisting of (safely) non-stationary verifying that A is true (updating with A), respectively.

Moreover, the following functions are introduced:

1. μ_t is the function $\mu_t : \{R_\alpha : \alpha \in \mathcal{A}\} \times \mathrm{Pow}(S) \to \mathrm{Pow}(S)$ such that

$$s \in \mu_t(R_\alpha, X) \quad \text{iff} \quad s \in X \Rightarrow (\{t : \langle s,t \rangle \in R_\alpha\} \subseteq X).$$

Note that $\mu_t(R_\alpha, X) = \mu(R_\alpha, X)$ for *atomic* action terms $\alpha \in \mathcal{A}_{at}$.

2. μ_c is the function $\mu_c : \{R_\alpha : \alpha \in \mathcal{A}\} \times \mathrm{Pow}(S) \to \mathrm{Pow}(S)$ such that

$$\mu_c(R_\alpha, X) = \mu_t(R_\alpha, X)$$

for all *elementary* action terms $\alpha \in \mathcal{A}_{el}$ where $\mathcal{A}_{el}$ is defined as $\mathcal{A}_{at} \cup \{A?, \tau A, \tau^\circ A, \diamond A : A \in \mathcal{L}(PDAL)\}$. For compound action terms, μ_c is inductively defined as follows:

$$\mu_c(R_{\alpha;\beta}, X) = \mu_c(R_\alpha, X) \cap \{s : \{t : \langle s,t \rangle \in R_\alpha\} \subseteq \mu_c(R_\beta, X)\}$$
$$\mu_c(R_{\alpha+\beta}, X) = \mu_c(R_\alpha, X) \cap \mu_c(R_\beta, X)$$
$$\mu_c(R_{\alpha^*}, X) = \bigcap_{n \geq 0} \mu_c(R_{\alpha^n}, X)$$

Finally, v is a function $v : \mathcal{P}_0 \to \mathrm{Pow}(S)$.

We define the meanings of formulas and terms by adding the following clauses:

- $\mathcal{M}, s \models \mathbf{tpres}(\alpha, A)$ iff $s \in \mu_t(R_\alpha, \|A\|^{\mathcal{M}})$.
- $\mathcal{M}, s \models \mathbf{cpres}(\alpha, A)$ iff $s \in \mu_c(R_\alpha, \|A\|^{\mathcal{M}})$.

So the following conditions are satisfied by $\mathcal{M}$ (the reference to models is left tacit): For all $s, t \in S$, $\alpha \in \mathcal{A}$:

$$\|\mathbf{cpres}(\alpha;\beta, A)\| = \|\mathbf{cpres}(\alpha, A)\| \cap \{s : \{t : \langle s,t \rangle \in R_\alpha\} \subseteq \|\mathbf{cpres}(\beta, A)\|\}$$
$$\|\mathbf{cpres}(\alpha+\beta, A)\| = \|\mathbf{cpres}(\alpha, A)\| \cap \|\mathbf{cpres}(\beta, A)\|$$
$$\|\mathbf{cpres}(\alpha^*, A)\| = \bigcap_{n \geq 0} \|\mathbf{cpres}(\alpha^n, A)\|$$

Logics and axiomatization. In addition to *PDL* axioms and rules, *PDAL* contains all instances of the axiom schemes for *(safe) nonstationary test, update, the 'any' action,* and the concepts of terminal and chronological *preservation:*

nTest.1	$< \tau A > \top \supset A$
nTest.2	$[\tau A]A$
snTest.1	$< \tau^\circ A > \top \supset A$
snTest.2	$[\tau^\circ A]A$
snTest.3	$A \supset < \tau^\circ A > \top$

Update.1	$[\diamond A]A$
Update.2	$[\diamond(A \vee B)]C \equiv ([\diamond A]C \wedge [\diamond B]C)$
Any.1	$[\nu]A \supset A$
Any.2	$[\nu]A \supset [\nu][\nu]A$
Any.3	$<\nu>[\nu]A \supset A$
Any.4	$[\nu]A \supset [\alpha]A$
tPres	$\mathbf{tpres}(\alpha, A) \equiv (A \supset [\alpha]A)$
cPres	$\mathbf{cpres}(\alpha, A) \equiv \left(A \supset \bigwedge_{\beta \in I(\alpha)} [\beta]A \right)$

The last mentioned axiom scheme makes use of the concept of *initial strings* of R_α. Define a function $I : \mathcal{A} \to \mathrm{Pow}(\mathcal{A})$ that assigns to each action term $\alpha \in \mathcal{A}$ the *set* of all action terms that denote *initial strings* of R_α. For *elementary* action terms, $\alpha \in \mathcal{A}_{el}$, we have $I(\alpha) = \{\alpha\}$, that is, an elementary action term has itself as its only initial string. For compound action terms, the auxiliary definition $(\alpha; I(\beta)) \stackrel{\mathrm{def}}{=} \{(\alpha; \gamma) : \gamma \in I(\beta)\}$ is needed. Then define $I(\alpha; \beta) = I(\alpha) \cup (\alpha; I(\beta))$, $I(\alpha + \beta) = I(\alpha) \cup I(\beta)$ and $I(\alpha^*) = (\alpha^*; I(\alpha))$. For instance, the initial strings of $R_{\alpha;\beta}$ are all those of R_α *plus* all those obtained from concatenating initial strings of R_β to R_α. Pratt [21] introduces formulas $\mathbf{throughout}(\alpha, A)$ similar to $\mathbf{cpres}(\alpha, A)$ which mean that formula A is true throughout (at every state of) the execution of action α. It turned out that the preservation concept of Pratt leads to a fairly complicated logic, so-called *process* logic [14]. In *PDAL*, a *definitional extension* of action models (the function μ_c) is of essentially the same expressibility.

Correctness and completeness. The following theorem can be proved by building upon the proof of the determination result for *PDL* and the fact that models for the language of *PDAL* are definitional extensions of the models for this language without the preservation operators (for details, see the extended version of the paper [22]).

Theorem 1. *(1) PDAL has the finite model property, (2) PDAL is (weakly) determined by the class of finite standard agent models and (3) PDAL is decidable.*

Example 1 (manufacturing example). We illustrate the expressibility of our framework with an example from the manufacturing domain. Consider an agent (robot) working at a car-manufacturing plant. The agent is supplied with a driver and a camera. The agent's task consists in turning screws until they are flush with the car-body. The manufacturing domain is described as follows: `screw` stands for "the action consisting of turning the screw", FULL for "the batteries are fully charged" and S_n for "the screwhead is n units apart from the car-body". Observe that, if $n = 0$, the screwhead is flush with the car-body. Of course, we assume that the S_i's are mutually exclusive, that is, $\neg(S_i \wedge S_j)$ whenever $i \neq j$. We define $\tau^{\diamond>} \stackrel{\mathrm{def}}{=} \tau^{\diamond}(\neg S_0)$ and $\tau^{\diamond=} \stackrel{\mathrm{def}}{=} \tau^{\diamond}(S_0)$. So we assume that the camera's operations of testing whether the screwhead is still apart from the car-body are safe. The entire plan our robot has to execute is $((\tau^{\circ>};\mathbf{screw})^*;\tau^{\circ=})$. If this plan terminates, it will terminate in S_0 (else $\tau^{\circ=}$ will abort). The crucial point is to find adequate conditions which ensure termination of the plan. For convenience, we define $\mathbf{bound}(\alpha, A) \overset{\text{def}}{=} A \wedge \mathbf{cpres}(\alpha, A)$, that is, A is a *boundary* condition during the performance of α.

(1) $\mathbf{bound}((\tau^{\circ>};\mathbf{screw})^*;\tau^{\circ=}, \mathrm{FULL})$

(2) $[\nu]((\mathrm{FULL} \wedge S_n) \supset (<\mathbf{screw}>\top \wedge [\mathbf{screw}]S_{n-1}))$ (for all $n > 0$)

(3) $[\nu]\mathbf{tpres}(\tau^{\circ>}, S_n)$ (for all $n > 0$)

The first assumption has the effect to ensure that for both the driver and the camera to work the batteries must be loaded during the *whole* performance of the plan. We also assume (2) as a premise, that is, the action $\mathbf{screw}$ applied to a state where the batteries are full and the screwhead is n units from the car-body will terminate and after termination the screwhead is $n - 1$ units from the car-body. Since our test operation $\tau^{\circ>}$ is non-stationary, we must add the further assumption (3) that guarantees that the test operation performed by the camera has no (destructive) effect on the position of the screw.

Within *PDAL*, we may prove the desired result ($\vdash$ is the deducibility relation of *PDAL*)

$$(1), (2), (3), S_n \vdash [(\tau^{\circ>};\mathbf{screw})^n]S_0 \wedge < (\tau^{\circ>};\mathbf{screw})^n;\tau^{\circ=} >\top,$$

that is, (i) whenever the action consisting of n times verifying $n > 0$ and then turning the screwhead terminates, the screwhead is fixed (flush with the car-body) and, (ii) the entire plan of performing this action *some number* of times and then verifying that the screwhead is fixed, will terminate if this number is n (for the proof, see [22]).

By way of example, we show how the *combinatorial* (frame) problem is solved in the dynamic logic setting. It is the problem of having to state frame axioms of format $p \supset [\alpha]p$ for *each* fact-action pair p-α. Nonmonotonic logics [13] circumvent this computational problem by introducing a 'blanket' frame axiom which covers all (atomic) facts and (elementary) actions. If one abstracts from the different syntactical appearance of the blanket frame axiom in nonmonotonic logics, an informal rendering of the frame assertion might read "if a fact p is true in a situation s and the action α is not abnormal wrt (with respect to) p when performed in s, then p is still true after termination of α" [9]. An action α is called *abnormal wrt p in s* if α reverses the truth-value of p. For instance, according to the policy of *causal* minimization (of abnormalities), a fact changes its truth-value if *and only if* a terminating action (performed by the agent) causes it to do so [16]. Our approach does not depend on normality assumptions of this sort. As said in the Introduction, these assumptions seem unnatural to us when multiagent domains are considered. On the contrary, we only force certain facts to persist, generally those which are preconditions to ensuing actions or effects of actions while other facts may *vary* due to activity of other agents. This counts as our solution to the *overcommitment problem* which is the problem of being overcommitted to a universe where actions of other agents have not occurred. For example, the boundary condition (1) states that FULL is chronologically preserved wrt the plan $(\tau^{\circ >}; \texttt{screw})^*; \tau^{\circ =}$, including the initial state of execution. Here, condition (1) is the weakest frame axiom which allows to deduce the result from conditions (1)–(3). Note that we do not need to apply a blanket frame axiom over and over again if a set of facts is to be propagated through a long sequence of actions. All we have to do is to derive obvious (monotonic) consequences of a single frame axiom like condition (1).

More recently, Etherington et al. [4] introduced a methodology of *scoped* nonmonotonic reasoning which does not suffer from the overcommitment problem. Instead of minimizing 'globally' (minimize changes of *all* facts if not forced otherwise), they propose to restrict the scope of reasoning to some pre-defined set of properties (for a thorough comparision with our account, see [22]).

4 The Update Problem in Dynamic Logic

Katsuno and Mendelzon [15] argue that a database can be modified in two fundamentally different ways. The first kind of modifying a database is called *revision*. When revising a database we obtain new information about a *static* world. For example, if we try to solve a crime, we may come to the conclusion that the butler is guilty (which we did not conclude before). The second type of database adaption, called *update*, is used when we want to bring the database up to date because the world described by it has *changed*. The fact that the master of the house has been murdered is certainly a case of update (the crime example is borrowed from [11]). The update problem is one of reasoning about action and change [26]. As opposed to revision, the minimization strategy—changing the database *as little as possible*—is not sensible for updates.

Example 2 (book and magazine example). Consider the book and magazine example similar to that in [15]. A room has three objects in it: a book, a magazine and a table. Suppose the database φ says either the book or the magazine is on the floor, but not both. Let B denote "only the book is on the floor" and M "the magazine is on the floor." We have $\varphi \equiv ((B \wedge \neg M) \vee (\neg B \wedge M))$. Now we order a robot to put the book on the floor. According to the revision postulate

(R2) If $\phi \wedge \mu$ is satisfiable then $\phi \circ \mu \equiv \phi \wedge \mu$.

of Katsuno and Mendelzon [15], the result of $\varphi \circ B$, the revision of φ by B, is $B \wedge \neg M$. In this way, the database is changed minimally. But the world described by φ has undergone a change due to an action. So we rather *update* φ with B, written $\varphi \diamond B$, in this situation. There is no argument why the magazine should *not* be on the floor after the performance of the update action.

In dynamic logic, one cannot mix up revisions with updates since only the latter adaption involves a state-transition. As said before, revision presumes a static world. More precisely, the propositions used to describe the state are static, that is, they are not implicitly dependent on the current state like "the book is on the floor." On the other hand, "at time t_0, the book is on the floor" is a static proposition (the distinction is due to [7]). Concerning the update operation, we introduced an operator ⋄ into dynamic logic which takes formulas to terms. Recall that the relation corresponding to ⋄A does not restrict $R_{\diamond A}$ such that only 'minimal' transitions $\langle s, t \rangle \in R_{\diamond A}$ are considered where t differs from s 'as little as possible'. If certain facts are to be maintained after all, this may be accomplished in the usual fashion, with the preservation operators.

Example 3 (disjunctive updates). The next example shows that the update postulates proposed by Katsuno and Mendelzon [15] assume that the world changes minimally after update. The example is borrowed from Goldszmidt and Pearl and reported in [26]. Here a robot is ordered to paint a wall in blue or white. We write BLUE for "the wall is painted in blue" and WHITE for "the wall is painted in white." Let $\varphi \equiv$ WHITE. Observe that the disjunctive update is already satisfied by the database. Application of the update postulate [15]

(U2)   If ϕ implies μ then $\phi \diamond \mu$ is equivalent to ϕ.

entails that the wall remains unchanged (white) after the update action. This conclusion seems too strong. All we know is that the wall is blue or white as a result of the update BLUE ∨ WHITE. It is easily seen that the modelling of updates in dynamic logic does not fall prey to the conclusion endorsed by (U2).

5   Combining *PDAL* with a Logic of Belief

We show how dynamic logic can be added externally to a modal logic of belief. In other words, the epistemic part of the logic is embedded inside dynamic logic. We get the following picture as illustrated in Fig. 1: the logic of belief is the *local* logic while dynamic logic is the *global* logic which describes the evolution of the agent's beliefs as a result of the execution of actions. Note that states $s \in S$ which are primitives in the modelling of actions in dynamic logic are now better termed *information states* (or structured states).

Concerning the logic of belief B, we decided for the popular modal system **KD45** (or *weak* **S5**) which is considered an appropriate logic to capture the propositional attitude of *belief*, for instance, by Halpern and Moses [12]. **KD45** contains both the 'positive introspection' axiom **4** "the agent believes what it believes" and the 'negative introspection' axiom **5** "the agent believes what it does not believe." Moreover, by axiom **D**, the agent is assumed to have no inconsistent beliefs, although the agent is not required to only believe true facts (a distinguished feature of belief as opposed to knowledge, see [12]). Semantically, we assume as given a set $W = \{w, w', ...\}$ of (possible) *worlds*. The relation $R \subseteq W \times W$ picks out those worlds $\{w' : \langle w, w' \rangle \in R\}$ which are possible (consistent) wrt what the agent believes in world w. An agent believes A, written

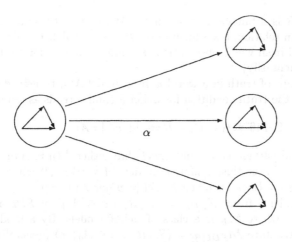

Fig. 1. *Relation R_α on* information *states corresponding to action α.*

BA, if A is true in all worlds the agent considers possible relative to the world it inhabits.

We proceed with the more formal developement which closely resembles the approach in [6] but adds a *dynamic* dimension to the logic of belief B.

Language. Let $\mathcal{P}_0$ be the set of propositional variables as above.

$$A ::= p \mid \neg A \mid A_1 \vee A_2 \mid \mathbf{B}A$$

where $p \in \mathcal{P}_0$ and $A \in \mathcal{L}(B)$. The definition of well-formed expressions of the logic resulting from externally adding dynamic logic, say *PDL*, to a logic of belief B simultaneously defines the notions of formula and term. Here, $ML(B)$ denotes the set of *monolithic formulas* of $\mathcal{L}(B)$, that is, formulas A that are not built up by any of the Boolean connectives. As shown in [6], the double parsing problem is thereby avoided. For instance, $\mathbf{B}A$ is in $ML(B)$ while $A \wedge B$ is not. For clarity, we give an inductive definition to define the syntax of the language rather than using the notation of Backus-Naur form as before. The sets $\mathcal{L}(PDL[B])$ and $\mathcal{A}$ are the smallest sets such that:

1. $A \in ML(B) \Rightarrow A \in \mathcal{L}(PDL[B])$;
2. $A, B \in \mathcal{L}(PDL[B]) \Rightarrow \neg A, (A \vee B) \in \mathcal{L}(PDL[B])$;
3. $\alpha \in \mathcal{A}, A \in \mathcal{L}(PDL[B]) \Rightarrow [\alpha]A \in \mathcal{L}(PDL[B])$;
4. $\alpha, \beta \in \mathcal{A} \Rightarrow \alpha;\beta, \alpha + \beta, \alpha^* \in \mathcal{A}$;
5. $A \in \mathcal{L}(PDL[B]) \Rightarrow A? \in \mathcal{A}$.

It is a straightforward exercise to set up the definition for $\mathcal{L}(PDAL[B])$.

Semantics. We start with the semantics of the underlying logic of belief B. A *belief model* $\mathcal{M}_B$ is a structure $\mathcal{M}_B = \langle W, R, l, w_c \rangle$ such that W is a set of worlds and R is a binary relation on W. We take R to be a serial, transitive and Euclidean relation. l is a function $l : \mathcal{P}_0 \to \mathrm{Pow}(W)$. Finally, $w_c \in W$ is the current world from which observations are made (motivation for this parameter is given momentarily).

The concept of truth in a world w in a model $\mathcal{M}_B$ is defined inductively as usual, where the truth condition for belief formulas reads as follows.

- $\mathcal{M}_B, w \models \mathbf{B}A$ iff $\forall t\, (\langle w, w' \rangle \in R \Rightarrow M, w' \models A)$.

The additional parameter w_c guarantees that either $\mathcal{M}_B \models A$ or $\mathcal{M}_B \models \neg A$ for all $A \in \mathcal{L}(B)$. To see this, assume a model $\mathcal{M} = \langle W, R, l \rangle$ where $W = \{w, w'\}$, $R = \emptyset$ and $l(p) = \{w\}$. Then neither $\mathcal{M} \models p$ nor $\mathcal{M} \models \neg p$.

Now consider a frame $\mathcal{F}_{PDL} = \langle S, \{R_a : a \in \mathcal{A}_0\} \rangle$ of PDL and a function $g : S \to \mathbf{M}_B$ where $\mathbf{M}_B$ is a class of belief models. By a model of $PDL[B]$ we mean a structure $\mathcal{M}_{PDL[B]} = \langle S, \{R_a : a \in \mathcal{A}_0\}, g \rangle$ (generalization to 'full' $PDAL$ is straightforward). The truth conditions are inductively defined by the following conditions:

- $\mathcal{M}_{PDL[B]}, s \models A$ ($A \in ML(B)$) iff $g(s) = \mathcal{M}_B$ and $\mathcal{M}_B \models A$.
- $\mathcal{M}_{PDL[B]}, s \models \neg A$ iff $\mathcal{M}_{PDL[B]}, s \not\models A$.
- $\mathcal{M}_{PDL[B]}, s \models A \vee B$ iff $\mathcal{M}_{PDL[B]}, s \models A$ or $\mathcal{M}_{FDL[B]}, s \models B$.
- $\mathcal{M}_{PDL[B]}, s \models [\alpha]A$ iff $\forall t\, (\langle s, t \rangle \in R_\alpha \Rightarrow \mathcal{M}_{PDL[B]}, t \models A)$.

Logics and axiomatization. B is the smallest subset $\mathbf{L} \subseteq \mathcal{L}(B)$ containing all instances of the following axiom schemes and is closed under the rules of inference **MP** and **NEC$_B$** (from $\vdash_B A$ infer $\vdash_B \mathbf{B}A$).

Taut	all classical propositional tautologies
K$_B$	$\mathbf{B}(A \supset B) \supset (\mathbf{B}A \supset \mathbf{B}B)$
D	$\neg \mathbf{B}\bot$
4	$\mathbf{B}A \supset \mathbf{B}\mathbf{B}A$
5	$\neg \mathbf{B}A \supset \mathbf{B}\neg \mathbf{B}A$

$PDAL[B]$ contains all instances of axiom schemes and rules of $PDAL$ and the additional rule of inference (called **Preserve** in [6])

LIFT   from $\vdash_R A$ infer $\vdash_{PDL[B]} A$ (for all $A \in \mathcal{L}(B)$).

Correctness and completeness. Concerning the logic of belief B (**KD45**), it is well known that B has the finite model property, is determined by the class of belief models and is decidable (see, for example, [12]). The following statement can be proved by means of the techniques developed in the paper of Finger and Gabbay [6].

Theorem 2. *(1) $PDAL[B]$ is (weakly) determined by the class of finite standard agent models and (2) $PDAL[B]$ is decidable.*

Example 4 (acidity test). We give an alternative solution to an example of Moore [19]. Consider an agent which is supposed to find out about the acidity of a solution by means of litmus paper. The point is that the agent is able to *deduce* the unobservable precondition, whether the solution is acid or alkaline, from the observable result, whether the paper turned red or blue. We write ACID for "the solution is acid" and RED for "the litmus paper is red". Assume the following three statements as premises: (1) ACID $\supset$ [τACID]BRED; (2) B(RED $\supset$ ACID); (3) ACID. First, we assume that if the solution is acid then after verifying that the solution is acid the agent believes that the paper is red. The second premise says that the agent believes that there is a connection between the color of the paper and the the solution being acid. Finally, we suppose that the solution is initially acid. With the proof theory of $PDAL[B]$ we get ($\vdash$ the deducibility relation of $PDAL[B]$)

$$(1),(2),(3) \vdash [\tau\text{ACID}]\mathbf{B}\text{ACID}$$

and, by axiom **D** and the same premise set, the weaker statement [τACID]$\neg$**B**$\neg$ACID. Of course, we cannot deduce [τACID]ACID, so destructive tests are covered as well. For nondestructive tests we may add the condition **bound**(τACID, ACID). In this case, the agent will even *know* that the solution is acid, by way of the generally accepted definition $\mathbf{K}A \stackrel{\text{def}}{=} \mathbf{B}A \wedge A$, where **K** is a knowledge operator.

6 Discussion

By extending standard *PDL* with natural action constructs, we have shown that dynamic logic provides a formally sound basis for exploring problems which are related to the frame problem and the update problem. We conceive of updates as a kind of 'change recording' actions and thereby capture a simple form of database modification without (so far) experiencing counter-intuitive results. Yet, our approach suggests no way to handle *revisions* which are intimately connected with some kind of minimization.

Interestingly, domain and plan constraints are easily encoded in *PDAL*. *Plan constraints* ensure the 'clean' behavior of a plan by imposing restrictions on accessible states (compare to premise (1) in Example 1). In *temporal plan theory* [17] this is achieved by stating temporal properties. For instance, the property guarantee states that always during the performance of a plan $\alpha_1, ..., \alpha_n$ where A holds, there exists a state which is attainable via β and satisfies B. Within *PDAL*, guarantee is expressed as **bound**($\alpha_1, ..., \alpha_n, A \supset <\beta> B$). For a dynamic rendering of other temporal properties, see [22]. *Domain constraints* make assertions about facts that are intended to be invariant over every performance of every action. Here, the 'any' action ν applies (recall premises (2) and (3) in Example 1).

In the context of conceptual modelling of databases, Dignum and Meyer [3] observe that (integrity) constraints often possess a *deontic* flavor. An undesirable state is denoted by the propositional constant V. Then $[\alpha]V$ is motivated as the dynamic logic interpretation of the prescriptive sentence $\mathbf{F}\alpha$ with informal

538

reading "it is forbidden to perform α" (**F** a deontic modality). In the parlance of [3], we could express the sentence "if the agent is in an undesirable state then executing a plan $\alpha_1, ..., \alpha_n$ is forbidden and so is the execution of all initial strings of the plan" by $\mathbf{cpres}(\alpha_1, ..., \alpha_n, V)$.

Admittedly, our theory of agents is still in a very preliminary state. Within our setting, the mental state of an agent consists of a single component, belief. In addition to belief and time, Shoham [25] takes into account *obligation* and *capability* as modalities, and Cohen and Levesque [2] *goals* and *intention*. It remains to be seen whether we can integrate mental categories different from belief to our framework while retaining the restriction that mental modalities are embedded within dynamic logic. This is not even true for Moore's [19] theory of knowledge and action which has the equivalents of both $[\alpha]KA$ and $K[\alpha]A$ in the logic. In the case of Example 4 (acidity test) we could provide a solution without admitting formulas $\mathbf{B}[\alpha]A$ while Moore's solution relies on formulas of this format.

References

1. Craig Boutilier and Nir Friedman. Nondeterministic actions and the frame problem. Unpublished manuscript, 1995.
2. Philip R. Cohen and Hector J. Levesque. Intention is choice with commitment. *Artificial Intelligence*, 42:213–261, 1990.
3. F. P. M. Dignum and J.-J. Ch. Meyer. Negations of transactions and their use in the specification of dynamic and deontic integrity constraints. In M. Z. Kwiatkowska, M. W. Shields, and R. M. Thomas, editors, *Proceedings of the International BCS-FACS Workshop on Semantics for Concurrency*, pages 61–80, University of Leicester, UK, 1990.
4. David W. Etherington, Sarit Kraus, and Donald Perlis. Nonmonotonicity and the scope of reasoning. *Artificial Intelligence*, 52:221–261, 1991.
5. Kit Fine and Gerhard Schurz. Transfer theorems for stratified multimodal logics. In J. Copland, editor, *Logic and Reality. Essays in Pure and Applied Logic. In Memory of Arthur Prior*. Oxford University Press, 1995.
6. Marcelo Finger and Dov M. Gabbay. Adding a temporal dimension to a logic system. *Journal of Logic, Language and Information*, 1:203–233, 1992.
7. Nir Friedman and Joseph Y. Halpern. A knowledge-based framework for belief change, part II: Revision and update. In *Proceedings Fourth International Conference on Principles of Knowledge Representation and Reasoning (KR-94)*, pages 190–201, 1994.
8. Michael P. Georgeff. Many agents are better than one. In F. M. Brown, editor, *Proceedings of the 1987 Workshop on the Frame Problem in Artificial Intelligence*, pages 59–75, Lawrence, Kansas, 1987.
9. Matthew L. Ginsberg. Computational considerations in reasoning about action. In James Allen, Richard Fikes, and Erik Sandevall, editors, *Proceedings of the Second International Conference On Principles of Knowledge Representation and Reasoning*, pages 250–261, Cambridge, Massachusetts, 1991.
10. R. Goldblatt. *Logics of Time and Computation*. Center for the Study of Language and Information Lecture Notes 7. Leland Stanford Junior University, Stanford, 1987.

11. Gösta Grahne. Updates and counterfactuals. In James Allen, Richard Fikes, and Erik Sandevall, editors, *Proceedings of the Second International Conference On Principles of Knowledge Representation and Reasoning*, pages 269–276, Cambridge, Massachusetts, 1991.

12. Joseph Y. Halpern and Yoram Moses. A guide to the modal logics of knowledge and belief: preliminary draft. In *Proceedings of the Ninth International Joint Conference on Artificial Intelligence*, pages 480–490, 1985.

13. Steve Hanks and Drew McDermott. Nonmonotonic logic and temporal projection. *Artificial Intelligence*, 33:379–412, 1987.

14. David Harel, Dexter Kozen, and Rohit Parikh. Process logic: expressiveness, decidability, completeness. *Journal of Computer and System Sciences*, 25:144–170, 1982.

15. Hirofumi Katsuno and Alberto O. Mendelzon. On the difference between updating a knowledge base and revising it. In James Allen, Richard Fikes, and Erik Sandevall, editors, *Proceedings of the Second International Conference On Principles of Knowledge Representation and Reasoning*, pages 387–394, Cambridge, Massachusetts, 1991.

16. Vladimir Lifschitz. Formal theories of action. In Matthew L. Ginsberg, editor, *Readings in Nonmonotonic Reasoning*, pages 410–432. Morgan Kaufmann Publishers, Inc., Los Altos, California, 1987.

17. Zohar Manna, Massimo Paltrinieri, and Richard Waldinger. Temporal planning. In *Workshop on Reasoning about Action and Change at IJCAI-93*, pages 37–41, Chambéry, France, 1993.

18. John McCarthy and Patrick J. Hayes. Some philosophical problems from the standpoint of Artificial Intelligence. In B. L. Webber and N.J. Nilsson, editors, *Readings in Artificial Intelligence*, pages 431–450. Morgan Kaufmann Publishers, Inc., Los Altos, California, 1981. Originally published in 1969.

19. Robert C. Moore. A formal theory of knowledge and action. In Jerry R. Hobbs and Robert C. Moore, editors, *Formal Theories of the Commonsense World*, pages 317–358. Ablex Publishing Corporation, Norwood, New Jersey, 1985.

20. Vaughan R. Pratt. Semantical considerations on Floyd-Hoare logic. In *Proceedings of the 17th Annual IEEE Symposium on Foundations of Computer Science, 1976*, pages 109–121, 1976.

21. Vaughan R. Pratt. A practical decision method for propositional dynamic logic. In *Proceedings 10th Annual Association for Computing Machinery (ACM) Symposium on Theory of Computing*, pages 326–337, 1978.

22. Helmut Prendinger and Gerhard Schurz. Reasoning about action and change: A dynamic logic approach. To appear in *Journal of Logic, Language and Information*.

23. Helmut Prendinger and Gerhard Schurz. Reasoning about action in dynamic logic. In *Proceedings Second World Conference on the Fundamentals of Artificial Intelligence (WOCFAI 95)*, pages 355–366, Paris, France, 1995.

24. Krister Segerberg. A completeness theorem in the modal logic of programs. In Tadeusz Traczyk, editor, *Universal Algebra and Applications*, volume 9 of *Banach Center Publications*, pages 31–46, Warsaw, 1982.

25. Yoav Shoham. Agent-oriented programming. *Artificial Intelligence*, 60:51–92, 1993.

26. Alvaro Del Val and Yoav Shoham. Deriving properties of belief update from theories of action. *Journal of Logic, Language, and Information*, 3:81–119, 1994.

Formalization of Reasoning about Default Action (Preliminary Report)*

Anna Radzikowska

Institute of Mathematics, Warsaw University of Technology
Plac Politechniki 1, 00–661 Warsaw, Poland
Email: annrad@im.pw.edu.pl

Abstract. In this paper we consider domains of actions involving causal chains of actions and default actions. In order to represent these domains, we define a language $\mathcal{AL}_0$ that extends the language $\mathcal{A}$ introduced by Gelfond and Lifschitz. We provide a translation from $\mathcal{AL}_0$ into circumscriptive theories and show that this translation is sound and complete relative to the semantics of $\mathcal{AL}_0$. Our approach is related to Sandewall's PMON logic and the occlusion concept. The analysis assumes a discrete linear model of time.

1 Introduction

Reasoning about action and change is one of the central problems in the theory of knowledge representation. In this area of research, it turned out to be extremely difficult to discuss the possibilities and limitations of the available methods in a precise and general way. Some of the recent works (see, for instance, [16], [12], [7] and [8]) attempt to overcome this problem by discussing action domains in a methodical and theoretically sound way. Sandewall ([16]) proposed a general framework to characterize various classes of dynamic systems. As part of this framework, several logics of preferential entailment are introduced and assessed for particular classes of action domains. The broadest class he investigates, $\mathcal{K}$-IA (and its associated reasoning method PMON), covers many of the most problematic examples traditionally studied in the AI literature.

Gelfond and Lifschitz ([4]) presented another approach to this issue. They proposed a declarative high-level action language $\mathcal{A}$ and defined a provably sound and complete translation of domains represented in this language into extended logic programs. The language $\mathcal{A}$ was later extended into the languages $\mathcal{A}_C$ ([1]), $\mathcal{A}_C^+$ ([17]), $\mathcal{AR}$ ([5]) and recently $\mathcal{ARD}$ ([6]), $\mathcal{AC}$ ([18]), and appropriate sound and complete translations into logic-based formalisms were provided. As it has been recently pointed out ([11]), action languages can be of interest in their own right, useful for better understanding of various aspects of commonsense knowledge and reasoning problems related to actions.

---

* This paper has been supported by the KBN Grant 3 P406 019 06.

Most of the existing approaches to action and change attract no much interest in default aspects of actions. In our previous papers ([14],[15]) we considered domains where actions with default effects occur.[2] In our everyday life we often deal with causal sequences of actions where one action, typically or necessarily, invokes another one (e.g. preparing tea usually follows boiling water). Also, we often face situations which (necessarily or typically) trigger some actions (e.g. when I found my car stolen, I immediately call the police). Finally, it is quite natural to anticipate some action at a particular point in time (e.g. the train from London typically arrives at 2pm). Then, in order to admit more realistic descriptions of dynamically changing words, it seems reasonable to consider domains where both causal chains of actions and default actions occur.

In this paper we consider a twofold extension of Sandewall's $\mathcal{K}$-IA class of action domains. First, causal chains of actions are permitted. Second, default actions may occur. In order to represent these domains, we introduce an action language $\mathcal{AL}_0$ that is another extension of the language $\mathcal{A}$ (the subscript 0 is used to distinguish this language from a more expressive language $\mathcal{AL}$ defined in the full version of the paper). Our aim is to define a translation from $\mathcal{AL}_0$ into a circumscriptive scheme and show soundness and completeness of this translation relative to the semantics of $\mathcal{AL}_0$. The proposed method of reasoning is based on Sandewall's PMON logic and the occlusion concept. A simple linear discrete model of time is assumed.

This paper is organized as follows. In Section 2 we characterize action domains under consideration and propose a reasoning method, GMD, adequate for these domains. Section 3 describes the action language $\mathcal{AL}_0$. In Section 4 we briefly present a fluent logic PFL. Next, in Section 5, we provide a circumscriptive scheme that can be viewed as a syntactic characterization of the GMD reasoning method. In Section 6 we show a translation from $\mathcal{AL}_0$ into circumscriptive theories and provide theorems stating soundness and completeness of this translation. The paper is completed with concluding remarks and some options for future work.

Due to the limited amount of space, proofs of theorems are relegated to the full version of the paper.

2 Domains of Actions

First, let us recall the key characteristics of the $\mathcal{K}$-IA class of action domains.
1. the commonsense law of inertia applies to all fluents;
2. all actions are known, together with their effects;
3. nondeterministic actions and actions with duration are allowed;
4. order of actions and any state of a system may be partially specified only.

In this paper we consider a twofold extension of the $\mathcal{K}$-IA family of action domains. First, we involve *causal chains of actions*.[3] This notion reflects a sequence

---

[2] Formalization of actions with typical effects within a dynamic/epistemic multi-agent system is presented in [3].

[3] A specific case of causal chains of actions has been considered in [15].

$<A_1, A_2, \ldots >$ (not necessarily finite) of actions such that an action A_i results from the performance of some previous action(s) A_j, $(1 \leq j < i)$. Specifically, an action A_j may either directly invoke A_i or it may cause a situation which triggers an execution of A_i.[4] Therefore A_i may be thought of as a *delayed, dynamic* effect of the causal action A_j.

When a performance of one action results from executing of another one, the time of its beginning may not be specified explicitly. For instance, it may be a consequence of some domain constraints. To simplify the current discussion we assume that the beginning of the resulting action is fixed relative only to a timepoint when the causal action starts. Similarly, a specific situation may trigger an action some time later. Also, for the purpose of simplicity, we assume that the triggered action is to be performed immediately, i.e. it starts in the causal situation.

Causal chains of actions may lead to concurrent actions. Due to the complexity of this problem we consider sequential scenarios only, i.e. domains where at each point in time at most one action may be performed.

The novelty of this approach is involving *default actions* into domains. This notion reflects an action which is *usually* performed either in particular situation or it is a typical consequence of some other action(s). An execution of an action of this kind may be viewed as dependent on a state of a system when the action begins. If the state is *normal* with respect to the particular action then the action may be executed; otherwise not. In fact, it is a system decision whether a particular state is normal relative to a particular action or not and it is not a matter of the action itself (i.e. its precondition, the adequacy of its qualification or its behaviour). Sometimes such a decision may be enforced by some later state of the system, i.e. the state contradicting effects of the action. Also, the system decides which action is executed when concurrency occurs. In general, the system adopts some strategy while deciding whether a particular default action may take place. This strategy reflects preferences of the system and guides the process of reasoning.

Formalizing default actions we do not specify what happens in abnormal states (relative to these actions) and simply assume that in such situations the actions are not executed.[5]

Since in our approach it is inessential whether an action has duration or not, we restrict ourselves to single-step actions only and omit temporal characteristics of their duration.

Finally, we assume that, unless otherwise is specified, an action has no effects if its precondition does not hold when the action begins.

2.1 The Reasoning Process

For a given domain of actions, our reasoning task is to conclude as much as possible about values of fluents at particular points in time. What we are also

---

[4] While considering causal chains of actions, we actually deal with some kind of the ramification problem.

[5] It is not difficult to extend our formalism and admit such specifications.

interested in is which action (and when) is actually executed. Involving default actions requires to define some preferential policy to decide which default actions are actually executed. However, the following idea seems to be intuitively justified: *in the absence of evidence on the contrary, assume that a state is normal with respect to a particular action.*

Since this policy determines as many as possible performances of default actions, we call it *General Maximization of Default Actions* (GMD, for short).

In general, the process of reasoning about domains under consideration proceeds in the following two stages. First, for every configuration of normalities of states, we determine all possible courses of events constrained by the inertia law. Next, we reject "histories" violating the adopted preferential policy.

The first step may be viewed as a process of simulating of all possible histories, while the next one selects histories relevant to the system preferences.

3 An Action Language $\mathcal{AL}_0$

In this section we define the action language $\mathcal{AL}_0$ capable of representing both causal chains of actions and default actions.

3.1 Syntax of $\mathcal{AL}_0$

Likewise the language $\mathcal{A}$, $\mathcal{AL}_0$ is rather a class of languages. A particular language of this class is characterized by:

- a nonempty set $\mathcal{F}$ of symbols, called *fluent names* (*fluents*, for short);
- a nonempty set $\mathcal{A}$ of symbols, called *action names* (*actions*);
- a nonempty set $\mathcal{T}$ of symbols, called *timepoints*.

In this paper we identify the set $\mathcal{T}$ with the set $\mathbb{N}$ of all natural numbers.
A *formula* is a propositional combination of fluents.
For any action $A \in \mathcal{A}$ we write $\overline{A}$ to denote either A or $\neg A$.
There are six types of statements in $\mathcal{AL}_0$.

A *value statement* is an expression of the form

$$\alpha \text{ at } t \tag{1}$$

with the intuitive meaning that a formula α holds at time t. For instance,

$$OnTable \lor InBox \text{ at } 3.$$

If $t = 0$ then (1) is written as

$$\text{initially } \alpha.$$

An *action statement* is an expression of the form

$$\overline{A} \text{ occurs at } t. \tag{2}$$

If $\overline{A}$ stands for an action A then (2) says that A is to be performed at time t, whereas if $\overline{A}$ is $\neg A$ then (2) means that an action A is not executed at time t.

Value statements and action statements will be referred to as *propositions*.

544

In order to express that an action A typically takes place at time t, a *typical action statement* is introduced. This is an expression of the form

$$\text{typically } A \text{ occurs at } t. \tag{3}$$

An *effect statement* is an expression of the form

$$A \text{ causes } \alpha \text{ if } \pi, \tag{4}$$

which informally asserts that an action A makes a formula α true, provided that a precondition π holds when A begins. For example,

$$Buy \text{ causes } HasTicket \lor HasPaper \text{ if } HasMoney.^6$$

If $\pi \equiv \top$ ($\top$ stands for the truth constant *True*) then we drop the part **if**.

A *release statement* is an expression of the form

$$A \text{ releases } f \text{ if } \pi. \tag{5}$$

Informally, (5) states that a fluent f is exempt from the inertia law whenever an action A is performed, provided that its precondition π holds at the beginning of A. Similarly, if $\pi \equiv \top$ then the part **if** is omitted. For example,

$$Toss \text{ releases } Heads.$$

Also, we define a *trigger statement* of the following two forms:

$$\pi \text{ triggers } \overline{A} \tag{6}$$
$$\text{typically } \pi \text{ triggers } A. \tag{7}$$

If $\overline{A}$ is an action A then (6) means that any situation satisfying π immediately causes an execution of an action A. For example,

$$HasTicket \land Arrived \text{ triggers } GetInTrain.$$

If $\overline{A}$ stands for $\neg A$ then (6) says that in any situation satisfying π, an action A cannot be performed. Then it has the same meaning as A **causes** $\bot$ **if** π and will be abbreviated to

$$\text{impossible } A \text{ if } \pi.$$

(7) is a default version of (6), that is it says that A is typically performed in any situation satisfying π.

Finally, we introduce an *invoke statement* of the following two forms

$$A \text{ invokes } \overline{B} \text{ after } t \text{ if } \pi \tag{8}$$
$$\text{typically } A \text{ invokes } B \text{ after } t \text{ if } \pi. \tag{9}$$

If $\overline{B}$ stands for an action B then (8) intuitively says that an action A causes a performance of B after t units of time, provided that π holds when A begins. If $t = 1$ (i.e. B immediately follows A) then the part **after** is omitted (similarly, we drop the part **if**, provided that $\pi \equiv \top$). For instance,

---

[6] Note that *Buy* is the nondeterministic action with three different outcomes.

$GetInRoom$ **invokes** $TurnOnLight$ **if** $IsDark$.

If $\overline{B}$ stands for $\neg B$ then (8) means that A makes the performance of B impossible after time t, provided that π is satisfied. We use the abbreviation

$$A \text{ disables } B \text{ after } t \text{ if } \pi.$$

For example,

$$CloseDoor \text{ disables } GetIn.$$

(9) represents a default version of (8).

A set D (not necessarily finite) of statements is called a *domain description* iff for every action A occurring in D there is an effect statement (4) (where $\alpha \not\equiv \bot$) or a release statement (5) in D describing effects of A.

The language of D will be denoted by $\mathcal{L}_D$. For a given D, we write $\mathcal{F}_D$ and $\mathcal{A}_D$ to denote the set of all fluents and the set of all actions occurring in D, respectively.

We say that a domain description D is *finite* iff it is a finite set of statements and the sets $\mathcal{F}_D$ and $\mathcal{A}_D$ are finite also.

Example 3.1 Consider a person, say Fred, who is at a railway station waiting for his train. Whenever he has no ticket, he normally buys it immediately. Having bought a ticket Fred usually gets in a train a single unit of time later, provided that the train is at the platform. Initially the train is at the platform and Fred has no ticket. The train leaves at time 3.

This domain can be encoded (below h, p and i stand for fluents $HasTicket$, $AtPlatform$ and $InTrain$, respectively)

> **initially** $\neg h \wedge p \wedge \neg i$;
> **typically** $\neg h$ **triggers** Buy;
> **typically** Buy **invokes** $GetIn$ **after** 2 **if** p;
> Buy **causes** h;
> $GetIn$ **causes** i **if** h;
> $Leave$ **causes** $\neg p$;
> $Leave$ **occurs at** 3. □

3.2 Semantics of $\mathcal{AL}_0$

In this section we provide the semantics of the action language $\mathcal{AL}_0$. To begin with, let us start with some preliminary terminology and notation.

A relation $\mathcal{R} \subseteq X \times Y$ is said *univalent on* X iff for every $x \in X$, $y_1, y_2 \in Y$, the conditions $(x, y_1) \in \mathcal{R}$ and $(x, y_2) \in \mathcal{R}$ imply $y_1 = y_2$.

Given a formula α, by $fl(\alpha)$ we denote the set of all fluents occurring in α.

For a relation $Rel \subseteq \mathcal{A} \times \mathbb{N}$, we write $(\overline{A}, t) \in Rel$ to denote that $(A, t) \in Rel$ if $\overline{A}$ stands for an action A, and $(A, t) \notin Rel$ if $\overline{A}$ is $\neg A$.

Given two functions $I_1, I_2 : X \rightarrow 2^Y$, we write $I_1 \prec I_2$ to denote that for every $x \in X$, $I_1(x) \subseteq I_2(x)$ and there is $x' \in X$ such that $I_1(x') \subset I_2(x')$.

A *valuation val* is a truth-valued function defined on the set of fluents $\mathcal{F}$ and the set of timepoints $\mathbb{N}$. For any timepoint t, $val(t)$ will be referred to as a *state*.

The function *val* can be easily extended for the set of all formulae $Form_{\mathcal{F}}$. A function $H : Form_{\mathcal{F}} \times \mathbb{N} \rightarrow \{0, 1\}$, called a *history function*, is defined according to the truth tables in propositional logic.

The semantics of the language $\mathcal{AL}_0$ is centered upon the following notion.

Definition 3.2 (Structure) A *structure* is a tuple $\Sigma = (H, I, E, N)$, where

- $H : Form_{\mathcal{F}} \times \mathbb{N} \rightarrow \{0, 1\}$ is a history function;

- $I : \mathcal{A} \times \mathbb{N} \rightarrow 2^{\mathcal{F}}$ is an *influence function* such that $I(A, t)$ yields a set of fluents influenced by executing of an action A at time t;

- $E \subseteq \mathcal{A} \times \mathbb{N}$ is a univalent relation on $\mathbb{N}$, called a *performance relation*, such that (A, t) yields that an action A is executed at time t;

- $N \subseteq \mathcal{A} \times \mathbb{N}$ is a *normality relation* such that (A, t) yields that at time t the state is normal with respect to an action A. □

Note that the univalence property of the relation E in a structure Σ guarantees that at every timepoint at most one action may be executed. It may seem strange why no demands are imposed on interconnections between the influence function and the performance relation (clearly, when an action is not executed, then it affects no fluent). However, our further discussion shows that this requirement is inessential. From similar reasons we impose no requirements on the relation N.

For a given structure $\Sigma = (H, I, E, N)$ and a statement s of $\mathcal{AL}_0$, we define that Σ is a *structure for s*, written $\Sigma \bowtie s$, as follows:

- $\Sigma \bowtie \alpha$ **at** t iff $H(\alpha, t) = 1$

- $\Sigma \bowtie \overline{A}$ **occurs at** t iff $(\overline{A}, t) \in E$

- $\Sigma \bowtie$ **typically** A **occurs at** t iff $(A, t) \in N \Rightarrow (A, t) \in E$

- $\Sigma \bowtie A$ **causes** α **if** π iff $\forall t \in \mathbb{N} [H(\pi, t) = 1 \ \& \ (A, t) \in E \Rightarrow H(\alpha, t+1) = 1$
$\& \ fl(\alpha) \subseteq I(A, t)]$

- $\Sigma \bowtie A$ **releases** f **if** π iff $\forall t \in \mathbb{N} [H(\pi, t) = 1 \ \& \ (A, t) \in E \Rightarrow f \in I(A, t)]$

- $\Sigma \bowtie \pi$ **triggers** $\overline{A}$ iff $\forall t \in \mathbb{N} [H(\pi, t) = 1 \Rightarrow (\overline{A}, t) \in E]$

- $\Sigma \bowtie$ **typically** π **triggers** A iff $\forall t \in \mathbb{N} [H(\pi, t) = 1 \ \& \ (A, t) \in N \Rightarrow (A, t) \in E]$

- $\Sigma \bowtie A$ **invokes** $\overline{B}$ **after** t **if** π iff $\forall t' \in \mathbb{N} [H(\pi, t') = 1 \ \& \ (A, t') \in E \Rightarrow$
$(\overline{B}, t+t') \in E]$

- $\Sigma \bowtie$ **typically** A **invokes** B **after** t **if** π iff $\forall t' \in \mathbb{N} [H(\pi, t') = 1 \ \&$
$(A, t') \in E \ \& \ (B, t+t') \in N \Rightarrow (B, t+t') \in E].$

We say that a proposition φ (i.e. a value statement or an action statement) is *true* in Σ iff Σ is a structure for φ.

Given a domain description D, a structure Σ is called a *structure for D* iff Σ is a structure for every statement in D.

What we are actually interested in are structures for D where the inertia law holds. Given a structure $\Sigma = (H, I, E, N)$, the function I indicates fluents the corresponding action affects. It seems then reasonable to make sets $I(A, t)$, for any action A and any timepoint t, as minimal as possible (relative to set inclusion).

We say that a structure $\Sigma = (H, I, E, N)$ for a domain description D is an *I-minimal structure for D* iff there is no structure $\Sigma' = (H', I', E', N')$ for D such that $H = H'$, $N = N'$ and $I' \prec I$.

An I-minimal structure for D determines minimal sets of fluents affected by actions at particular timepoints for a fixed history function and a normality relation. It amounts to determine such sets of fluents for all possible histories and all possible configurations of normalities of states (relative to particular actions).

However, I-minimal structures still admit spurious changes. In order to preserve the inertia law, we demand that all changes occuring in a history must result from action executions only. To this end, the class of I-minimal structures is further restricted.

Definition 3.3 (Model) Let $\Sigma = (H, I, E, N)$ be an I-minimal structure for a domain description D. We say that Σ is a *model of D* iff for every $t \in \mathbb{N}$ there is an action $A \in \mathcal{A}$ such that $\{f \in \mathcal{F} : H(f, t) \neq H(f, t+1)\} \subseteq I(A, t)$. $\square$

In other words, a fluent f may change at time t *only* if it is affected by some action A executed at $t-1$. However, at some timepoint t a fluent f may not actually change though there is an action performed at $t-1$ that influences f. From this reason we require that the set of fluents changing at time t is only a *subset* of the set of fluents affected by A at $t-1$.

A domain description D is called *consistent* if it has a model; otherwise it is called *inconsistent*.

We say that a domain description D *entails* a proposition φ, written $D \models \varphi$, iff φ is true in every model of D.

One may observe that a set of executable actions in a model of D (determined by the performance relation E) need not be minimal. Specifically, the relation E of the model may contain any pair (A, t) such that the execution of an action A at time t, though unexplained by D, has actually no effect since the precondition of A does not hold at time t. Since we are interested in propositions entailed by D, this requirement is inessential. However, it is easy to check that in a class of models of D that coincide on H, I and N, there is a model of D where E is minimal (relative to set inclusion).

Remark 3.4 It is worth noting that if no default action occurs in a domain description D then adding a value statement s to D restricts the class of models of D. Consequently, a set of propositions entailed by $D \cup \{s\}$ is made larger. This property, called *restricted monotonicity* ([9]), is then satisfied by the mentioned subclass of domains represented in $\mathcal{AL}_0$. $\square$

For a given consistent domain description D, the class of its models includes structures where the normality relation is defined arbitrary. Hence, a particular default action A is executed at a particular timepoint in one model, but is not performed (at the same time) in another. This class should be further restricted in order to select so-called *preferred models*. Adopting the preferential strategy GMD we select models Σ of D where as many as possible default actions are executed. Consequently, we prefer models of D where the normality relation is maximal (relative to set inclusion). This leads to the following definition.

Definition 3.5 (GMD-Preferred Model) Let D be a consistent domain description and let $\Sigma = (H, I, E, N)$ be a model of D. We say that Σ is a *GMD-preferred model of D* iff there is no model $\Sigma' = (H', I', E', N')$ such that $N \subset N'$.
□

We say that a proposition φ is *GMD-preferentially entailed by D*, written $D \approx \varphi$, iff φ is true in every GMD-preferred model.

It is easily observed that a new value statement, when added to D, may cause some default action(s) impossible to be performed. Then a set of propositions GMD-preferentially entailed by D often changes in a nonmonotonic fashion. Hence, in general case, the property of restricted monotonicity is not preserved.

Example 3.1 (continued) This domain description D has three classes of models. The first class involves models where Fred fails to buy a ticket and, consequently, cannot get in the train (i.e. the normality relation in these models does not contain a pair $(Buy, 0)$). The second class includes models where Fred buys a ticket but fails to take the train (i.e. the normality relation does not contain $(GetIn, 2)$). The last class covers models where Fred buys a ticket and gets in the train (i.e. the normality relation contains both $(Buy, 0)$ and $(GetIn, 2)$). Maximization of normalities leads to infinitely many GMD-preferred models of D with the normality relation $N = \mathcal{A} \times \mathbb{N}$ and the influence function such that $I(Buy, 0) = \{h\}$, $I(GetIn, 2) = \{i\}$, $I(Leave, 3) = \{p\}$ and $I(A, t) = \emptyset$ otherwise. Fig.1 depicts the course of events determined by the only GMD-preferred model of D with the minimal set of executable actions (states at timepoints $t > 4$ are the same as the state at time 4).

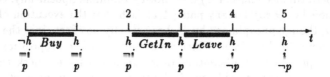

Fig. 1.

Suppose now that the train leaves at time 2. Intuitively, Fred either resigns to buy a ticket (e.g. he expects not to catch a train) or he buys it but fails to take his train. Since both default actions cannot be executed simultaneously, we have two classes of models of D: models where Fred fails to buy a ticket (i.e.

(Buy, 0) does not belong to their normality relation), and models where he buys
a ticket but fails to catch the train (i.e. their normality relation contains (Buy, 0)
but not ($GetIn$, 2)). Maximization of normalities leads to models where the only
abnormality occurs at time 0 with respect to the action Buy, and those where the
only abnormality occurs at time 2 relative to the action $GetIn$. Fig. 2 illustrates
courses of events determined by two GMD-preferred models with minimal sets
of executable actions.

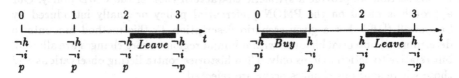

Fig. 2.

4 Propositional Fluent Logic PFL

In this section we briefly present the propositional fluent logic PFL based on the
logic FL introduced in [2] and extended in [14] and [15]. PFL is a three-sorted
FOPC with equality including a sort for temporal entities, a sort for truth-valued
fluents and a sort for actions. The language of PFL is denoted by $\mathcal{L}$(PFL). We
assume a linear discrete model of time containing all natural numbers.

The language $\mathcal{L}$(PFL) involves three sets of object constants: a set $\mathcal{F}$ of *fluent
constants*, a set $\mathcal{A}$ of *action constants* and the set $\mathbb{N}$ of timepoints. $\mathcal{L}$(PFL) has
the binary predicate symbol $<$ and the function symbols $+$ and $-$, interpreted on
natural numbers in the usual way. Also, we have two binary predicate symbols,
Holds and *Occlude*, which take a timepoint and a fluent as arguments, and the
predicate symbol *Action* which takes an action and a timepoint as arguments.
Finally, we introduce the unary predicates *Ab* with a timepoint as the only
argument.

Action(A, t) states that an action A occurs at time t (i.e. it is to be performed
during the period t to $t+1$). *Holds*(t, f) states that a fluent f is true at time t.
The predicate *Occlude* was originally introduced by Sandewall ([16]) in order
to explicitly indicate timepoints where changes are allowed. *Occlude*(t, f) states
that a fluent f is *occluded* at time t with the intended meaning that at time t
this fluent is exempt from the inertia law. Finally, $Ab(t)$ will be used to represent
an abnormality of a state at time t with respect to some action.

The *set of formulae* is constructed in the usual way using logical connectives
and quantifiers over temporal and fluent variables.

An action domain represented in $\mathcal{L}(PFL)$ will be referred to as a *scenario
description* and denoted by Γ. To distinguish between different types of formu-
lae in a scenario description Γ we write $\Gamma = \Gamma_{OBS} \cup \Gamma_{ACS} \cup \Gamma_{UNA}$ to state that
Γ consists of the set Γ_{OBS} of observation axioms representing (usually partial)

states of a system at particular timepoints, the set Γ_{ACS} of action axioms representing either action occurrences or action descriptions and the set Γ_{UNA} of appropriate unique name axioms for fluents and actions.

5   Circumscriptive policy

In this section we provide a syntactic characterization of the GMD policy. Our approach is based on the PMON preferential policy originally introduced by Sandewall ([16]). For a given scenario description Γ, this method amounts to determine all potential histories with minimal regions where changes are allowed, but relative to action axioms only. Then histories contradicting observations and those where spurious changes occur are rejected.

Following Sandewall's approach, we introduce the following *nochange axiom*:

$$\Gamma_{NCH} = \forall t, f. \neg Occlude(t+1, f) \rightarrow [Holds(t+1, f) \equiv Holds(t, f)]$$

in order to explicitly specify that a fluent may change its value when it is *occluded* only.

For a given scenario description Γ, the PMON minimization method can be then realize by the following circumscription, called a *PMON-Circumscription of Γ*[7]:

$$\text{CIRC}_{PMON}(\Gamma) = \Gamma_{NCH} \wedge \Gamma \wedge \text{CIRC}_{SO}(\Gamma_{ACS}; Occlude; Action),$$

where $\text{CIRC}_{SO}(\Gamma_{ACS}; Occlude; Action)$ is the second-order circumscription formula equivalent to

$$\Gamma_{ACS}(Occlude, Action) \wedge \forall \Phi, \Psi. \neg[\Gamma_{ACS}(\Phi, \Psi) \wedge \Phi < Occlude].$$

The PMON policy is provably applied to reason about action domains where causal chains of actions are not allowed. In order to reject cases where concurrent actions occur, the following *non-concurrency axiom* is addditionally introduced:[8]

$$\Gamma_{NCA} = \forall a, b \, \forall t. \, Action(a, t) \wedge Action(b, t) \rightarrow (a = b).$$

Given a scenario description Γ, the following circumscriptive scheme

$$\text{CIRC}_{EPMON}(\Gamma) = \Gamma_{NCA} \wedge \text{CIRC}_{PMON}(\Gamma).$$

is called an *Extended PMON-Circumscription of Γ* (EPMON-Circumscription of Γ, for short).

Note that a model of $\text{CIRC}_{EPMON}(\Gamma)$ does not determine the extension of the predicate *Action*. Therefore there are models where a particular action A is executed at some timepoint t, and models where A is not performed then.

---

[7] A simplified form of this scheme was proposed by Doherty in [2].

[8] This axiom was previously proposed in [15].

However, it concerns only those actions that actually lead to no effects since their preconditions are not satisfied when these actions (formally) begin.

The GMD policy amounts to select models of Γ, where as many as possible default actions are performed. Then we are to minimize abnormalities representing by Ab predicates. To this end, we propose the following circumscription.

Definition 5.1 (GMD-Circumscription) Let Γ be a scenario description and $\overline{Ab}$ be a tuple of all Ab predicates occurring in Γ. The *GMD-Circumscription of Γ* is the following formula

$$\text{CIRC}_{GMD}(\Gamma) = \text{CIRC}_{SO}(\text{CIRC}_{EPMON}(\Gamma); \overline{Ab}; Action, Holds, Occlude). \quad \square$$

For a given scenario description Γ, models of the GMD-Circumscription of Γ are called *GMD-intended models* of Γ.

Example 3.1 (continued) Below is the corresponding scenario description Γ

$$\neg Holds(0, h) \wedge \neg Holds(0, i) \wedge Holds(0, p)$$
$$\forall t. \neg Holds(t, h) \wedge \neg Ab_1(t) \rightarrow Action(Buy, t)$$
$$\forall t. Action(Buy, t) \wedge Holds(t, p) \wedge \neg Ab_2(t+2) \rightarrow Action(GetIn, t+2)$$
$$\forall t. Action(Buy, t) \rightarrow Holds(t+1, h) \wedge Occlude(t+1, h)$$
$$\forall t. Action(GetIn, t) \wedge Holds(t, h) \rightarrow Holds(t+1, i) \wedge Occlude(t+1, i)$$
$$\forall t. Action(Leave, t) \rightarrow \neg Holds(t+1, p) \wedge Occlude(t+1, p)$$
$$Action(Leave, 3).$$

For every combination of abnormalities, the EPMON-Circumscription determines models of Γ satisfying the inertia law where no concurrent actions are performed. Next, minimization of Ab-predicates rejects all models where any abnormality occurs. Consequently, we obtain infinitely many GMD-intended models of Γ that differ by extensions of the predicate $Action$ only. It is easily noted that a model $\mathcal{M}$, shown in the table below, is the only GMD-intended one where this extension is minimal.

Table 1: GMD-intended models[9]

	$\mathcal{M}$	$\mathcal{M}_1$	$\mathcal{M}_2$
0	$\neg h, p, \neg i$	$\neg h, p, \neg i$	$\neg h, p, \neg i$
1	$h^*, p, \neg i$	$\neg h, p, \neg i$	$h^*, p, \neg i$
2	$h, p, \neg i$	$\neg h, p, \neg i$	$h, p, \neg i$
3	h, p, i^*	$\neg h, \neg p^*, \neg i$	$h, \neg p^*, \neg i$
4	$h, \neg p^*, i$	$\neg h, \neg p, \neg i$	$h, \neg p, \neg i$
$\vdots$	$h, \neg p, i$	$\neg h, \neg p, \neg i$	$h, \neg p, \neg i$
Ab_1	$\emptyset$	$\{0\}$	$\emptyset$
Ab_2	$\emptyset$	$\emptyset$	$\{2\}$
$Action$	$\{(Buy, 0), (GetIn, 2), (Leave, 3)\}$	$\{(Leave, 2)\}$	$\{(Buy, 0), (Leave, 2)\}$

---

[9] In this table asterisks indicate occluded fluents at particular timepoints.

Suppose that the train leaves at time 2. Accordingly, the last axiom is replaced by $Action(Leaves, 2)$.

The PMON-Circumscription of Γ determines all models of Γ constrained by the inertia law. Note that it leads to models where the initial state, together with the state at time 2, are normal (relative to actions Buy and $GetIn$, respectively). Then the actions $GetIn$ and $Leaves$ take place simultaneously. However, the non-concurrency axiom makes these models rejected. By minimizing Ab predicates we get infinitely many GMD-intended models that differ by extensions of the predicate $Action$. Table 1 shows the only two $Action$-minimal GMD-intended models, namely $\mathcal{M}_1$ and $\mathcal{M}_2$. □

6 From $\mathcal{AL}_0$ into Circumscription

In this section we provide a two-step translation from $\mathcal{AL}_0$ into circumscriptive theories. We restrict attention to finite domain descriptions. For a given domain description D, the first step determines the corresponding scenario description Γ^D in $\mathcal{L}(PFL)$, whereas the next one amounts to apply GMD-Circumscription of the theory obtained in the first step.

Let D be a finite, consistent domain description and let $\mathcal{F}_D$ be the set of all fluents occurring in D and $\mathcal{A}_D$ be the set of all actions mentioned in D. Consider a language $\mathcal{L}(PFL)$, where $\mathcal{F}_D$ and $\mathcal{A}_D$ are sets of fluent constants and action constants, respectively.

For every statement s in D we define a corresponding first-order formula Θs of the language $\mathcal{L}(PFL)$ in the following manner.

If s is a value statement (1) then Θs is

$$Holds(t, \alpha),$$

where $Holds(t, \neg f)$ stands for $\neg Holds(t, f)$ and $Holds(t, \beta_1 \wedge \beta_2)$ stands for $Holds(t, \beta_1) \wedge Holds(t, \beta_2)$.

If s is an action statement of the form (2) then Θs is

$$Action(\overline{A}, t),$$

where $Action(\neg A, t)$ stands for $\neg Action(A, t)$.

If s is a typical action statement of the form (3) then Θs is

$$\neg Ab(t) \rightarrow Action(A, t).$$

If s is an effect statement (4) then Θs is

$$\forall t.\, Action(A, t) \wedge Holds(t, \pi) \rightarrow Holds(t+1, \alpha) \wedge \bigwedge_{f \in fl(\alpha)} Occlude(t+1, f).$$

If s is a release statement (5) then Θs is

$$\forall t.\, Action(A, t) \wedge Holds(t, \pi) \rightarrow Occlude(t+1, f).$$

If s is an invoke statement of the form (8) or (9) then Θs is

$$\forall t'.\ Action(A, t') \wedge Holds(t', \pi) \rightarrow Action(\overline{B}, t'+t)$$
$$\forall t'.\ Action(A, t') \wedge Holds(t', \pi) \wedge \neg Ab(t'+t) \rightarrow Action(B, t'+t),$$

respectively.

Finally, if s is a trigger statement of the form (6) or (7) then Θs is

$$\forall t.\ Holds(t, \pi) \rightarrow Action(\overline{A}, t)$$
$$\forall t.\ Holds(t, \pi) \wedge \neg Ab(t) \rightarrow Action(A, t).$$

Let Γ_{OBS}^{D} and Γ_{ACS}^{D} denote a set of first-order formulae in $\mathcal{L}(PFL)$ resulting from this translation of all value statements in D and the remaining statements in D, respectively. Therefore the theory $\Gamma^{D} = \Gamma_{OBS}^{D} \cup \Gamma_{ACS}^{D} \cup \Gamma_{UNA}^{D}$ is a scenario description corresponding to D.

6.1 Soundness and Completeness

Let Γ^{D} be a scenario description corresponding to a finite, consistent domain description D. We write ΘD to denote $\mathrm{CIRC}_{EPMON}(\Gamma^{D})$ and $\tilde{\Theta} D$ to denote $\mathrm{CIRC}_{GMD}(\Gamma^{D})$.

The following theorems state soundness and completeness of the translations Θ and $\tilde{\Theta}$ relative to the semantics of $\mathcal{AL}_0$.

Theorem 6.1 Let D be a consistent, finite domain description and let ΘD be defined as before. For any proposition φ in $\mathcal{L}_D$,

$$D \models\!\!\!\mid \varphi \quad \text{iff} \quad \Theta D \models \Theta \varphi. \qquad \Box$$

Theorem 6.2 Let D be a consistent, finite domain description and let $\tilde{\Theta} D$ be defined as before. For any proposition φ in $\mathcal{L}_D$,

$$D \models\!\!\!\approx \varphi \quad \text{iff} \quad \tilde{\Theta} D \models \Theta \varphi. \qquad \Box$$

7 Conclusion and Future Work

In this paper we have considered action domains with causal sequences of actions and default actions. We have defined the action language $\mathcal{AL}_0$ based on the language $\mathcal{A}$ proposed in [4] and have shown the translations from $\mathcal{AL}_0$ into circumscriptive theories, sound and complete relative to the semantics of $\mathcal{AL}_0$.

The problems of ramifications and qualifications of actions have not been studied here. Promising results recently obtained in [6], [13] and [18] inspire to futher studies. It is undoubtedly worthwhile to expand our formalism for cases where static and dynamic domain constraints occur.

Another problem deals with circumscription we apply as a syntactic characterization of the reasoning method. In particular, it is worth comparing our results with nested abnormality theories recently proposed by Lifschitz [10].

Finally, we wish to indicate the problem of so-called dependent actions (i.e. actions affecting the same fluents). In this paper we reject any form of concurrency. Nevertheless, there is no reason to admit performances of several independent actions at the same time. Moreover, it needs investigation how the mutual exclusion problem can be solved within our framework.

These topics will be discussed in our ongoing work.

References

1. C. Baral, M. Gelfond: Representing Concurrent Actions in Extended Logic Programming, in *Proc IJCAI-93*, Chambery, France, 1993, pp. 866–871.
2. P. Doherty: Reasoning about Action and Change using Occlusion, in *Proc. of ECAI-94*, Amsterdam, 1994, pp. 401–405.
3. B. Dunin-Kęplicz, A. Radzikowska: Epistemnic Approach to Actions with Typical Effects, in *Proc. of ECSQARU-95*, Fribourg, Switzerland, 1995, pp. 180–189.
4. M. Gelfond, V. Lifschitz: Representing Action and Change by Logic Programs, *The Journal of Logic Programming*, 17, 1993, pp. 301–322.
5. E. Giunghilia, G. N. Kartha, V. Lifschitz: Actions with Indirect Effects, in *Working Notes of the AAAI Spring Symposium on Extending Theories of Action*, 1994.
6. E. Giunghilia, V. Lifschitz: Dependent Fluents, in *Proc. of IJCAI-95*, Montreal, Canada, 1995, pp. 1964–1969.
7. G. N. Kartha: Soundness and Completeness Theorems for Three Formalizations of Actions, in *Proc. of IJCAI-93*, Chambery, France, 1993, pp. 724–729.
8. G. N. Kartha, V. Lifschitz: Actions with Indirect Effects, in *Proc. of 4th KR-94*, Bonn, Germany, 1994, pp. 341–350.
9. V. Lifschitz: Restricted monotonicity, in *Proc. of AAAI-93*, 1993, pp. 432–437.
10. V. Lifschitz: Nested Abnormality Theories, *Artificial Intelligence* 74, 1995, pp. 351–365.
11. V. Lifschitz: Two Components of An Action Language, in *Proc. of Third Symposium of Logical Formalizations of Commonsense Reasoning*, Stanford University, 1996.
12. F. Lin, Y. Shoham: Provably correct theories of actions (preliminary report), in *Proc. AAAI-91*, 1991, pp. 349–354.
13. N. McCain, H. Turner: A Causal Theory of Ramifications and Qualifications, in *Proc. of IJCAI-95*, Montreal, Canada, 1995, pp. 1978–1984.
14. A. Radzikowska: Circumscribing Features and Fluents: Reasoning about Action with Default Effects, in *Proc. of ECSQARU-95*, Fribourg, 1995, pp. 344–352.
15. A. Radzikowska: Reasoning about Action with Typical and Atypical Effects, in *Proc. of 19th German Annual Conference on AI*, Bielefeld, 1995, pp. 197–209.
16. E. Sandewall: Features and fluents: A systematic approach to the representation of knowledge about dynamical systems, Oxford University Press, 1994.
17. S. E. Bornscheuer, M. Thielscher: Representing Concurrent Actions and Solving Conflicts, in *Proc. of 18th German Annual Conference on AI*, Saarbrücken, 1994, pp. 16–27.
18. H. Turner: Representing Actions in Default Logic: A Situation Calculus Approach in *Proc. of Third Symposium of Logical Formalizations of Commonsense Reasoning*, Stanford University, USA, 1996.

An Architecture for Argumentative Dialogue Planning[*]

Chris Reed[1], Derek Long[2] and Maria Fox[2]

[1] Department of Computer Science,
University College London,
Gower St.,
London
WC1E 6BT
C.Reed@cs.ucl.ac.uk
http://www.cs.ucl.ac.uk/staff/C.Reed

[2] Department of Computer Science,
Durham University,
South Road,
Durham
D.P.Long@dur.ac.uk
M.Fox@dur.ac.uk

Topics: Argumentation theory; Fallacies and their role in practical reasoning

Abstract. Argument represents an opportunity for a system to convince a possibly sceptical or resistant audience of the veracity of its own beliefs. This ability is a vital component of rich communication, facilitating explanation, instruction, cooperation and conflict resolution. In this paper, a proposal is presented for the architecture of a system capable of constructing arguments. The design of the architecture has made use of the wealth of naturally occurring argument, which, unlike much natural language, is particularly suited to analysis due to its clear aims and structure. The proposed framework is based upon a core hierarchical planner conceptually split into four levels of processing, the highest being responsible for abstract, intentional and pragmatic guidance, and the lowest handling realisation into natural language. The higher levels will have control over not just the logical form of the argument, but also over matters of style and rhetoric, in order to produce as cogent and convincing an argument as possible.

1. Introduction

In this paper an architecture is presented for the autonomous construction of argument, outlining the components required for persuasive communication. Argument plays a crucial role for a communicative system as a means of persuading the human interlocutor to adopt some belief which is of interest to the system. This might be simply to improve the coherence between their beliefs or to obtain a specific effect such as an adoption of some new goal by the hearer, elicitation of particular information, or instruction upon some particular topic.

It is assumed throughout that the argument will ultimately be rendered in natural language, but this assumption does not form a cornerstone of the theory: the framework employs intentional structures which are sufficiently abstract to remain unchanged by a shift to some more simple and restrictive artificial language.

[*]This work has been funded by EPSRC grant no. 94313824.

After a brief introduction to argument as a form of natural language, the overview of the proposed architecture in §3 is followed by more detailed analyses of the five major components.

2. Argument

In order to model argument and design a rubric for its automatic synthesis, it is necessary to have a clear understanding of its nature. Unlike much natural discourse, arguments always have clearly defined goals, and in particular, to persuade a particular audience of a particular proposition. This perhaps oversimplifies the issue, since rarely will an argument be instigated between two parties who believe thesis and antithesis, and even more rarely does argument terminate with both parties believing the same proposition. This fuzziness concerning the beliefs of the participants is much more an intrinsic property of belief, rather than one of argument, since the actual process of argumentation can more accurately be perceived as the facilitation to the adoption of a belief on the part of the hearer. The process itself is the same, regardless of the myriad situations in which it can occur - public oration, debating house discourse, legal argument, newspaper commentary, scientific papers, etc.

3. Components of an Argumentative System

The architecture proposed in this paper rests in a hierarchical framework, reflecting the distinct though inter-related levels of structure within arguments identified through their analysis. There is a part of argument synthesis which is concerned with the resolution of syntax, expression and morphology, as with any natural language synthesis. This comprises the lowest level. Above this, there sits a level responsible for the coordination of sentence and intersentence relations, imposing structure at a more abstract level. This intermediate level is also responsible for handling aspects of communication such as focusing (Grosz and Sidner 1990). Finally, at the higher pragmatic levels, new machinery and techniques are required to produce cogent argument. This generation can be accomplished through two complimentary levels, one of which handles the structural aspects of argument, and the other effecting complex modifications to that structure and controlling stylistic devices. The former, the Argument Structure (AS) level, is at the highest level of abstraction, since it produces the logical form of the argument employing purely intentional data structures, by using logical operators, fallacy[1] operators and inductive operators. This form is then augmented and modified by the subordinate Eloquence Generation (EG) level, which employs heuristics based upon rhetoric and contextual parameters, and is concerned with such properties of a speech as its length, detail, meter, ordering of subarguments, grouping of subarguments,

---

[1]Lists of informal fallacies are common, though rarely identical: see, for example, Whately (1855) or (Johnson 1992).

enthymeme contraction, use of repetition, alliteration and so on. This architecture is summarised below in Fig.1.

Note that the 'intermediate level' has, in the figure, been labelled 'RST level': the functionality proposed for this level is in line with much of the work carried out on Rhetorical Structure Theory (Mann and Thompson 1986). (The inclusion of RST has also led to the Eloquence Generation level being so named: the more appropriate 'Rhetoric level' is open to confusion with RST).

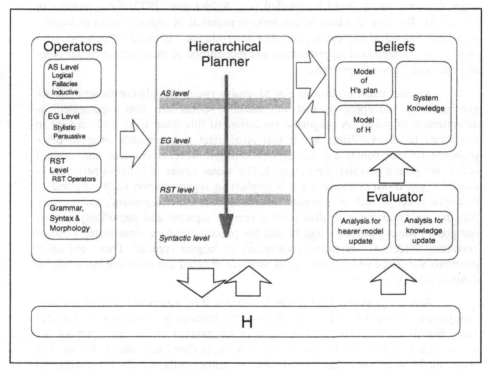

Fig. 1. System architecture overview

4. Supporting Components

4.1. Belief modelling

It has become clear that both AS and EG levels require access to the belief model of the audience, in addition, of course, to the beliefs of the system. The AS level needs to be able to assess in advance how different structures are likely to be received, and to take account of the beliefs of the audience in producing structure. The EG level employs the belief model to pitch an argument at the right level of detail, and to track the saliency of beliefs used during argumentation. A more detailed examination of the relationship between the higher levels and belief modelling is described below in §4.3 and §4.4.

There are a number of issues involved in analysing belief. There is more than one kind of belief - factual beliefs, which are either testable (at least in theory) or are definitional; opinions, which are based on morals and are thus ultimately personal and unprovable (since there is - and can be - no universally accepted and provably correct moral framework); and cultural beliefs, which are based upon sociocultural maxims (such as that which states that living into old age is desirable). This tripartite distinction has been based upon the views expressed implicitly by Blair (1838), when talking about general types of argument, but there are also numerous other divisions which could be detailed (see Ackermann (1972) for a number of examples). The way in which beliefs become manifest in argumentative dialogue - both their individual expression, and their interrelation - is chiefly dependent upon their class, so competent representation and recognition of these distinct belief types represents an important problem.

Another major problem is how to resolve two seemingly contradictory views generated by introspection, that of whether beliefs are best represented as dichotomous or scalar. A pragmatic resolution to this issue is in itself crucial to competent belief modelling, but it will also affect the approach taken to other problems (factual beliefs are usually dichotomous, whereas opinions and cultural beliefs are generally scalar, for example). The scalar nature of belief in particular is crucial to pin down precisely, for it is implicit in argumentation itself - arguments rely upon concepts such as 'persuasiveness' and 'strength of argument', and the very fact that an argument will often involve several separate and identifiably distinct component sub-arguments, suggests that the process involves some scalar value of 'mounting evidence', 'increasing conviction', or 'cogent context'. These and other problems associated with the concept of 'strength of belief' are discussed more fully in Galliers (1992).

Another property of belief which is manifest in argument is that of saliency. Enthymeme contraction (where an unnecessary premise or conclusion is omitted) relies heavily upon saliency and so too does the process of focusing: keeping the argument to the point (Cf. the lower level focusing of Grosz and Sidner (1990)). The belief model used must therefore be able to competently handle the concept of saliency.

Argument modelling also relies on a treatment of the complex phenomenon of mutual belief: an argument is based upon common ground - a set of mutual beliefs (ie. which both parties hold, and which both parties also know the other to hold). Mutual belief is defined in terms of an infinite regress of nested beliefs: the problem is to pragmatically choose a level of nesting beyond which 'mutual' belief is to be assumed. In making this choice, it is understood that no matter how many levels a system can cope with, it is always possible to construct a (highly convoluted) example which exceeds the capabilities of that system. From a psychological (and intuitive) point of view, choosing some arbitrary level of nesting by which to define mutuality seems rather implausible. In humans, it would appear that belief nesting is a resource bounded operation with no known limit, and it is possible to construct deeply nested examples which present remarkably little difficulty (such as the example in Fig. 2, below), though handling further complexity rapidly becomes extremely difficult and time consuming. It may be possible to utilise this evidence and allow for a similar process in an implementation, such that some fundamental operator (say, BMB, following Cohen and Levesque (1990)) is assumed at a naively shallow level of nesting, and then to allow the operator to be quite corrigible in the light of new evidence.

In the classic Hitchcock film "North by Northwest", Cary Grant is the central character in a case of mistaken identity - he is mistaken for a goodguy agent by the evil James Mason. Towards the end, a scene occurs in which CG masquerades as the goodguy agent he is thought to be. As part of a goodguy ploy, he informs JM of his wish to defect, at which he is shot by EveMarie Saint, a goodguy agent working undercover as accomplice to JM. In her role as a villain, she needs to stop the defection, believing him to be a traitor to the goodguys. Thus with the proposition P that 'ES is a goodguy',

```
BEL(ES, P)                                    She knows she's a goodguy
BEL(CG, P)                                    He knows she's a goodguy ...
BEL(ES, BEL(CG, ~P))                          ... but she doesn't realise that ...
BEL(CG, BEL(ES, BEL(CG, ~P)))                 ... and he knows she doesn't realise ...
BEL(Audience, BEL(CG, BEL(ES, BEL(CG, ~P))))  ... or at least that's what the audience thinks!
```

After JM has left however, CG gets up - the shooting had been faked. CG and ES must have been in league with each other after all. At this point the belief held must be changed to

```
BEL(Audience, BMB(CG, ES, P))))               They both know she's a goodguy, and
                                              both know that they both know it.
```

Hitchcock envisaged the whole situation and realised that it was unusual and interesting. There were in total, five levels of nested beliefs.

Fig. 2. Deep nesting of beliefs

4.2. Hierarchical planning

The role of planning in the construction of arguments is significant: arguments are constructed to achieve goals and are built out of primitive argument structures. this construction process includes consideration of the effects of the various components on the hearer's beliefs, the flow of the argument, its focus and force and the extent to which each part is supported by what precedes it. These issues are all interpretations, in the argument synthesis domain, of traditional planning problems, so it is no surprise that traditional planning has a role to play in this domain.

A considerable quantity of work already exists on the use of planners in the construction of monologue (Hovy (1993) contains a thorough overview). This work rests almost exclusively on planners employing a form of abstraction-based planning based on NOAH, a planner developed by Sacerdoti (1977) which is, by the standards of modern planning technology, rather dated. It is proposed that a more advanced style of abstraction-based hierarchical planning technology be used to support the architecture discussed in this paper, employing encapsulation-based abstraction (Fox and Long 1995) to exploit an intuitive structuring of the argument-planning domain. This approach involves the use of abstract operators which encapsulate bodies containing goals, the achievement of which sets the context for and then brings about the effects at a detailed level which are described at an abstract level in the abstract operator shells. The refinement of abstract operators into their more detailed bodies is carried out as an explicit and controlled stage in the planning process, rather than (as with the looser and less-powerful abstraction mechanism of NOAH) as the operators are applied. This allows a genuine hiding of detail until it is appropriate for the planner to consider it.

4.3.   Argument Structure

Through an analysis of a corpus of arguments drawn from a number of the sources mentioned in §2, the following structure has been identified:

(1)   An argument consists of one or more premises and exactly one conclusion

(2)   A premise can be a subargument (which itself consists of one or more premises and exactly one conclusion: the conclusion then stands as the premise in the superargument)

(3)   A subargument is an integral unit whose components cannot be referred to from elsewhere, nor can the conclusion of a subargument rest upon premises extraneous to that subargument

(4)   The only exception to (3) is where a conclusion in a distant subargument is restated locally as a premise

Analyses based on similar theories are performed in many texts: see for example, Johnson (1992) and Wilson (1980). It is this structure which the AS level constructs, linking premises to conclusions through the use of three groups of operators: standard logical relations, rhetorical fallacies, and inductively reasoned implications.

The first group comprises Modus Ponens, Modus Tollens, Conjunction, Disjunctive Syllogism, Hypothetical Syllogism and Constructive Dilemma (MP being by far the most common). The second group comprises around two dozen fallacies which, though illogical, are very common in natural language. Some are, without a doubt, tricks of the rhetorician, designed to mislead the audience (Red Herring and Straw Man, for example), whilst most are used quite inadvertently, but nevertheless assist in strengthening an argument (at least in the perceptions of some audiences) without recourse to factual support. To implement the use of fallacies could be seen as a violation of Grice's first maxim for successful communication, the maxim of quality, which demands sincerity and truthfulness (if the system employs a rhetorical trick, it is in some ways 'cheating'), but since humans generally utilise fallacies unwittingly, there would appear to be no lack of sincerity unless there is devious intent behind their application. Their use would be heavily constrained by tight preconditions specifying contextual factors in addition to both system knowledge and assumed audience beliefs. (Incidentally, if the preconditions were to include some analysis of the hearer's susceptibility to the fallacy, and its consequent success, it could then be argued that there is some measure of devious intent).

Finally, there are the inductive operators (inductive strength being defined as "...more probable than not that the conclusion follows [from the premises]."

(Johnson 1992)). The inductive operators are of three types: inductive generalisation, causal (based on Mill's methods) and analogical (eg. Long and Garigliano (1994)). All three will have preconditions that are again tightly specified for context and belief, but with the additional constraint of the 'criteria of inductive strength', ie. the requirements for their application to produce an inductively strong argument.

4.4. Eloquence Generation

Although the EG level also employs a number of operators, the bulk of its functionality is based on the application of heuristics. The first task of the EG level is to control stylistic presentation, comprising a number of relatively low level 'tweaks' such as the vocabulary range, syntactic construction (eg. the ratio of active to passive constructions) and the frequency of specific devices such as alliteration, repetition, metaphor and analogy. There are a large number of such factors detailed in the stylistic literature - see Sandell (1977) for an overview. To date, such factors have been controlled by quite artificial and simplistic means, typically, the user setting a number of variables to particular values (eg. Vreeswijk (1995)). For the instantiations to be made and altered 'on-the-fly', the EG level must refer to a body of parameters, in addition to the belief model of the audience and context, all of which are modified dynamically.

One vitally important parameter affecting the argument is the relationship which the speaker wishes to create or maintain with the hearer. This relationship is established through stylistic rather than structural means, and is not necessarily divorced from other aims: if a hearer accepts the speaker's authoritative stance, for example, the speaker may be able to use the relationship to reinforce his statements. Attempts at instigating different relationships between speaker and hearer account (at least in part) for a number of complex phenomena - humour, for example, is frequently used to establish the speaker as a friend (and therefore reliable and truthful). Such phenomena are not intended to fall inside the scope of this work.

The technical and general competence of the hearer are also important parameters to be considered at the outset. General competence determines the hearer's ability to understand complex argumentation (and to some extent, complex grammar); technical competence enables the argument to pitched at the right level, and affects the choice of appropriate vocabulary.

The initial goals of the argument are also best viewed as a parameter: as mentioned above, the primary goal in different situations (debating house, soap box, law courts, etc.) is subtly different: convincing an opponent is probably the most usual goal, but arguments are also used to impress, to instruct, to confound, to counter, to deceive and to provoke. Importantly, there are often several concurrent goals, such that a number of goal-dependent heuristics could be active at any one stage. Many of the EG level's suggestions are affected - vocabulary, grammar, gross structure and content all depend, to some extent, upon the goals of the system.

A number of other parameters have also been identified, including the speaker's potential gain from 'winning' the argument (in addition to the speaker's

562

gain as perceived by the hearer), the speaker's investment, or potential losses, suffered in losing the argument, the medium in which the argument is expressed and aspects of the speaker's model of the hearer's beliefs such as possible scepticism or bias. For example, advertisements usually have the aim of convincing the hearer to buy the product. In addition though, there are aims connected with the product identity, the particular advertising campaign, and often the speaker-hearer relationship. An attempt is frequently made to hide or play down the speaker's involvement and gain, whilst very careful analysis is made of the audience at which the advertisement is aimed, and the consequent weaknesses and beliefs of that audience.

Some of the EG heuristics are much less dependent upon these parameters: those aiding in the premise-conclusion ordering, for example, rely almost entirely upon general principles (though do also make use of factors such as the general tone of the argument). There are three possible statement orderings in an argument: Conclusion-First (C P*), Conclusion-Last (P* C), and Conclusion-Sandwich (P* C P*). The first is usually used where the P* are examples (in the case of factual conclusions: for opinions, the P* would often be analogies), especially when the hearer is being led to a hasty generalisation from the P* to the C. Conclusion-First is also used when the initial conclusion is deliberately provocative - the construction being used to draw attention to the argument (and consequently it is unusual to find a weak argument structured with the Conclusion-First ordering). It can also be forced should a premise not be accepted by the hearer, and require a sub-argument in its support. Conclusion-Last is the usual choice for longer, more complex or less convincing arguments (indeed some arguments are less convincing precisely because they are longer or more complex). It is also used for 'thin end of the wedge' argument and for grouping together premises which individually lend only very weak support to the conclusion. The Conclusion-Sandwich construction is rarer, usually occurring when the speaker has completed an Conclusion-Last argument which has not been accepted and therefore requires further support.

Finally, there are a number of features controlled by the EG level which lie somewhere between the stylistic and the rhetoric: well known artifices such as repetition and alliteration which can prove extremely effective in constructing eloquent and compelling argument. It is interesting to note that almost all of EG heuristics are devices listed by classical texts as figures and tropes. Indeed this is a uniquely interesting property of the EG level: the emphasis it puts in utilisation of ideas posited in precomputational treatises, especially the classical texts of Cicero and Aristotle (Freese (1926), for example) and the ideas developed in the late middle ages and renaissance (Whately (1855) and Blair (1838) have been used as source material for much of the analysis). This fact poses unique challenges as well as affording unique advantages: on the one hand, the ideas are clear and unbiased, whilst on the other, may be difficult to transcribe to implementation.

One of the key functions of both the AS and EG levels is to maintain a representation of the intention behind utterances. RST has been criticised for its restrictive handling of intention (Moore and Pollack 1992), (Young and Moore

1994), an undeniably vital ingredient in discourse. Thus the AS and EG levels have to manipulate rather different data structures from those employed at lower levels (which would appear quite reasonable, given the difference in data structures at the RST level compared with lower levels). Due to these important differences, the current work is concentrating specifically on the higher, more abstract functionality. The remaining, lower level functionality required to realise the abstract intentional structures into natural language utterances will be provided by a system such as LOLITA (Smith et al. 1994), a large scale, domain independent natural language system, in which a natural language generation algorithm is already implemented which subsumes responsibility for solving certain low-level text generation planning problems.

4.5. Interaction

It is intended that the system should be fully interactive, and enter in to dynamic natural language dialogue with a human interlocutor. There are a number of issues specific to this goal, most obviously, the differences between monologue and dialogue. Both monologue and dialogue require belief modelling of the audience (at the very least for argument, and quite possibly for all natural language); during production of a monologue however, no opportunity is presented for checking, revising and refining its audience model, which makes extended arguments simpler to construct (with no need to modify the plan according to new information regarding the beliefs of the hearer), but as a consequence, rather less likely to succeed at convincing a hearer of the proposition in hand. That is not to say that monologue is of no interest: quite to the contrary, since Debating House argument could be seen as the monologues of the two opposing interlocutors as opposed to true dialogue. Work on monologue-formed argument would probably be easier to tackle in the short term, but it ignores many critical aspects of modelling which are of longer term interest, in particular, in monologue the classical planning assumptions of perfect knowledge of effects and environment state can be employed, while with dialogue it is necessary to consider the problems of uncertainty, imperfect knowledge (and the need to acquire knowledge) and plan failure.

During true dialogue, then, competent belief revision is essential. This revision process is nontrivial and forms an important part of the functioning of the AS and EG levels, drawing in particular from the work of Gardenfors (1992) and Galliers (1992) for implementation.

Dialogue also affords an opportunity for the application of argument to the areas in which it is most useful: negotiation and conflict resolution (where the interlocutors may not already be disposed to helping one another). The literature details much work on negotiatory communication (eg. Zlotkin and Rosenschein (1990), emphasise conflict resolution in non-cooperative domains, ie. those in which the negotiation set may be empty). Much of this work has assumed agent-agent communication, whereas in this paper it is proposed that a similar approach could be adopted in modelling natural language human-computer interaction. There are situations in which the user has differing goals to that of the system, especially where the system has predefined aims. For example, a system which provides an interface to an online information provider might have goals of persuading the user that it has the required information, that it is the cheapest or most reliable source, and, therefore, that the user should buy from it immediately.

Another benefit (or complexity) of dialogue is that communicative failure can be detected and recovery instigated (a vital ability if misunderstandings are to be avoided: see for example, Moore and Swartout (1991)). It would appear that failure can occur at any of the levels of processing, each level producing characteristic failures: at the lowest level, lexical ambiguities such as homonyms can induce failure. At the RST level too, ambiguities in the hearer's RST analysis of an utterance can lead to serious failure. The EG level's enthymemes could misfire (a conclusion may be obvious and omitted by the speaker, but may be non-obvious to the hearer who then misunderstands the argument). Finally, the structural operators of the AS level could lead to misunderstanding in one of two ways: either the hearer fails to see the applicability of an induction or fallacy in trying to prove the speaker's point (a naive failure), or else the hearer identifies and draws attention to the rhetoric trick (an informed failure).

There would seem to be two possible methods by which failure could be recognised and dealt with. Firstly, there is the intuitively appealing idea of having the recognition and recovery occur at the level at which the error belongs. For example, if the failure was syntactic, it would become evident at the syntactic level that communication had failed, and there would then be immediate recovery. This may at first sight seem appealing, but it does not appear to be either computationally or psychologically very plausible: computationally, it is very difficult to see how the syntactic level could possibly recognise that a communication failure had occurred, and psychologically, from analysing argument, it is clear that even something as trivial as homonymity can rupture an argument to the point of stepping outside the entire argument and performing a little meta-conversation to rectify the disparity in the definitions employed. So it would appear that regardless of the type of the error, recovery is carried out at the highest level of planning. This would seem to suggest that perhaps recognition only occurs here too. This is supported by the fact that the major means of discovering failure (checking for disparity between the intentional structure behind the hearer's utterances and the system's belief model of the hearer) is only available to the higher planning levels. It is also only the higher level which can check the intentional structure to decide whether or not a communicative failure needs to be replanned because it is an intentional, core part of the argument, rather than some extraneous side effect (this is an important distinction, recognised in the fields of both communication recovery and of planning - see, for example, McCoy (1986) and Young and Moore (1994), respectively).

The reactivity inherent to such an approach to dialogue forces a style of planning in which plan construction is interleaved with plan execution, following the work of, for example, Ambros-Ingerson and Steel (1990).

5. Conclusion

This paper has presented an outline of the architecture required for constructing extended arguments from the highest level of pragmatic, intention-rich goals to a string of utterances, using a core hierarchical planner employing structural and rhetorical operators and heuristics. The proposed architecture is the first clear framework which presents a computational view of the rhetorical means used by humans to create cogent argument.

Work is currently underway to characterise the parameters and operators involved in the early stages of argument synthesis, and to relate these to aspects of argument unique to dialogue, such as turn taking, checking moves, etc., building upon the work of Levin and Moore (1977).

The architecture proposed will facilitate implementation of an HCI system capable of dealing with situations in which its own goals conflict with those of the user, or in which persuading the user to adopt new beliefs is a nontrivial task, whose success cannot be guaranteed.

References

Ackermann, R.J. "Belief and Knowledge", Anchor, New York (1972)

Ambros-Ingerson, J.A., Steel, S. "Integrated Planning, Execution and Monitoring", in Readings in Planning, (1990) 735-740.

Blair, H. "Lectures on Rhetoric and Belles Lettres", Charles Daly, London (1838)

Cohen, P.R., Levesque, H.J. "Rational Interaction as the Basis for Communication" in Cohen, P.R., Morgan, J., Pollack, M.E., (eds), Intentions in Communication, MIT Press, Boston (1990) 221-255

Fox, M., Long., D. P. "Hierarchical Planning using Abstraction", IEE Proceedings on Control Theory and Applications (1995)

Freese, J.H. (trans), Aristotle "The Art of Rhetoric", Heinmann, London (1926)

Galliers, J.R. "Autonomous belief revision and communication" in Gardenfors, P., (ed), Belief Revision, Cambridge University Press, Cambridge (1992) 220-246

Gardenfors, P. "Belief Revision: An Introduction" in Gardenfors, P., (eds), Belief Revision, Cambridge University Press, Cambridge (1992) 1-28

Grosz, B., Sidner, C.L "Plans for Discourse" in Cohen, P.R., Morgan, J. & Pollack, M.E., (eds), Intentions in Communication, MIT Press, Boston (1990) 418-444

Hovy, E. H. "Automated Discourse Generation Using Discourse Structure Relations", Artificial Intelligence 63,(1993) 341-385

Johnson, R.M. "A Logic Book", Wadsworth, Belmont CA (1992)

Levin, J.A & Moore, J.A. "Dialogue Games: Metacommunication Structures for Natural Language Interaction", Cognitive Science 1, (1977) 395-420

Long, D., Garigliano, R. "Reasoning by Analogy and Causality: A Model and Application", Ellis Horwood (1994)

McCoy, K.F. "Contextual Effects on Responses to Misconceptions", in Kempen, G., (ed), Natural Language Generation: New Results in Artificial Intelligence, Psychology and Linguistics, Kluwer (1986) 43-54

Mann, W.C., Thompson, S.A. "Rhetorical structure theory: description and construction of text structures" in Kempen, G., (ed), Natural Language Generation: New Results in Artificial Intelligence, Psychology and Linguistics, Kluwer (1986) 279-300

Moore, J.D., Pollack, M.E. "A Problem for RST: The Need for Multi-Level Discourse Analysis", Computational Linguistics 18 (4), (1992) 537-544

Moore, J.D., Swartout, W.R. "A Reactive Approach to Explanation: Taking the User's Feedback into Account", in Paris, C.L., Swartout, W.R. & Mann, W.C. (eds), Natural Language Generation in AI & Computational Linguistics, Kluwer, (1991) 3-48

Sacerdoti, E.D. "A Structure for Plans and Behaviour", Elsevier, Amsterdam (1977)

Sandell, R. "Linguistic Style and Persuasion", Academic Press, London (1977)

Smith, M.H., Garigliano, R., Morgan, R.C. "Generation in the LOLITA system: An engineering approach" in Proceedings of the 7th International Workshop on Natural Language Generation (INLG'94), Kennebunkport, Maine (1994)

Vreeswijk, G. "IACAS: an implementation of Chisolm's Principles of Knowledge", Proceedings of the 2nd Dutch/German Workshop on Nonmonotonic Reasoning, Utrecht, Witteveen, C. et al. (eds) (1995) 225-234

Whately, R. "Logic", Richard Griffin, London (1855)

Wilson, B.A. "The Anatomy of Argument", University Press of America, Washington (1980)

Young, R.M., Moore, J.D. "DPOCL: A principled approach to discourse planning" in Proceedings of the 7th International Workshop on Natural Language Generation, Kennebunkport, Maine, (1994) 13-20

Zlotkin, G. & Rosenchein, J.S. "Negotiation and Conflict Resolution in Non-Cooperative Domains", in Proceedings of the National Conference on AI (AAAI'90), (1990) 100-105

Skeptical Query-Answering
in Constrained Default Logic

Torsten Schaub[1] and Michael Thielscher[2]*

[1] Université d'Angers, 49045 Angers Cedex 01, France
[2] International Computer Science Institute, Berkeley, CA 94704-1198, USA

Abstract: An approach to skeptical query-answering in Constrained Default Logic based on the Connection Method is presented. We adapt a recently proposed general method to skeptical reasoning in Default Logics—a method which does neither strictly require the inspection of all extensions nor the computation of entire extensions to decide whether a formula is skeptically entailed. We combine this method with a credulous reasoner which uses the Connection Method as the underlying calculus for classical logic. Furthermore, we develop the notion of a skeptical default proof and show how such a proof can be extracted whenever our calculus proves skeptical entailment of a particular query.

1 Introduction

Nonmonotonic Logics in general, and approaches like Autoepistemic [10] or Default Logic [14] in particular, aim at extending an underlying classical logical system in order to provide conclusions that go beyond this system. For this, they induce one or several so-called *extensions* of a given world description, each of which represents a reasonable set of beliefs. This phenomenon of multiple extensions suggests two natural approaches to query-answering: A *credulous* one, in which a query is said to be derivable if it belongs to a *single* extension, and a *skeptical* one, in which one stipulates that a query lies in *all* extensions.

So far, computational approaches to nonmonotonic logics have mainly focused on the computation of entire extensions, like [4, 19, 7, 22, 11], or credulous query-answering, like [14, 18]. [8] compute intersections of extensions in Autoepistemic Logic. Skeptical query-answering has up to now been primarily studied in restricted nonmonotonic reasoning frameworks, like Theorist [12] (corresponding to so-called prerequisite-free default theories in Default Logic) [13, 20]. From the perspective of Default Logic, implementations of Circumscription, like [5], fall into the same category since they use roughly the same restricted fragment of Default Logic. Finally, a major category of implementations for fragments of Default Logic is given by the wide body of implementations of Logic Programming.

In what follows, we develop a method for skeptical query-answering in Default Logics. This work builds on [21], where a general framework to skeptical reasoning in (semi-monotonic) Default Logics was proposed. There, we have given a high-level description of skeptical query-answering that abstracts from an underlying credulous reasoner. In this paper, we make the aforementioned meta-algorithm precise and employ it to extend an existing approach to credulous query-answering [18] based on the Connection Method [1].

The reader may wonder why we have chosen Constrained Default Logic [16, 3] rather than Reiter's original approach. In fact, Constrained Default Logic serves us as an exemplary Default Logic enjoying the property of *semi-monotonicity*, which stipulates that

---

* On leave from FG Intellektik, TH Darmstadt

the addition of default rules to a theory does not invalidate the application of previously applied default rules. As pointed out in [14], semi-monotonicity is indispensable for feasible query-answering in Default Logics. This is so because it allows for local proof procedures focusing on default rules relevant for answering a query; otherwise the *whole* set of default rules has to be taken into account. Now, semi-monotonicity is only enjoyed by so-called *normal* default theories in Reiter's Default Logic. However, since both aforementioned Default Logics coincide on normal default theories, our exposition applies to this fragment of Reiter's Default Logic, too. As concerns other variants of Default Logic, we note that Constrained Default Logic yields the same conclusions as Cumulative Default Logic [2]; both variants differ in representational issues only. Moreover, Constrained Default Logic coincides with Rational Default Logic on the large fragment of so-called semi-normal default theories [9]. All these interrelations render our exemplar, Constrained Default Logic, a prime candidate for exposing our approach.

Of course, a similar question may arise concerning the choice of the Connection Method [1]. Unlike resolution-based methods that decompose formulas in order to derive a contradiction, the Connection Method analyses the structure of formulas for proving their unsatisfiability. In fact, we will see that skeptical query-answering requires numerous variants of similar subproofs. In such a case, it is advisable to reuse information gathered on similar structures. This approach is supported by the structure-sensitive nature of the Connection Method. As a result, we obtain a homogeneous characterization of skeptical default proofs at the level of the underlying deduction method.

The paper is organized as follows. After recapitulating the basic concepts of Constrained Default Logic in Section 2, we elaborate in Section 3 on the general framework to skeptical query answering proposed in [21]. The latter provides an abstract decision procedure for skeptical query-answering in (semi-monotonic) Default Logics; it has its roots in [13, 20], which address skeptical reasoning in Theorist [12]. In fact, we formalize the general ideas presented in [21] by appeal to sequences of default rules. As a result, we obtain in Section 3 an algorithm instantiating the general framework in [21]. The resulting formal underpinnings given in Theorem 3, 4, and 5 are obtained as corollaries to results in [21]. In addition, Section 3 offers a general definition of a *skeptical default proof*—a point left open in [21]. Such a skeptical default proof is returned by the algorithm developed in the same section. Section 4 describes our underlying method for credulous query-answering based on the Connection Method [18]. The major contribution of this paper is presented in Section 5: We develop an analytic calculus for skeptical query-answering by combing the approaches described in Section 3 and 4. We give a soundness and completeness result of this approach and illustrate how skeptical default proofs can be extracted whenever a query has been successfully proven. Our results are summarized in Section 6.

2 Constrained Default Logic

We consider a straightforward yet powerful extension of Constrained Default Logic [3], called *Pre-Constrained Default Logic* [17]. The idea is to supplement some initial consistency constraints that direct the subsequent reasoning process. This is a well-known technique, also used in Theorist [12], in which the "context of reasoning" is predetermined and subsequently dominated by some initial consistency requirements. These additional constraints play an important role in our approach to skeptical query-answering, as we will see below.

A *pre-constrained default theory* $(\mathcal{D}, \mathcal{W}, \mathcal{C})$ (*default theory*, for short) consists of a set of formulas $\mathcal{W}$, a set of default rules $\mathcal{D}$, and a set of formulas $\mathcal{C}$ representing some initial constraints. A *default rule* is any expression of the form $\frac{\alpha \,:\, \beta}{\gamma}$ where α, β, γ are formulas. For convenience, we denote the prerequisite α of a default rule by $Prereq(\delta)$, its justification β by $Justif(\delta)$, and its consequent γ by $Conseq(\delta)$.[3] A *normal default theory* is restricted to *normal default rules*, whose justification is equivalent to the consequent. For simplicity, we deal with a propositional language over a finite alphabet and assume that $\mathcal{W} \cup \mathcal{C}$ is satisfiable. A *constrained extension* is defined as follows.

Definition 1. Let $(\mathcal{D}, \mathcal{W}, \mathcal{C})$ be a default theory and let E and C be sets of formulas. Define $E_0 = \mathcal{W}$ and $C_0 = \mathcal{W} \cup \mathcal{C}$ and for $i \geq 0$

$$E_{i+1} = Th(E_i) \cup \{\ \gamma\ \mid \tfrac{\alpha \,:\, \beta}{\gamma} \in \mathcal{D}, \alpha \in E_i, C \cup \{\beta\} \cup \{\gamma\} \not\vdash \bot\}$$

$$C_{i+1} = Th(C_i) \cup \{\beta \wedge \gamma \mid \tfrac{\alpha \,:\, \beta}{\gamma} \in \mathcal{D}, \alpha \in E_i, C \cup \{\beta\} \cup \{\gamma\} \not\vdash \bot\}$$

(E, C) is a constrained extension of $(\mathcal{D}, \mathcal{W}, \mathcal{C})$ if $(E, C) = (\bigcup_{i=0}^{\infty} E_i, \bigcup_{i=0}^{\infty} C_i)$.

Observe that the initial constraints, $\mathcal{C}$, enter merely the final constraints C (at C_0) but not the extension E. Thus, the initial constraints $\mathcal{C}$ direct the reasoning process without actually becoming a part of it. In particular, they usually decrease the number of applicable default rules.

Let us guide the formal development of our approach by means of the following example. Consider the default statements "quakers are doves if they are no anti-pacifists," "republicans are hawks if they are no pacifists," "doves as well as hawks are traditionalists," along with the strict knowledge telling us that we have a "republican quaker." This is formalized in the following default theory:

$$\left(\left\{\tfrac{Q\,:\,P}{D}, \tfrac{R\,:\,\neg P}{H}, \tfrac{D\,:\,T}{T}, \tfrac{H\,:\,T}{T}\right\}, \{Q, R\}, \emptyset\right) \tag{1}$$

The first and second default cannot be combined in a single constrained extension due to their mutually exclusive justifications, P and $\neg P$. Hence, this theory has two constrained extensions, one containing D and T and another one containing H and T.

In the sequel, we follow [18] in dealing with default theories in atomic format in the following sense: For a default theory $\Delta = (\mathcal{D}, \mathcal{W}, \mathcal{C})$ in some language $\mathcal{L}_\Sigma$, let $\mathcal{L}_{\Sigma'}$ be the language obtained by adding, for each $\delta \in \mathcal{D}$, three new propositions, named $\alpha_\delta, \beta_\delta, \gamma_\delta$, which do not occur elsewhere. Then, Δ is mapped into a default theory $\Delta' = (\mathcal{D}', \mathcal{W}', \mathcal{C}')$ in $\mathcal{L}_{\Sigma'}$ where

$$\mathcal{D}' = \left\{\tfrac{\alpha_\delta \,:\, \beta_\delta}{\gamma_\delta} \ \middle|\ \delta \in \mathcal{D}\right\}$$

$$\mathcal{W}' = \mathcal{W} \cup \{Prereq(\delta) \to \alpha_\delta, \beta_\delta \to Justif(\delta), \gamma_\delta \to Conseq(\delta) \mid \delta \in \mathcal{D}\}$$

$$\mathcal{C}' = \mathcal{C}\,.$$

The resulting default theory Δ' is called the *atomic format* of the original default theory, Δ. As shown in [15], this transformation is a conservative extension of the formalism.

---

[3] These projections extend to sets and sequences of default rules in the obvious way.

Hence, it does not affect the computation of queries to the original default theory. We can therefore restrict our attention to atomic default rules without losing generality. The advantages of atomic default rules over arbitrary ones are, first, that their constituents are not spread over several clauses while transforming them into clausal format and, second, that these constituents, e.g. the consequents, are uniquely referable to. The motivations for this format are detailed in [18] and they are somehow similar to the ones for clausal form in automated theorem proving.

3 Skeptical reasoning in constrained default logic

In classical logic, we can say that a formula φ is derivable from a set of facts $\mathcal{W}$ iff it belongs to the deductive closure of $\mathcal{W}$, that is if $\varphi \in Th(\mathcal{W})$. Due to the possible existence of multiple extensions, this notion of derivability is not directly applicable to Default Logic. Rather we obtain two different notions of derivability: A formula φ is *credulously derivable* from $(\mathcal{D}, \mathcal{W}, \mathcal{C})$ iff $\varphi \in E$ for some constrained extension (E, C) of $(\mathcal{D}, \mathcal{W}, \mathcal{C})$.[4] And a formula φ is *skeptically derivable* from $(\mathcal{D}, \mathcal{W}, \mathcal{C})$ iff φ belongs to all such extensions of $(\mathcal{D}, \mathcal{W}, \mathcal{C})$. In our example, (1), D and H are only credulously derivable while T is also skeptically derivable.

In order to furnish a corresponding proof-theory, we need the following concepts. A *default proof segment* in a default theory $\Delta = (\mathcal{D}, \mathcal{W}, \mathcal{C})$ (or Δ-*segment*, for short) is a (finite) sequence of default rules $\langle \delta_i \rangle_{i \in I}$ such that

$$W \cup Conseq(\{\delta_0, \ldots, \delta_{i-1}\}) \vdash Prereq(\delta_i) \text{ for } i \in I \quad \text{and} \tag{2}$$

$$W \cup C \cup \{Conseq(\delta_i), Justif(\delta_i) \mid i \in I\} \text{ is satisfiable.} \tag{3}$$

A *credulous default proof*, or CDP for short, for a formula φ from Δ is a Δ-segment $\langle \delta_i \rangle_{i \in I}$ such that $W \cup \{Conseq(\delta_i) \mid i \in I\} \vdash \varphi$. Furthermore, we say that a formula φ is *provable from a* Δ-*segment* $\langle \delta_i \rangle_{i \in I}$ iff there is a CDP $\langle \delta_j \rangle_{j \in J}$ for φ such that $I \subseteq J$, that is, if the segment is extendible to a CDP for φ. In our example, there are five Δ-segments, each of which is extendible to one of the two CDPs of T, namely

$$\left\langle \frac{Q:P}{D}, \frac{D:T}{T} \right\rangle \quad \text{and} \quad \left\langle \frac{R:\neg P}{H}, \frac{H:T}{T} \right\rangle. \tag{4}$$

Clearly, a formula is credulously derivable iff it is provable from some Δ-segment, since in this case it has a CDP. Accordingly, a formula is skeptically derivable iff it is provable from all Δ-segments.[5]

A basic question is that on the concept of a default proof in skeptical reasoning. Since in general there is not a single CDP valid in each extension, it is however natural to view a skeptical default proof as being compound of multiple CDPs. In fact, we take a skeptical default proof of a formula to be a set $\mathcal{P}$ of CDPs such that $\mathcal{P}$ is *complete*, that is, for each constrained extension (E, C), $\mathcal{P}$ includes a proof which is valid in (E, C):

Definition 2 Skeptical Default Proof. Let $\Delta = (\mathcal{D}, \mathcal{W}, \mathcal{C})$ be a default theory and φ a formula. A <u>skeptical default proof</u> of φ from Δ is a set $\mathcal{P}$ of CDPs for φ such that for each constrained extension (E, C) of Δ there is some $\langle \delta_i \rangle_{i \in I} \in \mathcal{P}$ such that $C \cup \{Conseq(\delta_i), Justif(\delta_i) \mid i \in I\}$ is satisfiable.

---

[4] For brevity, we sometimes simply say φ "belongs to" (or "is contained in") an extension (E, C), which always means $\varphi \in E$.

[5] This is formally shown in [21].

This view relies heavily on the notion of CDPs. In fact, we keep this fundamental idea and base our method for skeptical reasoning on credulous reasoning, too [21]: The idea is to start with an arbitrary CDP of a given query. Then, we determine in some way a representative selection of Δ-segments incompatible with our initial CDP. These Δ-segments indicate extensions in which our initial default proof is invalid. Intuitively, they can be thought of as putative counterarguments challenging our initial CDP. Next, we verify in turn whether our query is derivable from each such Δ-segment. If this is indeed the case, then our initial query is skeptically derivable.

In order to illustrate this approach, let us verify that T is skeptically derivable in our example. We start with an arbitrary CDP of T from Default Theory (1). Consider the second CDP in (4): $\left\langle \frac{R:\neg P}{H}, \frac{H:T}{T} \right\rangle$. This proof takes place in the extension of (1) containing H and T. Next, we regard all Δ-segments 'challenging' default rules in $\left\langle \frac{R:\neg P}{H}, \frac{H:T}{T} \right\rangle$. This notion is captured formally by the property of *orthogonality*:[6] Two default proof segments $\langle \delta_i \rangle_{i \in I}$ and $\langle \delta_j \rangle_{j \in J}$ in a default theory $(\mathcal{D}, \mathcal{W}, \mathcal{C})$ are called $\mathcal{C}$-*orthogonal* iff $\mathcal{W} \cup \mathcal{C} \cup \{Conseq(\delta_k), Justif(\delta_k) \mid k \in I \cup J\}$ is unsatisfiable. That is, a Δ-segment is orthogonal to another one if its induced constraints, i.e., the set of justifications and consequents of its default rules, are incompatible with the same constraints of the other Δ-segment. Observe that each Δ-segment orthogonal to a given CDP indicates one or more extensions in which our CDP is not valid.

The formal basis for our approach is laid in the following theorem.

Theorem 3. *Let* $(\mathcal{D}, \mathcal{W}, \mathcal{C})$ *be a default theory and* φ *a formula. Then,* φ *is skeptically derivable from* $(\mathcal{D}, \mathcal{W}, \mathcal{C})$ *iff there is a* CDP $\langle \delta_i \rangle_{i \in I}$ *for* φ *and* φ *is provable from all default proof segments which are* $\mathcal{C}$-*orthogonal to* $\langle \delta_i \rangle_{i \in I}$.

In our example, there are two Δ-segments orthogonal to $\left\langle \frac{R:\neg P}{H}, \frac{H:T}{T} \right\rangle$, namely

$$\left\langle \frac{Q:P}{D} \right\rangle \text{ and } \left\langle \frac{Q:P}{D}, \frac{D:T}{T} \right\rangle.$$

This is so because the justification of the first default rule, $\frac{R:\neg P}{H}$, in our CDP is contradictory to the justification of default rule $\frac{Q:P}{D}$. There is no Δ-segment orthogonal to the second default rule in our default proof. As will be shown in Theorem 4, we can restrict our attention to minimal orthogonal Δ-segments. Accordingly, it is sufficient to consider the orthogonal Δ-segment $\left\langle \frac{Q:P}{D} \right\rangle$.

Intuitively, we then focus on all extensions of the initial default theory to which the Δ-segment $\left\langle \frac{Q:P}{D} \right\rangle$ contributes and check whether our initial query T belongs to these extensions too. Importantly, this is accomplished by using only default rules relevant for deriving T and hence without computing any extensions. We achieve this by checking whether T is skeptically derivable from the default theory obtained by 'applying' the default rules in our orthogonal Δ-segment. In this way, we try to prove our query under the restrictions imposed by the Δ-segment contesting our initial CDP. To this end, we add the consequence of the default rule $\frac{Q:P}{D}$ to the facts of Default Theory (1) while deleting the default rule itself. Furthermore, we have to add its justification to the set of

---

[6] Orthogonality usually refers to distinct extensions (c.f. [14]). In Constrained Default Logic, two constrained extensions (E, C) and (E', C') are orthogonal iff $C \cup C'$ is unsatisfiable [3].

initial constraints. This yields the following modified default theory:

$$\left(\left\{\frac{R:\neg P}{H}, \frac{D:T}{T}, \frac{H:T}{T}\right\}, \{Q, R, D\}, \{P\}\right) \tag{5}$$

Now, it remains to be shown that T be skeptically derivable from Default Theory (5). Proceeding recursively, we check first whether T is credulously derivable from this default theory. In fact, T can be proven by means of the CDP $\left\langle\frac{D:T}{T}\right\rangle$. Next, we have to proceed as above and in turn find all minimal Δ-segments orthogonal to $\frac{D:T}{T}$. However, there are no such segments since Default Theory (5) has a single extension containing D and T, in which, moreover, default rule $\frac{R:\neg P}{H}$ is blocked due to initial constraint P.

Importantly, the previous step supplies us with an alternative CDP of T from our original default theory in (1). This is obtained by appending Δ-segment $\left\langle\frac{Q:P}{D}\right\rangle$ and the CDP of T from default theory (5), viz. $\left\langle\frac{D:T}{T}\right\rangle$. This results in the first CDP given in (4). Since there are no other Δ-segments orthogonal to our initial default proof $\left\langle\frac{R:\neg P}{H}, \frac{H:T}{T}\right\rangle$, we are done. As a result, we have obtained a skeptical default proof consisting of two CDPs supporting the skeptical conclusion T. This approach is justified by the following theorem [21]:

Theorem 4. *Let $\langle\delta_i\rangle_{i\in I}$ be a default proof segment in a default theory $(\mathcal{D}, \mathcal{W}, \mathcal{C})$ and φ a formula. Then, φ is provable from all default proof segments $\langle\delta_j\rangle_{j\in J}$ where $I \subseteq J$ iff φ is skeptically derivable from*

$$(\mathcal{D} \setminus \{\delta_i \mid i \in I\}, \mathcal{W} \cup \{Conseq(\delta_i) \mid i \in I\}, \mathcal{C} \cup \{Justif(\delta_i) \mid i \in I\}).$$

Let us summarize our approach to checking whether a query φ is skeptically derivable: We start with a CDP of φ. Then, we determine all minimal orthogonal Δ-segments contesting our initial CDP. Next, we check in turn whether φ is skeptically derivable under the restrictions imposed by each such Δ-segment. In this way, we check whether φ belongs to all extensions orthogonal to the one containing our initial CDP of φ. It is important to note that the choice of the initial CDP is a *"don't care"*-choice in so far that deciding whether φ is skeptically entailed is independent of which CDP is initially chosen. The design of the resulting skeptical default proof, on the other hand, depends on the latter choice.

The approach we have outlined so far can be put together to a concrete algorithm as follows. Let $(\mathcal{D}, \mathcal{W}, \mathcal{C})$ be a default theory and φ be a formula. We assume a function $cred(\mathcal{D}, \mathcal{W}, \mathcal{C}, \varphi)$ such that

$$cred(\mathcal{D}, \mathcal{W}, \mathcal{C}, \varphi) = \begin{cases} \langle\delta_i\rangle_{i\in I} & \text{if } \langle\delta_i\rangle_{i\in I} \text{ is a CDP of } \varphi \text{ from } (\mathcal{D}, \mathcal{W}, \mathcal{C}) \\ \bot & \text{otherwise} \end{cases}$$

Moreover, given a CDP $\langle\delta_i\rangle_{i\in I}$, let $orth(\mathcal{D}, \mathcal{W}, \mathcal{C}, \langle\delta_i\rangle_{i\in I})$ yield the set of all minimal Δ-segments $\mathcal{C}$-orthogonal to $\langle\delta_i\rangle_{i\in I}$.[7] These two functions yield the following algorithm for skeptical query-answering. Similar to its credulous source, it returns $\bot$ if φ is not skeptically derivable; otherwise it returns a set of CDPs forming a skeptical default proof. The function "∘" concatenates two Δ-segments.

---

[7] Note that this set is finite in case of a finite propositional alphabet. The procedure *orth* will be designed below.

$skep(\mathcal{D}, \mathcal{W}, \mathcal{C}, \varphi) =$
if $cred(\mathcal{D}, \mathcal{W}, \mathcal{C}, \varphi) = \perp$
  then return $\perp$
  else let $\langle \delta_i \rangle_{i \in I} = cred(\mathcal{D}, \mathcal{W}, \mathcal{C}, \varphi)$ **in**
    let $\mathcal{O} = orth(\mathcal{D}, \mathcal{W}, \mathcal{C}, \langle \delta_i \rangle_{i \in I})$ **in**
    $\mathcal{P} := \emptyset$;
      while $\mathcal{O} \neq \emptyset$ **do**
        select $\langle \delta_j \rangle_{j \in J} \in \mathcal{O}$;
        let $\mathcal{D}' = \mathcal{D} \setminus \{\delta_j \mid j \in J\}$ **in**
        let $\mathcal{W}' = \mathcal{W} \cup \{Conseq(\delta_j) \mid j \in J\}$ **in**
        let $\mathcal{C}' = \mathcal{C} \cup \{Justif(\delta_j) \mid j \in J\}$ **in**
          if $skep(\mathcal{D}', \mathcal{W}', \mathcal{C}', \varphi) = \perp$
          then return $\perp$
          else $\mathcal{P} := \mathcal{P} \cup \{\langle \delta_j \rangle_{j \in J} \circ \langle \delta_l \rangle_{l \in L} \mid \langle \delta_l \rangle_{l \in L} \in skep(\mathcal{D}', \mathcal{W}', \mathcal{C}', \varphi)\}$
        fi ;
        $\mathcal{O} := \mathcal{O} \setminus \{\langle \delta_j \rangle_{j \in J}\}$
      od ;
      return $\{\langle \delta_i \rangle_{i \in I}\} \cup \mathcal{P}$
fi

The variable $\mathcal{P}$ accumulates the set of resulting CDPs. Whenever the procedure ends up with success, the CDPs in $\mathcal{P}$ form a skeptical default proof of φ from $(\mathcal{D}, \mathcal{W}, \mathcal{C})$.

Finally, we have to address the determination of minimal Δ-segments orthogonal to a CDP at hand. Again, this is accomplishable by appeal to credulous reasoning. Note that the notion of orthogonality refers to the consistency constraints induced by a CDP. For this, we have to consider additionally a default rule's justification any time the default rule applies: For a set of atomic default rules $\mathcal{D}$, we define the set of *normalization rules* as $N(\mathcal{D}) = \{\gamma_\delta \rightarrow \beta_\delta \mid \frac{\alpha_\delta : \beta_\delta}{\gamma_\delta} \in \mathcal{D}\}$. Intuitively, the addition of normalization rules to the facts of a default theory in atomic format turns each atomic default rule $\frac{\alpha_\delta : \beta_\delta}{\gamma_\delta}$ into a default rule of the form $\frac{\alpha_\delta : \beta_\delta}{\beta_\delta \wedge \gamma_\delta}$. With this, the following theorem tells us how to determine Δ-segments orthogonal to a CDP at hand:[8]

Theorem 5. *Let* $\langle \delta_i \rangle_{i \in I}$ *and* $\langle \delta_j \rangle_{j \in J}$ *be default proof segments in a default theory* $(\mathcal{D}, \mathcal{W}, \mathcal{C})$ *in atomic format. Then,* $\langle \delta_i \rangle_{i \in I}$ *and* $\langle \delta_j \rangle_{j \in J}$ *are C-orthogonal iff there is some* $i \in I$ *such that* $\langle \delta_j \rangle_{j \in J}$ *is a CDP for* $\neg Conseq(\delta_i) \vee \neg Justif(\delta_i)$ *in this default theory:*

$$(\mathcal{D} \setminus \{\delta_k \mid k < i\}, \mathcal{W} \cup N(\mathcal{D} \setminus \{\delta_k \mid k < i\}) \cup \{Conseq(\delta_k), Justif(\delta_k) \mid k < i\}, \mathcal{C}) \quad (6)$$

That is, in order to find Δ-segments orthogonal to our CDP $\langle \delta_1 = \frac{R : \neg P}{H}, \delta_2 = \frac{H : T}{T} \rangle$, we consider the default theories[9]

$$\left(\left\{ \frac{Q:P}{D}, \frac{R:\neg P}{H}, \frac{D:T}{T}, \frac{H:T}{T} \right\}, \{Q, R\} \cup \{D \rightarrow P, H \rightarrow \neg P\}, \emptyset \right) \quad (7)$$

---

[8] The following is a variant of Theorem 5.7 in [21], where a more complex normalization procedure is employed. By using $N(\mathcal{D})$ instead, we exploit the fact that we deal with default theories in atomic format only (c.f. Section 2).

[9] For sake of readability, we refrain from presenting Theory (7) and (8) in atomic format. Moreover, we discard normalization rules for normal default rules since they are tautological.

$$\left(\left\{ \tfrac{Q:P}{D}, \tfrac{D:T}{T}, \tfrac{H:T}{T} \right\}, \{Q,R\} \cup \{D \to P\} \cup \{H, \neg P\}, \emptyset \right) \qquad (8)$$

In turn, we must determine all minimal CDPs of the negated consequences or the negated justifications of $\frac{R:\neg P}{H}$ and $\frac{H:T}{T}$, respectively. That is, first we search for a proof for $\neg H \vee P$ from Theory (7) and then for a proof of $\neg T$ from Theory (8). While $\neg T$ is not provable from (8), $\neg H \vee P$ is provable from (7) yielding a single orthogonal Δ-segment, $\left\langle \frac{Q:P}{D} \right\rangle$.

Theorem 5 leads to the following algorithm for the function $orth$:

$orth(\mathcal{D}, \mathcal{W}, \mathcal{C}, \langle \delta_i \rangle_{i \in I}) =$

for $i \in I$ **do**

    let $\mathcal{D}' = \mathcal{D} \setminus \{ \delta_k \mid k < i \}$ **in**

    let $\mathcal{W}' = \mathcal{W} \cup N(\mathcal{D}') \cup \{ Conseq(\delta_k), Justif(\delta_k) \mid k < i \}$ **in**

    let $\mathcal{O} = \emptyset$ **in**

    let $\varphi = \neg Conseq(\delta_i) \vee \neg Justif(\delta_i)$ **in**

    $\mathcal{O} := \mathcal{O} \cup \{ \langle \delta_j \rangle_{j \in J} \mid \langle \delta_j \rangle_{j \in J} = cred(\mathcal{D}', \mathcal{W}', \mathcal{C}, \varphi)$ and there is no $J' \subset J$

                     such that $\langle \delta_{j'} \rangle_{j' \in J'} = cred(\mathcal{D}', \mathcal{W}', \mathcal{C}, \varphi) \}$

od ;

return $\mathcal{O}$

This procedure yields a set containing all minimal Δ-segments orthogonal to a given CDP $\langle \delta_i \rangle_{i \in I}$.

In all, our approach has the following advantages. First, it avoids the computation of entire extensions. Second, it is goal-directed and thus restricted to default rules relevant to proving a given query. Third, it takes advantage of basic techniques developed for credulous reasoning. Clearly, it is necessary to consider all mutually orthogonal CDPs belonging to distinct extensions. In this way, we cannot get around the exponential factor present in worst-case, where there is an exponential number of extensions each of which comprises a CDP orthogonal to all CDPs in all other extensions. In fact, skeptical reasoning is Π_2^P-complete [6]. While this theoretical threshold is inevitable in the worst-case, our local proof theory avoids investigating all or even entire extensions whenever a large domain is 'locally structured,' that is, if the query under consideration is deductively connected with merely a small fraction of the entire theory. This advantage becomes obvious by looking at some more examples. Suppose that we extend Default Theory (1) with a default rule like $\frac{R:W}{W}$ saying that "republicans are Western fans." Proving that W is skeptically derivable is doable by a single CDP, $\left\langle \frac{R:W}{W} \right\rangle$, since this CDP is not contested by any orthogonal Δ-segments. For another example, suppose that we extend Default Theory (1) with default rules like $\frac{Q:V}{V}$ and $\frac{R:\neg V}{\neg V}$ saying that "quakers are vegetarians" and "republicans aren't vegetarians." This leads to four distinct extensions. Proving that T is skeptically derivable however is doable with the same steps as described above. That is, the two additional extensions do not increase computational efforts. This is so because the new default rules are irrelevant to proving T.

4 Credulous query-answering

In what follows, we extend the approach for query-answering in Default Logics developed in [18] to Pre-Constrained Default Logic. This approach is based on the Connection Method [1], which allows for testing unsatisfiability of formulas in conjunctive normal form (CNF). Unlike resolution-based methods that decompose formulas in order to derive a contradiction, the Connection Method analyses the structure of formulas for proving their unsatisfiability.

In the Connection Method, formulas in CNF are displayed two-dimensionally in the form of *matrices* (see (9) for an exemplar). A matrix is a set of sets of literals (literal occurrences, to be precise).[10] Each column of a matrix represents a *clause* of the CNF of a formula. In order to show that a sentence φ is entailed by a sentence $\mathcal{W}$, we prove $\mathcal{W} \wedge \neg\varphi$ be unsatisfiable. In the Connection Method this is accomplished by path checking: A *path* through a matrix is a set of literals, one from each clause. A *connection* is an unordered pair of literals which are identical except for the negation sign (and possible indices). A *mating* is a set of connections. A mating *spans* a matrix if each path through the matrix contains a connection from the mating. Finally, a formula, like $\mathcal{W} \wedge \neg\varphi$, is unsatisfiable iff there is a spanning mating for its matrix.

The approach of [18] relies on the idea that a default rule can be decomposed into a classical implication along with two qualifying conditions, one accounting for the character of an inference rule and another one enforcing the respective consistency conditions. The computational counterparts of these qualifying conditions are given by the proof-oriented concepts of *admissibility* and *compatibility*, which we introduce in the sequel.

In order to find out whether a formula φ is in some extension of a default theory $(\mathcal{D}, \mathcal{W}, \mathcal{C})$, we first transform the default rules in $\mathcal{D}$ into their sentential counterparts. This yields a set of indexed implications:

$$W_{\mathcal{D}} = \left\{ \alpha_\delta \to \gamma_\delta \;\middle|\; \frac{\alpha_\delta : \beta_\delta}{\gamma_\delta} \in \mathcal{D} \right\}$$

Second, we transform both $\mathcal{W}$ and $W_{\mathcal{D}}$ into their clausal forms, $C_{\mathcal{W}}$ and $C_{\mathcal{D}}$. The clauses in $C_{\mathcal{D}}$, like $\{\neg\alpha_\delta, \gamma_\delta\}$, are called δ-*clauses*; all other clauses like those in $C_{\mathcal{W}}$ are referred to as ω-*clauses*. Finally, a query φ is derivable from $(\mathcal{D}, \mathcal{W}, \mathcal{C})$ iff there is a spanning mating for the matrix $C_{\mathcal{W}} \cup C_{\mathcal{D}} \cup \{\neg\varphi\}$ agreeing with the concepts of admissibility and compatibility.[11]

A useful concept is that of a *core* of a matrix M wrt a mating Π, which allows for isolating the clauses relevant to the underlying proof. [18] defines the core of M wrt Π as[12]

$$\kappa(M, \Pi) = \{c \in M \mid \exists \pi \in \Pi \,.\, c \cap \pi \neq \emptyset\}.$$

For instance, the core of Matrix (9) below wrt the mating drawn is given by all clauses connected by arcs. Then, the proof-theoretic counterpart of condition (2), also called groundedness, can be captured as follows [18]:

Definition 6 Admissibility. Let $C_{\mathcal{W}}$ be a set of ω-clauses and $C_{\mathcal{D}}$ be a set of δ-clauses and let Π be a mating for $C_{\mathcal{W}} \cup C_{\mathcal{D}}$. Then, $(C_{\mathcal{W}} \cup C_{\mathcal{D}}, \Pi)$ is <u>admissible</u> iff there is an enumeration $\langle\{\neg\alpha_{\delta_i}, \gamma_{\delta_i}\}\rangle_{i \in I}$ of $\kappa(C_{\mathcal{D}}, \Pi)$ such that for each $i \in I$, Π is a spanning mating for $C_{\mathcal{W}} \cup \left(\bigcup_{j<i}\{\{\neg\alpha_{\delta_j}, \gamma_{\delta_j}\}\}\right) \cup \{\{\neg\alpha_{\delta_i}\}\}$.

---

[10] In the sequel, we simply say literal instead of literal occurrences; the latter allow for distinguishing between identical literals in different clauses.

[11] Without loss of generality, we deal with atomic queries only, since any query can be transformed into 'atomic format.'

[12] Recall that we deal with literal occurrences.

Note that sometimes not all connections in Π are needed for showing the unsatisfiability of the previous submatrices.

As regards compatibility, we have to extend the corresponding notion found in [18] in order to deal with a set of pre-constraints C.

Definition 7 C-compatibility. Let C_W and C_C be sets of ω-clauses and let C_D be a set of δ-clauses. Let Π be a mating for $C_W \cup C_D$ and let $\langle\{\neg\alpha_{\delta_i}, \gamma_{\delta_i}\}\rangle_{i \in I}$ be an enumeration of $\kappa(C_D, \Pi)$. Then, $(C_W \cup C_D, \Pi)$ is $\underline{C\text{-compatible wrt } I}$ iff there is no spanning mating for $C_W \cup C_C \cup \left(\bigcup_{i \in I}\{\{\neg\alpha_{\delta_i}, \gamma_{\delta_i}\}, \{\beta_{\delta_i}\}\}\right)$.

The following theorem shows that our extended method is sound and complete:

Theorem 8. *Let (D, W, C) be a default theory in atomic format and φ an atomic formula. Then, $\varphi \in E$ for some constrained extension (E, C) of (D, W, C) iff there is a spanning mating Π for the matrix $M = C_W \cup C_D \cup \{\{\neg\varphi\}\}$ and an enumeration $\langle\{\neg\alpha_{\delta_i}, \gamma_{\delta_i}\}\rangle_{i \in I}$ of $\kappa(C_D, \Pi)$ which verifies $(C_W \cup C_D, \Pi)$ be admissible and C-compatible (wrt I).*

Finally, (M, Π) represents the CDP $\langle\delta_i\rangle_{i \in I}$ for φ from (D, W, C).

For illustration, let us verify that T is credulously derivable according to the recipe given above. The encoding of the set of default rules yields the set W_D of implications: $\{Q_{\delta_1} \to D_{\delta_1}, R_{\delta_2} \to H_{\delta_2}, D_{\delta_3} \to T_{\delta_3}, H_{\delta_4} \to T_{\delta_4}\}$. The indexes denote the respective default rules in (1) from left to right. In order to verify that a republican quaker is traditionalist, T, we first transform the facts in Default Theory (1) and the implications in W_D into their clausal form. The resulting clauses are given two-dimensionally as the first six columns of the matrix in (9). The full matrix is obtained by adding the clause containing the negated query, $\neg T$. In fact, the matrix has a spanning mating, viz $\{\{R, \neg R_{\delta_2}\}, \{H_{\delta_2}, \neg H_{\delta_4}\}, \{T_{\delta_4}, \neg T\}\}$. We have indicated these connections in (9) as arcs linking the respective literals.

$$
\begin{bmatrix}
& \neg Q_{\delta_1} & \neg R_{\delta_2} & \neg D_{\delta_3} & \neg H_{\delta_4} & \neg T \\
Q & R & D_{\delta_1} & H_{\delta_2} & T_{\delta_3} & T_{\delta_4}
\end{bmatrix} \tag{9}
$$

This proof corresponds to the second one in (4) and yields the following enumeration:

$$
\langle\{\neg R_{\delta_2}, H_{\delta_2}\}, \{\neg H_{\delta_4}, T_{\delta_4}\}\rangle \tag{10}
$$

For admissibility, we must therefore consider the following two submatrices of Matrix (9):

$$
\begin{bmatrix}
& \neg R_{\delta_2} \\
Q & R
\end{bmatrix}
\begin{bmatrix}
& \neg R_{\delta_4} & \neg H_{\delta_4} \\
Q & R & H_{\delta_2}
\end{bmatrix} \tag{11}
$$

Observe that each of these submatrices has a spanning mating, so the original matrix and its mating, given in (9), constitute an admissible proof.

For compatibility (or $\emptyset$-compatibility, to be precise), we have to verify that the following matrix has no spanning mating:[13]

$$
\begin{bmatrix}
& \neg R_{\delta_2} & & \neg H_{\delta_4} \\
Q & R & H_{\delta_2} & \neg P_{\delta_2} & T_{\delta_4} & T_{\delta_4}
\end{bmatrix} \tag{12}
$$

---

[13] The clauses $\{\neg P_{\delta_2}\}$ and $\{T_{\delta_4}\}$ represent the justifications of δ_2 and δ_4, respectively.

This is indeed the case since the matrix contains a non-complementary path, viz. $\{Q, R, H_{\delta_2}, T_{\delta_4}, \neg P_{\delta_2}\}$. We thus obtain an admissible and compatible proof for the original query, T, asking whether a republican quaker is traditionalist.

5 Skeptical query-answering

Let us now return to skeptical query-answering. The basic idea is to extend the given method for credulous query-answering by adding another specific condition on proofs ensuring that a query is skeptically derivable. This extra condition is motivated by the general idea described in Section 3.

At first, we account for the proof-theoretic counterpart of Δ-segments orthogonal to a given CDP. Let $C_{N(\mathcal{D})}$ be the clausal representation of $N(\mathcal{D})$, i.e., $C_{N(\mathcal{D})} = \{\{\neg\gamma_\delta, \beta_\delta\} \mid \delta \in \mathcal{D}\}$. These clauses are needed for adding the justifications of default rules while determining Δ-segments orthogonal to a CDP at hand. Clearly, this is obsolete for normal default theories.

Definition 9 Challenge. Let C_W and C_C be sets of ω-clauses and let C_D be a set of δ-clauses. Let Π be a mating for $C_W \cup C_D \cup \{\{\neg\varphi\}\}$ (for some φ) and let $\langle\{\neg\alpha_{\delta_i}, \gamma_{\delta_i}\}\rangle_{i\in I}$ be an enumeration of $\kappa(C_D, \Pi)$. Then, a <u>challenge</u> Λ at $i \in I$ is a minimal (wrt set inclusion) set of default rules $\Lambda = \{\delta \mid \{\neg\alpha_\delta, \gamma_\delta\} \in \kappa(C_D, \Pi_i)\}$ for some spanning mating Π_i for the matrix

$$M_i = C_W \cup C_{N(\mathcal{D})} \cup \{\{\gamma_{\delta_k}\}, \{\beta_{\delta_k}\} \mid k < i\} \cup C_D \cup \{\{\gamma_{\delta_i}\}, \{\beta_{\delta_i}\}\} \quad (13)$$

such that (M_i, Π_i) is admissible and C-compatible (wrt some index set I_i).

Observe that the matrix representation allows us to simplify (13) by replacing C_D and $C_{N(\mathcal{D})}$ by $(C_D \setminus \{\{\neg\alpha_{\delta_k}, \gamma_{\delta_k}\} \mid k \leq i\})$ and $(C_{N(\mathcal{D})} \setminus \{\{\neg\gamma_{\delta_k}, \beta_{\delta_k}\} \mid k \leq i\})$, respectively, since the subtracted clauses are subsumed by $\{\{\gamma_{\delta_k}\}, \{\beta_{\delta_k}\} \mid k < i\}$ and the query clauses $\{\{\gamma_{\delta_i}\}, \{\beta_{\delta_i}\}\}$. Now, the overall idea is that $(C_W \cup C_D, \Pi)$ represents a CDP $\langle\delta_i\rangle_{i\in I}$ for the considered query. Then, each default proof (M_i, Π_i) induces a challenge Λ that corresponds to a minimal Δ-segment C-orthogonal to $\langle\delta_i\rangle_{i\in I}$. This is so because (M_i, Π_i) represents a CDP of $\neg Conseq(\delta_i) \vee \neg Justif(\delta_i)$ from Default Theory (6) (c.f. Theorem 5). Of course, any matrix M_i may have several spanning matings Π_i and hence may induce different challenges.

Now, we are ready to formulate our additional condition on default proofs for skeptical reasoning:

Definition 10 Protection & Stability. Let C_W and C_C be sets of ω-clauses, C_D be a set of δ-clauses and φ an atomic formula. Let Π be a mating for $C_W \cup C_D$ and let $\langle\{\neg\alpha_{\delta_i}, \gamma_{\delta_i}\}\rangle_{i\in I}$ be an enumeration of $\kappa(C_D, \Pi)$. Let $\langle\Lambda_{ij}\rangle_{j\in J_i}$ be the family of all challenges at $i \in I$.

We say that $(C_W \cup C_D, \Pi)$ is <u>protected under C against Λ_{ij}</u> by (M_{ij}, Π_{ij}) iff Π_{ij} is a spanning mating for the matrix

$$M_{ij} = C_W \cup \{\{\gamma_\delta\} \mid \delta \in \Lambda_{ij}\} \cup C_D \cup \{\{\neg\varphi\}\}$$

such that (M_{ij}, Π_{ij}) is admissible, $C \cup Justif(\Lambda_{ij})$-compatible, and stable for φ under $C \cup Justif(\Lambda_{ij})$. We say that $(C_W \cup C_D, \Pi)$ is <u>stable for φ under C</u> iff $(C_W \cup C_D, \Pi)$ is protected under C against all challenges Λ_{ij}.

The idea is that $(C_W \cup C_D, \Pi)$ represents a CDP for query φ. In order to verify whether φ is skeptically derivable, we need to show that $(C_W \cup C_D, \Pi)$ in addition satisfies the stability criterion. For this, we proceed as follows. First, we isolate all challenges Λ against our CDP $(C_W \cup C_D, \Pi)$. In turn we verify whether φ is also skeptically derivable from the matrices obtained by adding the consequents of the default rules in Λ to C_W and taking the justifications of the default rules in Λ as additional constraints on the compatibility check. This amounts to verifying whether φ is in all extensions to which the default rules in Λ contribute. Accordingly, a skeptical default proof is given by a stable credulous default proof $(C_W \cup C_D, \Pi)$ along with all its protecting default proofs (M_{ij}, Π_{ij}). Of course, there may be several such skeptical proofs depending on the initial choice.

Now, let us examine whether our default proof in (9) is stable and thus renders T a skeptical conclusion of Default Theory (1). For this, we consider the obtained enumeration $\langle \{\neg R_{\delta_2}, H_{\delta_2}\}, \{\neg H_{\delta_4}, T_{\delta_4}\} \rangle$. In turn, we determine all emerging challenges. That is, we consider all minimal default proofs of $\neg H_{\delta_2} \vee P_{\delta_2}$ and $\neg T_{\delta_4}$. These formulas represent the negated consequents (and justifications) of the used default rules, δ_2 and δ_4, respectively.

In the first case, we consider the matrix obtained from our original matrix, (9), by replacing query clause $\{\neg T\}$ by clauses $\{H_{\delta_2}\}$ and $\{\neg P_{\delta_2}\}$. This allows us moreover to eliminate the δ-clause $\{\neg R_{\delta_2}, H_{\delta_2}\}$ from (9) since it is subsumed by $\{H_{\delta_2}\}$. Analogously, we can omit the normalization clause $\{\neg H_{\delta_2}, \neg P_{\delta_2}\}$ due to the presence of $\{\neg P_{\delta_2}\}$. Hence, we have to add only the normalization clause $\{\neg D_{\delta_1}, P_{\delta_1}\}$ since δ_3 and δ_4 are normal default rules. The modifications to our initial matrix in (9) are indicated as dashed boxes.[14] This results in the following derivative of Matrix (9):[15]

$$
M_2 = \begin{bmatrix} & \neg Q_{\delta_1} & \neg D_{\delta_3} & \neg H_{\delta_4} & \neg D_{\delta_1} & H_{\delta_2} & \neg P_{\delta_2} \\ Q & R & D_{\delta_1} & T_{\delta_3} & T_{\delta_4} & P_{\delta_1} & \end{bmatrix} \tag{14}
$$

By discarding the two query clauses, Matrix[16] M_2 can be seen as the proof-theoretic counterpart of Default Theory (7). In fact, M_2 admits the spanning mating $\Pi_2 = \{\{Q, \neg Q_{\delta_1}\}, \{D_{\delta_1}, \neg D_{\delta_1}\}, \{P_{\delta_1}, \neg P_{\delta_2}\}\}$. This CDP involves a single default rule, viz. δ_1. We thus obtain the singleton enumeration $\langle \{\neg Q_{\delta_1}, D_{\delta_1}\} \rangle$ inducing the following two matrices for verifying admissibility and compatibility, respectively:

$$
\begin{bmatrix} & \neg Q_{\delta_1} & \neg D_{\delta_1} & \neg H_{\delta_2} \\ Q & R & & P_{\delta_1} & \neg P_{\delta_2} \end{bmatrix} \begin{bmatrix} & \neg Q_{\delta_1} & \neg D_{\delta_1} & \neg H_{\delta_2} \\ Q & R & D_{\delta_1} & P_{\delta_1} & \neg P_{\delta_2} \end{bmatrix} \tag{15}
$$

Both matrices contain the normalization clauses $\{\neg D_{\delta_1}, P_{\delta_1}\}$ and $\{\neg H_{\delta_2}, \neg P_{\delta_2}\}$. The left matrix is complementary and thus confirms admissibility, while the right matrix has open path $\{Q, R, D_{\delta_1}, P_{\delta_1}, \neg H_{\delta_2}\}$ establishing compatibility. Consequently, (M_2, Π_2)

---

[14] This is done to underline the utility of structure-oriented theorem proving.

[15] For simplicity, we have refrained from turning the *two* query clauses $\{H_{\delta_2}\}$ and $\{\neg P_{\delta_2}\}$ into $\{H_{\delta_2}, \varphi\}, \{\neg P_{\delta_2}, \varphi\}$ along with the *single atomic* query clause $\{\neg \varphi\}$ (as stipulated in Theorem 8).

[16] The index 2 of M_2 and Π_2 reflects the index of the query $\neg H_{\delta_2} \vee P_{\delta_2}$.

provides us with a CDP of $\neg H_{\delta_2} \vee P_{\delta_2}$. This CDP is orthogonal to our initial proof in (9). As a result, (M_2, Π_2) induces the challenge $\Lambda_{21} = \{\delta_1\}$. There is no other challenge induced by M_2.

Now, let us first verify whether our CDP in (9) is protected against Λ_{21} by some other CDP before we determine more challenges: For establishing stability, intuitively, we verify whether our initial query, T, belongs to the extensions formed (among others) by the default rules in $\Lambda_{21} = \{\delta_1\}$. For this, we consider the matrix M_{21} obtained from our initial matrix in (9) by adding the consequents of all default rules in Λ_{21}. This amounts to replacing δ-clause $\{\neg Q_{\delta_1}, D_{\delta_1}\}$ in (9) by ω-clause $\{D_{\delta_1}\}$:

$$
M_{21} = \begin{bmatrix} & & & \neg R_{\delta_2} & \neg D_{\delta_2} & \neg H_{\delta_4} & \neg T \\ & Q & R & D_{\delta_1} & H_{\delta_2} & T_{\delta_3} & T_{\delta_4} \end{bmatrix} \tag{16}
$$

According to Definition 10 we then verify whether there is a spanning mating Π_{21} for M_{21} such that (M_{21}, Π_{21}) is admissible, $\{P_{\delta_1}\}$-compatible (since $Justif(\Lambda_{21}) = \{P_{\delta_1}\}$), and stable. Admissibility and compatibility of (M_{21}, Π_{21}) are easily verified by checking the two following matrices induced by the only δ-clause used in (16), i.e., $\{\neg D_{\delta_3}, T_{\delta_3}\}$:

$$
\begin{bmatrix} & & \neg D_{\delta_3} \\ Q & R & D_{\delta_1} \end{bmatrix} \quad \begin{bmatrix} & & & \neg D_{\delta_3} \\ Q & R & D_{\delta_1} & T_{\delta_3} & P_{\delta_1} \end{bmatrix}
$$

Complementarity of the left matrix establishes admissibility while non-complementarity (initiated by the open path $\{Q, R, D_{\delta_1}, T_{\delta_3}, P_{\delta_1}\}$) of the right matrix confirms $\{P_{\delta_1}\}$-compatibility. Now, it remains to be shown that (M_{21}, Π_{21}) along with its induced enumeration $\langle\{\neg D_{\delta_3}, T_{\delta_3}\}\rangle$ is stable for T under $\{P_{\delta_1}\}$. However, there is no challenge to $\{\neg D_{\delta_3}, T_{\delta_3}\}$ since $\neg T_{\delta_3}$ is not provable from the matrix obtained by replacing query clause $\{\neg T\}$ in (16) by $\{T_{\delta_3}\}$.[17] As a result, we obtain that our CDP in (9) is protected against Λ_{21} by (M_{21}, Π_{21}).

Next, we must consider all challenges of the second δ-clause in (10). For this, we determine all CDPs of $\neg T_{\delta_4}$ (the consequent of δ_4) from the matrix obtained in the following way. First, we add the clauses $\{H_{\delta_2}\}$ and $\{\neg P_{\delta_2}\}$ representing the 'consequent' and the 'justification' of the first δ-clause in (10) to our initial matrix, (9). The first addition is accomplished by replacing δ-clause $\{\neg R_{\delta_2}, H_{\delta_2}\}$ by $\{H_{\delta_2}\}$. Second, we add the normalization clause $\{\neg D_{\delta_1}, P_{\delta_1}\}$. As with Matrix (14), the second normalization clause $\{\neg H_{\delta_2}, \neg P_{\delta_2}\}$ can be omitted since it is subsumed by $\{\neg P_{\delta_2}\}$. Again, no normalization clauses are added for the normal default rules δ_3 and δ_4. Finally, we replace the original query clause $\{\neg T\}$ of (9) by $\{T_{\delta_4}\}$. This allows us to eliminate δ-clause $\{\neg H_{\delta_4}, T_{\delta_4}\}$ since it is subsumed by $\{T_{\delta_4}\}$:

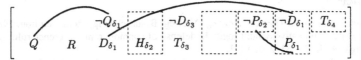

---

[17] In fact, there is no compatible default proof.

580

Observe that this matrix can be regarded as the proof-theoretic counterpart of Default Theory (8) if we discard the query clause. The above matrix has a spanning mating inducing the enumeration $\langle\{\neg Q_{\delta_1}, D_{\delta_1}\}\rangle$. Admissibility of the previous proof can be verified in a straightforward way, and to test compatibility, we have to consider the following matrix:

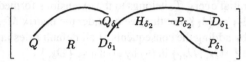

Obviously, this matrix has no open path so that our proof is not compatible. Accordingly, $\neg T_{\delta_4}$ is not derivable (c.f. Default Theory (8)) and therefore there are no more challenges, apart from Λ_{21}. In this way, we have shown that our initial CDP in (9) is stable for T (in addition to its admissibility and compatibility verified in Section 4). This is so because it is protected against its only challenge Λ_{21} by (M_{21}, Π_{21}). This tells us that T is skeptically derivable from Default Theory (9).

In general, we have the following result stating the adequacy of our proof method:

Theorem 11. *Let $(\mathcal{D}, \mathcal{W}, \mathcal{C})$ be a default theory in atomic format and φ an atomic formula. Then, $\varphi \in E$ for all constrained extension (E, C) of $(\mathcal{D}, \mathcal{W}, \mathcal{C})$ iff there is a spanning mating Π for the matrix $M = C_{\mathcal{W}} \cup C_{\mathcal{D}} \cup \{\{\neg\varphi\}\}$ and an enumeration $\langle\{\neg\alpha_{\delta_i}, \gamma_{\delta_i}\}\rangle_{i\in I}$ of $\kappa(C_{\mathcal{D}}, \Pi)$ such that $(C_{\mathcal{W}} \cup C_{\mathcal{D}}, \Pi)$ is admissible wrt I, C-compatible wrt I, and stable for φ under C.*

6 Conclusion

We have developed an approach to skeptical query-answering in Constrained Default Logic based on the Connection Method. This has been accomplished by elaborating on a recently proposed, general idea for skeptical reasoning in (semi-monotonic) Default Logics [21]. As a result, we have obtained a precise algorithm that returns a skeptical default proof if the query is contained in all extensions of the underlying default theory. The approach has then been combined with a method for credulous query-answering based on the Connection Method. This was accomplished by employing a further restriction on credulous default proofs, expressed by the *stability* criterion. This has led to a homogeneous characterization of skeptical default proofs at the level of the underlying deduction method. This approach was supported by the structure-sensitive nature of the Connection Method. The value of this for structure-sharing among the diverse subproofs involved is detailed for credulous query-answering in [18]. Even though we have not discussed it here, it should be obvious that the utility of structure-sharing applies to skeptical query-answering, too. We have tried to indicate this by stressing the common structures involved in the skeptical default proof carried out in the previous section.

References

1. W. Bibel. *Automated Theorem Proving*. Vieweg, Braunschweig, second edition, 1987.
2. G. Brewka. Cumulative default logic: In defense of nonmonotonic inference rules. *Artificial Intelligence*, 50(2):183–205, 1991.
3. J. Delgrande, T. Schaub, and W. Jackson. Alternative approaches to default logic. *Artificial Intelligence*, 70(1–2):167–237, 1994.

4. D. Etherington. *Reasoning with Incomplete Information*. Research Notes in Artificial Intelligence. Pitman, London, 1988.
5. M. Ginsberg. A circumscriprive theorem prover. In M. Reinfrank et al., editors, *Proceedings of the Second International Workshop on Non-Monotonic Reasoning*, pages 100–114. Springer, 1989.
6. G. Gottlob. Complexity results for nonmonotonic logics. *Journal of Logic and Computation*, 2(3):397–425, June 1992.
7. U. Junker and K. Konolige. Computing the extensions of autoepistemic and default logic with a TMS. In *Proceedings of the National Conference on Artificial Intelligence*, 1990.
8. W. Marek and M. Truszczyński. Computing intersection of autoepistemic expansions. In W. Marek et al., editors, *Proceedings of the First International Workshop on Logic Programming and Nonmonotonic Reasoning*, pages 37–50. MIT Press, 1991.
9. A. Mikitiuk and M. Truszczyński. Rational versus constrained default logic. In C. Mellish, editor, *Proceedings of the International Joint Conference on Artificial Intelligence*, pages 1509–1515. Morgan Kaufmann, 1995.
10. R. Moore. Semantical considerations on nonmonotonic logics. *Artificial Intelligence*, 25:75–94, 1985.
11. I. Niemelä. A decision method for nonmonotonic reasoning based on autoepistemic reasoning. In J. Doyle et al., editors, *Proceedings of the 4th International Conference on the Principles of Knowledge Representation and Reasoning*, pages 473–484. Morgan Kaufmann, 1994.
12. D. Poole. A logical framework for default reasoning. *Artificial Intelligence*, 36:27–47, 1988.
13. D. Poole. Compiling a default reasoning system into prolog. *New Generation Computing*, 9(1):3–38, 1991.
14. R. Reiter. A logic for default reasoning. *Artificial Intelligence*, 13(1–2):81–132, 1980.
15. A. Rothschild. Algorithmische Untersuchungen zu Defaultlogiken. Diplomarbeit, FG Intellektik, FB Informatik, TH Darmstadt, Germany, 1993.
16. T. Schaub. On commitment and cumulativity in default logics. In R. Kruse and P. Siegel, editors, *Proceedings of European Conference on Symbolic and Quantitative Approaches to Uncertainty*, pages 304–309. Springer, 1991.
17. T. Schaub. Variations of constrained default logic. In M. Clarke, R. Kruse, and S. Moral, editors, *Proceedings of European Conference on Symbolic and Quantitative Approaches to Reasoning and Uncertainty*, pages 312–317. Springer, 1993.
18. T. Schaub. A new methodology for query-answering in default logics via structure-oriented theorem proving. *Journal of Automated Reasoning*, 15(1):95–165, 1995.
19. C. Schwind. A tableaux–based theorem prover for a decidable subset of default logic. In M. Stickel, editor, *Proceedings of the Conference on Automated Deduction*. Springer, 1990.
20. M. Thielscher. On prediction in Theorist. *Artificial Intelligence*, 60(2):283–292, 1993.
21. M. Thielscher and T. Schaub. Default reasoning by deductive planning. *Journal of Automated Reasoning*, 15(1):1–40, 1995.
22. A. Zhang and W. Marek. On the classification and existence of structures in default logic. *Fundamenta Informaticae*, 8(4):485–499, 1990.

Type Theoretic Semantics for SemNet

Shiu S., Luo Z., Garigliano R.

Laboratory for Natural Language Engineering
School of Computer Science
University of Durham
South Road
Durham, DH1 3LE, United Kingdom.
email: s.k.shiu@durham.ac.uk

Abstract. Semantic Networks have long been recognised as an important tool for modelling human type reasoning. This paper describes an attempt to give a formal semantics of a semantic network in constructive type theory.

The particular net studied is SemNet, the internal knowledge representation for LOLITA[1]: a large scale natural language engineering system. SemNet has been designed with large scale, efficiency, integration and expressiveness in mind. It supports many different forms of plausible and valid reasoning, including: analogy, epistemic reasoning and inheritance. Type theory is used to define the syntactic and semantic models of Sem-Net. Because of the notion of an internal logic, which follows from the 'propositions as types' principle, both of these models can be reasoned about in the same framework. Once formal semantics have been defined they can then be used to analyse the different reasoning mechanisms.

A further advantage is that (because of applications to formal methods for software engineering) type checkers/proof assistants have been built. These tools are ideal for organising and managing the analysis of formal models.

The models are shown to be useful in analysing correctness of implementation of the algorithms, proving consistency and highlighting the assumptions and meaning of the valid and plausible reasoning.

1 Introduction

Semantic networks represent knowledge as a graph. The nodes correspond to facts or concepts and the arcs to relations between concepts. Within this framework there is a wide variety of definitions [AS93] [Sow84] [WS92]. Much of the research has been done in the area of natural language understanding. This has lead to models of intension, belief, hypothetical reasoning and commonsense reasoning.

Semantic networks are often criticised for lacking formal semantics [Woo91]. This is important to properly understand the associated reasoning mechanisms. The main problem is of moving from a graphical representation to a linear mathematical theory. There have been many approaches to formally modelling the

---

[1] Large-scale, Object based, Linguistic Interactor, Translator and Analyser

reasoning mechanisms of semantic networks, including: mapping to first order predicate logic [Sow84], using non-monotonic logics [FK88], and using set theoretic semantics [WS92]. These all succeed in capturing or demonstrating a particular aspect of a semantic network. What they do not do is provide a general framework for reasoning about a semantic network and its associated reasoning mechanisms.

Type theory can be used as a computational and a logical language. The key idea which makes this possible is that of 'propositions as types', discovered by Curry [CF58] and Howard [How80]. This allows for the notion of an internal logic which can be used to reason directly about type theoretic objects. There are different versions of type theory, including predicative type theories such as Martin-Löf's type theory [ML84] [NPS90], and impredicative systems such as F^ω and the Calculus of Constructions [Gir72] [CH88]. This work uses the unified theory of dependent types (UTT) [Luo94]. Type theory, through its inductive type schema gives us a computational understanding of a theory. The internal logic allows us to reason directly about such an inductive theory.

SemNet is a semantic network used as the internal knowledge base for LOLITA: a large scale Natural Language Engineering (NLE) system [M+95]. This paper describes work done to provide formal semantics for SemNet in UTT. A formalisation/definition of the syntax and (type theoretic) semantics of SemNet is given. An advantage of type theory is that this can all be done in a single, managable framework, moreover, in many circumstances, type theory offers 'natural' solutions to some otherwise awkward interpretation problems. The main contribution however is that the formalisation provides a basis for understanding the different reasoning mechanisms of SemNet.

There are a priori reasons for believing that constructive type theory is a good tool for this formalisation task.

- Intensional type theory is used as a tool to capture aspects of natural language [Can93]. Since SemNet represents knowledge as part of a NLE system it would be convenient (to allow comparisons between natural language and SemNet) if a similar language is used to formalise its meaning.
- Event nodes in SemNet correspond to propositions. This causes difficulties in set theory, as propositions cannot naturally be interpreted as sets. However in type theory the event nodes can naturally be treated as proposition types.
- SemNet nodes are understood as intensional concepts. This can be reflected by using different equality relations, as described in section 4.
- SemNet is a representation designed to work in a computer program. Type theory offers the ability to give a computational interpretation. This leads to the ability to verify the behaviour of the algorithms that operate on SemNet.
- Type checking proof assistants are available, providing goal directed proof development and machine checked proofs for propositions.
- In SemNet the meaning of any node is defined in terms of its relationship with other nodes (i.e. its location), so ultimately each node is only fully defined by the whole network. Knowledge is retrieved/inferred by path traversal, an event is recognised as entailed when a path is 'successful'. It seems possible

to identify the successful paths with proofs of the proposition, thus capturing this notion of meaning by location. This idea is not further developed in this paper, but shows a possible, more ambitious, aim of the formalisation.

The paper is organised as follows. Section 2 describes SemNet giving emphasis to those aspects relevant to the formalisation but also pointing out some of its features and uses. Section 3 describes constructive type theory and Lego (the proof assistant used) and how they are used to form the models. Section 4 describes some features of SemNet which are problematic for formal semantics and how type theory could offer a natural solution. Section 5 describes the type theoretic models. In addition an extension to the core is given showing how epistemic reasoning can be formalised. Section 6 draws conclusions and describes current work.

2 SemNet

SemNet is used as the internal knowledge base for LOLITA. LOLITA is a large scale NLE system. LOLITA is designed as a core system supplemented with a set of (natural language) applications, the former supplying basic natural language facilities to the latter. The core consists of analysis (transforming natural language text into SemNet), generation (transforming SemNet into natural language text), and various inference mechanisms. Prototype applications built include contents scanning, simple meaning based translation, natural language query, dialogue, and intelligent language tutoring.

2.1 General SemNet Principles

In common with semantic networks SemNet is a graph based representation, where concepts and relationships are represented by nodes and arcs. "Knowledge" is elicited by graph traversal. SemNet has been designed specifically for large scale NLE, in particular it needs to be expressive, efficient and easily integratable with other modules. SemNet supports many forms of reasoning as well as fully exploiting inheritance. There are, for example, models of epistemic reasoning, time and location [Sho96], reasoning by analogy [LG94] and standard logical connective reasoning. This section describes some of the main aspects needed for this discussion.

There are 3 types of nodes: entities, events (assertions) and actions. There are 3 types of directed arcs: subject, object and action which can be read/traversed in either direction. Only event nodes can have a subject, object or action arc attached. Only action nodes can be an action for an event node.

Figure 1 shows a section of SemNet graph. Event nodes correspond to statements. The event node E1 states that "Every FARMER1 OWNS a DONKEY1". The two 'spec' links are a shorthand for events with 'specialisation' as action. Important points to note are that the subject/object arcs give the statement direction. Events state that the referenced concepts are involved in a relation (labelled by the action). On each arc there is a quantification tag which makes explicit the way in which the pointed at concept is referenced.

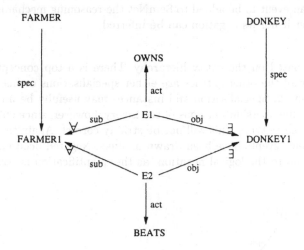

Fig. 1. A section of SemNet graph

The meaning of the tags are as follows:

- **Universal** ∀ refers to "instances" of the concept and says that all the "instances" of the concept are involved in relationship specified by the event.
- **Individual I** (not used in the diagram) refers to the concept as a "whole" and says that it is involved in the relationship specified by the event.
- **Existential** ∃ refers to the "instances" of the concept, but the "instance" involved depends on the particular "instance" of some other universally quantified concept involved in the event.

An important concept not mentioned is negation. An action can be negated, so that an event states that the referenced concepts are explicitly not in the labelled relationship. Therefore to negate an event, as well as negating the action the quantification tags may also need to be changed. It should be noted that the event nodes can be the subject or object of another event so that SemNet is 'propositional' [KC93].

A fundamental principle of the design is that concepts are not reduced to primitives. The meaning of any node is defined in terms of its relationship with other nodes, so ultimately each node is only fully defined by the whole semantic network.

2.2 Reasoning Mechanisms on SemNet

As stated there are many inference algorithms that have been designed and implemented for SemNet. This discussion will consider only reasoning by inheritance and epistemic reasoning. All reasoning proceeds by passing an event (intuitively a proposition) to SemNet and an algorithm determines whether the event (proposition) or its negation is entailed (validly or plausibly) by SemNet. Before allowing an event to be added to SemNet the reasoning mechanisms first check that neither it nor its negation can be inferred.

Inheritance is based on the entity hierarchy. There is a top concept (called "ENTITY") and all the other entities are either specialisations or instances of "Entity", see figure 2. Specialisation and instances may usefully be interpreted as the subset and membership relations of set theory, however, since entities are formally interpreted as types, this will not be strictly correct. Again the specialisation and instance events have been drawn as links. Negated occurences here actually correspond to the logical negation, as the quantification is straightforward.

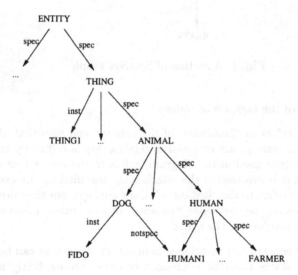

Fig. 2. Section of the Entity hierarchy of SemNet, the notspec relation is not inferrable from absence of a spec link.

To list the inheritance rules, a textual form for writing events will be used:

((Act,Bool) (A, Q1) (B, Q2))

where Act is an action node, A and B are entity nodes, Q1 and Q2 are quantification tags and Bool is a boolean value. For example, ((OWNS,true) (FARMER1,∀) (DONKEY1,∃)), represents E1 in figure 1. The quantification on spec and inst events is always 'I', and so they can be written as:

$A \succeq_s B$ informally, B is a subset of A

$\neg\,(A \succeq_s B)$ informally B is not a subset of A

$a \in A$ informally, a is a member of A

$a \notin A$ informally, a is not a member of A

The first rule of inference is that any event which is explicitly present in SemNet can be inferred. There are two rules for inferring spec events:

$$\frac{A \succeq_s B, B \succeq_s C}{A \succeq_s C} \qquad \frac{A \succeq_s B, \neg(A \succeq_s C)}{\neg(B \succeq_s C)}$$

where A, B, and C are entity nodes. At this stage there is no closed world assumption, i.e. the second rule is the only way in which a notspec relation can be inferred. Inheritance rules for the remaining (non epistemic) events are:

$$\frac{b \in B, A \succeq_s B}{b \in A} \qquad \frac{a \notin A, A \succeq_s B}{a \notin B}$$

$$\frac{A \succeq_s B, ((Act, true)(A, \forall)(O, Q))}{((Act, true)(B, \forall)(O, Q))} \qquad \frac{a \in A, ((Act, true)(A, \forall)(O, Q))}{((Act, true)(a, I)(O, Q))}$$

$$\frac{A \succeq_s B, ((Act, true)(A, \forall)(O, \forall)), b \in B}{((Act, true)(B, \exists)(O, \forall))} \qquad \frac{A \succeq_s B, ((Act, true)(B, \exists)(O, Q))}{((Act, true)(A, \exists)(O, Q))}$$

where 'Act' stands for an arbitrary action concept and '(O,Q)' stands for an arbitrary object and its quantification. There are an equivalent set of rules for the objects of events and a set of rules to handle negated actions (i.e. explicit negative facts, which should be distinguished from positive facts which cannot be inferred). These rules allow for very efficient inference by searching up and down the hierarchy.

Epistemic reasoning is based on epistemic events. Epistemic events have an agent[2] as subject, an event as object and an epistemic relation (for example, know, believe or think) as action, see figure 3.

It plausibly follows that a man that knows that "all farmers own a donkey" also knows that "all small farmers own a donkey". Informally this assumes that all men share the same hierarchy as LOLITA, and have the ability to make the inheritance. LOLITA achieves this by inducing a partial order on the event concepts from the hierarchy. Put informally, if Event2 follows from Event1 and the hierarchy then

Event1 $\succeq_e$ Event2

The inference rule can then be stated as,

$$\frac{((EpiAct, true)(A, Q)(Event1, I)), Event1 \succeq_e Event2}{((EpiAct, true)(A, Q)(Event2, I))}$$

where EpiAct is an epistemic action. There is a belief function associated with this rule, see [LG94] for more details.

---

[2] to which a personal world (in modal logic) can be ascribed

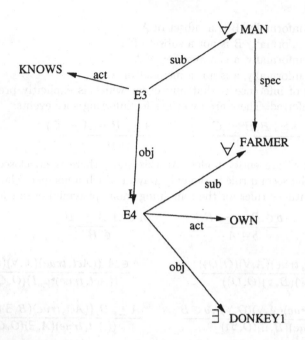

Fig. 3. Example of Epistemic Event, E3 represents "All men know that all farmers own a donkey".

3   Constructive Type Theory

Type theory is first understood in terms of objects and types. The judgement:

   a:A

in type theory is understood as "the object 'a' is of type 'A' ". Next more complex types can be defined using type constructors. Two important ones are the dependent product types and the strong sum types. The dependent product type has functions as objects. For a type A and any family of types B[x] indexed by arbitrary objects x of type A, $\Pi x{:}A.B(x)$ is the type of functions 'f' such that for any object a of type A, applying f to a yields an object of type B[a]. Intuitively it represents the set of (dependent) functions from A to B[x]:

$$\{f \mid f(a) : B[a], \forall a : A\}$$

Strong sum types are types of pairs of objects. For any type A and any family of types B[x] indexed by arbitrary object x of type A, $\Sigma x{:}A.B(x)$ is the type of pairs (a,b) where a is an object of type A and b is of type B[a]. Intuitively it represents the set of (dependent) pairs of elements of A and B[x]:

$$\{(a,b) \mid a : A, b : B[a]\}$$

The projection functions

$$\pi 1 : \Sigma \, x{:}A.B(x) \to A$$
$$\pi 2 : \Sigma \, x{:}A.B(x) \to B(x)$$

extract the first or second entry of the pair. For example, if $g{:}(\Sigma x{:}A.B(x))$, then $(\pi 1 \, g){:}A$. Because of the inherent dependency they are useful for describing complex types.

More generally types can be defined inductively. This is done by giving constructors to say how objects of this new type can be formed. For example, the type N of natural numbers can be specified by giving the introduction rules:

$$0 : N$$
$$succ : N \to N$$

Since, by definition, these rules exhaustively define how natural numbers may be constructed, an associated elimination rule can be inferred

$$C : N \to Type$$
$$(C \, 0) \to (\Pi x{:}N.C \, x \to C \, (succ \, x)) \to \Pi n{:}N.(C \, n)$$

which gives us a method for defining functions that operate on all objects of an inductive type. This method can be used to define standard functions on the natural numbers (such as addition and multiplication) and to prove properties about them (such as the Peano axioms). See, e.g., [NPS90] [Luo94] for details.

Various proof development systems have been developed based on type theories. These include NuPRL [Con86] ALF [MN94], Coq [Dow90] and Lego [LP92]. Lego is a proof assistant that implements UTT and which supports goal directed proof development. It has been used extensively in this project.

There are different versions of constructive type theory. The main differences are reflected in the different structures of their conceptual universe of types. Perhaps the best known is Martin-Löf's type theory [ML84] [NPS90]. This can be understood in a hierarchical way: one starts by introducing various basic types, and using type constructors, builds up more complex types, until finally, one may introduce type universes (names of types). There is not a type of all types, instead there is an infinite sequence of universes Type(0) : Type(1) : Type(2) : ..., which approximate the type of all types. In this theory types are not distinguished from propositions.

This work uses the unified theory of dependent types (UTT) [Luo94]. Here there is a distinction between the notions of data-types and logical propositions.

In UTT, propositions are logical formulas used to represent and reason about properties of the computational objects in the language. The universe 'Prop'is the logical universe where propositions reside; in other words , a proposition is an object of type 'Prop'. By the principle of 'proposition as types', these propositions are proof types, provable if and only if it is inhabited. In UTT there are internal notions of predicate and relations, essentially, if A is a type then a predicate over A is a function of type A $\to$ Prop and a (binary) relation is a function of type A $\to$ A $\to$ Prop.

4 Motivations for the type theoretic approach

This section describes some of the problematic features of SemNet, and how type theory offers a natural solution.

The intension problem: Many intensionally distinct concepts refer to the same object or extension. For example, consider the concept of "computer science students at Durham". Specialisations of this concept may be "AI students at Durham" and the "Computer Science football team". The extension of these concepts are (coincidentally) the same. However intensionally they are clearly different. This may lead to a problem when handling, say, explanations of causal reasoning. For example it may be explained that the football team are fit (since they play football), but the explanation that the AI students are fit (since they study AI) should not be inferred. Any formalisation should make this distinction. In type theory the relevant node/concept would be interpreted as an object of some type, say Entity. Different equality relations (i.e. functions with type "Entity → Entity → Prop") can be defined to represent extensional equivalence and intensional equivalence.

The quantification problem: SemNet contains knowledge that corresponds to natural language. It is therefore to be expected that any logical form or structure of the natural language utterance will have been preserved. To check this the meta-language should have the ability to provide formal semantics for natural language. Intensional type theory is a useful tool for formalising natural language. In particular constructive type theory has been shown to be useful for categorising some complex quantifications in natural language [Ran94].

SemNet handles complex quantifications by building up complex concepts and then referring to them. This is exactly how the problem is handled in [Ran94] using constructive type theory.

For example consider the sentence "Every farmer that owns a donkey beats it". To state this the sub-concepts 'donkey owning farmers', 'farmer owned donkeys' and 'donkey owned by particular farmer' are required. These are all available in the object event1 with dependent type:

event1 : Πx:F1.Σy:D1.(own x y)

x:F1 means x is a donkey owning farmer, and (π1 (event1 x)) is the donkey which 'x' beats. Thus:

event2 : Πx:F1.beat x (π1 (event1 x))

captures the required statement. This mirrors the way in which SemNet solves the problem. E1 in figure 1 is a defining event and builds the required concepts (corresponding to event1). E2 is an observed event which makes the required statement (corresponding to event2). See [LG94] for a further exposition of defined and observed events.

The Epistemic problem: Epistemic events of SemNet make propositions about relationships between entities and events (i.e. propositions). In a set theoretic formulation this corresponds to treating propositions (i.e. set theoretic statements) as sets. This is not a normal extension to set theory. An approximation can be made by relating the statements to sets which represent them. But by doing this some of the essence/meaning of the statement is clearly lost. In UTT propositions are types, and so event nodes can naturally be interpreted as propositions. Section 5.3 expands this idea and shows how it can be used to analyse the epistemic reasoning mechanism.

5   The Model

The previous section has outlined some of the problems of providing semantics for SemNet and given small local examples of how type theory may provide a more natural solution. This section describes the core of the formalisation of SemNet. The eventual aim is to incorporate the ad-hoc examples from above in the appropriate place of a core model.

Before providing semantics, type theory can be used to formalise the syntax of SemNet. This is done by defining SemNet as a (structured) type and defining rules of inference as inductive relations so as to give a computational understanding of SemNet. This 'computational' theory can be used to reason about and build 'correct' algorithms for SemNet. Semantics can then be provided by defining a function from SemNet (or its sub-structures) to propositions (objects of type 'Prop'). In this way, type theory allows for syntactic and semantic properties to be analysed all within UTT.

The core syntactic and semantic models are first described and then, to demonstrate the use of type theory, an extension of this core to provide semantics for epistemic events and reasoning is given.

5.1   The Syntactic Model

This section describes in overview the structure of the SemNet type and the functions that operate on it.

The SemNet type (i.e. the syntactic definition of SemNet) is intended to be as close as possible to the informal definition given in section 2. Nodes are defined as types. There are types Arc and Link. Arcs and Links are defined as objects of these types respectively. The Link type has a structure so that functions for extracting a 'source' node, 'target' node, arc and quantification can be defined. A set (actually implemented as a list) of links defines a type Graph. SemNet is defined as a strong sum type, whose objects are pairs with first element g of type Graph, and second element a (dependent) proposition of type P(g), where P is a predicate over type Graph. The predicate ensures all the constraints of SemNet are applied, (e.g. all events nodes to have a valid event structure, there must be a valid hierarchy etc...).

The inference engine of LOLITA works by passing a temporary event to SemNet, the engine returns a boolean value which depends on whether SemNet entails the event. So to define the inference a type Event must be defined. The Event type should have a structure so that functions for obtaining its subject, object, action, boolean and quantifications can be defined.

The inference rules are defined as an inductive relation

Infer : SemNet → Event → Prop

There is a function for adding events to SemNet. This function will only allow an object e, of type Event, to be added if neither e nor its negation can be inferred.

Consistency: to have a notion of consistency we must have the notion of negation. The negation of an event is defined by a function which inverts the boolean value of the event and changes the quantifications appropriately. Consistency can then be defined as a predicate on the type SemNet, which will be 'true' when for all objects e of type Event, not both e and its negation will be inferrable.

Rather than proving directly that SemNet is consistent, it has been proved (for the spec rules) that the function for adding events to SemNet preserves consistency.

Verification: The algorithm for performing inference can be proved correct. Basically the implementation can be shown to be sound by showing that it is inferrable when the implementation returns a 'true' boolean, and complete by showing that it returns a 'true' boolean the event is inferrable. These proofs have been outlined and are under construction in Lego.

5.2 Defining semantics for SemNet

The semantics are defined by functions from the event nodes of SemNet to propositions (i.e. objects of type 'Prop'). Entity nodes are mapped to types. Actions are defined as dependent (polymorphic) functions. The idea is that the action can be used by different events and entities to form propositions in a general way.

More formally

- MeaningEnt : Node → Type
- MeaningAct : Node → ΠT1,T2:Type.T1 → T2 → Prop
- MeaningEv : Node → Prop
  The meaning function for an event is more complex, as it covers all the quantification cases. It uses the sub-parts of the event to define a proposition. For example, the event E1 in figure 1, where the quantifications are $\forall$ and $\exists$, the interpretation is

  F1,D1 : Type as they are entity nodes,
  E1 : Πx:F1.$\exists$ y:D1.(Own F1 D1) x y

notice that (Own F1 D1) : F1 → D1 → Prop, so that the event E1 is interpreted as an object of type Prop.

The meaning of SemNet is then just defined as the conjunction of the meaning of the events in SemNet, so that MeaningSem : SemNet → Prop.

Soundness of semantics can now be defined as the proposition,

$$\Pi s\text{:SemNet.}\Pi e\text{:Event.(Infer s e)} \to (\text{MeaningSem s}) \to \text{MeaningEv e}$$

This proof has been outlined in LEGO and is under construction.

All modules which access SemNet can now be checked with respect to the semantics. It is worth noting that these analysis, as well as, showing that the rules of inference are "correct" also increase the likelihood that the formal semantics really do capture the intended meaning of SemNet.

5.3 The Epistemic Model

The discussion so far has presented the core of the syntactic and semantic models of SemNet. This section describes an extension to cover plausible epistemic reasoning.

The intuitive rules for this reasoning were given in section 2. They followed the idea that we are able to assume that agents have some knowledge and reasoning capability. In the example given we assume they have the same hierarchy as LOLITA and are able to infer by inheritance in the same way.

The Syntactic Epistemic Model: Informally the rules use a partial order on the events induced from the hierarchy. The partial order 'po' can be defined inductively as:

po : Event → Event → Prop
makepo : Πe1,e2:Event.Πs:SemNet.(Infer (Addtonet (h s) e1) e2)
→ po e1 e2

where h (: SemNet → SemNet) extracts the hierarchy from SemNet, and Addtonet (: SemNet → Event → SemNet) adds an event to SemNet.

The new rule of inference can now be defined as being that if an agent "knows" an event e1 then the agent can be inferred to know all events e2' such that "po e1 e2".

Although consistency is not relevant for such a rule, the syntactic model can still be used to show the implementation is correct.

The Semantic Epistemic Model: Semantics for epistemic events are similar to those above except that the action now takes objects of type Prop, rather than just types.

EpistemicAction : ΠA:Type.ΠP:Prop.A $\to$ P $\to$ Prop

and so the meaning of E3 in figure 3 is

E3 : Πx:Man.$\exists$ p:(Meaning E1).(Know Man (Meaning E4)) x p

Here we are making explicit use of the "proposition as types" principle. The meaning is defined as the proposition that there is a relationship (labelled know) between an agent (of type Man) and a proof of proposition (of type (Meaning E1)). With this interpretation it is easy to formally see the assumptions which are made in the inference. In particular it is assumed that each man knows the whole hierarchy and some (semantically) equivalent inference rule to LOLITA's inheritance.

6 Conclusions and further work

The first thing to note about this approach is that type theory is being used to define the syntax and the semantics of the language. Because of this the reasoning power of the internal logic can be used to prove both syntactic and semantic properties, all within a single framework.

Leaving aside the advantages of the syntactic model, this work has shown that the graphics based representation of meaning can be translated into type theory. It has also been shown that some of the type theory structures thought to be natural for describing SemNet are indeed suitable. Two different forms of reasoning have been usefully described and understood in terms of the semantics.

Work is in progress to extend the model to cover other aspects of reasoning, and also to show that the semantics can capture the complex quantifications described in section 3.

References

[AS93]   S. S. Ali and S. C. Shapiro. Natural language processing using a propositional semantic network with structured variables. *Minds and Machines*, 3, No 4, 1993.

[Can93] Ronnie Cann. *Formal Semantics, An Introduction*. Cambridge University Press, 1993.

[CF58]   H. B. Curry and R. Feys. *Cominatory Logic, volume 1*. North Holland Publishing Company, 1958.

[CH88]   Th. Coquand and G. Huet. The Calculus of Constructions. *Information and Computation*, 76(2/3), 1988.

[Con86] R. L. Constable. *Implementing Mathematics with the NuPRL Proof Development System*. Prentice Hall, 1986.

[Dow90]   G. Dowek. The coq proof assistant: User's guide (version 5.6). Technical report, INRIA-Rocquencourt and CNRS-ENS Lyon, 1990.

[FK88]   C. Froidevaux and D. Kayser. Inheritance in semantic networks and default logic. In P.Smets, E. H. Mamdani, D. Dubois, and H. Prade, editors, *Non-Standard Logics for Automated Reasoning*. Academic Press, Harcourt Brace Jovanovich Publishers, 1988.

[Gir72]   J. Y. Girard. *Interprétation fonctionelle et élimination des coupures de l'arithmetique*. PhD thesis, 1972.

[How80]   W. A. Howard. The formulae-as-types notion of construction. In J. Hindley and J. Seldin, editors, *To H. B. Curry: Essays on Combinatory Logic*. Academic Press, 1980.

[KC93]   D. Kumar and H. Chalupsky. Guest Editorial for Special Issue on Propositional Knowledge Representation. *Journal of Experimental and Theoretical Artificial Intelligence*, 5, No 2 and 3, 1993.

[LG94]   D. Long and R. Garigliano. *Reasoning By Analogy And Causality: A model and application*. Artificial Intelligence. Ellis Horwood, 1994.

[LP92]   Z. Luo and R. Pollack. LEGO Proof Development System: User's Manual. Technical report, University of Edinburgh LFCS report series, 1992.

[Luo94]   Z. Luo. *Computation and Reasoning: A Type Theory for Computer Science*. Oxford Science Publications, 1994.

[M+95]   R. Morgan et al. Description of the lolita system as used in muc-6. In *to appear in - Proceedings Sixth Message Understanding Conference (MUC-6)*, 1995.

[ML84]   P. Martin-Lof. *Intuitionistic Type Theory*. Bibliopolis, 1984.

[MN94]   L. Magnusson and B. Nordström. The ALF proof editor and its proof engine. In *Types for proof and programs, LNCS*, 1994.

[NPS90]   B. Nordstrom, K. Petersson, and J. Smith. *Programming in Martin-Lofs Type Theory: An Introduction*. Oxford University Press, 1990.

[Ran94]   A. Rante. *Type-Theoretical Grammar*. Oxford Science Publications, 1994.

[Sho96]   S. Short. *Knowledge Representation of Lolita*. PhD thesis, (to be submitted) Department of Computer Science, University of Durham, 1996.

[Sow84]   J. F. Sowa. *Conceptual Structures: Information Processing in Mind and Machine*. The Systems Programming Series. Addison Wesley, 1984.

[Woo91]   W. A. Woods. Understanding subsumption and taxonomy: A framework for progress. In J. F. Sowa, editor, *Principles of Semantic Networks: Explorations in the Representation of Knowledge*, chapter 1. Morgan Kauffman, 1991.

[WS92]   W. A. Woods and J. G. Schmolze. The KL-One family. *Computers Mathematics and Applications*, 23, No 2 and 3:133–177, 1992.

From Syllogisms to Audiences: The Prospects for Logic in a Rhetorical Model of Argumentation

Christopher W. Tindale
Department of Philosophy
Trent University, Ontario,
Canada

ABSTRACT: Central elements of formal and informal logics are examined in order to show that the product-oriented (logical) approach does not in itself constitute an adequate model of argumentation. Rather, the logical must be grounded in a rhetorical account of argumentation with its fuller treatment of context and richer notion of relevance. Ch. Perelman's model is a particularly suitable candidate for this. Contrary to the claims of some theorists, logical features are important to Perelman's overall treatment of argumentation.

This paper proceeds from two assumptions: (i) That an adequate model of argumentation must synthesize the logical, dialectical, and rhetorical approaches to the subject; and (ii) that, if the rhetorical is to form the groundwork for the synthesis, the best prospect is to be found in the account of Chaim Perelman.

In light of these assumptions, I develop the paper in the following way: (i) argue that product-oriented approaches to argumentation (as seen in formal and informal logics) cannot themselves constitute an adequate model of argumentation; and (ii) that Perelman's rhetorical model can accommodate the logical, product-oriented approach for a fuller treatment of argumentation. In proposing (ii), I disagree with the assessment of Perelman's New Rhetoric (with Olbrechts-Tyteca) put forward by van Eemeren & Grootendorst in the following:

> in the introduction to *La Nouvelle Rhetorique* [they] present logic as a completed whole which is no longer open to new developments and which will certainly not generate insights which may be practical to the development of argumentation theory. These authors link logic without further ado to the approach to reason *more geometrico*, and for that very same reason they automatically believe it to be inadequate and irrelevant, if not both. (1988:276-77)

In fact, a review of Perelman's ambivalent treatment of the logical shows this to overstate the case. While Perelman holds logic inadequate *as* a theory of argumentation, it is not irrelevant *to* such a theory.

Logic as the Product-Oriented Approach:

The logical approach focuses on argumentation as product. That is, its central concern is the collection of statements comprising a conclusion and one or more premises called an "argument," and the determination of such arguments as valid or invalid, strong or weak. Traditionally, the product approach is seen in the systems of formal logic.

Classical logic, and the theories based on it like modal logic or set theory, as well as rival logical theories, all find their bases in the idea of logical consequence (Read, 1995:36). On the traditional view, a good argument is a valid one, and validity is a matter of form. One proposition is a logical consequence of another only if together they match a valid pattern. Validity itself is determined in these forms according to a notion of truth-preservation. Valid arguments cannot have true premises and a false conclusion. The model implied in this truth-preserving system is that of mathematics.

The story against formal logic is that it is inadequate for the analysis of everyday argumentation. But two things in its defence merit noting here: (i) It's not clear that it *has* ever been advanced as an adequate model for treating everyday arguments (Cf. van Evra, 1985); and (ii) That it does not serve as a complete model for dealing with everyday arguments does not mean it cannot contribute to their analysis. Michael Scriven has observed: "The syllogism was probably nearer to reality (though not to comprehensiveness) than the propositional calculus, but not near enough to make it useful in handling the average editorial or columnist today" (1976:xv). The appropriateness of this comment may depend on what we understand "average" editorials to involve. Most everyday arguments do not fit patterns of the categorical syllogism or propositional logic; but some do. It overstates the case to propose that these argument forms are not useful. They are useful, if they are understood within a theory of argumentation that captures the full range of relations between arguers and audiences rather than just the products of those relations.

Stephen Toulmin's seminal text, *The Uses of Argument* (1958), is a precursor to Scriven's critique. And Toulmin's work is instructive for its detailing of a wide divergence between methods of professional (formal) logicians and those of every day arguers. While it may be the case that formal logicians have never claimed everyday arguments as their domain, Toulmin still accuses them of advancing a model of argument which they expect other types of argument to emulate (1958:126). In distinction to the kind of hierarchy that he sees proposed by formal logicians, with the formally valid argument at the pinnacle, Toulmin identifies a diverse range of arguments specific to different fields, which cannot be assessed by the same procedure and by appeal to the same standards (1958:14).

Toulmin calls the syllogism an unrepresentative and simple sort of argument and traces many of what he terms the "paradoxical commonplaces" of formal logic to the misapplication of this pattern to arguments of other sorts (1958:146). To his mind, different fields of argument will employ different standards of assessment as warrants, data, and backing come into play in different ways. The criteria of formal logic are, Toulmin claims, field-invariant. Thus they cannot deal adequately with the nuances of argument specific to different fields.

Toulmin's own account (which we cannot go into here) offers criteria that are field-dependent and will adapt themselves to the specifics of the field in question. In this way, at least, Toulmin represents a brand of more informal logic.

In 1980, Blair and Johnson identified an expansive range of components integral to the "informal logic point of view." These included a focus on natural language arguments and serious doubts whether deductive logic and standard inductive logic could model them; a view of argumentation as a dialectical process; and a conviction that there were standards and norms of argument evaluation beyond the categories of deductive validity and soundness (1980:x).

Each of these has been a recurring theme in subsequent models of informal logic, and the last has prompted a set of evaluative measures adopted by many informal logicians. There has also been a gradual equating of informal logic with the logic of argumentation (Habermas, 1984:22-3; Johnson and Blair, 1994:11; Walton, 1989:ix).

That 'informal logic' has achieved some kind of general account by 1995 can be seen in the confidence with which Johnson identifies four central characteristics. Informal logic is text-based (rather than speech-based), focuses on an argument (rather than a critical discussion), involves criteria (rather than rules), and is product-oriented (rather than procedure- or process-oriented) (1995:237).

It is this final point that most bears upon our current discussion. "IL," writes Johnson, "envisages a finished (to some degree) *product*, where the arguer is typically absent" (238-39). This leaves us with a set of premises supporting a conclusion. Other logicians, a myriad of them in fact, with an "informalist stripe" are said to share this view.

If both formal and informal logics are grounded in the same basic product-orientation, then what distinguishes them is the criteria by means of which arguments are evaluated.

In the classical tradition discussed above, the strongest claim that can be made about a premise's relation to its conclusion is that the premise entails the conclusion. But as a number of people have pointed out, formal validity is no guarantee of a good argument. Pinto makes the point particularly well when he argues that entailment is neither a necessary nor a sufficient condition for the premises and conclusion of an argument being suitably linked (1994; 1995).

> Not sufficient, because an argument of the form "P, therefore P" meets the criterion of entailment but is hopeless as an argument. Not necessary, because there are innumerable inductively strong arguments in which premises do not entail conclusions. The abstract structures that classical logic studies just don't coincide with the factors that make *arguments* logically good. (Pinto, 1995:277-78)

In place of validity and soundness, informal logicians speak of strength and cogency and evaluate arguments with criteria of relevance, sufficiency and acceptability (some accounts may add or substitutes criteria like truth or consistency). Furthermore, just as form and evaluation are related in formal logic, so evaluation in informal logic is related to the structure of arguments and informal logicians adopt diagramming techniques as a principal tool in their evaluations (Freeman, 1994). In these ideas, a common core of what constitutes informal logic has evolved, and the field has matured from inchoate confusions to a fully-formed discipline, with a recognized content and methodology.

Problems of the Product-Oriented Perspective:

Many of the problems of the product-oriented account as they arise in formal logic are remedied or ameliorated by the advances of informal logic. But two that still warrant attention can be discussed under the general headings: (i) adaptability; and (ii) relevance. To begin this, we return to Toulmin's critique.

(i) As we have seen, formal logic lacks adaptability to different fields. The seriousness of this failure is seen in Toulmin's charge that the field-dependence of logical categories is an essential feature because there are irreducible differences between the sorts of problem arguments can tackle. Having determined the kinds of problem appropriate to a particular case one can "determine what warrants, backing, and criteria of necessity are relevant to this case: there is no justification for applying analytic criteria in all fields of argument indiscriminately" (1958:176). Toulmin has expanded this notion of "field of argument" in terms of the problems that are said to be addressed by them. Thus geometrical argument is a field in which we are faced by geometrical problems. A moral argument is called for by a moral problem, and the need for a prediction calls for an argument with a predictive conclusion. Setting aside problems associated with defining fields in this way, Toulmin's point is to show how unadaptable formal arguments are when we attempt to apply them across the range of such problems in fields.

This criticism applies more to the strict sense of validity constraining formal reasoning than it does to its interest in arguments as products. But the types of arguments produced in such systems are not context-sensitive. Which is to say that they do not take account in their formulation of the diversity of situations from which such patterns are abstracted or the nuances of ordinary language statements that they translate.

Problems associated with translation are widely known and, for students of formal logic, experienced. But it bears repeating that even to test a simple argument of the *modus ponens* variety (if p then q, p /q), there is often indecision whether the 'p' of the first premise and the 'p' of the second premise (or the 'q' of the first premise and the 'q' of the conclusion) symbolize exactly the same expressed statements. Ordinary language arguments, rife with essential ambiguities and nuanced meanings, rarely lend themselves to such exact translations.

Meaning is just one feature of context that formal logic mistreats, if it deals with it at all. Generally, its treatment of the argument produced and the relations between statements within it is conducted without reference to the background -- the circumstances in which it arises, including the occasion and consequences; the arguer and her or his intentions in arguing; and the audience, with its background of beliefs and expectations. Toulmin anticipates many others when he criticizes the freezing of "statements into timeless propositions before admitting them into logic" (1958:182). Attention must be paid to the time and place of an utterance, and questions about the acceptability of an argument must be "understood and tackled *in a context*" (185). This the purely formal logician omits from the account even before beginning.

Informal logic fares better in these respects. While critiques of logic may tend to cast their nets around both formal and informal varieties, informal logic has the adaptability to respond to such critiques in a way its formal counterpart cannot.

Govier, for example, notes how the PPC (premise/premise/conclusion) structure of arguments "represents only the core of the argument" (1995:200), implying that other features beyond the premises and conclusion exist as *part of the argument*. This is an important, albeit vague, observation. Govier's surrounding discussion remarks on the social, practical and textual contexts and the backgrounds of non-argumentative discourse. Insofar as informal logicians address themselves to such features of the "argumentative context," they are expanding their accounts beyond the core of the product to accommodate dialectical and rhetorical aspects. Johnson exhibits just such an attitude by including the process of arguing and the arguers along with the product (Johnson, 1995:242). But, two points: Such accommodations must begin before the product is analysed. That is, the rhetorical (and dialectical) will underlie the logical. And, secondly, there is no direct mention of the central rhetorical feature of audience.

Informal logic's basic sets of criteria, like relevance, sufficiency and acceptability, allow for a more comprehensive assessment of the product than validity and soundness. Acceptability *can* involve considerations of the contexts in which arguments arise, and relevance, as we will see below, takes us beyond judgements of entailment.

Likewise, informal logicians' attention to the dynamics of argumentative discourse, to the ambiguities and vagueness of statements that serve as claims and premises, allows for a greater sensitivity to the "ordinariness" of everyday argumentation. Informal logic offers procedures for dealing with language problems in the standardizing of arguments (setting out the premises and conclusions) and further addresses them in the evaluation-criterion of acceptability.

On the other hand, this use of diagrams serves to emphasize the orientation on product. Reasoning is abstracted from its contexts and attention focused on supports within the argument. Some questions of evaluation, especially with respect to premise acceptability, may take us back to the context. But there is no guarantee that this will happen and premises can be assessed as acceptable according to whether or not they are known to be true to the evaluator. Despite this, however, we can conclude that, in general, informal logic offers a product-oriented model of argumentation with the prospect of some context-sensitivity and therefore richer in its evaluations and practical applications.

(ii) Relevance:

Formal entailment captures the idea of 'guaranteeing' or 'following from'. A premise entails a conclusion if, given the truth of the premise, the conclusion must also be true. But entailment is not strictly a notion of relevance in any way that might be useful. While truth, in these terms, is a *property of* statements, relevance is a *relation between* them. And entailment does not express the same type of relation. A typical textbook can give an inference like 'New York is in New York, therefore New York is in New York' to illustrate entailment. But for a premise to be relevant to a conclusion in any useful sense, it must act as *a reason* which increases our acceptance of, or convinces us to accept for the first time, the conclusion (Blair, 1989). Intuitively we recognize that a proposition may not be relevant to establishing a claim even though it is true. "Wayne Gretzky is clean-shaven" is not in any obvious sense relevant to the truth of the claim that "Wayne Gretzky is one of the premier hockey players in the world." However, in certain deductive systems, the rules governing logical implications (and all logical implications are also entailments) may well allow us to derive the one from the other.

When Read notes that truth-preservation in the classical account of formal logic endorses inferences in which the premises are irrelevant to the conclusion, he indicates a need to incorporate relevance into the criteria of logical consequence (Read, 1995:56).

What we need to support our intuitions about the Gretzky case is a more active sense of propositional relevance. To be relevant it is not enough for two propositions to be passively related; we require that one act upon the other such that it affects our beliefs about that other. Such an account is offered in many informal logic textbooks, which, in eschewing or supplementing details of formal consequence, avoid the problems associated with entailment. Still, when textbooks discuss relevance, they restrict their discussions to versions of what we will term *propositional relevance*. A premise is relevant to a conclusion if it increases (or decreases) our reasons for holding the conclusion. Trudy Govier offers a typical version of this: "Statement A is positively relevant to statement B if and only if the truth of A counts against the truth of B" (1992:146). Another sample account is offered by James Freeman:

If either the truth of the premise increases the likelihood that the
conclusion will be true or the falsity of the premise increases the
likelihood that the conclusion is false, then the premise is relevant to
the conclusion. If neither of these conditions holds, then the premise
is not relevant. (1993:199)

Both of these accounts offer a marked advance over entailment in that they are more suitable for dealing with arguments in natural language. But even strong accounts of propositional relevance are not sufficient to deal with the full range of situations in which questions of relevance arise. Consider, as one example here, the way Govier and Freeman treat the fallacy known as "Straw Man." Both logicians deal with it without making any modification to the accounts of propositional relevance they have provided (Govier, 1992:157; Freeman, 1993:210). That is, they present it as a fallacy of irrelevance even though it does not fit the notion of irrelevance in their accounts. Govier identifies straw man as a fallacy involving irrelevance and writes: "The straw man fallacy is committed when a person misrepresents an argument, theory, or claim..." (157). But previously she had defined 'irrelevance' as: "Statement A is irrelevant to statement B if and only if the truth of statement A counts neither for not against the truth of B" (146). How exactly is the straw man a case of irrelevant argument *in these terms*?

Similarly, Freeman's identification of the straw man as a special fallacy of irrelevance where "An opponent attacks an adversary's position by attributing to the adversary a [misrepresentation] statement S..." (210) does not appear to fit his definition of irrelevance, given above.

The problem arises here because the straw man is not a violation of propositional relevance at all but of *contextual relevance*. The arguer has constructed an argument which fails to be relevant to the context or background as it is constituted by his or her opponent's actual argument. As many instances of straw man indicate, an argument which commits this error may exhibit perfect propositional relevance *internally*. The informal logic account of relevance needs to be expanded to include features of the context, especially those involving the nature of the audience.

Our discussion has shown that both formal and informal logic (two models of the product-oriented perspective) fail to provide completely satisfactory accounts of relevance, although the latter offers advances over the former.

Rhetoric and Formal Logic:

Perelman begins his account by distinguishing argumentation from demonstration (1969:13-14). But his attitude toward demonstration is not straightforward and he will ultimately work with formal modes of reasoning in addressing what he calls "quasi-logical arguments."

The New Rhetoric gives the now-familiar contrast between argumentation and the classical concept of demonstration, and especially with formal logic which examines demonstrative methods of proof. The modern logician, building an axiomatic system, is not concerned with the origin of the axioms or the rules of deduction, or with the role the system plays in the elaboration of thought, or with the meaning of expressions. But when it comes to arguing -- using discourse to influence an audience's adherence to a thesis -- such things cannot be neglected (1969:13-14).

Effectively, such a contrast sets formal demonstration and argumentation into different spheres of influence. The point is emphasized elsewhere: "A purely formal identity is self-evident or posited by convention, but in any case it escapes controversy and hence argumentation" (1982:60). What cannot be argued is not pertinent to argumentation. Argumentation intervenes only where self-evidence is contested, where debate arises about the products of demonstration (1982:6). In this way Perelman escapes the consequences of Michel Meyer's observation that: "If everything is arguable, then nothing underlies argumentation" (Meyer, 1986:151). For Perelman, everything is not arguable. He assumes this when discussing arguments from authority (*argumentum ad vericundiam*). They belong within the realm of argumentation because they are of interest only in the absence of demonstrable proof, since "no authority can prevail against a demonstrable truth" (1982:94-95).

Unlike Descartes, who looked to build all knowledge on a foundation of indubitable self-evidence, Perelman holds such an enterprise to be an exception in the project of knowledge-acquisition, one appropriate only for scientists. In other fields, of philosophy or ethics or law, a quite different practice prevails. Here reasoning cannot be limited to deduction and induction (note, he does not exclude them: they are not adequate, but nor are they irrelevant), rather "a whole arsenal of arguments" should be used, along with a broader conception of reason which includes argumentative techniques and rhetoric as a theory of persuasive discourse (1982:160-161).

The remarks here are ambiguous. On the one hand, demonstration in the form of self-evident truths is outside of the domain of argumentation. On the other hand, deduction is *one of* the argumentative techniques to be employed. This relationship is clarified somewhat in Perelman's treatment of quasi-logical arguments (1969:193-260; 1982:53-80), which gain their force from similarities to formal reasoning. The claim is that, at root, quasi-logical argumentation is nonformal and considerable effort has been required to formalize it. The analyses provided work backwards from the formal scheme to the underlying argument. Quasi-logical arguments include those which depend on logical relations (contradiction, identity, transitivity), and those which depend on mathematical relations (connections between part and whole, smaller and larger, and frequency) (1969:194). For example, and significantly, argumentation makes "considerable use of the relation of logical consequence" (1969:230). Here the *enthymeme* is seen as a quasi-logical argument cast in syllogistic form, and Perelman further notes a wide usage of syllogistic chains in quasi-logical argumentation.

Such discussions indicate a need for studying logical form and formal techniques so as the see the quasi-logical argumentation underlying them and to understand how and when they can be employed. It also allows their adaptability to a wide range of cases. Formal logic is not being discarded as irrelevant here, but its role and relevance are being rethought in a different context. A rhetorical model of argumentation does not narrow the range of methods appropriate to it, but accommodates a wide range of methods, rethought in relation to the underlying rhetorical perspective (Willard, 1989).

The relationship between classical logic and rhetoric is seen finally in the somewhat idiosyncratic way that Perelman distinguishes the rational and the reasonable.

The distinction between formal and nonformal reasoning can be viewed as a distinction between what is rational and what is reasonable.

> The *rational* corresponds to mathematical reason ... which grasps
> necessary relations, which know *a priori* certain self-evident and
> immutable truths ... it owes nothing to experience or to dialogue, and
> depends neither on education nor on the culture of a milieu or an
> epoch. (1979:117)

The *reasonable*, by contrast, is that which is consistent with common sense or conventional wisdom. The 'reasonable person' is guided by the search for what is acceptable in their milieu. The vision of the 'rational person' separates the reason from other human faculties, inaugurates a being who functions as a machine, insensible to his or her humanity and the reactions of others. The 'reasonable person' locates reason as an essential component, but only one component, within the human project.

Importantly for the model of argumentation Perelman develops, the rational and the reasonable must coexist in a mutually supportive relationship. Should one dominate the other, we risk losing advances in thought based on scientific principles, or the guidance of reason to choose between systems (Laughlin and Hughes, 1986:188).

So there is a number of senses in which Perelman's texts include logic in his argumentative project. While it is not adequate for a theory of argumentation, it does contribute to one.

Perelman's model of argumentation is rooted in the rhetorical. Given that our arguments cannot be "proved" completely, they must be submitted for the judgement of those at whom they are directed. This is the audience, the focal point around which other features of context cohere. This means that, essentially, the audience determines the argument, and that an underlying, central sense of contextual relevance must be audience-relevance. This is the case for the arguer, who construct the argument in accordance with the audience's knowledge, background, etc. And this is the case for the evaluator, who critiques the argument in terms of its success in gaining the adherence of the audience for the thesis put forward.

I have only been able to begin to discuss the merits of Perelman's project in this paper. But I hope to have shown that he offers the prospect of a model of argumentation that both underlies and exceeds that of the product-oriented, logical approach.

---- Bibliography ----

Blair, J. A.: "Premise Relevance," *Norms in Argumentation*, Ed. R. Maier, Dordrecht, Foris Publications: (1989) 67--83.

Blair, J. A., Johnson, R.H.: "Introduction," *Informal Logic: The First International Symposium*. Ed. J. A. Blair and R. H. Johnson. Pt. Reyes. CA, Edgepress (1980).

Eemeren, F.H. van, Grootendorst, R.: "Rationale for a Pragma-Dialectical Perspective," *Argumentation* 2 (1988) 271--291.

Evra, J. van.: "Logic, the Liberal Science," *Teaching Philosophy* 8 (1985) 285--294.

Freeman, J. B.: "The Place of Informal Logic in Logic," in Johnson & Blair eds. *New Essays in Informal Logic* (1994) 36--49.

Freeman, J. B.: *Thinking Logically: Basic Concepts for Reasoning*, 2nd edition, New Jersey, Prentice-Hall Inc., (1993).

Govier, T.: "Non-adversarial Conceptions of Argument," in *Perspectives and Approaches: Proceedings of the Third ISSA Conference on Argumentation*. Vol.1. Eds. Frans H. van Eemeren, et. al. Amsterdam, Sicsat (1995) 196--206.

Govier, T.: *A Practical Study of Argument*, 3rd edition, Belmont, California, Wadsworth (1992).

Habermas, J.: *The Theory of Communicative Action: Reason and the Rationalization of Society*, Vol.1. Trans. Thomas McCarthy. Boston: Beacon Press (1984).

Johnson, R. H.: "Informal Logic and Pragma-Dialectics: Some Differences," in *Perspectives and Approaches: Proceedings of the Third ISSA Conference on Argumentation*. Vol.1. Eds. Frans H. van Eemeren, et. al. Amsterdam, Sicsat (1995) 237--245.

Laughlin, S. K., Hughes, D.T.: "The Rational and the Reasonable: Dialectic or Parallel Systems?" *Practical Reasoning in Human Affairs: Studies in Honor of Chaim Perelman*, Ed. J.L. Golden and J. J. Pilotta, Dordrecht, D. Reidel Publishing Company (1986) 187--205.

Meyer, M.: "Problematology and Rhetoric," *Practical Reasoning in Human Affairs: Studies in Honor of Chaim Perelman*, Ed. J.L. Golden and J.J. Pilotta, Dordrecht, D. Reidel Publishing Company (1986) 119--152.

Perelman, Ch.: *The Realm of Rhetoric*, Trans. William Kluback, Notre Dame, University of Notre Dame Press (1982).

Perelman, Ch.: "The Rational and the Reasonable" in *The New Rhetoric and the Humanities: Essays on Rhetoric and its Applications*, Dordrecht, D. Reidel Publishing Co., (1979).

Perelman, Ch., Olbrecht-Tyteca, L.: *The New Rhetoric: A Treatise on Argumentation*, Trans. John Wilkinson and Purcell Weaver, Notre Dame, University of Notre Dame Press (1969).

Pinto, R.C.: "The Relation of Argument to Inference," in Frans H. van Eemeren et. al. *Perspectives and Approaches: Proceedings of the Third ISSA Conference on Argumentation*. Vol.1 (1995) 271--286.

Pinto, R.C.: "Logic, Epistemology and Argument Appraisal," in Johnson & Blair eds. *New Essays in Informal Logic* (1994) 118--124.

Read, S.: *Thinking About Logic: An Introduction to the Philosophy of Logic*, Oxford, Oxford University Press (1995).

Scriven, M.: *Reasoning*, New York, McGraw-Hill Ryerson (1976).

Toulmin, S.: *The Uses of Argument*, Cambridge, Cambridge University Press (1958).

Walton, D. *Informal Logic: A Handbook for Critical Argumentation*, Cambridge, Cambridge University Press (1989).

Willard, C.A.: *A Theory of Argumentation*, Tuscaloosa, Alabama, The University of Alabama Press (1989).

Human Reasoning with Negative Defaults

Carl Vogel*

Institute for Computational Linguistics
Azenbergstr 12
University of Stuttgart
D-70174 Stuttgart
Germany
vogel@ims.uni-stuttgart.de

Abstract. This paper examines psychological data on human reasoning with sets of negative defaults. A negative default is a statement of the form: *Xs are typically not Ys*. While there is pragmatic motivation for chaining positive defaults, chaining negative defaults (concluding from, *As are typically not Bs* and *Bs are typically Cs* that *As are typically not Cs*) is far less reasonable. Default inheritance reasoners universally prohibit 'negative chaining'. However, examination of the psychological plausibility of various conflicting proof theories for default inheritance has demonstrated that some fundamental assumptions of the inheritance literature do not actually hold. This work has also revealed reasoning strategies which do describe human behaviors. In an effort to define inheritance reasoners that are more predictive of human reasoning with defaults, it is important to attend to these findings. This paper focuses on the fact that many people do in fact chain negative defaults. The paper identifies a group of subjects who consistently do so, and evaluates reasoning strategies which are predictive of the behavior of those subjects with respect to other 'benchmark' problems that have been addressed in the literature. Corroboration is found for 'most-path' reasoning.

1 Introduction

Path-based default inheritance reasoning is interesting as a form of nonmonotonic inference because of its attractive computational properties. For example, the skeptical theory of inference proposed by Horty, Thomason, and Touretzky (1990) is polynomial. This is not always true of inheritance (Selman & Levesque, 1989, 1993) and certainly not of nonmonotonic inference at large (although for restricted subsets this can be the case (Niemelä & Rintanen, 1994)). Inheritance can be viewed as a variant of the monadic predicate calculus, with a guarantee of inferential acyclicity, and for this reason should always be decidable, sometimes tractable. However, complexity is introduced in the varying ways that

---

* Thanks to Robin Cooper, Claire Hewson, Jon Oberlander and Jeff Pelletier for encouragement and feedback. I am grateful to the Marshall Aid Commemoration Commission for funding my Ph.D. research at the Centre for Cognitive Science at the University of Edinburgh, and to the SFB-340 B-9 for taking me to Stuttgart.

inference can be defined on the acyclic graphs isomorphic to inheritance theories. There are, of course, any number of ways of defining this inference, and competing methods appeal to 'intuitions' about which methods are most appropriate (Touretzky, Horty, & Thomason, 1987). The intuitions at stake are those which define the most plausible conclusions on the basis of particular interpretations of graphs for those graphs which different conclusions depending upon the reasoner applied. Arguments about plausibility, rather than validity, are made because inheritance is designed to be a psychologically plausible model of human reasoning with defaults.

Inheritance reasoners purport to provide a psychologically plausible model of reasoning with defaults, partially motivated by the idea that tangled hierarchies are ubiquitous in the organization of information, partly for the descriptive power they provide in the semantic analysis of natural language generics, and partly because the (sometimes) efficient decision procedures associated with the representation make it seem a reasonable descriptor of human reasoning, which is relatively efficient. It is surprising that until very recently there have been no psychological investigations designed to elucidate the semantics of reasoning with generics with respect to the idealizations of inheritance theory. Elio and Pelletier (1993) present results about the way people classify exceptional objects in light of default theories in relation to the way general default logics classify the same exceptional objects. They also present the first pilot study applying similar scrutiny to inheritance reasoners, but they do not consider other foundational claims of inheritance reasoning. Hewson and Vogel (1994) and Vogel (1995b) present work directly testing the plausibility of inheritance reasoning, attempting to assess the degree of fit that popular inheritance reasoners (like that of Horty et al., 1990) have with the data supplied by human reasoning.

Vogel (1995a) extends that work by analyzing the pooled data from the experiments, a substantial body of evidence on human reasoning with sets of defaults. That paper examined within-subject patterns of reasoning and identified three basic reasoning strategies: affirmation, shortest path reasoning, explicit link acceptance. However, out of 162 subjects, and after very stringent tests designed to identify subjects who consistently over the entire 40 problem questionnaire[2] answered as predicted by very foundational assumptions about reasoning, there were only 8 subjects left under analysis. One of the largest sources of elimination was invalid negative reasoning.[3] Of course, default reasoning in general is invalid in relation to classical first order logic, but invalid negative reasoning is invalid even in default systems. An example of invalid negative reasoning would be concluding from statements like *Cs are typically not Ds* and *Ds are typically Es* that *Cs are typically not Es*; that is, incorporating negative defaults as a nonfinal link in an argument. For convenience, such an argument will be termed a *negative chain* as distinct from a *negative path* which is the usual notion from inheritance. It is a very basic assumption of path-based inheritance reasoning

---

[2] As will be seen, the task was to a limited extent open ended; thus, fully consistent response patterns over the entire questionnaire are not generally to be expected.

[3] Padgham (1989) outlines valid methods negative default inheritance reasoning.

that a negative link can occur only at the end of a path. This fits with the intuition that if one knows that something (C) is in the complement set (non-D) of another set $(D)^4$, then one has no information about the relationship between C and anything else that D is related to. However, one of the surprising findings of the inheritance experiments was that a great many people reason incorrectly in these cases, generally propagating an initial negative link down a path, rendering the whole thing negative. What is not clear is whether two negative links in an argument are deemed to intensify or to cancel, although the evidence that seems to preclude cancellation.

Since it should not really be a surprise that people reason invalidly with negative information (for example, Wason and Johnson-Laird (1972) demonstrate the human propensity to seek only confirming evidence for hypotheses when disconfirming evidence would correctly resolve the issue, although here negative reasoning is used more weakly than in the case just discussed), and since the general task at stake here is psychologically plausible modeling of human reasoning with defaults[5], it is worth reconsidering the data with less stringent inclusion criteria so that more information can be obtained about reasoning strategies that people use. This paper develops a consistency criterion which allows 'negative argument chains'. This includes data from more subjects (32) and gives evidence for a forth basic default reasoning strategy: argument counting.

2 Defining Plausibility

The psychological plausibility of a logic can be defined as the degree to which it captures the reasoning patterns that people ordinarily use. Thus, the modeling task is closer to that of theoretical linguistics in finding the right level of grammatical representation to capture linguistic phenomena as it occurs, rather than as it is prescribed. That is, just as for natural language where a prescriptive grammar is no more 'correct' than a descriptive one, neither is there a 'correct' logic of human reasoning. One can of course define logics of 'ideally rational agents', however those logics are nearly always undecidable, and therefore are perhaps even farther from the ideal for a rational agent to be driven by them. Goldman (1986) also argues that the concept of rationality should be dictated more by what human behavior demonstrates, instead of rating human reasoning as defective. This paper is not concerned with whether logics provide a closer morphism to the processes that actually govern human reasoning than some other formal framework (like mental models, for instance); rather, its interest is the sort of logic that provides the best description language for expressing exactly those sentences that people are likely to agree are true or worth acting

---

[4] A negative default is assumed to be a statement of the form *As are typically not Ds*, and not the denial of a positive default: *It is false that As are typically Ds*.

[5] The next section defends this as an equal task to identifying 'correct' methods of default reasoning. Essentially, any computer system that reasons correctly, will also have to be able to understand how the people it interacts with reason, even if that reasoning is invalid.

upon. This position can of course be assailed for methodological reasons by people with different goals: if the goal is to build a machine that's a 'better' or more reliable reasoner than humans are then there is little reason to develop logics with an eye on psychological plausibility. On the other hand, if the goal is to describe human reasoning in a generative formal system, then this is a reasonable way of proceeding. Moreover, even if the goal is to develop a machine that reasons more correctly than humans, *if it is to interact with humans it will have to have a model of human reasoning.* From this perspective, plausibility can be measured in terms of the degree of fit between the conclusions licensed by a logic about a set of premises, and the conclusions reached by people. Depending upon the logic, there may be a correspondence between proof theoretic parameters and reasoning strategies that people use. This does not presume that people use the same reasoning strategy all the time. It is appropriate to determine sets of strategies, and to identify to the extent empirically possible that these strategies are employed by and across human reasoners.

3   Testing Plausibility

Hewson and Vogel (1994) addressed inheritance reasoning directly by attempting to determine what sorts of conclusions people find in sets of abstract defaults. The statements in the sets were abstract in the sense that uninterpreted Roman letters were used as labels for concepts (*Bs are typically Cs*) rather than given concrete interpretations (*Marines are typically short haired*). Their initial pilot studies gave subjects sets of statements either in sentential form or using the network syntax of inheritance reasoning and asked subjects to state what they felt reasonable to conclude on the basis of the given information about two of the concepts named in the set. These pilots produced little data, because task was too open ended. They decided instead to constrain answers more by supplying five possible answers with appropriate instantiations for X and Y depending on the exact problem: *Xs are normally Ys; Xs are normally not Ys; Xs are normally Ys and Xs are normally not Ys, it is indeterminate whether Xs are normally Ys or normally not Ys, I don't know.* The materials were questionnaires comprised of 40 problems, each problem containing a set of statements constituting a default theory about abstract concepts. Abstract concepts rather than concrete ones were used to allow control over the influence of other knowledge, belief or opinion (Kaufmann & Goldstein, 1967) — subjects were asked to give the answer they thought appropriate on the basis of the information given in each problem alone. Materials were supplied with the default theories in three modes of presentation: sentence only, graph only, sentence and graph. Each of 72 subjects (in a range of ages and generally with non-technical backgrounds) received all 40 questions (in a random order or the reverse) in one mode of presentation. The 40 problems were designed to encompass a number of inheritance networks specifically disputed in the literature in relation to the conclusions they should 'intuitively' license. An example problem from the graph+sentence mode of presentation is given in Figure 1. The problems were comprised strictly of defaults. In the graph-only condition, subjects were told that the arrow meant 'are typically', and thus had half as much priming of 'typicality' as the other subjects (which means they could have occasionally interpreted arrows as meaning 'strict is-a'). It is convenient to index each problem by its graph presentation. In each problem, the respondent is asked to characterize the relationship between the 'leftmost' and 'rightmost' (relative to the corresponding graph) class mentioned in the premises. The questionnaire contained full instructions to subjects on how to answer the problems; it was stressed that there were no right or wrong answers, and subjects were to say what *they* thought could be concluded from the information given.

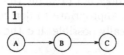

- As are normally Bs.
- Bs are normally Cs.

What can you conclude from these statements? Asterisk (*) the appropriate answer.

(a) As are normally Cs.
(b) As are normally not Cs.
(c) As are normally Cs and normally not Cs.
(d) It isn't definite whether As are normally Cs or normally not Cs.
(e) I don't know.

If you wish, explain why you reach this conclusion.

Fig. 1. An Example Question

The full list of problems is given in Table 1 in the random order that half the subjects were presented with (the other half had the reverse order). Herein, problems will be referred to either by graph or by the number next to the graph as it appears in Table 1. As mentioned, the multiple choice question was appropriately adapted to each problem, which was presented to subjects in exactly one of the three modes of presentation.

The problems were designed to test the more foundational assumptions of the inheritance literature: transitivity, negativity, redundancy and preemption. Others were there in the hope of clarifying high-level conflicting intuitions about what conclusions should follow from an inheritance network (Touretzky et al., 1987). The remaining problems were there as controls. Hewson and Vogel (1994)

611

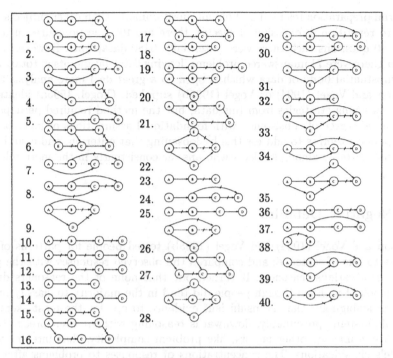

Table 1. Graphic Representation of the Problem Set

reported surprising findings. Firstly, most of the problems were rated indeterminate. Of the determinate response, while there was considerable support for transitivity, and some support for simple cases of preemption, there was little evidence for redundancy (redundant statements/links seemed to be interpreted as providing additional information rather than redundant and disregardable information), and people seemed to reason with negative information in a way wholly unpredicted by the literature. Hewson and Vogel (1994) did not report evidence about the various specific forms of preemption or subpath ambiguity. There was no clear finding with respect to mode of presentation, except that graphic presentation seemed to polarize responses more, suggesting that the graphic syntax of inheritance reasoning has much to do with the strength of 'intuitions' that have been reported as controversial in the proof theoretic literature (which includes equivocation over the use of defeasible links to represent strict relationships — the graph-only polarization may have picked this up as well). Vogel (1995b) partially replicated the experiment of (Hewson & Vogel, 1994), dropping the graph+sentence condition and testing the materials on an additional 98 subjects (undergraduates in English literature and composition courses); the results corroborated the earlier findings.

This paper examines the pooled data from the two experiments. After removal of unattempted questionnaires, and those for which collation errors during material preparation resulted in different questionnaires than other subjects had (due to repetition & omission of sheets), there are 162 questionnaires in total. With 40 questions each there were potentially 6480 datapoints. However, altogether there were 71 unanswered problems, leaving 6409 datapoints in total. This is a substantial body of data which is open to a great deal more analysis than Hewson and Vogel (1994) or Vogel (1995b) supplied. (Vogel, 1995a) identified three reasoning strategies from this data, but the method required perhaps an unrealistic degree of consistency with foundational assumptions of inheritance. This paper attempts to address this by loosening that criterion, and examining response patterns among subjects who answer consistently in the more relaxed setting.

4 Negative Defaults

Hewson and Vogel (1994) and Vogel (1995b) tested people for patterns of response to sets of sentences and compared the observed responses with the predictions of various reasoners. It turned out that none of the reasoners did a perfect job of it. Even though people did tend in the majority to classify problems as ambiguous, that is insufficient evidence to opt for the most skeptical possible system (presumably, downwards reasoning without even preemption) since there may be other factors, like problem complexity, that interact with people's classifications. The concentrations of responses to problems after the indeterminate[6] classifications are set aside yield evidence for less skeptical systems, or at least a stronger foothold from which to explore the factors that affect their classification. It is equally interesting to define reasoners which yield a closer approximation to observed behavior. Vogel (1995a) pooled the data from the two experiments, and examined within-subject patterns of reasoning (the previously mentioned analyses considered only between-subject systematicity for individual problems; the current work attempts to assess between subject systematicity over *sets* of related problems). Data obtained from subjects who did not conform to predictions of positive and negative transitivity (measured by problems containing no conflicts) were eliminated; this left 88 questionnaires. This means that subjects were removed from consideration who did not give answer category 'a' to each of, ⊙—⊙—ⓒ, ⊙—⊙—ⓒ—ⓓ, ⊙⌒⊙⌒ⓒ⌒ⓓ , and ⊙—⊙—ⓒ—ⓓ, as well as answer category 'b' to each of, ⊙—⊙—ⓒ, ⊙—⊙—ⓒ—ⓓ, ⊙⌒⊙⌒ⓒ⌒ⓓ , and ⊙ ⊙—ⓒ—ⓓ. This is motivated by the consideration that the reasoning of those subjects whose responses were not predicted by transitivity on graphs containing no conflicts would be better predicted by classical statistical models (or were otherwise relatively inconsistent[7] in response patterns), and are thus outside the scope of default inheritance. Additional subjects whose responses to

---

[6] Answers in categories 'c' or 'd' are referred to as *indeterminate classification* of a problem. An answer in category 'a' is *positive classification*; 'b', *negative classification*. An answer of either 'a' or 'b' is a *definite classification*.

[7] Not necessarily in the sense meaning contradictory, though.

problems containing non-final negative links (e.g. ⊙⁻⁺⊙⁻⊙) propagated negativity (i.e., concluded definitely that *As* are typically not *Cs*) were also eliminated. This left the materials of just 8 subjects. While this is a rather small resulting dataset, it should be pointed out eliminations were taken to be stringent: in the transitivity case, subjects were eliminated if they did not respond in accord with transitivity in each of the graphs without conflicts; in the negativity case, subjects were eliminated if they propagated negativity with any of the graphs involving nonfinal links. Degrees of satisfaction of these two properties also occur, but extreme stringency illustrates the main points allowing the identification of 3 clear credulous reasoning strategies. Most of the subjects were skeptical, preferring to classify a graph as indeterminate unless there were no conflicts at all. Of the remainer were credulous in the sense of being willing to make determinate classifications in the presence of potential conflicts. One strategy was *affirmation* (making a positive classification if an affirmative argument existed), another was *shortest path* reasoning, and the other was *explicit link acceptance.*

This paper conducts a similar analysis, but on the basis of less stringent subject selection criteria. Noting that a great many people reason with arguments with nonfinal negative links as if they were properly negative arguments, the data was filtered again to include subjects who consistently deemed such paths to be negative. That is, in addition to the respondingly consistently to the problems representative of regular positive and negative transitivity judgements, this paper also admitted each subject who gave a definite negative classification to all of the following problems: ⊙⁻⁺⊙⁻⊙⁻⊙, ⊙⁻⊙⁻⁺⊙⁻⊙, ⊙⁻⊙⁻⁺⊙⁻⊙, ⊙⁻⁺⊙⁻⁺⊙, ⊙⁻⊙⁻⁺⊙⁻⊙, ⊙⁻⊙⁻⊙⁻⊙, and ⊙⁻⁺⊙⁻⊙. Before examining within-subject patterns of response, it is useful to consider the overall response patterns of subjects in this group. Table 2 summarizes the results. For each problem, the corresponding number in the response column indicates the number of subjects in the subgroup who chose the classification in the column heading. The response category 'c/d' sums over both sorts of indeterminate classification (why this is done will be more clear shortly). Tables (2.a and 2.b) show the response patterns over the full questionnaire, and for convenience Table (2.c) repeats the same information, without the entries for the problems used as tests to identify this particular subgroup of subjects (those were the questions that sub-

jects answered in the way specified above). Note that problem 16 (⊙⁻⊙⁻⊙) could also have been used as a subject selection test. It is useful to consider how subjects who evaluated arguments with nonfinal negative links as valid negative arguments also evaluated more complicated problems involving more than one argument.

4.1 Between-Subject Findings

From Table 2 it is easy to see that the overwhelming majority of problems were classified as indeterminate. This corresponds to skeptical reasoning, since indeterminacy is basically the conclusion that no definite classification can be made,

Table 2.a

Problem	a	b	c/d
1	4	2	26
2	7	3	22
3	3	13	15
4	5	1	26
5	12	2	18
6	5	12	15
7	15	3	14
8	3	14	14
9	5	8	19
10	0	32	0
11	0	32	0
12	1	31	0
13	32	0	0
14	32	0	0
15	0	32	0
16	0	32	0
17	15	1	16
18	4	5	23
19	0	32	0
20	1	5	26

(2.a)

Table 2.b

Problem	a	b	c
21	1	3	28
22	1	10	20
23	0	32	0
24	0	32	0
25	0	32	0
26	1	13	18
27	1	7	23
28	7	4	21
29	1	6	24
30	32	0	0
31	17	0	15
32	1	31	0
33	1	11	19
34	1	29	2
35	1	12	16
36	0	31	0
37	31	0	0
38	1	1	28
39	8	10	13
40	10	10	12

(2.b)

Table 2.c

Problem	a	b	c/d
1	4	2	26
2	7	3	22
3	3	13	15
4	5	1	26
5	12	2	18
6	5	12	15
7	15	3	14
8	3	14	14
9	5	8	19
16	0	32	0
17	15	1	16
18	4	5	23
20	1	5	26
21	1	3	28
22	1	10	20
26	1	13	18
27	1	7	23
28	7	4	21
29	1	6	24
31	17	0	15
33	1	11	19
35	1	12	16
38	1	1	28
39	8	10	13
40	10	10	12

(2.c)

Table 2. Observed Frequency of Responses among Negative Reasoners

positive or negative (more on this will be said below). Thus, it is interesting to look at the set of definite classifications, and mainly for those where there was a significant difference between 'a' and 'b' responses[8]. In these terms, the problems which were given a positive definite classification are: (5) 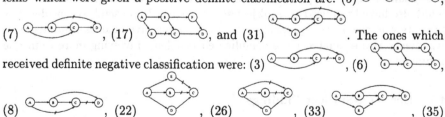,

(7) , (17) , and (31) . The ones which

received definite negative classification were: (3) , (6) ,

(8) , (22) , (26) , (33) , (35)

---
[8] In general, this paper does not report significance values.

. With one exception, both of these patterns are best described as being decided by the polarity of the greatest number of arguments in the set of defaults: *most-paths* reasoning. The exception is in the first positive case (5,) since there there are no more positive paths than negative ones (moreover, the shortest path is negative). In general, the inheritance literature would predict that this problem should get a definite negative classification. Note that its complement (29,) was given a negative classification (albeit by a smaller margin; among subjects who gave it a definite classification), symmetrically contrary to the literature's predictions. Thus, it would seem that some independent explanation should account for the definite classifications in those two cases.[9]

Problem 35 is topologically identical to problem 27 ( ), which received the same basic classification. This is actually one of the networks that has received attention in the literature on inheritance proof theory with respect to whether subpath ambiguity should propagate to containing paths (restricted versus full skepticism; see Touretzky et al. (1987)). On the basis of ambiguity about whether *As* are *Cs*, a restricted skeptical reasoner would determine that there is no positive path for the negative link between C and D to extend, and therefore that there is no negative path conflicting with the positive path through A, B and F to D. A fully skeptical reasoner would point at the conflict between that positive path and the negative one through A, B and C, yielding an indeterminate classification since it isn't clear whether *As are typically Ds* or *As are typically not Ds*. In either case, the assumption is that the negative link between C and D cannot extend the negative path from A to C. However, under the assumption that negative arguments can be defined that way, there are actually two negative arguments conflicting with the single positive argument. And indeed, subjects gave answer 'b' in response to both of these problems.

The only problem for which use of an argument with a nonfinal negative link as a negative argument would have made a difference, but didn't, is problem 40: . Here, curiously, there is effectively a random pattern of response across subjects in the subgroup (the problem received a statistically significant negative classification in the full sample). The same is true of problem 39 ( ) which has two genuinely negative paths. At the present I have no good explanation for this exception. However, it is evident that the main claim of this section are otherwise rather robust. Analysis of patterns of response among the entire subsample under consideration suggest that those

---

[9] Responses to cases where direct links connect the endpoints of a (single) path of opposite polarity (2, ; 18,) exhibit less systematicity and in the opposite direction (of the definite responses problem 2 was taken as mainly positive), thus indicating that preemption is not at work in 5 and 29.

subjects whose evaluation of use negative chains as proper negative paths in isolation also make the same evaluation in the context of complex problems and use this information as a way of counting negative arguments as evidence for a definite negative classification rather than a positive classification.

4.2 Within-Subject Findings

So far the analysis has been quantified over the entire subgroup under consideration. This is a methodologically sound approach, but it is also informative to look at within-subject patterns of response in order to obtain more specific information about potential reasoning strategies. The data is encapsulated in Table 4 (see Appendix A). The responses given by each of the 32 subjects in the subgroup to each of the answered questions is shown by listing the problem numbers appropriate to each answer category. Answers in category 'e' (don't know) were ignored. Table 2 used a response category 'c/d' for indeterminate responses which are given here in separate columns. This makes clear why it is reasonable to collapse the two categories into a indeterminacy. Answer category 'c' referred to *Xs are normally Ys and Xs are normally not Ys* which would generally be considered logically inconsistent. Subjects were asked to reply strictly on the basis of the information given, the way they thought the answers should be. However, while it has been shown that people entertain logically inconsistent beliefs (Braisby, Franks, & Hampton, 1994), there was only one problem for which the expected answer was 'c' (38, ⬡). It was expected that the graphs deemed truly ambiguous would be given answer category 'd': *it is indeterminate whether Xs are normally Ys or normally not Ys*. Given the surprising amount of answers in category 'c', seemingly with the interpretation intended for 'd', it is reasonable to coalesce the responses in these categories as was done in the preceding section (and by Hewson and Vogel (1994)).

One thing that is clear from this breakdown is that some of the subjects listed were distinctly less affirmative than others. Consider the subset which gave 5 or 4 affirmative answers (by construction of this sample, these subjects all gave affirmative classification to problems 13, 14, 30 and 37). This set has 11 members (subjects 48, 73, 91, 94, 100, 101, 104, 107, 115, 116 and 130). It can be observed that the only other problems that they answered with category 'a' were 5 (⊙—⊙—⊙—⊙) and 7 (⊙⊙—⊙⊙) (each by one subject) and 39 (⊙—⊙—⊙—⊙) and 40 (⊙⊙—⊙—⊙) (by three and four subjects, respectively). It is interesting that there was such agreement on those last two graphs among these subjects who tended to give few affirmative classifications. These were pointed to above as counter-examples to the most-paths reasoning strategy. However, the same subjects who gave few affirmative classifications also tended to give few negative classifications. All 11 gave answer 'b' for problem 16, yet it has already been pointed out that this problem could as well have been used to to discriminate subjects who reason within the predictions of transitivity applied to negative paths. Including problem 16 as a test, three of the 11 gave answer 'b' to only the problems that determined the subgroup of 32. This subset can be deemed the most skeptical since in all other cases they classified problems as indeterminate. The other seven of this 11 subjects gave negative classifications to ten other problems; four of that ten (6, ⟨graph⟩, 35, ⟨graph⟩,

39, ⟨graph⟩, 40, ⟨graph⟩) were each given answer 'b' by three subjects. However, there was agreement in definite classification on more than one of these problems only between two subjects: both subject 115 and 130 gave answer 'b' to both 6 and 35 (one rated 39 negative and 40 positive and the other rated 40 negative and 39 indeterminate). On each of these four problems, a most-path strategy using negative chains as negative paths would also predict negative classification. But this strategy is not predictive of these subjects' responses in general since so many other graphs for which it would have yielded negative classification were rated indeterminate. Moreover, as many of these 11 who rated 39 and 40 negative also rated both definitely positive. Therefore there seems to be little systematicity for those two problems in this subject pool. In sum, the behavior of these 11 subjects rise to a slightly broader range of skeptical reasoners: the least credulous of the subjects who tended to give definite answers were the ones who gave definite responses to problems consisting of only one argument chain or with multiple but nonconflicting argument chains. In the case of negative conclusions, they gave definite classifications even to arguments that included nonfinal negative links.

The section which reported between-subject systematicity noted that there were four problems which received significant numbers of positive classifications: (5) ⟨graph⟩, (7) ⟨graph⟩, (17) ⟨graph⟩, and (31) ⟨graph⟩. There were more problems in the set which received definite negative classification: (3) ⟨graph⟩, (6) ⟨graph⟩, (8) ⟨graph⟩, (22) ⟨graph⟩, (26) ⟨graph⟩, (29) ⟨graph⟩, (33) ⟨graph⟩ and (35) ⟨graph⟩. The advantage of doing a within-subject analysis is that it makes it possible to identify exactly which subjects respond in a consistent pattern. Five of the 21 subjects (72, 78, 89, 108 and 138) in the subgroup that complements the group discussed in the preceding paragraph gave answer 'a' to all four of problems 5, 7, 17, and 31, in accord with most-path reasoning in the positive case. But, in the negative case, only three of these subjects (89, 108 and 138) gave answer 'b' for at least one of the problems which tended to receive negative classifications. The other two gave answer 'b' primarily to just the problems which were used as tests to define the group used throughout this paper. However, six different

618

subjects (87, 105, 112, 127, 133, and 136) gave answer 'a' to three of the 'affirmative' problems and answer 'b' to at least three of the 'negative' problems (five of those subjects gave 'b' to as many as five of the 'negative problems). Of those, subjects 105 and 127 did not classify problem 5 (⊙—⊙—⊙—⊙, which does not have a most-paths explanation) affirmatively. Subjects 105, 127 and 138 all gave answer 'a' to 7, 17, and 31 and answer 'b' at least 2 of the 4 distinct problems for which the most reasonable description is most-path reasoning making crucial use of negative chains as proper negative paths. Subject 127 answered all 4 of those consistently. Thus, while one clear subset of the more credulous reasoners was credulous (and consistent with most-path reasoning) only in the case of affirmative classifications, here we have another group of approximately the same size which answered fairly consistently in accord with most-path reasoning in both the positive and negative instances where negative chains are counted as negative paths. Another two subjects (2 and 144) gave positive classification to three of the 'affirmative' problems and negative classification to one or two of the 'negative' ones.

5 Discussion

This paper has demonstrated that human subjects tend to be consistent in patterns of responses to sets of abstract generics. It seems that a subset consistently make use of arguments which have nonfinal negative links. It appears that subjects partition into: those who are completely skeptical of default arguments; those who do not see negative chains as negative paths (and whose behavior is predicted by either affirmation, shortest path reasoning or direct-link acceptance (Vogel, 1995a)); those who are skeptical unless there is an unopposed positive or negative argument (where a negative argument can be formed by a negative chain); those who reason with positive paths and negative paths and chains and use most-path reasoning.[10] This paper found specific support for most-path reasoning. The use of negative chains in the latter two cases is invalid in classical and default systems, but perhaps has rational explanation through the decontextualization involved in the experiments. It would be expected that the same results would not obtain given actual contexts instead of defaults over abstract contexts. The criteria for deciding which sets of responses to consider in this analysis were stringent: subjects were required to behave consistently with respect both the transitivity test and the negative chaining test.[11] Ongoing analysis of the data involves using less stringent inclusion criteria (degrees of satisfaction of the consistency tests) and further examination of the degree to which the five reasoning strategies are consistently predictive in general.

---

[10] Loosely speaking, of course—it's just easier to express the pretense that subjects behave according to these rules than that these generalizations are predictive of their behavior.

[11] While Vogel (1995a) *excluded* subjects who gave a determinate answer to *any* problem involving unopposed negative chains, the analysis here *included* only those subjects who gave determinate, negative, answers to *all* of those problems.

619

A Encapsulation of Subject Responses

For convenience in determining the structure of the problem corresponding to the problem numbers listed in the various answer categories in Table 4, Table 1 from above is reproduced here.

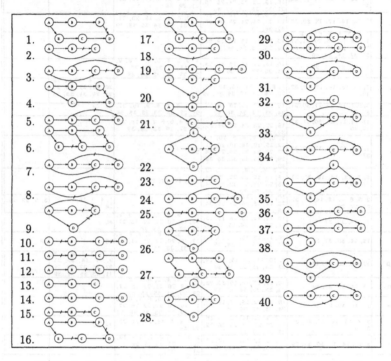

Table 3. Graphic Representation of the Problem Set (Table 1)

subject	Response 'a'	'b'	'c'	'd'	'e'
2	5, 7, 9, 13, 14, 28, 30, 31, 37	1, 8, 10, 11, 12, 15, 16, 18, 19, 23, 24, 25, 26, 32, 34, 36	2, 4, 17, 20, 21, 22, 29, 33, 38, 40	3, 6	27, 35, 39
48	7, 13, 14, 30, 37	10, 11, 12, 15, 16, 19, 23, 24, 25, 32, 34, 36	1, 2, 3, 4, 5, 6, 8, 9, 17, 18, 20, 21, 22, 26, 27, 28, 29, 31, 33, 35, 38, 39, 40		
49	2, 9, 13, 14, 30, 31, 37	3, 5, 6, 7, 8, 10, 11, 12, 15, 16, 17, 18, 19, 20, 21, 22, 23, 24, 25, 26, 27, 28, 29, 32, 33, 34, 35, 36, 39, 40		1, 4, 38	
52	13, 14, 28, 30, 31, 32, 34, 37	3, 6, 8, 9, 10, 11, 12, 15, 16, 19, 22, 23, 24, 25, 26, 27, 33, 35, 36, 39, 40	1, 2, 4, 5, 7, 17, 18, 20, 21, 29, 38		
65	7, 13, 14, 17, 28, 30, 37	6, 8, 9, 10, 11, 12, 15, 16, 19, 22, 23, 24, 25, 26, 27, 32, 36, 40	1, 2, 3, 4, 5, 18, 20, 21, 29, 31, 33, 34, 35, 38, 39		
72	1, 2, 3, 4, 5, 6, 7, 8, 9, 13, 14, 17, 18, 20, 21, 22, 26, 27, 28, 29, 30, 31, 33, 35, 37, 38, 39, 40	10, 11, 12, 15, 16, 19, 23, 24, 25, 32, 34, 36			
73	13, 14, 30, 37, 39	10, 11, 12, 15, 16, 19, 23, 24, 25, 32, 34, 36, 40	1, 2, 3, 4, 5, 6, 7, 8, 9, 17, 18, 20, 21, 22, 26, 27, 28, 29, 31, 33, 35, 38		
77	7, 12, 13, 14, 17, 30, 37, 40	3, 8, 10, 11, 15, 16, 19, 23, 24, 25, 32, 34, 36, 39	9	1, 2, 4, 5, 6, 18, 20, 21, 22, 26, 27, 28, 29, 31, 33, 35, 38	
78	3, 5, 7, 13, 14, 17, 30, 31, 37, 39	10, 11, 12, 15, 16, 19, 23, 24, 25, 32, 34, 36, 40	1, 2, 4, 6, 8, 9, 18, 21, 26, 27, 28, 29, 33, 35, 38	20, 22	
84	13, 14, 17, 28, 30, 31, 37, 39, 40	3, 9, 10, 11, 12, 15, 16, 19, 22, 23, 24, 25, 26, 32, 34, 36	1, 2, 4, 5, 6, 7, 8, 18, 20, 21, 27, 29, 33, 35, 38		
85	1, 2, 4, 13, 14, 17, 30, 31, 37	3, 5, 6, 7, 8, 9, 10, 11, 12, 15, 16, 18, 19, 20, 21, 22, 23, 24, 25, 26, 27, 28, 29, 32, 33, 34, 35, 36, 39, 40	38		
87	5, 7, 13, 14, 18, 30, 31, 37, 40	2, 3, 6, 8, 9, 10, 11, 12, 15, 16, 19, 20, 22, 23, 24, 25, 26, 28, 29, 32, 33, 34, 35, 36, 38, 39	1, 4, 17, 21, 27		
89	2, 5, 6, 7, 13, 14, 17, 30, 31, 37, 39	10, 11, 12, 15, 16, 19, 20, 23, 24, 25, 26, 32, 33, 34, 36		1, 4, 9, 18, 21, 27, 28, 38, 40	3, 8, 22, 29, 35
91	13, 14, 30, 37	10, 11, 12, 15, 16, 19, 23, 24, 25, 32, 34, 35, 36, 39	1, 2, 3, 4, 5, 6, 7, 8, 9, 17, 18, 20, 21, 22, 26, 27, 28, 29, 31, 33, 38	40	
94	13, 14, 30, 37, 39	10, 11, 12, 15, 16, 19, 21, 23, 24, 25, 32, 34, 36	1, 2, 3, 4, 5, 6, 7, 8, 9, 17, 18, 20, 22, 26, 27, 28, 29, 31, 33, 35	40	38
100	13, 14, 30, 37, 40	6, 10, 11, 12, 15, 16, 19, 23, 24, 25, 32, 34, 36		1, 2, 3, 4, 5, 7, 9, 17, 18, 20, 21, 22, 26, 27, 28, 29, 31, 33, 35, 38, 39	
101	13, 14, 30, 37, 40	10, 11, 12, 15, 16, 19, 23, 24, 25, 32, 34, 36, 39	1, 2, 3, 4, 5, 6, 7, 8, 9, 17, 18, 20, 21, 22, 26, 27, 28, 29, 31, 33, 35, 38		
104	13, 14, 30, 37, 40	10, 11, 12, 15, 16, 19, 23, 24, 25, 32, 34, 36	1, 2, 3, 4, 5, 6, 7, 8, 9, 17, 18, 20, 21, 22, 26, 27, 28, 29, 31, 33, 35, 38, 39		
105	3, 7, 13, 14, 17, 28, 30, 31, 37	6, 8, 9, 10, 11, 12, 15, 16, 19, 22, 23, 24, 25, 26, 32, 33, 34, 35, 36	1, 2, 4, 5, 18, 20, 21, 27, 29, 38	39, 40	
106	8, 13, 14, 17, 30, 37, 39	10, 11, 12, 15, 16, 19, 23, 24, 25, 32, 34, 36, 40	1, 2, 3, 4, 5, 6, 7, 9, 18, 20, 21, 22, 26, 27, 28, 29, 31, 33, 35, 38		
107	13, 14, 30, 37	10, 11, 12, 15, 16, 19, 23, 24, 25, 32, 34, 36	1, 2, 3, 4, 5, 6, 7, 8, 9, 17, 18, 20, 21, 22, 26, 27, 28, 29, 31, 33, 35, 38, 39, 40		
108	5, 6, 7, 13, 14, 17, 18, 30, 31, 37, 40	1, 2, 3, 4, 10, 11, 12, 15, 16, 19, 23, 24, 25, 26, 32, 34, 36, 39	8, 9, 20, 21, 22, 27, 28, 29, 35, 38		33
112	1, 4, 5, 6, 13, 14, 17, 18, 30, 31, 37	2, 3, 8, 10, 11, 12, 15, 16, 19, 20, 22, 23, 24, 25, 26, 28, 29, 32, 33, 34, 36	7, 9, 21, 35, 38, 39, 40	27	
115	13, 14, 30, 37, 40	6, 10, 11, 12, 15, 16, 19, 23, 24, 25, 29, 32, 33, 34, 35, 36, 39	1, 2, 3, 4, 5, 7, 8, 9, 18, 20, 21, 22, 26, 27, 28, 31, 38		
116	13, 14, 30, 37, 39	10, 11, 12, 15, 16, 19, 23, 24, 25, 32, 34, 36, 40	1, 2, 3, 4, 5, 6, 7, 8, 9, 17, 18, 20, 21, 22, 26, 27, 28, 29, 31, 33, 35, 38		
127	7, 13, 14, 17, 28, 30, 31, 37	3, 6, 8, 9, 10, 11, 12, 15, 16, 19, 22, 23, 24, 25, 26, 27, 32, 33, 34, 35, 36, 40	2, 4, 5, 18, 20, 21, 29, 38, 39	1	
130	5, 13, 14, 30, 37	6, 8, 9, 10, 11, 12, 15, 16, 18, 19, 22, 23, 24, 25, 32, 33, 34, 35, 36, 40	2, 3, 7, 20, 21, 26, 28, 29, 38	1, 4, 17, 27, 31, 39	
133	2, 4, 5, 9, 13, 14, 17, 30, 31, 37	3, 6, 7, 8, 10, 11, 12, 15, 16, 18, 19, 23, 24, 25, 26, 29, 32, 33, 34, 35, 36, 39		1, 20, 21, 22, 27, 28, 38, 40	
136	2, 5, 7, 9, 13, 14, 30, 31	3, 6, 8, 10, 11, 12, 15, 16, 19, 23, 24, 25, 32, 34	4, 17, 18, 20, 21, 22, 26, 27, 28, 29, 33, 39, 40	1	
138	2, 5, 6, 7, 13, 14, 17, 30, 31, 37, 40	3, 8, 10, 11, 12, 15, 16, 19, 23, 24, 25, 27, 32, 35, 36	1, 4, 9, 18, 20, 29, 38, 39	21, 22, 26, 28, 33, 34	
144	1, 4, 5, 7, 13, 14, 17, 30, 37	3, 10, 11, 12, 15, 16, 19, 23, 24, 25, 32, 34, 36	2, 6, 8, 9, 18, 20, 21, 22, 26, 27, 28, 29, 31, 33, 35, 38, 39, 40		
161	7, 8, 13, 14, 30, 31, 37	10, 11, 12, 15, 16, 19, 23, 24, 25, 27, 32, 34, 35, 36		1, 2, 3, 4, 5, 6, 9, 17, 18, 20, 21, 22, 26, 28, 29, 33, 38, 39, 40	

Table 4. Individual Subject Response Details

References

Braisby, N., Franks, B., & Hampton, J. (1994). On the Psychological Basis for Rigid Designation. In *Proceedings of the Sixteenth Annual Conference of the Cognitive Science Society*, pp. 56–65. Atlanta, Georgia.

Elio, R. & Pelletier, F. J. (1993). Human Benchmarks on AI's Benchmark Problems. In *Proceedings of the Fifteenth Annual Conference of the Cognitive Science Society*, pp. 406–411. June 18-21, 1993. Boulder, Colorado.

Goldman, Alvin, I. (1986). *Epistemology and Cognition.* Cambridge: Harvard University Press.

Hewson, C. & Vogel, C. (1994). Psychological Evidence for Assumptions of Path-Based Inheritance Reasoning. In *Proceedings of the Sixteenth Annual Conference of the Cognitive Science Society*, pp. 409–14. Atlanta, Georgia.

Horty, J., Thomason, R., & Touretzky, D. (1990). A Skeptical Theory of Inheritance in Nonmonotonic Semantic Networks. *Artificial Intelligence, 42*(2-3), 311–48.

Kaufmann, H. & Goldstein, S. (1967). The Effects of Emotional Value of Conclusions upon Distortions in Syllogistic Reasoning. *Psychonomic Science, 7*, 367–8.

Niemelä, I. & Rintanen, J. (1994). On the Impact of Stratification on the Complexity of Nonmonotonic Reasoning. *Journal of Applied Non-Classical Logics, 4*(2), 141–79.

Padgham, L. (1989). Negative Reasoning Using Inheritance. In *Proceedings of the 11th International Joint Conference on Artificial Intelligence*, pp. 1086–93. Detroit, Michigan.

Selman, B. & Levesque, H. (1989). The Tractability of Path-Based Inheritance. In *Proceedings of the 11th International Joint Conference on Artificial Intelligence*, pp. 102–9. Detroit, Michigan.

Selman, B. & Levesque, H. (1993). The Complexity of Path-Based Defeasible Inheritance. *Artificial Intelligence, 62*, 303–39.

Touretzky, D., Horty, J., & Thomason, R. (1987). A Clash of Intuitions: The Current State of Non-Monotonic Inheritance Systems. In *Proceedings of the 10th International Joint Conference on Artificial Intelligence*, pp. 476–82. Milan, Italy.

Vogel, C. M. (1995a). Defining Psychologically Plausible Default Inheritance Reasoners. Unpublished manuscript, Institute for Computational Linguistics, University of Stuttgart.

Vogel, C. M. (1995b). *Inheritance Reasoning: Psychological Plausibility, Proof Theory and Semantics.* Ph.D. thesis, Centre for Cognitive Science, University of Edinburgh.

Wason, P. C. & Johnson-Laird, P. N. (1972). *Psychology of Reasoing: Structure and Content.* Cambridge, Mass.: Harvard University Press.

On the Semantics of the Unknown

STEFFO WEBER
UNIVERSITÄT DORTMUND
INFORMATIK VIII

`steffo@kibosh.informatik.uni-dortmund.de`

ABSTRACT. How can a database express and conclude that it has information on some sentence Φ but not on Ψ? We develop a formal semantics for an operator $\Box$, to be read as "It is entailed by the database that ...". We define a consequence relation based on a preference semantics and analyze under which circumstances this relation fulfils the condition that $\neg\Box\Phi$ is entailed by a set X of formulas whenever $\Phi \notin Cn(X)$. It turns out that this condition holds if the elements of X are modal Horn clauses or propositional sentences. We further relate our logic to some well known nonmonotonic logics. We show that it is located between autoepistemic logic and McDermott's nonmonotonic S5; moreover, all S5 theorems are valid. But unlike McDermott's nonmonotonic S5 it does not collapse to the monotonic case.

1. MOTIVATION

Suppose we have a database and a certain deduction mechanism. The database could be a set of propositional sentences and the deduction mechanism could be the provability relation of classical propositional logic. Imagine that a user begins to fill the database by telling it that A holds, $B \lor C$ does not hold, $A \to B$ holds etc. After a certain period of time the user wants to find out whether the database contains information on A. The easiest way would be to put the query A? If this query fails (i.e. if the database answers 'No.'), we could ask for $\neg A$? If both queries fail, then we know that the database does not have information on A.

This procedure enables us to *check* whether there is information on A or not. But can we say that the database itself has information on whether A is known or unknown? The answer is 'No', because there is no sentence like 'unknown A' contained in the database. The database cannot even represent a fact like 'unknown A' because there is no operator 'unknown', if we have a propositional or any other classical language.

The author would like to thank Gerd Brewka and Joachim Hertzberg for invaluable comments on a draft version of this paper.

The task of this paper is to develop a logic which enables a database to reason about its own content or data. Assume that we have enriched our language by an operator $\Box$ with the intended meaning that $\Box A$ should represent something like 'it is known to the database that A' or 'A can be derived from the database' (by means of its deduction mechanism). The intended meaning of $\Box A$ is similar to that of A if none of them is negated: there should be no difference between telling the database that 'A holds' or telling it 'you *know* that A holds'. If, however, the database has *no* information on A, for example when it is empty, then $\neg\Box A$ should be entailed. This is different from a database which entails $\neg A$, because in general such a database could not be empty.

1.1. Potential impact. A logic which enables a database to reason about its own content and identifies the sentences which are *not* known has a great impact on AI applications. For example, machine learning algorithms, could really improve when having the explicit information that A is unknown. This has been reported by, e.g. Hirsh [Hirsh, 1990] (cf. also the transcript of the discussion on meta-reasoning in [Brazdil and Konolige, 1990]). Moreover, there are machine learning systems and knowledge-acquisition systems whose inference engine provide an unknown operator (cf. [Morik et al.,1993], [Emde, 1991]). Unfortunately there is a severe theoretical drawback of these systems: they lack a clear semantics. One part of the semantics is algebraical or set theoretical while the semantics for the unknown operator is given in a proof theoretical manner. For example, the semantics for the unknown-operator in [Morik et al.,1993] is described as (. . .)unknown *[A] evaluates to true if and only if the proposition A (. . .) cannot be proved*. This is a reading rather than a semantics. The logic presented in this paper will fix this problem.

The plot is as follows: after having presented some preliminaries we will investigate what is meant by 'A is unknown', thus developing a formal semantics which eliminates the aforementioned problems with earlier approaches. In the ensuing section we will study fundamental properties of our logic. Section 4 is devoted to a comparison with Moore's autoepistemic logic and the nonmonotonic logic NML-2 of McDermott. It will turn out that our logic which was originally intended to model the semantics of an unknown-operator used by some AI-systems, is located between autoepistemic logic and McDermott's nonmonotonic S5.

2. Semantical Investigations

2.1. Terminology. A (propositional) *signature* Σ is a denumerable set of propositional variables. The language based on Σ is defined as the smallest set $\mathcal{L}(\Sigma)$ such that $\Sigma \subset \mathcal{L}(\Sigma)$ and if $\Phi, \Psi \in \mathcal{L}(\Sigma)$ then $\neg\Phi, \Phi \rightarrow \Psi, \Phi \wedge \Psi, \Phi \vee \Psi, \Phi \oplus \Psi, \Diamond\Phi, \Box\Phi \in \mathcal{L}(\Sigma)$. The elements of Σ are also called *atomic formulas*; a formula preceded by $\Diamond$ or $\Box$ is called a *modal formula*. A *nonmodal formula* is a formula which has no modal subformula. We will omit the reference to a special signature Σ when it is clear from the context or not important and then just talk about $\mathcal{L}$. Further, let $\mathcal{L}_0 \subset \mathcal{L}$ denote the classical nonmodal propositional language. The elements of $\mathcal{L}_0$ are usually denoted by capital latin letters, whereas the elements of $\mathcal{L}$ are denoted by greek capital letters (it is often important for us to distinguish between both sets).

As usual, the *degree* $d(\Phi)$ of a formula Φ is the number of connectives and operators occurring in Φ. The notion of degree will only be needed in inductive proofs.

A *propositional interpretation function* of a signature Σ is a mapping $V : \Sigma \rightarrow \{t, f\}$. A *Kripke structure* $\mathfrak{M}$ is a tuple (M, V)[1] where M is an index set (also called set of possible worlds or states) and $V = \{V_1, \ldots, V_{|M|}\}$ is a set of propositional interpretation functions. The set of all Kripke structures interpreting a signature Σ is denoted by STRUCT(Σ).

We will make the convention that '$\subset$' means proper containment.

2.2. Unknown means satisfiability.

Let us return to our database. We say that the database $X \subseteq \mathcal{L}$ does not have any information about a sentence $\Phi \in \mathcal{L}$ if and only if neither Φ nor $\neg\Phi$ can be proved from the database[2]. An uncareful transformation of this idea into the definition of validity could yield the potpourri semantics mentioned in the Introduction. The key idea to obtain an appropriate semantics is that if neither Φ nor $\neg\Phi$ is provable from X, then Φ as well as $\neg\Phi$ is satisfiable with respect to X. Thus, Φ is unknown w.r.t. the database X if both, $\{\Phi\} \cup X$ and $\{\neg\Phi\} \cup X$ are satisfiable. Such an unknown-operator could be modeled by using Kripke structures as interpreting structures. The validity relation $\models$ is identical to the validity relation of modal S5.

Definition 1 (Validity, Kripke-model). *Let $\Phi, \Psi \in \mathcal{L}$ be a formula, $\mathfrak{M} = (M, V)$ a Kripke structure, $\alpha \in M$*

---

[1] Normally, Kripke structures have an accessibility relation $R \subseteq M \times M$. Throughout this paper we assume R to be complete (i.e. $R = M \times M$); it follows that M is also universal (i.e. reflexive, symmetric and transitive).

[2] An alternative would be to restrict the non-provability condition to Φ, i.e. to say that we don't have any information on Φ if and only if we cannot prove Φ. We judge this to be a matter of taste.

$$\mathfrak{M} \models_\alpha \Phi \qquad \text{for an atomic } \Phi \text{ if } V_\alpha(\Phi) = t$$
$$\mathfrak{M} \models_\alpha \Phi \wedge \Psi \qquad \text{iff } \mathfrak{M} \models_\alpha \Phi \text{ and } \mathfrak{M} \models_\alpha \Psi$$
$$\mathfrak{M} \models_\alpha \Phi \to \Psi \qquad \text{iff } \mathfrak{M} \not\models_\alpha \Phi \text{ or } \mathfrak{M} \models_\alpha \Psi$$
$$\mathfrak{M} \models_\alpha \Diamond\Phi \qquad \text{iff there is } \beta \in M \text{ such that } \mathfrak{M} \models_\beta \Phi$$
$$\mathfrak{M} \models_\alpha \neg\Phi \qquad \text{iff } \mathfrak{M} \not\models_\alpha \Phi$$

The connectives $\vee$ (disjunction), $\oplus$ (exclusive or) and the operator $\Box$ are defined as abbreviations in the usual way.

$\mathfrak{M}$ *is a* (Kripke-) *model for Φ, alternatively Φ is valid in $\mathfrak{M}$, ($\mathfrak{M} \models \Phi$) if for all $\alpha \in M$, we have $\mathfrak{M} \models_\alpha \Phi$. We extend the relation $\models$ in the usual way to sets of formulas.*

The next step is to define the concept of *entailment* (denoted by the relational symbol $\Vdash$). As said before, the goal is to have an entailment relation $\Vdash$ such that $X \Vdash \neg\Box\neg\Phi$ (or shorter, $X \Vdash \Diamond\Phi$), whenever Φ is satisfiable w.r.t. X or, in other terms, whenever $X \cup \{\Phi\}$ has a model. However, if we allow Φ to be an arbitrary sentence, this could yield counter-intuitive results: let $X = \{B\}$; there is a Kripke-structure $\mathfrak{A}$ such that $\mathfrak{A} \models X \cup \{\Box C\}$. Thus, $X \cup \{\Box C\}$ is satisfiable and we should have $X \Vdash \Diamond\Box C$. But $\Diamond\Box C$ is semantically equivalent to $\Box C$. Hence, $X \Vdash \Box C$. This contradicts our intuition because the conclusion 'C is known' is entailed by a database which had no information about C. We will therefore restrict Φ to be a nonmodal sentence[3].

We have to find out whether there is a model $\mathfrak{A}$ of X such that $\Phi \in \mathcal{L}_0$ is true in at least one of $\mathfrak{A}$'s possible worlds. Fortunately, we can restrict the search for such a model to those models of X which contain a *maximal* set of possible worlds. To see why, note that if $X \cup \Diamond\Phi$ has a model then $X \cup \Diamond\Phi$ is valid in some model $\mathfrak{A}_{max}$ of X which contains a maximal set of possible worlds (remember that Φ is nonmodal). Generally,

X entails Ψ if and only if Ψ is valid in each structure $\mathfrak{A} = (M, V)$ with

1. $\mathfrak{A} \models X$ and
2. M is maximal, that is V contains as many two valued interpretations as possible, that is $\mathfrak{A}$ makes as many formulas of the form $\Diamond A$ valid as possible, where A is a nonmodal formula.

To ensure that we consider only maximal models (in the above sense), we must be able to compare arbitrary structures to find out which structure is a maximal model. This can be done by correlating all structures via the substructure relation. The idea behind substructures is that, $\mathfrak{M} = (M, V)$ is a substructure of $\mathfrak{N} = (N, W)$ if $\mathfrak{N}$ 'extends' $\mathfrak{M}$, i.e if

---

[3] This restriction can practically be justified by the fact that in the aforementioned AI systems the operator unknown can only be applied to propositions which do not already contain this operator.

$M \subseteq N$ (and $V \subseteq W$). However, since M and N could arbitrary index sets (i.e. M could be a set of natural numbers, while N could be set of characters), we have to ensure, that they are 'comparable'. This will be guaranteed by existence of an isomorphism.

Definition 2 (Substructure). *Two structures* $\mathfrak{M} = (M, V)$, $\mathfrak{N} = (N, W)$ *are* isomorphic *(denoted by* $\mathfrak{M} \cong \mathfrak{N}$*) if and only if there is an isomorphism* $I : M \to N$ *such that* $V_\alpha = W_{I(\alpha)}$, *for every* $\alpha \in M$.

We say that $\mathfrak{M} = (M, V)$ *is a* substructure *of* $\mathfrak{N} = (N, W)$, *denoted by* $\mathfrak{M} \preccurlyeq \mathfrak{N}$, *if and only if there is* $\mathfrak{N}' = (N', W')$ *such that* $N' \subseteq N$, $W' \subseteq W$ *and* $\mathfrak{N}'$ *is isomorphic to* $\mathfrak{M}$. *We say that* $\mathfrak{M}$ *is a* strict substructure *of* $\mathfrak{N}$, *denoted by* $\mathfrak{M} \prec \mathfrak{N}$, *if and only if there is* $\mathfrak{N}' = (N', W')$ *such that* $N' \subset N$, $W' \subset W$ *and* $\mathfrak{N}'$ *is isomorphic to* $\mathfrak{M}$.

We can use $\prec$ as a preference relation in the sense of [Shoham, 1988] and define preferred models on the basis of this preference relation.

Definition 3 (Preferred model). *Let* $\Phi \in \mathcal{L}$, $X \subseteq \mathcal{L}$. *Define a relation* $\models^\pm \subseteq \models$ *such that*

1. $\mathfrak{M} \models^\pm \Phi$ *if and only if there is no* $\mathfrak{M}'$ *such that* $\mathfrak{M} \prec \mathfrak{M}'$ *and* $\mathfrak{M}' \models \Phi$; *we say that* $\mathfrak{M}$ *is a* preferred *or* maximal *model for* Φ.
2. $\mathfrak{M} \models^\pm X$ *if and only if there is no* $\mathfrak{M}'$ *such that* $\mathfrak{M} \prec \mathfrak{M}'$ *and* $\mathfrak{M}' \models X$; *we say that* $\mathfrak{M}$ *is a* preferred *or* maximal *model for* X.

Please note that the extension from formulas to sets of formulas in the above definition is not equivalent to '$\mathfrak{M} \models^\pm X$ if and only if $\mathfrak{M} \models^\pm \Phi$, for every $\Phi \in X$'.

Following Shoham [Shoham, 1988] we define an entailment relation $\Vdash$ as follows:

Definition 4 ($\Vdash, Cn$). $X \Vdash \Phi$ *if and only if,* $\mathfrak{M} \models^\pm X$ *implies* $\mathfrak{M} \models \Phi$. $Cn(X)$ *is the set of all formulas entailed by* X, *i.e.* $Cn(X) =_{\text{def}} \{\Phi \mid X \Vdash \Phi\}$.

The following examples illustrate the above definitions

Example 1. *Let* $\Sigma = \{A, B\}$ *be a signature,* $X = \{\neg \Box A \to A\}$. *There are two preferred models* $\mathfrak{M}_1 = (\{\alpha, \beta\}, V_1)$ *and* $\mathfrak{M}_2 = (\{\gamma, \delta\}, V_2)$ *for* X:

$$V_{1_\alpha}(A) = t \mid V_{1_\beta}(A) = t \parallel V_{2_\gamma}(A) = f \mid V_{2_\delta}(A) = f$$
$$V_{1_\alpha}(B) = t \mid V_{1_\beta}(B) = f \parallel V_{2_\gamma}(B) = t \mid V_{2_\delta}(B) = f$$

Thus, we have $X \Vdash \Diamond B \wedge \Diamond \neg B$ *but* $X \not\Vdash \Diamond A$ *and* $X \not\Vdash \Diamond \neg A$.

Example 2. *Let* $\Sigma = \{A, B\}$ *be a propositional signature,* $X = \emptyset$. *Then* $\{\Diamond A, \Diamond B, \Diamond \neg A, \Diamond \neg B\} \subset Cn(X)$.

2.3. **Properties.** In this Section we will discuss some properties of our logic. The main questions we have in mind are:

1. Is the entailment relation $\Vdash$ appropriate w.r.t. the analysis of the meaning of unknown, i.e. does it hold for all nonmodal A that

$$X \not\Vdash A \text{ and } X \not\Vdash \neg A \text{ implies } X \Vdash \Diamond A \text{ and } X \Vdash \Diamond \neg A$$

2. What general properties, which we know from classical logic, e.g. closure properties, do hold for our logic?

2.3.1. *Is $\Vdash$ appropriate?* The answer is 'yes' if X has a unique preferred model and 'no' otherwise. The reason that, in general, we could have $X \not\Vdash \Diamond A$ even if $X \not\Vdash A$ and $X \not\Vdash \neg A$ is that the concept of satisfiability can be expressed in the object language itself: let $X = \{\Diamond A \oplus \Diamond B\}$ (literally, either A is satisfiable or B is satisfiable). X has two preferred models, one satisfying $\Diamond A$ and $\neg B$ and the other one satisfying $\Diamond B$ and $\neg A$. Clearly, we have $X \not\Vdash A$ and $X \not\Vdash \neg A$ but also $X \not\Vdash \Diamond A$.

These issues are related to Default theories or autoepistemic theories which have multiple extensions and we will discuss the relationship between multiple extensions and theories which have more than one preferred model in Section 4. However, we can identify restrictions $\mathcal{L}'$ of our language $\mathcal{L}$ such that every $X \subset \mathcal{L}'$ has at most one preferred model. One of these restrictions is based on an extension of the well-known notion of Horn-theories to a modal language.

Definition 5 (M-literal, modal Horn). *Let $\Phi \in \mathcal{L}$ be a formula. Φ is said to be an M-literal if and only if Φ has the form $\Diamond A$ or $\neg \Diamond A$ where A is any nonmodal formula*

An M-literal is said to be positive *if and only if it has the form $\Diamond A$. A negative M-literal is an M-literal of the form $\neg \Diamond A$.*

Φ is said to be modal Horn *if and only if Φ has the form*

$$\Phi_1 \vee \Phi_2 \vee \ldots \vee \Phi_n$$

where each Φ_i is an M-literal and at most one Φ_i is a positive M-literal.

Proposition 1. *The following subsets X of $\mathcal{L}$ have at most one unique model:*

1. $X \subset \mathcal{L}_0$
2. $X \subset \mathcal{L}_{Horn} =_{def} \{\Phi \mid \Phi \text{ is modal Horn}\}$

Proof. Part (1) is simple. For part (2) assume that $X \subset \mathcal{L}_{Horn}$ has two preferred models, say $\mathfrak{A}$ and $\mathfrak{A}'$. We will first prove the following

Lemma 1. *Let X be a set of formulas such that X has two preferred models $\mathfrak{M} = (M, V)$, $\mathfrak{N} = (N, W)$ and $\mathfrak{M} \not\cong \mathfrak{N}$ ($\mathfrak{M}$ is not isomorphic to*

$\mathfrak{N}$). *Then there are formulas* Φ, Ψ *such that* Φ *has the form* $\Diamond A$ *and* Ψ *has the form* $\Diamond B$ *(where* $A, B \in \mathcal{L}_0$*) and*

$$\mathfrak{M} \models \{\Phi\} \cup X \qquad \mathfrak{N} \models \{\Psi\} \cup X$$
$$\mathfrak{M} \not\models \{\Psi\} \cup X \qquad \mathfrak{N} \not\models \{\Phi\} \cup X$$

Proof. Assume to the contrary that for all nonmodal A we have

$$\mathfrak{M} \models \Diamond A \Leftrightarrow \mathfrak{N} \models \Diamond A$$

This implies that for all nonmodal C we have

$$\mathfrak{M} \models C \Leftrightarrow \mathfrak{N} \models C$$

and thus, for all $\Phi \in \mathcal{L}$

(*) $$\mathfrak{M} \models \Phi \Leftrightarrow \mathfrak{N} \models \Phi$$

From $\mathfrak{M} \not\cong \mathfrak{N}$ we can conclude that

1. there is no isomorphism from M to N or
2. there is at least one $\alpha \in M$ such that for all $\beta \in N$ it holds that $V_\alpha \neq W_\beta$

But since (1) implies (2), we can assume without loss of generality that there is at least one such $\alpha \in M$ and that $\alpha \notin N$. Define, $\mathfrak{N}' =_{\text{def}} (N \cup \{\alpha\}, W \cup \{V_\alpha\})$. Clearly, $\mathfrak{N} \prec \mathfrak{N}'$ and by (*) and the construction of $\mathfrak{N}'$ we have that

$$\mathfrak{N} \models \Phi \Leftrightarrow \mathfrak{N}' \models \Phi \qquad \text{for all } \Phi \in \mathcal{L}$$

Thus, $\mathfrak{N}$ cannot be a preferred model of X. ∎

Let Φ, Ψ be the formulas of the preceding Lemma; assume w.l.g that $\mathfrak{A} \models \Phi$ and $\mathfrak{A}' \models \Psi$. By the preceding Lemma we can conclude that

$$\mathfrak{A} \models \{\Phi \vee \Psi\} \cup X \text{ and } \mathfrak{A}' \models \{\Phi \vee \Psi\} \cup X$$

Thus, there must be $\Theta \in X$ such that Θ is semantically equivalent to $\Phi \vee \Psi$, i.e. for all structures $\mathfrak{M}$ it holds that $\mathfrak{M} \models \Theta$ if and only if $\mathfrak{M} \models \Phi \vee \Psi$. Since Φ has the form $\Diamond A$ and Ψ has the form $\Diamond B$, there must be a sentence semantically equivalent to $\Diamond A \vee \Diamond B$ in X. But this is impossible since $X \subset \mathcal{L}_{\text{Horn}}$ ∎

2.3.2. *Other properties.* Let us know state some closure properties.

Proposition 2. *Define* $\text{Th}_{S5}(X)$ *to be the smallest set containing* X*, all classical axioms, as well as all instances of the axiom schemes*

$$\Box(\Phi \rightarrow \Psi) \rightarrow (\Box\Phi \rightarrow \Box\Psi) \quad \textbf{K}$$
$$\Box\Phi \rightarrow \Phi \quad \textbf{T}$$
$$\Box\Phi \rightarrow \Box\Box\Phi \quad \textbf{4}$$
$$\Diamond\Phi \rightarrow \Box\Diamond\Phi \quad \textbf{5}$$

which is closed under application of modus ponens and the rule of necessitation, i.e. $\Phi/\Box\Phi$.

Let $\mathfrak{A}$ *be a structure and* $T_{\mathfrak{A}} =_{\text{def}} \{\Phi \mid \mathfrak{A} \models \Phi\}$. *Then,* $T_{\mathfrak{A}}$ *is closed under* Th_{S5}

Proof. Easy (omitted). ∎

Observation 1. *The nonmonotonic consequence operator* Cn *enjoys the properties of inclusion, closure, nonmonotonicity, cumulativity.*

Proof. Easy (omitted). ∎

Let us now turn to an important property, namely compactness. The fact that $\models$ is compact follows from [Chellas, 1980], who calls it compactness of consistency. If Cn were monotonic, then the compactness of entailment would follow immediately. However, in the presence of non-monotonicity, compactness of entailment is a nontrivial property.

Theorem 1 (Compactness of consistency). *Let* $X \subseteq \mathcal{L}$ *be a possibly infinite set of formulas.* X *has a model if and only if every finite subset of* X *has a model.*

Proof. See e.g. [Chellas, 1980] ∎

Theorem 2 (Compactness of entailment). *Let* $X \Vdash \Phi$ *for some* X *and some* Φ. *Then there is a finite set* $X_{fin} \subseteq X$ *such that* $X_{fin} \Vdash \Phi$.

Proof. By induction on the degree $d(\Phi)$ of Φ.

$d(\Phi) = 0$: In this case Φ is atomic. Assume to the contrary that for all $X_{\text{fin}} \subseteq X$ we have $X_{\text{fin}} \not\Vdash \Phi$. Hence, there is $\mathfrak{M} \models X_{\text{fin}} \cup \{\neg\Phi\}$, for all X_{fin}. Thus, by Theorem 1, there is $\mathfrak{A} = (M, V)$ such that $\mathfrak{A} \models X \cup \{\neg\Phi\}$. Now, either $\mathfrak{A}$ is a preferred model of X, or there is $\mathfrak{A}'$, $\mathfrak{A} \preccurlyeq \mathfrak{A}'$ such that $\mathfrak{A}'$ is a preferred model of X. But neither $\mathfrak{A}$ nor $\mathfrak{A}'$ can be a model of Φ. This contradicts the assumption that $X \Vdash \Phi$.

$d(\Phi) = n + 1$: We have to distinguish several cases:

(i) Φ has the form $\Diamond\Psi$. We have to find X_{fin} such that $X_{\text{fin}} \Vdash \Diamond\Psi$. There are two possibilities

(a) $X \Vdash \Psi$. By induction hypothesis, there is X_{fin}' such that $X_{\text{fin}}' \Vdash \Psi$. Then, define $X_{\text{fin}} = X_{\text{fin}}'$. Clearly, $X_{\text{fin}} \Vdash \Diamond\Psi$.

(b) $X \not\Vdash \Psi$. Define, $X_{\text{fin}} = \emptyset$. Since $X \Vdash \Diamond\Psi$, we have $X_{\text{fin}} \Vdash \Diamond\Psi$.

(ii) Φ has the form $\Psi \wedge \Theta$. By induction hypothesis there is $X_{\text{fin}}' \Vdash \Psi$ and $X_{\text{fin}}'' \Vdash \Theta$. Define $X_{\text{fin}} = X_{\text{fin}}' \cup X_{\text{fin}}''$. Clearly, $X_{\text{fin}} \Vdash \Psi \wedge \Theta$.

(iii) Φ has the form $\neg\Psi$. Again, we have several subcases analyzing the form of Ψ.

(a) Ψ has the form $\neg\Theta$. But then Φ is semantically equivalent to Θ and by induction hypothesis we have, ex. $X_{\mathrm{fin}}' \Vdash \Theta$. Thus, define $X_{\mathrm{fin}} = X_{\mathrm{fin}}'$.

(b) Ψ has the form $\Diamond\Theta$. Thus, Φ is equivalent to $\neg\Diamond\Theta$ which is equivalent to $\neg\Theta$. By induction hypothesis and since $d(\neg\Theta) = n$ there is $X_{\mathrm{fin}}' \Vdash \neg\Theta$. Thus, define $X_{\mathrm{fin}} = X_{\mathrm{fin}}'$.

(c) Ψ has the form $\Theta \wedge \Upsilon$. Thus, Φ is equivalent to $\neg\Theta \vee \neg\Upsilon$. By induction hypothesis there is $X_{\mathrm{fin}}' \Vdash \neg\Theta$ and $X_{\mathrm{fin}}'' \Vdash \neg\Upsilon$. Then, define $X_{\mathrm{fin}} = X_{\mathrm{fin}}'$ or define $X_{\mathrm{fin}} = X_{\mathrm{fin}}''$ (both works).

■

From a pragmatical point of view it is reasonable to ask what impact compactness has, since a database is always considered to be finite. However, due to its general nature, the compactness theorem is a useful tool like completeness theorems or interpolation lemmas. The main reason for including the compactness theorem is that in [Kraus et al., 1990] compactness property is assumed.

Let me briefly state what we have got so far: we have defined a cumulative, nonmonotonic entailment relation $\Vdash$ and an operator $\Diamond$ such that for any nonmodal sentence A the database X entails $\Diamond A$ if and only if $X \not\Vdash A$ (provided that X has a unique preferred model). Moreover, the consequence operator Cn is inclusive, idempotent and compact (in algebraical terms: a compact preclosure operator).

3. A PRACTICAL APPLICATION

As already mentioned in the introduction, the knowledge about which sentences are unknown can improve the efficiency of machine learning algorithms. But there are currently more applications. Consider the database $X = \{\Diamond A \to A\}$. This set has two preferred models, yielding $\Diamond A \notin Cn(X)$ and $\Diamond\neg A \notin Cn(X)$. This is a consequence of treating $\to$ as material implication. Nevertheless, sometimes formulas like $\Diamond A \to A$ are used to express "If A is consistent with your formulas, then assert A" (cf. [Emde, 1991]). The point is that even though $\to$ is given the semantics of material implication, as in [Emde, 1991], [Morik et al.,1993], it is *used* or to be read as an inference rule.

Lukaszewicz described a cunning trick to rule out $\mathfrak{A}_1 = (\{1\}, \{V_1(A) = f\}$ as a preferred model for X. Thus, the only preferred model which remains is $\mathfrak{A}_2 = (\{1\}, \{V_1(A) = t\}$ (cf. [Lukaszewicz, 1990]). We will adapt this trick for our needs. In order to make it work, we have to ensure that each formula in our database has a certain normal form.

Definition 6 (Ordered MCNF). *A formula* $\Phi_1 \wedge \Phi_2 \wedge \ldots \wedge \Phi_n$ *is said to be in* ordered conjunctive normal form *if and only if each* Φ_i *is of the form*

$$\neg \Box B \vee \Box C_1 \vee \ldots \vee \Box C_k \vee A$$

where $B, C_1, \ldots, C_k, A \in \mathcal{L}_0$.

It is known that each modal formula can be reduced in S5 to ordered MCNF (cf. [Hughes and Cresswell, 1968]). Please, note that we can write each conjunct Φ_i as $(\Box B \wedge \neg \Box C_1 \wedge \ldots \wedge \neg \Box C_k) \to A$.

Definition 7 (Applicable [Lukaszewicz, 1990]). *Let* $\Phi \in \mathcal{L}$ *in ordered MCNF,* $T = \mathrm{Th}_{S5}(T)$ *a set of formulas. A conjunct* $\Phi_i = (\Box B \wedge \neg \Box C_1 \wedge \ldots \wedge \neg \Box C_n) \to A$ *of* Φ *is said to be* applicable *w.r.t.* T *if and only if* $B \in T,\ C_1, \ldots, C_n \notin T$; *otherwise,* Φ_i *is said to be inapplicable w.r.t* T.

Definition 8 (Strongly preferred model). *Let* $X \subset \mathcal{L}$ *be a set of formulas,* $\mathfrak{A}$ *a structure and* $T_{\mathfrak{A}} =_{\mathrm{def}} \{\Phi \mid \mathfrak{A} \models \Phi\}$. *Define*

$$X' =_{\mathrm{def}} X \backslash \{\Phi \mid \Phi \text{ is inapplicable w.r.t } T_{\mathfrak{A}}\}$$

$\mathfrak{A}$ *is a* strongly preferred model *for* X *if and only if*

Cond. A $\qquad\qquad\qquad \mathfrak{A} \stackrel{\bullet}{\models} X$

Cond. B $\qquad\qquad\qquad \mathfrak{A} \not\stackrel{\bullet}{\models} X'$

We say that $X \Vdash_{\mathrm{S}} \Phi$ if and only if every strongly preferred model of X is a model of Φ. Clearly, $X \Vdash_{\mathrm{S}} \Phi$ implies $X \Vdash \Phi$.

Example 3. *Let* $\Sigma = \{A\}$, $X = \{\Diamond A \to A\}$. *There are two preferred models* $\mathfrak{A}_1, \mathfrak{A}_2$ *for* X *but only one them is strongly preferred. Let* $\mathfrak{A}_1 = (\{1\}, \{I_1(A) = t\})$ *and* $\mathfrak{A}_2 = (\{1\}, \{I_1(A) = f\})$. $\Diamond A \to A$ *is applicable w.r.t* $T_{\mathfrak{A}_1}$ *but not w.r.t* $T_{\mathfrak{A}_2}$. *Thus we have to check whether Cond. B holds.*

1. $\mathfrak{A}_1 \stackrel{\bullet}{\models} X \backslash \emptyset$ *(there are no inapplicable formulas w.r.t* $T_{\mathfrak{A}_1}$*) and*
2. $\mathfrak{A}_2 \not\stackrel{\bullet}{\models} X \backslash X$ *(every formula of* X *is inapplicable w.r.t* $T_{\mathfrak{A}_2}$*).*

Hence, $X \Vdash_{\mathrm{S}} A$.

We close this Section with the observation that the strong consequence operator $Cn_S(X) =_{\mathrm{def}} \{\Phi \mid X \Vdash_{\mathrm{S}} \Phi\}$ inherits all closure properties as well as compactness of Cn.

4. RELATIONSHIP TO NONMONOTONIC MODAL LOGICS

We will now relate our logic to some nonmonotonic modal formalisms. One of the earliest attempts to attack the problem of nonmonotonic reasoning was by means of modal logic. In 1982 McDermott introduced some nonmonotonic versions of well-known monotonic modal logics such as

S4 and S5. But unfortunately, it turned out that the most promising formalization of nonmonotonic reasoning based on modal S5 collapses to monotonic S5. Then, in 1983, Moore decided to develop a new formalism called autoepistemic logic, which is a reconstruction of an earlier proposal by Doyle and McDermott. Autoepistemic logics are nowadays the most prominent nonmonotonic modal formalisms.

From a syntactical point of view, autoepistemic logic can be regarded to be a weaker[4] system than a nonmonotonic formalism based on modal S5, because autoepistemic logic is based on a system which is called *weak S5* (i.e. K45). The main result of this Section is that our logic lies between nonmonotonic K45 and S5, that is, it can be seen as an extension of autoepistemic logic towards McDermott's ideas. Thus, we will first compare our logic with Moore's autoepistemic logic and then with McDermott's Nonmonotonic Logic II (NML-2).

4.1. Extensional entailment. Many nonmonotonic formalisms for example, Default Logics and Autoepistemic Logics do not discuss consequence operators. They rather make use of the notion of an *extension*. The main difference between the set of (nonmonotonic) consequences and an extension is that if we are given a set X of sentences, then the set of (nonmonotonic) consequences is unique while X may have different extensions.

In order to provide a basis for a comparison between logics based on the notion of extension and our logic, we have to say what corresponds to an extension in our logic. According to Stalnaker, an extension can be regarded as a final belief set; thus, containing a maximal set of beliefs[5]. One property which reflects the intuition about final beliefs sets is that they should be *stable* (cf.[Stalnaker, 1993]).

Definition 9 (Stable). *A set $X \in \mathcal{L}$ of sentences is stable if and only if it meets the following requirements:*

1. X *is closed under classical (nonmodal) propositional consequence* Cn_{cl}
2. *if* $\Phi \in X$ *then* $\neg \Diamond \neg \Phi \in X$
3. *if* $\neg \Phi \notin X$ *then* $\Diamond \Phi \in X$

---

[4]Please note that the terms stronger (containing more axioms) and weaker (containing less axioms) are somewhat meaningless within nonmonotonic logics, because the addition of axioms does not guarantee that we get more theorems.

[5]Stalnaker talks about autoepistemic extensions, but I think his interpretation does also apply to every other nonmonotonic formalism which makes use of the term 'extension'

We propose to regard every preferred model of X as a final belief state. The set of formulas which are valid in a preferred model of X is a final belief set. This motivates the following definition.

Definition 10 (Extensional entailment). *Let $X, Y \subseteq \mathcal{L}$; we say X extensionally entails Y (denoted by $X \Vdash_{\mathbb{E}} Y$ if and only if there is a structure $\mathfrak{A}$ such that $\mathfrak{A} \models X$ and $Y = \{\Phi \mid \mathfrak{A} \models \Phi\}$. Y is said to be an S5-based extension of X (because the validity relation $\models$ is that of modal S5).*

Please note that the relation $\Vdash_{\mathbb{E}}$ is only defined between sets of formulas. Further, $Cn(X) = \bigcap \{T \mid X \Vdash_{\mathbb{E}} T\}$.

It has already been observed by Konolige, Moore and Fitting (cf. [Konolige, 1988]) that if we have the validity relation of modal S5, then the set of all sentences valid in an S5 structure is stable.

Proposition 3 (Konolige, Moore, Fitting). *Let $\mathfrak{A}$ be a structure. Then the set $T_{\mathfrak{A}} =_{\text{def}} \{\Phi \mid \mathfrak{A} \models \Phi\}$ is stable.*

Extensions can be characterized by self-referential equations. We will now give such an equation for S5-based extensions.

Theorem 3. $X \Vdash_{\mathbb{E}} T$ *if and only if* $T = Th_{S5}(X \cup \{\Diamond A \mid \neg A \in \mathcal{L}_0 \backslash T\})$.

Proof. Let RHS denote the right hand side (i.e $Th_{S5}(X \cup \{\Diamond A \mid \neg A \in \mathcal{L}_0 \backslash T\})$) of the above equation.

'$\Rightarrow$'

$RHS \subseteq T$: $X \Vdash_{\mathbb{E}} T$ implies that T is stable and consistent. From stability we can conclude that $\neg A \notin T$ implies $\Diamond A \in T$, thus T contains all formulas which are added to the right hand side (RHS) of the above equation via the condition $\{\Diamond A \mid A \in \mathcal{L}_0$ and $\neg A \notin T\}$. Moreover, T is closed under Th_{S5}.

$T \subseteq RHS$: It holds that $Th_{S5}(X) \subseteq$ RHS and since $X \subseteq T$ and T is closed under Th_{S5} we have $Th_{S5}(X) \subseteq T$. It remains to show that $T \backslash Th_{S5}(X) \subset$ RHS. But this can be reduced to show that each formula of the form $\Diamond \Phi \in T \backslash Th_{S5}(X)$ is in RHS. The proof can be carried out by induction on the degree of Φ; we have to distinguish the following cases:

$\Phi \in \mathcal{L}_0$: In this case $\Diamond \Phi$ is added to the RHS via the condition $\{\Diamond A \mid A \in \mathcal{L}_0$ and $\neg A \notin T\}$

$\Phi \equiv \Diamond \Phi'$: In this case we must ensure that $\Diamond \Diamond \Phi'$ is added to the RHS. By induction hypothesis we have $\Diamond \Phi' \in$ RHS, thus by $\Diamond \Psi \rightarrow \Box \Diamond \Psi$ (Axiom **5**) we have that $\Box \Diamond \Phi' \in$ RHS. Since RHS is closed under Th_{S5} we have $\Diamond \Diamond \Phi' \in$ RHS.

$\Phi \equiv \Box \Phi'$: We have to show that $\Diamond \Box \Phi' \in$ RHS. By induction hypothesis, $\Box \Phi' \in$ RHS and, again, since $\Psi \rightarrow \Diamond \Psi$ is a theorem of S5 and RHS is closed under Th_{S5}, we have $\Diamond \Box \Phi' \in$ RHS.

'⇐' We have to show that there is $\mathfrak{A}$ such that $\mathfrak{A} \models_{\!\!\underline{\cdot}} X$ and $T = \{\Phi \mid \mathfrak{A} \models \Phi\}$. It is sufficient to prove that $Th_{S5}(T) \neq \mathcal{L}$ because then such $\mathfrak{A}$ exists. It is clear that for any T we have: $\lozenge A \in T$ iff $\neg A \notin \mathcal{L}_0 \backslash T$. Thus T is consistent or it does not exist. ∎

The notion of extensional entailment as well as the syntactical characterization of extensions provide the basis for a comparison between our logic and two other nonmonotonic modal formalisms.

4.2. **Autoepistemic logics.** Autoepistemic logics (AEL) can be considered as a semantical approach to common sense reasoning (in opposite to syntactical approaches like Default Logic or the logics by Doyle and McDermott).

As already mentioned, autoepistemic logic uses the concept of extension to characterize the possible belief states. A set of sentences $T \subset \mathcal{L}$ is an AE-extension of a set (of initial premises) X if and only if

$$T = Cn_{\mathrm{cl}}(X \cup \{\Box\Phi \mid \Phi \in T\} \cup \{\neg\Box\Phi \mid \Phi \notin T\})$$

Cn_{cl} denotes the consequence operator of classical propositional logic.

The main result of this subsection is that if T is an autoepistemic extension of X and T contains all instances of the modal axiom scheme **T**, then $X \Vdash_{\mathrm{E}} T$ (Corollary 1). We already know by Proposition 3 that the set of all sentences which are valid in a structure $\mathfrak{A}$ is stable. Note, that the converse of Proposition 3 does not hold because there are stable sets $T \subset \mathcal{L}(\Sigma)$ (e.g. an AE-extension of $\{A, \lozenge\neg A\}$) which do not have a model from STRUCT(Σ). However, if T contains all instances of the modal axiom scheme **T** and is closed under the application of modus ponens and the rule of necessitation, then the converse of Proposition 3 does hold.

Definition 11 ($T_{\mathfrak{A}}, T_X$). *Let Σ be a signature, $\mathfrak{A} \in STRUCT(\Sigma)$, $X \subseteq \mathcal{L}(\Sigma)$. Define*

$$T_{\mathfrak{A}} =_{\mathrm{def}} \{\Phi \mid \mathfrak{A} \models \Phi\} \text{ and}$$
$$T_X =_{\mathrm{def}} \{T_{\mathfrak{A}} \mid \mathfrak{A} \in STRUCT(\Sigma) \text{ and } \mathfrak{A} \models X\}$$

Proposition 4. *Let Σ be a signature, $X \subset \mathcal{L}(\Sigma)$ and S_X be the set of all consistent stable theories T which contain X, every instance of the axiom scheme **T** and which are closed under the rule of necessitation. Then, $T_X = S_X$.*

Proof. $T_X \subseteq S_X$ is immediately clear, since every $T_{\mathfrak{A}} \in T_X$ is a stable theory containing X. Assume that there is $T \in S_X$ such that $T \notin T_X$. Thus, for all $\mathfrak{A}$ such that $\mathfrak{A} \models X$ we have $\mathfrak{A} \not\models T$. This means that $T \cup X$ does not have a model. Hence, T cannot be consistent, because $X \subset T$ – a contradiction. ∎

Corollary 1. *Let $X, T \subset \mathcal{L}$. T is an AE-extension of X which contains all instances of the axiom scheme* **T** *and which is closed under the rule of necessitation if and only if $X \Vdash_E T$.*

Proposition 4 and its Corollary relate AEL to our logic via semantical terms like extensional entailment. Let us now look at the syntactical aspects which become clearer when looking at a syntactical characterization (like the one of Theorem 3) of autoepistemic logic. Konolige showed that there is a close correspondence between AEL and the modal system K45. Let Th_{K45} be the syntactical consequence operator of modal K45 (cf. Proposition 2). We say that Φ ($\Phi \in \mathcal{L}$) is *strongly* K45-provable from a set $X \subseteq \mathcal{L}$ if and only if there are formulas $\Phi_1, \ldots, \Phi_n \in X$ such that $\Phi_1 \wedge \ldots \Phi_n \to \Phi \in Th_{K45}(X)$; denote the set of all sentences which are strongly K45-provable from X by $Th^S_{K45}(X)$. T is an autoepistemic extension of X if and only if

$$T = Th^S_{K45}(X \cup \{\Diamond A \mid \neg A \in \mathcal{L}_0 \backslash T\} \cup \{\Box A \mid A \in T \cap \mathcal{L}_0\})$$

If we compare the above characterization with Theorem 3 we can see that the main difference between our logic and autoepistemic logic is the absence of the axiom scheme **T**. Interestingly, both logics add only all *nonmodal* sentences which are consistent with the initial set X.

4.3. **Nonmonotonic logic II.** McDermott's notion of an NML2-S5-Extension can be written as

Definition 12 (NML2-S5-Extension). *Let $X, T \subset \mathcal{L}$. T is an* NML2-S5-*extension of X if and only if*

$$T = Th_{S5}(X \cup \{\Diamond \Phi \mid \Phi \in \mathcal{L} \text{ and } \neg \Phi \notin T\})$$

In the above definition of an extension, $\Diamond$ plays the role of something like 'it is S5-consistent, that ...' where Φ is S5-consistent with X means that $\neg \Phi \notin Th_{S5}(X)$. This is exactly the drawback we discussed when trying to find a definition of the entailment relation in Section 2.2. This deficiency leads to the collapse of nonmonotonic S5 to monotonic S5:

$$\bigcap \{S \mid S \text{ is an NML2-S5-extension of } X\} = Th_{S5}(X),$$

If we compare the syntactical characterization of NML2-S5-extensions of X with syntactical characterization by Theorem 3, then we see that the only difference is that McDermott forces *all* sentences which are consistent with an extension to enter the extension, i.e. $\Phi \in \mathcal{L}$ and $\neg \Phi \notin T$, whereas we do only require that T contains all *nonmodal* sentences which are consistent with T.

Theorem 4. *Let $X, T \subset \mathcal{L}$. If $X \Vdash_E T$ then T is as NML2-S5-extension.*

Proof. Follows from Theorem 3 and by the fact that $\mathcal{L}_0 \subset \mathcal{L}$. ∎

5. CONCLUSION

We showed that the unknown-operator, as used for example in practical AI systems, can be given a semantics which is based on a subset $\models$ of S5's validity relation $\models$. In addition, our logic is – at least from a syntactical point of view – located between autoepistemic logic and McDermott's NML-2 based on S5. We thus have a reasonable formal basis for a semantics of reasoning systems which supply an unknown-operator.

A matter of further research is to incorporate some aspects of paraconsistency, because it could happen easily that different users assert contradicting information to the database. We are currently extending the valuation functions V_α of Kripke structures to a mapping $V \to \{t, f, \top\}$ where $\top$ is a second designated truth-value.

REFERENCES

[Brazdil and Konolige, 1990] P. Brazdil and K. Konolige (Editors) *Machine Learning, Meta-Reasoning and Logics*, Kluwer, Norwell, 1990.

[Chellas, 1980] Brian F. Chellas *Modal Logic*, Cambridge University Press, Cambridge, 1980.

[Emde, 1991] Werner Emde *Modellbildung, Wissensrevision und Wissensrepräsentation im maschinellen Lernen*, Springer, Berlin, 1991.

[Doyle and McDermott, 1980] J. Doyle and D. McDermott *Nonmonotonic Logic I*, ARTIFICIAL INTELLIGENCE, Vol.13, 1980.

[Hirsh, 1990] H. Hirsh *Overgenerality in EBG*, in MACHINE LEARNING, META-REASONING AND LOGICS edited by Brazdil and Konolige, Kluwer, Norwell, 1990.

[Hughes and Cresswell, 1968] Hughes and Cresswell *Introduction to modal logic*, Routledge, London, 1990.

[Konolige, 1988] Kurt Konolige *On the Relation between Default and Autoepistemic Logic*, ARTIFICIAL INTELLIGENCE, Vol.35, 1988.

[Kraus et al., 1990] Kraus, Lehmann and Magidor *Nonmonotonic Reasoning, Preferential models and Cumulative Logics*, ARTIFICIAL INTELLIGENCE, Vol.44, 1990.

[Lukaszewicz, 1990] W.Lukaszewicz *Non-Monotonic Reasoning*, Ellis Horwood Ltd., Chichester, 1990.

[McDermott, 1982] D. McDermott *Nonmonotonic Logic II*, JOURNAL OF THE ACM, Vol.29, 1982.

[Moore, 1985] R. Moore *Semantical considerations on Nonmonotonic Logic*, ARTIFICIAL INTELLIGENCE, Vol.25, 1985.

[Morik et al.,1993] K. Morik, S. Wrobel, J. Kietz and W. Emde *Knowledge Acquisition and Machine Learning*, Academic Press, London, 1933.

[Reiter, 1980] R. Reiter *A Logic for Default Reasoning*, ARTIFICIAL INTELLIGENCE, Vol.13, 1980.

[Shoham, 1988] Y. Shoham *Reasoning about change*, MIT Press, Cambridge MA, 1988.

[Stalnaker, 1993] Robert C. Stalnaker *A Note on Nonmonotonic Modal Logic*, ARTIFICIAL INTELLIGENCE, Vol.64 1993.

System J - Revision Entailment
Default reasoning through ranking measure updates

Emil Weydert

Max-Planck-Institute for Computer Science
Im Stadtwald, D-66123 Saarbrücken, Germany
emil@mpi-sb.mpg.de

Abstract

This paper introduces a new default reasoning paradigm based on Spohn's revision methodology for ranking measures, i.e. Jeffrey-conditionalization. Its main features are representation independence, robustness, strong inheritance properties together with a quasi-probabilistic justification and the satisfaction of Lehmann's postulates for preferential consequence relations.

1. Introduction

In recent years, the rationality postulates for nonmonotonic reasoning first proposed and investigated by Dov Gabbay [Gabbay 85], Daniel Lehmann [Kraus *et al.* 90] and David Makinson [Makinson 94] have become a powerful tool for evaluating default formalisms. The resulting paradigm shift in default reasoning has promoted several new accounts based on default prioritization, e.g. rational closure [Lehmann and Magidor 92], system Z [Pearl 90] or conditional entailment [Geffner and Pearl 92], and (quasi-)probabilistic approaches, e.g. entropy-maximization [Goldszmidt *et al.* 90, Weydert 95] or belief functions [Benferhat *et al.* 95]. However, all these proposals continue to violate some basic desiderata concerning semantic foundations, representation independence, intuitive justifications, inheritance features, computational properties and user-friendliness. Preferential construction entailment [Weydert 95a] was an attempt to tackle the first four problems in a novel way within a quasi-probabilistic preferential framework based on the comparison of constructible κ-ranking measure models. It might be helpful to recall the basic ideas behind this approach.

Ranking measure semantics. *Defaults induce constraints on rough, order-of-magnitude-probability-like distributions called ranking measures.*

Explicitation strategy. *Implicit information which could be reflected by the syntactic structure of the default knowledge base is made explicit and exploited.*

Preferential entailment. *Default reasoning is based on a preferential consequence relation obtained from a suitable preference order on ranking measures (e.g. implementing normality maximization principles).*

Construction philosophy. *Ranking measures should be seen as constructions resulting from an information aggregation process controlled by the given defaults.*

Preferential construction entailment, the original implementation of these principles, has two major strengths. First its natural quasi-probabilistic background which is substantial for reasoning with and about independencies. Secondly the reconciliation of inheritance to exceptional subclasses and representation independence, based among others on a suitable language extension. Unfortunately, this very pleasant behaviour is overshadowed by a certain overinterpretation of default knowledge. Furthermore, the whole approach still appears to offer room for considerable simplifications which would not only smoothen technical investigations but also positively affect the credibility, computability and applicability of the corresponding default reasoning philosophy.

Generally speaking, what we want to achieve in this paper is a major reformulation of the preferential construction paradigm which guarantees not only representation-independence, reasonable inheritance properties and results backed by quasi-probabilistic justifications, but also more transparency and robustness. More concretely, we shall be interested in extended ranking measure models of default knowledge which can be obtained through iterated, variable-strength revision of a noninformative prior with the material implications induced by our defaults. They will allow us to define a more realistic preferential entailment relation backed by epistemic dynamics.

To start, we recall our semantic framework based on ranking measures and go on by discussing general principles and problems of nonmonotonic entailment relations. Next we motivate and define an extended language and model concept allowing us to avoid conflicts between inheritance principles and conditional logic. On a descriptive level, this means splitting up default knowledge into default conditionals encoding ranking measure constraints and explicit belief statements fixing revision items. This determines our monotonic logic. In a further step, we introduce a Spohn-type revision methodology which implements a minimal change strategy through Jeffrey-conditionalization on ranking measures and gives us an appropriate constructibility notion. Based on the concept of ranking measure constructions, we can then introduce our new default inference relation, J-entailment. To conclude, we discuss some examples illustrating the different features of our approach and relate it to other proposals.

2. Monotonic entailment

In our framework, the basic notion is that of a ranking measure. These are plausibility measures which associate with each proposition a value in a suitable linearly ordered valuation structure $\mathcal{V}$. We may think of these values as degrees of plausibility, normality or disbelief, but also as subjective semi-qualitative or order-of-magnitude probabilities. The most general definition still in line with our needs and aspirations for a powerful quasi-probabilistic approach to default reasoning is the following one.

Definition 2.1 $\mathcal{R}$ is called a *ranking measure* over $\mathcal{B}$ with values in $\mathcal{V}$ iff

1. $\mathcal{B} = (B, \cap, \cup, -, \emptyset, S)$ is a *boolean set-algebra* on S (in particular $B \subseteq 2^S$),

2. $\mathcal{V} = (V, -\infty, o, *, «)$ is a *ranking algebra* : $(V\backslash\{-\infty\}, o, *, «)$ is the negative half of a nontrivial ordered commutative group with unit o, o is the «-maximal and $-\infty$ is the «-minimal and absorptive element for $*$, i.e. for all $v \in V$, $-\infty * v = v * -\infty = -\infty$,

3. $\mathcal{R} : B \to V$ is a function s.t. for all $A, B \in B$,

$\mathcal{R}(A \cup B) = \max_{«}\{\mathcal{R}(A), \mathcal{R}(B)\}, \mathcal{R}(\emptyset) = -\infty$ and $\mathcal{R}(S) = o,$

4. If $A = \cup\{A_i \mid i \in I\} \in B$ and for all $i \in I, \mathcal{R}(A_i) = -\infty$, then $\mathcal{R}(A) = -\infty$ *(coherence).*

The *conditional ranking measure* $\mathcal{R}(\mid) : B \times B \to V$ corresponding to $\mathcal{R}$ is now defined by the equations

5. $\mathcal{R}(A \cap B) = \mathcal{R}(B \mid A) * \mathcal{R}(A)$, for $\mathcal{R}(A) \neq -\infty$, and $\mathcal{R}(B \mid A) = -\infty$, for $\mathcal{R}(A) = -\infty$.

Let $\mathcal{R}_0$ be the uniform ranking measure over B, i.e. $\mathcal{R}_0(A) = o$ for all $A \in B\setminus\{\emptyset\}$.

There is an obvious analogy with probability distributions in the sense that here the ordinary probabilistic valuation algebra ($[0, 1]$, 0, 1, +, x, <) is replaced by (V, $-\infty$, o, $\max_{«}$, *, «) - in **2.1.3**, we may assume that A, B are disjoint - and σ-additivity is changed into coherence. $\mathcal{R}(A)$ « $\mathcal{R}(B)$ is supposed to mean that A is in some way negligible w.r.t. B. The main difference with probability measures is that less-or-equal ranking measure values don't add up, i.e. for $\mathcal{R}(A)$ « $\mathcal{R}(B)$ or $\mathcal{R}(A) = \mathcal{R}(B)$, we always have $\mathcal{R}(A) = \mathcal{R}(A \cup B)$. The multiplicative connective * on V is needed to get a reasonable notion of conditional ranking measures and independency, whereas coherence ensures that $\mathcal{R}(A) = -\infty$ really expresses the impossibility of A. If B is compact, coherence automatically holds.

There are two major examples of ranking measures. First, the *κ-ranking measures,* whose ranking algebra is derived from the ordered additive group of integers Z, i.e. $V = V(Z) = \{0, -1, -2, ..., -\infty\}$, o = 0, * = + and « = <. They constitute a generalization of Spohn's natural conditional functions [Spohn 90] which have been used for modeling semi-qualitative belief strength and revision and have been introduced as κ-rankings to default reasoning by Judea Pearl. Secondly, the *π-ranking measures,* whose valuation structure is obtained from the odered multiplicative group of positive reals R^+, i.e. $V = V(R^+) = [0, 1]$, o = 1, * = x and « = <. They generalize Dubois and Prade's [Dubois and Prade 88] possibility distributions which are prominent tools in reasoning with uncertainty. Our ranking measure approach is slightly more general because we don't require continuity, that is

• $\mathcal{R}(\cup\{A_i \mid i \in I\}) = \sup_{«}\{\mathcal{R}(A_i) \mid i \in I\}$ for $\cup\{A_i \mid i \in I\}, A_i \in B$.

The ranking measure framework was first advocated in [Weydert 91] and further developed in [Weydert 94] as a natural and powerful semantic tool for interpreting default conditionals. Independently, other authors have made similar observations for the more restricted notions mentioned above. The basic idea is that a default $\varphi \to \psi$ may be seen as telling us – possibly among others – that the ranking measure value of $\varphi \& \neg\psi$, that is its degree of plausibility, is either smaller than that of $\varphi \& \psi$ or simply vanishing, if $\varphi \& \neg\psi$ is considered impossible. However, defaults may well carry additional information which cannot be summarized in this way. Therefore, it is very important to make a distinction between arbitrary defaults $\varphi \to \psi$ and those interpreted exclusively as ranking measure constraints, which are called default conditionals and denoted by $\varphi \Rightarrow \psi$. So, accepting the conditional Bird(*Tweety*) $\Rightarrow$ Fly(*Tweety*) simply means attributing a higher plausibility to Bird(*Tweety*)&Fly(*Tweety*) than to Bird(*Tweety*)&¬Fly(*Tweety*) or rejecting Bird(*Tweety*) altogether.

Now, let us proceed to the formal definition of our ranking measure semantics for default conditionals. Let L be a standard language of propositional logic based on an infinite set of propositional variables $Var = \{P_i \mid i \in I\}$, $L°(\Rightarrow) = \{\varphi \Rightarrow \psi \mid \varphi, \psi \in L\}$, and $L(\Rightarrow)$ be the closure of $L°(\Rightarrow)$ under the usual propositional connectives F, T, ¬, &, v, → and ↔. Furthermore, let $Mod_L(\varphi) = \{\omega \in \{0, 1\}^{Var} \mid \omega$ verifies $\varphi\}$, $\mathcal{B}_L$ be the set-algebra on $\mathcal{B}_L = \{Mod_L(\varphi) \mid \varphi \in L\}$ and ⊢ be the classical propositional inference relation on L. In the obvious way, we can then define a satisfaction relation ⊨ between ranking measures $\mathcal{R} : \mathcal{B}_L \rightarrow \mathcal{V_R} = (V_\mathcal{R}, -\infty, o, *, «)$ and $L(\Rightarrow)$-formulas. To simplify notation, we write $\mathcal{R}(Mod_L(\varphi)) = \mathcal{R}(\varphi)$. The propositional connectives are handled as usual.

- $\mathcal{R} \models \varphi \Rightarrow \psi$ iff $\mathcal{R}(\neg\psi \mid \varphi) « o$ iff $\mathcal{R}(\varphi\&\neg\psi) « \mathcal{R}(\varphi\&\psi)$ or $\mathcal{R}(\varphi\&\neg\psi) = -\infty$.

Given these truth-conditions, ⇒ represents normal or plausible implication. The satisfaction relation ⊨ induces a monotonic entailment relation ‖– on $L(\Rightarrow)$ defined by $\Sigma \Vdash \psi$ iff for all $\mathcal{R}$ s.t. $\mathcal{R} \models \varphi$ for every $\varphi \in \Sigma$, we have $\mathcal{R} \models \psi$. It can be axiomatized by the rules for rational conditional logic, i.e. the full object-level version of Lehmann's rationality postulates [Kraus *et al.* 90] over $L(\Rightarrow)$. Note however that the richness of our model structures cannot be completely grasped by this conditional language. In particular, if $\Vdash_\nu$ is the monotonic consequence relation obtained by considering only ranking measures $\mathcal{R}$ with $V_\mathcal{R} = V$, then $\Vdash_\nu = \Vdash$ for finite premise sets Σ. In this context, only the availability of the infinite total ordering « counts. But for infinite Σ, the situation changes. If $\Sigma = \{P_1 v...vP_{2n} \Rightarrow P_0, P_1 v...vP_{2n+1} \Rightarrow \neg P_0 \neg (P_1 \Rightarrow F) \mid n \in \mathbf{Nat}\}$, we have $\Sigma \Vdash_{\nu(Z)} F$ although $\Sigma \nVdash F$, because $V(\mathbf{Z})$ doesn't allow infinitely ascending sequences. On the other hand, for countable languages we get $\Vdash_{\nu(\mathbf{R}+)} = \Vdash$. This might be interpreted as an argument for π-ranking measures. However, there are also discrete ranking algebras V with $\Vdash_\nu = \Vdash$. For convenience, we introduce the comparative negligibility modality < from [Weydert 91, 94] as a notational tool, with $\varphi < \psi$ being an abbreviation for $\varphi v \psi \Rightarrow \neg\varphi$. It is easy to see that $\mathcal{R} \models \varphi < \psi$ iff $\mathcal{R}(\varphi) « \mathcal{R}(\psi)$ or $\mathcal{R}(\varphi) = -\infty$. So, we may read $\varphi < \psi$ as "φ *is negligible w.r.t.* ψ". Observe that $\Vdash \varphi \Rightarrow \psi \leftrightarrow \varphi\&\neg\psi < \varphi\&\psi$ and $\Vdash F < F$ (F is absolutely negligible).

3. Nonmonotonic entailment

The role of default knowledge Δ is to sanction and control plausible and therefore possibly defeasible inferences from given sets of facts Σ ($\Sigma \subseteq L$). In the following, we are going to concentrate ourselves on finite fact sets Σ and finite default knowledge bases $\Delta = \{\varphi_i \dashrightarrow \psi_i \mid i \leq n\}$ with $\varphi_i, \psi_i \in L$. What are the minimal requirements for a plausible inference relation ⊨ if $\Sigma \cup \Delta \models \psi$ is intended to express that ψ ($\in L$) can be reasonably assumed given $\Sigma \cup \Delta$? First of all, we have to guarantee the canonical link between $\dashrightarrow$ and ⊨ which is at the heart of any reasoning with defaults.

(1) $\{\varphi\} \cup \{\varphi \dashrightarrow \psi\} \models \psi$ if $\varphi\&\psi \nVdash F$ (*Defeasible modus ponens*).

Furthermore, $\models$ should satisfy reasonable structural properties, notably the core of Lehmann's rationality postulates formulated for arbitrary but fixed default knowledge sets Δ. Let's abbreviate $\Sigma \cup \Delta \models \psi$ by $\Sigma \models_\Delta \psi$.

(2) $\Sigma \vdash \varphi$ implies $\Sigma \models_\Delta \varphi$ (*Supraclassicality w.r.t.* $\vdash$),

(3) $\Sigma \models_\Delta \varphi$ implies ($\Sigma \models_\Delta \psi$ iff $\Sigma \cup \{\varphi\} \models_\Delta \psi$) (*Cumulativity*),

(4) $\Sigma \cup \{\varphi\} \models_\Delta \psi$ and $\Sigma \cup \{\varphi'\} \models_\Delta \psi$ implies $\Sigma \cup \{\varphi \vee \varphi'\} \models_\Delta \psi$ (*Or*).

This list is equivalent to the rules for *preferential consequence relations* [Kraus *et al.* 90]. There is another principle which certainly holds for normal implication $\Rightarrow$ (object-level) but appears to be too restrictive for nonmonotonic inference (meta-level).

• $\Sigma \models_\Delta \psi$ and $\Sigma \not\models_\Delta \neg \varphi$ implies $\Sigma \cup \{\varphi\} \models_\Delta \psi$ (*Rational monotony*).

Preferential consequence relations verifying this postulate are called *rational*. There are several reasons why we should not insist on rational monotony for defeasible reasoning. First of all, rational monotony sometimes forces us to accept shaky inferences which do not well reflect the incompleteness and weakness of our knowledge. Consider for instance $\Delta = \{\mathbf{T} \dashrightarrow \neg\alpha, \mathbf{T} \dashrightarrow \neg\beta, \alpha \vee \gamma \dashrightarrow \neg\gamma\}$, where α, β, γ are independent propositional variables. Assuming rationality we would have to support either $\{\beta \vee \gamma\} \models_\Delta \neg\gamma$ or $\{\beta \vee \alpha\} \models_\Delta \neg\beta$, which may seem quite arbitrary. Another major problem resulting from its non-Horn character is that rational monotony is not conserved under intersections. This is undesirable because the combination of reasonable inference relations shouldn't be much worse than the relations themselves. These examples suggest that rational monotony, at least, shouldn't be a conditio sine qua non for defeasible inference.

A very fundamental issue is syntax independence. Nonmonotonic conclusions should not depend on purely syntactic choices. We therefore propose two related postulates concerned with the relations between syntactic variations of the same default set. To begin with, we introduce a rudimentary inference notion $\vdash'$ on $L°(\dashrightarrow)$ axiomatized by

• $\vdash \varphi \leftrightarrow \varphi'$ and $\vdash \psi \leftrightarrow \psi'$ implies $\varphi \dashrightarrow \psi \vdash' \varphi' \dashrightarrow \psi'$ (*Extensionality for* $\dashrightarrow$).

It says that defaults whose premises and conclusions have the same propositional logical content are equivalent. The above requirement for $\models$ therefore suggests

(5) $\Delta' \dashv\vdash' \Delta'$ implies $\Sigma \models_\Delta \psi$ iff $\Sigma \models_{\Delta'} \psi$ (*Left equivalence w.r.t.* $\vdash°$).

Furthermore, we would like to have invariance w.r.t. variable renaming. For each permutation f of the propositional variables in *Var*, let f* be a corresponding function on $L°(\dashrightarrow)$ inductively defined by

• $f^*(P_i) = f(P_i)$, $f^*(\varphi \# \psi) = f^*(\varphi) \# f^*(\psi)$ for every propositional connective # and $\dashrightarrow$.

Certainly, we are hardly willing to consider inference relations violating

(6) $\Sigma \cup \Delta \models \psi$ iff $f^{*"}(\Sigma \cup \Delta) \models f^*(\psi)$ (*Renaming invariance*).

Of course, these principles only constitute a first step. Taken together, they are still much too weak to guarantee even a moderately reasonable behaviour of $\vDash$. In particular, defeasible inheritance to irrelevant subclasses or defeasible chaining may fail. For instance, we may define an elementary preferential consequence relation $\vDash_0$ which verifies defeasible modus ponens, left equivalence w.r.t. $\vdash^\circ$ and renaming invariance by setting $-\!\!» = \Rightarrow$ and

- $\{\varphi_i \mid i{\leq}n\} \cup \Delta \vDash_0 \psi$ iff $\Delta \Vdash \varphi_0 \&...\&\varphi_n \Rightarrow \psi$.

However, even for independent propositional variables α, β, γ, $\vDash_0$ violates very basic default inference patterns.

- $\{\alpha \& \gamma\} \cup \{\alpha \Rightarrow \beta\} \nvDash_0 \beta$ *(No irrelevance reasoning)*.

- $\{\alpha\} \cup \{\alpha \Rightarrow \gamma, \gamma \Rightarrow \beta\} \nvDash_0 \beta$ *(No defeasible chaining)*.

But there are many nonmonotonic inference relations which support these plausible inference steps, in particular those implementing normality maximization like rational closure or system Z. On the other hand, there are also desiderata which pose much bigger problems. Consider for instance the following important scheme for inheritance through exceptional subclasses.

- $\{\alpha \& \neg \gamma\} \cup \{\alpha -\!\!» \beta, \alpha -\!\!» \gamma\} \vDash \beta$ for logically independent α, β, $\gamma \in L$ **(EI)**.

A major drawback of otherwise well-behaved and semantically interpretable popular formalisms like system Z is that they violate this fundamental requirement and its relatives. Note that the pattern doesn't present any problems for most traditional, rule-oriented default formalisms, like Reiter's default logic (haunted by other serious difficulties). Unfortunately, this failure is not only a local inadequacy of some particular account but a slingshot for all purely conditional approaches to default reasoning. In fact, there is an impossibility theorem which sets narrow limits to our defeasible inference dreams about the possible coexistence of basic inheritance features and naturally interacting defaults.

Fact 3.1 Let $\vDash$ be a finitary preferential inference relation for $L \cup L^\circ(\Rightarrow)$ s.t. $\Delta \dashv\Vdash \Delta'$ implies $\vDash_\Delta = \vDash_{\Delta'}$ *(left equivalence w.r.t. $\Vdash$)* and $\vDash$ satisfies **EI**. Then we have

- $\{\alpha \& \neg \gamma\} \cup \{\alpha \Rightarrow \beta, \alpha \Rightarrow \gamma\} \vDash \mathbf{F}$ (thus $\vDash \gamma$) for logically independent α, β, $\gamma \in L$.

This is quite disturbing because we cannot accept the trivialization of plausible inferences from exceptional premises like $\alpha \& \neg \gamma$. Of course, one might think of ad hoc solutions to fix this problem. For instance, we could introduce syntactic restrictions like considering only propositional variables. Or we could require that implicit independence assumptions have to be made explicit by the user. But all these proposals strongly conflict with basic intentions and intuitions of defeasible inference, which should exploit implicit assumptions by itself and, among others, support inheritance through exceptional subclasses. On the other hand, the impossibility result clearly demonstrates that sometimes, default knowledge $\Delta = \{\varphi_i -\!\!» \psi_i \mid i{\leq}n\}$ cannot be comprehensively represented by sets of default conditionals $\{\varphi_i \Rightarrow \psi_i \mid i{\leq}n\}$, i.e. ranking measure constraints. Additional informations, possibly conveyed by the choice of defaults used to encode the ranking constraints, may be necessary. And it should be possible to make them explicit. That is, we want to translate $\Delta = \{\varphi_i \dashrightarrow \psi_i \mid i \leq n\}$ by $\{\varphi_i \Rightarrow \psi_i \mid i \leq n\} \cup X$, where X represents some knowledgeable, semantically meaningful information extracted from Δ but not graspable by default conditionals alone.

The strategy adopted in [Weydert 95a] is to mark the material implications $\varphi \rightarrow \psi$, alternatively the abnormality parts $\varphi \& \neg \psi$, corresponding to the defaults $\varphi \dashrightarrow \psi \in \Delta$. This is done through so-called *default-declarations* $\mathbf{I}(\varphi \rightarrow \psi)$. The unary modality $\mathbf{I}$ over L should here be seen as expressing some form of explicit belief. It should be closed under substitution by logically equivalent formulas but inert w.r.t. more sophisticated logical manipulations. So, we may read $\mathbf{I}(\varphi)$ as "*the proposition φ is explicitly believed* ". On a general level, these I-expressions are used to reflect the individual, independent character of the given defaults and to structure what is normal or exceptional. Consequently, in our framework, every default knowledge base Δ over $L^\circ(\dashrightarrow)$ is translated into a corresponding $L^\circ(\Rightarrow, \mathbf{I})$-subset

- $\Delta' = \Delta^{\mathbf{I}} \cup \Delta^{\Rightarrow} = \{\mathbf{I}(\varphi \rightarrow \psi) \mid \varphi \dashrightarrow \psi \in \Delta\} \cup \{\varphi \Rightarrow \psi \mid \varphi \dashrightarrow \psi \in \Delta\}$,

where $L^\circ(\Rightarrow, \mathbf{I}) = L^\circ(\Rightarrow) \cup L^\circ(\mathbf{I}) = \{\varphi \Rightarrow \psi \mid \varphi, \psi \in L\} \cup \{\mathbf{I}(\varphi) \mid \varphi \in L\}$. Of course, we may also consider arbitrary knowledge bases $\Phi = \Phi^{\mathbf{I}} \cup \Psi^{\Rightarrow}$ with $L^\circ(\mathbf{I}) \supseteq \Phi^{\mathbf{I}}$ and $L^\circ(\Rightarrow) \supseteq \Psi^{\Rightarrow}$. What could be an appropriate semantics for $L^\circ(\Rightarrow, \mathbf{I})$? A straightforward possibility would be to consider pairs $(\mathcal{R}, \mathfrak{l})$, where $\mathcal{R}$ is a ranking measure over $\mathcal{B}_L$ and $\mathfrak{l}$ a subset of $\mathcal{B}_L$. The truth condition for $\mathbf{I}(\varphi)$ then would be

- $(\mathcal{R}, \mathfrak{l}) \models \mathbf{I}(\varphi)$ iff $\text{Mod}_L(\varphi) \in \mathfrak{l}$.

However, the particular use we are going to make of I-statements in the context of our revised preferential construction paradigm suggests a slightly different approach.

4. Extended monotonic entailment

As we noted before, $\mathbf{I}$ can be seen as describing indvidual explicit beliefs, up to logical equivalence. Default conditionals, on the other hand, represent constraints on ranking measures, determining which ones are to be considered admissible. In particular, we may interpret implicit belief in φ by $\mathcal{R}(\neg\varphi) \ll o$ and encode it with $\top \Rightarrow \varphi$. Note that implicit belief is correlated with explicit belief because $\varphi \dashrightarrow \psi$ induces $\mathbf{I}(\varphi \rightarrow \psi)$ as well as $\varphi \Rightarrow \psi$ which $\Vdash$—entails $\top \Rightarrow (\varphi \rightarrow \psi)$. This is the static viewpoint. But we may also adopt a more dynamic perspective and exploit the possible history of these beliefs. A natural approach is to see them as resulting from an update process which starts with the informationless uniform state $\mathcal{R}_0$ and proceeds by iterated, variable-strength revision with propositions in the scope of $\mathbf{I}$, e.g. material implications obtained from a default set. Hence, the role of I-statements would be to single out possible building blocks for the incremental construction of epistemic ranking measure models of the conditional constraints $\Delta^{\Rightarrow}$ through an update process.

What would be a suitable revision procedure ? Adopting the minimal change paradigm, the canonical strategy for updating ranking measures with constraints of the form $\mathcal{R}(\neg\varphi) \underset{\ll}{\leq} r$ (i.e. revision by φ if $r \ll o$) is Jeffrey-conditionalization-as-needed-as-possible, adapted to our quasi-probabilistic context. This approach is close to what Spohn has done [Spohn 90].

Definition 4.1 Let $\mathcal{R}$ be a ranking measure over $\mathcal{B}_L$ with algebra $\mathcal{V} = (V, -\infty, o, *, \ll)$, $A \in B_L$ and $r \in V$. If $\mathcal{R}(\neg A) = -\infty$, we set $\mathcal{R}[A : r] = \mathcal{R}$. If $\mathcal{R}(\neg A) \neq -\infty$, then $\mathcal{R}[A : r]$ is the unique ranking measure $\mathcal{R}^* : \mathcal{B}_L \to \mathcal{V}$ s.t.

- for all $B \in B_L$, $\mathcal{R}^*(B|A) = \mathcal{R}(B|A)$ and $\mathcal{R}^*(B|\neg A) = \mathcal{R}(B|\neg A)$,

- if $r \ll o$, $\mathcal{R}^*(A) = \min_\ll\{r, \mathcal{R}(A)\}$ and $\mathcal{R}^*(\neg A) = o$,

- if $r = o$, $\mathcal{R}^* = \mathcal{R}$.

Hence, if $r \ll o$, we obtain $\mathcal{R}[A : r]$ by uniformly shifting A downwards until we reach $\underset{\ll}{\leq} r$, and by uniformly shifting $\neg A$ upwards to the top, i.e. o. That is, we change only what has to be changed to ensure that $\mathcal{R}^*(A) \underset{\ll}{\leq} r$, if possible. If the constraint is already verified by $\mathcal{R}$, in particular if it is tautological ($r = o$), then we stay with the original ranking measure. Also note that if $\mathcal{R}(A) = -\infty$, $\mathcal{R}$ is not revisable with A.

We are now interested in those ranking measures which can be constructed from $\mathcal{R}_o$ by iterated variable-strength (no restrictions on r) revision with given propositions.

Definition 4.2 Let $\mathcal{L} = \{A_i \mid i \leq n\} \subseteq B_L$ and $\mathcal{R} : \mathcal{B}_L \to \mathcal{V}$ be a ranking measure. Then $\mathcal{R}$ is called $\mathcal{L}$-*constructible* iff there are $r_i \in V$ s.t. $\mathcal{R} = \mathcal{R}_o[A_0 : r_0] \dots [A_n : r_n]$, also written as $\mathcal{R}_o[A_0 : r_0, \dots, A_n : r_n]$.

It is easy to see that a ranking measure $\mathcal{R}$ is constructible iff $\mathcal{R}$ is *finitary*, i.e. iff there is a partition $\{A_0, \dots, A_p\} \subseteq B_L$ s.t. for all $A \in B_L$, $\mathcal{R}(A) = \max_\ll\{\mathcal{R}(A_i) \mid i \leq p, A \cap A_i \neq \emptyset\} \cup \{-\infty\}$. Then $\mathcal{R}$ is $\{A_0, \dots, A_p\}$-constructible. All this suggests a possible refinement of ranking measures taking into account not only the static but also the dynamic or historic identity of these epistemic valuations. Our semantics for $L^\circ(\Rightarrow, I)$ is based on these more fine-grained concepts.

Definition 4.3 If $B_L \supseteq \{A_i \mid i \leq n\}$ and $V \supseteq \{r_i \mid i \leq n\}$, then $A = (A_0 : r_0, \dots, A_n : r_n)_{BL,\nu}$ is called a *ranking measure construction* over B_L with algebra $\mathcal{V}$. $L^\circ(\Rightarrow, I)$-*structures* are ranking measure constructions over B_L (with arbitrary ranking algebras). The satisfaction relation between $L^\circ(\Rightarrow, I)$-structures and $L^\circ(\Rightarrow, I)$-formulas is denoted by $\Vert=$. For propositional connectives, satisfaction is defined as usual. For conditional and explicit belief formulas, we set

- $(A_0 : r_0, \dots, A_n : r_n)_{BL,\nu} \Vert= \varphi \Rightarrow \psi$ iff $\mathcal{R}_o[A_0 : r_0, \dots, A_n : r_n] \models \varphi \Rightarrow \psi$,

- $(A_0 : r_0, \dots, A_n : r_n)_{BL,\nu} \Vert= I(\varphi)$ iff $\text{Mod}_L(\neg\varphi) \in \{A_i \mid i \leq n\}$.

A ranking measure construction satisfies a conditional iff it holds in the constructed ranking measure. As we know, the sentences in the scope of **I** may be seen as explicit beliefs obtained through revision steps. Because believing φ presupposes $\mathcal{R}(\neg\varphi) \ll o$, we should certainly be allowed to shift $\neg\varphi$ downwards. This is achieved through the second condition. Note that we do not require $r_i \neq o$ (the belief A_i may be redundant). The corresponding monotonic entailment relation ‖-' can be axiomatized by the rules for rational conditional logic and extensionality for I, i.e.

- $\vdash \psi \leftrightarrow \varphi$ implies $I(\psi)$ ‖-' $I(\varphi)$.

In particular, ‖-' $\cap$ $L°(\Rightarrow)$ = ‖-. This gives us a monotonic logic for defaults as long as we adopt the following interpretation and translation principle.

- $\varphi \dashrightarrow \psi = \varphi \Rightarrow \psi$ & $I(\varphi \rightarrow \psi)$.

As ‖-, ‖-' only reflects a small fragment of the semantic richness of ranking measure constructions which will only be exploited at the level of nonmonotonic reasoning.

5. Revision entailment

How should we proceed to arrive at a reasonable nonmonotonic entailment relation for $L°(\Rightarrow, I)$? Because we are looking for a system which verifies Lehmann's core postulates, we may try to find a suitable preference order over ranking measure cons-tructions. Traditionally, for ranking measures, this has been achieved through some type of normality maximization or minimum specificity ordering [Weydert 95a, Ben-ferhat *et al.* 95]. But, as we know, this is not enough to give us inheritance through exceptional subclasses. We have to take into account the update structure.

The meaning of individual ranking measure values is difficult to grasp. In fact, they may be seen as an auxiliary technical notion without a clear-cut objective meaning. For instance, how should we decide whether a proposition P deserves a π-ranking value of 0.5 or 0.6 ? Therefore, it seems reasonable to concentrate ourselves on the set of revisable propositions $\mathcal{V}(A) = \{A_i \mid i{\le}n\}$ of $L°(\Rightarrow, I)$-structures $A = (A_0 : r_0, ..., A_n : r_n)_{BL,\nu}$. An obvious choice in line with the minimality philosophy would be to prefer those $L°(\Rightarrow, I)$-models of the given default knowledge base $\Phi^I \cup \Psi^\Rightarrow$, where this set becomes minimal. Because in the present framework $\{A_i \mid i{\le}n\}$ is always finite, minima necessarily exist. In fact, this simple idea backs a very powerful entailment notion which is orthogonal to the classical normality maximization strategy and does not depend on complex ranking measure orderings.

Definition 5.1 J-entailment, denoted by $\approx_J$, is a finitary default inference relation defined for $\Sigma = \{\varphi_i \mid i{\le}n\}\cup\{\psi\} \subseteq L$, $\Phi \subseteq L°(I)$, $\Psi \subseteq L°(\Rightarrow)$ and $\varphi_\Sigma = \varphi_0\&...\&\varphi_n$ by

- $\Sigma \cup \Phi \cup \Psi \approx_J \psi$ iff $\{A \mid \mathcal{V}(A)$ minimal with A ‖= $\Phi \cup \Psi\}$ ‖= $\varphi_\Sigma \Rightarrow \psi$.

If there are A ‖= $\Phi \cup \Psi$ with $\mathcal{V}(A) = \{Mod_L(\neg\varphi) \mid I(\varphi){\in}\Phi\}$, this is equivalent to say that $\varphi_0\&...\&\varphi_n \Rightarrow \psi$ is verified by each of them. Note that this requirement always holds for $\Phi \cup \Psi = \Delta^I \cup \Delta^\Rightarrow$. It is not difficult to see that $\approx_J$ is well-behaved as far as our general principles (1) - (6) for defeasible entailment are concerned.

Theorem 5.1 For fixed $\Phi \cup \Psi$, $\vDash_J$ is a finitary preferential consequence relation (w.r.t. $\vdash$) which satisfies defeasible modus ponens, left equivalence w.r.t. $\Vdash'$ and renaming invariance. Furthermore, it is invariant w.r.t. arbitrary automorphisms of $\mathfrak{B}_L$.

Rational monotony, however, which some authors reject as a an excessively strong nonmonotonic inference postulate [Geffner and Pearl 92] (cf. our discussion above) - not to be confused with rational monotony for object-level default conditionals which appears to be not only acceptable but indispensable - is violated by $\vDash_J$. These are good news if you reject extremely speculative inferences and care about robustness.

On the other hand, J-entailment validates all the standard patterns for most-specific-subclass and irrelevancereasoning as well as defeasible transitivity (chaining). The following example addresses all these issues. Let α, β, γ, φ, ψ, ϕ be logically independent L-formulas, e.g. α = Course-participant(John), β = Black-haired(John), φ = Student(John), ψ = Adult(John) and ϕ = Employed(John).

- $\{\alpha \& \beta\} \cup \{\alpha \dashrightarrow \varphi, \varphi \dashrightarrow \psi, \psi \dashrightarrow \phi, \varphi \dashrightarrow \neg\phi\} \vDash_J \neg\phi$ (*Extended specificity*).

Of course, there are many formalisms getting this extended specificity example right. What is more interesting is the validity of inheritance through exceptional subclasses.

- $\{\alpha \& \neg\gamma\} \cup \{\alpha \dashrightarrow \beta, \alpha \dashrightarrow \gamma\} \vDash_J \beta$ (*Exceptional inheritance*).

The reason why $\vDash_J$ succeeds is that the impossibility result mentioned above mainly depends on the $\Vdash$-equivalence between the conditional premise sets $\{\alpha \Rightarrow \beta, \alpha \Rightarrow \gamma\}$ and $\{\alpha \Rightarrow (\beta \leftrightarrow \gamma), \alpha \Rightarrow \gamma\}$. But this relation is blocked by the fact that $\Delta = \{\alpha \Rightarrow \beta, \alpha \Rightarrow \gamma\} \cup \{I(\alpha \rightarrow \beta), I(\alpha \rightarrow \gamma)\}$ and $\{\alpha \Rightarrow (\beta \leftrightarrow \gamma), \alpha \Rightarrow \gamma\} \cup \{I(\alpha \rightarrow (\beta \leftrightarrow \gamma)), I(\alpha \rightarrow \gamma)\}$ are no longer equivalent w.r.t. $\Vdash'$. Let's have a look at the preferred models of Δ. Because we are only interested in the resulting ranking measures and the set of revisable propositions, it is enough to consider representative structures of the form $(\alpha \& \neg\beta : \mathfrak{r}_0, \alpha \& \neg\gamma : r_1)_{BL,\nu}$ with $r_j < o$. For $\mathcal{R} = \mathcal{R}_0[\alpha \& \neg\beta : \mathfrak{r}_0, \alpha \& \neg\gamma : r_1]$, we have $\mathcal{R}(\alpha \& \neg\gamma \& \neg\beta) = \mathcal{R}(\alpha \& \neg\gamma) * \mathcal{R}(\alpha \& \neg\beta) \ll \mathcal{R}(\alpha \& \neg\gamma) * \mathcal{R}(\alpha \& \beta) = \mathcal{R}(\alpha \& \neg\gamma \& \beta)$. This holds by Jeffrey-conditionalization, the algebraic properties of ν and the enforcement of $\mathcal{R}(\alpha \& \neg\beta) \ll \mathcal{R}(\alpha \& \beta)$ by $\alpha \Rightarrow \beta$. A further desirable property is illustrated by

- $\{\neg\varphi\} \cup \{T \dashrightarrow \varphi, T \dashrightarrow \varphi\vee\psi\} \nvDash_J \psi$ if $\neg\varphi \nVdash \psi$ (*Redundancy-tolerance*).

Because the second default might well encode redundant information - recalling that $T \Rightarrow \varphi \Vdash' T \Rightarrow \varphi\vee\psi$ - we should not infer anything which does not already follow from $\{\neg\varphi\} \cup \{T \dashrightarrow \varphi\}$. A falsifying ranking measure construction here is $(\neg\varphi : r_0, \neg\varphi \& \neg\psi : o)$ with $\mathfrak{r}_0 < o$. Another important feature of $\vDash_J$ is its robustness, e.g. its cautious stance w.r.t. default counting which reflects well the rather shaky character of default information, often ignored by rational inference relations. This conservativism is exemplified by the rejection of several questionable reasoning patterns sanctioned by less robust approoaches. Consider for instance this variation of the Nixon diamond discussed in [Benferhat *et al.* 95].

- $\{\alpha \& \beta \& \gamma\} \cup \{\alpha \dashrightarrow \varphi, \beta \dashrightarrow \varphi, \gamma \dashrightarrow \neg\varphi\} \nvDash_J \varphi, \neg\varphi$ (*Assymmetric Nixon diamond*).

Staying ambiguous seems to be the right answer here. Falsifying the first two defaults does not necessarily weigh more heavily than violating the third one. Another prominent example correctly handled by our approach goes back to Geffner.

- $\{\alpha\&\beta\&\gamma\} \cup \{ \alpha -\!\!»-\varphi, \gamma -\!\!»-\varphi, \alpha\&\beta -\!\!» \varphi\} \; |\!/\!\approx_J \varphi, -\varphi$ *(Geffner's example)*.

Because $\alpha\&\beta$ and γ are unrelated, there is no real reason to prefer either φ or $\neg\varphi$. Our next example exhibits well the differences between quasi-probabilistic approaches correctly handling independence and syntax-oriented prioritization formalisms.

- $\{\alpha\vee(\beta\&\gamma)\} \cup \{\mathbf{T} -\!\!»-\beta, \mathbf{T} -\!\!»-\gamma, \alpha\vee\beta -\!\!»-\alpha, \alpha\vee\gamma -\!\!»-\alpha\} \; |\!/\!\approx_J -\alpha$ *(Anti-prioritization)*.

To start, we note that the corresponding default conditionals can be written as $\beta < \mathbf{T}$, $\gamma < \mathbf{T}$, $\alpha < \beta$ and $\alpha < \gamma$ where $<$ is the negligibility modality previously introduced as an abbreviation. If all we know is that β and γ are negligible w.r.t. $\mathbf{T}$ and that α is so w.r.t. β and γ, then putting α automatically below $\beta\&\gamma$ is hardly justifiable given that $\beta\&\gamma$ is also exceptional w.r.t. β and γ.

What is the relationship between the preferential construction framework described in [Weydert 95a] and our present approach ? Let V be an arbitrary ranking algebra. For all $A \in B_L$, $r \in V$, we have $\mathcal{R}_{A,r} = \mathcal{R}_0[A : r]$. Let $\mathcal{R}$ and $\mathcal{R}'$ be two finitary ranking measures with partitions $\{A_0, ..., A_n\}$ resp. $\{P_0, ..., P_{n'}\}$ which are compatible with each other, i.e. without an $A \in B_L$ s.t. $\mathcal{R}(\neg A) \ll o$ and $\mathcal{R}'(A) \ll o$. We can then define a ranking measure $\mathcal{R}*\mathcal{R}' : B_L \to V$ with partition $\{A_i \cap P_j \mid A_i \cap P_j \neq \emptyset, i \leq n, j \leq n'\}$ by setting $\mathcal{R}*\mathcal{R}'(A_i \cap P_j) = \mathcal{R}(A_i \cap P_j)*\mathcal{R}'(A_i \cap P_j)$ for all $i \leq n, j \leq n'$. For each $\Phi \subseteq L°(I)$, let $\mathbf{S}_{BL,V}(\Phi)$ be the set including $\mathcal{R}_0$ and all those ranking measures obtained by combining through $*$ finitely many compatible $\mathcal{R}_{\mathrm{Mod}(\varphi),r}$ with $I(\neg\varphi) \in \Phi$ and $r \in V$.

Theorem 5.2 Let $\Sigma \cup \Delta^I \cup \Delta^{\Rightarrow}$ be a finite knowledge base where $\Sigma = \{\varphi_i \mid i \leq n\} \cup \{\psi\} \subseteq L$, $\Delta \subseteq L°(-\!\!»)$ and $\varphi_\Sigma = \varphi_0\&...\&\varphi_n$. Then

- $\Sigma \cup \Delta^I \cup \Delta^{\Rightarrow} \approx_J \psi$ iff $\{\mathcal{R} : B_L \to V \mid \mathcal{R} \models \Delta^{\Rightarrow}\} \cap \mathbf{S}_{BL,V}(\Delta^I) \models \varphi_\Sigma \Rightarrow \psi$.

The preferential construction entailment notion $\approx_{PC}$ [Weydert 95a] is different from $\approx_J$ insofar as it assumes $V = V(\mathbf{Z})$, considers ranking measures constructible from $\mathcal{R}_{A,r}$ with $*$ and $\max_\ll$, exploits normality maximization and satisfies rational monotony. This causes a more speculative but less differentiated reasoning behaviour which affects robustness and promotes some counterintuitive results.

To round up the picture, let's compare **J**-entailment with other related approaches. Its advantages may be summarized in a placative way as follows.

- It has better inheritance features than rational closure $\approx_R$ [Lehmann 92] or system Z [Pearl 90].

- It is less syntax-dependent and anti-probabilistic than lexicographic closure $\approx_{LC}$ [Lehmann, Benferhat *et al.* 93] or conditional entailment $\approx_{CE}$ [Geffner and Pearl 92].

648

- It is less speculative and more easy to handle than maximum entropy entailment $\models_{ME}$ [Goldszmidt et al. 90, Weydert 95].

- It is more redundancy-tolerant and robust than LCD-inference $\models_{LCD}$ [Benferhat et al. 95] which exploits the belief function paradigm.

In addition, one of the strengths of our account is that we allow inheritance through exceptional subclasses together with a powerful monotonic object-level logic based on the default conditional $\Rightarrow$ and our explicit belief modality I. The following figure shows the handling of the above examples by different default formalisms. The main purpose of this small (and slightly biased) feature list is to illustrate some basic characteristics of the main conditional logical approaches and to provide a general impression of their strengths and presumable problems. A detailed analysis of their relationships remains still to be done

Principles :	RM	EI	GE	AP	RE	AN
$\models_J$	0	1	1	1	1	1
$\models_{PC}$	1	1	1	1	1	0
$\models_{LCD}$	0	1	1	0	0	0
$\models_{RC}$	1	0	0	0	1	1
$\models_{ME}$	1	1	1	1	1	0
$\models_{CE}$	0	1	1	0	0	1
$\models_{LC}$	1	1	0	0	0	0

RM : rational monotony, EI : inheritance through exceptional subclasses, GE : Geffner's example, AP : anti-prioritization, RE : redundancy-tolerance, AN : assymmetric Nixon diamond.

References

[Benferhat et al. 93] S. Benferhart, C. Cayrol, D. Dubois, J. Lang, H. Prade. Inconsistency management and prioritized syntax-based entailment. In *Proceedings of IJCAI 93*, Morgan Kaufmann, 1993.

[Benferhat et al. 95] S. Benferhart, A Saffiotti, P. Smets. In *Proceedings of UAI 95*, Morgan Kaufmann, 1995.

[Bacchus et al. 93] F. Bacchus, A.J. Grove, J. Y. Halpern. Statistical foundations for default reasoning. In *Proceedings of IJCAI 93*, Morgan Kaufmann, 1993.

[Dubois and Prade 88] D. Dubois, H. Prade. *Possibility Theory*. Plenum Press, New York 1988.

[Gabbay 85] D. Gabbay. Theoretical foundations for nonmonotonic reasoning in expert systems. In Logics and models of concurrent systems, ed. K.R. Apt. Berlin, Springer-Verlag, 1985.

[Geffner and Pearl 92] H. Geffner, J. Pearl. Conditional entailment : bridging two approaches to default reasoning. *Artificial Intelligence*, 53: 209 - 244, 1992.

[Goldszmidt *et al.* 90] M. Goldszmidt, P. Morris, J. Pearl. A maximum entropy approach to nonmonotonic reasoning. In *Proceedings of AAAI 90*. Morgan Kaufmann 1990.

[Kraus *et al.* 90] S. Kraus, D. Lehmann, M. Magidor. Nonmonotonic reasoning, preferential models and cumulative logics. *Artificial Intelligence*, 44: 167-207, 1990.

[Lehmann 92] D. Lehmann. Another perspective on default reasoning. Technical report TR-92-12, Hebrew University, Jerusalem, 1992.

[Lehmann and Magidor 92] D. Lehmann, M. Magidor. What does a conditional knowledge base entail ? *Artificial Intelligence*, 55:1-60, 1992.

[Makinson 94] D. Makinson. General patterns in nonmonotonic reasoning. In *Handbook of Logic in Artificial Intelligence and Logic Programming (vol II)*, eds. D.Gabbay, C. Hogger. Oxford University Press, 1994.

[Pearl 90] J. Pearl. System Z: a natural ordering of defaults with tractable applications to nonmonotonic reasoning. Proc. of the Third Conference on Theoretical Aspects of Reasoning about Knowledge. Morgan Kaufmann.

[Spohn 90] W. Spohn. A general non-probabilistic theory of inductive reasoning. In R.D. Shachter *et al.* (eds.), *Uncertainty in Artificial Intelligence 4*, North-Holland, Amsterdam 1990.

[Weydert 91] E. Weydert. Qualitative magnitude reasoning. Towards a new semantics for default reasoning. In J. Dix *et al.* (eds.), *Nonmonotonic and Inductive Reasoning*. Springer-Verlag 1991.

[Weydert 93] E. Weydert. Plausible inference for default conditionals. In *Proceedings of ECSQARU 93*. Springer, Berlin, 1993.

[Weydert 94] E. Weydert. General belief measures. In *Tenth Conference on Uncertainty in Artificial Intelligence*. Morgan Kaufmann, 1994.

[Weydert 95] E. Weydert. Defaults and infinitesimals. Defeasible inference by non-archimedean entropy maximization. In *Eleventh Conference on Uncertainty in Artificial Intelligence*. Morgan Kaufmann, 1995.

[Weydert 95a] E. Weydert. Default entailment. A preferential construction semantics for defeasible inference. In Dreschler-Fischer, Pribbenow (eds.) *Proceedings of KI-95* . Springer-Verlag 1995.

Deep Disagreements and Public Demoralization

John Woods

University of Lethbridge Dept. of Philosophy
4401 University Drive
Lethbridge, Alberta TIK 3M4
Canada
and University of Amsterdam

1 Standoffs

One of the attractions of the theory of argument in the present day is the care it takes with practical affairs. It is nothing but good that more or less sophisticated techniques now exist for the display and appraisal of arguments about things as important as income tax policy, auto repairs and the upkeep of public parks. Such are practical matters calling for practical reasoning. One of the hallmarks of practical life, unlike the purely theoretical domain, is that its disagreements can be truly awful things, contentions that cry out for techniques of resolution.

It is with this connection in mind that I propose to develop an analytic model addressing a particularly intractable kind of disagreement and to demonstrate an applied analysis supports a limitation theorem for conflict resolution by rational means. The analysis is then made to predict stabilization strategies that lie on the further side of our limitation theorem. The limitations in question are broadly logical in character and they are made interesting because, among other things, they create a paralysis in public discourse. This paralysis arises from argumentational blockages which I propose to call **standoffs of force five**. Their analysis will be the principal business of the present section of this paper. The concept of demoralization will occupy us in section two.

I shall say that an argument is in a basic standoff or a standoff of *force one* just in case the participants disagree on some point at issue and there is no agreement about procedures which would or might lead to agreement. Standoffs of *force two* are standoffs of force one which satisfy a further condition: there is no consensus to agree to disagree, to let sleeping dogs lie, so to speak.

If a further condition is fulfilled, then we have a standoff of *force three*: there is no agreement to send the dispute to third-party determination. As we begin to see, standoffs are "bad things", logically speaking, never mind that they are also all too human. The natural mark of a standoff is its susceptibility to the fallacies of irrelevance and question- begging. Standoffs of force three are, like the others, of shifting degrees of importance, and their importance will vary in turn with the interests of the protagonists. There are, however, cases galore in which the following things are true:

A: the standoff is a standoff *in a community* and thus answers to certain demographic criteria. Whatever these might be in fine, they suffice to make it true not just that Harry and Charlie are in a standoff, but that the country[1] is;

B: the standoff involves a claim about what to do or should be done, and so is of an essentially normative character;

C: the matter in dispute is of a type that places it within the legitimate authority of a third party to settle without the direct consent of the contenders. In many such cases, the standoff includes a disagreement about what the government, say, should do about the matter in dispute, concerning which it is empowered to act; and

D: the matter is frequently such as to present a third party - the government or the courts - with what William James called a *forced option*.[2] So not taking any action is equivalent to acting for one of the contending parties and against the other(s).

It is important to be clear about the normative character of standoffs of the sort we propose here to examine. I shall say that a subject S holds a *deontic* belief B *proactively*, if and only if,

(a) B is a belief in the form "Nobody should do action A"

and

(b) holding B commits S to the truth of the proposition, P, "Everybody should take appropriate steps to prevent the performance of A by anyone."

The proactive commitments of S's subscription to a deontic belief B can, in turn, be characterized as follows: Anyone who satisfied (a) but who did not act in fulfilment of P is

pragmatically inconsistent[3]

and

guilty of wrongful omission by his or her own lights.

---

[1] Or the trade union or the university, or the political party. The demographic conditions for the attribution of standoffs to groups include group-percentages involved in the disagreement, the standing of the disputants in the group, statistical distribution of the issues in contention, the momentousness of the issue, and so on. The concept of group is a particularly intractable problem for the social sciences generally, as Margaret Gilbert's *On Social Facts* (Princeton: Princeton University Press, 1993) ably attests.

[2] William James, "The Will to Believe", in *Essays in Pragmatism*, New York: Hafner Publishing Company, 1957, p. 89.

[3] At a gently intuitive level, a charge of pragmatic inconsistency echoes the familiar complaint against "not practising what you preach". A deeper analysis, which I shall not pursue here, reveals how surprisingly pernicious this breed of inconsistency actually is. See John Woods, "Dialectical Blindspots", *Philosophy and Rhetoric*, 26 (1993), 251-265.

Conspicuous by its imprecision is the requirement of the proactive deontic believer to take "appropriate steps". I shall not attempt to subdue that imprecision beyond pointing out that in a rough and ready way preventative measures are held to conditions of proportionality and procedural normalcy. If I think that students should not wear baseball caps in my classes, it may be enough of a discouragement that I merely say so. If I see that my neighbour's house is being burgled, phoning the police will suffice in the general case. If a toddler is being physically savaged by a drunken brute, and I am the only third party at the scene, it may be that I will have to intervene directly. But if I am ninety-three years old and much reduced by arthritis, my intervention may have to be more honorific than actual.

A further imprecision affects what logicians call the *quantifiers* embedded in clauses (a) and (b) of our definition of proactively-held deontic beliefs. These are the expressions 'nobody', 'everyone' and 'anyone'. Quantifiers have *scopes* or ranges of applicability. They "range over" classes of individuals to whom their attached predicates are presumed to apply. If I think that everyone should try to discourage government corruption, whether I myself am bound to take such steps will vary as between my own government and, say, the government of France. Whatever the details of such variation we may take it that I myself am excluded from the range of quantifiers in the proposition "Everybody should try to prevent government corruption" in its application to France.

In what follows, the standoffs I shall be considering will be understood to satisfy these four additional conditions, A to D, as well as the definition of deontic beliefs proactively held. It will facilitate the exposition we call to mind the recent career of the abortion issue in countries such as Canada.

Condition C is especially important. Although a standoff is not a standoff of force three unless the disputants fail to agree to volunteer their differences to a third party for his or her (or its) resolution, there is nothing to say that the impacted issue might be of a kind that makes it legitimate for a third party, the government say, to *appropriate* the issue and enforce a resolution of it. I shall say something about just such a circumstance in a moment; for the present it would be well to bring out further indications of the flavour of force three intractability. Force three standoffs resist two common strategies for reconciliation. They resist the device of reconciliation by way of *analogies*; disputants are routinely disposed to judge proffered analogies as false analogies.[4] They also resist the application of what might be called, in a flexibly metaphorical way, the Fundamental Law on Collective Bargaining, in a form that extends to negotiated acceptance in a quite general sense. The Fundamental Law is an instrument designed to discourage extremism. It presumes that negotiable disagreement are those whose rational settlement involves a splitting of differences. This may well be true of any

---

[4] Concerning which see John Woods and Brent Hudak, "By Parity of Reasoning", *Informal Logic*, XI (1989), 125-134, especially pp. 128-130, in which the most celebrated, to date, analogical defence of the permissibility of abortion is found wanting. This is the argument proposed by Judith Jarvis Thomson in "A Defence of Abortion", *Philosophy and Pubic Affairs* 1, (1971), 47-66.

disagreement properly judged to be negotiable, but it is characteristic of force three disagreement that at least one party regards the issue as non-negotiable. Consider again the anti-abortion protagonist who regards abortion as murder at any pre-natal stage. It is a distinctive feature of such a conviction that for anyone holding it there are no differences to split, and for anyone opposing it the presumption of negotiability is fatally and ludicrously flawed.

Even so, for a good many standoffs of force three, although disputants will not themselves negotiate resolutions, they will submit to settlements imposed by lawfully constituted third parties who (or which) appropriate the issues as their own. To the extent that this is so, we may suppose that disputants honour the presumption of prior consent embedded in the mythic device of the social contract. In saying so, a definition presses for recognition.

> *Def. 1*: A force three standoff fulfilling conditions A to D is a *political* standoff in a population **P** to the extent that **P** acknowledges the presumption of prior consent to a resolution imposed by a lawfully constituted third party recognized as such in **P**.

A significant contrast announces itself straightaway.

> *Def. 2*: A force three standoff fulfilling condition A to D is a *purely moral* standoff in a population **P** to the extent that **P** refuses this presumption of prior content.

It is well to note that I am employing the terms "political" and "purely moral" in a somewhat technical way. A disagreement qualifies as political or as moral not by virtue of the subject matter of the issue in question but rather by way of procedural qualifications and limitations on its resolution.

Standoffs of *force four* can now be recognized . They are standoffs of force three which also satisfy *Def. 2*. They are thus purely moral disagreements, and so disagreements of such a kind that lawfully enforced settlements will not be consented to. Whereas political disagreements are arguments that disputants are prepared to lose, moral disagreements of force four are arguments people are not prepared to lose.

Crucial to the distinction at hand is the idea of withheld consent. If S is a party to a standoff of force four concerning some contested issue, say the practice of A-ing, then S proactively holds a deontic belief with respect to A-ing, namely, "Nobody should do A". By the conditions that qualify S's deontic belief as proactive, S is committed to take steps to prevent the practice of A-ing. If S's government acts in such a way as to make A-ing lawfully permissible, S must take *some* steps to undo the government's approval, and whatever those steps are they must be such a character as to imply S's conviction that the government's action is not morally authoritative. It is not enough that a "loser" in a government-resolved force four contention is not happy about having "lost". If is rather that he is not happy about having "lost" by virtue of wrongful actions by his government, hence actions by which S judges himself to be morally unbound.

It is here that the idea of proportionality of preventative measures can be seen to bear some real theoretical weight. If S believes that not even his government is morally entitled to decide a force four contention "against" him, then his own proactive commitments by his own lights may obligate him to a course of civil disobedience or rebellion. In this, he is not always wrong, unless it can be shown that civil rebellion is never morally justified.

Standoffs of force four are dangerous things, to say the least, but they are also sometimes resolvable in principle by what is known as a Pascalian minimax strategy. The abortion issue provides a handy illustration. X and Y are in a standoff of force four about abortion on demand at any time during pregnancy. X is *pro* and Y *contra*. If X and Y are willing to admit the possibility that they might be mistaken is their respective views, then costs can be reckoned. The cost of the mistake if Y is mistaken is that he or she would have scores of thousands of Canadian women a year encumbered with pregnancies they did not want and need not have endured. It is clearly a non-trivial cost. On the other hand, the cost of the mistake if X is mistaken in the unjustified killing of that same number of unborn each year. If X and Y are able to agree on which would be the greater cost, then a resolution rule drops out.

> **RR**: Settle the issue in such a way as minimizes the realization of
> the greater possible cost. The higher the cost the more it is mandatory
> to minimize the possibility of its exaction.

It may strike us as astonishing that in the recent history of the abortion debate in Canada, resolution rule **RR** is utterly conspicuous by its utter absence. It can safely be conjectured that the issue failed a condition which makes **RR** an applicable rule. **RR** is an applicable rule only if each party to the disagreement is prepared to admit the possibility that he or she is mistaken. On something as morally and metaphysically complex as the issue of when a member of the biological species homo sapiens acquires protection against termination of its life, it could only be expected that those having views about the matter would readily admit the possibility of mistake. Yet precisely this seems to have been missing from the Canadian record.

The idea of the possibility of mistake is a theoretically elusive one. For present purposes an economical characterization will have to do. Let us say that a disputant recognizes the possibility that his own position is mistaken precisely to the extent that he thinks that **RR** is an applicable rule in the argument at hand. With that said, our final category of standoffs can be specified. An argument is a standoff of *force five* when it is an argument of force four concerning which disputants are closed-minded. Closed-mindedness is not, here, a term of abuse. A person is closed-minded about an issue precisely to the extent to which he is unprepared to submit the issue to **RR**-determination. Most people are close-minded about whether the gunning down of patrons at arbitrarily selected fast food outlets is all right as a form of recreation. It is certain that **RR** would decide the issue in their favour, but that they would even consider submitting it to such a strategy is lunatic.

Governments, the dominant élites (to borrow Walter Lipman's term) and the whole citizenry have a large stake in discouraging force five standoffs. Force five standoffs are the non-meterological counterparts of force ten hurricanes. They require one to batten down the hatches and to head inland. And clearly it would have been better had they never cropped up.[5] But crop up they do, more in some societies than in others. It might be supposed that the propensity to standoff is greater the more a society is liberal and pluralistic in its dominant cultural arrangements. Whether such a culture will be beset by standoffs of force five, rather than by disagreements of lesser rank, will in large measure be function of what it is pluralistic about and so will turn on whether it fosters differences in value appropriate to the greater intractability of force five. It is sometimes apparent that certain kinds of multicultural pluralism are ripe occasions of force five turmoil, but it is by no means always so. The recent deadlock in Canada is in no obvious way a uniquely ethnic or even religious matter. The question now before us is that of how force five standoffs are to be managed in societies such as our own.

Something lies ready to hand and it is available to the entire network of the country's rulers, its dominant élites, and it enters the spiritual marrow from a pervasive cultural mist. It is called education. Education here is the education appropriate to the normative socialization of a liberal, pluralistic population. It is, or involves, the inculcation or transformation of values by means of the social technology of persuasion. Logic now defers to rhetoric.[6] In so saying, we have our *Limitation Theorem*: Standoffs of force five are logically irresolvable.

There is a huge literature regretting the social technology of persuasion. Much of the worry has to do with its concealedness, its manipulation and its sheer effi-

---

[5] Some naïfs celebrate pluralism with respect to values as occasion for the promotion and intensification of normative conflict. Cf. Isaiah Berlin, *Concepts and Categories*, H. Hardy (ed.), Harmondsworth, Middlesex: *Penguin Books* (1978). Bernard Williams counsels against attempts to eliminate normative conflict. Doing so, in the general case at least, incurs "the loss of a sense of loss" and a flattening of human experience: Bernard Williams, *Moral Luck*, Cambridge: Cambridge University Press (1981); p.80, p.82. This does seem rather precious. In fairness to Williams, he does say that normative conflict is "not necessarily pathological"; and this leaves plenty of room to regret it when (contingently) it *is* pathological (as it clearly often is in the case of force five standoffs).

[6] Logicians tend to be dismissive of rhetoric, legatees all of Socrates' hostility to the Sophists of yore. I am not one of them. If concerning any issue there should happen to be a fact of the matter, then it would be oddly illogical not to want to present the truth persuasively, that is, in ways that maximize the chance of getting others to see that it is true. More interesting are those cases concerning which no consensus exists with regard to the question of truth. For wide ranges of such issues, giving up on critical discussion and going fishing is not a realistic option, and we are left with the hard question of what to do in the wake of intractable disagreement. Some issues require the fixation of belief, never mind that the matter at hand is underdetermined by the agreed-upon evidence. (Thus every scientific theory worthy of the name). So a central task for rhetoric is to specify measures for the fixation of belief about matters underdetermined by the agreed-upon evidence and to establish that (and in what sense) such measures are acceptable. It is a very hard question, as is evidenced by the frequency with which belief-fiction is accomplished by measures thought to be seriously unacceptable, as we will now see.

cacy. People complain that the technology (i.e., the mass media), is concentrated
in too few or in any case the wrong hands. Others fret about the domination of
economic considerations over the selection, duration and mode of transmission of
mass media signals. Some say that the majority of such transmissions are continuous with mass advertising which, in turn, is continuous with propaganda, and
that the continuity-relation is transitive. Still others are troubled by the individual's inaccessibility to the media or by the fact, if it is a fact, that televisual
argumentation is structurally illucid for the ordinary viewer and thus beyond
his competence to assess.[7] Of two things I have no doubt. One is that much of
the mass media bashing of recent years is half-baked, paranoid and politically
self-serving. The other is that there is evidence enough of thought-control to
warrant our concern and chagrin.

I want, in any case, to return to the business at hand and to concentrate
on what proved to have been the rhetorical targets of the dominant élites in
the recent course of the education of the Canadian public on the matter of
abortion. In the early 1990's the abortion issue had been giving the appearance
of becoming, if it were not already one, a nasty force five standoff in Canada.
Objective: neutralize it, reduce it to nothing of higher rank than a political
standoff. And so,

1. Discredit the leaders as extremists. Show them in disagreeable, noisy, uncivil
   and, if possible, criminal contention. Hope for early boredom.
2. Demoralize the population. Restrict the national discussion to shrill set
   pieces from the partis pris. Do nothing to encourage the notion that this is
   an issue of complexity and perhaps of central national importance. (Merely
   saying that it is, is another matter entirely).
3. Marginalize the visible protagonists. Take care to identify Ms. X as a member
   of the Feminist Caucus and Mr. Y as a Mormon bishop. For all its dialectical limitations, endorse and promote Ramsey's Maxim:[8] the most sensible

---

[7] Here is a very small sample of the literature. H. Cantril, *The Invasion from
Mars: A Study in the Psychology of Panic*, Princeton: Princeton University Press
(1940); Harold Innis, *The Bias of Communication*, Toronto: University of Toronto
Press (1951); George Grant, *Technology and Empire*, Toronto: House of Anansi
(1969); Marshall McLuhan, *Understanding Media: The Extension of Man*, New York:
McGraw- Hill (1964); Jacques Ellul, *Propaganda: The Formation of Men's Attitudes*,
New York: Knopf (1965); D. Altheide, *Creating Reality: How TV News Distorts
Events*, Newbury Park, CA: Sage (1976); Michael Geis, *The Language of Televi-
sion Advertising*, New York: Academic Press (1982); Trudy Govier (ed.), *Selected
Issues in Logic and Communication*, Belmont, CA: Wadsworth (1988); Edward S.
Herman and Noam Chomsky, *Manufacturing Consent: The Political Economy of the
Mass Media*, New York: Pantheon Books (1988); and papers by John D. May, John
McMurtry, Lenore Langsdorf and John Dolan, all in *Informal Logic*, X (1988).

[8] Regarding issues about fundamental issues that seem insusceptible of consensus,
F.P. Ramsey proposed that "in such cases it is a heuristic maxim that the truth lies
not in one of the two disputed views, but in some third possibility which has not
yet been thought of, which we can only discover by rejecting something assumed
as obvious by both disputants." F.R. Ramsey, *The Foundations of Mathematics*,
London: Routledge and Keagan Paul 1931, pp. 115-116.

solution is a compromise. It sounds so right.

4. Trivialize the contending values, not only as extreme but as moral. Thus moral choices are private, a matter of personal opinion only, are not to be taken up evasively ("forcing your personal values upon me".) Invoke the non-cognitivism of moral principles. Emphasize their relativity.

5. Adjust the taxonomy. These extreme positions are also religious positions, fine as long as they do not intrude and, in any event, subject by implicit prior consent to the sanctity of the separation of Church and State. Since it certainly is not up to the Church to decide for all, the right of decision must be the State's to take or not.

6. Saturate communications with euphemisms, the more vapid the better. (Thus "pro-choice", "pro-life"). Keep disclosure of clinical details to a minimum. (No footage of an abortion; no footage of a thirteen year old's labour). Be "tasteful".

7. Guilt by association. President Ceauscescu ran a monstrous tyranny. He imposed absurd and burdensome reproduction targets on Romanian women, in an effort to increase the population substantially. Abortions were also forbidden. Now all that nonsense is over.[9]

In a conference in Amsterdam in 1990[10] I said that my own best guess was that the Canadian population was getting weary of this business, and that the drift of feeling was toward some kind of negotiated settlement by way of the Fundamental Law on collective bargaining; although it must also be recognized that the issue might just go away under press of dialectical fatigue.

As events have turned, we have seen that the first strategy failed, for Parliament has abandoned all efforts to legislate on this issue. The second thing happened, an utter astonishment. A standoff of force five plummetted not to a political standoff of force three, but to a standoff of *force one*. With respect to abortion, Canadians have in effect agreed to disagree, a particularly striking outcome given the issues involved.

We see in this a consequential overdoing of the rhetorical technologies. The target was to downgrade the abortion controversy from the dangerous perch of force fivedom to a level on which Parliament could work; that is, the objective was to convert a closed-minded moral disagreement into a resolvable political one. But something else happened. Parliament tried to treat abortion as a political problem; it openly connived in the conversion of a moral irresolvability

---

[9] "Of all Nicolae Ceaucescu's legacies, perhaps the saddest can be found in Romania's maternity wards. Visitors are shocked at the sight of abandoned babies, sick pregnant women and doctors working 24-hour shifts to perform abortions - all because of a mad plan to increase Romania's population", *The Economist*, 314, 20 January 1990, p.52.

[10] The Royal Netherlands Academy Conference on Logic and Politics, February 1990. The idea of force five standoffs is given more technical expression in my "Public Policy and Standoffs of Force Five", in E.M. Barth and E.C.W. Krabbe (eds.), *Logic and Political Culture*, Amsterdam: North-Holland 1992, 97-108.

into a political resolvability. Parliament failed. And this, it may be said, was provocation of further degradation to level one. It is degradation in two senses. The standoff-grade is the lowest possible; and a sizeable percent of the Canadian electorate is seized of a fundamental matter of conscience on which it is now prepared to be inert. In the steep fall to force one, millions of Canadians have decided to do nothing about what they believe with unshaken conviction to be sixty thousand murders a year. In thinking so, they may well be wrong. But they do think so, and this means that the technology of persuasion has, under our very noses, transformed a noisy and even dangerously fractious population into a profoundly demoralized one. It is not a felicitous outcome.

2 Demoralization

There is plenty of evidence that the people of Canada face the end of the millenium with lowered spirits, but none that their lack of morale derives at all significantly from the present situation in the country as with regard to abortion. The claim with which we ended the preceding section would then appear to be empirically quite untrue.

By "demoralization" I do not here mean a lowering of spirits, or any particular variety of lassitude. Such a meaning is comparatively new, ensuing from the past century. In the century before it, demoralization was the corruption of morals. *The Oxford Dictionary of English Etymology* (1966), reminds us that the French equivalent *demoraliser*, a word of the French Revolution, provoked La-Harpe's scorn: *"Si Demoraliser pouvait être français, il signifierait cesser parler de morale"*. This, the eighteenth century sense, is closer to what I myself intend.

It is one thing to state the case you intend, it is another thing to make it. Doing so is the task of the present section. Its development will be facilitated if, from time to time, we replace the example of abortion in Canada of the 1990's with the example of the treatment of German Jews in the 1930's and 1940's. A great many people are infuriated by the abortion-holocaust analogy. In this they may be right. I myself intend no such analogy. I want rather to ask about a society in which the question of whether Jews should be exterminated is a question that divided opinion in ways that qualify it as a standoff of force five. I want then to consider, if the Canadian "solution" of the abortion question were applied to the Jewish question, how that state of affairs should best be understood.

Consider then a community K of B-holders, where B is a proactive deontic belief. Suppose that in the broader community, U, of which K is a part, B is disbelieved and resisted by an opposite proactive belief. As it stands, in our definition of proactive deontic beliefs the quantifiers 'nobody', 'everybody' and 'anyone' range over the members of U. But if the disagreement between K and U with respect to B is downgraded from one of force five to one of force one then, in effect, the quantifier of B is itself restricted to K; that is, its scope is cutdown from U to K and is a perfect illustration of the ancient wisdom of "minding your own business" and "not forcing your own personal values on others". B, recall, is a proposition in the form "Nobody should do A". And since it is held proactively, it commits its holders to proposition P, "Everybody should take appropriate steps to prevent anyone's performing A". The quantifier cutdown already noted for B, may also apply to P. That is, if the issue in question has been downgraded from a standoff of force five to a standoff of force one, proactive holders of B are, in effect, committed to the proposition that everyone *in K* is obligated to take steps to prevent anyone *in K* from doing A. We then see that there are two subcases to take note of. One is the case in which

I. Members of K shrink the scope of the quantifier in B to K itself. (This is the case in which, for example, Dutch Reformed Christians come to believe that *they themselves* should not participate in abortion.)

In contrast is a second case, in which

II. the quantifier of B remains fixed over U, but the quantifier of P is restricted to K. (Thus, if a Dutch Reformed Christian believes that no one should participate in abortion, he is in a situation of type II if he also holds that *all Dutch Reformed Christians* should take steps to prevent *any Dutch Reformed Christian* from participating in abortions).

In judging situations of types I and II, perspective is everything. From where a pro-abortionist stands, these are exactly the situations to be hoped for in a pluralistically tolerant society. But from the point of view of the anti-abortionist, both I and II are troubling. In situation I, the anti-abortionist decides to hold his strong belief against abortion in so "cutdown" a form as to approach to contradiction. For although he believes that no one should participate in abortions, the claim of situation I is the lesser one, that no K-member should be thus involved. The anti-abortionist has settled for a good deal less than what he actually believes. Situations of type II avoid the problem with respect to the scope of the belief that abortion is wrong (viz., for everyone), but they reduce the scope of preventative action. A B-holders' position becomes one in which, although he believes that no one should perform or receive abortions, he is not prepared to take preventative measures beyond those who already share that belief. We see, then, that a situation-II anti-abortionist is in violation of the definition of his own proactive belief. In restricting the range of preventative action in the ways here noted, he stands convicted of pragmatic inconsistency and moral wrongdoing, by his own lights.[11]

---

[11] It is worth noting how structurally intractable a type-II situation is. It is natural to suppose that members of K would aspire to a condition in which their role as "losers" of the contention in question is reversed, with their opponents now bearing that mantle. On the substance of the issue they would be quite right, again by their own lights, so to aspire, but structurally speaking it is no improvement. The pattern of alienation would merely have shifted from K to those who disagree with K. Either way there would be a significant chunk of the larger society which, in its own judgement, would be pragmatically inconsistent and guilty of non-trivial wrongdoing.

This being so, it cannot be an unmixed blessing for the pro-abortionist that his opponent's position has been downgraded from one of force five standing to one of force three standing or lower. Such is a transition that condemns his U-compatriot to pragmatic inconsistency and to the conviction that he is guilty of moral wrongdoing precisely for his having accommodated the pro-abortionists' own position. It is an unsatisfactory situation whatever the specific content of the disagreement at hand, provided that it was once a disagreement of higher than force three. It is not in general a good thing to have K's in U's that are false to their own beliefs in ways that convict them of inconsistency and serious wrongdoing.

A K-community in a situation of type-II is one that is alienated from the larger community U and whose membership in U is nontrivially compromised. People might, by and large, be prepared to tolerate such alienation where K is small and the issue is question sufficiently *recherché;*, if not downright nutty. One might even venture to say that alienation and compromised U-citizenship (where U is a state) are the structurally inescapable constituents of normative pluralism in U, and (with Isaiah Berlin) that a multicultural U is in general a better thing to have than a normatively monolithic U.[12]

So it might be proposed that to the extent that multiculturalism is a defensible ideal for polity, that situations of type II themselves must be defensible. Perspective being all, it depends on where the judges of this question stand on the issue whose situation-II "resolution" constitutes the alienation. We have no hesitation in saying that a situation of type-II with respect to the Jewish question in Germany was monstrous. Germany's demoralization was radically complete owing to the collapse of resistance to genocide. No doubt it will be said that the policy towards Jews was self-evidently outrageous and that to think otherwise one has to be quite mad. This is the pathologization of normative disagreement, and from where I stand this is the right judgement in that case. But from where the racists of Germany stood any such judgement is dismissible out of hand. The relevant thing to say for present purposes has a peculiarly structured texture. It generalizes as follows.

> *Demoralization Rule:* For any group K, in a society U, and for any proactively held belief B with regard to an action A, which satisfy conditions on a situation of type II, U is in a state of demoralization approximating to that of Germany in the 1930's and 1940's to the extent that members of K attach to action (or policy) A the same moral weight as opponents of the Jewish policy originally attached to it.

It is of little moment that members of K may be objectively wrong in their assessment of the moral gravamen of action (or policy) A. What signifies is that

---

[12] So we are recognizing that a normatively pluralistic society is not a matter of some people liking garlic and others hating the stuff, or of some people preferring to spend Saturday evenings at the Polish League's weekly dances and others opting for a night at the opera. Pluralism is a social condition with normative bite only to the extent that significant groups are significantly alienated from the whole.

K, and U too, are in a situation of type II and that they could not be so unless, in particular, the K-community believed unshakably that its assessment of the moral gravamen of A *was* objectively correct. Thus the relevant question to ask is whether it is a good thing for societies to be in a state of affairs that models our Demoralization Rule? The answer is clearly "No" to the extent to which it is a corollary of the Demoralization Rule that

> *Demoralization Corollary:* For any K satisfying the structural conditions of the Demoralization Rule, members of K will judges themselves and their larger society to be guilty of moral cowardice and political impotence with regard to a matter of great moral significance.

That a group such as K could well satisfy the Demoralization Corollary in a merely *tacit* way only adds to its ills. For now it is a state of bad faith or self-deception.

What is to be done in a society[13] in which a demoralized group K takes the issue on which it has "lost" to have a gravamen approximating to the killing of Jews in Germany? Clearly the broader society of which K is a part has a huge stake in promoting a shift from the type-II situation in which it finds itself. There are three options available for its consideration:

> *Option 1:* It can try to move itself into a situation of type I. That is, it can try to convince members of K that its abhorrence of action (or policy) A is applicable to the performance of A only by members of K themselves.

> *Option 2:* It can try to re-energize the members of K, and to encourage them to resume strategies of intervention with respect to the performance of A by anyone in the broader community U.

> *Option 3:* It can try to reduce the standoff of force five with respect to the issue in question to a standoff of force four. That is, it can try to persuade disputants to be open-minded, rather than close-minded about its disagreements.

Of the three, the second option has the least to recommend it, in as much as it risks the encouragement of civil strife. When abstractly considered, option one has its attractions. For it is tantamount to attempting to use the technologies of persuasion to convince members of K that their harsh judgement of A is valid only as applied to themselves. But we need only remind ourselves that option one was as available to the Nazis as it is to our hypothetical state U to see that it is something that members of K would resist to the uttermost (and should).

This makes option three something of a standout. In it, disputants are encouraged to be open-minded about the object of their disagreement. For this to

---

[13] The idea of what a society might "do" about a problem is, of course, a huge expository convenience. I shall leave this notion to function intuitively, beyond saying that as here applied it is the relevant totality of a society's varied devices for forming public opinion and conforming public behaviour to it.

happen, each must recognize that his own position might be mistaken, that is, that it is a real possibility that what his opponent says is true. Recognizing that one's being mistaken is a real possibility is consistent with one's total confidence that one is not in fact mistaken. Acknowledging the real possibility that one is mistaken is not a *probative* matter, that is, it is no evidence that one *is* mistaken. Its significance, then, is procedural. Parties locked in a standoff of force four dispose themselves to the application of the resolution rule **RR**, of section 1.

> **RR**: Settle the [force four] issue in such a way as minimizes the
> realization of the greater possible cost. The higher the cost the more it
> is mandatory to minimize the possibility of its exaction.

It is easy to see how option three and the **RR** rule which attends it would have settled things in Germany. Even the most ardent Nazi, if he recognized the real possibility that Jews have the full moral protections of human personhood could be made to see that if he were mistaken in thinking otherwise, the cost of the mistake would be mass murder; and, further, that this is a cost that outweighs any cost attaching clemency towards the Jews even if it chanced that they did not have the full moral protection of human personhood.

What is so deeply striking about **RR**-resolutions is that *they are arrangements that regulate behaviour without changing belief.* The Nazi who acquiesces in a **RR**-resolution of the Jewish question can go to his grave with the untroubled conviction that Jews are subhuman and wholly without the moral protections of life, and yet in full consistency with that conviction have quit killing Jews. It is a salutary arrangement.[14]

Standoffs of force five are a structural embarrassment for any polity in which public policy is shaped significantly by any mechanism that cuts across the grain of popular disagreement. They are an embarrassment virtually guaranteed to eventuate in any polity that is significantly multicultural in its makeup.

In characterizating such embarrassments as structural, it is intended that standoffs of force five persist at instrumental cost. They are disagreements that disturb the workings of polities in which they occur enduringly; that is to say, they have the potential to threaten the very conditions under which a social arrangement even qualifies as a polity. And, as we have seen, there is no modern, technologically mature polity which lacks mechanisms of social persuasion whose function is to discourage, contain, marginalize, trivialize or subdue these *crises de conscience.*

After consideration of the instrumental or structural debilities of standoffs of force five which are left standing, so to speak, it becomes clear that the central question is one of the quality of life in societies in which standoffs of force five are dealt with in the ways lately noted.

---

[14] As for abortion in Canada, under option three, I shall leave it to the reader to ascertain its providence with respect to rule **RR**.

Practical Reasoning with Procedural Knowledge

(A LOGIC OF BDI AGENTS WITH KNOW-HOW)

Michael Wooldridge

Department of Computing
Manchester Metropolitan University
Chester Street, Manchester M1 5GD
United Kingdom
M.Wooldridge@doc.mmu.ac.uk

Abstract. In this paper, we present a new logic for specifying the behaviour of multi-agent systems. In this logic, agents are viewed as *BDI* systems, in that their state is characterised in terms of *beliefs*, *desires*, and *intentions*: the semantics of the BDI component of the logic are based on the well-known system of Rao and Georgeff. In addition, agents have available to them a library of plans, representing their 'know-how': procedural knowledge about how to achieve their intentions. These plans are, in effect, programs, that specify how a group of agents can work in parallel to achieve certain ends. The logic provides a rich set of constructs for describing the structure and execution of plans. Some properties of the logic are investigated, (in particular, those relating to plans), and some comments on future work are presented.

1 Introduction

There is currently much international interest in computer systems that go under the banner of *intelligent agents* [16]. Crudely, an intelligent agent is a system that is situated in a dynamic environment, of which it has an incomplete view, and over which it can exert partial control through the performance of actions. Agents will typically be allocated several (possibly conflicting) tasks, and will be required to make decisions about how to achieve these tasks in time for these decisions to have useful consequences [8].

An obvious research problem is to devise software architectures that are capable of satisfying these requirements. Various solutions have been proposed, many of which are reviewed in [16]. One solution in particular, that is currently the subject of much ongoing research, is the *belief-desire-intention* (BDI) architecture [10]. A representative BDI architecture, (the PRS [4]), is illustrated in Figure 1. As this figure shows, a BDI architecture typically contains four key data structures. An agent's *beliefs* correspond to information the agent has about the world, which may be incomplete or incorrect. Beliefs may be as simple as variables, (in the sense of, e.g., PASCAL programs), but implemented BDI agents typically represent beliefs symbolically (e.g., as PROLOG-like facts [4]). An agent's *desires* intuitively correspond to the tasks allocated to it. (Implemented BDI agents require that desires be consistent, although *human* desires often fail in this respect.)

An agent's *intentions* represent desires that it has committed to achieving. The intuition is that an agent will not, in general, be able to achieve *all* its desires, even if these

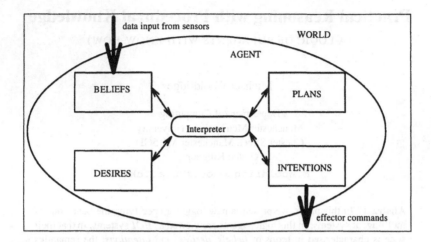

Fig. 1. A BDI Agent Architecture

desires *are* consistent. Agents must therefore fix upon some subset of available desires and commit resources to achieving them. These chosen desires are *intentions*. An agent will typically continue to try to achieve an intention until either it believes the intention is satisfied, or else it believes the intention is no longer achievable [2].

The final data structure in a BDI agent is a *plan library*. A plan library is a set of plans (a.k.a. *recipes*) that specify courses of action that may be followed by an agent in order to achieve its intentions. An agent's plan library represents its *procedural knowledge*, or *know-how*. A plan contains two parts: a *body*, or *program*, which defines a course of action; and a *descriptor*, which states both the circumstances under which the plan can be used (i.e., its pre-condition), and what intentions the plan may be used in order to achieve (i.e., its post-condition).

The *interpreter* in Figure 1 is responsible for updating beliefs from observations made of the world, generating new desires (tasks) on the basis of new beliefs, and selecting from the set of currently active desires some subset to act as intentions. Finally, the interpreter must select an action to perform on the basis of the agent's current intentions and procedural knowledge.

In order to give a formal semantics to BDI architectures, a range of *BDI logics* have been developed by Rao and Georgeff [9, 11]. These logics are extensions to the branching time logic CTL' [3], which also contain normal modal connectives for representing beliefs, desires, and intentions. Most work on BDI logics has focussed on possible relationships between the three 'mental states' [9], and more recently, on developing proof methods for restricted forms of the logics [11].

In short, the aim of this paper is to extend the basic BDI framework [9, 11] with an apparatus that allows us to represent the plans (options) that agents have available to them, and how these plans can be executed. As such, this paper builds on earlier attempts to represent BDI agents with plan libraries [12, 7], as well as more general attempts to represent agents that can act in complex ways [2]. We begin, in the following subsection, with a brief rationale for our work, and then in section 2, we introduce the basic semantic objects that underpin our new logic, and then formally define plans and the semantics of plan execution. In section 3, we present the new logic itself, which we shall call $\mathcal{L}$. Some properties of the logic are investigated, and some conclusions are presented in section 4.

Motivation: Some previous attempts have been made to graft a logic of plans onto the basic BDI framework [12, 7]. However, these logics treat plans as *syntactic* objects. This makes it unclear what plans *denote*, and also makes the semantics of quantifying over plans somewhat complex. For these reasons, we here introduce a new BDI logic of planning agents: $\mathcal{L}$. The new logic is similar in many respects to those defined in [12, 7], (in particular, the BDI semantics are based on [9]). However, we give a *semantic* account of plans, that has the flavour of dynamic logic [5].

2 Plans and Plan Execution

As noted above, we intend our logic $\mathcal{L}$ to let us represent the properties of reasoning agents, each of which has associated with it a *plan library*, containing plans that it can use in order to achieve its intentions. In this section, we formally define plans and the semantics of plan execution. We begin by introducing the basic semantic objects of our logic.

2.1 Worlds, Situations, and Paths

The logic $\mathcal{L}$ that we develop in section 3 allows us to represent the properties of a system that may evolve in different ways, depending upon the choices made by *agents* within it. We let D_{Ag} be the set of all agents, and use a binary *branching time relation*, R, to model all possible courses of system history. The relation R holds over a set T of *time-points*, i.e., $R \subseteq T \times T$. Any time-point may be transformed into another through the execution of a *primitive action* by some agent: arcs in R thus correspond to the performance of such actions. We let D_{Ac} be the set of all primitive actions, and assume an arc labeling function Act that gives the action associated with every arc in R. Similarly, we assume a function Agt, which gives the agent associated with every primitive action.

Definition 1. A *world* is a pair (T', R'), where $T' \subseteq T$ is a non-empty set of time points, and $R' \subseteq T' \times T'$ is a total, backwards linear branching time relation on T'. Let $W = \{w, w', \ldots\}$ be the set of all worlds (over T). If $w \in W$, then we write T_w for the set of time points in w, and R_w for the branching time relation in w.

Definition 2. A pair (w, t), where $w \in W$ and $t \in T_w$, is known as a *situation*. If $w \in W$, then the set of all situations in w is denoted by S_w, i.e., $S_w = \{(w, t) \mid t \in T_w\}$. Let $S = \bigcup_{w \in W} S_w$ be the set of all situations. We use s (with decorations: $s', s_1, \ldots$) to stand for members of S.

We now present some technical apparatus for manipulating branching time structures.

⟨sit-set⟩ ::= any element of ℘(S)	(conditions)
⟨plan-body⟩ ::= any element of D_{Ac}	(primitive actions)
| ⟨plan-body⟩; ⟨plan-body⟩	(sequential composition)
| ⟨plan-body⟩ '|' ⟨plan-body⟩	(non-deterministic choice)
| ⟨plan-body⟩ || ⟨plan-body⟩	(parallel composition)
| ⟨plan-body⟩*	(iteration)
| ⟨sit-set⟩?	(test actions)

Fig. 2. Plan Body Structure

Definition 3. Let $w \in W$ be a world. Then a *finite path* through w is a sequence

$$(t_0, t_1, \ldots, t_k)$$

of time points, such that $\forall u \in \{0, \ldots, k-1\}$, we have $(t_u, t_{u+1}) \in R_w$. Let *fpaths*(w) denote the set of finite paths through w. An *infinite path* (or just 'path') through w is a sequence $(t_u | u \in I\!N)$, such that $\forall u \in I\!N$, we have $(t_u, t_{u+1}) \in R_w$. Let *paths*(w) denote the set of paths through w. If p is a (finite or infinite) path and $u \in I\!N$, then $p(u)$ denotes the $u+1$'th element of p (where this is defined). Thus $p(0)$ is the first time-point in p, $p(1)$ is the second, and so on. If p is a (finite or infinite) path and $u \in I\!N$, then the path obtained from p by removing its first u time-points is denoted by $p^{(u)}$ (where this is defined).

2.2 Plan Structure

An agent's plan library is a set of 'recipes', which the agent can use use in order to bring about its intentions. Plans are actually *multi-agent* plans, which closely resemble parallel programs. A plan contains a *plan body*, which represents the 'program' part of the plan, and a *plan descriptor*, which characterizes the pre- and post-conditions of the plan. The atomic components of plan bodies are *actions*, i.e., elements of the set D_{Ac}. Actions are composed into plan bodies by the use of *plan constructors*: these are precisely the kind of constructs that one would expect to find in a parallel programming language[1], allowing for sequential and parallel composition, iteration, and choice. Formally, the set D_B, of all plan bodies, is defined by the grammar in Figure 2. We use β (with decorations: $\beta', \beta_1, \ldots$) to stand for members of D_B.

There are two points to note about this definition. First, although we have used a grammar to define plan bodies, and they thus appear to be syntactic objects, they are in fact *semantic* objects, built up from other semantic objects (actions and sets of situations). There are good reasons for emphasizing this point. In our language, $\mathcal{L}$, we will need to be able to quantify over plans (and plan bodies). Hence there must be terms in the language which stand for plans (and plan bodies), and both plans and plan bodies

---

[1] It is worth noting that we do not have any of the CSP-like primitives for communication and synchronization that one finds in parallel languages like occam [6].

must appear in the domain of the language. Thus plans are not syntactic constructs: they are semantic objects. The second point to note is that a test action takes as its argument a *set of situations*: the test action c? will succeed if the current situation is a member of the set c. A more natural representation for conditions might appear to be formulae of the language, so a test action φ? would succeed if the formula φ was satisfied in the current situation. But this would confuse syntax and semantics: test actions are semantic objects, since they are part of plan bodies, which in turn are contained within the domain of the language. In contrast, formulae are syntactic constructs. Putting formulae of the object-language into the domain of the object-language would, in effect, make $\mathcal{L}$ a kind of *self-referential meta-language*, and such languages have a number of difficulties associated with them [16].

A plan body is not, in itself, of much use to an agent, as it specifies neither the circumstances under which the plan may be used, nor what it is good for. A plan body is thus very much like an undocumented fragment of program code. For this reason, we introduce *plan descriptors*, which characterize the *pre-* and *post-conditions* associated with plans. The plan descriptor associated with a plan body represents both when the plan body can be executed, and what execution of the plan body will achieve.

Definition 4. A *plan descriptor*, δ, is a binary relation $\delta \subseteq S \times S$, with the constraint that if $((w, t), (w', t')) \in \delta$, then $w = w'$. Let Δ be the set of all plan descriptors.

Plan descriptors are interpreted as follows. If $\delta \in \Delta$ is intended to characterize the behaviour of a plan body $\beta \in D_B$, then: (i) dom δ represents the set of situations from which execution of β may legally commence — intuitively, dom δ represents the *pre-condition* of β; (ii) ran δ represents the set of situations that may arise as a result of executing β from one of the situations in dom δ — intuitively, ran δ represents the *post-condition* of β; and (iii) if $(s, s') \in \delta$, then s' is a situation that could possibly arise as a result of executing β starting in situation s.

The constraint on plan descriptors, (that if $((w, t), (w', t')) \in \delta$ then $w = w'$), ensures that plan execution always happens *within* worlds, rather than *between* worlds.

This method for characterizing the pre- and post-conditions of a plan might at first sight appear to be somewhat roundabout — a more obvious approach would be to characterize these conditions as formulae of $\mathcal{L}$. However, this approach would run into exactly the same difficulties that we outlined above with respect to test actions, in that putting formulae into the domain of the language is problematic. It is worth noting that the approach we have adopted is essentially identical to the way that programs are represented in dynamic logic, where the behaviour of a program is represented as a binary relation over program states [5].

Definition 5. A *plan* is a pair (β, δ), where $\beta \in D_B$ is a plan body, and $\delta \in \Delta$ is a plan descriptor, intended to represent the behaviour of β. Let $D_\Pi = D_B \times \Delta$ be the set of all plans; we use π (with decorations: $\pi', \pi_1, \ldots$) to stand for members of D_Π. If $\pi \in D_\Pi$, then let $\hat{\beta}(\pi) \in D_B$ denote the body of π, and $\hat{\delta}(\pi) \in \Delta$ denote the descriptor in π. Thus dom $\hat{\delta}(\pi)$ represents the pre-condition of π, and ran $\hat{\delta}(\pi)$ represents the post-condition. If $s \in$ dom $\hat{\delta}(\pi)$, then let $\hat{\delta}(\pi)(s)$ denote the image of s through $\hat{\delta}(\pi)$, i.e., $\hat{\delta}(\pi)(s) = \{s' \mid (s, s') \in \hat{\delta}(\pi)\}$.

If $\beta \in D_B$ is a plan body, then we denote by $agents(\beta)$ the set of all agents that could possibly be required to perform the actions in β:

$$agents(\alpha) \stackrel{\text{def}}{=} \{Agt(\alpha)\} \qquad\qquad (\text{where } \alpha \in D_{Ac})$$
$$agents(\beta \oplus \beta') \stackrel{\text{def}}{=} agents(\beta) \cup agents(\beta') \quad (\text{where } \oplus \in \{;,|,\|\})$$
$$agents(\beta*) \stackrel{\text{def}}{=} agents(\beta)$$
$$agents(c?) \stackrel{\text{def}}{=} \emptyset.$$

2.3 Plan Execution

We now turn to the semantics of plan execution. We define a 4-place meta-level predicate $exec$, such that $exec(\beta, p, u, v)$ holds just in case the plan body $\beta \in D_B$ is executed on path p between times $u, v \in I\!N$. Formally, the $exec$ predicate is defined inductively by six equations: one each for the plan body constructors, and one for the execution of primitive actions. The first equation represents the base case, where a primitive action is executed.

$$exec(\alpha, p, u, v) \text{ iff } v = u + 1 \text{ and } Act(p(u), p(u+1)) = \alpha \quad (\text{where } \alpha \in D_{Ac}) \quad (1)$$

The second equation captures the semantics of sequential composition: $\beta; \beta'$ will be executed between times u and v iff there is some time point n between u and v such that β is executed between u and n, and β' is executed between n and v.

$$exec(\beta; \beta', p, u, v) \text{ iff } \exists n \in \{u, \ldots, v\} \text{ s.t. } exec(\beta, p, u, n) \text{ and } exec(\beta', p, n, v) \quad (2)$$

The semantics of non-deterministic choice are even simpler: $\beta \mid \beta'$ will be executed between times u and v iff either β or β' is executed between those times.

$$exec(\beta \mid \beta', p, u, v) \text{ iff } exec(\beta, p, u, v) \text{ or } exec(\beta', p, u, v) \quad (3)$$

For the execution of parallel plan bodies $\beta \parallel \beta'$, we require that both β and β' are executed over the path, with the same start and end times. The semantics of concurrency clearly represent a simplification, which we make in order to prevent the formalism becoming complicated by tangential side issues.

$$exec(\beta \parallel \beta', p, u, v) \text{ iff } exec(\beta, p, u, v) \text{ and } exec(\beta', p, u, v) \quad (4)$$

The semantics of iteration rely upon the fact that executing $\beta*$ is the same as either (i) doing nothing, or (ii) executing β once and then executing $\beta*$. This leads to the following *fixed point* equation, where the right hand side is defined in terms of the left hand side.

$$exec(\beta*, p, u, v) \text{ iff } u = v \text{ or } exec(\beta; (\beta*), p, u, v) \quad (5)$$

(The reader may like to compare this equation with the fixed-point semantics given to loops in imperative programming languages [13].) Finally, we have an equation that defines the semantics of test actions (the free variable w, that appears on the right hand side of this equation, is the world through which p is a path, and in practice this variable will always be bound).

$$exec(c?, p, u, v) \text{ iff } (w, p(u)) \in c \tag{6}$$

Finally, we mention some assumptions relating to plans. First, the notion of *soundness*. Intuitively, a plan is sound if its plan body is completely correct with respect to its plan descriptor. That is, a plan $\pi \in D_\Pi$ is sound iff whenever its body $\hat{\beta}(\pi)$ is executed from a situation (w, t) such that $(w, t) \in \text{dom } \hat{\delta}(\pi)$, it will terminate in some situation (w, t') such that $((w, t), (w, t')) \in \hat{\delta}(\pi)$. For simplicity, we shall assume that (i) all plans are sound; and (ii) plan bodies are only executed when their pre-condition holds. Intuitively, condition (ii) requires that agents only execute a plan when they *know* the pre-condition of the plan is satisfied, i.e., they are *competent* with respect to plan pre-conditions. Formally, soundness is expressed as follows: $\forall \pi \in D_\Pi$, $\forall w \in W$, $\forall p \in paths(w)$, $\forall u, v \in I\!N$, if $exec(\hat{\beta}(\pi), p, u, v)$ then $(w, p(u)) \in \text{dom } \hat{\delta}(\pi)$ and $(w, p(v)) \in \hat{\delta}(\pi)((w, p(u)))$.

3   A Logic of BDI Agents with Procedural Knowledge

In this section, we formally define our logic $\mathcal{L}$, which is an extension to the expressive branching time logic CTL* [3]. The logic builds on the work of Rao and Georgeff [9], and our own previous work in agent theory [15].

$\mathcal{L}$ contains the usual connectives and quantifiers of sorted first-order logic: we take as primitive the connectives $\neg$ (not) and $\vee$ (or), and the universal quantifier $\forall$ (for all), and define the remaining classical connectives and existential quantifier in terms of these. As $\mathcal{L}$ is based on CTL*, a distinction is made between *state formulae* and *path formulae*. The idea is that $\mathcal{L}$ is interpreted over a tree-like branching time structure. Formulae that express a property of nodes in this structure are known as *state formulae*, whereas formulae that express a property of paths through the structure are known as *path formulae*. State formulae can be ordinary first-order formulae, but various other additional modal connectives are also provided for making state formulae. Thus (Bel i φ) is intended to express the fact that the agent denoted by i believes φ (where φ is some state formula). The semantics of belief are given in terms of an accessibility relation over possible worlds, in much the standard modal logic tradition [1], with the properties required of belief accessibility relations ensuring that the logic of belief corresponds to the normal modal system KD45 (weak-S5). The state formulae (Goal i φ) and (Int i φ) mean that agent i has a desire or intention of φ, respectively: the logics of desire and intention correspond to the normal modal system KD. (Note that worlds in $\mathcal{L}$ are not instantaneous states (as in [15]), but are themselves branching time structures: the intuition is that belief accessible worlds represent an agent's uncertainty not only about how the world actually is, but also about its past and future; similarly for desires and intentions [9].)

In addition, $\mathcal{L}$ contains various connectives for representing the plans possessed by agents. The state formula (Has i π) is used to represent the fact that in the current state, agent i is in possession of the plan denoted by π. The state formulae (Pre π) and (Post π) represent the fact that the pre- and post-conditions of the plan π respectively are satisfied in the current world-state. The formula (Body π β) is used to represent the fact that β is the body of the plan denoted by π. We also have a connective (Holds c), which means that the *condition* denoted by c is satisfied in the current world state.

Turning to path formulae, (Exec β) means that the plan body denoted by β is executed on the current path. State formulae may be related to path formulae by using the CTL* *path quantifier* A. This connective means 'on all paths'. It has a dual, existential connective E, meaning 'on some path'. Thus Aφ means that the path formula φ is satisfied on all histories originating from the current world state, and Eφ means that φ is satisfied on at least one history that originates from the current world state. Path formulae may be built up from state formulae (or other path formulae) by using two *temporal connectives*: the U connectives means 'until', and so a formula $\varphi U \psi$ means 'φ is satisfied until ψ is satisfied'. The $\bigcirc$ connective means 'next', and so $\bigcirc \varphi$ means that φ will be satisfied in the next state.

3.1 Syntax

$\mathcal{L}$ is a *many sorted* logic, which permits quantification over various types of individuals: agents, actions, plans, plan bodies, sets of agents (groups), sets of situations (conditions), and other individuals in the world. All of these sorts must have a corresponding set of terms in the alphabet of the language.

Definition 6. The alphabet of $\mathcal{L}$ contains the following symbols:

1. A denumerable set *Pred* of *predicate symbols*;
2. A denumerable set *Fun* of *function symbols*, the union of the following mutually disjoint sets:
   - Fun_{Ag} — functions that return agents;
   - Fun_{Ac} — functions that return actions;
   - Fun_{Π} — functions that return plans;
   - Fun_B — functions that return plan bodies;
   - Fun_{Gr} — functions that return sets of agents (groups);
   - Fun_C — functions that return sets of situations (conditions);
   - Fun_U — functions that return other individuals.
3. A denumerable set *Var* of *variable symbols*, the union of the mutually disjoint sets Var_{Ag}, Var_{Ac}, Var_{Π}, Var_B, Var_{Gr}, Var_C, and Var_U.
4. The *operator symbols* true, Bel, Goal, Int, Agts, $=$, $\in$, A, Pre, Post, Body, Has, Holds, Exec, U, and $\bigcirc$.
5. The *classical connectives* $\vee$ (or) and $\neg$ (not), and the *universal quantifier*, $\forall$.
6. The *punctuation symbols*), (, and $\cdot$.

Associated with each predicate and function symbol is a natural number called its *arity*, given by the function *arity* : *Pred* $\cup$ *Fun* $\rightarrow$ $I\!N$. Predicates of arity 0 are known as *proposition symbols*, and functions of arity 0 are known as *constants*.

$$
\begin{array}{ll}
\langle ag\text{-}term\rangle ::= \text{any element of } Term_{Ag} & \langle \Pi\text{-}term\rangle ::= \text{any element of } Term_{\Pi} \\
\langle \beta\text{-}term\rangle ::= \text{any element of } Term_{B} & \langle gr\text{-}term\rangle ::= \text{any element of } Term_{Gr} \\
\langle c\text{-}term\rangle ::= \text{any element of } Term_{C} & \langle term\rangle ::= \text{any element of } Term \\
\langle pred\text{-}sym\rangle ::= \text{any element of } Pred & \langle var\rangle ::= \text{any element of } Var
\end{array}
$$

$\langle state\text{-}fmla\rangle ::=$

true	$\mid \langle pred\text{-}sym\rangle(\langle term\rangle,\dots,\langle term\rangle) \mid$
(Bel $\langle ag\text{-}term\rangle$ $\langle state\text{-}fmla\rangle$)	$\mid$ (Goal $\langle ag\text{-}term\rangle$ $\langle state\text{-}fmla\rangle$) $\mid$
(Int $\langle ag\text{-}term\rangle$ $\langle state\text{-}fmla\rangle$)	$\mid$ (Agts $\langle \beta\text{-}term\rangle$ $\langle gr\text{-}term\rangle$) $\mid$
($\langle term\rangle = \langle term\rangle$)	$\mid$ ($\langle ag\text{-}term\rangle \in \langle gr\text{-}term\rangle$) $\mid$
(Pre $\langle \Pi\text{-}term\rangle$)	$\mid$ (Post $\langle \Pi\text{-}term\rangle$) $\mid$
(Body $\langle \Pi\text{-}term\rangle$ $\langle \beta\text{-}term\rangle$)	$\mid$ (Has $\langle ag\text{-}term\rangle$ $\langle \Pi\text{-}term\rangle$) $\mid$
(Holds $\langle c\text{-}term\rangle$)	$\mid$ A$\langle path\text{-}fmla\rangle$ $\mid$
$\neg\langle state\text{-}fmla\rangle$	$\mid \langle state\text{-}fmla\rangle \vee \langle state\text{-}fmla\rangle$ $\mid$
$\forall\langle var\rangle \cdot \langle state\text{-}fmla\rangle$	

$\langle path\text{-}fmla\rangle ::=$

(Exec $\langle \beta\text{-}term\rangle$)	$\mid \langle state\text{-}fmla\rangle$ $\mid$
$\langle path\text{-}fmla\rangle U\langle path\text{-}fmla\rangle$	$\mid \bigcirc \langle path\text{-}fmla\rangle$ $\mid$
$\neg\langle path\text{-}fmla\rangle$	$\mid \langle path\text{-}fmla\rangle \vee \langle path\text{-}fmla\rangle$ $\mid$
$\forall\langle var\rangle \cdot \langle path\text{-}fmla\rangle$	

$\langle fmla\rangle ::= \langle state\text{-}fmla\rangle$

Fig. 3. Syntax

Definition 7. A *sort* is either *Ag*, *Ac*, Π, *B*, *Gr*, *C*, or *U*. If σ is a sort, then the set $Term_\sigma$, of *terms of sort* σ, is defined as follows:

1. if $x \in Var_\sigma$, then $x \in Term_\sigma$;
2. if $f \in Fun_\sigma$, $arity(f) = n$, and $\{\tau_1,\dots,\tau_n\} \subseteq Term$, then $f(\tau_1,\dots,\tau_n) \in Term_\sigma$

where the set *Term*, of all terms, is defined by

$$Term = \bigcup\{Term_\sigma \mid \sigma \in \{Ag, Ac, \Pi, B, Gr, C, U\}\}.$$

We use τ (with decorations: $\tau', \tau_1, \dots$) to stand for members of *Term*.

The syntax of the language is then defined by the grammar in Figure 3 (it is assumed that predicate and function symbols are applied to the appropriate number of arguments).

3.2 Semantics

In addition to the various semantic sets discussed above, the world may contain other objects (such as, for example, blocks and tables), given by the set D_U. The objects over which we can quantify in $\mathcal{L}$ together constitute a *domain*.

Definition 8. A *domain* is a structure: $D = (D_{Ag}, D_{Ac}, D_\Pi, D_B, D_{Gr}, D_C, D_U)$ where:

- $D_{Ag} = \{1, \ldots, n\}$ is a non-empty set of agents;
- $D_{Ac} = \{\alpha, \alpha', \ldots\}$ is a non-empty set of actions;
- $D_\Pi = \{\pi, \pi', \ldots\}$ is a non-empty set of plans;
- D_B is a set of plan bodies;
- $D_{Gr} = \wp(D_{Ag}) - \{\emptyset\}$ is the set of non-empty subsets of D_{Ag}, i.e., the set of agent groups over D_{Ag};
- D_C is a non-empty set of situations; and
- D_U is a non-empty set of other individuals

such that (i) all actions in elements of D_B are members of D_{Ac}; (ii) all plan bodies in elements of D_Π must be in D_B; and (iii) any plan bodies contained in elements of D_B are also in D_B. If D is a domain, then we denote by $\bar{D}$ the set $\bigcup \{D_\sigma \mid \sigma \in \{Ag, Ac, \Pi, B, Gr, C, U\}\}$. If D is a domain and $u \in I\!N$, then by $\bar{D}^u$ we mean the set of u-tuples over $\bar{D}$.

In order to interpret $\mathcal{L}$, we need various functions that associate symbols of the language with semantic objects. The first of these is an *interpretation for predicates*.

Definition 9. A *predicate interpretation*, Φ, is a function

$$\Phi : Pred \times W \times T \to \wp(\bigcup_{u \in I\!N} \bar{D}^u)$$

such that $\forall Q \in Pred, \forall n \in I\!N, \forall w \in W, \forall t \in T_w$, if $arity(Q) = n$ then $\Phi(Q, w, t) \subseteq \bar{D}^n$ (i.e., predicate interpretations preserve arity).

Definition 10. An *interpretation for functions*, F, is a second-order function

$$F : Fun \to (\bigcup_{u \in I\!N} \bar{D}^u \to \bar{D})$$

such that (i) $\forall f \in Fun, \forall n \in I\!N$, if $arity(f) = n$ then dom $F(f) \subseteq \bar{D}^n$ (i.e., function interpretations preserve arity), and (ii) F preserves sorts.

Similarly, a variable assignment associates variables with elements of the domain.

Definition 11. A *variable assignment*, V, is a function $V : Var \to \bar{D}$, such that if $x \in Var_\sigma$, then $V(x) \in D_\sigma$, (i.e., variable assignments preserve sorts).

We now introduce a derived function $[\![\ldots]\!]_{V,F}$, which gives the *denotation* of an arbitrary term.

Definition 12. If V is a variable assignment and F is a function interpretation, then by $[\![\ldots]\!]_{V,F}$, we mean the function $[\![\ldots]\!]_{V,F} : Term \to \bar{D}$, which interprets arbitrary terms relative to V and F:

$$[\![\tau]\!]_{V,F} \stackrel{\text{def}}{=} \begin{cases} F(f)([\![\tau_1]\!]_{V,F}, \ldots, [\![\tau_n]\!]_{V,F}) & \text{where } \tau \text{ is } f(\tau_1, \ldots, \tau_n) \\ V(\tau) & \text{otherwise.} \end{cases}$$

Since V and F will always be clear from context, reference to them will be suppressed. We can now define models for $\mathcal{L}$.

Definition 13. A *model*, M, for $\mathcal{L}$, is a structure

$$M = (T, R, W, D, Act, Agt, P, BR, DR, IR, F, \Phi)$$

where:

- T is the set of all time points;
- $R \subseteq T \times T$ is a total, backwards-linear branching time relation over T;
- W is a set of worlds, such that $\forall w \in W$, we have:
  1. $T_w \subseteq T$;
  2. R_w is the relation obtained from R by removing from it any arcs that contain components not in T_w;
- $D = (D_{Ag}, D_{Ac}, D_{\Pi}, D_B, D_{Gr}, D_C, D_U)$ is a domain;
- $Act : R \to D_{Ac}$ associates an action with every arc in R;
- $Agt : D_{Ac} \to D_{Ag}$ associates an agent with every action;
- $P : D_{Ag} \times W \times T \to \wp(D_{\Pi})$ gives the *plan library* of every agent in every situation;
- $BR : D_{Ag} \to \wp(W \times T \times W)$ associates with every agent a serial, transitive, euclidean *belief accessibility relation*;
- $DR : D_{Ag} \to \wp(W \times T \times W)$ associates with every agent a serial *desire accessibility relation*;
- $IR : D_{Ag} \to \wp(W \times T \times W)$ associates with every agent a serial *intention accessibility relation*;
- $F : Fun \to (\bigcup_{u \in \mathbb{N}} \bar{D}^u \to \bar{D})$ interprets functions;
- $\Phi : Pred \times W \times T \to \wp(\bigcup_{u \in \mathbb{N}} \bar{D}^u)$ interprets predicates.

The formal semantics of the language are defined in two parts, for path formulae and state formulae respectively. The semantics of path formulae are given via the path formula satisfaction relation, '$\models$', which holds between structures of the form (M, V, w, p), (where M is a model, V is a variable assignment, w is a world in M, and p is a path through w), and path formulae. The rules defining this relation are given in Figure 4. The semantics of state formulae are given via the state formula satisfaction relation, which for convenience we also write as '$\models$': context will always make it clear which relation is intended. The state formula satisfaction relation holds between structures of the form (M, V, w, t), (where M is a model, V is a variable assignment, w is a world in M, and $t \in T_w$ is a time-point in w) and state formulae. The rules defining this relation are also given in Figure 4. We assume the standard interpretation for validity. Thus a path formula φ is valid, (notation: $\models_{\mathcal{P}} \varphi$), iff for all (M, V, w, p), we have $(M, V, w, p) \models \varphi$. Similarly, a state formula φ is valid iff for all (M, V, w, t) we have $(M, V, w, t) \models \varphi$. We write $\models_S \varphi$ to indicate that the state formula φ is valid. Satisfiability for path and state formulae are defined in the obvious way.

Path Formulae Semantics

$(M, V, w, p) \models \varphi$ iff $(M, V, w, p(0)) \models \varphi$ (where φ is a state formula)

$(M, V, w, p) \models \neg\varphi$ iff $(M, V, w, p) \not\models \varphi$

$(M, V, w, p) \models \varphi \vee \psi$ iff $(M, V, w, p) \models \varphi$ or $(M, V, w, p) \models \psi$

$(M, V, w, p) \models \forall x \cdot \varphi$ iff $(M, V \dagger \{x \mapsto d\}, w, p) \models \varphi$
                for all $d \in \bar{D}$ s.t. x and d are of the same sort

$(M, V, w, p) \models \varphi U \psi$ iff $\exists u \in I\!N$ such that $(M, V, w, p^{(u)}) \models \psi$ and
                $\forall v \in I\!N$, if $(0 \leq v < u)$, then $(M, V, w, p^{(v)}) \models \varphi$

$(M, V, w, p) \models \bigcirc\varphi$ iff $(M, V, w, p^{(1)}) \models \varphi$

$(M, V, w, p) \models (\text{Exec } \beta)$ iff $\exists u \in I\!N$ such that $exec([\![\beta]\!], p, 0, u)$

State Formulae Semantics

$(M, V, w, t) \models \text{true}$

$(M, V, w, t) \models Q(\tau_1, \ldots, \tau_n)$ iff $\langle [\![\tau_1]\!], \ldots, [\![\tau_n]\!]\rangle \in \Phi(Q, w, t)$

$(M, V, w, t) \models \neg\varphi$ iff $(M, V, w, t) \not\models \varphi$

$(M, V, w, t) \models \varphi \vee \psi$ iff $(M, V, w, t) \models \varphi$ or $(M, V, w, t) \models \psi$

$(M, V, w, t) \models \forall x \cdot \varphi$ iff $(M, V \dagger \{x \mapsto d\}, w, t) \models \varphi$
                for all $d \in \bar{D}$ s.t. x and d are of the same sort

$(M, V, w, t) \models (\text{Bel } i\ \varphi)$ iff $\forall w' \in W$, if $(w, t, w') \in BR([\![i]\!])$, then $(M, V, w', t) \models \varphi$

$(M, V, w, t) \models (\text{Goal } i\ \varphi)$ iff $\forall w' \in W$, if $(w, t, w') \in DR([\![i]\!])$, then $(M, V, w', t) \models \varphi$

$(M, V, w, t) \models (\text{Int } i\ \varphi)$ iff $\forall w' \in W$, if $(w, t, w') \in IR([\![i]\!])$, then $(M, V, w', t) \models \varphi$

$(M, V, w, t) \models (\text{Agts } \beta\ g)$ iff $agents([\![\beta]\!]) = [\![g]\!]$

$(M, V, w, t) \models (\tau = \tau')$ iff $[\![\tau]\!] = [\![\tau']\!]$

$(M, V, w, t) \models A\varphi$ iff $\forall p \in paths(w)$, if $p(0) = t$, then $(M, V, w, p) \models \varphi$

$(M, V, w, t) \models (\text{Pre } \pi)$ iff $(w, t) \in \text{dom } \hat{\delta}([\![\pi]\!])$

$(M, V, w, t) \models (\text{Post } \pi)$ iff $(w, t) \in \text{ran } \hat{\delta}([\![\pi]\!])$

$(M, V, w, t) \models (\text{Body } \pi\ \beta)$ iff $\hat{\beta}([\![\pi]\!]) = [\![\beta]\!]$

$(M, V, w, t) \models (\text{Has } i\ \pi)$ iff $[\![\pi]\!] \in P([\![i]\!], w, t)$

$(M, V, w, t) \models (\text{Holds } c)$ iff $(w, t) \in [\![c]\!]$

Fig. 4. Semantics of $\mathcal{L}$

3.3 Derived Connectives

In addition to the basic connectives defined above, it is useful to introduce some *derived* constructs. These derived connectives do not add to the expressive power of the language, but are intended to make formulae more concise and readable. First, we assume that the remaining connectives of classical logic, (i.e., $\wedge$ — 'and', $\Rightarrow$ — 'if... then...', and $\Leftrightarrow$ — 'if, and only if') have been defined as normal, in terms of $\neg$ and $\vee$. Similarly, we assume that the existential quantifier, $\exists$, has been defined as the dual of $\forall$. Next, we introduce the *existential path quantifier*, E, which is defined as the dual of the universal path quantifier A. Thus a formula $E\varphi$ is interpreted as 'on some path, φ', or 'optionally, φ':

$$E\varphi \stackrel{\text{def}}{=} \neg A\neg\varphi.$$

It is also convenient to introduce further temporal connectives. The unary connective ◇ means 'sometimes'. Thus the path formula ◇φ will be satisfied on some path if φ is satisfied at some point along the path. The unary □ connective means 'now, and always'. Thus □φ will be satisfied on some path if φ is satisfied at all points along the path. We also have a weak version of the U connective: φWψ is read 'φ *unless* ψ'.

$$◇\varphi \stackrel{\text{def}}{=} \text{true}\text{U}\varphi \qquad □\varphi \stackrel{\text{def}}{=} \neg◇\neg\varphi \qquad \varphi\text{W}\psi \stackrel{\text{def}}{=} (\varphi\text{U}\psi) \lor □\varphi.$$

Thus φWψ means that either: (i) φ is satisfied until ψ is satisfied, or else (ii) φ is always satisfied. It is *weak* because it does not require that ψ be eventually satisfied.

Talking about groups: The language $\mathcal{L}$ provides us with the ability to use simple (typed) set theory to relate the properties of agents and groups of agents. The operators ⊆ and ⊂ relate groups together, and have the obvious set-theoretic interpretation; (Singleton g i) means g is a singleton group with i as the only member; (Singleton g) simply means g is a singleton.

$$(g \subseteq g') \stackrel{\text{def}}{=} \forall i \cdot (i \in g) \Rightarrow (i \in g') \quad (\text{Singleton } g \; i) \stackrel{\text{def}}{=} \forall j \cdot (j \in g) \Rightarrow (j = i)$$
$$(g \subset g') \stackrel{\text{def}}{=} (g \subseteq g') \land \neg(g = g') \quad (\text{Singleton } g) \stackrel{\text{def}}{=} \exists i \cdot (\text{Singleton } g \; i)$$

(Agt β i) means that i is the only agent required to perform plan body β.

$$(\text{Agt } \beta \; i) \stackrel{\text{def}}{=} \forall g \cdot (\text{Agts } \beta \; g) \Rightarrow (\text{Singleton } g \; i)$$

Talking about plans: Next, we introduce some operators that will allow us to conveniently represent the structure and properties of plans. First, we introduce two constructs, (Pre π φ) and (Post π φ), that allow us to represent the pre- and post-conditions of plans as formulae of $\mathcal{L}$. Thus (Pre π φ) means that φ corresponds to the pre-condition of π — that φ is satisfied in just those situations where the pre-condition of π is satisfied:

$$(\text{Pre } \pi \; \varphi) \stackrel{\text{def}}{=} \text{A}□((\text{Pre } \pi) \Leftrightarrow \varphi).$$

Similarly, (Post π φ) means that φ is satisfied in just those situations in which the post-condition of π is satisfied:

$$(\text{Post } \pi \; \varphi) \stackrel{\text{def}}{=} \text{A}□((\text{Post } \pi) \Leftrightarrow \varphi).$$

These definitions say that if (Pre π φ), then $(M, V, w, t) \models \varphi$ iff $(w, t) \in \text{dom } \hat{\delta}([\![\pi]\!])$, and if (Post π φ), then $(M, V, w, t) \models \varphi$ iff $(w, t) \in \text{ran } \hat{\delta}([\![\pi]\!])$. We write (Plan π φ ψ β) to express the fact that plan π has pre-condition φ, post-condition ψ, and body β:

$$(\text{Plan } \pi \; \varphi \; \psi \; \beta) \stackrel{\text{def}}{=} (\text{Pre } \pi \; \varphi) \land (\text{Post } \pi \; \psi) \land (\text{Body } \pi \; \beta).$$

It is often useful to be able to talk about the *structure* of plans: how their bodies are put together, in terms of the constructors ;, |, ||, and so on. In order to do this, we introduce some logical functions (i.e., functions denoted by elements of the set *Fun*). We introduce one function for each of the plan constructors:

$$seq \text{ for } ; \quad par \text{ for } \| \quad test \text{ for } ? \quad or \text{ for } | \quad iter \text{ for } *.$$

We require that these functions satisfy certain properties. For example, for all $\beta, \beta' \in Term_B$, we require that $seq(\beta, \beta')$ returns the plan body $[\![\beta]\!]; [\![\beta']\!]$, i.e., that $seq(\beta, \beta')$ returns the plan body obtained by conjoining the plan bodies denoted by β and β' with the sequential composition constructor. Similarly, we require than $par(\beta, \beta')$ returns $[\![\beta]\!] \| [\![\beta']\!]$, that $or(\beta, \beta')$ returns $[\![\beta]\!] | [\![\beta']\!]$, that $iter(\beta)$ returns $[\![\beta]\!]*$, and finally, that $test(c)$ returns $[\![c]\!]?$, for all $c \in Term_C$. These functions allow us to construct plan bodies within our language. However, complex plan bodies written out in full using these functions become hard to read. To make such expressions more readable, we introduce a *quoting convention*. The idea is best illustrated by example. We write

$$\ulcorner \beta; \beta' \urcorner \qquad \text{to abbreviate } seq(\beta, \beta')$$
$$\ulcorner \beta; (\beta' \| \beta'') \urcorner \quad \text{to abbreviate } seq(\beta, par(\beta', \beta''))$$
$$\ulcorner \beta; (\beta' \| \beta'')* \urcorner \text{ to abbreviate } seq(\beta, iter(par(\beta', \beta'')))$$

and so on. In the interests of consistency, we shall generally use quotes even where they are not strictly required.

Next, we introduce a construct that makes test actions more readable. Let $c \in Term_C$ be a term denoting a situation set, (i.e., a condition), and let φ be a state formula. Then $(c \equiv \varphi)$ represents the fact that φ is satisfied in just those situations denoted by c:

$$(c \equiv \varphi) \stackrel{\text{def}}{=} \mathsf{A} \Box ((\mathsf{Holds}\ c) \Leftrightarrow \varphi).$$

Hereafter, instead of writing c? we write φ?, where it is understood that $(c \equiv \varphi)$. Thus, when we write $(\mathsf{Exec}\ \ulcorner (\mathsf{Bel}\ i\ p)? \urcorner)$, it should be understood that this abbreviates $\forall c \cdot (c \equiv (\mathsf{Bel}\ i\ p)) \Rightarrow (\mathsf{Exec}\ \ulcorner c? \urcorner)$. Any formula abbreviated in this way may be systematically rewritten into the fully expanded form. Note that one property of this style of abbreviation is $\models_{\mathcal{P}} (\mathsf{Exec}\ \ulcorner \varphi? \urcorner) \Leftrightarrow \varphi$.

The readability of plan body expressions may be further improved by the introduction of derived constructs corresponding to the high-level statement-types one would expect to find in a standard imperative language such as PASCAL. First, the *if... then...* construct:

$$\ulcorner \mathtt{if}\ \varphi\ \mathtt{then}\ \beta\ \mathtt{else}\ \beta' \urcorner \stackrel{\text{def}}{=} \ulcorner (\varphi?; \beta) | (\neg\varphi; \beta') \urcorner.$$

While and *repeat* loops are similarly easy to define:

$$\ulcorner \mathtt{while}\ \varphi\ \mathtt{do}\ \beta \urcorner \stackrel{\text{def}}{=} \ulcorner (\varphi?; \beta)*; \neg\varphi? \urcorner$$
$$\ulcorner \mathtt{repeat}\ \beta\ \mathtt{until}\ \varphi \urcorner \stackrel{\text{def}}{=} \ulcorner \beta; \mathtt{while}\ \neg\varphi\ \mathtt{do}\ \beta \urcorner.$$

Finally, we define an await construct:

$$\ulcorner \mathtt{await}\ \varphi \urcorner \stackrel{\text{def}}{=} \ulcorner \mathtt{repeat}\ \mathsf{true}?\ \mathtt{until}\ \varphi \urcorner.$$

Thus await φ will be executed on a path p if there is some point on p at which φ is true. There is thus a close relationship between await and the temporal 'sometimes' connective: $\models_{\mathcal{P}} (\mathsf{Exec}\ \ulcorner \mathtt{await}\ \varphi \urcorner) \Leftrightarrow \Diamond\varphi$.

3.4 Some Properties of $\mathcal{L}$

After introducing a new logic by means of its syntax and semantics, it is usual to illustrate its properties, typically by means of a Hilbert-style axiom system. However, no complete axiomatization is currently known for CTL*, the logic that underpins $\mathcal{L}^2$. For this reason, instead of attempting a complete axiomatization, we simply identify some valid formulae of $\mathcal{L}$, focusing in particular on plan execution. First, notice that the semantics of $\mathcal{L}$ generalize those of sorted first-order logic, and hence, in turn, propositional logic. Thus $\mathcal{L}$ admits propositional and sorted first-order reasoning, as one might expect. In addition, the semantics of the BDI component of $\mathcal{L}$ ensure that axioms corresponding to the normal modal system KD45 (weak-S5) are valid for the Bel modalities, and axioms corresponding to the normal modal system KD are valid for Goal and Int modalities. Rao and Georgeff prove that these axioms together constitute a sound and complete axiomatization of this 'basic BDI system' [11]. With respect to the CTL* component of the logic, it is not difficult to see that the axioms one would expect of CTL* are valid in $\mathcal{L}$ [14].

Turning to plans, it is not difficult to see that the logic of plan execution is similar to that of many program logics. For example, one can show that if the plan body $\ulcorner\beta \mid \beta'\urcorner$ is executed, then the pre-condition of β or the pre-condition of β' must hold prior to execution:

$$\models_{\mathcal{P}} (\text{Plan}\,\pi\,\varphi\,\psi\,\beta) \wedge (\text{Plan}\,\pi'\,\varphi'\,\psi'\,\beta') \wedge (\text{Exec}\,\ulcorner\beta \mid \beta'\urcorner) \Rightarrow (\text{Exec}\,\ulcorner(\varphi\vee\varphi')?; (\beta \mid \beta')\urcorner)$$

As a corollary, one can show that agents believe that plans behave in this way:

$$\models_{\mathcal{S}} (\text{Bel}\,i\,(\text{Plan}\,\pi\,\varphi\,\psi\,\beta)) \wedge (\text{Bel}\,i\,(\text{Plan}\,\pi'\,\varphi'\,\psi'\,\beta')) \wedge (\text{Bel}\,i\,\text{A}(\text{Exec}\,\ulcorner\beta \mid \beta'\urcorner)) \Rightarrow$$
$$(\text{Bel}\,i\,\varphi \vee \varphi').$$

In a similar way, one can prove various properties of derived constructs such as `while` loops and `if` statements.

4 Concluding Remarks

It is now widely accepted that the technology of multi-agent systems will play a key role in the development of future distributed systems. As the use of multi-agent technology becomes more commonplace, so the need for a firm theoretical foundations for it will grow. In this paper, we hope to have contributed to such a foundation, by presenting a new logic that can be used to give an abstract semantics to a significant class of intelligent agent architectures. In these so-called BDI architectures, the internal state of an agent is characterised by symbolic data structures loosely corresponding to beliefs, desires, and intentions. In addition, such agents have available to them a library of plans, representing their 'know-how': procedural knowledge about how to achieve goals.

Acknowledgments: This work was carried out while I was a visiting researcher at Daimler-Benz research institute in Berlin. I would like to thank Kurt Sundermeyer for arranging the visit, and in particular, Stefan Bussmann, Michael Georgeff, Afsaneh Haddadi, and Anand Rao, who all read and carefully commented on earlier drafts of this paper.

---

[2] The system presented in [14] reportedly contains an error in the proof of completeness.

References

1. B. Chellas. *Modal Logic: An Introduction*. Cambridge University Press: Cambridge, England, 1980.
2. P. R. Cohen and H. J. Levesque. Intention is choice with commitment. *Artificial Intelligence*, 42:213–261, 1990.
3. E. A. Emerson and J. Y. Halpern. 'Sometimes' and 'not never' revisited: on branching time versus linear time temporal logic. *Journal of the ACM*, 33(1):151–178, 1986.
4. M. P. Georgeff and A. L. Lansky. Reactive reasoning and planning. In *Proceedings of the Sixth National Conference on Artificial Intelligence (AAAI-87)*, pages 677–682, Seattle, WA, 1987.
5. D. Harel. Dynamic logic. In D. Gabbay and F. Guenther, editors, *Handbook of Philosophical Logic Volume II — Extensions of Classical Logic*, pages 497–604. D. Reidel Publishing Company: Dordrecht, The Netherlands, 1984. (Synthese library Volume 164).
6. C. A. R. Hoare. Communicating sequential processes. *Communications of the ACM*, 21:666–677, 1978.
7. D. Kinny, M. Ljungberg, A. S. Rao, E. Sonenberg, G. Tidhar, and E. Werner. Planned team activity. In C. Castelfranchi and E. Werner, editors, *Artificial Social Systems — Selected Papers from the Fourth European Workshop on Modelling Autonomous Agents in a Multi-Agent World, MAAMAW-92 (LNAI Volume 830)*, pages 226–256. Springer-Verlag: Heidelberg, Germany, 1992.
8. A. S. Rao and M. Georgeff. BDI Agents: from theory to practice. In *Proceedings of the First International Conference on Multi-Agent Systems (ICMAS-95)*, pages 312–319, San Francisco, CA, June 1995.
9. A. S. Rao and M. P. Georgeff. Modeling rational agents within a BDI-architecture. In R. Fikes and E. Sandewall, editors, *Proceedings of Knowledge Representation and Reasoning (KR&R-91)*, pages 473–484. Morgan Kaufmann Publishers: San Mateo, CA, April 1991.
10. A. S. Rao and M. P. Georgeff. An abstract architecture for rational agents. In C. Rich, W. Swartout, and B. Nebel, editors, *Proceedings of Knowledge Representation and Reasoning (KR&R-92)*, pages 439–449, 1992.
11. A. S. Rao and M. P. Georgeff. Formal models and decision procedures for multi-agent systems. Technical Note 61, Australian AI Institute, Level 6, 171 La Trobe Street, Melbourne, Australia, June 1995.
12. A. S. Rao, M. P. Georgeff, and E. A. Sonenberg. Social plans: A preliminary report. In E. Werner and Y. Demazeau, editors, *Decentralized AI 3 — Proceedings of the Third European Workshop on Modelling Autonomous Agents in a Multi-Agent World (MAAMAW-91)*, pages 57–76. Elsevier Science Publishers B.V.: Amsterdam, The Netherlands, 1992.
13. D. A. Schmidt. *Denotational Semantics*. Allyn and Bacon: Newton, MA, 1986.
14. C. Stirling. Completeness results for full branching time logic. In *REX School-Workshop on Linear Time, Branching Time, and Partial Order in Logics and Models for Concurrency*, Noordwijkerhout, Netherlands, 1988.
15. M. Wooldridge. Coherent social action. In *Proceedings of the Eleventh European Conference on Artificial Intelligence (ECAI-94)*, pages 279–283, Amsterdam, The Netherlands, 1994.
16. M. Wooldridge and N. R. Jennings. Intelligent agents: Theory and practice. *The Knowledge Engineering Review*, 10(2):115–152, 1995.

Towards the Assessment of Logics for Concurrent Actions

Choong-Ho Yi
Department of Computer Science
University of Karlstad
S-651 88 Karlstad, Sweden
E-mail: Choong-ho.Yi@hks.se

Abstract. We have introduced concurrency into the framework of Sandewall. The resulting formalism is capable of reasoning about interdependent as well as independent concurrent actions. Following Sandewall's systematical method, we have then applied the entailment criterion PCM to selecting intended models of common sense theories where concurrent actions are allowed, and proved that the criterion leads to only intended models for a subset of such theories.

1 Introduction

With restriction to the case where actions are assumed to occur sequentially, a number of nonmonotonic logics have been proposed in AI. In the meanwhile, research has advanced and one began to investigate model-theoretically whether a logic at hand produces conclusions correctly for a given theory. Sandewall [San94] introduces a new approach in this context. For each of the major logics presented by then he identified a corresponding class of reasoning problems for which the logic is *proved* to obtain exactly the intended conclusions, i.e. the *range of applicability of the logic*.

On the other hand, logics have been suggested by, e.g. Kowalski & Sergot [KS86], Allen [All91], Pelavin [Pel91], Lansky [Lan90], Georgeff [Geo86], and more recently, by Grosse & Khalil [GKar] and Thielscher [Thi95] which, directly or indirectly, allow concurrency. As was the case in reasoning with sequential actions, the importance and the need to identify the range of applicability of a given logic could not be emphasized too much even when concurrent actions are allowed. By this time, however, there has not been reported any systematic result in that direction. Actually, in contrast to the case for sequential actions where several entailment criteria, e.g. chronologically maximal ignorance [Sho88], have been proposed for selecting intended models, no such method has been tested for dealing with concurrent actions.

The work presented in this paper is an approach "in that direction" which has been done based directly on Sandewall. By making necessary generalizations we have introduced concurrent actions into his framework which was restricted to sequential actions. The resulting formalism is capable of reasoning about interdependent as well as independent concurrent actions. Then as a first step, we picked out the simplest entailment criterion PCM (prototypical chronological minimization of change) of Sandewall, allowed independent concurrent actions into the class of reasoning problems for which he had proven PCM to be correct, and proved that PCM is still correct for this extended class.

2 Sandewall's Systematical Approach

In Sandewall's approach[1], common-sense reasoning is understood on the basis of an *underlying semantics* which views the interaction between the ego of an intelligent agent and a world as a *game*, and which characterizes the actions the agent may evoke during the game in terms of a *trajectory semantics*. The *inertia problem* is approached by building inertia into the underlying semantics, i.e. the world has inertia so that features remain unchanged unless actions which override the inertia are performed.

2.1 The Game

The game is made in terms of a *finite development*

$$\langle \mathcal{B}, M, R, \mathcal{A}, \mathcal{C} \rangle.$$

$\mathcal{B}$ is a set of integers representing the time points at which the ego and the world alternate in the game and the largest member n of the set is "now". M assigns values to temporal constants and object constants. R is a mapping from a set $\{0, \ldots, n\}$ of time points to a set $\mathcal{R}$ of states, i.e. R is a history of the world up to n. The pair $\langle M, R \rangle$ then constitutes an interpretation for a given object domain $\mathcal{O}$. $\mathcal{A}$ is a set of tuples $\langle s, E, t \rangle$ where s and t are start respective end time of the action E and $s < t \leq n$, i.e. a set of actions which have been terminated at time n. $\mathcal{C}$ is a set of tuples $\langle s, E \rangle$ where s is start time and $s \leq n$, i.e. a set of actions which have been started but not terminated yet at time n. The tuple $\langle \mathcal{B}, M, R, \mathcal{A}, \mathcal{C} \rangle$ works as a "game board" in the game; the ego and the world alternate and extend it such that, roughly, the world executes the actions which are evoked by the ego.

2.2 The Trajectory Semantics

The *trajectory semantics* characterizes actions in terms of two functions. The function $\mathtt{Infl}(E, r)$ represents a set of features which may be affected if the action E is performed in the state r. The function $\mathtt{Trajs}(E, r)$ represents a set of possible *trajectories* of E initiated in r, where a trajectory, written as v, is expressed as a finite sequence

$$\langle r'_1, \ldots, r'_k \rangle$$

of partial states r'_i ($1 \leq i \leq k$) each of which assigns values to exactly those features appearing in $\mathtt{Infl}(E, r)$. This sequence is a trajectory of the action of the

---

[1] The presentation in this section is mainly based on [San93].

form $[s, s+k]E$, where s is start point in time and $s+k$ end point, and describes the effects of E successively for ecah time point during the execution period. Therefore, in the trajectory semantics one cares not only about the results of an action, but also about its trajectories. Since it is the world which performs actions, the pair $\langle \text{Infl}, \text{Trajs} \rangle$ characterizes a world. Later, ego and world will be defined exactly in a trajectory semantics generalized for concurrency.

2.3 Commonsense Scenarios

A commonsense theory is expressed as a tuple

$$\langle \mathcal{O}, \Pi, \text{SCD}, \text{OBS} \rangle.$$

$\mathcal{O}$ is an object domain. Π is a set of formulae describing the effects of actions, e.g.

$$[s, t] \, OpenWindow \Rightarrow [t] \, Window \doteq Open$$

which means if the *OpenWindow* action happens over the time interval s to t, then the feature *Window* representing the openness of the window has value *Open* at time t. Actually, this set is an exhaustive description of Trajs in logical formulae. SCD represents the actions scheduled to be performed, and is a set of action statements, e.g. $[3, 5] \, OpenWindow$, and time statements, e.g. $t_1 < s_2$. The effect $[5] \, Window \doteq Open$ of performing $[3, 5] \, OpenWindow$ is then obtained by applying the action statement to Π. OBS is a set of observation statements, i.e. any formulae not containing action statements.

2.4 Intended Models

If a scenario $\Upsilon = \langle \mathcal{O}, \Pi, \text{SCD}, \text{OBS} \rangle$ is given, then the set of *intended models* of Υ is defined as follows. First, select an arbitrary world which is exactly characterized by Π, select an arbitrary ego, an arbitrary initial state and an arbitrary initial mapping for temporal and object constants. Let $Mod(\Upsilon)$ be a set of completed developments $\langle \mathcal{B}, M, R, \mathcal{A}, \mathcal{C} \rangle$ obtained from games between them over Υ such that there is a 1:1 correspondence between members of the set $\mathcal{A}$ and those of SCD (i.e. all of the scheduled actions have been performed successfully), and all formulae in SCD $\cup$ OBS are true in $\langle M, R \rangle$ having $\mathcal{O}$ as object domain. Then,

$$\{\langle M, R \rangle | \langle \mathcal{B}, M, R, \mathcal{A}, \mathcal{C} \rangle \in Mod(\Upsilon)\}$$

is the set of intended models of Υ.

2.5 Taxonomy of Reasoning Problems

One of the characteristics of Sandewall's systematic approach is the use of *taxonomy* of reasoning problems. The taxonomy is obtained by making explicitly *ontological assumptions* about actions and world and *epistemological assumptions* about knowledge about the actions and the world to be reasoned with. For example, the ontological characteristic **I** represents that inertia holds; **A** represents "alternative results", i.e. the effects of an action are conditional on the starting state; **C** represents that concurrent actions are allowed; **D** represents dependencies between features, i.e. change in one feature implies possibility of immediate change in another feature; and so on. The classical frame problem is then denoted as **IA**, and the ramification problem is in the **IAD** ontological family.

In addition, for a more precise specification, he provides *sub-characteristics* which are additional constraints within characteristics and which are written with small letters. For example, **Is** represents the subfamily of **I** where all actions take a single time step; **An** denotes the subfamily of **A** where all features which are allowed to be influenced as result of an action in a given state should change their value if the action is performed in that state: if a feature with three possible values red, yellow, green is influenced by an action, then the action is allowed to nondeterministically change the value from red to yellow or red to green, but it is not allowed to choose between switching from red to green or keeping it red; and so on. All of the sub-characteristics can be defined precisely in terms of the trajectory semantics.

In order to characterize the epistemological assumptions, a list of epistemological characteristics is provided. For example, $\mathcal{K}$ denotes complete and correct knowledge about actions. $\mathcal{K}$p represents that in addition there are no observations about any time point after the initial one. Therefore $\mathcal{K}$p denotes pure prediction problem.

The ontological and the epistemological descriptors are then combined and characterize a class of systems or reasoning problems. For example, the combination $\mathcal{K}$p−**IsAn** represents a set of reasoning problems satisfying the restrictions **IsAn** and $\mathcal{K}$p. Such combinations are used for identifying the applicability of different logics. That is, the correctness of a logic is defined for a class of reasoning problems in terms of equality between the set of intended models and the set of preferred models.

2.6 Assessment of PCM

The entailment criteria PCM has been formalized by Sandewall as follows. Let $I = \langle M, R \rangle$ be an interpretation, then the *breakset* of I at time t is defined as a set of features which change value from time $t - 1$ to t; formally

$$breakset(I, t) = \{f_i \mid R(f_i, t - 1) \neq R(f_i, t)\}.$$

Definition [San94] Let $I = \langle M, R \rangle$ and $I' = \langle M', R' \rangle$ be interpretations, then I is said to be *PCM-preferred* over I', written as $I \ll_{pcm} I'$, iff $M = M'$ and there is some time point t such that

- $R(f, t) = R'(f, t)$ for all features f in ν and for all time points $t < \mathrm{t}$, and

- $breakset(I, \mathrm{t}) \subset breakset(I', \mathrm{t})$. □

Sandewall has shown that PCM guarantees only intended models for reasoning problems within the class $\mathcal{K}p-\textbf{IsAn}$ described above. For the detailed discussion and the full proofs, please refer to [San94].

3 Concurrency in the Trajectory Semantics

In Sandewall [San94] the trajectory semantics was defined with the restriction that only sequential actions are allowed, i.e. at most one action is considered at a time. However, in dealing with concurrent actions new problems arise which were not there for sequential actions. Concurrent actions imply that at least two actions are involved at a given time, and, consequently, that interactions may arise between them. Therefore, the semantics must be modified.

3.1 Concurrent Interactions

In a broad sense, concurrent actions may be *interacting* or *noninteracting*, and, if they interact, they may interact *interferingly* or *noninterferingly*. Given two or more actions, one cannot say unconditionally whether they interact or not, and if they do, whether they interfere or not. The behaviour of individual actions in these respects is dependent on their *start states* and the *trajectories chosen* for them. The set of features influenced by executing an action in a state may be different if the action occurs in another state. What it means is that any two overlapping actions E_1 and E_2 which influence no feature in common if they start in state r_1 and r_2 respectively, i.e.

$$\texttt{Infl}(E_1, r_1) \cap \texttt{Infl}(E_2, r_2) = \emptyset,$$

can easily show different effects if performed in different states, e.g. such that

$$\texttt{Infl}(E_1, r_1) \cap \texttt{Infl}(E_2, r_3) \neq \emptyset.$$

For another example, let

$$\{v_1, v_2\} \subseteq \texttt{Trajs}(E_1, r_i)$$
$$\{v_3, v_4\} \subseteq \texttt{Trajs}(E_2, r_j).$$

Then it may be the case that the trajectory v_1 interacts with v_3, and, also with v_4 but differently than with v_3, while there is no interaction between v_2 and v_4.

However, as will be discussed when we define concurrent interactions formally, the interactions are relative to the *start time points* of actions as well.

3.2 Trajectory Preserving Condition

As mentioned in our previous discussion, the function $\texttt{Trajs}(E, r)$ captures the set of possible trajectories of the action E w.r.t. its starting state r. In the case of sequential actions, it was enough to say merely that there are several ways for a given action to go. In discussing about concurrent actions, one also needs to know the *conditions under which each trajectory proceeds as such*, since, unless these conditions are available, interactions between trajectories of concurrent actions cannot be represented and reasoned about effectively. What is missing in this function is to represent such conditions for every trajectory $v \in \texttt{Trajs}(E, r)$.

Definition Let E be an action and r a state, then each trajectory $v \in \texttt{Trajs}(E, r)$ is defined now as a pair

$$\langle\langle r'_1, \ldots, r'_k \rangle, \langle r''_1, \ldots, r''_k \rangle\rangle$$

of finite sequences of partial states where $\langle r'_1, \ldots, r'_k \rangle$ is a *trajectory description* and is our "old" trajectory, and $\langle r''_1, \ldots, r''_k \rangle$ is *trajectory preserving condition*. If the action E is started at time point s, each r''_i ($1 \leq i \leq k$) in the trajectory preserving condition specifies conditions to hold at time point $s + i$ in order for the action E to proceed as described by $\langle r'_1, \ldots, r'_k \rangle$. □

The trajectory description will be written as d, and the trajectory preserving condition as pc. The trajectory preserving condition is a generalization of the *prevail condition* of Sandewall & Rnnquist [SR86] in that both two refer to conditions that should hold during an action performance. The difference is that their prevail condition only represents conditions that should hold during the *whole duration* of an action, while the trajectory preserving condition can freely refer to conditions not solely for the whole duration but also for some parts of it, or even for a time point.

3.3 Concurrent Interactions in terms of Infl and pc

In our formalism, concurrent interactions are considered at the level of *concurrent trajectories*, i.e. the trajectories of concurrent actions. Concurrent trajectories can interact in two ways, namely, by influencing some feature in common, or by influencing a feature which appears in the preserving condition of the other trajectory. The formal definition follows. For the forthcoming discussion, we introduce some notations first. For a given trajectory $\langle d, pc \rangle = \langle\langle r'_1, \ldots, r'_m \rangle, \langle r''_1, \ldots, r''_m \rangle\rangle$, $d(k)$ and $pc(k)$ shall be the k:th ($1 \leq k \leq m$) element of the trajectory description d and that of the trajectory preserving condition pc, respectively. Similarly, $d(f, k)$ and $pc(f, k)$ will be used to express the value of a feature f defined in $d(k)$ respectively $pc(k)$. Let $v = \langle d, pc \rangle$, then $length(v)$ shall be the length of time period over which the trajectory v proceeds, i.e. m. In addition, by $\mathcal{F}(r)$ we denote the set of features which are defined in a given state r.

Definition Given two arbitrary actions of the form $[s_1, t_1]E_1$ and $[s_2, t_2]E_2$ such that $max(s_1, s_2) < min(t_1, t_2)$, i.e. they are concurrent actions, let R be an arbitrary history defined over $[0, s]$ where $s \geq max(s_1, s_2)$, and, for $1 \leq i \leq 2$, let $R(s_i) = r_i$, let $v_i = \langle d_i, pc_i \rangle$ be a member of $\texttt{Trajs}(E_i, r_i)$ of length $t_i - s_i$, let $x_i = max(s_1, s_2) - s_i + 1$ and $y_i = min(t_1, t_2) - s_i$, i.e. x_i and y_i are intended to represent the first respectively the last moment at which v_i might interact with the other trajectory. Then v_1 and v_2 are said to

- Infl-*interact* iff
  - $\text{Infl}(E_1, r_1) \cap \text{Infl}(E_2, r_2) \neq \emptyset$;

- *pc-interact* iff
  - $\text{Infl}(E_1, r_1) \cap \mathcal{F}(pc_2(k_2)) \neq \emptyset$
    for some k_2 where $x_2 \leq k_2 \leq y_2$ or
  - $\text{Infl}(E_2, r_2) \cap \mathcal{F}(pc_1(k_1)) \neq \emptyset$
    for some k_1 where $x_1 \leq k_1 \leq y_1$. □

Notice that $\text{Infl}(E_1, r_1)$ and $\text{Infl}(E_2, r_2)$ represent the set of features defined in the elements of d_1 respectively d_2, and that the interactions in the above definition are relative to the choice of s_1 and s_2. Additionally, let

$$f_1 \in \text{Infl}(E_1, r_1) \cap \text{Infl}(E_2, r_2),$$
$$f_2 \in (\text{Infl}(E_1, r_1) \cap \mathcal{F}(pc_2(k_2))) \cup$$
$$(\text{Infl}(E_2, r_2) \cap \mathcal{F}(pc_1(k_1))).$$

Then, for any f_1 and f_2, v_1 and v_2 are said to Infl-interact through f_1 and *pc*-interact through f_2, respectively. Therefore, concurrent trajectories may Infl-, or *pc*-interact, or both of the two.

Based on these concepts, we can go on and identify clearly interferences too. Infl-interacting trajectories interfere iff they assign at some overlapping time point different values to some feature through which they Infl-interact. And the *pc*-interacting interfere iff one trajectory assigns at some overlapping time point a different value from the trajectory preserving condition of the other to some feature through which they *pc*-interact. The precise definition is as follows.

Definition Let the same assumptions be given as in the previous definition, and let k_1 and k_2 be any time points satisfying $x_1 \leq k_1 \leq y_1$, $x_2 \leq k_2 \leq y_2$ and $s_1 + k_1 = s_2 + k_2$. Then the trajectories v_1 and v_2 are said to

- Infl-*interfere* iff
  - they Infl-interact through some feature f
    and
  - $d_1(f, k_1) \neq d_2(f, k_2)$ for some k_1 and k_2;

- *pc-interfere* iff
  - they *pc*-interact through some feature f and
  - $d_1(f, k_1) \neq pc_2(f, k_2) \lor d_2(f, k_2) \neq pc_1(f, k_1)$
    for some k_1 and k_2. □

In addition, independent actions are defined trivially such that they neither Infl-interact nor *pc*-interact.

3.4 An Example

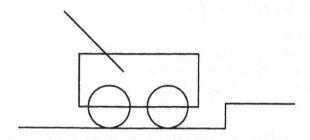

Fig. 1. A cart in front of the curb.

As an example, consider a situation shown in Figure 1 where a cart is standing in front of the curb, and we want to move it over the curb. In order to do that, you should first press down the handle while the front wheels go over the curb and then lift the handle while the back wheels go over. The vertical position of the front wheels and the back wheels is represented in relation to the curb by the features VpF repectively VpB which have as their value domain

$$\{on, lifted\}.$$

Therefore, the wheels may be *on* the ground, under or over the curb, or *lifted* higher than the curb so as to pass it freely. Similarly, HpF and HpB whose value domain is

$$\{before, passed, elsew\},$$

state the horizontal position of the front and the back wheels in relation to the curb. *before* means that the front or the back wheels are directly before the curb and ready to go over, *passed* they passed it, and *elsew* the wheels are elsewhere before the curb but not directly in front of it. Let PH denote the action "press handle" whose intended effect is to lift the front wheels. The action PC, "push cart", has the effect of moving the cart over the curb while its wheels are lifted. By restricting to these actions and some states which would characterize the actions well, let us briefly consider about concurrent interactions in the trajectory semantics.

Let r_1 be any state which satisfies

$$r_1 \supset \{VpF \hat{=} on, VpB \hat{=} on\},$$

and let

$$\mathrm{Infl}(PH, r_1) = \{VpF\}.$$

It means, pressing down the handle of the cart when the wheels are on the ground, can influence the vertical position of the front wheels. Then consider a trajectory $v_1 = \langle d_1, pc_1 \rangle \in \text{Trajs}(PH, r_1)$ where

$$d_1 = \langle \{ VpF \hat{=} lifted \}, \{ VpF \hat{=} lifted \},$$
$$\{ VpF \hat{=} lifted \}, \{ VpF \hat{=} lifted \} \rangle$$
$$pc_1 = \langle \emptyset, \emptyset, \emptyset, \emptyset \rangle.$$

That is, a possible trajectory of the action PH initiated in r_1 is that it proceeds over 4 time units and holds the front wheels lifted over the interval, i.e. $d_1(VpF, 1) = \ldots = d_1(VpF, 4) = lifted$. No trajectory preserving condition is required. Of course, $\text{Trajs}(PH, r_1)$ may also contain other trajectories. For convenience, however, we assume that v_1 is its only member.

Next, suppose a state r_2 such that

$$\{ VpF \hat{=} lifted, HpF \hat{=} before, VpB \hat{=} on, HpB \hat{=} elsew \}.$$

Then, for the action PC we consider in a similar way a trajectory $v_2 = \langle d_2, pc_2 \rangle$ of $\text{Trajs}(PC, r_2)$ where

$$\text{Infl}(PC, r_2) = \{ HpF, HpB \}$$
$$d_2 = \langle \{ HpF \hat{=} before, HpB \hat{=} elsew \},$$
$$\{ HpF \hat{=} passed, HpB \hat{=} elsew \},$$
$$\{ HpF \hat{=} passed, HpB \hat{=} before \} \rangle,$$
$$pc_2 = \langle \{ VpF \hat{=} lifted \}, \{ VpF \hat{=} lifted \}, \emptyset \rangle.$$

According to the trajectory v_2, pushing the cart in r_2 would proceed as follows; if the front wheels continue to be held lifted over the curb over $[s+1, s+2]$ where s is start time of PC, i.e. $pc_2(VpF, 1) = pc_2(VpF, 2) = lifted$, then the cart rolls on back wheels so that the front wheels pass the curb at $[s+2]$, i.e. $d_2(HpF, 2) = passed$, and the back wheels reach the curb at $[s+3]$, i.e. $d_2(HpB, 3) = before$. Here, too, v_2 is assumed to be the only member of $\text{Trajs}(PC, r_2)$.

Now, let r be the initial state

$$\{ VpF \hat{=} on, HpF \hat{=} before, VpB \hat{=} on, HpB \hat{=} elsew \}$$

pictured in Figure 1, and let 0 be initial time point. In addition suppose that we press the handle of the cart over the interval $[0, 4]$ and push the cart concurrently over $[1, 4]$. Since r satisfies the condition of r_1, $v_1 \in \text{Trajs}(PC, r)$. Let v_1 be chosen for $[0, 4] PH$. (We need to and will discuss in detail about choosing trajectories for given actions in next section.) By starting the trajectory v_1 from time 0, the front wheels are lifted at succeeding time point 1, and this is the only change caused by v_i at time 1; recall $d_1(1) = \{ VpF \hat{=} lifted \}$. Therefore the state of the world is changed from r to r_2 over $[0, 1]$. And so, let the trajectory v_2 be selected for $[1, 4] PC$. The concurrent trajectories v_1 and v_2 interact, namely pc-interact through the feature VpF, but not interfere. Actually, v_1 enables v_2 such that the trajectory preserving condition pc_2, i.e. $pc_2(1)$ and $pc_2(2)$, is satisfied by $d_1(2)$ and $d_1(3)$.

4 Trajectory Semantics World and Ego

For dealing with concurrency in the trajectory semantics, the *single-timestep ego-world game* where the world advances time by exactly one time step at a time, is adopted. It offers a clear and simple underlying semantics, and reduces the technical complexity. Following Sandewall, trajectory semantics world and ego are defined as follows. Formal definitions are given in [Yi95].

4.1 Trajectory Semantics World in a Single-timestep Ego-World Game

As mentioned previously, the ego-world game is performed in terms of a finite development $\langle \mathcal{B}, M, R, \mathcal{A}, \mathcal{C} \rangle$. Let a world description $\langle \mathtt{Trajs}, \mathtt{Infl} \rangle$ be given, and let $\langle \mathcal{B}, M, R, \mathcal{A}, \mathcal{C} \rangle$ be a development given for a single-timestep game between a trajectory semantics world and a trajectory semantics ego where the "now" time, i.e. $max(\mathcal{B})$, is n, and the history R is defined over $[0, n]$. Assume that the world takes over the control now, and that the world modifies the development into $\langle \mathcal{B}', M', R', \mathcal{A}', \mathcal{C}' \rangle$. As we will see, the modification is made differently according to whether the current action set $\mathcal{C}$ is empty or not. However, the following hold irrespective of it; $\mathcal{B}' = \mathcal{B} \cup n+1$, i.e. the now-time is increased by one time point, $M \subseteq M'$, and the restriction of R' to the period $[0, n]$ equals R.

If $\mathcal{C} = \emptyset$, then it means that no action is going on at time n. It's because all actions being processed have "died out", and because the ego has decided not to start any new action at this moment. In this case, the world extends history such that $R'(n + 1) = R(n)$, and $\mathcal{A}'$ and $\mathcal{C}'$ are set to $\mathcal{A}$ and $\emptyset$ respectively.

On the other hand, at n there may be an arbitrary number of actions to be considered, i.e. $\mathcal{C}$ has an arbitrary number of members $\langle s_i, E_i \rangle$ where s_i is start time of E_i and $s_i \leq n$. Here, it's not sure whether all members of $\mathcal{C}$ can be performed concurrently. For example, it may be that some actions are evoked at n but interfere with other actions which have been started previously. When conflict arises between concurrent actions, one may choose to perform as many actions as possible, or break the game there, or abandon all interfering ones, or save earlier actions first, and so on. However, rather than choose a specific "policy" among them, we leave our underlying semantics open and more general on the question, and simply find some combination of compatible actions containing some or all members of $\mathcal{C}$. Actually, it's a combination of mutually compatible trajectories which are in accordance with the history R. There may be more than one such combination. The precise definition follows.

Definition Let a development $\langle \mathcal{B}, M, R, \mathcal{A}, \mathcal{C} \rangle$ be given where $max(\mathcal{B}) = n$ and R is defined over $[0, n]$. Then a *compatible-trajectories combination*, written as c, for $\mathcal{C}$ is defined as any set of trajectories $v_i = \langle d_i, pc_i \rangle \in \mathtt{Trajs}(E_i, R(s_i))$ for some or all members $\langle s_i, E_i \rangle$ of $\mathcal{C}$ which satisfy

- $length(v_i) > n - s_i$,

- if $n > s_i$, then $d_i(k) \cup pc_i(k) \subseteq R(s_i + k)$ for all $1 \leq k \leq n - s_i$, i.e. v_i agrees with the previous history,

- v_i interferes in no way with any other member of c and

- the trajectory preserving condition pc_i is satisfied by other trajectories in c or by applying inertia. □

Let c be a compatible-trajectories combination selected by the world. Then the world extends the history as follows.

$$R'(n + 1) = R(n) \oplus \bigcup_{\langle d_i, pc_i \rangle \in c} d_i(n - s_i + 1)$$

where $\oplus$ is Sandewall's "override" operation over states such that the value of a feature f in $[r \oplus r']$ equals that in state r' if f is defined there, otherwise that in state r. Using this $\oplus$ operation inertia is interwoven into the semantics. Notice that, since the trajectories in c do not interfere, the partial states at time $n + 1$, $d_i(n - s_i + 1)$, which are obtained from the trajectories, can be "put together" into a union without causing any conflict. On the other hand, the trajectory preserving conditions at that time point, $pc_i(n - s_i + 1)$, do not participate in the history extension. They are expected to be satisfied by other trajectories $d_j(n - s_j + 1)$ or inertia.

For some trajectory $\langle d_j, pc_j \rangle \in$ c, if $d_j(n - s_j + 1)$ is the last element of d_j, then it means that the trajectory v_j has been performed successfully and is terminated at $n + 1$. Therefore C' is obtained by first making a set of corresponding tuples $\langle s_i, E_i \rangle$ for each member v_i of c and then removing from the set all of the "terminated" members $\langle s_j, E_j \rangle$. On the other hand, tuples of the form $\langle s_j, E_j, n + 1 \rangle$ are added to the past action set $\mathcal{A}$ for the completed actions E_j, and $\mathcal{A}'$ is set to the resulting set.

4.2 Trajectory Semantics Ego

In the game it is the ego that activates one or more actions in its turn. Let a development $\langle \mathcal{B}, M, R, \mathcal{A}, C \rangle$ be given where $max(\mathcal{B}) = n$, then for each action E_i which is started by the ego at time n, a corresponding tuple $\langle n, E_i \rangle$ shall be added to the C component. If the ego passes on the control to the world without evoking any new actions, then no change is made for C. This definition does not need to be restricted to the single-timestep games.

5 Reapplying PCM to Concurrent Actions

As a subset of concurrent actions, we have defined into Sandewall's taxonomy the class **Ci** where all trajectories of concurrent actions are mutually independent. Then we have extended the class $\mathcal{K}p-$**IsAn** to $\mathcal{K}p-$**IsAnCi** and analyzed the applicability of PCM on the new class. By generalizing Sandewall's proof of the correctness of PCM for $\mathcal{K}p-$**IsAn**, we have proven that PCM still obtains only intended models for the scenarios within the new class. For want of space, the full proofs and details are reserved in [Yi95].

6 Conclusion

This work gives a base for analyzing the range of applicability of logics for concurrency. The result of our work implies that Sandewall's systematic approach can easily be extended to concurrency, and that most of the results shown by him for the case of sequential actions may be reobtained similarly for concurrent actions as well after necessary modifications.

References

[All91] James F. Allen. Temporal reasoning and planning. In J. F. Allen, H. A. Kautz, R. N. Pelavin, and J. D. Tenenberg, editors, *Reasoning about Plans*, pages 1–68. Morgan Kaufmann, 1991.

[Geo86] Michael P. Georgeff. The representation of events in multigent domains. In *AAAI*, pages 70–75, 1986.

[GKar] G. Groose and H. Khalil. State event logic. In *MEDLAR Special Issue of the Bulletin of the Interest Group in Pure and Applied Logics (IGPL)*, To appear.

[KS86] Robert Kowalski and Marek Sergot. A logic-based calculus of events. *New Generation Computing*, 4:67–95, 1986.

[Lan90] Amy L. Lansky. Localised representation and planning. In J. Allen, J. Hendler, and A. Tate, editors, *Readings in Planning*, pages 670–674. Morgan Kaufmann, 1990.

[Pel91] Richard N. Pelavin. Planning with simultaneous actions and external events. In J. F. Allen, H. A. Kautz, R. N. Pelavin, and J. D. Tenenberg, editors, *Reasoning about Plans*, pages 127–211. Morgan Kaufmann, 1991.

[San93] Erik Sandewall. The range of applicability of nonmonotonic logics for the inertia problem. In *IJCAI*, 1993.

[San94] Erik Sandewall. *Features and Fluents, A Systematic Approach to the Representation of Knowledge about Dynamical Systems*. Oxford University Press, 1994.

[Sho88] Y. Shoham. *Reasoning about Change: Time and Causation from the Standpoint of Artificial Intelligence*. MIT Press, Massachusetts, 1988.

[SR86] Erik Sandewall and Ralph Rönnquist. A representation of action structures. In *AAAI*, pages 89–97, 1986.

[Thi95] Michael Thielscher. On the logic of dynamic systems. In *AAAI Spring Symposium: Extending Theories of Action*, Stanford University, 1995.

[Yi95] Choong-Ho Yi. Reasoning about concurrent actions in the trajectory semantics. Licentiate Thesis, 1995. Department of Computer and Information Science, Linköping University.

Default Reasoning and Belief Revision in the CIN Project

G. Antoniou[1] and M.A. Williams[2]

[1] Griffith University, CIT
Nathan, QLD 4111, Australia
ga@cit.gu.edu.au
[2] Dept. of Management, The University of Newcastle
Callaghan, NSW 2308, Australia
maryanne@frey.newcastle.edu.au

1 Motivation and overview of the CIN Project

Most information systems are faced with incomplete information, even simple database management systems, therefore they must make plausible conjectures in order to operate in a satisfactory way. A simple example is the Closed World Assumption which is used extensively in database applications. Nonmonotonic reasoning (NMR) provides formal methods which support such a behaviour; in default logic, for example, the plausible conjectures are based on 'rules of thumb'.

Information is subject to change due to the inherent uncertainty of information or because the environment is volatile and dynamic. Current nonmonotonic reasoning systems neglect the problems raised by change. Belief revision (BR) is the research area that has developed techniques capable of dealing with changing information.

The CIN Project (Changing and Incomplete Information) aims to develop an *Intelligent Information Management Toolkit* which provides the user with a suite of methods for dealing with incomplete and changing information to solve real–world problems. It is designed to be an open system because of our contention that finding *the* right method for NMR and BR is an elusive dream, and that we should instead seek to determine the most appropriate method for the specific problem at hand. Innovative aspects of the project include the following:

- Full integration of default reasoning and belief revision: Current nonmonotonic reasoning methods can only cope with the acquisition of new information by recomputing extensions from scratch, which is computationally prohibitive. Instead we propose the use of belief revision techniques to preserve big portions of the conclusions previously drawn (details can be found in a forthcoming paper)
- Focus on engineering issues such as modularity and integration with object–oriented methods on the knowledge representation level.

2 The current state of the system

2.1 Default reasoning

Currently the system is capable of computing *extensions* of (finite) default theories according to Reiter's original default logic, Lukaszewicz' Justified Default Logic and Constrained Default Logic; also, a partial priority ordering on defaults may be introduced.

The implementation of default logic builds upon the tableau–based theorem prover *Bluegum* [3] which performs classical deduction and consistency checks. The theorem prover is implemented as a set of C++ objects that model all the various proof entities, ranging from variables and clauses up to a fully–featured tableau encapsulation and proof generation class capable of generating traces of proof constructions; it also makes extensive use of object–oriented techniques such as inheritance, polymorphism, dynamic binding and parameterized typing.

Tableau–based theorem provers are particularly well–suited for default logic implementations because the same tableau may be used both for checks of deducibility and for consistency checks associated with default application.

The implementation computes extensions according to the operational process model described in [1], which amounts to traversing the so–called process tree that gives an overview of all possible processes. *Tree–pruning techniques* are used to make the search more efficient. They are based on the property that for two sets of generating defaults leading to different extensions, one set cannot be a proper subset of the other. Applying this observation to tree-pruning leads to sophisticated methods that can cut down the size of the process tree significantly [2]. Further improvement of the performance is achieved by the use of stratification techniques that have been proposed in the literature.

2.2 Belief revision

Belief revision in the AGM paradigm was originally developed by Alchourron, Gärdenfors and Makinson and it has become one of the standard frameworks for modeling changes to repositories of information. In particular, it provides operators for modeling the revision and contraction of information. The logical properties of a body of information are not strong enough to uniquely determine a revision or contraction operator, therefore their principal constructions rely on some form of underlying preference relation, such as an *epistemic entrenchment ordering* or *partial entrenchment ranking*. The latter introduces degrees of acceptance of beliefs. Intuitively, the higher the ordinal assigned to a sentence the more important it is with respect to change.

Adjustments were proposed by Williams [4] as a way of realizing the principle of an *absolute* measure of minimal change. The computational model for adjustments described in [5] forms the basis of the Belief Revision component of the CIN system.

The Belief Revision System [6] was designed using object-oriented techniques, and makes aggressive use of parameterized typing, algorithm and container abstraction techniques, and iterators and adaptors. The main emphasis is on producing lean, efficient abstractions that will be of wider use in the development of the CIN Project's Intelligent Information Management Toolkit.

The basic functionality of the belief revision component of the CIN system is to recompute the degrees of belief acceptance once a modification has to be made (that means, once the degree of acceptance of a sentence has changed).

The underlying idea of the implementation is to assign to each rank, which represents a degree of acceptance, a layer object which encapsulates the tableau corresponding to the sentences included in that layer. Two of the entry points offered provide a means to locate the sentences entailed, and whether a particular sentence is entailed.

The basic operation of an adjustement (see [5]) requires that a sentence α is entailed by the layer mapping to ranks $\geq i$, and to compute the set of sentences entailed by the same sequence of layers. To determine the rank of a sentence the tableaux corresponding to the layers are built up in a stepwise manner, this additive property reduces the amount of computation substantially. In this way the system's architecture facilitates the process of adjustment efficiently (see [6] for details).

3 Current and future work

The next steps in the development of the CIN system include the implementations for goal–driven query evaluation in default logic and its variants. Also we are working on the integration of default reasoning with reason–based belief revision operators. Future features of the CIN system include modular concepts, object–oriented design and approximative default reasoning (restricting the consistency checks to parts of the knowledge base).

References

1. G. Antoniou. *Nonmonotonic Reasoning.* MIT Press 1996 (in press).
2. A. Courtney. *Towards a Default Logic Workbench: Computation of Extensions.* Honours Thesis. Department of Computer Science, University of Sydney.
3. K. Wallace (1994). *Proof Truncation Techniques in Model Elimination Tableau-Based Theorem Provers.* PhD Dissertation, University of Newcastle.
4. M.A. Williams, *Transmutations of Knowledge Systems*, in J. Doyle, E. Sandewall, and P. Torasso (eds), Principles of Knowledge Representation and Reasoning: Proceedings of the Fourth International Conference, Morgan Kaufmann, San Mateo, CA, 619 - 629, 1994.
5. M.A. Williams, *Iterated Theory Base Change: A Computational Model*, in the Proceedings of the Fourteenth International Joint Conference on Artificial Intelligence, Montréal, 1541 - 1550, 1995.
6. M.A. Williams, K. Wallace, G. Antoniou, *An Object-Oriented Implementation of Belief Revision*, in the Proceedings of the 8th Australian Joint Conference on Artificial Intelligence, World Scientific 1995, 259-266.

Mechanizing Multi-Agent Reasoning with Belief Contexts*

Alessandro Cimatti and Luciano Serafini

IRST — Istituto per la Ricerca Scientifica e Tecnologica
I-38050 Povo, Trento, Italy
email:{cx,serafini}@irst.itc.it

A system for the mechanization of multi-agent reasoning (see for instance [4, 10]) has to cope with the representation of mutual belief, common and nested belief, ignorance and ignorance ascription. These forms of reasoning can be uniformly represented in the framework of belief contexts [5, 7, 8]. The basic feature of belief contexts is *modularity*. A multi-agent scenario can be described by different, separated modules, called *contexts*, connected together by means of suitable interaction mechanisms, called *bridge rules*. Intuitively, different contexts can represent the sets of beliefs of different agents, the beliefs of an agent in different situations, the view of an external observer, or the beliefs ascribed by an agent to another agent. Bridge rules represent the relations between contexts. Some examples are the introduction of new beliefs as effect of communication, learning or belief revision.

The modular structure of the formalism gives a number of representational and implmentational advantages (these issues are thoroughly discussed in [3] and [1]):

Incrementality. Belief contexts allow for incremental and independent development of different parts of the knowledge base.

Locality. Reasoning in a multi-agent scenario is intrinsically local. This is formalized by reasoning within a single context, and allows for the application of general purpose inference techniques to very small subsets of the global system. This is substantially different from other approaches, where *ad hoc* reasoning techniques try to isolate the relevant information out of a global, unstructured theory.

Compositionality. The overall reasoning process is formalized by composing local reasoning steps by means of bridge rules. As a result, the process of inference can take advantage of the structure of the system, and hence be much more efficient.

Belief contexts have been mechanized in GETFOL. GETFOL [0] is an interactive system for the mechanization of multi-context systems[1], developed on top of a

---

* Fausto Giunchiglia has provided basic motivations, feedback and encouragement. We thank Massimo Benerecetti, Paolo Bouquet, Enrico Giunchiglia, Kurt Konolige, and John McCarthy for useful feedback and discussions. Lorenzo Galvagni has developed an implementation of multi-context systems in GETFOL.

[1] Belief contexts are a particular class of multi-context systems.

reimplementation [9] of **FOL** [12]. In the following, for lack of space, we do not go into the details of this mechanization. We give instead an overview of the peculiar features of belief contexts. For a thorough description of the mechanization of multi-agent reasoning in **GETFOL**, the reader is referred to [2].

Formally, a context is a theory which we present as a formal system $\langle L, \Omega, \Delta \rangle$, where L is a logical language, $\Omega \subseteq L$ is the set of axioms (basic facts of the view), and Δ is a deductive machinery. L contains formulas of the form $B_i(``A")$, for each wff $A$ and for each agent $i$, used to express belief. Intuitively, $B_i(``A")$ means that agent i believes the proposition expressed by A. To express that A is a common belief [11], L contains formulas of the form $CB(``A")$. Ignorance is formalized in belief contexts as non belief. Therefore, $\neg B_i(``A")$ expresses the fact that i does not believe A. In order to deal with ignorance, however, we explicitly represent in the formal language the assumptions which qualify ignorance ascription. In particular, L contains formulas of the form $ARF_i(``A_1, \ldots, A_n", ``A")$, meaning that $A_1, \ldots, A_n$ are all the relevant facts available to i to infer the conclusion A.

Contexts are connected together by means of bridge rules. Formally, a bridge rule is an inference rule with premises and conclusion in different contexts. Bridge rules can be thought of as constraining the interpretation of different formulas in different contexts. Consider the contexts ϵ and i, representing the view of an external observer and of agent i, respectively. The following bridge rules enforce the mutual relations between i and ϵ:

$$\frac{\epsilon : B_i(``A")}{i : A} \; \mathcal{R}_{dn.} \quad \frac{i : A}{\epsilon : B_i(``A")} \; \mathcal{R}_{up.} \quad \frac{\epsilon : CB(``A")}{i : A} \; CB_{inst} \quad \frac{\epsilon : CB(``A")}{i : CB(``A")} \; CB_{prop}$$

$$\frac{i : A_1 \cdots i : A_n \quad \epsilon : ARF_i(``A_1, \ldots, A_n", ``A")}{\epsilon : \neg B_i(``A")} \; \text{Bel-Clo}$$

Consider the rule $\mathcal{R}_{dn.}$. $\mathcal{R}_{dn.}$ states that the formula A of context i (written $i : A$ to stress the context dependence) is derivable from $\epsilon : B_i(``A")$. Conversely, $\mathcal{R}_{up.}$ states that $\epsilon : B_i(``A")$ is derivable from $i : A$ ($\mathcal{R}_{up.}$ is applicable only if $i : A$ does not depend on any assumption in i). The rules CB_{prop} and CB_{inst} enforce the fact that if A is a common belief, then i believes it, and also believes that it is a common belief. Bel-Clo is applicable only if $A_1, \ldots, A_n \nvdash_i A$, and $i : A_1, \ldots, i : A_n$ do not depend on any assumption in i. Bel-Clo allows to infer ignorance of agent i.

The bridge rules above formalize the fact that context i is a model of agent i from the point of view of the external observer. Therefore, in order to derive in context ϵ properties about the knowledge of agent i, context i can be called into play in a systematic manner. The intuition is that, rather than reasoning about i in context ϵ (e.g. by axiomatizing i's inference ability), it is possible reason "as i", locally in context i, and export the results in ϵ. This general mechanism can be iterated to different depths. For instance, in order to formalize i reasoning about agent j, a different context ij can be defined which represents the belief set ascribed by i to j.

696

The procedures implementing the basic operations of building the model of an agent, reasoning inside it, and exporting the result, are the kernel of the mechanization of belief contexts in GETFOL (see [2]). The formal proofs representing reasoning in the multi-agent scenario can be mechanically built by uniform application of such procedures, and share a common structure close to the reasoning being formalized.

References

1. A. Cimatti and L. Serafini. Multi-Agent Reasoning with Belief Contexts II: Elaboration Tolerance. In *Proc. 1st Int. Conference on Multi-Agent Systems (ICMAS-95)*, pages 57–64, 1995. Presented at *Commonsense-96*, Third Symposium on Logical Formalizations of Commonsense Reasoning, Stanford University, 1996. URL: ftp://ftp.mrg.dist.unige.it/pub/mrg-ftp/9412-09.ps.Z
2. A. Cimatti and L. Serafini. Multi-Agent Reasoning with Belief Contexts III: Towards the Mechanization. In P. Brezillon and S. Abu-Hakima, editors, *Proc. of the IJCAI-95 Workshop on "Modelling Context in Knowledge Representation and Reasoning"*, pages 35–45, 1995. URL: ftp://ftp.mrg.dist.unige.it/pub/mrg-ftp/9506-02.ps.Z
3. A. Cimatti and L. Serafini. Multi-Agent Reasoning with Belief Contexts: the Approach and a Case Study. In M. Wooldridge and N. R. Jennings, editors, *Intelligent Agents: Proceedings of 1994 Workshop on Agent Theories, Architectures, and Languages*, number 890 in Lecture Notes in Computer Science, pages 71–85. Springer Verlag, 1995. URL: ftp://ftp.mrg.dist.unige.it/pub/mrg-ftp/9312-01.ps.Z
4. L. G. Creary. Propositional Attitudes: Fregean representation and simulative reasoning. In *Proc. of the 6th International Joint Conference on Artificial Intelligence*, pages 176–181, 1979.
5. F. Giunchiglia. Contextual reasoning. *Epistemologia, special issue on I Linguaggi e le Macchine*, XVI:345–364, 1993. Short version in Proceedings IJCAI'93 Workshop on Using Knowledge in its Context, Chambery, France, 1993, pp. 39–49.
6. F. Giunchiglia. GETFOL Manual - GETFOL version 2.0. Technical Report 92-0010, DIST - University of Genoa, Genoa, Italy, March 1994. URL: ftp://ftp.mrg.dist.unige.it/pub/mrg-ftp/92-0010.ps.Z
7. F. Giunchiglia and L. Serafini. Multilanguage hierarchical logics (or: how we can do without modal logics). *Artificial Intelligence*, 65:29–70, 1994. URL: ftp://ftp.mrg.dist.unige.it/pub/mrg-ftp/9110-07.ps.Z
8. F. Giunchiglia, L. Serafini, E. Giunchiglia, and M. Frixione. Non-Omniscient Belief as Context-Based Reasoning. In *Proc. of the 13th International Joint Conference on Artificial Intelligence*, pages 548–554, Chambery, France, 1993. URL: ftp://ftp.mrg.dist.unige.it/pub/mrg-ftp/9206-03.ps.Z
9. F. Giunchiglia and R.W. Weyhrauch. FOL User Manual - FOL version 2. Manual 9109-08, IRST, Trento, Italy, 1991.
10. A. R. Haas. A Syntactic Theory of Belief and Action. *Artificial Intelligence*, 28:245–292, 1986.
11. R.C. Moore. The role of logic in knowledge representation and commonsense reasoning. In *National Conference on Artificial Intelligence*. AAAI, 1982.
12. R.W. Weyhrauch. Prolegomena to a Theory of Mechanized Formal Reasoning. *Artificial Intelligence*, 13(1):133–176, 1980.

Arguments and Mental Models: A Position Paper

David W. Green
Centre for Cognitive Science, Department of Psychology, University College
London, Gower Street London WC1E 6BT, U.K.

Abstract This paper explores the relationship between arguments and mental
models in the context of practical reasoning. Argument-models are
distinguished from other kinds of mental model, in particular causal models.
Examples of both kinds are presented. In addition, the process of
constructing a mental model is distinguished from the process of resolving
the model in order to reach a decision. It is suggested that argumentation-
models are especially important in social situations whereas causal models
may be more relevant when an individual has to make a number of
interrelated decisions. The representation relevant to decision and action
differs in these two cases as does the nature of the resolution process. Some
experimental and empirical work is discussed which illustrates the proposal.

1. Mental models and social agents

We start from two fundamental assumptions: 1) individuals construct mental models
in order to understand the world around them and 2) human beings are social agents.
We consider both these assumptions first of all.

The concept of a mental model One critical function of mind is to create models of
reality so as to allow effective action and decision-making [7]. A mental model
symbolizes a situation or entity in the world by letting mental tokens, their properties
and relations correspond to actual entities and their properties and relations [e.g.
19,20]. Such models may be constructed from perceptual information or from verbal
descriptions.

One theory based on the concept of a mental model has been applied successfully
to the various domains of deduction [22] and to induction [21]. Essentially, the more
models needed for a deduction the more difficult individuals find it. In the case of
induction, the addition of information to the model covers a variety of operations of
linguistic generalization. However, the theory remains to be extended to the domain
of practical reasoning. This paper suggests one approach.

The cognition of social agents The ability to model the social world is important to
us. We need to understand how our close relations, friends and peers think and feel
and might respond. Byrne & Whitten [5] supposed that pressures to understand and
to predict the behaviour of others in the social group was critical to the evolution of
the human mind [see also 6,9,18]. Humans are embedded in a social setting and the
biological bases of our cognitive and affective worlds reflects this [4].

A critical ability in social groups is the ability to argue. Groups reach decisions
by arguing the merits of different courses of action. In order to participate in such
discussions individuals need to be able to keep track of the various arguments and be
able to contribute to them. Further, when advocating a particular course of action,
individuals may take account of the likely objections that others may propose.
Mental models need to represent social situations and the way in which actions are

achieved through talk [12]. But since we are born into a social group we learn its practices [36]. How we think may then be usefully viewed as internal argumentation [3]. Thinking as internal argument provides a natural way in which to link the individual to the social context. We propose then that thinking and decision-making can be viewed as an internal process of argumentation in which the arguments for one course of action are weighed against those of another. Decisions can be reached by constructing and resolving an argumentation-model. In some cases, though, argument is used to build a causal model of some phenomenon. We consider the nature of the argumentation-model first of all.

2. The concept of an argumentation-model

The present proposal is that when individuals contribute to a discussion or consider a possible decision they establish an argument-model in which mental tokens symbolize arguments for action or for decision. Adapting a scheme proposed by Toulmin [35, p. 104] we suppose that a mental token comprises a possible claim, decision or action, any data and a warrant. Unlike Toulmin's notion, the tokens are conceptual-intentional objects (rather than linguistic objects). They are however interlinked in various ways. For instance, an argument may support or rebut another argument. We illustrate Toulmin's scheme:

Jack has normal blood cholesterol

Therefore, other things being equal he is unlikely to die of a heart attack

Here, a claim or conclusion follows to the extent that normal blood cholesterol is consistent with a healthy heart. But this claim is hedged, other things may not equal for various reasons. Toulmin proposed that individuals establish certain claims (e.g. Jack is unlikely to die of heart disease) on the basis of certain *data* (e.g. test results indicating that Jack has normal blood cholesterol or claims from other arguments) and a *warrant* (that normal blood cholesterol is consistent with a healthy heart). This claim can be challenged in various ways - it is defeasible [32,33]. The data might be disputed or the warrant might be queried. If the warrant is challenged, an argument to support the warrant might refer to various pieces of medical evidence or to some causal mechanism (Toulmin termed such an argument the backing). But the claim can be challenged in another way. It ignores other relevant pieces of information. It may turn out that Jack smokes, takes little exercise and has high blood pressure. These data are consistent with an alternative claim that Jack may get heart disease. This would *rebut* the earlier claim. The particular view of argument recruited here can be traced as Loui [27] showed to the work of Keynes in the 1920's [24].

An exposition of experimental psychological work on argumentation is beyond the scope of this paper but we note that warrants can be of different types [17,29]. In particular, the arguments involving them are not necessarily deductively valid. In addition, in everyday talk individuals do not always mention relevant data [1] and so arguments are incomplete. Further, the kinds of things they say depends on what effects they are trying to achieve [10]. A particular kind of argument will be developed with a view to countering particular objections. On social and political issues individuals tend to construct arguments in favour of just one side - "myside bias" [31; see also 25], though they may know of arguments on the other side. Baron [2] found that on social issues most individuals rate arguments with arguments only one side more highly than those with arguments on both sides though a critical factor according to Baron is the extent to which individuals consider that open-mindedness is a virtue. This affects how many other side arguments they give. Such research does not of course imply that individuals only consider one kind argument when they make decisions. Toulmin's scheme has also been applied to the protocols of individuals estimating probabilities [8]. The scheme has also been deployed in the AI literature on risk and probability [e.g. 11]. Other technically sophisticated approaches also exist - see, for instance, Pollock [33] where the notion of warrant is used more restrictively.

Warrants and decisions: model theory predictions and experimental data The notion of a mental token is general and so we can use existing properties of model theory to derive predictions about argument-models. If arguments are necessary then we expect that the nature of the decisions individuals reach will depend on the kinds of arguments (especially warrants) they invoke and the nature of the resolution process. A fundamental property of model theory is that individuals minimize what is explicitly represented mentally. Decisions may then be based on what is initially represented. Effects attributable to the initial representation are termed focus effects [26]. However, precisely what decision is reached will depend on the extent to which alternatives have been considered and on the extent to which the relationship between these alternatives has been made explicit. That is, it will depend on the extent to which the model has been fleshed-out. Evidence that individuals consider alternative possibilities may be more forthcoming in circumstances where a decision can be expressed on a scale. The presence of an alternative may bias the decision. We have observed such an effect in a task that allows subjects to express a decision as a scale value [16].

We adapted a scenario from Macrae [28]. Individuals were told about Lucy who normally eats in a particular restaurant. On the Saturday in question she follows habit but gets food poisoning and is ill in hospital for four days. Individuals are asked to decide what level of fine the restaurant should pay from £0 to £4000 in steps of £500. The basic idea underlying these experiments is that individuals reach a decision by considering the data and relevant warrants. They construct an argument-model. Where there is more than one argument token then individuals must weight or endorse one of them in order to resolve the model. However, although a single fine value must be selected, any other active token will bias the decision. The amount of bias will be less if the resolution process involves the explicit rebuttal of that token. Unless, in other words, the individual elaborates the argument-model.

In a series of experiments, based on the above scenario, we have begun to explore these expectations. We established that the level of fine (holding the data constant) depends on the warrant. So, for instance, individuals warrant their decision in one of two main ways. They either write that the food-poisoning was a one-off accident since Lucy would not keep visiting a restaurant which gave her food poisoning or they state that a restaurant has a duty of care towards its customers. In one study the mean fine increased from £400 for those stating that the event was a one-off accident to £2570 for those stating that the restaurant has a duty of care. If individuals need a warrant for a decision then presenting them with a warrant should lead many to adopt it - if it is at least plausible. In terms of the theory of mental model we predicted a focus effect. We found just such an effect with these materials. In addition, we found an effect of rebuttal.

In order to examine these questions we presented participants either with the one-off accident warrant or the duty of care warrant. They were free to select it or to write down an alternative warrant. Individuals who rejected a presented warrant (e.g. the one-off accident warrant) invariably generated the alternative warrant (in this case, the duty of care warrant). This meant that we could compare the difference in fine level for the one-off accident warrant and the duty of care warrant for participants who selected the presented warrant with that of participants who generated an alternative. In one study participants could write down an alternative without explicitly stating why they rejected the presented warrant. In this case, the difference in the amount of fine for the two warrants was the same for participants who selected the presented warrant compared to those who did not. In either case the difference averaged £1080. In a second study, participants who generated an alternative warrant had to write down why they rejected the presented warrant: that is, they had to rebut the presented warrant explicitly. As predicted, the difference in the amount of fine for these two warrants was greater for those participants who explicitly rebutted the presented alternative. For those accepting a presented warrant the difference was £923 whereas for those generating the warrants the difference was £2446. When individuals make decisions without explicitly rebutting an alternative warrant they seem to be affected by implicit alternatives making participants less willing to choose more extreme values. When they explicitly rebut a presented alternative, and hence flesh-out their mental model, then their judgements are more extreme.

3. Argumentation-models and causal-models

So far we have assumed that the argumentation-model is critical to decision-making. The structure of the argumentation-model corresponds to the relationship between arguments - such as whether they support or rebut one another and not to the order or form in which these arguments are expressed. In a social situation, such arguments may, in addition, be indexed by the names of the persons who advocate them. But we need to enquire as to the relationship between this kind of model and other mental models, in particular causal models.

Individuals seek to understand their worlds. In explaining someone's actions, for instance, we may construct a causal model based around possible motives and opportunities [30]. In the case of physical phenomena (e.g. the incidence of coronary heart disease or unemployment) we may construct a causal mental model that shows the influence of various factors on the effect in question. This model corresponds to the putative mechanisms in the world. Such a model can be constructed on basis of argument. An argument, for instance, might be used to support a connection between one kind of factor and another in causal model. That is the outcome of argumentation is a model of a certain type in which the tokens correspond to those factors that are causally relevant. Decisions can be then be based on the manipulations of this model rather than on the arguments that led to its construction. It is the causal model which is functionally relevant to performance.

Recently [13], we have explored the way individuals understand an ecological phenomenon. Individuals construct a mental model based on prior beliefs about predators and prey and mentally simulate their causal model in order either to explain or to generate population changes in one or the other [see 23 for a different but related example]. Here participants' predictions derive from the mental simulation of the causal model. The arguments that lead to the construction of the causal model are implicit rather than explicit in its structure. In explaining the nature of the population changes individuals describe the history of the simulation. The reason why the simulation takes the form it does stems from the nature of the model and the values chosen. These in turn can be accounted for in terms of the arguments used to construct the model in the first place. But in terms of the predictions individuals make they derive from the causal model.

Consider another example. Individuals can be asked to construct a causal model of some of the risk factors for coronary heart disease - CHD [15]. Here arguments can be used to propose specific links (e.g. high cholesterol increases risk of CHD; exercise can reduce risk of high cholesterol). Such arguments may be based on remembered information. After representing their model as a diagram, individuals were asked to assess the strength (out of 100) of the paths they have drawn. We have found that if individuals are then asked to rate the effectiveness of different actions designed to reduce the risk of CHD (e.g. take more exercise; eat less fat) then these ratings are well predicted by the paths strengths in the causal model.

Imagine that individuals have to decide how effective exercise will be in reducing the risk of CHD. We suppose that they trace paths connecting exercise with the risk of CHD. Given, for example, a direct path from exercise to CHD and indirect path via cholesterol, total path strength is equal to the strength of the direct path (e.g. 30/100 or 0.3) plus the strength of the indirect path. If the path from exercise to cholesterol has a strength of 20/100 (i.e. 0.2) and that from cholesterol to CHD has strength of 80/100 (i.e. 0.8) then the strength of the indirect path is 0.16 (0.2 * 0.80) and so total path strength is 0.46. We found that total path strength, rather than number of paths, predicts the ratings of effectiveness. Of course, these judgements of effectiveness can be warranted ultimately by appealing to the arguments used in constructing the causal model. Once the model is constructed though there is no need to consider the arguments that went into constructing the paths and the path strengths in order to reach a decision. The argument-model is now implicit in the structure of the causal model.

An interesting related example of this duality occurs in the context of persuasive communications. Green [14] examined the advertising copy of a well known breakfast cereal. This copy included claims such as the following: Britain has one of the highest death rates from heart disease in the Western world (data; no warrant); cholesterol is a major risk factor for heart disease (the warrant was that this is common knowledge) and oat bran may help reduce cholesterol (warrant: recent studies). When asked what kind of benefit they could perceive from eating the product, over 30% of young adults reported that it could reduce the risk of CHD. The suggestion here is that individuals seek to find a way in which consumption of the product can lead solving the problem raised by the copy, namely the risk of CHD. They find a path in their causal model that connects the product to the problem.

We have supposed that individuals use arguments to construct a causal model. But, the relationship is not one-way. Individuals can use a causal model to construct an argument. Thus, the presence of an established mechanism (e.g. a chemical reaction) can be used to argue that some event was caused by some combination of circumstances; on the other hand, the absence of such a mechanism can be used to refute a proposal.

4. Agents' activities

The nature of the decision individuals reach depends on the extent to which they flesh-out their argumentation-model of the problem. When they do so the influence of alternative but unselected arguments for a decision, is reduced. In some cases individuals, we suggest, rely on the causal model constructed from arguments. In reaching a decision, individuals use the model to compute the impact of various factors on a target factor. In such activity, the arguments that led to construction of the model remain implicit.

If a causal model can be constructed when do participants rely on it to reach a decision rather than on the argumentation-model involved in its construction? In a social setting individuals may need to account for, or to explain, their decision-making and so the argumentation-model provides an available conceptual resource for such accounts. In a social setting too, individuals may need to deliberate with others in order to reach a decision. On the other hand, when individuals have to make a series of interlinked decisions (e.g. how to change their lifestyle in order to reduce the risk of CHD) then the causal model might provide a more effective representation since it acts as a reduced representation in which only paths and their strengths need to be consulted.

A number of further questions are raised by the idea of a dual representation in reasoning about practical decisions. We consider just one. It follows from the present suggestion that once a causal model is constructed then any uncertainties concerning a particular link in the causal model or the complexity of argument invoked in establishing it will be impounded in terms of some simple variable (e.g. path strength). The ease of reaching a decision involving that causal path should then be independent of the argumentation-model. On the other hand, when individuals come to justify or account for that decision, complexity of the argumentation-model should affect the ease of providing an account.

References

Antaki, C. (1994). *Explaining and arguing: The Social Organization of Accounts*. London: Sage Publications.

Baron, J. Myside bias in thinking about abortion. *Thinking and Reasoning, 1*, 221-235.

Billig, M. (1987). *Arguing and Thinking: A Rhetorical Approach to Social Psychology*. Cambridge: Cambridge University Press.

Brothers, L. (1990). The social brain: A project for integrating primate behavior and neurophysiology in a new domain. *Concepts in NeuroScience, 1*, 27-51.

Byrne, R., & Whitten, A. (1988). *Machiavellian Intelligence: social expertise and the evolution of intellect in monkeys, apes and humans*. Oxford: Oxford University Press.

Cosmides, L., Tooby, J., & Barkow, J. (1992). Introduction:evolutionary psychology and conceptual integration. In J.Barkow et al. *The Adapted Mind*. Oxford: Oxford University Press.

Craik, K. (1943). *The Nature of Explanation*. Cambridge, England: Cambridge University Press.

Curley, S.P., Browne, G.J., Smith, G.F., & Benson, P.G. (1995) Arguments in practical reasoning underlying constructed probability responses. *Journal of Behavioral Decision-Making, 8*, 1-20.

Dunbar, R. (1993). Coevolution of neocortical size, group size, and language in humans. *Behavioral and Brain Sciences, 16*, 681-735.

Edwards, D., & Potter, J. (1993). Language and causation: a discursive action model of description and attribution. *Psychological Review, 100*, 23-41.

Fox, J. (1994). On the necessity of probability: reasons to believe and grounds for doubt. In G.Wright & P. Ayton (Eds.). *Subjective Probability*. John Wiley & Sons.

Green, D.W. (1994). Induction: representation, strategy and argument. *International Studies in the Philosophy of Science, 8*, 45-50.

Green, D.W. (1995). Explaining and envisaging an ecological phenomenon. *British Journal of Psychology* (in press).

Green, D.W. (1996) Inferring health claims: a case study. *Forensic Linguistics* (in press).

Green, D.W., & McManus, I.C. (1995). Cognitive structural models: the perception of risk and prevention in coronary heart disease. *British Journal of Psychology, 86*, 321-336.

Green, D.W., Muckli, L., & McClelland, A. (1995). Argument and decision. Paper presented to the British Psychological Society Cognitive Psychology Section Meeting, Bristol, September 7-9, 1995.

Huff, A.S. (1991) Ed. *Mapping strategic thought*. New York: John Wiley & Sons

Humphrey, N. (1984). *Consciousness Regained*. Oxford: Oxford University Press.

Johnson-Laird, P.N. (1983). *Mental models: Towards a Cognitive Science of Language, Inference and Consciousness*. Cambridge, MA: Harvard University Press.

Johnson-Laird, P.N. (1993). *Human and Machine Thinking*. Hillsdale, NJ: Lawrence Erlbaum Associates.

Johnson-Laird, P.N. (1994). A model theory of induction. *International Studies in the Philosophy of Science, 8*, 5-29.

Johnson-Laird, P.N., Byrne, R.M.J., & Schaeken, W. (1992). Propositional reasoning by model. *Psychological Review, 99*, 418-439.

Jungermann, H., and Thuring, M. (1987). The use of mental models for generating scenarios. In G.Wright and P. Ayton (eds.). *Judgemental Forecasting*. London: John Wiley & Sons Ltd.

Keynes, J.M. (1921). *Treatise on probability*. London: Macmillan & Co.

Kuhn, D. (1991). *The Skills of Argument*. Cambridge, England: Cambridge University Press.

Legrenzi, P., Girotto, V., & Johnson-Laird, P.N. (1993). Focusing in reasoning and decision-making. *Cognition, 49*, 37-66.

Loui, R-P. (1991). Argument and belief: where we stand in the Keynesian tradition. *Minds and Machines, 1*, 357-365.

Macrae, C.N. (1992). A tale of two curries: counter-factual thinking and accident-related judgements. *Personality and Social Psychology Bulletin, 18*, 84-87.

Mason, R.O. & Mitroff, I.I. (1981). *Challenging strategic planning assumptions: theory, cases and techniques.* New York: John Wiley & Sons.

Pennington, N., & Hastie, R. (1993). Reasoning in explanation-based decision-making. *Cognition, 49,* 123-163.

Perkins, D.N. (1989). Reasoning as it is and could be: an empirical perspective. In D.M. Topping, D.C. Crowell, & V.N. Kobayashi (Eds.). *Thinking across cultures: the third international conference on thinking.* Hillsdale, NJ: Lawrence Erlbaum Associates, cited in Baron (1995).

Pollock, J. (1991). Self-defeating arguments. *Minds and Machines, 1,* 367-392.

Pollock, J. (1992). New foundations for practical reasoning. *Minds and Machines, 2,* 113-144.

Pollock, J. (1993). The phylogeny of rationality. *Cognitive Science, 17,* 563-588.

Toulmin, S. (1958). *The Uses of Argument.* Cambridge: Cambridge University Press.

Vygotsky, L.S. (1960). The development of higher mental functions. Quoted in J.V. Wertsch (1985). *Vygotsky and the Social Formation of Mind.* Cambridge, MA: Harvard University Press.

Argumentation and Decision Making: A Position Paper

Simon Parsons* and John Fox

Advanced Computation Laboratory,
Imperial Cancer Research Fund,
P.O. Box 123,
Lincoln's Inn Fields,
London WC2A 3PX,
United Kingdom.

Abstract. This paper summarises our position on the use of symbolic methods for reasoning under uncertainty, and argumentation in particular. Our view is that argumentation offers a complement to numerical methods for reasoning about belief, and a general framework within which many competing approaches can be understood. In applications we have found that argumentation offers a variety of benefits for practical reasoning systems. The presentation is historical, emphasising the reasons which motivated the development of the argumentation framework, drawing primarily on work carried out by researchers at the Imperial Cancer Research Fund since about 1980.

1 The need for symbolic decision making

Work on argumentation at the Imperial Cancer Research Fund arose out of a series of studies on medical decision making, including work on modelling human diagnostic reasoning [9] and comparisons of the relative merits of numerical and symbolic inference techniques in clinical decision making [11, 16, 22]. These studies strongly suggested, contrary to the assumption prevalent at the time, that numerical methods were not the only sound and practical means of making decisions under uncertainty.

The first line of work was in an empirical tradition, concerned with how people make decisions and the strengths and weaknesses of the decision making process. One study investigated diagnostic decision making by medical students and compared two computational models of decision making on a simulated medical diagnosis problem [10]. The models were implemented as sets of production rules, one implementing a statistical model and one a knowledge-based, semi-qualitative model. The results strongly suggested that the latter model gave a better account of human reasoning under uncertainty. This is consistent with a well established finding from psychological research that people do not manage uncertainty in ways which closely resemble normative probabilistic reasoning.

---

* Current address: Department of Electronic Engineering, Queen Mary and Westfield College, Mile End Road, London E1 4NS, United Kingdom. S.Parsons@qmw.ac.uk.

Classical decision theorists have strongly criticised human judgement precisely on the grounds that people do not comply with the requirements of probability theory. In contrast some cognitive scientists have questioned the force of this observation by emphasising the flexibility and other virtues of human reasoning (e.g. [24]). This observation was reinforced by a study which investigated the performance of an expert system based on a model of human reasoning [11]. In a realistic diagnostic problem in gastroenterology it was found that the diagnostic accuracy of the expert system was very similar to that of a Bayesian diagnostic system though the expert system achieved this level of performance using only half the information provided to the probabilistic system.

Other empirical studies suggest that, notwithstanding the quantitative precision of probabilistic evidence analysis, much practical medical decision making can be successfully carried out without precise numerical data. For example [16], a rule-based system for leukaemia diagnosis was developed using the EMYCIN expert system shell. The accuracy of the system, as compared with the decisions of a domain expert, was 64%. When the CFs were limited to just two values, 1.0 or 0.5 (loosely "certain" or "uncertain") a 5% *increase* in correct diagnoses was observed! In another study, decision making using a semiqualitative decision procedure was compared with a probabilistic procedure using data regarding the admission of 140 patients to a coronary care unit. It was found that the symbolic decision model performed at least as well as the probabilistic model as determined by an ROC analysis [22]. Other independent studies have confirmed the finding that practical decision making can be carried out successfully without depending upon precise quantitative data (e.g. [4]).

2   A model for reasoning under uncertainty

Having established these results, the motivation for further work was an interest in building decision support systems which would have a number of advantages over conventionally available technologies:

- They would not depend upon the availability of objective statistical data (which are frequently not available in complex domains like medicine)
- They could make use of other kinds of reasoning than statistical inference (e.g. causal, functional and temporal reasoning) which might be more intuitive than quantitative reasoning.
- They could support all phases of decision making, not just evidence analysis, such as recognising when a decision is needed, what the decision options are, what information is relevant to the choice and so forth.

An initial framework for "symbolic decision theory", using first-order logic for much of the deductive reasoning required and argumentation for the management of uncertainty, was proposed [12, 14], and evidence for the practicality of the theory came from its use in the Oxford System of Medicine (OSM) [13]. This is a decision support system aimed at general practitioners which provides a generic decision procedure for a range of medical decision tasks. Evidence that the symbolic decision theory is very versatile comes from the wide range of possible applications that have now been developed [8].

In addition, however, it was agreed, in accordance with the view generally held by decision theorists, that any decision procedure which is to be used for practical applications, particularly those like medical applications which have safety implications, must be given a sound theoretical underpinning. While the symbolic approach might be "inspired" by observations of human flexibility, people make mistakes and it is clearly not desirable to emulate those mistakes!

3 Formalising the model

The most contentious element of symbolic decision theory is the use of argumentation as the basic framework for reasoning under uncertainty. The central idea in argumentation is that of a *tentative proof* of a proposition. The fact that a proposition can have arguments for and against it suggests a divergence between argumentation and classical first order logic in which propositions are true or false. Furthermore, as pointed out in [2], argumentation has many commonalities with intuitionistic logic, suggesting that argumentation might be given a sound basis in category theory since this is possible for intuitionistic logic. The first steps in providing this basis are detailed in [2] which identifies the structure of the space of arguments, along with the kind of operations possible over them. The rest of the formalisation is provided in [1], which also highlights the link between argumentation and Dempster-Shafer theory.

With this semantics in mind, it is possible to define a logic LA in which the consequences of a database are a set of arguments, and this is the subject of [3] and [19]. In this work logical formulae are augmented with their proofs and when formulae are combined, the proofs are handled in an appropriate manner. This means that it is possible to determine the validity of formulae derived in the logic based upon the strength of the arguments for and against individual formulae, and that the way in which this is done is in accordance with the category-theoretic semantics. This process has been automated in the Argumentation Theorem Prover [20], and a summary of the formal model is recorded in [18].

4 A general model of reasoning

In addition to the arguments for argumentation as a symbolic model of decision making, we can argue [15] that it is a model of "practical reasoning" of the kind that humans indulge in every day. It captures many of the modes of commonsense reasoning—finding support for ideas, attacking other ideas, and trying to attack the support of other ideas. It handles contradictions, and should also enable the resolution of conflicting arguments at the meta-level. Furthermore, there is a strong case that argumentation provides a general framework for unifying many methods for reasoning under uncertainty, such as possibility and probability [23] theories. In this role argumentation is less a formalisation of human reasoning than a tool that can enable the use of other formalisms. Argumentation provides a general way of combining logical reasoning with Bayesian probability by using it to construct a network of influences between relevant variables. Indeed, argumentation is sufficiently general as to underlie symbolic as well as quantitative formalisms [21].

However, it is possible to do more with argumentation than just provide a framework for using established formalisms, instead, as is discussed in [6], it is possible to handle inconsistent information. That is, it is possible to have certain arguments for both a proposition p and its negation $\neg p$. This inconsistency enables LA to provide a ranking over the propositions for which arguments may be proposed. In particular, arguments for propositions are allocated different classes of acceptability [5, 7], the allocation depending on factors such as whether an argument is based on a consistent database or whether there are any counter-arguments. This approach can be used to give a purely logical approach to uncertainty that ranks propositions only on the structure of the arguments for them, and it can be augmented by the use of preference relations over subsets of the database.

5 Summary

To summarise, empirical evidence for the usefulness of a symbolic theory of decision making has led to the development of a formal model based on first-order logic combined with the use of argumentation for handling uncertainty. The versatility of the model is suggested by its wide practical applicability, while the justification for the use of argumentation is based upon its proven practical flexibility and its well-developed formal semantics.

References

1. Ambler, S. (1992) A categorical approach to the semantics of argumentation, Technical Report 606, Department of Computer Science, Queen Mary and Westfield College.
2. Ambler, S. and Krause, P. (1992) Enriched categories in the semantics of evidential reasoning, Technical Report 153 Advanced Computation Laboratory, Imperial Cancer Research Fund.
3. Ambler, S. and Krause, P. (1992) The development of a "Logic of Argumentation", *Proceedings of the International Conference on Information Processing and the Management of Uncertainty*, Palma.
4. Chard, T. (1991) Qualitative probability versus quantitative probability in clinical diagnosis: a study using a computer simulation, *Medical Decision Making*, 11, 38–41.
5. Elvang-Gøransson, M. and Hunter, A. (1993) Argumentative logics—reasoning with classically inconsistent information, *Data and Knowledge Engineering*, 16, 125–145.
6. Elvang-Gøransson, M., Krause, P. and Fox, J. (1993) A logical approach to handling uncertainty, *Proceedings of the Workshop on Modelling Problems in Control and Supervision of Complex Dynamic Systems*, Lyngby, Denmark.

7. Elvang-Gøransson, M., Krause, P. and Fox, J. (1993) Dialectic reasoning with inconsistent information, *Proceedings of 9th Conference on Uncertainty in Artificial Intelligence*, Washington, D. C.

8. Fox, J. (1996) A unified framework for hypothetical and practical reasoning (2): lessons from clinical medicine, *Proceedings of the International Conference on Formal and Applied Practical Reasoning*, (this volume).

9. Fox, J. (1987) Making decisions under the influence of knowledge, in *Modelling Cognition*, P. Morris ed., John Wiley & Sons Ltd.

10. Fox, J. (1980) Making decisions under the influence of memory, *Psychological Review*, **87**, 2, 190–211.

11. Fox, J., Barber D., and Bardhan, K. D. (1980) Alternatives to Bayes? A quantitative comparison with rule-based diagnostic inference, *Methods of Information in Medicine*, **19**, 210–215.

12. Fox, J., Clark, D. A., Glowinski, A. Gordon, C. and O'Neil, M. J. (1990) Using predicate logic to integrate qualitative reasoning and classical decision theory, *IEEE Transactions on Systems, Man and Cybernetics*, **20**, 347–357.

13. Fox, J., Glowinski, A. Gordon, C. Hajnal, S. and O'Neil, M. (1990) Logic engineering for knowledge engineering: design and implementation of the Oxford System of Medicine, *Artificial Intelligence in Medicine*, **2**, 323–339.

14. Fox, J., Glowinski, A. J., O'Neil, M. J. and Clark, D. A. (1988) Decision making as a logical process, *Proceedings of Expert Systems '88*, Cambridge.

15. Fox, J., Krause, P. and Ambler, S. (1992) Arguments, contradictions and practical reasoning, *Proceedings of the European Conference on Artificial Intelligence*, Vienna.

16. Fox, J., Myers, C. D., Greaves, M. F., and Pegram, S. (1985) Knowledge acquisition for expert systems: experience in leukemia diagnosis, *Methods of Information in Medicine*, **24**, 65–72.

17. Fox, J., O'Neil, M., Glowinski, A. J. and Clark, D. (1988) A logic of decision making, *Illinois Interdisciplinary Workshop on Decision Making*, Urbana, Illinois.

18. Krause, P., Ambler, S., Elvang-Gørannson, M. and Fox, J. (1994) A logic of argumentation for reasoning under uncertainty, *Computational Intelligence*, **11**, 113–131.

19. Krause, P., Ambler, S. and Fox, J. (1993) The development of a "Logic of Argumentation", in *Advanced Methods in Artificial Intelligence*, B. Bouchon-Meunier, L. Valverde and R. R. Yager eds., Springer-Verlag, Berlin.

20. Krause, P., Ambler, S. and Fox, J. (1993) ATP user manual, Technical Report 187, Advanced Computation Laboratory, Imperial Cancer Research Fund.

21. Krause, P. J., Fox, J. and Ambler S. (1992) Argumentation as a unifying concept for reasoning under uncertainty, Technical Report 166, Advanced Computation Laboratory, Imperial Cancer Research Fund.

22. O'Neil, M. and Glowinski, A. (1990) Evaluating and validating very large knowledge-based systems, *Medical Informatics*, **3**, 237–251.

23. Parsons, S. (1996) Defining normative systems for qualitative argumentation, *Proceedings of the International Conference on Formal and Applied Practical Reasoning*, (this volume).

24. Shanteau, J. (1987) Psychological characteristics of expert decision makers, in *Expert Judgement and Expert Systems*, J. Mumpower ed., NATO ASI Series, vol F35.

The Implementation of LENA

Claudia M.G.M. Oliveira
Instituto Militar de Engenharia - IME, Rio de Janeiro, Brazil
e-mail: cmaria@ime.eb.br

1 Overview of LENA

LENA stands for Logic Engine for Natural Agent. It is meant as a system
which integrates mechanisms that when combined in certain ways will produce
a mode of human practical reasoning, supporting the following central concepts:
1. Declarative Unit, the basic information unit; **2. Databases**, structures of
declarative units; **3. Input**, a function associating with each database Δ and
a declarative unit δ, a new database $\Delta' = Input(\Delta, \delta)$, the result of adding
δ to A; **4. Integrity**, the qualification of databases as "acceptable" or "unacceptable" (consistency for instance); **5. Update**, *Input* with integrity criteria,
such that $\Delta' = Update(\Delta, \delta)$ is the input of δ to Δ maintaining integrity; **6.
Consequence**, a binary relation between databases presented algorithmically;
7. Abduction, a function that produces the necessary additions to a database
Δ such that the relation $Input(\Delta, \Delta') \vdash \delta$ holds for some goal δ; **8. Actions**,
the application of an update to the database provided that a precondition is
met with; **9. Goals, desires and wants**, the mechanisms through which the
user expresses her demands from the system, basically a query mechanism. The
author represents goals by a sequence of declarative units.

2 Overview of the Implementing System - Witty

In order to implement a LENA we will need a system in which there is a clear definition of two levels: one for knowledge representation and retrieval with inference
capabilities and another for programs. The natural candidate for a programming
language which quite naturally communicates with a knowledge based system
is Prolog, mostly because of the nature of its variable bindings. The knowledge
based level could also be Prolog as it is explored in [BK82]. Here we generalize
those ideas to a knowledge level based on general clauses so as to have the power
of full first order logic in the knowledge based level. Therefore, Witty [Oli95]
presents a great advantage in relation to Prolog, where programs as well as data
are represented in the same language and the same context and even this distinction - database × program - is not present in the implemented system, in
Witty here we treat them totally differently.

Witty is formed by two levels of representation: the logical level is a database
(collection of databases defined as pairs $< s, \Gamma >$, where s is a label and Γ is a
multiset of clauses and a classical first order inference procedure that with the
addition of mechanisms of **Filter** and **Oracle** it can become the implementation
of a scope of non-classical logics; the second level is a program interpreter.

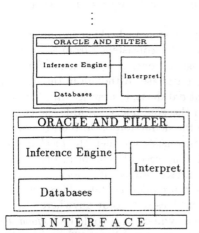

Basically, a **Filter** is program in the language interpreted by the control which restricts the derivations, by evaluating the labels of the formulas in one resolution step and validating/invalidating the step according to some theory of labels. If a filter $fil(s_1, s_2, s_3)$ is applied after a successful resolution step then s_1 is the label of the near parent clause, s_2 is the label of the far parent clause and s_3 is the resulting label. If fil_3 succeeds then the resolvents label is s_3 and the derivation proceeds; otherwise the derivation backtracks.

Likewise, an **Oracle** is a program which extends resolution such that an alternative direction will be given to the proof-search whenever an unsuccessful path is reached, seeking to solve the problem outside the current setting, using extra data. If an oracle $op(L, s_1, s_2)$ is applied after a failed resolution step then L is the literal which failed to be resolved and s_1 is the current resolvent's label and s_2 is the resulting label. If op_3 succeeds then the resolvents label is s_3 and the derivation proceeds; otherwise the derivation backtracks.

Consider the basic syntax of Prolog programs. Instead of having atomic facts which are the base case for these programs we have primitives executed by the system with respect mostly to the inference system. These primitives are: $PROVE(x_1, \cdots, x_n, x)$ - executes the inference procedure for sentence x, using databases $x_1, \cdots, x_n$, and x_1 is the set of support; $ASSERT(x_1, x_2)$ - introduces sentence x_2 into database x_1; $DELETE(x_1, x_2)$ - eliminates sentence x_2 from database x_1; $DEFBASE(x)$ - creates empty database x; $GETVAR(x_1, x_2, x_3)$ - instantiates variable x_3 with the x_2-th value obtained for x_1 in a derivation.

3 The language LENA

LENA is intended to be an integrated reasoner based on Logic Programming. The approach to building a reasoner is to take a reasonable deductive database language, it with *mechanisms*, namely: 1. structure of databases; 2. input mechanisms; 3. consistency checking procedures; 4. updating mechanisms; 5. query mechanisms. We build the "the basic five" in our system as follows:

1. Structure of databases - What we call a database $< s, \Gamma >$ we will use as a declarative unit where s is a label and Γ is the set of clauses resulting from the normalization of a formula. Our collection of databases W is one database in LENA.

2. Input mechanisms - Quite naturally, this is basically $ASSERT(t, A)$. The different structures of databases are built via the generation of t. Therefore one would have a meta-theory for generating these labels, represented as a label generating program.

3. Consistency checking procedures - Consistency checking in our system is programmable. There are two distinct types of inconsistency from a practical reasoner's point of view. The first one is logical consistency, when a formula and its negation are both derivable from the database. Although in theory, any query can be derived from such an inconsistent database, in practice this does not happen because most provers will work from the goal query towards the empty clause and the inconsistency might not be in the deduction path. If on the other hand the user is trying to spot this type of inconsistency then the mechanisms he must build are fully described in [Oli95]. The second type of inconsistency is a more useful one. In order to represent unacceptable situations, a set (or multiset, or list, etc) of integrity constraints is provided as data. These clauses all have the same head, some predicate which does not occur anywhere else in the database. It could be ⊥, to give it a familiar symbol, or it could be *nogood*, to make a distinction from the usual falsehood constant predicate. Thus, in order to verify whether a database is inconsistent, *nogood* would be queried.

4. Updating mechanisms - Updating is a combination of the consistency checking mechanism with the input mechanism, in the sense that if a database is to be updated with a formula, the latter must be either consistent with the former or there must be a way to revise the database and restore consistency. This mechanism is part of the underlying language Gabbay proposes for LENA, which also has an implementation in Witty.

5. Query mechanisms - These are basically realized by the *PROVE* primitive. The modes in which a query can be answered are altered by means of defining variations of the basic *PROVE* primitive which change the underlying reasoning pattern. These variations include: 1. preparation of the goal formula; 2. setting the initial label; 3. setting the filter and the oracle.

It is clear so far that LENA is not a system but an outline of a system, or a collection of requirements for a practical reasoning system. It becomes a system though, given a logical language upon which to define database structure, consequence relation, etc. Gabbay exemplifies a LENA for a choice of logical language called CondLP [GGMO94a].

References

[BK82]   Kenneth A. Bowen and R. Kowalski. Amalgamating language and met-alanguage in logic programming. *A.P.I.C. Studies in Data Processing*, 153–172, 1982.

[GGMO94a] Dov Gabbay, L. Giordano, A. Martelli, and N. Olivetti. Conditional logic programming. 1994.

[GGMO94b] Dov Gabbay, L. Giordano, A. Martelli, and N. Olivetti. Lena (a logic engine for natural agent). 1994.

[Oli95]   Claudia Oliveira. *An Architecture for Labelled Theorem Proving*. PhD thesis, University of London - Imperial College, 1995.

The Implementation of CondLP

Claudia M.G.M. Oliveira
Instituto Militar de Engenharia - IME, Rio de Janeiro, Brazil
e-mail: cmaria@ime.eb.br

Introduction The language CondLP [GGMO94a] is an extension of NProlog [GR84] where the Horn fragment is extended to allow embedded implications both in goals and in clause bodies. Consistency checking is automatic for all goals containing implications: some of the database is checked for consistency against the hypothesis in the implication. Embedded implications work as hypothetical reasoning. Moreover, the database is divided into protected clauses and removable clauses. So whenever a goal $A \Rightarrow B$ is to be derived the hypothesis A is tested for consistency against the whole database and only removable clauses can be removed to obtain consistency.

In the Witty environment [Oli95], there is a general clauses database system and a general clauses theorem prover which can be programmed by a Horn language enriched with primitives. These primitives can be used as meta-predicates to the first order theories in the database system. We define the system CondLP in terms of Witty primitives.

The syntax of CondLP distinguishes three types of formulas: clauses (D), goals (G) and integrity constraints (I). Following the definition of the language, we present four other elements that will accomplish the summarized operational semantics: 1. a derivability relation; 2. a consistency check procedure; 3. an updating function; 4. an ordering relation.

Database Organization A database in Witty is a pair $< l, \Gamma >$ where l is a given name and Γ is a multiset of clauses. What is considered in CondLP as the database $\Delta \| L$ we will represent as a collection of Witty databases. We consider all protected clause in one database $< delta, \Delta >$. The removable clauses are stored each in a separate database and these will be referred to by: $< l_1, C_1 >$ $\cdots < l_n, C_n >$ so that we can activate/ deactivate the clauses separately. Also $l_1, \cdots, l_n$ can be give an ordering prioritizing the removable clauses.

In order to implement NProlog's derivability, goals will be translated in terms of sequences of $PROVEs$, defined as: $PROVE_{n+m+1}$, where $\overline{B}_i$ is a sequence of labels, $PROVE(\overline{B}_i, H_1, \cdots, H_m, G) \equiv PROVE(\overline{B}_i, H_1 \rightarrow, \cdots, \rightarrow H_m \rightarrow G)$ with the difference that only G is considered as in the set of support. In this way the contrapositive is not obtained since derivation will always start from one of the clauses generated by G and moreover the problem of having the hypothesis added to the database, universally quantified, is overcome.

Database Clauses These are Horn clauses which might have implications embedded in the body. The syntax of clauses is the language that will be used with the *ASSERT* primitive: *If G is a goal and A is an atom then $G \to A$ is a clause.* The implication in CondLP is intuitionistic. For each implication we define a program on the variables of the implicational subformula and a literal with the same name and on the same variables. For instance: $(P(x) \Rightarrow Q(x)) \to R$ is asserted to the database as $cond1(x) \vee R$ with the creation of the program: $cond_1(x) \leftarrow provecond([P(x)], Q(x))$. (*provecond* is later defined).

Definition 1. We define τ_A as a translation algorithm from CondLP database clauses into a set of programs P_i and a formula F as follows: 1. input $G \to H$, where G is a goal and H is an atom; 2. substitute all subformulas of G, $A \Rightarrow (G_1 \wedge G_2)$ by $(A \Rightarrow G_1) \wedge (A \Rightarrow G_2)$ recursively. The resulting G is a conjunction of implications $A_1 \Rightarrow (A_2 \Rightarrow (\cdots \Rightarrow A_n) \cdots)$ where A_i are atoms; 3. for each implication of size > 1, $A_{i1} \Rightarrow (A_{i2} \Rightarrow (\cdots \Rightarrow A_{in_i}) \cdots)$ create a program *condi* which reads: $cond_i(x_{i1}, \cdots, x_{ik}) \leftarrow provecond([A_{i1}, \cdots, A_{in_{i-1}}], A_{in_i})$., where $x_{i1}, \cdots, x_{ik}$ are the variables in the implication subformula. Substitute the implication subformula in question by the literal $cond_i(x_{i1}, \cdots, x_{ik})$. The resulting formula is $F = L_1 \wedge L_2 \wedge \cdots \wedge L_n \to H$ where some L_j are $cond_i$ literals and some are atoms in the original formula. This function is used as follows: if $\tau_A(t : F_1) = F$ where $t : F_1$ is a labelled CondLP sentence, add F to the database t in the system with $ASSERT(t, F)$.

Goals In CondLP a goal is either: 1. an atom A ($\top$ is an atom); 2. the conjunction of two goals $G_1 \wedge G_2$; 3. the conditional $A \Rightarrow G$ where A is an atom.

Definition 2. We define τ_P as a translation algorithm from CondLP goal formulas into a program P as follows: 1. input G; 2. substitute all subformulas of G of the form $A \Rightarrow (G_1 \wedge G_2)$ by $(A \Rightarrow G_1) \wedge (A \Rightarrow G_2)$ recursively. The resulting formula is still a CondLP formula which is a conjunction of implications of the form $A_1 \Rightarrow (A_2 \Rightarrow (\cdots \Rightarrow A_n) \cdots)$, where A_i are atoms; 3. substitute all implications $A_1 \Rightarrow (A_2 \Rightarrow (\cdots \Rightarrow A_n) \cdots)$ by sequences $provecond([A_1, \cdots, A_{n-1}], A_n)$ thus obtaining $\bigwedge_{i=1}^{m} provecond([A_{i1}, \cdots, A_{in_i-1}], A_{in_i})$; 4. the resulting Witty program P is obtained by substituting all signs $\wedge$ by &.

Database Revision $< delta, \Delta >$ may also include integrity constraints with the following syntax: *If G is a goal and $\perp$ is an arbitrary predicate then $G \to \perp$ is an intergrity constraint.* They are translated into the database as ordinary database clauses via the translation procedure τ_A (definition 1).

Derivations As previously mentioned, derivations are performed by the interplay of 4 connected mechanisms: 1. a derivability of a goal from a database of clauses; 2. a consistency check procedure; 3. an updating function which will revise a program $\Delta \| L$ with respect to an atom A; 4. an ordering relation between subsets of L with respect to which maximal elements are taken. These definitions are found in [GGMO94a]. The following is the implementation of the whole consequence relation in Witty.

Procedure 0.1 *Given: 1. the set of databases $W = \{< gamma, \Gamma >\}$, where Γ contains database clauses obtained from CondLP formulas by translation via τ_A (definition 1);*

2. the set of programs $\Pi = \left\{ \begin{array}{l} condi(\cdots) \leftarrow provecond(\cdots). \\ orac(L, l_1, l_2) \leftarrow L. \end{array} \right\}$

where $condi_{ki}$ are the programs obtained in the translation via τ_A and $orac_3$ is the oracle then the verification of $\Gamma \vdash_c \perp$ is computed by $PROVE(gamma, \perp)$.

The updating function is implemented by the following Witty programs.

Procedure 0.2 *Given the following setting: 1. the set of databases $W = \{< delta, \Delta >, < l_i, C_i >\}$, for $1 \le i \le k$, where Δ and C_i contains clauses obtained from CondLP formulas by translation via τ_A (definition 1); 2. $\Pi =$*

$$\left\{ \begin{array}{l} condi(\cdots) \leftarrow provecond(\cdots). \\ orac(L, l_1, l_2) \leftarrow L. \\ provecond([A_1, \cdots, A_n], B) \leftarrow listRevise([A_1, \cdots, A_n]) \& \\ \quad PROVE(delta, A_1, \cdots, A_n, B). \\ provecond([], B) \leftarrow PROVE(delta, B). \\ listRevise([A_1|T]) \leftarrow \cdots \& Revise(l_k, A_1) \& listRevise(T). \\ listRevise([]) \leftarrow \top. \\ Revise(x, y) \leftarrow \cdots \& PROVE(x, delta, y, \perp) \& \cdots \\ tau\_A(x, y) \leftarrow \cdots \\ tau\_P(x, y) \leftarrow \cdots \\ goal(x) \leftarrow tau\_P(x, P) \& P. \end{array} \right\}$$

where: 1. $cond\_i_{ki}$ are the programs obtained in the translation via τ_A; 2. $orac_3$ is the oracle; 3. $provecond_2$ is the variant of the $PROVE$ program, which revises databases with respect to the list of hypothesis then derives the conclusion from the revised database; 4. $listRevise_2$ goes through the list of hypothesis and it is used in conjunction with $Revise_2$, which in turn verifies inconsistencies in each databases according to algorithm $MAX(P, A)$, using database delta; 5. tau_A_2 translates database clauses and tau_P_2 translates goals; 6. $goal\_1$ is the actual query program; it obtains the translation of the goal, which is a program, and executes it.

References

[GGMO94a] Dov Gabbay, L. Giordano, A. Martelli, and N. Olivetti. Conditional logic programming. 1994.

[GGMO94b] Dov Gabbay, L. Giordano, A. Martelli, and N. Olivetti. Lena (a logic engine for natural agent). 1994.

[GR84] Dov Gabbay and U. Reyle. N-prolog: an extension of prolog with hypothetical implications. *Journal of Logic Programming*, 4:319–355, 1984.

[Oli95] Claudia Oliveira. *An Architecture for Labelled Theorem Proving*. PhD thesis, University of London - Imperial College, 1995.

How To Reason About *Akratic* Action Practically?

(position paper: theory of agent, theory of action, inconsistency and actions)

Valentyn Omelyanchyk, Institute of philosophy,
Triochsviatytelska,4. 252001 Kiev -1, UKRAINE
E-mail: vo @ cpi.freenet.kiev.ua

1. *Akrasia* According to Aristotle.

Akratic or incontinent action is the action which an agent freely, knowingly and intentionally performs against his better judgement that an incompatible action is the better thing to do. According to Aristotle, who had formulated the problem of *akratic* action, the incontinent people differ from the wicked in that they think that they ought to live according to reason, whereas the wicked do not presupposed such moral requirement. As a result the first ones regret their incontinent actions, the second ones do not (*EN* 1150b 30 ff.). Thus the first ones have an ability to change their character while the second ones do not.

How *akrasia* is possible? The key of Aristotle's representation of the problem is his conception of "practical syllogism", so that Aristotle represents this problem referring to the idea of a reasoning about *akratic* action. In practical syllogism the major premise indicates that something ought to be done or avoided, the minor premise is a factual premise which informs the person whether the particular relevant facts fall under the general rule of the major premise. The conclusion of the practical syllogism is the action itself.

How to reason about *akratic* action? It is evident that practical syllogism presupposes a view of the relation between knowledge and action, so that knowing what is best "implies" that we *do* what is best. In other words, the existence of practical syllogism is a strong argument for the rationality of all kinds of human behaviour.

This problem is analogous to the problem of the future contingents in (*De Int.*, Ch. IX). Here Aristotle formulates the puzzle that very fact of reasoning about the future "implies" deterministic ontological position. In the our case, Aristotle also formulates the puzzle that the very fact of reasoning about actions "implies" rationalistic position in the practical philosophy.

It is difficult to say how Aristotle understood the solution of this puzzle (but see 3 below). Traditionally it has been suggested that minor premise of the practical syllogism, informing about particular fact, somehow becomes ignored or forgotten during the state of passion (cf. *EN* 1147 b 9-18). Anyway, for us it is important to notice that in *Nicomachean Ethics* the idea of practical syllogism appears *only* in the context of the problem of *akratic* action.

2. *Akratic* Action in the Mentalistic Action Theories.

Being eliminated from this context, the idea of the practical reasoning ("practical syllogism") is collapsed into the idea of the theoretical reasoning about the practical domain: as in the theoretical domain we find reasons for our beliefs, so as in the practical domain we find reasons for our actions. But this position implies the claim of the rationality of all human behaviour.

What is action in this case? According to Mill's paradigm, an action is 'the state of mind called a volition, followed by an effect'. In the frame of this paradigm the reduction of the *final causality*-clauses, like "S does A to bring about B", to the *cause-effect*-clauses, like "S does A *because* S wants to bring about B", is the standard move. Reasoning theoretically about practical domain, Action Theory considers as his central object the causally connected sequence of events that are involved in the performance of a human action with the central question where to locate the action in this causal chain.

Causal ontology in Action Theory necessarily leads to *mentalistic stance* in Action Theory according to which a notion of action is

• reducible notion,
• reducible to mentalistic notions.

Most prominent Mentalistic Action Theories today are Causal Action Theory (Davidson) and Agency Action Theory (Chisholm)[1] In the first one the notion of action is reduced to the notion of mental event, being identified with event, but only with one which is caused in the right way by a mental state. In the second case an action is reduced to the notion of a substantial self or agent: an action is an event caused by a person who realises a purpose or an aim or who fits a norm.

Trying to solve the *akrasia* problem, Mentalistic Action Theories reduce it to the *psychological problem* about the imbalance between evaluational and motivational structures of an agent, so that the very possibility of *akrasia* is based on the possibility of imbalance between reason and motivation[2]

That is why to solve the *akrasia* problem means in the frame of Mentalistic Action to demonstrate consistency of the assertions like followings (*pace* Davidson):

• (P1): if an agent wants to do x more than he wants to do y and he believe himself free to do either x or y, then he will intentionally do x if he does either x or y.
• (P2): if an agent judges that it would be better to do x then to do y, then he wants to do x more then he wants to do y.
• (P3) There are incontinent actions where incontinent action is the action x such that (a) the agent does x intentionally, (b) the agent believes there is an alternative action y open to him, (c) the agent judges that, all things considered, it would be better to do y than to do x.

These assertions *prima facie* are inconsistent, so that one cannot reason about *akratic* action. But bearing in mind that (P2) fixes the connection between evaluation and motivation, (P2) is the natural point for making a distinction which make it possible to avoid this inconsistency. For example, one can distinguish between wants as desires and wants as rational preferences, or between person's better judgement as evaluative reason and person's better judgement as practical reason (a mental state). As a result this solution of the *akrasia* problem permits to reason about *akratic* action, but it has to do

---

[1] See: M. Brand 'Intending and Acting (Toward Naturalised Action Theory), MIT Press, Camb.,1984.

[2] A. Walker 'The Problem of Weakness of Will' - Nous 23 (1989) 6, p.653-76.

with an agent that is *not fully well-put-together* [3]. But just this conclusion can be attacked by sceptic: Mentalistic Action Theories attribute to *akratic* person a kind of schizophrenia. This is the price of reasoning about *akratic* action in Mentalistic Action Theories.

3. *Akratic* Action in Non-Mentalistic Action Theory.

According to Aristotelian Action Theory [4] the *akrasia* problem is not a psychological, but an *ontological* problem. Resolution of this problem presupposes postulating in our ontology the new kind of *irreducible* entities, namely actions. [5]

From this it follows that in Aristotelian Action Theory the notion of action is *more fundamental* then the notion of self or person, then the other mentalistic notions like purpose, norm etc. Thus in the frame of Aristotelian Action Theory one can speak about an *impersonal* action, an action *without* purpose, intention etc. [6]

Such kind of actions constitutes the fist level of Aristotelian Action Theory. At this level the main question is: what kind of process an action is? With Aristotle we can understand an action as an *ability-process* (a process where some ability is realised). Thus the first level of Aristotelian Action Theory presupposes a *modal* ontology. The logic which describes this ontology is the logic of Aristotelian modality which is based on the analogy:

$$\text{necessity} / \text{possibility} = \text{particular} / \text{general.} \, [7]$$

The most simple logic which fits this analogy is J.Lukasiewicz's L-modal logic [8]. Aristotelian necessity, which is necessity *de re*, has the meaning in the intended model of this logic as "to be in a relation to a sample", so that the most simple action is the ability-process of *following a sample of behaviour*.

The second level of Aristotelian Action Theory deals with the problem of *constituting or mustering of self or agent*. At this level one deals with the practical reasoning properly speaking. It is the process of attribution of agents, goals, intentions etc. to actions. Only at this level we have non-schizophrenic resolution of the akrasia problem, bearing in mind Aristotelian thesis that a person is constituted by actions (*EN* 1103 b 31f). [9]

At this level the *akrasia* problem can be naturally posed as the problem of genesis of self or person, so that the distinguishing trait of the *akratic* action is the following: to be an

---

[3] Cf. 'What is special in incontinence is that the actor cannot understand himself: he recognises in his own intentional behaviour, something essentially surd' (D.Davidson 'How is Weakness of the Will Possible' - In: Essays on Action and Events, 1980,p.42).

[4] See my book 'Possibility, Structure, Action (An Introduction to Modal Realism)', Kiev,1991.

[5] Cf. D.-H. Ruben 'Three Theories of Action' - In: Abstracts of X Congress of LMPS, Florence, 1995,p.599.

[6] Cf. Aristotelian distinction *praxis / poiesis*.

[7] V.Omelyantchik 'Ghost of Aristotelian Modality'. - In: Abstracts of X Congress of LMPS, Florence, 1995,p. 356.

[8] Cf. D.Walton 'Time and Modality in the *Can* of Opportunity' - In: Action Theory (Ed. M.Brand, D.Walton),1976,p.285.

[9] More accurately: 'by free deeds (*praxeis*)', but I ignore this distinction in the exposition.

action with the agent in change. Thus one can reason about *akratic* actions without the rationality presupposition about a human behaviour, from the one side, and without schizophrenic presupposition about an agent, from the other side, considering changeable agents. Reasoning about *akratic* action means mustering an agent who constituted by this kind of actions, so that practical reasoning is not descriptive, but *constructive* process here.[10]

Action like this one can be classified as a *passive action.* It cannot be reduced to two consequent states (active and passive, respectively) with the unchangeable agent(s). Being understand in this positive way, the problem of *akrasia* is assimilated, generally speaking, to the problem of training and cultivating one's character.

---

[10] Cf. E. von Glaserfeld 'Radical Constructivisme' - In: Power, Autonomy, Utopia (Ed. D.Trappl), 1989, p.107-116.

Authors Index

Lecture Notes in Artificial Intelligence (LNAI)

Lecture Notes in Computer Science